ARMORED SCALE INSECT PESTS

of Trees and Shrubs

ARMORED SCALE INSECT PESTS

of Trees and Shrubs
(Hemiptera: Diaspididae)

DOUGLASS R. MILLER

and

JOHN A. DAVIDSON

COMSTOCK PUBLISHING ASSOCIATES

a division of

Cornell University Press, Ithaca and London

Douglass R. Miller is Research Entomologist with the Systematic
Entomology Laboratory, Agricultural Research Service, U.S. Department
of Agriculture at the Beltsville Agricultural Research Center, MD. He
received B.S., M.S., and Ph.D. degrees in entomology from the University
of California at Davis. He, Y. Ben-Dov, and G. A. P. Gibson developed
'ScaleNet,' an online information system on the scale insects of the
world.

John A. Davidson is Professor Emeritus in the Department of
Entomology at the University of Maryland, College Park. A graduate of
Columbia Union College, Takoma Park, MD, he received M.S. and Ph.D.
degrees in entomology from the University of Maryland and is coauthor
of Landscape IPM with Michael J. Raupp.

Publication of this book was made possible, in part, by the financial
support of the University of Maryland and the United States Department
of Agriculture

First published 2005 by Cornell University Press

Printed in Hong Kong

Library of Congress Cataloging-in-Publication Data

Miller, Douglass R., 1942–
    Armored scale insect pests of trees and shrubs / by Douglass R. Miller
and John A. Davidson.
        p. cm.
    Includes bibliographical references and index.
    ISBN 0-8014-4279-6 (cloth : alk. paper)
  1. Trees—Diseases and pests. 2. Shrubs—Diseases and pests. 3. Scale
insects—Identification. I. Davidson, John A. II. Title.
SB761.M55 2005
634.9'6752—dc22   2005002689

Cornell University Press strives to use environmentally responsible
suppliers and materials to the fullest extent possible in the publishing
of its books. Such materials include vegetable-based, low-VOC inks
and acid-free papers that are recycled, totally chlorine-free, or partly
composed of nonwood fibers. For further information, visit our website
at www.cornellpress.cornell.edu.

Cloth printing   10 9 8 7 6 5 4 3 2 1

We dedicate this volume to those who stimulated and encouraged our interests in the fascinating world of nature, including Lester E. Harris Jr. (teacher and mentor), Robert House (teacher), John W. Johnson (teacher), Betsy Miller (mother), and Gayle H. Nelson (mentor). The first author dedicates this volume to the late Judith F. Miller for her encouragement and moral support over more than forty years.

Contents

Preface   xi

Introduction   1
  Biology   *1*
    General life history   *1*
    Female cover formation   *5*
    Male cover formation   *5*
  Ecology   *5*
  Management   *6*
    Detection   *7*
    Identification   *7*
    Monitoring   *7*
    Chemical control   *7*
      Oils   *7*
      Soaps   *7*
      Synthetic organic insecticides   *8*
      Growth regulators   *8*
      Pheromones   *8*
      Scale resistance to insecticides   *8*
    Biological control   *8*
    Cultural control   *12*
    Host-plant resistance   *12*
    Integrated pest management   *13*
  Economic Importance   *13*
  Morphology   *13*
    Glossary of morphological terms   *14*
  Materials and Methods   *14*
    Collection and dry preservation   *18*
    Liquid preservation   *18*
    Slide mounting   *18*
      Temporary mounts   *18*
      Permanent mounts   *18*

Key to Adult Females (Microscopic Characters)   20

Field Key to Economic Armored Scales   29

Treatment of 110 Species   37
  Abgrallaspis cyanophylli (Signoret)—Cyanophyllum scale   *38*
  Abgrallaspis degenerata (Leonardi)—Degenerate scale   *42*
  Abgrallaspis ithacae (Ferris)—Hemlock scale   *44*
  Andaspis punicae (Laing)—Litchi scale   *48*
  Aonidiella aurantii (Maskell)—California red scale   *50*
  Aonidiella citrina (Coquillett)—Yellow scale   *54*
  Aonidiella orientalis (Newstead)—Oriental armored scale   *58*
  Aonidiella taxus Leonardi—Asiatic red scale   *62*
  Aspidiella sacchari (Cockerell)—Sugarcane scale   *66*
  Aspidiotus cryptomeriae Kuwana—Cryptomeria scale   *68*

Aspidiotus destructor Signoret—Coconut scale   72

Aspidiotus excisus Green—Aglaonema scale   76

Aspidiotus nerii Bouché—Oleander scale   78

Aspidiotus spinosus Comstock—Spinose scale   82

Aulacaspis rosae (Bouché)—Rose scale   86

Aulacaspis tubercularis Newstead—White mango scale   90

Aulacaspis yasumatsui Takagi—Cycad aulacaspis scale   94

Carulaspis juniperi (Bouché)—Juniper scale   96

Carulaspis minima (Targioni Tozzetti)—Minute cypress scale   100

Chionaspis americana Johnson—Elm scurfy scale   102

Chionaspis corni Cooley—Dogwood scale   106

Chionaspis furfura (Fitch)—Scurfy scale   110

Chionaspis heterophyllae Cooley—Pine scale   112

Chionaspis pinifoliae (Fitch)—Pine needle scale   116

Chionaspis salicis (Linnaeus)—Willow scale   118

Chrysomphalus aonidum (Linnaeus)—Florida red scale   122

Chrysomphalus bifasciculatus (Ferris)—Bifasciculate scale   126

Chrysomyphalus dictyospermi (Morgan)—Dictyospermum scale   130

Clavaspis herculeana (Cockerell and Hadden)—Herculeana scale   134

Clavaspis ulmi (Johnson)—Elm armored scale   136

Comstockiella sabalis (Comstock)—Palmetto scale   140

Cupressaspis shastae (Coleman)—Redwood scale   142

Diaspidiotus ancylus (Putnam)—Putnam scale   146

Diaspidiotus forbesi (Johnson)—Forbes scale   152

Diaspidiotus gigas (Thiem and Gerneck)—Poplar scale   154

Diaspidiotus juglansregiae (Comstock)—Walnut scale   158

Diaspidiotus liquidambaris (Kotinsky)—Sweetgum scale   162

Diaspidiotus osborni (Newell and Cockerell)—Osborn scale   166

Diaspidiotus ostreaeformis (Curtis)—European fruit scale   168

Diaspidiotus perniciosus (Comstock)—San Jose scale   172

Diaspidiotus uvae (Comstock)—Grape scale   176

Diaspis boisduvalii Signoret—Boisduval scale   180

Diaspis bromeliae (Kerner)—Pineapple scale   182

Diaspis echinocacti (Bouché)—Cactus scale   186

Duplaspidiotus claviger (Cockerell)—Camellia mining scale   188

Duplaspidiotus tesseratus (D'Emmerez de Charmoy)—Tesserate scale   192

Dynaspidiotus britannicus (Newstead)—Holly scale   194

Epidiaspis leperii (Signoret)—Italian pear scale   198

Fiorinia externa Ferris—Elongate hemlock scale   200

Fiorinia fioriniae (Targioni Tozzetti)—Palm fiorinia scale   204

Fiorinia japonica Kuwana—Coniferous fiorinia scale   206

Fiorinia theae Green—Tea scale   210

Froggattiella penicillata (Green)—Penicillate scale   212

Furcaspis biformis (Cockerell)—Orchid scale   216

Furchadaspis zamiae (Morgan)—Zamia scale   218

Gymnaspis aechmeae Newstead—Flyspeck scale   222

Hemiberlesia lataniae (Signoret)—Latania scale   224

Hemiberlesia neodiffinis Miller and Davidson—False diffinis scale   228

Hemiberlesia rapax (Comstock)—Greedy scale   232

Howardia biclavis (Comstock)—Mining scale   234

Ischnaspis longirostris (Signoret)—Black thread scale   238

Kuwanaspis pseudoleucaspis (Kuwana)—Bamboo diaspidid   240

Lepidosaphes beckii (Newman)—Purple scale   244

Lepidosaphes camelliae Hoke—Camellia scale   248

Lepidosaphes conchiformis (Gmelin)—Fig scale   250

Lepidosaphes gloverii (Packard)—Glover scale   254

Lepidosaphes pallida (Maskell)—Maskell scale   256

Lepidosaphes pini (Maskell)—Pine oystershell scale   260

Lepidosaphes pinnaeformis (Bouché)—Cymbidium scale   262
Lepidosaphes ulmi (Linnaeus)—Oystershell scale   266
Lepidosaphes yanagicola (Kuwana)—Fire bush scale   270
Lindingaspis rossi (Maskell)—Black araucaria scale   272
Lopholeucaspis japonica (Cockerell)—Japanese maple scale   276
Melanaspis lilacina (Cockerell)—Dark oak scale   278
Melanaspis obscura (Comstock)—Obscure scale   282
Melanaspis tenebricosa (Comstock)—Gloomy scale   284
Mercetaspis halli (Green)—Hall scale   288
Morganella longispina (Morgan)—Plumose scale   290
Neopinnaspis harperi McKenzie—Harper scale   294
Nuculaspis californica (Coleman)—Black pineleaf scale   296
Nuculaspis pseudomeyeri (Kuwana)—False Meyer scale   300
Nuculaspis tsugae (Marlatt)—Shortneedle conifer scale   304
Odonaspis ruthae (Kotinsky)—Bermuda grass scale   306
Parlatoreopsis chinensis (Marlatt)—Chinese obscure scale   310
Parlatoria blanchardi (Targioni Tozzetti)—Parlatoria date scale   312
Parlatoria camelliae (Comstock)—Camellia parlatoria scale   316
Parlatoria oleae (Colvee)—Olive scale   320
Parlatoria pergandii (Comstock)—Chaff scale   324
Parlatoria pittospori Maskell—Pittosporum scale   328
Parlatoria proteus (Curtis)—Proteus scale   330
Parlatoria theae Cockerell—Tea parlatoria scale   334
Parlatoria ziziphi (Lucas)—Black parlatoria scale   336
Pinnaspis aspidistrae (Signoret)—Fern scale   340
Pinnaspis strachani (Cooley)—Lesser snow scale   344
Pseudaonidia duplex (Cockerell)—Camphor scale   346
Pseudaonidia paeoniae (Cockerell)—Peony scale   350
Pseudaonidia trilobitiformis (Green)—Trilobe scale   354
Pseudaulacaspis cockerelli (Cooley)—False oleander scale   356
Pseudaulacaspis pentagona (Targioni Tozzetti)—White peach scale   360
Pseudaulacaspis prunicola (Maskell)—White prunicola scale   364
Pseudischnaspis bowreyi (Cockerell)—Bowrey scale   368
Pseudoparlatoria ostreata (Cockerell)—Gray scale   370
Pseudoparlatoria parlatorioides (Comstock)—False parlatoria scale   374
Quernaspis quercus (Comstock)—Oak scale   376
Rhizaspidiotus dearnessi (Cockerell)—Dearness scale   380
Selenaspidus albus (McKenzie)—White euphorbia scale   382
Selenaspidus articulatus (Morgan)—Rufous scale   386
Situlaspis yuccae (Cockerell)—Yucca scale   388
Unaspis citri (Comstock)—Citrus snow scale   392
Unaspis euonymi (Comstock)—Euonymus scale   396

References Cited   399

Index of Host Plants of Armored Scales   419

Index of Armored Scales, Natural Enemies, and General Subjects   437

Preface

Armored scales are members of the family Diaspididae and are highly specialized plant parasites. The presence of a cover that is not attached to the body makes armored scales easily recognizable in the field. Species occur in all areas of the world with the exception of the polar regions and are common plant pests wherever they are found. Most armored scale pests are invasive in the countries where they cause damage. With increased international trade, these often concealed and cryptic pests pose a serious threat to the world agricultural economy. In the United States alone they cost more than $1 billion annually in damage and control costs.

This volume has been in preparation for more than 20 years. It began as a syllabus used for part of a short course on the economic scale insects of the United States. Over the years the course was attended by more than 100 students and was taught by colleagues from numerous state and federal organizations. Although the participants were primarily from the United States, attendees came from as far away as Africa and New Zealand.

The original concept of this project was to write a three-volume handbook that would cover all groups of scale insects including armored scale, mealybugs, soft scales, felt scales, margarodids, and other scale families that contain pest species. Unfortunately, our ambitions exceeded our abilities, and our contribution will be restricted to the armored scales. We should point out, however, that this family includes the most damaging and speciose of all of the major scale insects groups.

The primary objectives of this book are to provide sufficient information on the economic armored scales in the United States so that readers can: (1) recognize undetermined or unfamiliar species based on their field appearance; (2) validate or improve field identifications by examining slide-mounted specimens with a compound microscope; (3) understand the general life history, economic importance, host range, and distribution of destructive armored scales; (4) maintain armored scale populations below damaging thresholds; (5) obtain more detailed information through the bibliography and references; and (6) enjoy the diversity, adaptability, and splendor of this interesting group of unusual insects.

We give special acknowledgment to our key contributors: Ray Gill, California Department of Food and Agriculture, Sacramento, CA; Gary Miller, Systematic Entomology Laboratory, Beltsville, MD; and Avas Hamon, Division of Plant Industry (DPI), Gainesville, FL, for their important roles in helping finish this book. Without their assistance this work would provide far less information and may never have been finished. Ray provided numerous magnificent slides that make this book a unique tool for field identification of armored scales. He also has shared his encyclopedic knowledge of the armored scales of the western United States and was a great help by reviewing the manuscript in its entirety. He assisted us in locating infestations of species that we could not have found on our own and introduced us to many of the knowledgeable county agents in California. Gary Miller has been instrumental in developing the graphic portions of the manuscript, and with his wonderful knack with computers has made the illustrations and plates of a high quality. It is only because of him that the color plates and line drawings are organized and presented in a way that makes them an important part of the information content of this book. Avas Hamon also provided color slides of Florida pest species. He was instrumental in organizing two collecting trips in his state and introduced us to a wonderful group of DPI inspectors who were extremely helpful. His assistance and encouragement are greatly appreciated.

For the past 20 to 25 years colleagues from all over the United States have assisted us in many ways. Some, we are afraid, have been forgotten and for this we apologize. Those who have assisted us in locating specimens for photography are as follows: Donn Johnson and Barbara Lewis, Department of Entomology, University of Arkansas, Fayetteville, AK; Chris Baptista, David Mills, and Scott Soby, State Agricultural Laboratory, Arizona Department of Agriculture, Phoenix, AZ; Jerry Davidson, Santa Barbara County, Agricultural Commissioners Office, CA; David Callum, County of San Diego, CA; Rosier Garrison, Los Angeles County, Department of Agricultural Commissioner, South Gate, CA; Lorenzo Fernandez and Nick Nisson, Orange County Agricultural Commissioner's Office, Anaheim, CA; Andy Blanton, Instant Jungle International, Anaheim, CA; Scala Nursery, Santa Ana, CA; Quail Botanical Garden, Encinitas, CA; Norman Smith, Fresno County Department of Agriculture, Fresno, CA; Penny Gullan, Department of Entomology, University of California, Davis, CA; Terrence Williams, DPI, Orlando, FL; Harry P. Leu Botanical Gardens, Orlando, FL; Jonathan Crane, Waldy Klassen, Jorge Peña, Tropical Research and Education Center, University of Florida, Homestead, FL; Louis Loydga and Gwen Myres, DPI, Coral Gables, FL; Don Evans and Craig Allen, Fairchild Tropical Gardens, Coral Gables, FL; Maria Quintanilla, DPI, Ft. Lauderdale, FL; Matt Brodie, Scott Krueger, DPI, Naples, FL; Freda Long, Susan Gallagher, Elsie Page in private gardens in Naples, FL; Jan Abernathie and David Tetzlaff, Caribbean Gardens: The Zoo in Naples, FL; F. William Howard, University of Florida, Ft. Lauderdale, FL; James F. Stimmel, Pennsylvania Department of Agriculture, Harrisburg, PA; JoAnne Lund, USDA Forestry Sciences Laboratory, Rhinelander, WI; Pedro Millan, Mildred Sosa, Julio Navarro, Miguel Gonzalez, Edda Martinez, Lorimar Figueroa, Manuel Matos, Shara Madera, and Teresita Alemany, Animal and Plant Inspection Service, Plant Protection and Quarantine, Ponce and Mayaguez, Puerto Rico; Stanton S. Gill, University of Maryland Extension Service, Ellicott City, MD; Maren E. Gimpel, Systematic Entomology Laboratory, Beltsville, MD; William F. Gimpel, Maryland Department of Agriculture, Annapolis, MD; Gale E. Ridge, Department of Entomology, Connecticut Agriculture Experiment Station, New Haven, CT.

Pamela J. Hollyoak was hired as a contractor by the Systematic Entomology Laboratory to prepare some of the early illustrations for the book. She mostly did a series of blue line drawings that were corrected and inked by Davidson (with signature PJH and JAD); a few were corrected and inked by Pamela (PJH). We are grateful for her assistance.

Those who read the manuscript and made many important and useful comments include Cliff Sadof, Purdue University, West Lafayette, IN; Ray Gill, California Department of Food and Agriculture, Sacramento; James Stimmel, Pennsylvania Department of Agriculture, Harrisburg; Ronald Ochoa, Systematic Entomology Laboratory (SEL), Plant Sciences Institute, Agricultural Research Service, U. S. Department of Agriculture, Washington, DC, and Beltsville, MD; Jerry Davidson, Santa Barbara County Commissioner's Office, Santa Barbara, CA; Rosier Garrison, Los Angeles County, Department of Agricultural Commissioner, South Gate, CA; Nick Nisson, Orange County Agricultural Commissioner's Office, Anaheim, CA. Special thanks to John Brown (SEL) who read the entire manuscript and put considerable effort into its improvement.

We are most grateful to those individuals who have allowed us to use their color images including: Ray Gill, California Department of Food and Agriculture, Sacramento; Avas Hamon, previously from Division of Plant Industry, Gainesville, FL; Richard Casagrande, Department of Plant Sciences, University of Rhode Island, Kingston; Don Alstad, Department of Ecology, Evolution, and Behavior, University of Minnesota, St. Paul; Michael J. Raupp, Department of Entomology, University of Maryland, College Park; Michael Rose, Department of Entomology, Montana State University, Bozeman; Michael L. Williams, Department of Entomology/Plant Pathology, Auburn University, AL; Manya B. Stoetzel, Systematic Entomology Laboratory, USDA/ARS, Beltsville, MD; Jorge Peña, Tropical Research and Education Center, University of Florida, Homestead, FL; and the late Warren T. Johnson, Department of Entomology, Cornell University, Ithaca, NY.

Finally we are grateful to Systematic Entomology Laboratory technical staff, Debra Creel and Nit Malikul for preparation of important specimens and to Erica Limones, an intern in the Laboratory, for spending endless hours scanning the line drawings.

ARMORED SCALE INSECT PESTS

of Trees and Shrubs

Introduction

One-third of the more than 7300 species of scale insects (superfamily Coccoidea) are armored scales (family Diaspididae). They are the most speciose family of scale insects including about 2400 species in 380 genera. Other common groups of scale insects include mealybugs (Pseudococcidae), soft scales (Coccidae), pit scales (Asterolecaniidae), giant scales (Margarodidae), and felt scales (Eriococcidae) to name a few of the 20 or more extant families. Scale insects are part of the order Hemiptera (previously considered as part of the Homoptera but this order is no longer considered valid) and are worldwide in distribution. Although there are several classifications of the Diaspididae, there are two groups that contain a majority of the species and are relatively easy to recognize. They often are used as informal groups and are referred to as diaspidines and aspidiotines; they are based on two of the major subfamilies of armored scales, the Diaspidinae and Aspidiotinae. There are no obvious characters that separate groups all of the time, but most species are consistent with the following combination of characters. Diaspidines produce an elongate scale cover and have two-barred macroducts, more than one seta on each antenna, gland spines between the pygidial lobes, and pores near the spiracles. Aspidiotines produce an oval or round cover and have one-barred macroducts, one seta on each antenna, plates between the pygidial lobes, and no pores near the spiracles.

Armored scales occur on a variety of host plants encompassing more than 1380 plant genera in 182 plant families (Borchsenius 1966). The most prevalent host families are: Leguminosae with about 230 species of armored scales, Gramineae with about 150 species, and Euphorbiaceae with 145 species. Armored scales usually are pests on plants that survive for more than a single year including fruit and nut crops, forest trees, and ornamentals, such as landscape perennials, shrubs, shade trees, and greenhouse plants. Miller and Davidson (1990) compiled a list of 199 species that are considered pests in at least some part of the world. This figure is only about 8% of the total number of described species, and their economic impact is quite significant. In the United States, all scale insects are estimated to cause millions (Kosztarab 1977) or even billions (Kosztarab 1990) of dollars in damage and control costs annually. Armored scales are responsible for major portions of these costs.

BIOLOGY

General life history

(Plate 1; Figs. 1 and 2) Armored scales have three female instars and five male instars including adults. Most are protected by a separate cover that consists of about 50% wax and 50% of a non-waxy component that may be a polyphenol or melanin-like compound (Ebstein and Gerson 1971) and shed skins. Flat strands of wax produced primarily from the pygidial wax glands form the basic wax component of the cover. It is continuously cemented with material from the anal opening (Foldi 1990). The structure of the cover varies primarily at the generic level. It is thin and transparent in some genera (Plate 14A and B) and dense and opaque (Plate 79A) in others. Females of many species produce a ventral cover that usually is thin (Plate 36I) but in some species is remarkably thick (Plate 57B). Thus, depending on the genus, the cover may be more or less resistant to water loss as well as to penetration by chemicals. Shed skins of the first two instars are incorporated in two ways. In armored scales with elongate or oyster-shell-shaped covers, the exuviae are usually marginal or submarginal (Plate 1J and K; Fig. 2H). In species that form round or oval covers, the shed skins are generally central or subcentral (Plate 51A; Fig. 1E). These terms are used in relation to the front and back orientation of the scale cover only. Occasionally an individual of a species that incorporates exuviae centrally may have shed skins positioned near the lateral margin of the cover; but because the shed skins are near the lateral margin of the cover, not the front, they are considered to be central; they are centrally located with regard to the front and back orientation of the scale cover. Cover formation and female morphology may vary dramatically depending on their feeding location on the host, that is, bark, leaves, petioles, or fruit. For example, Knipscher et al. (1976) demonstrated that *Chionaspis sylvatica* Sanders was the bark form of *C. nyssae* (sweetgum scale), which occurs on leaves. Intermediate forms were found on leaf petioles. The bark form is oyster-shell shaped, the wax is thick, and the scale is usually dingy gray. The leaf form is triangular or circular, thin, and pure white. The scale cover of *Diaspidiotus ancylus* (Putnam scale) on the leaves or fruit is conspicuously white (Plate 36C), whereas the cover on the bark is dark brown or gray (Plate 37A) (Polavarapu et al. 2000). Similar dimorphic forms have been reported in species of several armored scale genera. Females in some genera do not produce a typical waxy cover for protection. Instead they are completely or partially enclosed in an enlarged, robust, second-instar shed skin (Plate 53A and B). These kinds of scale insects are called pupillarial.

Most armored scale species are biparental and mating is required for egg production. Other species appear to consist of biparental and uniparental populations, while yet others never seem to produce males. Armored scales overwinter in many different life stages including eggs, second-instar males and females, or mated adult females. Most species produce less than 100 eggs per female, which is quite different from many soft scales that lay 1000 or more. Many species lay eggs under the scale cover (Plate 14B); a few lay eggs that hatch within minutes of being laid; and a few other species lay nymphs that hatched from the egg inside the body of the adult female. The color of mature eggs is consistent for each species and may be relatively constant for most genera. Species treated in this volume have eggs that are white (Plate 48H), yellow (Plate 13H), salmon (103H), red, brown (Plate 18F), pink (Plate 70G), purple (Plate

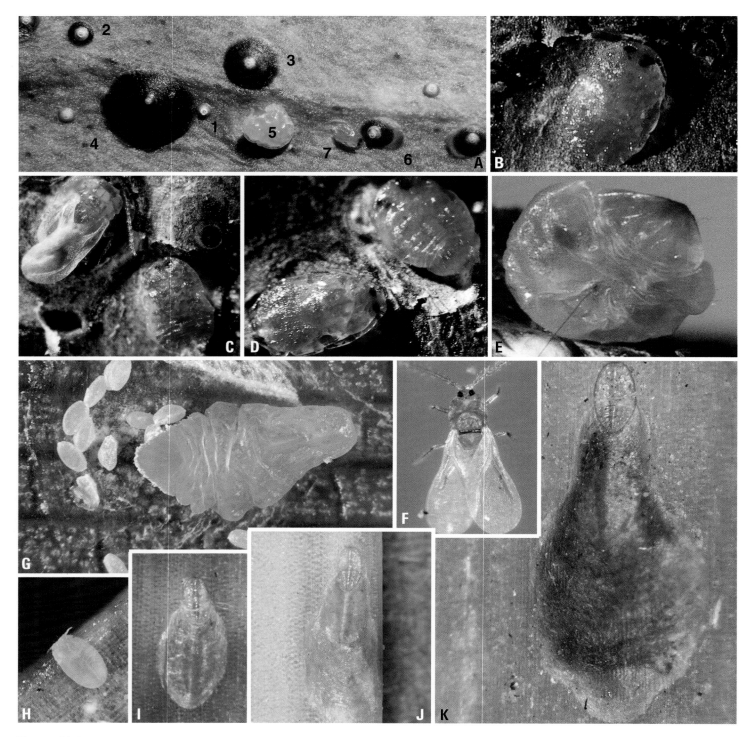

Plate 1. Biology

Aspidiotines:

A. Covers of *Chrysomphalus aonidum* (Florida red scale) (1) capped crawler, (2) second-instar female cover, (3) young third-instar (adult) female cover, (4) mature third-instar (adult) female cover, (5) body of young adult female, (6) cover of fourth-instar (pupa) male, and (7) body of pupal male (J. A. Davidson).

B. Third-instar (prepupa) male body of *Melanaspis obscura* (obscure scale) (J. A. Davidson).

C. Newly molted adult male and pupa of *M. obscura* (obscure scale) (J. A. Davidson).

D. Note eyes of prepupa inside body of second-instar male (right) and fourth-instar (pupa) male body (left) of *M. obscura* (obscure scale) (J. A. Davidson).

E. Adult female of *M. obscura* (obscure scale) raised off of host showing thin filamentous mouthparts (J. A. Davidson).

F. Adult male of *Diaspidiotus ancylus* (Putnam scale) (J. A. Davidson).

Diaspidines:

G. Adult female body with eggs of *Pseudaulacaspis cockerelli* (false oleander scale) (J. A. Davidson).

H. Active first-instar crawler of *Fiorinia japonica* (coniferous fiorinia scale) (J. A. Davidson).

I. Cover of second-instar female of *Pinnaspis aspidistrae* (fern scale) (J. A. Davidson).

J. Cover of young third-instar (adult) female of *P. aspidistrae* (fern scale) (J. A. Davidson).

K. Cover of mature third-instar (adult) female of *P. aspidistrae* (fern scale) (J. A. Davidson).

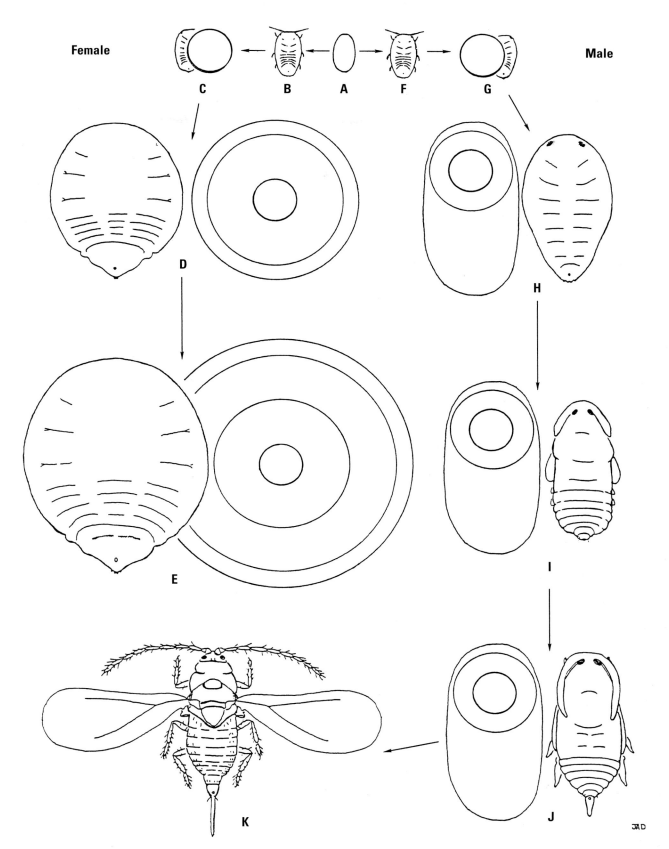

Figure 1. Aspidiotine life-history chart and scale-cover development. Female: **A**, egg; **B**, active crawler; **C**, settled crawler and white cap; **D**, second-instar female and partially formed cover; **E**, third-instar (adult) female and partially formed cover. Male: **A**, egg; **F**, active crawler; **G**, settled crawler and white cap; **H**, second-instar male and cover; **I**, third-instar (prepupa) male and cover; **J**, fourth-instar (pupa) male and cover; **K**, fifth-instar (adult) male.

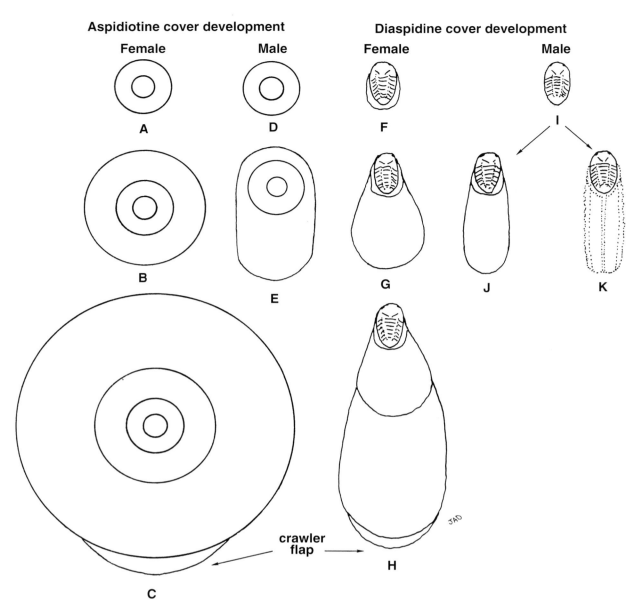

Aspidiotine cover development

Female   Male

Diaspidine cover development

Female   Male

A   D   F   I

B   E   G   J   K

crawler flap →

C   H

Figure 2. Scale-cover development comparisons. **A**, aspidiotine female crawler cover with white cap and added wax; **B**, fully formed aspidiotine second-instar female cover; **C**, fully formed aspidiotine third-instar (adult) female cover with crawler flap; **D**, aspidiotine male crawler cover with white cap and added wax (same appearance as in female); **E**, fully formed aspidiotine second-instar male cover (note elongate form); **F**, fully formed diaspidine female crawler cover; **G**, fully formed diaspidine second-instar female cover; **H**, fully formed diaspidine third-instar (adult) female cover with crawler flap; **I**, recently settled diaspidine male crawler cover; **J**, fully formed diaspidine second-instar male cover (not felted and without longitudinal carinae); **K**, fully formed diaspidine second-instar male cover (felted, with three longitudinal carinae).

28F), or black. Individual females of the *Pseudaulacaspis pentagona* (white peach scale), produce salmon female eggs first, followed by white male eggs later (Plate 103H) (Bennett and Brown 1958). Egg shells (called chorions) are actually white or transparent; hence, egg color reflects the pigment contained in the egg contents and developing embryo. Crawlers retain this color when they hatch but may change when they begin feeding. The color of gravid females tends to be the same as the newly hatched crawler, but recently molted females can be a different color. For example, the newly formed adult females of *Chionaspis pinifoliae* (pine needle scale) are yellow, but as eggs develop inside the female, the body color changes to pink or red (Plate 27E).

Crawlers, or first instars, are active or settled. The active phase lasts for several hours to several days (Koteja 1990; Willard 1973). While active, crawlers' extended antennae and legs are visible around the perimeter of the body (Plate 27G; Fig. 1B and F). Settled crawlers retract their antennae and legs under the body and thus these structures are not visible from above (Plate 27H; Fig. 1C and G).

Armored scales have a rather unusual biology in that they are enclosed in a scale cover that is constructed of wax and other compounds and is detached from the body of the female. The enclosed life style has mandated that armored scales no longer produce honeydew from the anus and has caused them to alter their feeding behavior to tap individual plant cells rather than feeding from the phloem sap like their other scale insect relatives. Glands within the body secrete substances through the anal opening that are used in cementing the waxy components of the cover together (Foldi 1990). Sex pheromones are produced by the adult female in pygidial glands near the anal opening and apparently exit the female body through the anal opening (Gieselmann 1990).

Female cover formation

Aspidiotine crawlers (those that usually produce circular covers) secrete a white waxy cover that encloses the top of the body within 24 to 48 hours of settling. This so-called cap (or white cap) (Plate 9E; Fig. 1C) is most conspicuous and durable in genera that form thick scale covers, for example, *Lindingaspis* and *Melanaspis,* and less obvious in those that form thin, fragile covers, for example, *Aonidiella* and *Aspidiotus.* Some diaspidine crawlers (those that primarily form elongate covers) produce a transparent, essentially invisible wax cover (Plate 1J; Fig. 2F) that does not include strands of wax and disintegrates as the scale matures, for example, *Carulaspis* and *Pseudaulacaspis.* Other diaspidines produce a series of strands of wax almost immediately after settling, and these strands become intertwined and form a thin mat over the crawler body (excluding the head), for example, *Chionaspis* and *Lepidosaphes.* After the initial cover is formed (Fig. 1C), an anal secretion is produced that eventually cements the filaments together into a true cover. Throughout this process the body of the crawler rotates back and forth around the inserted mouthparts to form the first part of the cover. As the crawler enlarges, it adds new scale material around the edge of the initial cover; this wax is denser and usually is of a different texture than the initial cover (Fig. 2A). This material is the same as the wax that is added to the cover by later instars. When the scale has nearly completed the crawler stage, the dorsum becomes hard and may blacken. The shed skin divides along the body margin, the dorsal surface is pushed up against the wax cover, and the ventral surface is pushed back and remains attached to the exuviae. The crawler shed skin is yellow in many genera including *Aspidiotus* (Plate 14C) and *Diaspidiotus;* black in *Melanaspis, Lindingaspis* (Plate 76C), and some species of *Parlatoria;* and black or yellow orange in *Hemiberlesia lataniae* (Plate 61B and C) (Stoetzel 1976). When the crawler cover is rubbed, the color of the exposed shed skin, can be a useful diagnostic character.

The second instar detaches itself from the upper surface of the crawler shed skin, which becomes an integral part of the scale cover. After further cover enlargement with wax produced by pygidial wax glands (Figs. 1D, 2B and G), the dorsal derm of the second instar hardens and darkens similar to the process described for the crawler. The shed skin of the second instar is located under the crawler shed skin and is incorporated in the wax cover. The adult female completes cover enlargement with wax produced by glands on the pygidium (Figs. 1E, 2C and H). The newly produced wax is molded by structures at the end of the female body similar to the way a trowel is used to form concrete. The body of the female pivots in characteristic ways around the inserted mouthparts to form elongate (Plate 1J; Fig. 2H) or rounded (Plate 1A; Figs. 1E and 2C) scale covers characteristic of the species. If mated, adult females of many species, such as *Abgrallaspis ithacae* (hemlock scale), *Melanaspis obscura* (obscure scale), *Nuculaspis californica* (black pineleaf scale), and *Furcaspis biformis* (orchid scale) (Plate 58A; Fig. 2C and H), produce a thin cover extension or flap several days before crawlers are ready to emerge from the female cover. The flap has a horizontal slitlike opening that allows the emerging

crawlers access to the outside. In pupillarial scales, a slit forms at the posterior end of the second-instar shed skin for the same purpose.

Male cover formation

Male covers are simpler in construction (Figs. 1H, 2J and K) than those of the female and can be made either from the same materials as the female (Plate 28C; Figs. 1H and 2J) or from a white felted wax (Plate 19C; Fig. 2K) that is very different in appearance from other armored scale covers. Male aspidiotine crawlers generally form a distinct white cap (Fig. 1G) and diaspidines produce the nearly transparent covering similar to that produced by the female crawlers. The process of scale enlargement and molting to the second instar also is similar. From this point, developmental and cover formation processes differ significantly from the female. Bodies of second-instar males are narrower and longer than those of the female and this difference is reflected in the more elongate scale cover of the second-instar male. Second-instar covers of many genera of armored scales are constructed of materials that are similar to those of the female and can be distinguished only by the shape, for example, *Lepidosaphes* and all aspidiotines. However, a few genera of diaspidines form felted white covers that are conspicuously different from the second-instar female cover—of the economic armored scales, felted wax covers occur in *Aulacaspis, Chionaspis, Diaspis, Fiorinia, Pinnaspis, Pseudaulacaspis,* and *Unaspis.* These often have either one or three ridges (sometimes called carinae) that run the length of the cover, but in some instances no ridge is clearly visible. After both kinds of second-instar covers are made, a small posterior wax flap is produced when the cover is near completion. This flap serves as a way for the adult male to exit. Third, fourth, and adult male instars do not add to the cover but develop within the structure constructed by first and second instars. Unlike the second-instar female, the shed skin of the second-, third-, and fourth-instar males is not incorporated in the scale cover. The process of development to the adult male from a mature second instar takes only a few days or weeks depending on the species.

ECOLOGY

McClure (1990b) reviewed the seasonal history and ecology of armored scales. The rate of development resulting in the number of generations per year (voltinism) exhibited by a species is influenced primarily by temperature and humidity. Thus, the same species may have 2 generations in the mountains and 3 or more in the valleys of the same region, for example, *Parlatoria oleae* (olive scale) (Applebaum and Rosen 1964). In the laboratory, *Pseudaulacaspis pentagona* (white peach scale) completes a generation in 110 days at 13 °C, and 40 days at 26 °C (Ball 1980). In Maryland this species has 3 generations per year, whereas in Florida there are up to 5 generations per year. Armored scales usually have 1 to 3 generations per year, but more are possible in the warm south.

Because plants also use temperature and humidity as developmental cues, the developmental stages of plants (phenology) can be used as predictors of scale insect development, particularly for first-generation activity in the spring. For example, Mussey and Potter (1997) showed that egg hatch occurred in *Chionaspis pinifoliae* (pine needle scale) when *Crataegus viridis* was in 50% bloom, and eggs hatched in *Unaspis euonymi* (euonymus scale) when *Crataegus viridis* was in 95% bloom. Physiological development time is calculated in day-degrees (DD), which is defined as the sum of the difference between the average daily temperature and a cited threshold, the zero development point of the species. Potter et al. (1989) found that first crawler hatch for *Melanaspis obscura* (obscure scale) in Lexington, Kentucky, corresponded to a mean accumulation of 1521 DD calculated from a threshold of 40°F. From 1984–1986 this ranged from 22 June to 6 July. Some other armored scales for which day-degree development is known include *Lepidosaphes ulmi* (oystershell scale) (Mussey and Potter 1997), *Diaspidiotus perniciosus* (San Jose scale) (Huba 1962), *Chionaspis pinifoliae* (pine needle scale) (Mussey and Potter 1997), *Aonidiella aurantii* (California red scale) (Hoffman and Kennett 1985), *Carulaspis juniperi* (juniper scale) (Mussey and Potter 1997), *Diaspidiotus uvae* (grape scale) (Johnson et al. 1999), *Diaspidiotus juglansregiae* (walnut scale) (Mussey and Potter 1997), *Pseudaulacaspis pentagona* (white peach scale) (Mazzoni and Cravedi 1999), *Hemiberlesia rapax* (greedy scale) (Blank et al. 1995), and *Unaspis euonymi* (euonymus scale) (Savopoulou-Soultani 1997; Mussey and Potter 1997).

The overwintering stage varies among species in genera such as *Diaspidiotus*, and even within a species in different geographic regions and climates. For example, *Diaspidiotus perniciosus* (San Jose scale) overwinters as crawlers in Oregon (Westigard 1979; Jorgensen et al. 1981) and second instars in New Zealand (Wearing 1976). Most U.S. species overwinter as second-instar males and females or fertilized females.

Host-plant specificity is highly variable among species. There are a few amazing generalists, such as *Aspidiotus nerii* (oleander scale), which has been recorded from more than 100 plant families (Beardsley and González 1975); Borchsenius (1966) and Dekle (1977) treat 5 scale species that are recorded from more than 200 genera of host plants including: *Aonidiella aurantii* (California red scale) (202 genera), *Aspidiotus nerii* (oleander scale) (235), *Chrysomphalus aonidum* (Florida red scale) (259), *Diaspidiotus perniciosus* (San Jose scale) (240), and *Hemiberlesia lataniae* (latania scale) (276 genera). Conversely, a surprisingly large number of species are restricted to a limited range of host plants. In the data presented here there are 38 economically important species that have been reported on fewer than 10 plant families: *Abgrallaspis degenerata* (degenerate scale), *A. ithacae* (hemlock scale), *Aonidiella taxus* (Asiatic red scale), *Aspidiella sacchari* (sugarcane scale), *Aspidiotus cryptomeriae* (cryptomeria scale), *Aulacaspis rosae* (rose scale), *Aulacaspis yasumatsui* (cycad aulacaspis scale), *Carulaspis juniperi* (juniper scale), *C. minima* (minute cypress scale), *Chionaspis corni* (dogwood scale), *C. heterophyllae* (pine scale), *C. pinifoliae* (pine needle scale), *Comstockiella sabalis* (palmetto scale), *Cupressaspis shastae* (redwood scale), *D. liquidambaris* (sweetgum scale), *Diaspidiotus osborni* (Osborn scale), *D. ostreaeformis* (European fruit scale), *Fiorinia externa* (elongate hemlock scale), *F. japonica* (coniferous fiorinia scale), *Froggattiella penicillata* (penicillate scale), *Furchadaspis zamiae* (zamia scale), *Kuwanaspis pseudoleucaspis* (bamboo diaspidid), *Lepidosaphes camelliae* (camellia scale), *L. pallida* (Maskell scale), *L. pini* (pine oystershell scale), *L. yanagicola* (fire bush scale), *Melanaspis lilacina* (dark oak scale), *M. obscura* (obscure scale), *Nilotaspis halli* (Hall scale), *Nuculaspis californica* (black pineleaf scale), *N. tsugae* (shortneedle conifer scale), *Odonaspis ruthae* (Bermuda grass scale), *Parlatoria blanchardii* (parlatoria date scale), *P. ziziphi* (black parlatoria scale), *Quernaspis quercus* (oak scale), *Selenaspidus albus* (white euphorbia scale), and *Unaspis citri* (citrus snow scale), *U. euonymi* (euonymus scale). Most conifer-, grass-, and palm-infesting species have a limited host diversity.

How 1- to 2-mm-long armored scale insects are able to obtain food through relatively thick-barked plants is an interesting phenomenon. The feeding mechanism has been reviewed by Beardsley and González (1975) and McClure (1990a). The threadlike mouthparts (stylets) are composed of two outer, hardened filament-like blades (mandibles) that saw into the plant tissue. These enclose two thicker filaments (maxillae) that lock together to form two canals. The salivary canal transmits salivary secretions into the plant, and the food canal carries plant-cell contents into the scale body. In honeydew-producing scales like soft scales and mealybugs, the food canal primarily carries large quantities of plant sap from the phloem. In armored scales the food canal contains the contents of various kinds of cells and therefore does not transport large quantities of liquid. Because they ingest small quantities of liquid, armored scales do not produce honeydew. It also is important to understand that only the first two instars of the male feed. The prepupa, pupa, and adult male lack mouthparts or they are vestigial.

Armored scales have very long stylets in relation to their body length; sometimes several times longer than the body. The reason for this may be that they need to exploit a large area of plant tissue to obtain sufficient nutrition to survive. The effect of scale insect feeding on plants is not well understood. Feeding effects are most obvious in some leaf-inhabiting species. *Pseudaulacaspis cockerelli* and *Parlatoria ziziphi* (Plates 102C and 96I) appear to probe plant tissues in a straight line causing a discolored stylet track that is longer than the body of the scale. Several other species inhabit leaves or green stems and appear to feed in a circular area around the body, such as *Diaspidiotus ancylus* and *Unaspis euonymi*. The conspicuous yellow or red areas that are produced on the host make it easy for Integrated Pest Management (IPM) specialists to detect pest scale infestations. In some leaf-feeding species no feeding tracks are apparent. When leaves become heavily infested, they usually become discolored or chlorotic or both and eventually die and fall from the host. Similarly, when bark-infesting species reach high population levels, affected leaves begin to drop and eventually dieback occurs, that is, both leaves and branches die.

Armored scales are particularly well adapted to survive in urban conditions. The reasons for this are not completely clear. In one study of *Pseudaulacaspis pentagona* (white peach scale), Hanks and Denno (1993c) demonstrated that the patchy distribution of the white peach scale was best explained by the combined effects of natural enemies and water stress. Scale infestations were lowest on water-stressed mulberry trees where natural enemies were common and highest on trees that grew on sites with adequate water where generalist predators were rare.

MANAGEMENT

Most species of trees and shrubs in the United States are susceptible to infestation by one or more species of armored scale. Scale infestations in natural sites generally remain below damaging levels. For example, we have observed *Chionaspis pinifoliae* (pine needle scale) on the same white pines and mugo pines for 30 years, and the scale populations have never reached pest levels. The interactions of parasites, predators, growing conditions, plant genetics, plant defenses, and pesticide use all are involved in maintaining scale insect populations below damaging levels. When plants are grown commercially or are managed in landscape settings, pesticides are commonly applied to control various pests such as caterpillars and leaf beetles. The results of these treatments can be unintended suppression of the natural enemies of scale insects. For example, on the College Park Campus of the University of Maryland we maintained a map of the distribution of more than 30 scale insect species on ornamental plantings. In the 1970s and early '80s, it was standard procedure to spray pesticides to control several insect pests. For the past 20 or more years, an integrated pest management program (IPM) has been implemented and residual pesticides are almost never used. Unfortunately, for those of us interested in scale insects, the populations of most scale species have virtually disappeared and only rarely can a few individuals be located. Implementation of IPM strategies using the integration of diverse control tactics and horticultural techniques generally can keep scale insect populations below damaging levels. To maintain low scale populations, the following management strategies and tactics should be understood.

Detection

Before plants are purchased they should be carefully examined for scales and other insects. This includes examining the trunk, pulling loose bark, studying wound areas such as pruning scars, and looking at both surfaces of the leaves; essentially inspecting any location where scale insects are found on their hosts. Established plants also should be inspected periodically. Low populations are difficult to detect, but high populations are easier to locate because they often cause leaf mottling and dieback.

Identification

Accurate identification of armored scale pests is a critical first step toward developing effective pest-management systems. Because the biology of most armored scale species differ, it is important to determine the correct identity of the scale so that pertinent information can be located in the literature. Such basic life-history differences as the timing of crawler emergence, number of generations per year, or overwintering stages can have dramatic influence on the development of effective management systems. The timing of application and susceptibility of natural enemies to certain chemical treatments can influence choice of control methods. But this information will not be available unless the correct name of the pest species is known. In some instances it is possible to make determinations of armored scale pests in the field by studying their external appearance, distribution, location on the host, and host preferences. In fact, one objective of this book is to provide tools that will allow this kind of identification system. We strongly urge, however, that field identifications be confirmed by mounting specimens on microscope slides and examining them with a compound microscope whenever possible. If this is not possible, it may be necessary to submit specimens to someone who can make these kinds of determinations.

Monitoring

Once low scale populations have been detected and identified, they should be visually monitored when crawlers are expected to appear to determine if they are under suppression by predators and parasites or if the population size is increasing and dispersing more widely (see discussion of locating evidence of the presence of natural enemies in the biological control section). If the scale population is increasing or is causing objectionable levels of feeding symptoms, then control tactics may be warranted.

Where insecticide sprays are the tactic of choice, crawler monitoring is necessary. This is done in two ways: (1) Two-sided sticky tape can be wrapped around an infested twig or branch, and wandering crawlers will become tangled in the glue on the margins of the tape, making them easy to observe using a hand lens (Dreistadt et al. 1994); (2) crawlers can be detected by the use of sticky card traps placed in the canopy of infested trees (Dreistadt et al. 1994) to capture crawlers blown by the wind. Tatara (1999) found that the optimum time to spray first-generation *Pseudaulacaspis pentagona* (white peach scale) on tea bushes was two to five days after the peak number of crawlers were captured on sticky traps placed inside the bushes. Detection by sticky traps is most effective for monitoring first-generation emergence of crawlers that hatch over a three- or four-week period, for example, *Aspidiotus cryptomeriae* (cryptomeria scale), *Diaspidiotus perniciosus* (San Jose scale), *Fiorinia externa* (elongate hemlock scale), *Hemiberlesia lataniae* (latania scale), *Pseudaulacaspis pentagona* (white peach scale), and *P. prunicola* (white prunicola scale) (Stoetzel and Davidson 1974). It is important to note that scale parasites can be abundant when crawlers are active (Stoetzel and Davidson 1974; Potter et al. 1989) and also are detectable in crawler sticky traps. Because of this, crawler sprays may be counterproductive when high levels of effective natural enemies are present.

Chemical control

Chemicals that are active only while they are wet, which generally dry in a few hours, are termed biorationals because they have only minor impact on parasite and predator populations. Horticultural oils and insecticidal soaps are two commonly used biorationals. They apparently work by suffocation and penetration of the respiratory system and interfere with various physiological processes (Davidson et al. 1991).

Oils. Some of the first chemicals developed for scale insect control included petroleum oils. Their use began in the late 1800s primarily as dormant bark sprays because they often caused toxic reactions to foliage (phytotoxicity). Relatively good bark-infesting scale pest control was achieved. Improvements were made in the 1920s when quick-breaking oil and water emulsion sprays became available. These new oils were lighter and gave much improved coverage. Oil sprays were largely ignored following World War II when the synthetic organic pesticide industry developed. Eventually, the harmful effects of some of these persistent pesticides, such as bioaccumulation in birds and rampant pesticide resistance in pest species, brought attention back to oils. Interest in environmentally safe oil sprays returned in the 1970s and further research was to have important benefits. The light horticultural oils available today can be sprayed on the bark and foliage of most plants at most times of the year when label directions are followed. Dormant and summer oils are among the most useful materials to reduce population levels of bark-infesting armored scales. Scales with thin and loosely attached covers, such as *Lopholeucaspis japonica* (Japanese maple scale), are easier to control than those with thick and strongly attached covers, such as *Melanaspis obscura* (obscure scale). Armored scales show little or no resistance to oils; beneficial insect populations usually survive oil sprays (Raupp et al. 2001) because they are lethal only when they are wet, and oils dry quickly (Davidson et al. 1991). Oils are inexpensive, easy to use, and safe to handle compared to residual chemicals. The major drawbacks to oil sprays during the growing season are: They cannot be mixed with certain insecticides and fungicides; coverage must be very thorough; they must be applied to dry plants under good drying conditions; and they should not be sprayed on plants stressed by heat, drought, wind, or pest infestation. Davidson et al. (1991) provide an excellent review of the use of oils to control pests on agricultural and ornamental crops.

Soaps. Insecticidal soaps are long-chain fatty acids specifically formulated for high insect mortality. They usually are sprayed on the leaves and bark of infested plants to control crawlers and other soft-bodied pests. As they are nonpersistent, they are active only while they are wet. Like oil sprays, they often are used to encourage predators and parasites and have minimal impact on these natural enemies. Thorough coverage is required and accurate timing at peak crawler emergence is essential. Soaps are believed to kill by disrupting membrane function. They can be inactivated by chemicals in hard water, which causes the fatty acids to precipitate. Buffering and conditioning agents are commercially available that can reduce this problem. A number of plants are sensitive to soaps, so it is important to know which scale hosts are harmed by contact with these chemicals. Soap sprays should be used only during the growing season (Davidson and Raupp 1999).

Synthetic organic insecticides. Several classes of pesticides have been developed with labels for armored scale crawler control. Those that act on the nervous system include organochlorines (e.g., endosulfan), organophosphates (e.g., acephate, dimethoate, azinphosmethyl, malathion, methidathion), carbamates (e.g., carbaryl), and pyrethroids (e.g., bifenthrin, cyfluthrin, fluvalinate, fenvalerate, lambda-cyhalothrin, pemethrin). Most of the compounds have long residual activity, which may depress the beneficial effects of natural enemies unless properly managed (Raupp et al. 2001). Some organophosphates have systemic activity in plants (e.g., acephate, dimethoate). Neonicotinoids (e.g., imidacloprid) are the newest class of insecticides, but they don't seem to work particularly well for armored scales (Shetlar 2002). In fact, they have been reported to suppress populations of scale parasitoids (Sadof and Sclar 2000; Rebek and Sadof 2003). This class of insecticide acts on the nicotinoid receptor sites in the nervous system resulting in the inhibition of feeding. Imidacloprid is a systemic insecticide with slow uptake but long activity time in woody plants when applied by soil drench or tree injection. Soil-drench applications of this insecticide are suggested to be less effective than crawler sprays because the low mobility of this material makes it difficult for scales to receive sufficient toxicant from the contents of damaged cells.

Armored scales are generally more difficult to control than other scale insects using classical contact or systemic insecticides. Contact spray may work well on crawlers and adult males providing that: Applications are accurately timed using an appropriate monitoring technique; a chemical is available that will have little or no impact on natural enemy populations; coverage is thorough; and the scale is not resistant to the chemical. Instars that are protected by the scale cover, particularly second instars and adult females, are relatively unaffected by contact sprays. When attempting to reduce armored scale populations in the older instars, the scale feeding site should be carefully considered. In our experience, species that primarily feed through the bark are not affected by most contact sprays, soil drenches, or bark injections of systemic insecticides. Armored scales that principally feed on leaves and have not begun to lay eggs, however, may be controlled with some systemic materials used as leaf sprays, bark injections, or soil drenches (Rebek and Sadof 2003).

It is interesting that Raupp et al. (2001) found that excessive use of insecticides actually increased the number of scale species in urban plantings. When ornamental landscapes were treated with persistent cover sprays for more than four consecutive years, nine species of armored scales were present at the end of the experiment; when similar sites were treated the same way but for less than four years they had only four species of armored scales.

Growth regulators. Darvas and Varjas (1990) reviewed growth-regulator effects on armored scales. Two types of insect growth regulators are used as insecticides and cause lethal effects on the developmental or reproductive physiology of the scale insect. They usually act in a specific window of time in a susceptible stage of the scale's life cycle. Chitin synthesis inhibitors, such as buprofezin, prevent molting and act best on crawlers; the effectiveness decreases with age. Adults that survive such treatments have been shown to produce fewer eggs. Synthetic compounds that mimic juvenile hormones, such as s-kinoprene and pyriproxyfen, also interfere with molting. Early instars suffer the highest mortality from treatment with this kind of compound. Properly used insect growth regulators do not have serious effects on some parasites but may interfere with the process of pupation in some lady beetle predators.

Pheromones. Adult female armored scales are sedentary and are enclosed within their covers for the duration of their existence. Because of the protection afforded by this scale cover, the short-lived but active adult males cannot visually or tactilely locate the female for mating. To overcome this impediment, virgin females produce airborne sex pheromones that attract males to their scale cover. Gieselmann and Rice (1990) reviewed the use of armored scale pheromones for managing pest species. They discuss four species for which the pheromones have been identified and synthesized: *Aonidiella aurantii* (California red scale), *A. citrina* (yellow scale), *Pseudaulacaspis pentagona* (white peach scale), and *Diaspidiotus perniciosus* (San Jose scale). Interestingly, late-instar male scales are sensitive to pesticide sprays, and pheromone traps can be used to time sprays to take advantage of this sensitivity. Rice et al. (1979) found that male sprays are effective for control of San Jose scale in the western United States but were not compatible with other IPM programs and pollinators. Downing and Logan (1977) found that male sprays for *D. perniciosus* (San Jose scale) were effective and feasible on apples in western Canada. Armored scale sex pheromones also function as kairomones that attract certain scale parasites.

Scale resistance to insecticides. Armored scales were among the first insects to exhibit insecticide resistance. One of the most serious pests in early American agriculture was *Diaspidiotus perniciosus* (San Jose scale). Among the materials first used to control this devastating pest were lime sulfur and hydrocyanic acid (HCN). It soon became apparent that the impact of these pesticides in controlling *D. perniciosus* (San Jose scale) lessened through time. Melander (1914) demonstrated resistance of this pest to lime sulfur and recognized that levels of coverage with lime sulfur and genetic factors in the scale were involved in causing this resistance. *Aonidiella aurantii* (California red scale) resistance to HCN was reported by Dickson (1941), who found that it was a sex-linked, incompletely dominant genetic trait. Following the 50-year use of organophosphate and carbamate insecticides in California, insecticide resistance in armored scales is deeply entrenched. Grafton-Cardwell et al. (2001) found that 40% of 163,000 acres of citrus contained *Aonidiella aurantii* (California red scale) and *A. citrina* (yellow scale) that were resistant to insecticides in these classes.

Biological control

Scale insects were among the most serious pests encountered by early American farmers. *Diaspidiotus perniciosus* (San Jose scale), introduced from the Orient, so devastated fruit tree crops in the 1800s that it became responsible for the creation of the first U.S. quarantine laws. Modern pesticides were yet to be invented, hence, the only recourse was to send entomologists to countries where scales were native and under control to search for natural enemies, primarily predaceous and parasitic insects. Fortunately, the first attempt at introducing a natural enemy succeeded magnificently and the vedalia ladybird beetle from Australia controlled the cottony cushion scale (a margarodid scale) and saved the California citrus industry. Thus began the field that we now call biological control.

In the intervening years many biological control projects have been initiated to control arthropod pests and weeds. Hall et al. (1980) analyzed data in Clausen (1978) who discussed a series of biological control projects. Hall et al. found that 16% of the projects produced effective control of the pest organism and 42% gave partial control. According to DeBach et al. (1971), one-fifth of all successful biological control projects have targeted armored scale insects. Based on this, it appears that armored scale pests are unusually susceptible to suppression by natural enemies.

The major natural enemies of armored scales are ectoparasites (primarily wasps in the aphelinid genus *Aphytis*), endoparasitic wasps in the families Aphelinidae and Encyrtidae, and predators including certain groups of beetles such as lady beetles, lacewings, gall flies, thrips, mites, and entomopathogenic fungi. Rosen and DeBach (1990) considered *Aphytis* species to be the most effective biological control agents of armored scale insects. Van Driesche and Bellows (1996) asserted that half of all successful biological control projects that employed introduction strategies were accomplished by using aphelinid and encyrtid parasites.

All known ectoparasitic wasps of armored scales are in the large genus *Aphytis* (Rosen and DeBach 1990). These wasps are about 1 mm long. Adult females search for armored scales primarily using chemical cues. Once a scale is located, the female lowers her ovipositor and drills a hole through the scale cover (Plate 2A) and into the body of the scale. The wounded scale then 'bleeds' body fluid through the opening (Plate 2B), which the female consumes (Plate 2C). Several kinds of parasites wound scales for the purpose of feeding on the scale body fluids. This habit is highly destructive to most scales as the puncture they sustain in this process is usually lethal. When the female is ready to lay an egg she does so by placing the egg on the body of the scale using an adhesive pad to keep it in place (Rosen and DeBach 1990). The wasp larva remains attached to the scale and grows as it extracts nutrients from the slowly deflating scale (Plate 2D). Results of this behavior are a mature *Aphytis* larva and a shriveled scale (Plate 2E). The larva pupates under the wax cover of the scale. When an adult emerges, it chews its way through the cover leaving a circular emergence hole (Plate 2I). These wasps are very effective biological control agents because they often have three generations to each single generation of the scale pest and the intrinsic rate of increase may be three to five times that of the scale (Abdelrahman 1974).

Female endoparasitic wasps locate host scales in the same manner but differ by inserting the egg into the body of the scale. The egg hatches inside the scale and the larva consumes the body contents (Plate 2F shows a cleared female *Diaspidiotus ancylus* [Putnam scale] containing a mature wasp larva). Scales that contain endoparasitic wasp larvae often can be distinguished by a translucent central area and a blackened or melanized marginal area (Plate 2G shows a parasitized adult female *Melanaspis obscura* [obscure scale]). The dark form of the wasp pupa is quite easy to distinguish through the skin of the dead scale body (Plate 2G). Plate 2H shows a wasp pupa

within the skin of a *Pinnaspis aspidistrae* (fern scale). Examples of adult wasps of three parasite species that attack armored scales are shown on Plate 2J, K, and L.

Lady beetles are major predators of scale insects. Feeding strategies among these beetles are highly varied, but some groups feed primarily on aphids, some concentrate on mites, while others prefer scale insects (Gordon 1985). Within the scale insect feeders, there are some that specialize on armored scales. Unfortunately, only a few of the more than 4000 described species of lady beetles have been studied in regard to host preferences (Drea and Gordon 1990). In North America *Axion*, *Chilocorus*, *Coccidophilus*, *Exochomus*, *Hyperaspis*, *Microweisea*, *Rhizobius*, *Scymnus*, *Zagloba*, and *Zilus* contain species that have been reported to feed on armored scales, with *Chilocorus* feeding on the most species (Drea and Gordon 1990). *Chilocorus* species are generally black with red spots and are difficult to determine to species. Plate 3E depicts *C. stigma* on the left and *C. kuwanae* on the right. The first is native to North America while the second was introduced to control *Unaspis euonymi* (euonymus scale). *Chilocorus kuwanae* eggs are placed under scale covers of their prey (Plate 3A) or in bark cracks. The early instar larva (Plate 3B) is tan with black spots that bear spines. Mature larvae fasten themselves to the substrate to pupate and the pupa remains in the split larval skin (Plate 3C). Once adults emerge, they seek and feed on armored scales such as *Unaspis euonymi* (euonymus scale) (Plate 3D).

Although *Chilocorus* species are among the largest lady beetles that feed on armored scales, *Rhizobius* contains several small species that are about the size of the scale upon which they feed. *Rhizobius lophanthae* is found throughout the southern United States and is commercially available for release in conservatories and greenhouses. Eggs (Plate 3F) are placed under and among the host scale covers. Larvae (Plate 3G) are small and cryptic while they feed in host colonies. Pupae (Plate 3H) are naked except for a few defensive secretory hairs. Adults (Plate 3I) are black with a reddish thorax and are covered with fine hairs.

Drea (1990a) reviewed the importance of seven beetle families that contain species that are occasional predators of armored scales. Of these, *Cybocephalus* species in the sap beetle family Nitidulidae is the most important. Species of this genus have been recorded on 30 different genera of armored scales and have been used effectively in the biological control of some pest species.

Sixteen species of neuropterans in 9 predaceous genera have been reported to feed on 21 species of armored scales in the United States. Only one, the dusty wing *Aleuropteryx juniperi* Ohm, which feeds on *Carulaspis juniperi* (juniper scale), is considered to be important (Drea 1990b).

Nine thrips species have been reported to prey on armored scales, but their impact on scale populations has never been seriously studied. Only one species is characterized as feeding primarily on scale insects: *Aleurodothrips fasciapennis* (Franklin) (Palmer and Mound 1990).

Approximately 10 families of mites contain species known to prey on armored scales. Only one, *Hemisarcoptes malus* (Shimer), is considered to be a significant biological control agent. Interestingly, this mite produces a special, non-feeding nymphal instar that attaches itself to *Chilocorus* lady beetles in order to move from one scale colony to another (Gerson et al. 1990).

Flies generally are not natural enemies of armored scales although 16 species of cecidomyiid flies or gall midges have been reported to feed on a diverse array of diaspidids. These flies are part of a group of gall midges that are predators on small

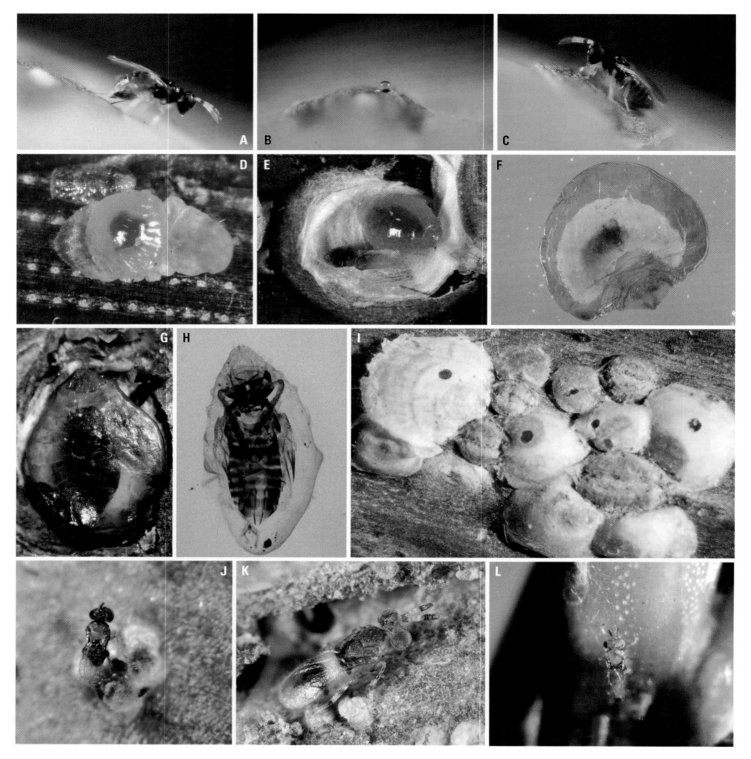

Plate 2. Biological Control (wasp parasites)

A. Female *Coccobius* wasp piercing *Aonidiella aurantii* (California red scale) cover with ovipositor (M. Rose).

B. Droplet of hemolymph exuding from wound caused by *Coccobius* ovipositor puncture (M. Rose).

C. Female *Coccobius* wasp feeding on hemolymph from ovipositor puncture (M. Rose).

D. Young ectoparasitic *Aphytis* larva feeding on *Chionaspis pinifoliae* (pine needle scale) (J. A. Davidson).

E. Mature ectoparasitic *Aphytis* larva beside consumed female of *Hemiberlesia lataniae* (latania scale) (J. A. Davidson).

F. Compound microscope view of female *Diaspidiotus ancylus* (Putnam scale) containing endoparasitic wasp larva (J. A. Davidson).

G. Endoparasitic wasp pupa in female *Melanaspis obscura* (obscure scale) (J. A. Davidson).

H. Compound microscopic view of endoparasitic wasp pupa shed skin inside female of *Pinnaspis aspidistrae* (fern scale) (J. A. Davidson).

I. Parasitic wasp emergence holes in scale covers of *Pseudaulacaspis pentagona* (white peach scale) (J. A. Davidson).

J. Parasitic wasp, *Encarsia berlesei* on *P. pentagona* (white peach scale)) (J. A. Davidson).

K. Parasitic wasp, *Coccobius varicornis* on *Diaspidiotus ancylus* (Putnam scale)) (J. A. Davidson).

L. Parasitic wasp, *Aphytis hispanicus* on *Lepidosaphes pallida* (Maskell scale)) (J. A. Davidson).

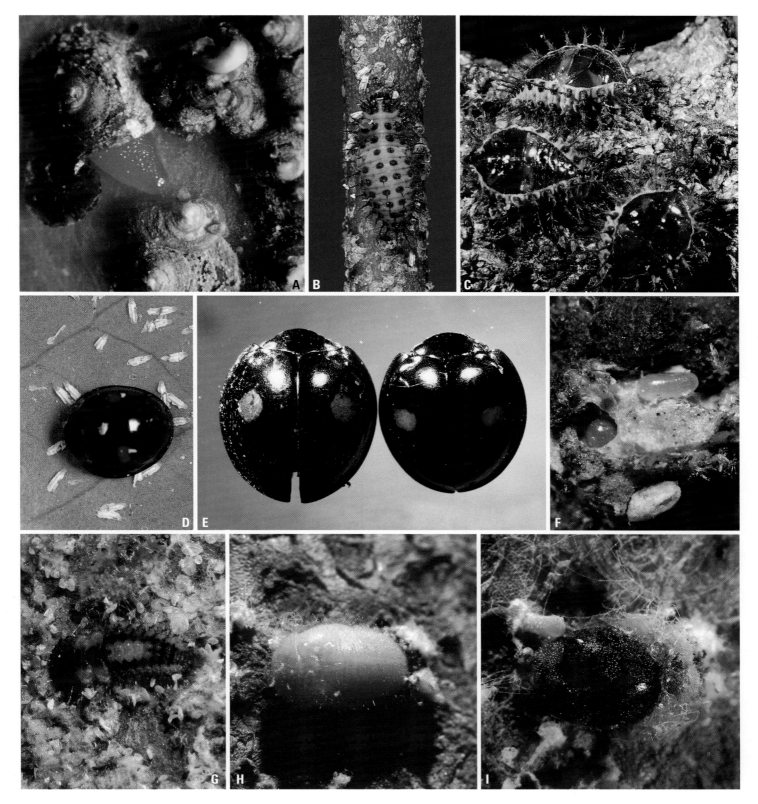

Plate 3. Biological Control (lady beetles)

A. Egg of *Chilocorus kuwanae* inserted under scale cover (M. J. Raupp).

B. Young larva of *C. kuwanae* feeding on *Unaspis euonymi* (euonymus scale) (M. J. Raupp).

C. Pupae of *C. kuwanae* in larval shed skins (J. A. Davidson).

D. Adult *C. kuwanae* feeding on *U. euonymi* (euonymus scale) (M. J. Raupp).

E. Adult *Chilocorus stigma* (left) compared to adult *C. kuwanae* (right) (J. A. Davidson).

F. Egg of *Rhizobius lophanthae* on cover of *Pseudaulacaspis pentagona* (white peach scale) (J. A. Davidson).

G. Larva of *R. lophanthae* feeding on *P. pentagona* (J. A. Davidson).

H. Pupa of *R. lophanthae* on *P. pentagona*; note droplets on ends of hairs (J. A. Davidson).

I. Adult of *R. lophanthae* feeding on *P. pentagona* (J. A. Davidson).

insects; it is not clear whether they are closely tied to armored scales or are more general predators (Harris 1990).

Three types of biological control strategies are recognized depending on the method of implementation: (1) Importation of exotic species to control introduced pests (classical) (Rose and DeBach 1990b); (2) enhancement of the impact of existing species by mass rearing or other augmentative techniques (augmentation) (Rose 1990a); and (3) conservation of existing species by minimizing pesticide usage and managing biotic factors (conservation) (Rose and DeBach 1990a). Classical methods have been the most effective but also have been used for the greatest duration. When Albert Koebele introduced the vedalia beetle to California for control of the cottony cushion scale on citrus, it took only two years to gain statewide control of the pest (Van Driesche and Bellows 1996). It is important to point out that when this famous biological control project had such great success, synthetic insecticides had not been invented. Augmentation strategies generally have been used with biological control agents that have been successful but their populations have been depleted by unfavorable climatic conditions or chemical controls. To control *Aonidiella aurantii* (California red scale) in citrus in California this strategy is regularly used to replenish the pesticide-depleted populations of parasitic wasps that keep this pest in check (Flint 1984). Conservation techniques are essential parts of most successful IPM programs and are most effective for native scale pests because the full complement of beneficial organisms are potentially available for suppression strategies (Van Driesche and Bellows 1996). Perhaps the most important conservation technique in armored scale management is pesticide selection and use. Biorational sprays, such as oils, soaps, and BT, should be used whenever possible rather than broad-spectrum residual sprays, and these should only be applied as spot sprays to infested plants rather than as cover sprays to all plants to protect reservoirs of beneficials. Spot sprays can reduce pesticide use by 90% in landscape (Holmes and Davidson 1984) and nursery systems (Davidson et al. 1988).

Cultural control

Flint and Gouveia (2001) reviewed cultural practices used in IPM programs; those that pertain to armored scale control are summarized as follows. (1) Proper site selection for planting ornamentals is an important aspect of cultural control because plants poorly adapted to growing sites can be more prone to serious insect attack. (2) Sanitation is an important cultural control practice. When branches of trees and shrubs are found to be heavily encrusted with scales, the contaminated portion of the host should be removed as quickly as possible so that it doesn't serve as a reservoir for new infestations in the area. Effects of oil sprays can be enhanced on lightly encrusted areas of some hosts, such as Japanese flowering cherries contaminated with *Pseudaulacaspis prunicola* (white prunicola scale), by abrading with a moderately stiff brush prior to oil applications. (3) Alternate host destruction generally is not considered practical for armored scale management since most infested plants are trees and shrubs that are prized in the landscape. Where *P. pentagona* (white peach scale) is a perennial pest of important ornamental and fruit crops, however, and where white mulberry serves as a reservoir for recontamination of these plants, it may be of value to rogue out the weedy infestation sources. (4) Habitat modification can have significant impact on scale infestations when causes of population buildup are understood. For example, scale-susceptible hosts should never be located near unpaved roads as dust particles have a significant, detrimental impact on parasite populations. Appar-

ently the dust particles kill the parasites but do not affect the scale insects protected by a scale cover (Rose and DeBach 1990b). It also is known that crawlers wander for a shorter period of time on dusty leaves than on clean leaves (Willard 1973). (5) Several management strategies that are purported to enhance the beauty and health of the landscape often cause more harm than benefit. The plant culture activity that seems to have the greatest benefit to armored scale population explosion is fertilization. McClure (1980b) demonstrated that when hemlocks were fertilized with ammonium nitrate, *Fiorinia externa* (elongate hemlock scale) developed faster, laid more eggs, and suffered less mortality. The same benefits pertain to other sucking pests, such as leafhoppers, aphids, and mites (Flint and Gouveia 2001). One of the problems is that commercial nursery growers are anxious to produce healthy-looking plants for the marketplace as rapidly as possible. To accomplish this, they often apply fertilizers far above the recommended rates (Gill 1996). The end product is that the nursery grower, the distributor, and the homeowner end up on the infamous 'pesticide treadmill.' High fertilizer rates encourage plants to produce luscious, nitrogen-rich foliar growth at the expense of valuable root growth and often plant defenses (Herms and Mattson 1992). Sucking insects increase at more rapid rates than normal, which causes increased application of insecticides.

Host-plant resistance

Resistance to pest infestation often is thought to be a genetic trait of the plant at the cultivar level (Flint and Gouveia 2001). Many resistant plants repel the scale or provide insufficient resources for survival. Others allow pest feeding and survival but show no obvious long-term effects and are considered to be tolerant.

Little is known about plant resistance or tolerance in armored scales. Williams and Greathead (1990) reviewed the effect that different growth habits of sugar cane varieties have on population densities of *Melanaspis glomerata* and *Aulacaspis tegalensis*. Crawlers settle in protected areas of the leaf sheaths that surround the main stalk of the plant. Cultivars that developed the heaviest scale infestations had leaf sheaths that were closely attached to the stalk even after death of the plant. The so-called free-thrashing cultivars have their leaf sheaths weakly attached to the stalks, allowing parasites and predators access to the scale infestations and better control of populations of the pests. Sadof and Neal (1993) showed that *Unaspis euonymi* (euonymus scale) that fed on yellow leaf areas of the variegated cultivar *Euonymus japonica* 'Aureus' produced 22% fewer oocytes as compared with individuals that fed on green leaf areas. They suggest that reduced oocyte production was due to the relatively poor nutritional quality of yellow leaves. Yet, because of a 17% increase in oocyte production by scales feeding on stems of variegated plants and a 33% reduction in plant growth, it is difficult to say that variegated cultivars are resistant to euonymus scale (Sadof and Raupp 1991). Johnson and Lyon (1988) mention several species of euonymus that are resistant to *U. euonymi* (euonymus scales) and tout *Euonymus kiautshovica* as highly tolerant to attack by this pest. Warner (1949) notes similar resistance in the Arnold Arboretum where plants of *Euonymus japonica* were heavily infested with the scale, but plants of *E. alata*, *E. sachalinensis*, and *E. sanguinea* that were interplanted with infested plants were completely free of the scale. In Maryland, Gill et al. (1982) found heavily infested *E. japonica* plants growing among *E. kiautshovica*. After 3 years several of the *E. japonica* plants were dead, but the tolerant *E. kiautshovica* plants had only light infestations, never showed damage symptoms, and grew on the site for 16 more years. We have examined specimens of *U.*

euonymi (euonymus scale) from 11 species of euonymus, but most are apparently tolerant because reports of damage on most of these species are rare. A clear method for alleviating problems from this scale is to avoid planting susceptible species like *E. japonica* and choosing tolerant or resistant species such as *E. alata* or *E. kiautshovica*. We are quick to mention, however, that *E. alata* may be resistant to *U. euonymi*, but it is susceptible to *Lepidosaphes yanagicola* (fire bush scale).

An interesting area of plant–scale interactions was first hypothesized by Edmunds (1973) and is called *demic adaptation*. More information on this theory is presented in the biology section of *Nuculaspis californica* (black pineleaf scale). Briefly, it appears that movement of genes among scale colonies is restricted by their limited mobility. As such, when host plants are separated in space, it is difficult for genes to flow among scale colonies because of their genetic isolation. Thus, it is likely for isolated scale colonies to develop unique genetic populations over time. Research results relating to this theory have been mixed (Alstad 1998; Boecklen and Mopper 1998; Hanks and Denno 1994; Wainhouse and Howell 1983).

Integrated Pest Management (IPM)

Concerns about the environmental impact of pesticides, pesticide resistance, and pesticide contaminated foods have forced control specialists to develop and implement an ecological approach to pest suppression called IPM. This technique seeks to maintain long-term control of pest species by integrating management strategies such as biological control, habitat modification, manipulation of cultural practices, and use of resistant varieties of affected plant species. Pesticides are used only when no other cost-effective method will work. Pesticides are chosen and applied to minimize their effects on beneficial organisms, human health, and the environment. Because of the relatively high susceptibility of armored scales to suppression strategies with natural enemies, they also are effectively controlled using IPM approaches. Flint and Gouveia (2001) present an excellent review of IPM methods and techniques for use by pest management specialists working in agricultural crops, urban landscapes, greenhouses, and other managed ecosystems. Methods that have been employed to manage pests in citrus using IPM techniques are described in detail by Flint (1984).

ECONOMIC IMPORTANCE

Armored scales are mainly pests of perennial plants in managed systems. They may cause considerable damage and economic loss in environments such as fruit and nut tree orchards, nurseries, greenhouses, landscapes, and forests (Miller and Kosztarab 1979).

Kosztarab (1990) reviewed the effects of scale insect feeding on plants and recognized 11 types of damage depending on the scale species involved and its preferred feeding site. Damage symptoms generally are most evident when feeding occurs on leaves or fruit. Species such as *Unaspis euonymi* (euonymus scale) on euonymus, *Aonidiella aurantii* (California red scale) on citrus, and *A. taxus* (Asiatic red scale) on podocarpus remove chlorophyll in a circle around the scale and create a yellow halo. Other species such as *Pseudaulacaspis cockerelli* (false oleander scale) on palm, *Aulacaspis tubercularis* (white mango scale) on mango, and *Fiorinia theae* (tea scale) on camellia feed in one direction on a leaf forming a narrow chlorotic area (called a stylet track). This track may become darker with age. Many scale species prefer to feed on the lower surface of leaves, for example, *Fiorinia externa* (elongate hemlock scale), *Aspidiotus*

cryptomeriae (cryptomeria scale), and *Abgrallaspis ithacae* (hemlock scale), but their presence is normally evidenced by the chlorotic blotches that are visible on the upper leaf surfaces. Some leaf feeders do not produce obvious yellow areas, for example, species such as *Carulaspis juniperi* (juniper scale) on juniper, *Diaspis bromeliae* (pineapple scale) on bromeliads, and *D. echinocacti* (cactus scale) on cactus. A few leaf-feeding species produce reddish halos, such as *Diaspidiotus perniciosus* (San Jose scale) on several hosts, *D. ancylus* (Putnam scale) on blueberry, and *Parlatoria oleae* (olive scale) on mahonia. Regardless of the scale species or the feeding symptoms, heavily infested leaves usually turn yellow or brown and die and fall from the host. Some armored scale species even produce red halos on fruit, including *Diaspidiotus perniciosus* (San Jose scale) on apple, *D. ancylus* (Putnam scale) on green blueberries, and *Hemiberlesia lataniae* (latania scale) on cotoneaster. This kind of damage can reduce the value of the fruit by rendering it unsaleable in fresh markets.

Armored scales that feed on and under bark do not usually cause obvious damage symptoms, but some species cause reddish discoloration of the wood under the bark where the scale has been feeding, evidently caused from a reaction to the scale's salivary secretions, for example, *Diaspidiotus perniciosus* (San Jose scale) on almond (Plate 44I) and *D. ancylus* (Putnam scale) on blueberry (Plate 37G). A few species, such as *Epidiaspis leperii* (Italian pear scale) and *Diaspidiotus gigas* (poplar scale), which feed primarily on the bark, cause depressions or pits under the body of the scale (Kosztarab and Kozár 1978). We often have observed this kind of host distortion on pin oaks in Maryland (Plate 79D). One Maryland nursery rogued out an entire planting of pin oak saplings because the bark on the young trees was heavily gnarled by *Melanaspis obscura* (obscure scale) (Cornell, personal communication 1988).

Heavy scale infestations can become so dense that they cover the bark of twigs and branches and frequently are associated with dieback. This kind of damage is commonly visible when branches of Japanese flowering cherry are heavily encrusted with the white covers of *Pseudaulacaspis prunicola* (white prunicola scale) (Plate 104H). We have observed healthy 20-foot trees that apparently died within 3 or 4 years because of heavy infestation of this pest (Plate 104I).

Perhaps the most damaging of all armored scales are those that attack all parts of the plant: fruit, leaves, stems, and branches. *Aonidiella aurantii* (California red scale) infests citrus in this manner and is considered the most potentially injurious insect pest of California citrus (Flint 1984).

Several authors have estimated economic losses due to scale insect pests on a particular crop system. For example, a loss of $3.85 million in Texas on citrus due to *Chrysomphalus aonidum* (Florida red scale) (Anonymous 1977); $7 million in Florida on citrus due to *Unaspis citri* (citrus snow scale) (Kosztarab 1990); $22.8 million in California on citrus due to several scale pests (Hawthorne 1975), and more than $37 million in Georgia due to all scales on ornamental plants, lawns, and turf (Ellis and Howell 1982). Kosztarab (1977) estimated $500 million in annual economic loss in the United States due to all scale insects and in 1990 he estimated that this figure had risen to as much as $5 billion annually. If these figures are correct, extrapolation to armored scales alone probably would exceed $2 billion annually.

MORPHOLOGY

The presence of a scale cover in armored scales has brought about the development of a unique set of morphological structures. Compared with most other scale insects, the body struc-tures that are present are greatly reduced. For example, in adult females, legs are absent and antennae are reduced to unseg-mented tubercles. Body segmentation is difficult to discern, and the posterior segments are sclerotized and partially coalesced into a structure called the pygidium. Remnants of wax-produc-ing structures are visible on slide-mounted specimens even when the glandular components have dissolved in the mount-ing process. Other specialized structures at the posterior end of the body include gland spines, setae, plates, and lobes. The form, shape, and size of these structures are important in iden-tifying species and are discussed in more detail below.

To see these structures, good slide mounts must be made using the techniques discussed under slide-mount preparations below. Newly molted adult females make the best mounts and were used for making the line illustrations whenever possible.

Glossary of morphological terms

For the most part, reference to Figures 3, 4, and 5 will be the quickest source of information about the identity, placement, shape, and structure of a diagnostic character. A brief descrip-tion of most major morphological features used in adult female keys and descriptions is included below.

Abdominal segments—the posterior 8 divisions in the body normally represented by a set of setae on each segment. Because the abdomen is modified into a sclerotized pygidium, abdominal segmentation is not always apparent. See Figures 3 and 4 for a representation of the arrangement and placement of body segments. There are slight dif-ferences between the aspidiotine and diaspidine arrangement of seg-ments. The distribution of macro- and microducts can be quite important identification characters, but in order for these to be useful it is important to determine their placement on particular abdominal segments.

Anal opening—a sclerotized opening on the dorsal surface of the pygid-ium. The size and distance of the anal opening from the median lobes are important taxonomic characters.

Antenna—the antenna is represented by an unsegmented, sclerotized projection that bears one or more setae. It generally occurs between the anterior margin of the mouthparts and the anterior margin of the head. The number of setae and rarely the placement of the antennae are useful taxonomic characters.

Basal sclerosis—the large sclerotization that is attached to the medial part of the median lobes and extending toward the anal opening (see Fig. 5). The presence or absence of this structure can be useful taxo-nomically.

Cicatrices—clear or slightly pigmented areas or spots on the dorsolat-eral surface of the body. When present they generally are on the sub-margins of the prothorax and abdominal segments 1 and 3. A few species have more than 3 pairs of cicatrices, and the presence and dis-tribution of these provide useful diagnostic characters. Some scien-tists distinguish between sclerotized bosses and cicatrices, but we treat them as synonymous.

Eye—these structures usually are inconspicuous and are represented by a small pigmented spot or dome, but in a few cases they are either spinelike or absent. Their position on the female body can be a useful taxonomic character.

Gland spines—dermal projections that contain at least one microduct, and usually with a simple apex (see Fig. 5). The gland spines that typ-ically occur along the pygidial margin between the lobes are narrow; those in the submarginal areas of the thorax and anterior abdominal segments are broad and short. Some scientists consider gland spines to be different from gland tubercles, but we consider them to be the same. The distribution of gland spines is occasionally important in species recognition.

Lobes—flat sclerotized projections that are usually lobular in shape and occur on the pygidial margin (see Fig. 5). Generally the lobes have small apical or lateral notches. The number, relative sizes, and shape of the lobes are important characters, as are the presence and distri-bution of notches on the lobes.

Macroducts—these are the internal tubular vestibules of wax glands. They look like cylinders or barrels and have sclerotized rings on the inner end that look like bars when viewed from the side. They are called two-barred or one-barred ducts depending on the number of rings (see Fig. 5). Macroducts generally are on the dorsal aspect of the pygidium but can be scattered over the rest of the dorsal surface; occa-sionally they are present near the submargin of the ventral thorax and anterior abdomen. The distribution of these ducts is important in rec-ognizing certain species.

Median lobes—the pair of lobes located at the posterior end of the pygidium.

Mesothorax—the second segment of the thorax located between the pro-thorax and mesothorax.

Metathorax—the third or posterior-most segment of the thorax located between the mesothorax and abdomen and containing the posterior spiracle.

Microducts—the thin tubular vestibules of wax glands (see Fig. 5). They look like narrow tubes and normally are scattered over the ventral surface, especially near the anterolateral margin of the pygidium, near the spiracles, and near the mouthparts. On the dorsum they usually are present along the body margin of the head, thorax, and anterior portion of the abdomen; a few may be present submedially on the thorax and anterior abdomen. Microducts sometimes are associated with plates and gland spines on the pygidium. Although the distri-bution of these structures can be diagnostic, perfect slide mounts are required or they are difficult to see and therefore are of limited use.

Paraphyses—the slender, sclerotized rods that most often arise from the margins of the lobes but occasionally also are found in the spaces between the lobes (see Fig. 5). They are most obvious in dorsal view and are not always included in ventral aspects of the line drawings. The presence or absence, size, shape, and distribution of these struc-tures are useful taxonomically.

Perispiracular pores—the wax pores that surround the spiracles and usually have 3 or 5 loculi. The presence or absence, numbers, and number of loculi are useful diagnostic characters.

Perivulvar pores—the wax pores that usually have 5 loculi and are in clusters around the vulva. The presence or absence, numbers, and dis-tribution of these pores are useful diagnostic characters.

Plates—the flat dermal projections that have their apices fringed or at least divided (see Fig. 5). Usually they occur between the pygidial lobes and can sometimes be very long and ornate. Some may contain microducts. The numbers, distribution, and shape of the plates can be important taxonomically.

Prothorax—the first or anterior-most segment of the thorax located between the head and mesothorax.

Pygidium—the partially fused and heavily sclerotized posterior 4 or 5 abdominal segments. We have arbitrarily selected the anterior edge of the pygidium as segment 4 for standardization purposes even though it varies between segments 4 and 5. The shape, dermal pattern, and composition of the pygidium are important taxonomic characters.

Second lobes—the pair of lobes located laterad of the median lobes (see Fig. 5). These are usually single in aspidiotines and bilobed in dias-pidines.

Spiracles—two pairs of sclerotized tubes that open on the ventral surface of the thorax. These are conduits of the respiratory system and are not currently considered to be important in diagnosing species.

Yoked median lobes—median lobes that have a sclerotization that ties them together (see Fig. 5).

MATERIALS AND METHODS

Measurements (in millimeters and microns) and counts were taken from 10 specimens collected from as many localities and hosts as possible; they are given in the text as the range followed by the mean in parentheses, for example, 4.1–7.0 (6.6). Micro-scopic examination was undertaken using Wild, Zeiss, and Leica compound microscopes at magnifications of 200X, 400X, and 1000X. Illustrations were made freehand without the use of a camera lucida, projector, or grid.

A few measurements require special mention. The length of the anal opening is the longest length of the anal opening itself, not the length including the sclerotized outside ring. The dis-tance from the anal opening to the base of the median lobes is the dimension from the posterior end of the sclerotized outside

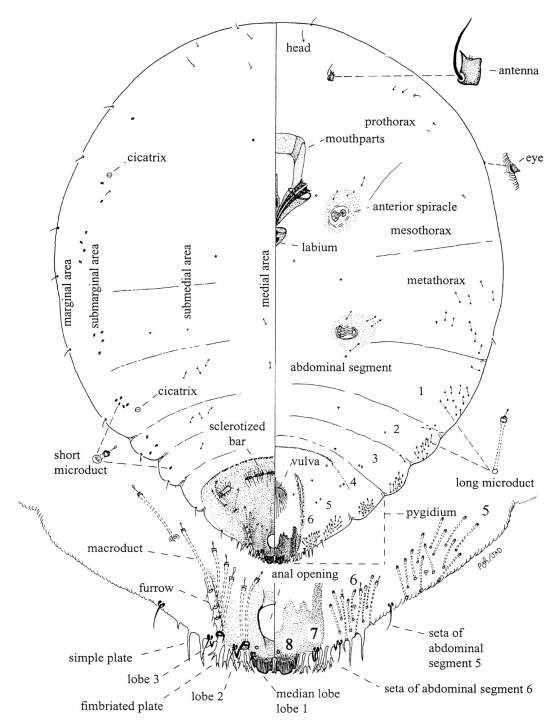

Figure 3. General morphology of slide-mounted adult female aspidiotine.

ring of the anal opening to the base of the dorsal seta located at the base of the median lobes. The length of the median lobes is the distance from the base of the dorsal seta located near the base of the median lobes to the posterior apex of the median lobe. The width of the median lobes is the greatest width of the lobe. The distance between the median lobes is the shortest distance between the lobes with the exception of diverging lobes, which was measured at the distance between the lobe midlengths. Some of these measurements are a bit subjective, but we have standardized them as much as possible.

Some other terminology needs explanation. The gland-spine formula or plate formula is given as the number of these structures that occur in the space between the median and second lobe, between the second and third lobe (or location where the third lobe would normally be if it were present as indicated by the dorsal segmental setae), and between the third and fourth lobe (using the dorsal segmental setae as markers). Thus, the formula 2-2-3 means there are 2 plates (or gland spines) between the median and second lobes, 2 between the second and third lobes, and 3 between the third and fourth lobes (or the location

Diaspidine morphology

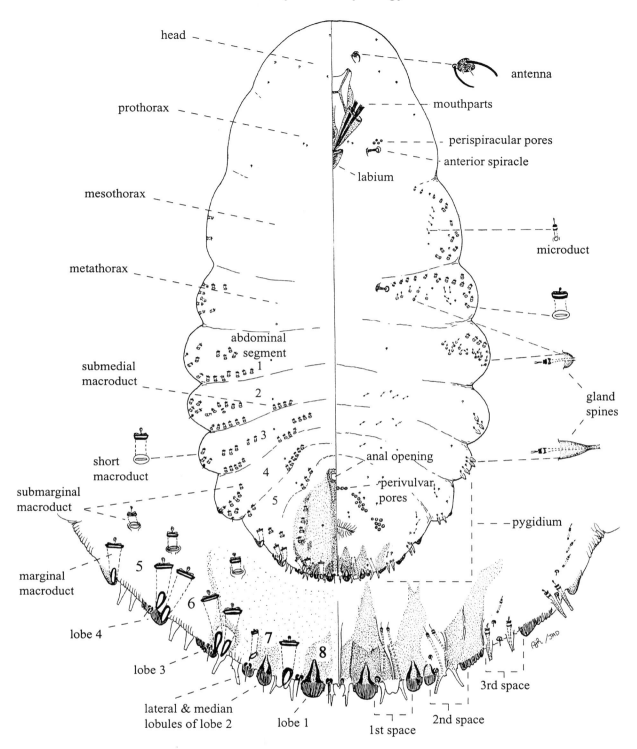

Figure 4. General morphology of slide-mounted adult female diaspidine.

where they would be if they were present). The paraphysis formula is similar to the plate formula but counts the paraphyses on the pygidial margin touching the lobes or spaced between the lobes rather than the plates. Thus, the paraphysis formula 3-5-2 means there are 2 paraphyses between the median and second lobes, 5 between the second and third lobes, and 2 between the third and fourth lobes (or the location where they would be if they were present).

The descriptions and illustrations of adult females are original works designed specifically for this publication. Most of the information on life history and economic importance is compiled and synthesized from the literature with occasional original observations. Host and distribution information is primarily based on our personal examination of specimens deposited in the United States National Entomological Collection, Smithsonian Institution, Beltsville, Maryland (USNM);

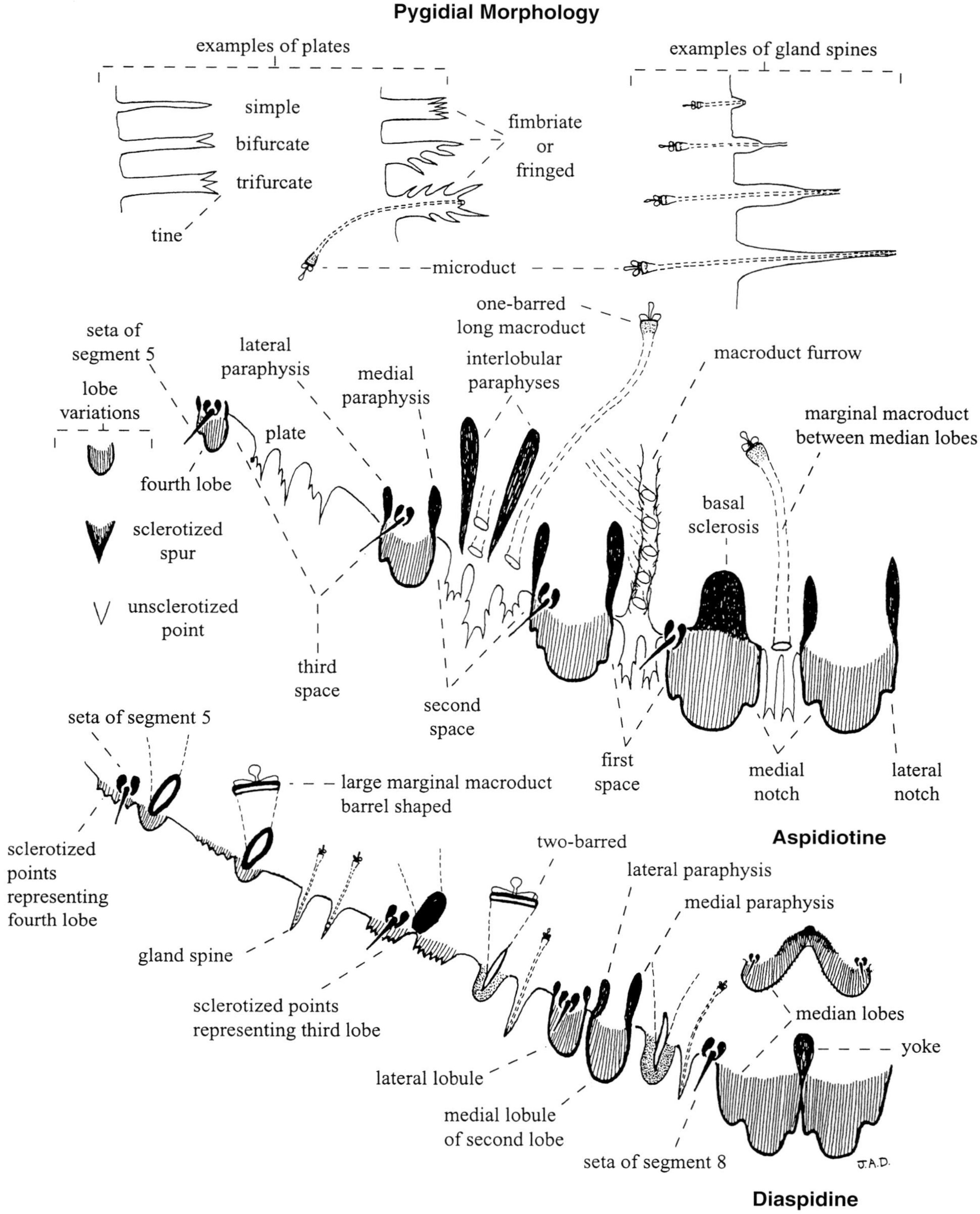

Pygidial Morphology

Figure 5. General morphology of pygidial margin.

these were supplemented from the literature when the published information was judged to be correct. Color photographs principally were taken from the slide collection of Ray Gill, California Department of Food and Agriculture, Sacramento, or from photographs taken by us. Acknowledgments are given in the legend with each color image.

The following discussion on methods and techniques for preserving and preparing armored scales is extracted from a syllabus chapter written by Richard Wilkey in 1991 and a handout developed by the staff of the Systematic Entomology Laboratory, most recently by Debra Creel and Nit Malikul (2002).

Collection and dry preservation

Plant material containing armored scales can be kept for years by simply placing the infested plant specimen in a coin-sized paper envelope or in a small cardboard box. In the USNM Collection at Beltsville, specimens collected as long ago as the late 1800s have been preserved in this manner. Surprisingly, they are easily and effectively mounted on microscope slides after more than 100 years. In the field, plant materials containing armored scales are best collected in bags for more careful examination in the laboratory. Often specimens that could not be seen in the field are easily collected while using a low-power dissecting microscope. Paper bags are best used for this purpose to avoid the accumulation of moisture and the development of mold, but plastic bags are convenient for short-term storage. Standard collection data, for example, host, date, location, etc., should accompany each sample.

Liquid preservation

There are many different kinds of collecting fluids, but alcohol is adequate for preserving most armored scales. Seventy or 75% ethanol (ETOH) or isopropyl works very well. An important step that few collectors use is to 'quick fix' the specimens so they are more easily processed later. When unfixed specimens are kept in alcohol for any length of time, they can become hard and stiff and are difficult or impossible to clear and mount on slides. The process of fixing is achieved by heating the alcohol. This is usually done by holding a match close to the vial with the cap removed. Fixing takes place when the alcohol comes to a slight boil. Care must be taken to avoid overheating the vial, which may cause the alcohol to boil away and the vial to ignite or the vial to crack. Always direct the opening of the vial away from yourself and other people to prevent burns. For preservation of specimens that will be used for molecular analysis it is best to place them in ETOH that is as close to absolute as possible. Specimens should be stored in a low-temperature freezer if possible. A standard freezer will suffice if it is the only option.

Slide mounting

In general, there are two types of mounts: temporary and permanent. Temporary mounts deteriorate over time while permanent mounts can be usable after more than 100 years. The advantage of temporary mounts is that they can be prepared in a few minutes whereas permanent mounts take hours or even days depending on the procedure followed. Specimens that are to be placed in a museum as vouchers must be mounted following the permanent protocol.

Equipment needed for preparation of specimens on microscope slides can be difficult to obtain commercially. Because of this, many coccidologists have learned to make their own tools. Equipment that is essential includes: needles, small spatulas, fine tweezers, small cutting tools, eye droppers, slide-warming trays or a slide oven, and slide storage boxes or slide trays. Glassware needs include: microscope slides, cover slips (usually 15 mm), Syracuse dishes or small watchglasses, polyethylene wash bottles, storage bottles, dropper bottles (polyethylene squeeze type), and a balsam dispensing bottle with a central glass rod. Chemicals include KOH pellets, balsam, distilled water, absolute alcohol, and stains such as lignin pink or acid fuchsin. Other materials include a small hot plate, slide labels, and a computer or pen and ink to print labels.

Temporary mounts. Specimens can be placed directly into temporary mounting media such as Hoyer's (a chloral hydrate and gum arabic combination) or polyvinyl alcohol (PVA). It is best to puncture or make an incision into the body margin opposite the posterior spiracle (see Fig. 3). Gently place a cover glass onto the mounting media containing the specimen and heat on a hot plate. Be careful not to allow the media to boil. If the specimens are large, it is best to tease out as much of the body contents as possible in alcohol before placing them on the slide. In general, young adult females that are not gravid and that are alive when collected make the best mounts.

The most commonly used temporary mounting medium is Hoyer's, with PVA a distant second. The problem with Hoyer's is that chloral hydrate, which is a primary component, can be used as a drug, so a special permit is required to purchase it. Temporary mounts generally are unpredictable. In some cases, crystallization occurs and the medium becomes cloudy, making the included specimens difficult or impossible to find and study. In other cases, bubbles form around the edges of the cover slip and the medium eventually turns brown or black. It is possible to retard medium deterioration by placing a thin layer of a ringing compound around the perimeter of the cover slip so that the mounting medium is not exposed to air.

Permanent mounts. There are nearly as many different techniques for permanently mounting armored scales as there are scientists studying the group, but most differences are slight modifications of a fairly generic procedure. In recent years, some of the more noxious chemicals that traditionally were used in this process have been replaced with compounds that are less likely to pose safety hazards. Chemicals like tetrahydrofuran, xylene, and Essig's aphid fluid are not being used as frequently by armored scale preparators as they were in the past. Substances like Histoclear and dish-washing detergents are commonly used as replacements.

The procedure used in the Systematic Entomology Laboratory (ARS, USDA) at Beltsville, Maryland, is as follows.

- Place armored scales in cold 10% solution of potassium hydroxide (KOH) for 24 hours to soften body contents. (Specimens should not be left in KOH any longer than necessary because it slowly dissolves components of the cuticle and obliterates important characters.)
- An alternative is to heat the KOH solution containing the armored scales on a hot plate set at 110°F. Make sure to cover the specimen dish to prevent evaporation.
- Using a sharp probe or needle, make a small incision on the side of the abdomen. If several specimens are being prepared, vary the location of the incision so that among the specimens there are some complete left and right body margins.
- With a small spatula or other tool, gently press down on body and tease out the contents. If some contents do not come out after pumping, place specimens back on the hot plate for another 5 minutes or place them into a solution of half dish-washing detergent and half distilled water for 3 minutes.
- With scale still in distilled water, lightly push down on specimens with spatula in a pumping action and leave for 5 minutes.
- Add a few drops of stain stock to the distilled water solution and allow it to stand for about 5 minutes.
- Pump in 70% ETOH and leave it for 5 to 10 minutes.
- Pump in 95% ETOH. If specimens are stained too darkly, add water a drop at a time to remove some of the stain. Leave in 95% ETOH for a minimum of 5 minutes.

- Transfer to clove oil for a minimum of 5 minutes. Gently pump specimens to remove excess ETOH. Specimens can be left overnight in clove oil if necessary.
- Place drop of thinned Canada balsam in the center of a slide. (Thin balsam with Histoclear if necessary.) Gently transfer specimens to slide.
- Carefully place coverslip over balsam. Place slide on hot plate until balsam is spread evenly.
- Cure slides in a 110 °F oven (40 °C) for one month so the balsam is no longer soft. If no oven is available, slides will dry after 2 months at room temperature.

- Place a slide label to the left of the coverslip that contains the following information if available: *Genus / species* Author / *Host* / date (7-IV-2000) / STATE or COUNTRY, City, County / Collector / specimen ID number.
- To remount a balsam mount, place the slide in Histoclear to dissolve the old balsam, then follow the preceding mounting procedure.

Key to Adult Females (Microscopic Characters)

These keys provide a useful feature to aid in navigation. The number found in parentheses indicates the step immediately preceding, allowing a reader to reverse or backtrack in the identification process. A simple example: the 2(1) indicates that the reader was at key choice 1 before being directed to choice 2. Numbers will not always be closely sequential in that manner, however, for instance, backtracking from 106 returns you to 93, or 69 refers you back to number 1.

The designation of '(in part)' means that the associated species appears in more than one place in the key.

1   With at least 1 pore near posterior or anterior spiracles (Fig. 54) 2
     Without pores near spiracles 69

2(1) With perivulvar pores (Fig. 21) 8
     Without perivulvar pores 3

3(2) Marginal macroducts about same size as other pygidial macroducts (Fig. 117) 4
     Marginal macroducts larger than at least some medial pygidial macroducts (Fig. 85) 7

4(3) Dorsum of pygidium without areolation pattern; without deep notch on body margin between
        spiracles 5
     Dorsum of pygidium with areolation pattern; with deep notch on body margin between spiracles (on
        many hosts; from Florida and Hawaii in the U.S.) (Fig. 54) . . . .
        . . . . tesserate scale, *Duplaspidiotus tesseratus* (D'Emmerez de Charmoy)

5(4) Without gland spines or with 1 pair of small gland spines between median lobes; with 2 or 3 pairs
        of obvious lobes (Fig. 117); on numerous woody hosts, not on bamboo 6
     With 3 or more conspicuous gland spines between median lobes (Fig. 61); without well-defined lobes
        on pygidium; occurring on bamboo (from 9 states) . . . .
        . . . . penicillate scale, *Froggattiella penicillata* (Green)

6(5) Body elongate, length greater than 2 times maximum width; with medial macroducts on segment 7
        (Fig. 117); without abdominal spurs (primarily on citrus; from Florida and Louisiana) . . . .
        . . . . citrus snow scale, *Unaspis citri* (Comstock) (in part)
     Body turbinate, length less than 2 times maximum width; with medial macroducts absent from
        segment 7; with abdominal spurs in lateral areas of segments 2 to 4 (Fig. 68) (on many hosts; occur-
        ring out-of-doors in Florida and Hawaii, occasionally in greenhouses elsewhere) . . . .
        . . . . mining scale, *Howardia biclavis* (Comstock)

7(3) Body elongate oval; macroduct absent between median lobes (occurring on rosaceous trees like pears,
        cherries, plums; eradicated from California) (Fig. 85) . . . .
        . . . . Hall scale, *Mercetaspis halli* (Green)
     Body turbinate; macroduct present between median lobes (occurring primarily on cycads; found out-
        of-doors in California and Hawaii) (Fig. 63) . . . . zamia scale, *Furchadaspis zamiae* (Morgan)

8(2) Body elongate, oval, or turbinate, not as described below 11
     Body elongate, head and/or anterior 2 thoracic segments forming rectangular structure with lateral
        margins at least partially parallel, rectangular structure wider than remainder of body (Fig. 20) . . .
        . . . . (*Aulacaspis*) . . . 9

9(8) Mouthparts without sclerotized projections on either side of labium; not on mango 10
     Mouthparts with sclerotized projections on each side of labium forming a tubular structure; on
        mango (from Florida) (Fig. 21) . . . . white mango scale, *Aulacaspis tubercularis* Newstead

10(9) Rectangular area comprising pro- and mesothoracic segments (on cycads; from Florida) (Fig. 22) . . .
        . . . . cycad aulacaspis scale, *Aulacaspis yasumatsui* Takagi (in part)
     Rectangular area comprising head, pro- and mesothoracic segments (on rosaceous hosts particularly
        roses and berries; widespread) (Fig. 20) . . . . rose scale, *Aulacaspis rosae* (Bouché)

11(8) Body elongate, length usually 2 times or more than greatest width; widest part of body usually located
        at metathorax or abdomen (Fig. 87) 12
     Body oval or turbinate, length usually less than 2 times greatest width; widest part of body usually
        located at head, prothorax or mesothorax (Fig. 62) 45

12(11) Median lobes separated by space at least one-quarter width of lobe (Fig. 81) or forming apical notch
        (Fig. 27); with or without gland spines between median lobes 16

Median lobes fused to their apex (Fig. 112) or separated by narrow space less than one-quarter width of lobe (Fig. 87), usually not forming apical notch; without gland spines between median lobes 13

13(12) Apex of pygidium rounded; with 15 or more perivulvar pores on each side of body; normally with perispiracular pores near posterior spiracle (Fig. 101) 14
Apex of pygidium acute; with 13 or fewer perivulvar pores on each side of body; without perispiracular pores near posterior spiracle (on many hosts; from California, Florida, Georgia, and Hawaii) (Fig. 87) . . . . Harper scale, *Neopinnaspis harperi* McKenzie

14(13) Median lobes with small space between them (Fig. 102); on many hosts, rarely on oaks . . . . (*Pinnaspis*) . . . 15
Median lobes fused, without space between them; on oaks (from 10 states) (Fig. 112) . . . . oak scale, *Quernaspis quercus* (Comstock)

15(14) Preanal sclerosis lacking or represented by light sclerotized patch; median lobes protrude less than or about same distance as second lobes; posterior spiracles each with 1–12 (4) pores (on many hosts; from many states, often in greenhouses) (Fig. 101) . . . . fern scale, *Pinnaspis aspidistrae* (Signoret)
Preanal sclerosis represented by sclerotized bar; median lobes protrude beyond or about same distance as second lobes; posterior spiracles each with 0–4 (2) pores (on many hosts; from 7 states) (Fig. 102) . . . . lesser snow scale, *Pinnaspis strachani* (Cooley)

16(12) Without perispiracular pores near posterior spiracles 25
With perispiracular pores near posterior spiracles (Fig. 118) 17

17(16) Macroducts absent from submarginal and medial areas of abdominal segment 7 (Fig. 31) 19
Macroduct present in submarginal and medial areas of abdominal segment 7 (Fig. 118) . . . . (*Unaspis*) . . . 18

18(17) With more than 4 perivulvar pores on each side of body (primarily on euonymus; from many states) (Fig. 118) . . . . euonymus scale, *Unaspis euonymi* (Comstock)
With 4 or fewer perivulvar pores on each side of body (primarily on citrus; from Florida and Louisiana) (Fig. 117) . . . . citrus snow scale, *Unaspis citri* (Comstock) (in part)

19(17) Median lobes closely appressed for about one-half of their length, without a distinct space between them (Fig. 26); not on pines 21
Median lobes separated by distinct space, not closely appressed basally (Fig. 29); on pines 20

20(19) Median lobes narrow, strongly divergent; median lobe apices widely separated, with short yoke (on pines; from many states) (Fig. 29) . . . . pine scale, *Chionaspis heterophyllae* Cooley
Median lobes thick, weakly divergent; median lobe apices narrowly separated, with anteriorly protruding yoke (on several needle-bearing conifers; from many states) (Fig. 30) . . . . pine needle scale, *Chionaspis pinifoliae* (Fitch)

21(19) Median lobes without deep notch on lateral margin; not on elm 22
Median lobes with deep notch on lateral margin; on elm (from many states) (Fig. 26) . . . . elm scurfy scale, *Chionaspis americana* Johnson

22(21) Median lobes broadly rounded 24
Median lobes with medial margins parallel half of length then strongly divergent 23

23(22) With sclerotized yoke joining median lobes anteriorly; widest part of body at metathorax or segment 1; on dogwood; from many states (Fig. 27) . . . . dogwood scale, *Chionaspis corni* Cooley
Without sclerotized yoke joining median lobes anteriorly; widest part of body at mesothorax; on cycads; from Florida (Fig. 22) . . . . cycad aulacaspis scale, *Aulacaspis yasumatsui* Takagi (in part)

24(22) Yoke between median lobes as long as median lobes; lateral sclerosis on median lobe conspicuous; polyphagous, primarily on rosaceous hosts (from many states) (Fig. 28) . . . . scurfy scale, *Chionaspis furfura* (Fitch)
Yoke between median lobes short or absent, less than length of median lobes; lateral sclerosis on median lobe inconspicuous; on poplar and willow (from many states) (Fig. 31) . . . . willow scale, *Chionaspis salicis* (Linnaeus)

25(16) Pygidium without reticulate pattern on dorsal surface 27
Pygidium with distinct reticulate pattern on dorsal surface 26

26(25) Second lobes bilobate, with 2 lobules; marginal macroducts barrel shaped (on many hosts; from many states, occurring out-of-doors in southern states) (Fig. 69) . . . . black thread scale, *Ischnaspis longirostris* (Signoret)
Second lobes simple, with 1 lobule; marginal macroducts elongate (on many hosts; from Florida) (Fig. 105) . . . . trilobe scale, *Pseudaonidia trilobitiformis* (Green) (in part)

27(25) With at least a few macroducts on medial or submarginal areas of pygidium (Fig. 81) 31
Without macroducts on medial or submarginal areas of pygidium (Fig. 58) . . . . (*Fiorinia*) . . . 28

28(27) Head without a tubercle between antennae 29
Head with conspicuous tubercle between antennae (common on camellia and holly; from many states) (Fig. 60) . . . . tea scale, *Fiorinia theae* Green

29(28) With 4 to 6 large macroducts on each side of abdomen (Fig. 59); on conifers 30
With 3 large macroducts on each side of pygidium; normally not on conifers, often on palms (out-of-doors in Alabama, California, Florida, Georgia, Mississippi, and Texas, occasionally in greenhouses elsewhere) (Fig. 58) .... palm fiorinia scale, *Fiorinia fioriniae* (Targioni Tozzetti)

30(29) With 1 size of macroduct along margin of pygidium (usually on hemlock; from northeastern U.S.) (Fig. 57)................................... elongate hemlock scale, *Fiorinia externa* Ferris
With 2 sizes of macroducts along margin of pygidium (on many conifers including hemlock; from the District of Columbia, Maryland, and Virginia) (Fig. 59) ....
.... coniferous fiorinia scale, *Fiorinia japonica* Kuwana

31(27) Gland spines or plates between median lobes absent or simple, without apical fringe (sometimes with 1 or 2 tines near base of gland spine) 33
Gland spines or plates between median lobes apically fringed (Fig. 70) 32

32(31) Perivulvar pores in clusters on abdominal segments 4 to 6; dorsum of pygidium with sclerotized spots; on many ornamental hosts, not on bamboo (from 10 eastern states) (Fig. 81) ....
.... Japanese maple scale, *Lopholeucaspis japonica* (Cockerell)
Perivulvar pores in clusters restricted to abdominal segment 6; dorsum of pygidium uniformly sclerotized; on bamboo (from 6 eastern states, California, and Hawaii) (Fig. 70) ....
.... bamboo diaspidid, *Kuwanaspis pseudoleucaspis* (Kuwana)

33(31) Median lobes rounded or lateral margin conspicuously longer than medial margin (Fig. 9); with 2 gland spines between median lobes 34
Median lobes pointed, medial margin of lobe conspicuously longer than lateral margin; without gland spines between median lobes (on many hosts; from Hawaii and 6 southern states out-of-doors, common on imported nursery stock in northern area) (Fig. 106) ....
.... false oleander scale, *Pseudaulacaspis cockerelli* (Cooley)

34(33) Median lobes round, medial margin of lobe about same length as lateral margin (Fig. 77) ....
.... (*Lepidosaphes*) ... 35
Median lobes with lateral margin conspicuously longer than medial margin (primarily on litchi fruit; from Florida) (Fig. 9) .... litchi scale, *Andaspis punicae* (Laing)

35(34) Sclerotized spurs absent from body margin of abdominal segments 2, 3, and 4 40
Sclerotized spurs present on body margin of 1 or more of abdominal segments 2, 3, and 4 (Fig. 78) .
.... 36

36(35) Eye absent or represented by a small dome; rarely on orchids 37
Eye represented by conspicuous sclerotized spur; on orchids mostly in greenhouses (from 10 states) (Fig. 77) .... cymbidium scale, *Lepidosaphes pinnaeformis* (Bouché) (in part)

37(36) Mature female without sclerotized pattern on thorax; without sclerotized dermal pockets on thorax
.... 38
Mature female with distinctive sclerotized pattern on thorax; with sclerotized dermal pockets on pro- and mesothorax (common on citrus; out-of-doors in Alabama, California, Florida, Georgia, Louisiana, Mississippi, South Carolina, and Texas, in grocery stores elsewhere) (Fig. 74) ....
.... Glover scale, *Lepidosaphes gloverii* (Packard)

38(37) Macroducts on segment 6 in continuous band from submargin to submedial area, with more than 4 ducts in each cluster; with cicatrices on dorsum of abdomen (Fig. 71) 39
Macroducts on segment 6 restricted to submedial cluster, with 4 or fewer ducts in each cluster; without cicatrices on dorsum of abdomen (often on *Euonymus alata*; from 11 eastern states and Oklahoma) (Fig. 79) .... fire bush scale, *Lepidosaphes yanagicola* Kuwana

39(38) Body margin with only 1 lateral spur on each side of abdomen (not illustrated) (common on citrus; out-of-doors in Alabama, California, Florida, Georgia, Louisiana, and North Carolina, in grocery stores elsewhere) (Fig. 71) .... purple scale, *Lepidosaphes beckii* (Newman) (in part)
Body margin with 2 or more lateral spurs on each side of abdomen (on many hosts; common in many states) (Fig. 78) .... oystershell scale, *Lepidosaphes ulmi* (Linnaeus)

40(35) With a small dorsal macroduct anterior of second lobes (Fig. 74) 43
Without a small dorsal macroduct anterior of second lobes 41

41(40) Perivulvar pores arranged in 5 groups, pores on segment 6 only (Fig. 73) 42
Perivulvar pores arranged in more than 5 groups, some pores on segments 5 and 6 (usually on pines; from Hawaii, Maryland, New Jersey, and Pennsylvania) (Fig. 76) ....
.... pine oystershell scale, *Lepidosaphes pini* (Maskell)

42(41) Cicatrices present on abdomen; common on citrus (out-of-doors in Alabama, California, Florida, Georgia, Louisiana, and North Carolina, in grocery stores elsewhere) (Fig. 71) ....
.... purple scale, *Lepidosaphes beckii* (Newman) (in part)
Cicatrices absent from abdomen; usually on figs and ornamental trees (from California) (Fig. 73) ..
.... fig scale, *Lepidosaphes conchiformis* (Gmelin)

43(40) Eye absent or represented by a small dome; rarely on orchids 44
Eye represented by conspicuous sclerotized spur; on orchids mostly in greenhouses (reported from 10 states) (Fig. 77) .... cymbidium scale, *Lepidosaphes pinnaeformis* (Bouché) (in part)

44(43) Small dorsal duct anterior of second lobe thin, about same thickness as microducts in gland spines; on conifers (from 15 states) (Fig. 75) ....
.... Maskell scale, *Lepidosaphes pallida* (Maskell)
Small dorsal duct anterior of second lobe thick, conspicuously wider than microducts in gland spines; common on camellias and holly (from 19 states) (Fig. 72) ....
.... camellia scale, *Lepidosaphes camelliae* Hoke

45(11) Marginal macroducts barrel shaped, length of duct usually less than 3 times width of inner end of duct (Fig. 24) 52
Marginal macroducts slender, length of duct usually more than 3 times width of inner end of duct (Fig. 62) 46

46(45) Dorsum of pygidium with reticulate pattern (Fig. 53) 49
Dorsum of pygidium evenly sclerotized, without reticulate pattern 47

47(46) With more than 11 perivulvar pores on each side of body (Fig. 91); without plates anterior of median lobes 48
With 11 or fewer perivulvar pores on each side of body; with distinct plates between median and second lobes and between second and third lobes (on orchids; occurring out-of-doors in Florida and Hawaii, occasionally in greenhouses elsewhere) (Fig. 62) ....
.... orchid scale, *Furcaspis biformis* (Cockerell)

48(47) With long setae along margin of head and thorax (longer than length of spiracle); perispiracular pores absent from posterior spiracles; on palms (from out-of-doors in California, Florida, Georgia, Louisiana, North Carolina, South Carolina, and Texas, rarely in greenhouses elsewhere) (Fig. 37) .
.... Palmetto scale, *Comstockiella sabalis* (Comstock)
With short setae along margin of head and thorax (shorter than length of spiracle); perispiracular pores present near posterior spiracles; on grass (from 11 southern states) (Fig. 91) ....
.... Bermuda grass scale, *Odonaspis ruthae* Kotinsky (in part)

49(46) Without clubbed-shaped paraphysis between median lobes and lobe 2 ...... (*Pseudaonidia*) ... 50
With clubbed-shaped paraphysis between median lobes and lobe 2 (on many woody hosts particularly camellia, ligustrum, and viburnum; from Florida and Hawaii) (Fig. 53) ....
.... camellia mining scale, *Duplaspidiotus claviger* (Cockerell)

50(49) Posterior spiracles without associated pores; pygidial macroducts long, about 10 times longer than width of dermal orifice (Fig. 103) 51
Posterior spiracles with at least 1 associated pore; pygidial macroducts short, about 5 times longer than width of dermal orifice (usually on camellia or rhododendron; widely distributed in U.S.) (Fig. 104) .... peony scale, *Pseudaonidia paeoniae* (Cockerell)

51(50) Second lobe protruding posteriorly less than or about same as median lobes; second lobes conspicuously thinner than median lobes, half or less of width of median lobes (on many hosts; from Alabama, Florida, Georgia, Louisiana, Mississippi, and Texas) (Fig. 103) ....
.... camphor scale, *Pseudaonidia duplex* (Cockerell)
Second lobe protruding posteriorly more than median lobes; second lobes about same width as median lobes (on many hosts; from Florida) (Fig. 105) ....
.... trilobe scale, *Pseudaonidia trilobitiformis* (Green)(in part)

52(45) Barrel-shaped macroduct present between median lobes (Fig. 23) 56
Without macroduct between median lobes or duct between median lobes long and slender 53

53(52) Perispiracular pores absent near posterior spiracles; without submedial macroducts on segment 7 .
.... 54
Perispiracular pores present near posterior spiracles; with submedial macroducts on segment 7 (on grass; from 11 southern states) (Fig. 91) ....
.... Bermuda grass scale, *Odonaspis ruthae* Kotinsky (in part)

54(53) With more than 5 perispiracular pores associated with each anterior spiracle; median lobes narrowly rounded or pointed (Fig. 107) .... (*Pseudaulacaspis*) ... 55
With less than 5 perispiracular pores associated with each anterior spiracle; median lobes broadly rounded (on conifers, especially juniper-like groups [Cupressaceae]; widespread in U.S.) (Fig. 24) .... minute cypress scale, *Carulaspis minima* (Targioni Tozzetti)

55(54) Third space usually with 1 gland spine; at least 1 bifurcate or trifurcate gland spine in second, third, or fourth spaces (on many hosts including mulberry; common in many states) (Fig. 107). . .
.... white peach scale, *Pseudaulacaspis pentagona* (Targioni Tozzetti)
Third space usually with 2 or more gland spines; gland spines rarely with bifurcate or trifurcate apex (on many host including *Prunus* and *Ligustrum*; common in many states) (Fig. 108) ....
.... white prunicola scale, *Pseudaulacaspis prunicola* (Maskell)

56(52) Plates or gland spines in space between median and second lobes with at least 2 apical fimbriations, usually more (Fig. 100) 62
Plates or gland spines between median and second lobes simple, without apical fimbriations (Fig. 23) 57

57(56) Submarginal areas of posterior pygidial segments with at least 2 barrel-shaped macroducts (Fig. 50) 59
Submarginal areas of posterior pygidial segments without macroducts 58

58(57) Median lobe broadly rounded, without notches, lateral and medial margins about equal; without globular sclerosis in space between median and second lobes (common on juniper and other juniper-like conifers [Cupressaceae]; widespread in U.S.) (Fig. 23) . . . . juniper scale, *Carulaspis juniperi* (Bouché)
Median lobe narrowly rounded or acute, with notches, lateral margin conspicuously longer than medial margin; with globular sclerosis in space between median and second lobes (on many hosts; from California, Florida, and Missouri) (Fig. 92) . . . . Chinese obscure scale, *Parlatoreopsis chinensis* (Marlatt)

59(57) Perispiracular pores absent near posterior spiracle; without conspicuous, comma-shaped paraphyses attached to lateral margins of lobes 2, 3, and 4 (Fig. 50) . . . . (*Diaspis*) . . . 60
Perispiracular pores present near posterior spiracle; with conspicuous comma-shaped paraphyses attached to lateral margins of lobes 2, 3, and 4 (on many hosts; in 9 states) (Fig. 56) . . . . Italian pear scale, *Epidiaspis leperii* (Signoret)

60(59) Without submedial macroducts on abdominal segments 2 to 5; with notches on medial margin of median lobes; rarely on cactus 61
With submedial macroducts on abdominal segments 2 to 5; without notches on medial margin of median lobes; primarily on cactus (widespread in U.S., often in greenhouses) (Fig. 52) . . . . cactus scale, *Diaspis echinocacti* (Bouché)

61(60) With 2 large submarginal macroducts on each side of the pygidium and many small macroducts; anterior spiracles with 1–7 (3) associated perispiracular pores (normally on orchids and palms; widespread in U.S., often in greenhouses) (Fig. 50) . . . . . . Boisduval scale, *Diaspis boisduvalii* Signoret
With more than 2 large submarginal macroducts on each side of the pygidium and few small macroducts; anterior spiracles with 12–16 (14) associated perispiracular pores; (normally on bromeliads; from 13 states, often in greenhouses) (Fig. 51) . . . . pineapple scale, *Diaspis bromeliae* (Kerner)

62(56) Without conspicuous ear-like lobes on body margin laterad of mouthparts (sometimes with spine-like projection in this area) 63
With conspicuous ear-like lobes on body margin laterad of mouthparts (on citrus; from Florida and Hawaii) (Fig. 100) . . . . black parlatoria scale, *Parlatoria ziziphi* (Lucas)

63(62) Median lobes with notches; prothoracic cicatrix absent or small and unsclerotized; with at least a few gland spines on head, prothorax, or mesothorax (Fig. 95) 64
Median lobes usually without notches, when present, with only 1 on lateral margin; prothoracic cicatrix large, conspicuous, and sclerotized; without gland spines on head, prothorax, or mesothorax (on palms; eradicated from Arizona, California, and Texas) (Fig. 93) . . . . parlatoria date scale, *Parlatoria blanchardi* (Targioni Tozzetti)

64(63) Prepygidial macroducts absent in submedial areas, sometimes with microducts in these areas 66
Prepygidial macroducts present in submedial areas (Fig. 95) 65

65(64) First 3 lobes usually with 1 lateral notch and no medial notch; plates between median and second lobes usually with 2 tines (on many hosts; from Arizona, California, Delaware, and Maryland) (Fig. 95) . . . . olive scale, *Parlatoria oleae* (Colvee)
First 3 lobes usually with at least 1 lateral notch and 1 medial notch; plates between median and second lobes normally with more than 2 tines (often on pittosporum and leptospermum; from California) (Fig. 97) . . . . pittosporum scale, *Parlatoria pittospori* Maskell

66(64) With small dermal pocket between posterior spiracle and body margin (Fig. 98) 67
Without dermal pocket between posterior spiracle and body margin (common on citrus; from many states) (Fig. 96) . . . . chaff scale, *Parlatoria pergandii* Comstock

67(66) Eye absent or dome shaped (Fig. 99) 68
Eye spurlike, apically pointed (on many hosts; found out-of-doors in Florida, Georgia, Louisiana, and Texas, frequent in greenhouses elsewhere) (Fig. 98) . . . . . proteus scale, *Parlatoria proteus* (Curtis)

68(67) With 22–41 (31) perivulvar pores on each side of body; most plates in spaces between first 3 lobes with 2 microducts; 17–27 (23) macroducts on each side of pygidium (on many hosts; from the District of Columbia, Hawaii, Maryland, Mississippi, North Carolina, Texas, and Virginia) (Fig. 99) . . . . tea parlatoria scale, *Parlatoria theae* Cockerell
With 10–17 (13) perivulvar pores on each side of body; most plates in spaces between first 3 lobes with 1 microduct; 14–17 (16) macroducts on each side of pygidium (on many hosts, common on camellia; widespread in U.S.) (Fig. 94) . . . . camellia parlatoria scale, *Parlatoria camelliae* Comstock

69(1) With perivulvar pores (Fig. 18) 87
Without perivulvar pores 70

70(69) Second and/or third lobes absent (Fig. 86) or conspicuously smaller than median lobes (Fig. 31) 75
Second and third lobes about same size as median lobes (Fig. 82) 71

71(70) Paraphyses less than 2 times length of median lobes; macroducts broad, orifices at least twice diameter of setal base near second lobe; body of mature adult female with pygidium partially or completely invaginated in lobes formed by anterior abdominal segments (Fig. 13) 73
At least 1 paraphysis greater than 3 times length of median lobes; macroducts thin, orifices about same diameter or smaller than setal base near second lobe; body of mature adult female not as above (Fig. 82) 72

72(71) Medial paraphysis of second lobes longer than paraphysis in space between median and second lobes; anal opening situated near apical third of pygidium (on numerous tree species, particularly maple; found primarily in eastern U.S.) (Fig. 84) . . . . . . gloomy scale, *Melanaspis tenebricosa* (Comstock)
Medial paraphysis attached to medial margin of second lobes shorter than paraphysis in space between median and second lobes; anal opening situated near center of pygidium (on oak; from Arizona, California, and New Mexico) (Fig. 82) . . . . dark oak scale, *Melanaspis lilacina* (Cockerell)

73(71) Without tubercle on lateral margin of segment 3; with sclerotized apophyses anterolaterad of vulva; usually on citrus (Fig. 10) 74
With tubercles on lateral margin of segment 3; without apophyses anterolaterad of vulva; on podocarpus and yew (out-of-doors from Alabama, Florida, and Louisiana, occasionally on imported nursery stock in northern areas) (Fig. 13) . . . . Asiatic red scale, *Aonidiella taxus* Leonardi

74(73) With 2 conspicuous scleroses associated with apophysis anterolaterad of vulva; longest macroduct in first space 62–89 (81) μ long; duct between median lobes 70–89 (78) μ long (polyphagous, usually on citrus; out-of-doors from Alabama, California, Florida, Georgia, Louisiana, and Texas, common in grocery stores elsewhere) (Fig. 10) . . . . California red scale, *Aonidiella aurantii* (Maskell)
Without scleroses associated with apophysis anterolaterad of vulva; longest macroduct in first space 82–110 (94) μ long; duct between the median lobes 70–102 (85) μ long (usually on citrus; out-of-doors from California, Florida, and Texas) (Fig. 11) . . . yellow scale, *Aonidiella citrina* (Coquillett)

75(70) Without conspicuous basal sclerosis attached to median lobes (disregard paraphysis-like sclerotization attached to medial margin of median lobes) 78
With conspicuous basal sclerosis attached to median lobes (basal sclerosis is contiguous with lateral and medial margins of median lobes not just medial margin as in a paraphysis-like sclerotization), sclerosis about equal in length or longer than median lobes (Fig. 113) 76

76(75) Pygidium with inconspicuous, simple plates shorter than length of median lobes (Fig. 113) 77
Pygidium with conspicuous, fringed plates exceeding length of median lobes (on many hosts; from Florida and Hawaii) (Fig. 86) . . . . plumose scale, *Morganella longispina* (Morgan)

77(76) With numerous macroducts along body from segments 1 to 8; without club-shaped paraphysis attached to lateral margin of median lobes; usually on Ericaceae or Compositae (widespread throughout U.S.) (Fig. 113) . . . . Dearness scale, *Rhizaspidiotus dearnessi* (Cockerell)
With macroducts restricted to pygidium, absent from segment 1 to segment 4; with conspicuous club-shaped paraphysis attached to lateral margin of median lobes; usually on trees such as elm (widespread throughout U.S.) (Fig. 36) . . . . . elm armored scale, *Clavaspis ulmi* (Johnson) (in part)

78(75) Pygidium with at least median lobes present 79
Pygidium without median lobes, second and third lobes absent or represented by small sclerotized nobs (on bromeliads; out-of-doors from Florida, common in greenhouses elsewhere) (Fig. 64) . . . . flyspeck scale, *Gymnaspis aechmeae* Newstead

79(78) Without indentation and associated spine between meso- and metathorax; without a spur-like third lobe 80
With conspicuous indentation and associated tubercle or spine between meso- and metathorax; with a spur-like third lobe (on *Euphorbia*; from California) (Fig. 114) . . . . white euphorbia scale, *Selenaspidus albus* (McKenzie)

80(79) Without a mushroom-shaped paraphysis in space between median and second lobe 81
With a mushroom-shaped paraphysis in space between median and second lobe (on many hosts; from Florida, Hawaii, and Texas) (Fig. 35) . . . . herculeana scale, *Clavaspis herculeana* (Cockerell and Hadden)

81(80) With marginal macroduct between median lobes (Fig. 38) 83
Without marginal macroduct between median lobes 82

82(81) Plates anterior of third lobe simple, without tines and associated microducts; anal opening located less than length of opening from base of median lobes (on numerous hosts; widespread in U.S.) (Fig. 67) . . . . greedy scale, *Hemiberlesia rapax* (Comstock)
Plates anterior of third lobe distinctive, with tines and associated microducts; anal opening located more than length of opening from base of median lobes (mostly on deciduous trees and shrubs; widespread in U.S.) (Fig. 66) . . . . false diffinis scale, *Hemiberlesia neodiffinis* Miller and Davidson

83(81) Medial margin of macroduct furrow between median and second lobes sclerotized and forming paraphysis; with plates anterior of seta marking segment 6 85
Medial margin of macroduct furrow between median and second lobes without paraphysis; without plates anterior of seta marking segment 6 (Fig. 38) 84

84(83) Macroducts short and barrel shaped, length about 2 times diameter of orifice; macroducts in marginal areas of segments 2 to 4 (on many hosts; from southwestern U.S.) (Fig. 116) . . . . yucca scale, *Situlaspis yuccae* (Cockerell) (in part)
Macroducts long and narrow, length about 4 or 5 times diameter of orifice; macroducts absent from marginal areas of segments 2 to 4 (on conifers; from many states) (Fig. 38) . . . . redwood scale, *Cupressaspis shastae* (Coleman)

85(83) Plates in third space simple, without tines, plate microducts restricted to body margin (Fig. 36) 86

Plates in third space of distinctive shape, with 1 or 2 lateral tines and central microduct (on many hosts especially Rosaceae; widespread in U.S.) (Fig. 48) . . . . San Jose scale, *Diaspidiotus perniciosus* (Comstock)

86(85) Plates in first and second spaces simple or with 1 or 2 small tines; second and sometimes third lobes represented by small sclerotized protrusions; without plates between median lobes; often found on trees, particularly elms (widespread in U.S.) (Fig. 36) . . . . elm armored scale, *Clavaspis ulmi* (Johnson) (in part)
Plates in first and second space with several conspicuous tines; second and third lobes absent or represented by thin unsclerotized points; 2 plates between median lobes; usually on sweetgum (widespread in U.S.) (Figs. 38 and 39) . . . . sweetgum scale, *Diaspidiotus liquidambaris* (Kotinsky)

87(69) Marginal macroducts on pygidium long, length greater than 3 times width of orifice (Fig. 43) . . . 91
Marginal macroducts on pygidium short, length less than 3 times width of orifice (Fig. 116) . . . 88

88(87) Perivulvar pores normal in size and shape; with more than 10 pores on each side of body (Fig. 111) 89
Perivulvar pores small and poorly formed; with less than 10 pores on each side of body (on many hosts; from southwestern U.S.) (Fig. 116) . . . . . yucca scale, *Situlaspis yuccae* (Cockerell) (in part)

89(88) Second lobes with 2 lobules; with simple gland spines between lobes (Fig. 111) . . . . (*Pseudoparlatoria*) . . . 90
Second lobes simple, with 1 lobule; with fringed plates between lobes (on many hosts; widespread in U.S.) (Fig. 18) . . . . oleander scale, *Aspidiotus nerii* Bouché

90(89) With 24–52 (35) macroducts on each side of body; usually having an anteromedial cluster of perivulvar pores (on many hosts; from Florida) (Fig. 110) . . . . gray scale, *Pseudoparlatoria ostreata* Cockerell
With 14–22 (17) macroducts on each side of body; usually lacking an anteromedial cluster of perivulvar pores (on many hosts; from Florida, Georgia, and Texas) (Fig. 111) . . . . false parlatoria scale, *Pseudoparlatoria parlatorioides* (Comstock)

91(87) Without indentation between meso- and metathorax (Fig. 33) 93
With distinct indentation between meso and metathorax (Fig. 43) 92

92(91) Third lobes absent or flat and rounded (on many tree species especially English walnut; widespread in U.S.) (Fig. 43) . . . . walnut scale, *Diaspidiotus juglansregiae* (Comstock) (in part)
Third lobes narrow and spurlike (on many hosts; from Florida) (Fig. 115) . . . . rufous scale, *Selenaspidus articulatus* (Morgan)

93(91) Third lobes absent (Fig. 13), represented by small sclerotized protrusion (Fig. 43), or by unsclerotized point (Fig. 19); distinctly smaller than second lobes (Fig. 5) 106
Third lobes large and sclerotized, usually about same size and shape as second lobes (Fig. 33) 94

94(93) Paraphyses absent or inconspicuous, all shorter than length of median lobes (disregard paraphysis-like sclerosis between median lobes) (Fig. 88) 100
Paraphyses conspicuous, at least 1 paraphysis as long as or longer than length of median lobes (disregard paraphysis-like sclerosis between median lobes) (Fig. 12) 95

95(94) Without cluster of macroducts on submarginal areas of prepygidial segments 97
With at least 1 cluster of macroducts on submarginal areas of prepygidial segments (Fig. 32) . . . 96

96(95) With cluster of macroducts on submarginal areas of abdominal segments 2 and 3 (on many hosts; from 10 states) (Fig. 33) . . . . bifasciculate scale, *Chrysomphalus bifasciculatus* Ferris
With cluster of macroducts on submarginal areas of abdominal segment 2 only (on many hosts; out-of-doors from Florida, Georgia, Hawaii, Louisiana, Mississippi, and Texas, common in greenhouses elsewhere) (Fig. 32) . . . . Florida red scale, *Chrysomphalus aonidum* (Linnaeus)

97(95) Row of small paraphyses absent from area marking fourth lobe 98
Row of small paraphyses present in area marking fourth lobe (on many hosts, common on araucaria; out-of-doors from California and Hawaii) (Fig. 80) . . . . black araucaria scale, *Lindingaspis rossi* (Maskell)

98(97) Without clavate processes on plates between lobes 3 and 4 99
With clavate processes on plates between lobes 3 and 4 (on many hosts; out-of-doors from 11 southern states) (Fig. 34) . . . . dictyospermum scale, *Chrysomphalus dictyospermi* (Morgan)

99(98) Body elongate in mature females; with large macroduct between median lobes; on many tropical hosts (out-of-doors from Florida, rare in greenhouses elsewhere) (Fig. 109) . . . . Bowrey scale, *Pseudischnaspis bowreyi* (Cockerell)
Body oval or turbinate in mature females; without macroduct between median lobes; primarily on oak (widespread in eastern U.S., introduced in California) (Fig. 83) . . . . obscure scale, *Melanaspis obscura* (Comstock)

100(94) Large macroducts absent from prepygidial segments 103
Large macroducts present on prepygidial segments (Fig. 12) 101

101(100) Plates between lobes 3 and 4 without sickle-shape apices and attached microduct 102
Plates between lobes 3 and 4 with sickle-shape apices and attached microduct (on many hosts, especially palms; out-of-doors from Florida, occasionally in greenhouses elsewhere) (Fig. 12) . . . . oriental armored scale, *Aonidiella orientalis* (Newstead)

102(101) Body shape broadly oval; plates unusually broad, as wide as long; body at maturity heavily sclerotized; on conifers (widespread in U.S.) (Fig. 88) . . . .
. . . . black pineleaf scale, *Nuculaspis californica* (Coleman)
Body shape pyriform; plates narrow, narrower than long; body at maturity not sclerotized; common on camellia (from California, Missouri, and Oregon) (Fig. 7) . . . .
. . . . degenerate scale, *Abgrallaspis degenerata* (Leonardi) (in part)

103(100) Macroduct between median lobes extending to posterior margin of anal opening (Fig. 55) 104
Macroduct between median lobes extending about halfway to posterior margin of anal opening (on several hosts including aglaonema; from Florida) (Fig. 17) . . . .
. . . . aglaonema scale, *Aspidiotus excisus* Green

104(103) Fringe plates present anterior of seta marking segment 5; without medial notch on second lobes (Fig. 15) 105
Fringe plates absent anterior of seta marking segment 5; with medial notch on second lobes (common on *Laurus* and holly; from 9 states) (Fig. 55) . . . .
. . . . holly scale, *Dynaspidiotus britannicus* (Newstead)

105(104) Second lobes not protruding beyond median lobes; posterior apex of anal opening located 2.5–4.3 (3.6) times length of anal opening from base of median lobes; occurring on conifers (temperate species, from Connecticut, Delaware, Indiana, Maryland, New York, and Pennsylvania) (Fig. 15) . . . .
. . . . cryptomeria scale, *Aspidiotus cryptomeriae* Kuwana
Second lobes normally protruding beyond median lobes; posterior apex of anal opening located 1.3–2.2 (1.5) times length of anal opening from base of median lobes; rarely on conifers (tropical species; from Florida, Hawaii, and Georgia) (Fig. 16) . . . .
. . . . coconut scale, *Aspidiotus destructor* Signoret

106(93) Median and second lobes distinctly dissimilar in shape, or second lobes absent (Fig. 49) 116
Median and second lobes similar in shape (Fig. 14) 107

107(106) Plates in first space between median lobe and second lobe with more than 2 tines (Fig. 6) 110
Plates in first space between median lobe and second lobe simple or bifurcate, with little or no fringing (Fig. 41) 108

108(107) Median lobes without basal sclerosis; with conspicuous paraphyses attached to lateral margin of median lobes; on trees and shrubs (Fig. 42) 109
Median lobes with conspicuous basal sclerosis; without definite paraphysis attached to lateral margin of median lobes; on sugarcane and other grasses (from Florida, Hawaii, and Texas) (Fig. 14)
. . . . sugarcane scale, *Aspidiella sacchari* (Cockerell) (in part)

109(108) With distinct indentations between prothorax and mesothorax and between mesothorax and metathorax (on many tree species especially English walnut; widespread in U.S.) (Fig. 43) . . . .
. . . . walnut scale, *Diaspidiotus juglansregiae* (Comstock) (in part)
Without distinct indentations between prothorax and mesothorax and between mesothorax and metathorax (on many trees and shrubs; widespread in U.S.) (Fig. 41) . . . .
. . . . Forbes scale, *Diaspidiotus forbesi* (Johnson)

110(107) Large macroducts present on any or all of prepygidial segments 2, 3, and 4 (Fig. 8) 113
Large macroducts restricted to pygidium, absent from prepygidial segments 2, 3, and 4 111

111(110) Median lobes without basal sclerosis; median lobes about as wide as long (Fig. 6) 112
Median lobes with noticeable basal sclerosis; median lobes longer than wide (on many hosts; out-of-doors from Alabama, California, Florida, Georgia, Hawaii, Louisiana, Mississippi, and Texas, occasional in greenhouses elsewhere) (Fig. 19) . . . .
. . . . spinose scale, *Aspidiotus spinosus* Comstock (in part)

112(111) Anal opening large, 17–23 (21) μ long; anal opening located 1.3–1.8 (1.5) times length of anal opening from base of median lobes; second lobe conspicuously narrower than median lobes (on many hosts; out-of-doors from Alabama, California, Florida, Louisiana, and Texas, sometimes in greenhouses elsewhere) (Fig. 6) . . . . cyanophyllum scale, *Abgrallaspis cyanophylli* (Signoret) (in part)
Anal opening small, 10–19 (13) μ long; anal opening located 2.0–2.2 (2.1) times length of anal opening from base of median lobes; second lobe about same width as median lobes (on many hosts; widespread in U.S.) (Fig. 40) . . . . Putnam scale, *Diaspidiotus ancylus* (Putnam) (in part)

113(110) With at least 1 macroduct between median lobes (Fig. 8) 114
Without macroduct between median lobes (on conifers; from Pennsylvania and New York) (Fig. 89) . . . . false Meyer scale, *Nuculaspis pseudomeyeri* (Kuwana)

114(113) Third lobes sclerotized, obvious (Fig. 90) 115
Third lobes absent or reduced to unsclerotized point (on a few conifers; from 10 eastern states) (Fig. 8) . . . . hemlock scale, *Abgrallaspis ithacae* (Ferris)

115(114) With 8–15 (12) perivulvar pores on each side of body; total of 36–88 (54) macroducts on each side of body; on conifers (from Connecticut, New Jersey, and Rhode Island) (Fig. 90) . . . .
. . . . shortneedle conifer scale, *Nuculaspis tsugae* (Marlatt)
With 1–6 (4) perivulvar pores on each side of body; total of 16–36 (30) macroducts on each side of body; common on camellia (from California, Missouri, and Oregon) (Fig. 7) . . . .
. . . . degenerate scale, *Abgrallaspis degenerata* (Leonardi) (in part)

116(106) Without plates anterior of seta marking segment 6 (Fig. 42) 123
With plates anterior of seta marking segment 6 (Fig. 49) 117

117(116) With macroduct between median lobes (Fig. 6) 119

Without macroduct between median lobes 118

118(117) Anal opening large, greater than 19μ long; axis of median lobes usually converging apically (on many hosts, rare on grape; widespread in U.S.) (Fig. 65) . . . . latania scale, *Hemiberlesia lataniae* (Signoret) (in part)

Anal opening small, less than 18μ long; axis of median lobes usually parallel or diverging apically (on grape; widespread in U.S.) (Fig. 49) . . . . grape scale, *Diaspidiotus uvae* (Comstock)

119(117) With more than 5 perivulvar pores on each side of body; without club-shaped paraphysis attached to lateral margin of median lobes 120

With less than 5 perivulvar pores on each side of body; conspicuous club-shaped paraphysis attached to lateral margin of median lobes (usually on trees such as elm; widespread in U.S.) (Fig. 36) . . . . elm armored scale, *Clavaspis ulmi* (Johnson) (in part)

120(119) Anal opening large, greater than 16μ long 121

Anal opening small, less than 15μ long (on many hosts; out-of-doors from Alabama, California, Florida, Georgia, Hawaii, Louisiana, Mississippi, and Texas, occasionally found in greenhouses) (Fig. 19) . . . . spinose scale, *Aspidiotus spinosus* Comstock (in part)

121(120) Plates anterior of seta marking segment 6 without fimbriations (Fig. 39) 122

Plates anterior of seta marking segment 6 fimbriate (on many hosts; out-of-doors from Alabama, California, Florida, Louisiana, and Texas, occasionally found in greenhouses elsewhere) (Fig. 6) . . . . cyanophyllum scale, *Abgrallaspis cyanophylli* (Signoret) (in part)

122(121) Anal opening large, 20μ long or more (on many hosts; widespread in U.S.) (Fig. 65) . . . . latania scale, *Hemiberlesia lataniae* (Signoret) (in part)

Anal opening small, 20μ long or less (on many hosts; widespread in U.S.) (Fig. 39) . . . . Putnam scale, *Diaspidiotus ancylus* (Putnam) (in part)

123(116) Large macroducts absent from dorsum of segment 3 126

Large macroducts present on dorsum of segment 3 (Fig. 47) 124

124(123) Dorsal macroducts numerous, with 50–109 (83) on each side of body 125

Dorsal macroducts few, with 25–40 (35) on each side of body (on many trees and shrubs; from 22 cooler northern states) (Fig. 47) . . . . European fruit scale, *Diaspidiotus ostreaeformis* (Curtis)

125(124) Median lobes without conspicuous basal sclerosis; with paraphysis attached to lateral margin of median lobes (on willow and poplar; from 11 cooler northern states) (Fig. 42) . . . . poplar scale, *Diaspidiotus gigas* (Thiem and Gerneck)

Median lobes with conspicuous basal sclerosis; without definite paraphysis attached to lateral margin of median lobes (on sugarcane and other grasses; from Florida, Hawaii, and Texas) (Fig. 14) . . . . sugarcane scale, *Aspidiella sacchari* (Cockerell) (in part)

126(123) Anal opening small, less than 20μ long 127

Anal opening large, more than 20μ long (on many hosts; widespread in U.S.) (Fig. 65) . . . . latania scale, *Hemiberlesia lataniae* (Signoret) (in part)

127(126) Second lobes absent or represented by unsclerotized point; medial margin of median lobes distinctly shorter than lateral margin (common on oak and walnut; widespread in U.S.) (Fig. 46) . . . . Osborn scale, *Diaspidiotus osborni* (Newell and Cockerell)

Second lobes represented by sclerotized swelling about same shape as median lobes, but smaller; medial margin of median lobes about same size as lateral margin (on many trees and shrubs; widespread in U.S.) (Fig. 41) . . . . Forbes scale, *Diaspidiotus forbesi* (Johnson) (in part)

Field Key to Economic Armored Scales

1   Not on conifers 15
    On conifers 2

2(1)   Scale cover of adult female on needles or fruit 3
    Scale cover of adult female on bark (rarely on conifers; widespread) (Plate 40) ....
    .... walnut scale, *Diaspidiotus juglansregiae* (Comstock) (in part)

3(2)   Adult female cover with 2 shed skins; cover not formed by second-instar skin; adult female visible
    when cover removed 6
    Adult female cover with only 1 shed skin; cover formed by enlarged second-instar skin; adult female
    enclosed in second-instar shed skin 4

4(3)   Shed skin marginal; male cover white 5
    Shed skin submarginal; male cover gray or brown (common on juniper, also on other similar conifers
    [Cupressaceae]; from many states) (Plate 35) ..... redwood scale, *Cupressaspis shastae* (Coleman)

5(4)   Central black blotch absent from cover of adult female (primarily on hemlock; eastern U.S.) (Plate
    53) .... elongate hemlock scale, *Fiorinia externa* Ferris
    Central black blotch present on cover of adult female (on many conifers; eastern U.S.) (Plate 55) ..
    .... coniferous fiorinia scale, *Fiorinia japonica* Kuwana

6(3)   Adult female cover brown or black 8
    Adult female cover white 7

7(6)   Mature female cover elongate; shed skins marginal ....
    1) pine needle scale, *Chionaspis pinifoliae* (Fitch) (Plate 27) (primarily on pines; widespread)
    2) pine scale, *C. heterophyllae* Cooley (Plate 26) (primarily on pines; widespread)
    Mature female cover circular or oval; shed skins central or subcentral ....
    1) juniper scale, *Carulaspis juniperi* (Bouché) (Plate 21) (on junipers or similar conifers [Cupres-
    saceae]; widespread)
    2) minute cypress scale, *Carulaspis minima* (Targioni Tozzetti) (Plate 20) (on junipers or similar
    conifers; widespread)

8(6)   Female cover circular, oval, or elongate oval, not oyster-shell shaped; shed skins central or
    subcentral 10
    Female oyster-shell shaped; shed skins marginal 9

9(8)   Mature female cover broadly oyster-shell shaped; often at base of needles in sheath well concealed
    (primarily on pine needles; from a few eastern states) (Plate 72) ....
    .... pine oystershell scale, *Lepidosaphes pini* (Maskell)
    Mature female cover narrowly oyster-shell shaped; often exposed on foliage, not concealed (on juniper
    and similar conifers [Cupressaceae]; from a few eastern states) (Plate 71) ....
    .... Maskell scale, *Lepidosaphes pallida* (Maskell)

10(8)   Adult female cover oval to elongate oval 12
    Adult female cover circular 11

11(10)   Adult female cover delicate, translucent, clear or red, without concentric rings; body yellow to brown;
    (primarily on yews and podocarpus; from a few southeastern states) (Plate 11) ....
    .... Asiatic red scale, *Aonidiella taxus* Leonardi
    Adult female cover thick, opaque, dark brown to black, often with light-colored concentric rings;
    body clear or light pink (polyphagous, common on araucaria; outside in California and Hawaii, rarely
    taken in greenhouses elsewhere) (Plate 76) ....
    .... black araucaria scale, *Lindingaspis rossi* (Maskell) (in part)

12(10)   Mature adult female body normally unsclerotized except on pygidium; without retracted pygidium
   13
    Mature adult female body heavily sclerotized; with pygidium retracted (primarily on pine needles;
    widespread) (Plate 84) .... black pineleaf scale, *Nuculaspis californica* (Coleman)

13(12)   Scale cover occurring under epidermis of leaf 14
    Scale cover occurring on surface of epidermis of leaf
    1) false Meyer scale, *Nuculaspis pseudomeyeri* (Kuwana) (Plate 85) (on juniper, chamaecyparis and
    similar conifers [Cupressaceae]; from the District of Columbia, New York, Pennsylvania)

2) shortneedle conifer scale, *Nuculaspis tsugae* (Marlatt) (Plate 86) (on 8 diverse coniferous genera; Connecticut, New Jersey, Rhode Island)

14(13) Newly formed adult female cover nearly flat, translucent (on many coniferous hosts; northeastern U.S.) (Plate 13) . . . . cryptomeria scale, *Aspidiotus cryptomeriae* Kuwana
Newly formed adult female cover convex, opaque . . . .
1) hemlock scale, *Abgrallaspis ithacae* (Ferris) (Plate 6) (on several coniferous hosts; eastern U.S.)
2) black araucaria scale, *Lindingaspis rossi* (Maskell) (in part) (Plate 76) (polyphagous; outside in California and Hawaii, rare in greenhouses elsewhere) (in part)

15(1) Not on grasses 19
On grasses, including bamboo 16

16(15) Cover oval or round, not oyster-shell shaped; on bamboo and other grasses 17
Cover elongate, oyster-shell shaped (on bamboo; several states) (Plate 66) . . . .
. . . . bamboo diaspidid, *Kuwanaspis pseudoleucaspis* (Kuwana)

17(16) Shed skins marginal; scale cover oval 18
Shed skins subcentral; scale cover circular (on many grass hosts, common on sugar cane; Florida, Hawaii, and Texas) (Plate 12) . . . . sugarcane scale, *Aspidiella sacchari* (Cockerell)

18(17) On bamboo only; scale cover light brown or tan (southern U.S.) (Plate 57) . . . .
. . . . penicillate scale, *Froggattiella penicillata* (Green)
Not on bamboo; scale cover white (southern U.S.) (Plate 87) . . . .
. . . . Bermuda grass scale, *Odonaspis ruthae* Kotinsky

19(15) Male cover of similar shape and texture as female cover, or male absent; shed skins on male cover not marginal 51
Male cover differently shaped than female cover; shed skins on male cover marginal 20

20(19) Shed skins on adult female cover marginal or submarginal; if in doubt, scale cover in most species elongate, oyster-shell shaped or oval 27
Shed skins on adult female cover central or subcentral; if in doubt, scale cover in most species circular 21

21(20) Occurring primarily on leaves or cactus pads 24
Occurring primarily on stems or bark 22

22(21) Body of adult female yellow, sometimes with pink tinge when laying eggs 23
Body of adult female pink or red (on fruit trees; from 9 states) (Plate 52) . . . .
. . . . Italian pear scale, *Epidiaspis leperii* (Signoret)

23(22) Eggs pink (polyphagous; primarily from Pennsylvania southward) (Plate 104) . . . .
. . . . white prunicola scale, *Pseudaulacaspis prunicola* (Maskell) (in part)
Eggs pink and white (common on prunus species; northeastern U.S.) (Plate 103) . . . .
. . . . white peach scale, *Pseudaulacaspis pentagona* (Targioni Tozzetti) (in part)

24(21) Rare on cactus 25
Common on cactus (outside in southern areas, common in greenhouses) (Plate 48) . . . .
. . . . cactus scale, *Diaspis echinocacti* (Bouché)

25(24) Scale cover large, 1.5–2.5 mm; common on orchids, bromeliads, and palms in greenhouses and conservatories throughout U.S. 26
Scale cover small, 1.0–1.5 mm; common on desert plants and ivy in dry southwestern U.S. (Plate 112)
. . . . yucca scale, *Situlaspis yuccae* (Cockerell)

26(25) Common on orchids and palms (widely distributed, especially in greenhouses) (Plate 46) . . . .
. . . . Boisduval scale, *Diaspis boisduvalii* Signoret
Common on bromeliads (widely distributed, especially in greenhouses) (Plate 47) . . . .
. . . . pineapple scale, *Diaspis bromeliae* (Kerner)

27(20) Eggs red, purple, or dark orange; when eggs absent, body of adult female red, brownish red, purple, or dark orange 38
Eggs yellow, white, pink, light orange, or light brown; when eggs absent, body of adult female yellow, yellow brown, or light orange 28

28(27) Scale cover of adult female other than white 32
Scale cover of adult female white 29

29(28) Scale cover of adult female round or oval 30
Scale cover of adult female oyster-shell shaped (polyphagous including cycads; from many southern states including Florida) (Plate 102) . . . . false oleander scale, *Pseudaulacaspis cockerelli* (Cooley)

30(29) Not commonly collected on cycads 31
On cycads (Florida) (Plate 20) . . . . cycad aulacaspis scale, *Aulacaspis yasumatsui* Takagi

31(30) Eggs pink (polyphagous; primarily from Maryland southward) (Plate 104) . . . .
. . . . white prunicola scale, *Pseudaulacaspis prunicola* (Maskell) (in part)
Eggs pink and white (common on prunus species; northeastern U.S.) (Plate 103) . . . .
. . . . white peach scale, *Pseudaulacaspis pentagona* (Targioni Tozzetti) (in part)

32(28) Scale cover of adult female not black with a white fringe; body of female without marginal lobe adjacent to mouthparts 33

Scale cover of adult female black with white fringe; body of female with marginal lobe adjacent to mouthparts (on citrus; Florida) (Plate 96) . . . . black parlatoria scale, *Parlatoria ziziphi* (Lucas)

33(32) Not pupillarial, with adult female directly under cover; cover with first- and second-instar shed skins 34
Pupillarial, with adult female inside second-instar shed skin; cover with only first-instar shed skin. . . . .
1) palm fiorinia scale, *Fiorinia fioriniae* (Targioni Tozzetti) (in part) (Plate 54) (polyphagous; primarily from eastern U.S., outside in southern areas)
2) tea scale, *Fiorinia theae* Green (Plate 56) (on many ornamental hosts especially camellia and holly; outside in southeastern U.S. sometimes overwintering as far north as Maryland)

34(33) Cover of adult female usually brown 35
Cover of adult female usually white or gray (polyphagous, often collected on hibiscus; outside in southern U.S.) (Plate 98) . . . . lesser snow scale, *Pinnaspis strachani* (Cooley)

35(34) Adult female cover large, 1.5–2.5 mm long; obvious on host 36
Adult female cover small, 0.8–1.2 mm long; difficult to locate, occurring in cracks of bark on host (primarily on *Prunus* species; eradicated from California) (Plate 82) . . . .
. . . . Hall scale, *Mercetaspis halli* (Green)

36(35) Normally not collected on euonymus 37
Frequently collected on euonymus (widespread in U.S.) (Plate 114) . . . .
. . . . euonymus scale, *Unaspis euonymi* (Comstock)

37(36) Widespread in U.S., in greenhouses in northern areas, outside in south as far north as Maryland (frequently collected on citrus, liriope, aspidistra, or various ferns; reported on about 70 host genera) (Plate 97) . . . . fern scale, *Pinnaspis aspidistrae* (Signoret)
Reported from Florida and Louisiana in U.S. (frequently collected on citrus, reported on about 10 host genera) (Plate 113) . . . . citrus snow scale, *Unaspis citri* (Comstock)

38(27) Adult female cover circular or oval 43
Adult female cover oyster-shell shaped 39

39(38) Not commonly collected on oaks . . . . (*Chionaspis*) . . . 40
On oaks (from 10 southern states) (Plate 108) . . . . oak scale, *Quernaspis quercus* (Comstock)

40(39) Not commonly collected on dogwood 41
Most frequently collected on dogwood (in northern U.S.) (Plate 24) . . . .
. . . . dogwood scale, *Chionaspis corni* Cooley

41(40) Rarely collected on willow or poplar 42
Commonly collected on willow and poplar (widespread in U.S.) (Plate 28) . . . .
. . . . willow scale, *Chionaspis salicis* (Linnaeus)

42(41) Most frequently collected on fruit trees such as apples and pears (widespread in U.S.) (Plate 25) . . .
. . . . scurfy scale, *Chionaspis furfura* (Fitch)
Most frequently collected on elm and hackberry (widespread in U.S.) (Plate 23) . . . .
. . . . elm scurfy scale, *Chionaspis americana* Johnson

43(38) Male cover without 3 distinct longitudinal ridges . . . . (*Parlatoria*) . . . 45
Male cover with 3 distinct longitudinal ridges 44

44(43) Occurring in Florida (commonly collected on mango, but polyphagous) (Plate 19) . . . .
. . . . white mango scale, *Aulacaspis tubercularis* Newstead
Occurring throughout U.S. including Florida (generally restricted to rosaceous hosts, most commonly collected on roses and *Rubus*) (Plate 18) . . . . rose scale, *Aulacaspis rosae* (Bouché)

45(43) Not commonly collected on camellia 46
Commonly collected on camellia, but polyphagous (occurring wherever camellias grow) (Plate 90) .
. . . . camellia parlatoria scale, *Parlatoria camelliae* Comstock (in part)

46(45) Not commonly collected on pittosporum 47
Commonly collected on pittosporum, but polyphagous (from California) (Plate 93) . . . .
. . . . pittosporum scale, *Parlatoria pittospori* Maskell (in part)

47(46) Not commonly collected on date palms; scale cover without white powder 48
Commonly collected on date palms; scale cover dusted with white powder (eradicated from Arizona, California, and Texas) (Plate 89) . . . .
. . . . parlatoria date scale, *Parlatoria blanchardi* (Targioni Tozzetti) (in part)

48(47) Not commonly collected on citrus 49
Commonly collected on citrus, but polyphagous (outside in southern citrus growing areas, in greenhouses in northern areas) (Plate 92) . . . . chaff scale, *Parlatoria pergandii* Comstock (in part)

49(48) Not commonly collected on orchids 50
Commonly collected on orchids, but polyphagous (usually in greenhouses) (Plate 94) . . . .
. . . . proteus scale, *Parlatoria proteus* (Curtis) (in part)

50(49) Commonly collected on fruit trees, especially olive, but polyphagous (from Arizona, California, Delaware, and Maryland) (Plate 91) . . . . olive scale, *Parlatoria oleae* (Colvee) (in part)
Commonly collected on maple, acuba, and euonymus, but polyphagous (from the District of Columbia, Hawaii, Maryland, Mississippi, North Carolina, Texas, and Virginia) (Plate 95) . . . .
. . . . tea parlatoria scale, *Parlatoria theae* Cockerell (in part)

51(19) Adult female cover with at least 1 shed skin (when covered with wax, probe shed skin area with pin to ascertain presence of shed skins) 52
Adult female cover without shed skin (when covered with wax, probe apparent shed skin area with pin to ascertain absence of shed skins) (on palms; from southern states outside, occasionally in other states indoors) (Plate 34) . . . . palmetto scale, *Comstockiella sabalis* (Comstock)

52(51) Shed skins on adult female cover central or subcentral; if in doubt, scale cover in most species circular 79
Shed skins on adult female cover marginal or submarginal; if in doubt, scale cover in most species elongate, oyster-shell shaped or oval 53

53(52) Not pupillarial, with female directly under cover; cover with first- and second-instar shed skins 55
Pupillarial, with adult female inside second-instar shed skin; cover with only first-instar shed skin 54

54(53) Body of adult female white; cover overlain by white wax (polyphagous; eastern U.S.) (Plate 77) . . . . Japanese maple scale *Lopholeucaspis japonica* (Cockerell)
Body of adult female brownish yellow or brownish orange; cover without white wax (polyphagous; primarily from eastern U.S., outside in southern areas) (Plate 54) . . . . palm fiorinia scale, *Fiorinia fioriniae* (Targioni Tozzetti) (in part)

55(53) Cover of adult female oval or circular 65
Cover of adult female elongate 56

56(55) Cover of adult female oyster-shell shaped, expanded posteriorly 60
Cover of adult female parallel sided, not expanded posteriorly 57

57(56) Scale cover of adult female brown 58
Scale cover of adult female black (polyphagous; outside in southern states and Hawaii, common in greenhouses) (Plate 65) . . . . black thread scale, *Ischnaspis longirostris* (Signoret)

58(57) Body small, 1.0–2.0 mm 59
Body large, 2.5–3.5 mm (most commonly found on citrus; from citrus-growing areas) (Plate 70) . . . . Glover scale, *Lepidosaphes gloverii* (Packard)

59(58) Commonly collected on stems and bark of litchi (also on other tropical hosts; Florida) (Plate 7) . . . . litchi scale, *Andaspis punicae* (Laing)
Not reported from litchi (most often found on fruit trees and holly; California, Florida, Georgia, and Hawaii) (Plate 83) . . . . Harper scale, *Neopinnaspis harperi* McKenzie

60(56) Not commonly collected on citrus 61
Commonly collected on citrus, but polyphagous (outside in citrus-growing areas, in greenhouses elsewhere) (Plate 67) . . . . purple scale, *Lepidosaphes beckii* (Newman)

61(60) Not commonly collected on camellia 62
Commonly collected on leaves of camellia (from about 12 other hosts; primarily from eastern U.S.) (Plate 68) . . . . camellia scale, *Lepidosaphes camelliae* Hoke

62(61) Not commonly collected on fig 63
Commonly collected on fig, but polyphagous (from California) (Plate 70) . . . . fig scale, *Lepidosaphes conchiformis* (Gmelin)

63(62) Not commonly collected on orchid 64
Commonly collected on orchids, especially cymbidiums (mostly on orchids; mostly in greenhouses) (Plate 73) . . . . cymbidium scale, *Lepidosaphes pinnaeformis* (Bouché)

64(63) Commonly collected on winged euonymus or firebush, on several other trees and shrubs (from eastern U.S.) (Plate 75) . . . . fire bush scale, *Lepidosaphes yanagicola* Kuwana
Polyphagous, frequently on lilac, beech, birch, ash, maple, willow, elm, boxwood, apple, pear, and prunus in U.S. (widely distributed, rare or absent in southern areas) (Plate 74) . . . . oystershell scale, *Lepidosaphes ulmi* (Linnaeus)

65(55) Scale cover of adult female not entirely black, sometimes with central black area 66
Scale cover of adult female entirely black (polyphagous, common on sea grape and agave; outside in Florida, rare in greenhouses elsewhere) (Plate 105) . . . . Bowrey scale, *Pseudischnaspis bowreyi* (Cockerell)

66(65) Body of adult female not yellow 68
Body of adult female yellow 67

67(66) Not found on cycads; newly formed adult female covers translucent . . . .
1) gray scale, *Pseudoparlatoria ostreata* Cockerell (Plate 106) (polyphagous; Florida)
2) false parlatoria scale, *Pseudoparlatoria parlatorioides* (Comstock) (Plate 107) (polyphagous; outside in Alabama, Florida, Georgia, Texas, rarely in greenhouses elsewhere)
Frequently found on cycads; newly formed adult female covers opaque (also common on bird-of-paradise; outside in California and Hawaii, occasionally found in greenhouses elsewhere) (Plate 59) . . . . zamia scale, *Furchadaspis zamiae* (Morgan)

68(66) Adult female body brown, pink, purple, or red when newly molted 69
Adult female body white when newly molted (polyphagous; Florida and Hawaii) Plate 64) . . . . mining scale, *Howardia biclavis* (Comstock) (in part)

69(68) Shed skins of adult female large; about one-third size of wax portion of scale cover 71
Shed skins of adult female small; much less than one-third size of wax portion of scale cover ....
.... 70

70(69) Not under flakes of bark, exposed on surface of host; on bark and leaves
    1) camphor scale, *Pseudaonidia duplex* (Cockerell) (in part) (Plate 99) (polyphagous; southeastern
    U.S.)
    2) trilobe scale, *Pseudaonidia trilobitiformis* (Green) (in part) (Plate 101) (polyphagous; Florida)
Often under flakes of bark; usually on trunks and stems (common on camellia and rhododendron,
polyphagous; occurring in 18 states including Florida) (Plate 100) ....
.... peony scale, *Pseudaonidia paeoniae* (Cockerell) (in part)

71(69) Not commonly collected on camellia 72
Commonly collected on camellia, but polyphagous (occurring wherever camellias grow) (Plate 90) .
.... camellia parlatoria scale, *Parlatoria camelliae* Comstock (in part)

72(71) Not commonly collected on pittosporum 73
Commonly collected on pittosporum, but polyphagous (from California) (Plate 93) ....
.... pittosporum scale, *Parlatoria pittospori* Maskell (in part)

73(72) Not commonly collected on date palms 74
Commonly collected on date palms (eradicated from Arizona, California, and Texas) (Plate 89) ....
.... parlatoria date scale, *Parlatoria blanchardi* (Targioni Tozzetti) (in part)

74(73) Not commonly collected on citrus 75
Commonly collected on citrus, but polyphagous (occurring outside in southern citrus-growing areas,
also in greenhouses in northern areas) (Plate 92) ....
.... chaff scale, *Parlatoria pergandii* Comstock (in part)

75(74) Not commonly collected on orchids 76
Commonly collected on orchids, but polyphagous (usually in greenhouses) (Plate 94) ....
.... proteus scale, *Parlatoria proteus* (Curtis) (in part)

76(75) Shed skins of adult female cover dark brown or black, rarely yellow 77
Shed skins of adult female cover light brown 78

77(76) Commonly collected on fruit trees, especially olive (from Arizona, California, Delaware, and Maryland) (Plate 91) .... olive scale, *Parlatoria oleae* (Colvee) (in part)
Commonly collected on maple, acuba, and euonymus, but polyphagous (from the District of Columbia, Hawaii, Maryland, Mississippi, North Carolina, Texas, and Virginia) (Plate 95) ....
.... tea parlatoria scale, *Parlatoria theae* Cockerell (in part)

78(76) Shed skins of adult females with black stripe or spot (polyphagous, common on maple, acuba, and
euonymus; from District of Columbia, Hawaii, Maryland, Mississippi, North Carolina, Texas, Virginia) (Plate 95) .... tea parlatoria scale, *Parlatoria theae* Cockerell (in part)
Shed skins of adult females without black stripe or spot (on many trees and shrubs; from California,
Florida, Missouri) (Plate 88) .... Chinese obscure scale, *Parlatoreopsis chinensis* (Marlatt)

79(52) Body of egg-laying females about same shape as newly matured females, without lateral margins of
body developed into lobes that surround pygidium making body kidney shaped 83
Body of egg-laying females of different shape than newly matured females, with lateral margins of
body developed into lobes that surround pygidium making body kidney shaped 80

80(79) Two shed skins present in adult female cover; not forming pits or galls 81
One shed skin present in adult female cover; forming pits or galls on leaves of liquidambar, also on
bark (widespread in U.S.) (Plate 41) ....
.... sweetgum scale, *Diaspidiotus liquidambaris* (Kotinsky) (in part)

81(80) Scale cover of newly matured adult female translucent, body and/or eggs clearly visible through cover
.... 82
Scale cover of newly matured adult female opaque white, body and/or eggs barely or not visible
through cover (polyphagous, common on palms; outside in Florida, occasionally in greenhouses elsewhere) (Plate 10) .... Oriental armored scale, *Aonidiella orientalis* (Newstead) (in part)

82(81) Scale cover of adult female translucent, showing red or orange body color in living specimens; occurring on trunks, stems, leaves, or fruit of host (polyphagous, common on citrus; outside in Alabama,
California, Florida, Georgia, Louisiana, Texas, common in grocery stores) (Plate 8) ....
.... California red scale, *Aonidiella aurantii* (Maskell)
Scale cover of adult female translucent, showing yellow body color in living specimens; usually on
leaves and fruit of host (polyphagous, common on citrus; outside in California, Florida, Texas) (Plate
9) .... yellow scale, *Aonidiella citrina* (Coquillett)

83(79) Body of adult female without notches on lateral margin of thorax 91
Body of adult female with at least 1 notch on lateral margin of thorax 84

84(83) Body of adult female not yellow 87
Body of adult female yellow 85

85(84) Scale cover of adult female not white, sometimes semitransparent 86
Scale cover of adult female white, not transparent (on *Euphorbia*; California) (Plate 110) ....
.... white euphorbia scale, *Selenaspidus albus* (McKenzie)

86(85) Body of adult female with single notch; normally on leaves and fruit of host (polyphagous; Florida) (Plate 111) . . . . rufous scale, *Selenaspidus articulatus* (Morgan)
Body of adult female with 2 notches; normally on bark and stems (polyphagous, common on walnut; widespread) (Plate 40) . . . . walnut scale, *Diaspidiotus juglansregiae* (Comstock) (in part)

87(84) Usually occurring under flakes of bark; on bark only 88
Not under flakes of bark, exposed on surface of host; on bark and leaves . . . .
1) camphor scale, *Pseudaonidia duplex* (Cockerell) (in part) (Plate 99) (polyphagous; southeastern U.S.)
2) trilobe scale, *Pseudaonidia trilobitiformis* (Green) (in part) (Plate 101) (polyphagous; Florida)

88(87) Body of adult female with mesothorax of same size or larger than other thoracic segments 89
Body of adult female with mesothorax conspicuously smaller than other thoracic segments (common on camellia, polyphagous; Florida and Hawaii) (Plate 50) . . . .
. . . . tesserate scale, *Duplaspidiotus tesseratus* (D'Emmerez de Charmoy)

89(88) Without circular areas on head and prothorax 90
Two circular areas on each side of head and prothorax (common on camellia, polyphagous; Florida and Hawaii) (Plate 49) . . . . camellia mining scale, *Duplaspidiotus claviger* (Cockerell)

90(89) Shed skins of adult female cover yellow or orange (common on camellia and rhododendron, polyphagous; occurring in 18 states including Florida) (Plate 100) . . . .
. . . . peony scale, *Pseudaonidia paeoniae* (Cockerell) (in part)
Shed skins of adult female cover brown (polyphagous; Florida and Hawaii) (Plate 64) . . . .
. . . . mining scale, *Howardia biclavis* (Comstock) (in part)

91(83) Not pupillarial; adult female body evident under scale cover 92
Pupillarial; adult female body inside second-instar shed skin (common on bromeliads; outside in Florida, in greenhouses elsewhere) (Plate 60) . . . .
. . . . flyspeck scale, *Gymnaspis aechmeae* Newstead

92(91) Two shed skins present in adult female cover; not forming pits or galls 93
One shed skin present in adult female cover; forming pits or galls (on leaves of liquidambar, also on bark, but not forming galls; widespread in U.S.) (Plate 41) . . . .
. . . . sweetgum scale, *Diaspidiotus liquidambaris* (Kotinsky)(in part)

93(92) Shed skins of adult female cover not black when rubbed 99
Shed skins of adult female cover black when rubbed 94

94(93) Body of mature adult female clear, light pink, or purple 96
Body of mature adult female white or yellow 95

95(94) Crawler flap inconspicuous, not curved upward at end of female cover; ventral cover thin . . . .
1) latania scale, *Hemiberlesia lataniae* (Signoret) (in part) (Plate 61) (polyphagous; widespread)
2) false diffinis scale, *Hemiberlesia neodiffinis* Miller and Davidson (Plate 62) (primarily on trees; eastern U.S. and Texas)
3) greedy scale, *Hemiberlesia rapax* (Comstock) (in part) (Plate 63) (polyphagous; widespread)
Crawler flap conspicuous, curved upward at end of female cover; ventral cover thick (polyphagous; Florida, Hawaii) (Plate 82) . . . . plumose scale, *Morganella longispina* (Morgan)

96(94) Normally found on oaks 97
Rarely collected on oaks 98

97(96) Restricted to eastern part of U.S. except for introduced population near Sacramento, California (Plate 79) . . . . obscure scale, *Melanaspis obscura* (Comstock)
Restricted to California, Arizona, and New Mexico (Plate 78) . . . .
. . . . dark oak scale, *Melanaspis lilacina* (Cockerell)

98(96) Present outside in California and Hawaii; often collected on conifers, especially araucaria (Plate 76)
. . . . black araucaria scale, *Lindingaspis rossi* (Maskell) (in part)
Present in eastern U.S.; usually collected on maple (Plate 80) . . . .
. . . . gloomy scale, *Melanaspis tenebricosa* (Comstock)

99(93) Rarely on stems and bark of grape 100
Often on bark of grape (primarily in eastern U.S.) (Plate 45) . . . .
. . . . grape scale *Diaspidiotus uvae* (Comstock)

100(99) Body of adult female not visible through scale cover 101
Body of adult female usually visible through scale cover . . . .
1) cyanophyllum scale, *Abgrallaspis cyanophylli* (Signoret) (Plate 4) (polyphagous; outside in southern U.S.)
2) coconut scale, *Aspidiotus destructor* Signoret (Plate 14) (polyphagous; outside in Florida, Hawaii, and Georgia)
3) aglaonema scale, *Aspidiotus excisus* Green (Plate 15) (polyphagous; outside in Florida)
4) oleander scale, *Aspidiotus nerii* Bouché (Plate 16) (polyphagous; outside in southern U.S.)
5) herculeana scale, *Clavaspis herculeana* (Cockerell and Hadden) (in part) (Plate 32) (polyphagous, especially tropical plants; outside in Florida, Hawaii, Texas)
6) holly scale, *Dynaspidiotus britannicus* (Newstead) (Plate 51) (northern U.S.; common on holly and laurel)

101(100) Adult females rarely, or not, occurring on leaves of host 112
Adult females predominantly on leaves of host 102

102(101) Scale cover of adult female not dark brown, black, or dark reddish brown 104
Scale cover of adult female dark brown, black, or dark reddish brown 103

103(102) Not commonly collected on orchids; posterior end of male cover same color as rest of cover ......
1) Florida red scale *Chrysomphalus aonidum* (Linnaeus) (Plate 29) (polyphagous; outside in Florida, Georgia, Hawaii, Louisiana, Mississippi, Texas, common in greenhouses elsewhere)
2) bifasciculate scale, *Chrysomphalus bifasciculatus* Ferris (Plate 30) (polyphagous, common on holly and citrus; Alabama, California, Georgia, Louisiana, North Carolina, South Carolina, Texas, Virginia)
3) dictyospermum scale, *Chrysomphalus dictyospermi* (Morgan) (Plate 31) (polyphagous; Alabama, Arizona, California, Florida, Georgia, Hawaii, Louisiana, Mississippi, New Mexico, South Carolina, Texas)
Commonly collected on orchids; posterior end of male cover conspicuously lighter than rest of cover (outside in Florida, Hawaii, greenhouses elsewhere) (Plate 58) ....
.... orchid scale, *Furcaspis biformis* (Cockerell)

104(102) Scale cover of adult female not white 108
Scale cover of adult female white 105

105(104) Adult female cover conspicuously convex ....
1) latania scale, *Hemiberlesia lataniae* (Signoret) (in part) (Plate 61) (polyphagous; widespread)
2) greedy scale, *Hemiberlesia rapax* (Comstock) (in part) (Plate 63) (polyphagous; widespread)
Adult female cover slightly convex 106

106(105) Thorax and head of old adult females brown and sclerotized 107
Thorax and head of old adult females yellow, not sclerotized (polyphagous, especially trees and shrubs; southern states) (Plate 17) .... spinose scale, *Aspidiotus spinosus* Comstock (in part)

107(106) Shed skins of adult female yellow or red (polyphagous, especially trees and shrubs; widespread) (Plate 36) .... Putnam scale, *Diaspidiotus ancylus* (Putnam) (in part)
Shed skins of adult female brown or transparent (polyphagous, especially palms; outside in Florida, occasionally found in greenhouses) (Plate 10) ....
.... Oriental armored scale, *Aonidiella orientalis* (Newstead) (in part)

108(104) Shed skins of adult female cover not yellow when rubbed 111
Shed skins of adult female cover yellow when rubbed 109

109(108) Cover of adult female light brown, cream, or opaque yellow 110
Cover of adult female usually gray (polyphagous, common on rosaceous hosts; widespread) (Plate 44)
.... San Jose scale, *Diaspidiotus perniciosus* (Comstock) (in part)

110(109) Cover of adult female light brown (polyphagous, especially trees and shrubs; southern states) (Plate 17) .... spinose scale, *Aspidiotus spinosus* Comstock (in part)
Cover of adult female cream or opaque yellow (common on camellia, also on other shrubs; California, Oregon, and Missouri) (Plate 5) .... degenerate scale, *Abgrallaspis degenerata* (Leonardi)

111(108) Cover of adult female caramel brown (polyphagous, especially tropical plants; Florida, Hawaii, and Texas) (Plate 32) .... herculeana scale, *Clavaspis herculeana* (Cockerell and Hadden) (in part)
Cover of adult female light brown (polyphagous, especially palms; outside in Florida, occasionally found in greenhouses) (Plate 10) ....
.... Oriental armored scale, *Aonidiella orientalis* (Newstead) (in part)

112(101) Body of old, living adult females without sclerotized and darkened marginal areas on head and thorax
.... 117
Body of old, living adult females with brown sclerotized area on at least marginal area of head and thorax; sometimes entire thorax and head sclerotized 113

113(112) Rarely collected on poplar and willow 114
Restricted to poplar and willow (northern U.S.) (Plate 39) ....
.... poplar scale, *Diaspidiotus gigas* (Thiem and Gerneck)

114(113) Rarely collected on oak 115
Most frequently collected on oak (also on other trees; widespread in U.S.) (Plate 42) ....
.... Osborn scale, *Diaspidiotus osborni* (Newell and Cockerell)

115(114) Rarely on composite and ericaceous hosts 116
Most frequently collected on composite and ericaceous hosts such as snake weed and cranberries (widespread) (Plate 109) .... Dearness scale, *Rhizaspidiotus dearnessi* (Cockerell)

116(115) Scale cover of adult female gray (when moved from under epidermis)......................
1) Putnam scale, *Diaspidiotus ancylus* (Putnam) (in part) (Plate 37) (polyphagous, especially trees and shrubs; widespread)
2) European fruit scale, *Diaspidiotus ostreaeformis* (Curtis) (Plate 43) (primarily on trees and large shrubs; northern states)
Scale cover of adult female light brown or cream (polyphagous, especially palms; outside in Florida, occasionally found in greenhouses) (Plate 10) ....
.... Oriental armored scale, *Aonidiella orientalis* (Newstead) (in part)

117(112) Cover of adult female flat or slightly convex 118
Cover of adult female strongly convex..
1) latania scale, *Hemiberlesia lataniae* (Signoret) (in part) (Plate 61) (polyphagous; widespread)

2) false diffinis scale, *Hemiberlesia neodiffinis* Miller and Davidson (Plate 62) (primarily on trees; eastern U.S. and Texas)

3) greedy scale, *Hemiberlesia rapax* (Comstock) (in part) (Plate 63) (polyphagous; widespread)

118(117) Shed skins of adult female cover yellow or orange when rubbed 119
Shed skins of adult female cover light to dark brown when rubbed (polyphagous, mostly tropical plants; Florida, Hawaii, and Texas) (Plate 32) ....
.... herculeana scale, *Clavaspis herculeana* (Cockerell and Hadden) (in part)

119(118) Rarely collected on elm 120
Commonly collected on elm (also on other trees and shrubs; widespread) (Plate 33) ....
.... elm armored scale, *Clavaspis ulmi* (Johnson)

120(119) Shed skins of adult female yellow when rubbed 121
Shed skins of adult female orange when rubbed (polyphagous, mainly on trees and shrubs; widespread) (Plate 38) .... Forbes scale, *Diaspidiotus forbesi* (Johnson)

121(120) Scale cover of adult female usually gray; second-instar female covers gray with subcentral white ring (polyphagous, common on rosaceous hosts; widespread) (Plate 44) ....
.... San Jose scale, *Diaspidiotus perniciosus* (Comstock) (in part)
Scale cover of adult female usually light brown; second-instar female covers brown without subcentral white ring (polyphagous, especially on trees and shrubs; southern states) (Plate 17).......
.... spinose scale, *Aspidiotus spinosus* Comstock (in part)

Abgrallaspis cyanophylli (Signoret)

Suggested Common Name Cyanophyllum scale (also called palm scale).

Common Synonyms and Combinations *Aspidiotus cyanophylli* Signoret, *Hemiberlesia cyanophylli* (Signoret), *Furcaspis cyanophylli* (Signoret), *Diaspidiotus cyanophylli* (Signoret), *Evaspidiotus cyanophylli* (Signoret).

Field Characters (Plate 4) Adult female cover variable, flat, irregularly circular to elongate oval, semitransparent, white to gray; shed skins central, yellow. Male cover elongate oval, same color and texture as female cover; shed skin submarginal. Body of adult female and eggs yellow. On leaves, stems, and fruit of host, normally on undersides of leaves.

Slide-mounted Characters Adult female (Fig. 6) with 3 pairs of well-developed lobes, usually with small fourth lobe; paraphyses weakly developed in first space, occasionally with 1 in second space. Median lobes separated by space 0.3–0.5 (0.4) times width of median lobe, with noticeable paraphysis-like sclerotization on medial margin of lobe, occasionally with smaller sclerotization on lateral margin, some specimens with heavily sclerotized area between lobes but no definite yoke, with medial and lateral margins parallel, with 1 lateral notch, with 1 medial notch; second lobes simple, noticeably narrower than median lobes, slightly shorter, with 1 lateral notch, 0–1 (1) medial notch; third lobes simple, slender, about one-half size of second lobe, without notches; fourth lobe usually represented by low series of sclerotized points. Plates with associated microduct on segment 5, microduct located near base, plates between first 3 pairs of lobes broad, apically fringed, plates anterior of lobe 3 variable, normally slightly branched, fimbriate on sides, occasionally simple, plate formula 2-3-3 or 2-3-4, with 2 broad, apically fringed plates between medial lobes, longer than median lobes. Macroducts of 1 size, 1 macroduct between median lobes, extending 1.2–1.7 (1.4) times distance between posterior apex of anal opening and base of median lobes, 34–54 (45) μ long, longest macroduct in first space 43–73 (55) μ long, with 8–18 (13) on each side of pygidium on segments 5 to 8, cactus-infesting forms normally with largest number of ducts, ducts in submarginal and marginal areas, some macroduct orifices anterior of anal opening; prepygidial macroducts usually absent. Pygidial microducts on venter in submarginal areas of segments 5 and 6, with 6–11 (8) ducts; prepygidial ducts of 1 size, submarginal ducts sometimes on head, scattered on thorax, usually present on each of segments 1 to 4, submedial ducts near each thoracic spiracle, anterior of mouthparts, and on segments 1 and 2; pygidial microducts absent from dorsum, of 3 sizes on prepygidial areas, short ducts present along body margin from head or prothorax to segment 2 or 3, long ducts present in same area from mesothorax or metathorax to segment 3 or 4, intermediate ducts in submedial areas of mesothorax or prothorax to segment 1, 2, or 3. Perivulvar pores in 4 or 5 groups, 7–12 (10) pores on each side of body. Perispiracular pores absent. Anal opening located 1.3–1.8 (1.5) times length of anal opening from base of median lobes, anal opening 17–23 (21) μ long. Dorsal seta laterad of median lobes 1.0–1.2 (1.1) times length of median lobe. Eyes represented by distinct spur on body margin at level of anterior spiracle. Antennae each with 1 seta. Cicatrices usually on prothorax and segment 1. Body pear shaped. Segment 4 without ventral marginal seta.

Affinities The cyanophyllum scale is most similar to *Hemiberlesia palmae*, the tropical palm scale, but is different in having the second lobes sclerotized, the plates anterior of lobe 3 only slightly branched, and the anal opening smaller and farther removed from the posterior apex than on *H. palmae*. The trop-

ical palm scale has unsclerotized second lobes and apically fringed plates anterior of lobe 3.

Hosts Polyphagous. Borchsenius (1966) records it from 75 genera in 18 families of host plants; Dekle (1977) reports it from 174 host genera. We have examined U.S. specimens from the following host genera: *Acacia, Annona, Begonia,* 'cactus,' *Castanea, Castanopsis, Cocos, Copernicia, Diospyros, Echinopsis, Ehretia, Eucalyptus, Eupritchardia, Ficus, Hepatica, Jasminum, Kleinhovia, Livistona, Malus, Mangifera,* 'palm,' *Pandanus, Persea, Phoenix, Porliera, Psidium, Randia, Tabebuia, Tamarindus,* and *Viola.* Common U.S. hosts are avocado, cactus, and palms. Based on material in the USNM Collection, cyanophyllum scale often is intercepted in quarantine on species of cactus, orchids, and palms, as well as crops such as citrus, banana, and mango.

Distribution We have examined U.S. specimens from AL, CA, CT, DC, FL, KS, LA, NH, NJ, NY, OH, PA, TX, VA. The species is reported from IN, MD, MS. The cyanophyllum scale occurs out-of-doors in the southern states only. It commonly is taken in quarantine from the Caribbean islands, Central and South America, Europe, Japan, and Mexico and is present in most tropical and subtropical areas of the world.

Biology The life history of this species has not been studied in the United States. Gerson and Zor (1973) examined the species on avocado in Israel. The species prefers the undersides of leaves of the host and settles near the main veins or the leaf edge. When it settles on branches it is found on smooth areas not on the roughened, corky bark. In Israel, crawlers are present year-round, and there are apparently four overlapping generations. Hadzibejli (1983) indicates that the species has two and a partial third generation in the western part of the Georgia (former Soviet Union). Both uniparental (Israel) and biparental (U.S.) strains are known (Dekle 1977; Brown 1965). According to He et al. (1998), in laboratory experiments the survival rate of this pest at 25 °C was high at 75.9%, whereas at 30 °C it was only 44.7%. The nymphal stages lasted from 37 to 64.5 days at 20 to 28 °C. The number of crawlers per female was greatest (142.1) at 28 °C. The average longevity of females decreased with increasing temperatures (200.2 days at 20 °C and 62.1 day at 30 °C). The survival rate at RH 75% was higher than that at either RH 60% or at 90%. The life history of this species also has been discussed by Shiao (1979, 1981). The scale completes about five generations a year. The female lays at least 60 eggs and about 93% hatch. The sex ratio (female : male) is approximately 3 : 5. In 1978 the nymphal and adult density reached its peak in August and September, respectively. The population was affected by temperature, rainfall, a parasitoid, and a fungus. The cyanophyllum scale apparently prefers warm humid areas; in Israel it is found along the coastal plain (Gerson and Zor 1973).

Economic Importance The cyanophyllum scale is a general pest in nurseries in the southern United States and in greenhouses in the North. Dekle (1977) records it as a serious pest of palms in Florida although it generally is controlled in Florida by 'wasp parasites.' According to Shiao (1979) this scale is one of the most important tea pests in northern Taiwan. It causes damage by producing chlorotic spots on the leaves and ultimately premature leaf drop. Wysoki (1997) records it as a pest of mango in Israel, and Silva (1950) reports it as a problem in cocoa in Brazil. Miller and Davidson (1990) consider this species to be a pest.

Selected References Davidson (1964); Gerson and Zor (1973); Komosinska (1969).

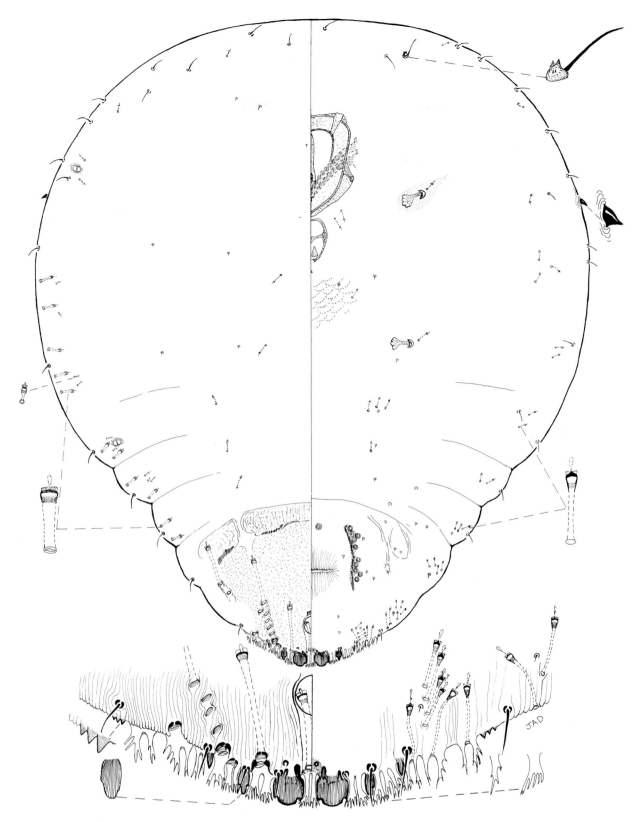

Figure 6. *Abgrallaspis cyanophylli* (Signoret), cyanophyllum scale, Daytona Beach, FL, on *Viola* sp., XII-1-1970.

Plate 4. *Abgrallaspis cyanophylli* (Signoret), Cyanophyllum scale

A. Five adult female covers and two adult male covers on cactus (R. J. Gill).

B. Newly molted adult female as seen through scale cover on *Crassula* (J. A. Davidson).

C. Mature covers of one adult female and several males on *Euphorbia* (R. J. Gill).

D. Heavy infestation on cactus (R. J. Gill).

E. Covers of dead scales on *Euphorbia* (R. J. Gill).

F. Male cover on *Crassula* (J. A. Davidson).

G. Body of young adult female on *Crassula* (J. A. Davidson).

40

Plate 5. *Abgrallaspis degenerata* (Leonardi), Degenerate scale

A. Unrubbed adult female cover on *Camellia* (J. A. Davidson).
B. Rubbed adult female cover on *Camellia* showing yellow crawler shed skin (J. A. Davidson).
C. Adult female cover on *Camellia* with crawler flap (D. R. Miller).
D. Old adult female covers on *Camellia* (D. R. Miller).

E. Male cover on *Camellia* (J. A. Davidson).
F. Male cover on *Camellia* turned over showing parasite larva in ruptured body (J. A. Davidson).
G. Distance view of female covers on *Camellia* (J. A. Davidson).
H. Body of adult female and crawler on *Camellia* (J. A. Davidson).

41

Abgrallaspis degenerata (Leonardi)

Suggested Common Name Degenerate scale.

Common Synonyms and Combinations *Chrysomphalus degeneratus* Leonardi, *Aspidiotus (Chrysomphalus) degeneratus* (Leonardi), *Aspidiotus degeneratus* (Leonardi), *Hemiberlesia degenerata* (Leonardi), *Diaspidiotus degeneratus* (Leonardi), *Dynaspidiotus degeneratus* (Leonardi).

Field Characters (Plate 5) Adult female cover slightly convex, circular, opaque yellow or cream colored; shed skins central or subcentral, yellow. Male cover elongate oval, same color and texture as female cover; shed skin submarginal. Body of adult female yellow. On leaves.

Slide-mounted Characters Adult female (Fig. 7) with 3 pairs of well-developed lobes, often with small fourth lobe; paraphyses weakly developed, when visible usually 2-2-0, occasionally 2-1-0, 3-3-0, 2-3-0. Median lobes separated by space 0.5–0.6 (0.6) times width of median lobe, with noticeable paraphysis-like sclerotization on medial margin of lobe, occasionally with smaller sclerotization on lateral margin, without yoke or basal sclerotization, with medial and lateral margins usually converging apically, sometimes parallel, with 1 lateral notch, with 1 medial notch; second lobes simple, noticeably narrower than median lobes, shorter, with 1 lateral notch, 0–1 (0) medial notch; third lobes simple, slender, about one-half size of second lobe, with 1–2 (2) lateral notches, 0–1 (0) medial notch; fourth lobe present in about half of specimens, when present usually bilobed, represented by low series of sclerotized points. Plates with associated microduct on segment 5, microduct located near base, plates between first 3 pairs of lobes broad, apically fringed, plates anterior of lobe 3 variable, usually simple with small basal fringing, plate formula usually 2-3-3, sometimes 2-3-4 or 2-3-2, with 2 broad, apically fringed plates between medial lobes, usually longer than median lobes. Macroducts of 2 sizes, large-size ducts in marginal and submarginal areas of segments 2 or 3 to 7 or 8, small-size ducts on venter in marginal or submarginal areas of meta- or mesothorax to segments 2 or 3, usually with 1 macroduct between median lobes, extending 1.2–1.6 (1.5) times distance between posterior apex of anal opening and base of median lobes, 38–45 (42) µ long, longest macroduct in first space 39–54 (45) µ long, with 10–16 (14) on each side of pygidium on segments 5 to 8, total of 16–36 (30) macroducts on each side of body, ducts in submarginal and marginal areas, some macroduct orifices anterior of anal opening; prepygidial macroducts present on meso- or metathorax to segment 4. Pygidial microducts on venter in submarginal areas of segment 5, with 2–4 (3) ducts; prepygidial ducts of 1 size, submarginal ducts on segments 2 or 3 to 4, submedial ducts near each thoracic spiracle and on any or all of segments 1 to 3; pygidial microducts absent from dorsum, of 1 or 2 sizes on prepygidial areas, short or long ducts present along body margin from head to metathorax or segment 1, long ducts present in submedial area of any or all of metathorax, segments 1, 2, or 3. Perivulvar pores in 4 groups, 1–6 (4) pores on each side of body. Perispiracular pores absent. Anal opening located 1.7–2.3 (2.0) times length of anal opening from base of median lobes, anal opening 13–15 (14) µ long. Dorsal seta laterad of median lobes 0.5–1.1 (0.7) times length of median lobe. Eyes variable, when present represented by distinct spur or small dome on body margin at level of anterior edge of clypeolabral shield. Antennae each with 1 seta. Cicatrices usually on prothorax and segment 1. Body pear shaped.

Affinities The degenerate scale is most similar to *Abgrallaspis fraxini*, the ash scale, but is different in lacking submedial macroducts on abdominal segments 1 to 3. These ducts are present in *A. fraxini*. For an additional comparison see affinities section of *Nuculaspis pseudomeyeri*.

Hosts Somewhat limited host range. Borchsenius (1966) records it from 8 genera in 5 families of host plants; McKenzie (1956) reports it from 6 genera in California. We have examined U.S. specimens from the following host genera: *Camellia, Cleyera, Euonymus, Eurya, Ilex, Schima,* and *Vaccinium*. The most common host of this scale seems to be camellia.

Distribution We have examined U.S. specimens from CA, MO, OR. We have seen material from China, England, Italy, and Japan, and it is recorded from Greece, Korea, and Portugal (Nakahara 1982). It commonly is taken in U.S. quarantine from Japan.

Biology The life history of this species apparently has not been studied.

Economic Importance At one time degenerate scale was considered to be a nursery pest in California, especially on camellias (McKenzie 1956), but in recent times it has become uncommon and of no economic importance (Gill 1997). Miller and Davidson (1990) consider this species to be a pest.

Selected References Gill (1997); McKenzie (1956).

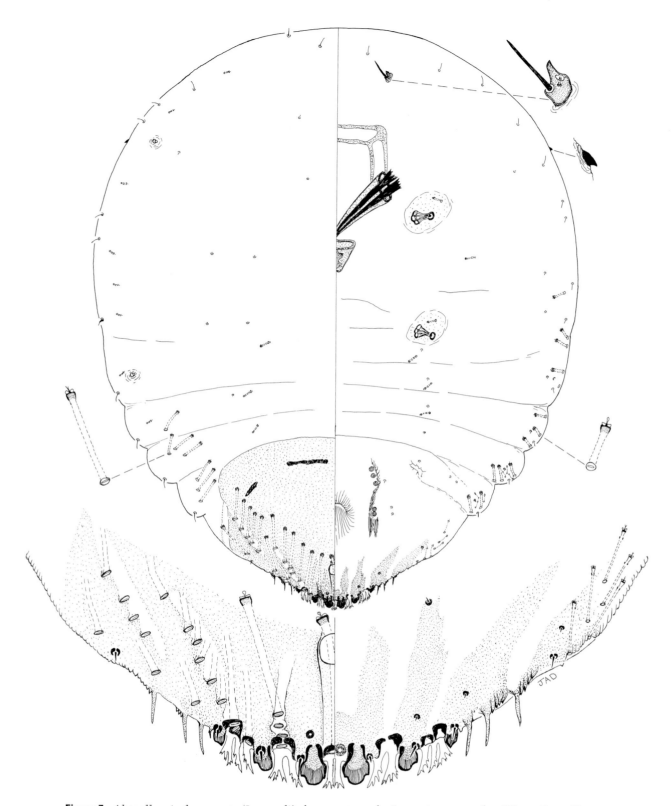

Figure 7. *Abgrallaspis degenerata* (Leonardi), degenerate scale, Japan, intercepted at HI, on *Camellia* sp., XII-2-1958.

Abgrallaspis ithacae (Ferris)

ESA Approved Common Name Hemlock scale.

Common Synonyms and Combinations *Aspidaspis ithacae* Ferris, *Gonaspidiotus ithacae* (Ferris).

Field Characters (Plate 6) Adult female cover slightly convex, circular to oval, gray to black with white margins; shed skins subcentral. Male cover elongate oval, blackish; shed skin submarginal. Body of adult female yellow green; eggs pale yellow. On undersides of needles and occasionally on petioles, usually under host epidermis.

Slide-mounted Characters Adult female (Fig. 8) with 2 pairs of well-developed lobes, usually with small third lobe; paraphyses absent. Median lobes separated by space 0.3–0.5 (0.4) times width of median lobe, with paraphysis-like sclerotization on medial and lateral margin of lobe, without yoke, with medial margins parallel or slightly converging apically, lateral margins parallel or diverging apically, with 0–1 (1) lateral notch, with 0–1 (1) medial notch; second lobes simple, variable in size, usually about same size and shape as median lobes, occasionally much smaller, with 0–1 (1) lateral notch, 0–1 (0) medial notch; third lobes simple, usually represented by inconspicuous, unsclerotized point. Plates with associated microduct on segment 5, microduct located near base, plates between first 3 pairs of lobes broad, apically fringed, plates anterior of lobe 3 variable, normally slightly branched, fimbriate on sides, occasionally simple, plate formula 2-3-3 or 2-3-4, with 2 apically fringed plates between medial lobes, about same length or slightly shorter than median lobes. Macroducts of 1 size, 1 macroduct between median lobes, extending 1.1–1.3 (1.2) times distance between posterior apex of anal opening and base of median lobes, 29–43 (35) μ long, longest macroduct in first space 34–48 (39) μ long, with 15–37 (25) on each side of pygidium on segments 5 to 8, ducts in submarginal and marginal areas, total macroducts 25–58 (38) on each side of body, some macroduct orifices anterior of anal opening; prepygidial macroducts present on segments 2 or 3 to 4. Pygidial microducts on venter in submarginal areas of segments 5 and sometimes 6, with 2–9 (5) ducts; prepygidial ducts of 2 sizes, larger size in submarginal areas of mesothorax and/or metathorax and segments 1 and/or 2, smaller size in submarginal areas of segments 2 or 3 to 4, submedial ducts near each thoracic spiracle, posterior of mouthparts, and on metathorax or segment 1 to segment 3; pygidial microducts absent from dorsum, of 2 sizes on prepygidial areas, short ducts present along body margin from head or prothorax to segment 2 or 3, long ducts present in same area from head or prothorax to segment 3 or 4, long ducts present in medial and submedial areas of mesothorax or metathorax to segment 1, 2, or 3. Perivulvar pores in 4 or 5 groups, 11–23 (17) pores on each side of body. Perispiracular pores absent. Anal opening located 1.2–1.7 (1.5) times length of anal opening from base of median lobes, anal opening 16–24 (20) μ long. Dorsal seta laterad of median lobes 1.0–1.3 (1.1) times length of median lobe. Eyes variable, when present represented by raised sclerotized area that may be rounded dome, single spur, or series or points, on body margin at slightly posterior of level of antenna. Antennae each with 1 seta. Cicatrices usually on prothorax and segment 1. Body pear shaped, thorax and head sclerotized in fully mature adult females.

Affinities The hemlock scale is most similar to *Nuculaspis abietus* but occurs in the eastern United States, has macroducts loosely distributed along the pygidial margin, has the space between the median lobes about 0.4 times the width of a median lobe, and has 25–58 (38) macroducts on each side of the body. *Nuculaspis abietus* occurs in the northwestern United States, has the macroducts restricted to the pygidial margin, has the space between the median lobes about equal to the width of a median lobe, and has more than 45 macroducts on each side of the body. The hemlock scale also is similar to *Nuculaspis californica*, the black pineleaf scale, but differs by having 3 pairs of lobes and by having the macroducts loosely distributed along the pygidial margin. *Nuculaspis californica* has 4 pairs of lobes and has the macroducts closely appressed to the pygidial margin.

Hosts Narrow range of hosts. Borchsenius (1966) records it from 4 genera in 1 family. We have examined material from *Abies* (fir) and *Tsuga* (hemlock). *Picea* (spruce), *Pinus* (pine), and *Pseudotsuga* (Douglas fir) have been reported; we suspect that spruce and Douglas fir are good hosts but are suspicious of the pine records.

Distribution We have examined U.S. specimens from IN, MD, NY, VA. The species is reported from CT, GA, OH, PA, TN, WV. Published records from ID, MT, OR, WA are apparently incorrect.

Biology The hemlock scale has been studied in detail by Stoetzel and Davidson (1974) on hemlock in Maryland. There are 2 generations per year and the overwintering stages are the second-instar male and female. Adults first appear in the middle of March and eggs and crawlers of the summer generation are present during June into early July. The summer generation develops rapidly, and by early August eggs and crawlers of the winter generation begin to appear. Crawlers are present from August to early October. The overwintering second instars are first observed in early September. The species has winged males. Feeding occurs primarily on the undersides of leaves.

Economic Importance Heavy infestations are reported to cause early leaf drop, dieback of branches (Pirone 1970), and defoliation of trees (Westcott 1973). Feeding females cause a yellow area on the upper surface of the needle; four to six individuals on a needle cause it to fall from the tree (Johnson and Lyon 1976). Miller and Davidson (1990) consider this species to be a pest.

Selected References Ferris (1938a); Kosztarab (1963); Stoetzel and Davidson (1974).

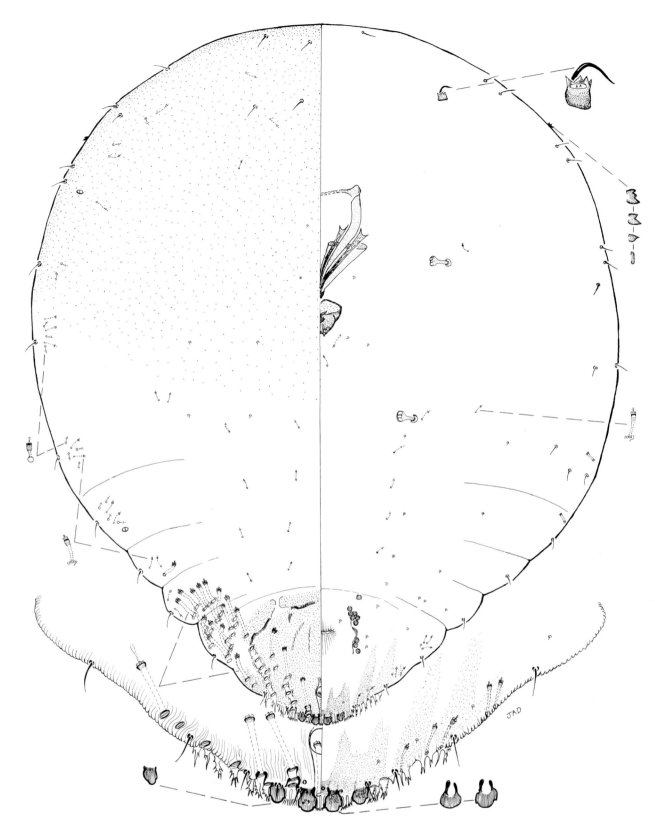

Figure 8. *Abgrallaspis ithacae* (Ferris), hemlock scale, Marion, VA, on *Tsuga canadensis*, VIII-5-1954.

Plate 6. *Abgrallaspis ithacae* (Ferris), **Hemlock scale**

A. Close-up of adult female cover with crawler flap on hemlock (J. A. Davidson).

B. Distance shot of adult female cover on hemlock (J. A. Davidson).

C. Male covers on hemlock (M. B. Stoetzel).

D. Male and crawler covers on hemlock (J. A. Davidson).

Plate 7. *Andaspis punicae* (Laing), Litchi scale

A. Adult female covers on litchi tree partially obscured by algae (J. A. Davidson).

B. Adult male cover on litchi (J. A. Davidson).

C. Adult male covers on litchi (J. A. Davidson).

D. Distance view of cryptic female covers on algae-covered litchi bark (J. A. Davidson).

E. Distance view of infestation on litchi (J. A. Davidson).

F. Distance view of heavy infestation on litchi (J. A. Davidson).

G. Distance view of old female bodies revealed after removal of overlying vegetative mass on litchi (J. A. Davidson).

H. Terminal dieback on heavily infested litchi tree (J. Peña).

I. Litchi tree suspected of dying as a result of scale infestation (J. Peña).

Andaspis punicae (Laing)

Suggested Common Name Litchi scale.

Common Synonyms and Combinations *Lepidosaphes punicae* Laing.

Field Characters (Plate 7) Adult female cover oyster-shell shaped, generally straight, usually brown, sometimes reddish brown or nearly black, with whitish powder along sides of cover; shed skins marginal, golden brown (Laing 1929). On stems and branches.

Slide-mounted Characters Adult female (Fig. 9) with 1 pair of distinct lobes, second, third, and fourth lobes represented by series of points; paraphyses incorporated in enlarged setal bases, present on medial and lateral margins of median lobes where enlarged, also present as narrow sclerotized band extending from dorsal setal base to medial margin of lobes 2, 3, and 4. Median lobes closely appressed, separated by space 0.1–0.2 (0.1) times width of median lobe, without basal sclerosis or yoke, medial margin parallel diverging about three-quarters of length, lateral margins converging, with 3–7 (5) lateral notches, 0–1 (1) medial notches; second lobes usually simple, rarely bilobed, about one-quarter size of median lobe, with 1–3 (2) lateral notches, absent medially; third lobes simple, wider than second lobes, with 1–6 lateral notches, without medial notches; fourth lobes about same size as third lobes, with 3–5 (4) lateral notches, with medial notches. Gland spines with associated ducts, gland spines short in first space, gland-spine formula 2-2-2, with 7–12 (10) gland spines near each body margin from segment 2 to segment 4, some having appearance of gland tubercles; median lobes with 2 gland spines between them, about same length or slightly shorter than median lobes. Macroducts of 2 sizes, larger size on margin only, short and wide, duct in first space 19–28 (22) μ long, 6 large ducts on each side of pygidium, present on segment 5 to 7, smaller size abundant, 1 or 2 submarginal ducts on segment 7, submarginal clusters present from mesothorax or metathorax to segment 6, increasing in size posteriorly, submedial clusters from segment 2 or 3 to 6, ducts on thorax anteriorly becoming more abundant on venter than on dorsum, 35–54 (46) macroducts on pygidium from segment 5 to 7, total of 72–123 (102) dorsal macroducts on each side of body, orifices of macroducts occurring anterior of anal opening. Pygidial microducts on venter about same size as small macroducts on dorsum, in submarginal clusters on segments 5, 6, and 7, with 7–26 (18) ducts on each side of body; prepygidial ducts in marginal cluster on head, scattered along body margin from mesothorax to segment 2 or 3, forming band between posterior spiracles; on dorsum pygidial ducts absent; prepygidial microducts undifferentiated from small macroducts, present on anterior margin of head. Perivulvar pores in 5 groups, with 11–15 (12) pores on each side of pygidium. Perispiracular pores with 3 loculi, anterior spiracles each with 1–2 (1) pores, posterior spiracles without pores. Anal opening located 11–17 (13) times length of anal opening from base of median lobes, anal opening 8–12 (10) μ long. Dorsal seta laterad of median lobe 0.6–0.8 (0.7) times length of lobe. Eyes normally absent, occasionally represented by small, flat, sclerotized area, at level of anterior edge of clypeolabral shield. Antennae each with 2, rarely 1, large setae. Prepygidium membranous. Pigmented cicatrices normally on segments 1, 2, and 4, non-pigmented cicatrix often on prothorax. Space between median lobes with 2 setae. Head often with anterior margin slightly sclerotized.

Affinities Of the economic species, litchi scale is most similar to *Lepidosaphes ulmi*, oystershell scale, but differs by lacking lateral spurs, by having pigmented cicatrices on segments 1, 2, and 4, and by having triangular-shaped lobes; whereas, *L. ulmi* has lateral spurs on several abdominal segments, has pigmented or non-pigmented cicatrices on a combination other than 1, 2, and 4, and has rounded median lobes.

Hosts Small number of hosts. Borchsenius (1966) records it from 2 genera in 2 families. We have examined specimens from hosts in the following genera: *Artocarpus, Erythrina, Ficus, Litchi, Mammea, Nephelium, Plumeria, Psittacanthus, Rosa, Solanum, Vanilla*, and *Yucca*.

Distribution This species is known only from FL and HI in the United States. It has been reported from the Caroline Islands, Guam, Palau, and Tanzania. We have examined specimens from other locations as follows: Barbados, Dominican Republic, Guatemala, Honduras, Philippines, and Thailand. This species commonly is taken at U.S. ports-of-entry from the Caribbean and Thailand.

Biology Little is known about the biology of this species other than it occurs on the stems and trunks of the host.

Economic Importance This species causes severe dieback on large litchi trees and even death of small trees in southern Florida and is causing serious problems in commercial orchards. Heavily infested trees have abnormal exfoliating bark, which may be associated with infestations of this pest (Miller 2002, personal observation).

Selected References Laing (1929); Rao and Ferris (1952).

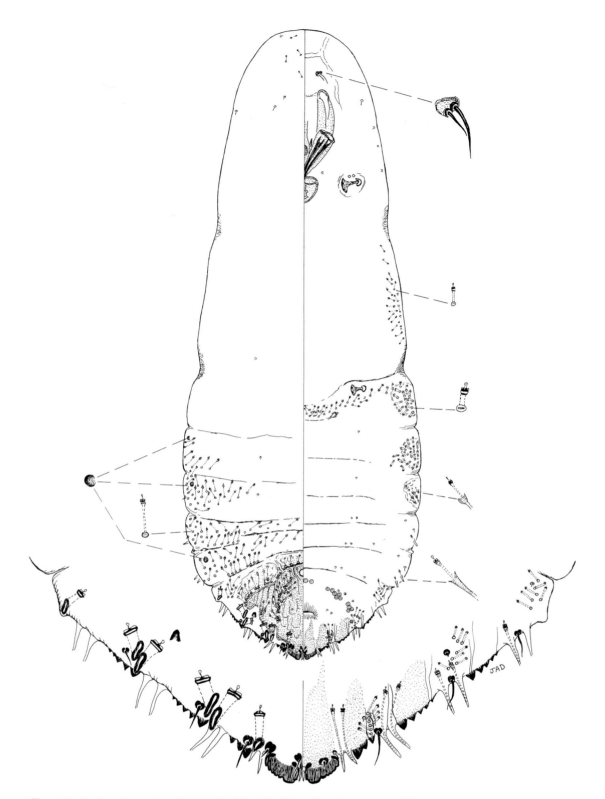

Figure 9. *Andaspis punicae* (Laing), litchi scale, Ocoa, Dominican Republic, on *Ficus carica*, XI-18-1957.

Aonidiella aurantii (Maskell)

ESA Approved Common Name California red scale (also called red scale, red orange scale, orange scale, and cochinella roja Australiana).

Common Synonyms and Combinations *Aspidiotus aurantii* Maskell, *Aspidiotus citri* Comstock, *Chrysomphalus aurantii* (Maskell), *Aspidiotus coccineus* Gennadius, *Aonidia gennadii* Targioni-Tozzetti, *Aonidia aonidum* Targioni-Tozzetti, *Aonidia aurantii* (Maskell).

Field Characters (Plate 8) Adult female cover flat, circular; shed skins central or subcentral, translucent. Male cover elongate oval, similar in color and texture to female cover; shed skin submarginal. Body of adult female usually red to reddish brown giving scale cover reddish-brown appearance; eggs hatch inside body. Body of mature adult female after mating reniform (kidney shaped), sclerotized submarginally. On trunk, stems, leaves, and fruit.

Slide-mounted Characters Adult female (Fig. 10) with 3 pairs of well-developed lobes, fourth lobe represented by series of sclerotized points; paraphyses usually weakly developed, difficult to discern, when present, paraphysis formula of 3-3-1 or 3-2-2. Median lobes separated by space 0.3–0.7 (0.4) times width of median lobe, with noticeable paraphysis-like sclerotization on medial and lateral margin of lobe, without yoke, usually with medial and lateral margins parallel, with 1 lateral notch, with 1 medial notch; second lobes simple, equal to or slightly smaller than median lobes, with 1–3 (2) lateral notches, 0–1 (1) medial notch; third lobes simple, about same size as second lobe, with 1–3 (2) lateral notches, 0–1 (0) medial notches; fourth lobe represented by series of sclerotized points. Plates fimbriate, those in third space deeply bifurcate, with microduct emptying between main branches; plate formula 2-2-3 or rarely 2-2-2, with plate in second space sometimes divided giving appearance of 3 plates; with 2 broad, apically fringed plates between median lobes, equal to or longer than median lobes. Macroducts of 1 variable size, all about same width, extending 1.8–2.6 (2.1) times distance between posterior apex of anal opening and base of median lobes, 70–89 (78) μ long, longest macroduct in first space 62–89 (81) μ long, with 20–26 (23) on each side of pygidium on segments 5 to 8, ducts in submarginal and marginal areas, some macroduct orifices anterior of anal opening; prepygidial macroducts absent. Pygidial microducts on venter in submarginal areas of segment 5, with 5–8 (6) ducts; prepygidial ducts of 1 size, submarginal ducts on each of segments 2 to 4, submedial ducts near mouthparts, rarely present near posterior spiracle, often present on segment 1 and/or 2, in distinct cluster on segment 1; pygidial microducts absent from dorsum, of 1 size on prepygidial areas, in submarginal areas of prothorax to segment 1 or 2. Perivulvar and perispiracular pores absent. Anal opening located 2.4–3.5 (3.1) times length of anal opening from base of median lobes, anal opening 10–15 (13) μ long. Dorsal seta laterad of median lobes 1.0–1.7 (1.3) times length of median lobe. Eyes absent or represented by small spur. Antennae each with 1 seta. Cicatrices usually on prothorax and segment 1. Body with lateral areas of thorax and segment 1 swollen to surround pygidium in mature specimens (reniform), marginal and submarginal areas heavily sclerotized. Normally each apophysis anterolaterad of vulva with 2 associated scleroses. Ventral cluster of microducts on submarginal area of segment 4 composed of 8–19 (12) ducts.

Affinities In the United States the California red scale is most similar to *Aonidiella citrina*, the yellow scale. It differs by generally having a red appearance in the field, by feeding on all parts of the host, and by inhabiting coastal and inland habitats. Yellow scale is normally yellow (when alive), prefers the leaves and fruit, and inhabits inland areas. Morphologically these species are very similar; no single character will differentiate them in all specimens, but we have found the following features to be effective when used in combination and when identifying a series of specimens. California red scale has 2 conspicuous scleroses associated with the apophysis anterolaterad of the vulva, has the plates with the fringing less pronounced than on yellow scale, has slightly shorter macroducts—the longest duct in the first space is 62–89 (81) μ long and the duct between the median lobes is 70–89 (78) μ long, and has a submarginal cluster of 8–19 (12) microducts on segment 4. Yellow scale has the scleroses anterior of the vulva absent or inconspicuous, has deeply grooved fringing on the posterior plates, has slightly longer macroducts—the longest duct in the first space is 82–110 (94) μ long and the duct between the median lobes is 70–102 (85) μ long, and has 4–11 (8) submarginal microducts on each side of pygidium on segment 4.

Hosts Polyphagous. Collections are most commonly made on *Citrus* spp. Borchsenius (1966) lists hosts in 75 families and 202 genera; Dekle (1977) reports it from 11 genera in Florida. We have examined specimens from *Annona, Barringtonia, Ceratonia, Clematis, Cocculus, Codiaeum, Colocasia, Cordyline, Cupressus, Cycas, Elaeagnus, Ephedra, Erythrina, Eucalyptus, Euonymus, Ficus, Fortunella, Fraxinus, Gardenia, Grammatophyllum, Guaiacum, Haworthia, Huernia, Ilex, Jasminum, Juglans, Laelia, Lagerstroemia, Laurus, Ligustrum, Lilium, Macadamia, Macrozamia, Mangifera, Melia, Moringa, Morus, Murraya, Nerium, Olea, Opuntia, Pandanus, Pelargonium, Persea, Philodendron, Phoenix, Plumeria, Polianthes, Protea, Psidium, Punica, Pyrus, Ravenala, Reseda, Ricinus, Rosa, Ruscus, Ruta, Salix, Spondias, Tamarix, Vitis,* and *Zizyphus.* California red scale often is intercepted in quarantine on citrus, rose, mango, avocado, and ficus.

Distribution Occurs out-of-doors in AL, CA, FL, GA, LA, TX and has been eradicated from AZ. It has been collected in greenhouses in DC, KS, MO, NJ, OR, PA. This species is widely distributed and may be taken in quarantine from most tropical and subtropical areas of the world. It apparently occurs in all the citrus-growing areas of the world, particularly between the latitudes 25° and 40° North and South (DeBach 1962). Nakahara (1982) records it from the New World (Antigua, Bahamas, Barbados, Bermuda, Dominica, Dominican Republic, Guadeloupe, Guyana, Jamaica, Mexico, Nicaragua, Martinique, Montserrat, Panama, St. Lucia, St. Vincent, Trinidad, South America); Europe (Mediterranean area); former Soviet Union (Black Sea Coast, Transcaucasia); Africa (North Africa, Angola, Canary Islands, Congo, Ethiopia, Guinea, Kenya, Madagascar, Madeira Island, Mauritius, Mozambique, Reunion, Rhodesia, Senegal, Seychelles, Sudan, South Africa, Tanzania, Togo, Uganda, Zambia); Asia (widespread); others (Australia, Fiji, Guam, New Britain, New Caledonia, New Zealand, Russell Island, Samoa, Society Islands, Solomon Islands, Tonga, Wallis Island). A distribution map of this species was published by CAB International (1996a).

Biology According to Ebeling (1959) the California red scale attacks all parts of the tree: trunk, branches, twigs, leaves, and fruit. On leaves, crawlers usually settle along the midrib and major veins and on fruit in the tiny depressions of the oil glands. Caswell (1962) indicated that feeding occurs in individual plant cells rather than vascular tissues. There may be as many as 150 crawlers on each citrus leaf, and nearly twice as many on fruit (Bodenheimer 1951). Crawlers can be transported by wind over considerable distances (Quayle 1938) and adult males have been shown to disperse widely in spite of their short dispersal period and delicate morphology (Rice and Moreno 1970). According to

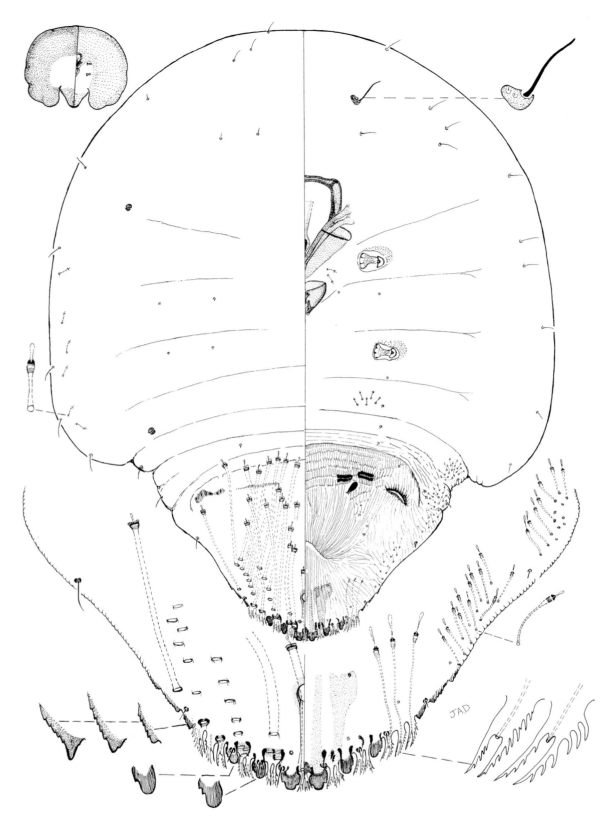

Figure 10. *Aonidiella aurantii* (Maskell), California red scale, Mexico, intercepted at San Antonio, TX, on *Citrus* sp., III-18-1969.

Yan and Isman (1986) the daily cycle of the adult males is influenced by temperature, light intensity, and humidity. High temperatures and light intensity tend to shift male emergence to mid-afternoon from early evening. These authors also showed that adult males rarely live longer than a day and their longevity is significantly shortened by high temperature, high light intensity, and low humidity. Generally, there are approximately equal numbers of males and females in the population (Caswell 1962) although this ratio may vary depending on the host and the temperature (Orphanides 1982). Males are attracted to virgin females by a sex pheromone (Roelofs et al. 1978) and only mated females reproduce. Mated adult females are attached to the scale cover, a habit that renders this stage impervious to attack from wasp parasites (Rosen and DeBach 1978). Eggs hatch inside the body of the female, and crawlers are laid on the host. In California, crawlers are present most of the year; there are 2 generations per year in the coastal areas and 3 in inland areas (Rosen and DeBach 1978). The number of annual generations reported from other countries include: 6 in Argentina (Costilla et al. 1970) and Australia (Bodenheimer 1951); 5 in Australia (Smith 1981; Snowball 1971); 4 to 5 in Israel (Avidov and Harpaz 1969); 4 and a partial 5th in South Africa (Bedford and Georgalia 1978); 4 in Sicily (Inserra 1969), Egypt (Habib et al. 1971), and Cyprus (Orphanides 1982); 4 in middle Egypt and 3 in north coastal areas (Habib et al. 1972); 3 in San Joaquin Valley of California, with a possible 4th on fruit (Carroll and Luck 1984); 3 in Nagasaki Prefecture of Japan (Ohgushi et al. 1967), Morocco (Bénassy and Euverte 1967; Delucchi 1965), Egypt (Abdel-Fattah et al. 1978), Turkey (Onder 1982), Italy (Battaglia and Viggiani 1982), New Zealand (Gellatley 1968), and Crete (Alexandrakis 1983). In most areas, the generations overlap significantly, making it difficult to discern distinct yearly cycles. California red scales reached the largest size when they were reared in cool autumn temperatures (Hare and Luck 1991). Karaca and Uygun (1992) showed that population densities varied on different citrus species and varieties.

Economic Importance The literature is full of reports of the impact of this species as a pest, particularly before the introduction of effective natural enemies. Examples of these reports follow: California red scale is probably the most important citrus pest in the world (Ebeling 1959); attacks are very severe on citrus in Chile (Ebeling 1945); an important pest of citrus in Uruguay (Asplanato and Garcia Mari 1998); one of the most important pests of citrus in Australia (McKeown 1945); causes greater damage than any other citrus pest in California (Ryan 1935); by far the worst pest of citrus in South Africa (Smit 1937; Bedford 1996); a serious pest of apples in New Zealand (Richards 1960); causes severe damage to mulberry in India (Chatterjee 1961); important pest in Queensland, Australia, on fig, walnut, apple, mulberry, peach, plum, and grape (Brimblecombe 1962); a pest of avocado in Israel (Gerson and Zor 1973); is a pest of guava (Butani 1974); causes damage to citrus, figs, and olive in Egypt (El-Minshawy et al. 1974); a pest of apple and pear in France (Geoffrion 1979); a pest of citrus in Yemen (Van Harten 1992); may cause damage when populations are heavy in California on nut crops, apples, avocados, bananas, citrus, coconuts, figs, grapes, mangoes, olives, pears, pistachios, and walnuts (Kenneth 1981); a serious, widespread pest of citrus and many ornamental plants (Gill 1982); a serious pest of citrus in Turkey (Gumus and Uygun 1992); a pest of roses in India (Hole and Salunkhe 1999). Infestations on young trees are especially damaging; newly planted trees experience set back and even death under certain circumstances (Ebeling 1959). Johnson and Lyon (1976) state that infestation by this pest leads to yellowing foliage, defoliation, and the death of twigs and branches. Young fruit also may be shed (Caswell 1962). A toxic substance may be produced by the scale that causes damage in addition to that caused by the feeding process (Ebeling 1959). According to Talhouk (1975) it is a very important pest in Morocco, Italy, Greece, Cyprus, Turkey, Lebanon, Egypt, Pakistan, India, China, Southeast Asia, California, Texas, Mexico, Argentina, Peru, and Australia. It is important in Tunisia, Israel, Arizona, Bolivia, South Africa, and Southern Rhodesia. It is of less importance, but still economic, in Algeria and Brazil. The California red scale is listed as one of the 43 principal armored scale pests of the world by Beardsley and González (1975) and is considered a serious pest by Miller and Davidson (1990). A generalized population model has been developed in California that allows prediction of population peaks and is important in pest management programs. Counts of adult males taken at pheromone traps serve as the monitoring data needed to make population forecasts and implement control strategies (Morse et al. 1985). A program called SCALEMAN was developed to help growers make decisions about controlling this pest (Clift and Beattie 1993). Some research has been conducted on host-plant resistance in citrus, and it has demonstrated that species with high hydrocarbon content are most susceptible, whereas those with high alcoholic terpene content are most resistant (Salama and Saleh 1984). In California, the California red scale is effectively controlled by *Aphytis melinus* in the warm interior areas (Rosen and DeBach 1978; Murdoch et al. 1996) and by *A. lingnanensis* in the cooler coastal areas (Rosen and DeBach 1978). Hare et al. (1993) discovered a kairomone produced by this scale that is highly attractive to *A. melinus*. The presence and concentration of this chemical in concert with the size of the cover are important initial stimuli for the investigating wasp (Morgan and Hare 1998). Ants of several species have been demonstrated to have an important negative effect on the natural enemies of California red scale (Steyn 1958; Bedford 1968; Milne 1974; Samways et al. 1983; James et al. 1997) and must be considered in any pest management system. Although the California red scale does not produce honeydew itself, small populations of soft scales or mealybugs may deposit enough honeydew to afford protection to coexisting armored scales.

Selected References Bodenheimer (1951); Ebeling (1959); McKenzie (1956).

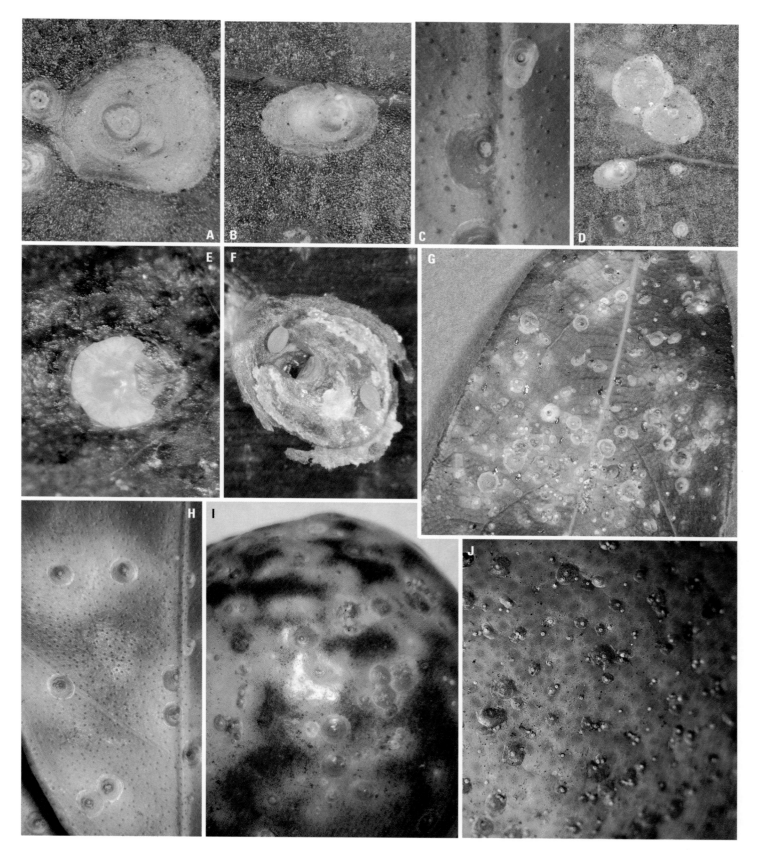

Plate 8. *Aonidiella aurantii* (Maskell), California red scale

A. Close-up of adult female cover and settled crawler on laurel (J. A. Davidson).
B. Close-up of adult male cover on laurel (J. A. Davidson).
C. Adult female and male covers on citrus (R. J. Gill).
D. Adult female and male covers on laurel (J. A. Davidson).
E. Body of young adult female on citrus (J. A. Davidson).
F. Two crawlers and body of older adult female showing characteristic retracted pygidium on umbrella pine (J. A. Davidson).
G. Damage to laurel leaf (J. A. Davidson).
H. Damage to citrus leaf (R. J. Gill).
I. Damage to olive fruit (R. J. Gill).
J. Damage to orange fruit (J. A. Davidson).

Aonidiella citrina (Coquillett)

ESA Approved Common Name Yellow scale (also called California yellow scale).

Common Synonyms and Combinations *Aspidiotus citrinus* Coquillett, *Aspidiotus aurantii* var. *citrinus* Coquillett, *Chrysomphalus citrinus* (Coquillett).

Field Characters (Plate 9) Same as for *A. aurantii* except body color usually yellow when alive and usually on leaves and fruit, rarely on trunk and stems.

Slide-mounted Characters Adult female (Fig. 11) with 3 pairs of well-developed lobes, fourth lobe represented by series of sclerotized points; paraphyses usually weakly developed, difficult to discern, when present, paraphysis formula of 2-2-0, 2-3-0, 3-3-0, 3-3-1 or 3-2-2. Median lobes separated by space 0.5–0.7 (0.6) times width of median lobe, with noticeable paraphysis-like sclerotization on medial and lateral margin of lobe, without yoke, usually with medial and lateral margins parallel, with 1 lateral notch, with 1 medial notch; second lobes simple, equal to or slightly smaller than median lobes, with 1–2 (1) lateral notches, 0–1 (1) medial notch; third lobes simple, about same size as second lobe, with 1–2 (1) lateral notches, 0–1 (1) medial notches; fourth lobe represented by series of sclerotized points. Plates fimbriate, those in third space deeply bifurcate, with microduct emptying between main branches; plate formula 2-2-3 or rarely 2-2-2, with plate in second space sometimes divided giving appearance of 3 plates; with 2 broad, apically fringed plates between median lobes, equal to or longer than median lobes. Macroducts of 1 variable size, all about same width, 1 macroduct between median lobes, extending 1.8–2.5 (2.2) times distance between posterior apex of anal opening and base of median lobes, 70–102 (85) μ long, longest macroduct in first space 82–110 (94) μ long, with 20–26 (23) on each side of pygidium on segments 5 to 8, ducts in submarginal and marginal areas, some macroduct orifices anterior of anal opening; prepygidial macroducts absent. Pygidial microducts on venter in submarginal areas of segment 5, with 5–8 (7) ducts; prepygidial ducts of 1 size, submarginal ducts on each of segments 2 to 4, submedial ducts near mouthparts, rarely present near posterior spiracle, often present on segment 1 and/or 2, in distinct cluster on segment 1; pygidial microducts absent from dorsum, of 1 size on prepygidial areas, in submarginal areas of prothorax to segment 1 or 2. Perivulvar and perispiracular pores absent. Anal opening located 2.6–3.4 (2.9) times length of anal opening from base of median lobes, anal opening 12–17 (13) μ long. Dorsal seta laterad of median lobes 1.0–1.7 (1.3) times length of median lobe. Eyes absent or represented by small spur. Antennae each with 1 seta. Cicatrices usually on prothorax and segment 1. Body with lateral areas of thorax and segment 1 swollen to surround pygidium in mature specimens (reniform), marginal and submarginal areas heavily sclerotized. Normally each apophysis anterolaterad of vulva without associated scleroses, rarely with faint sclerotized areas. Ventral cluster of microducts on submarginal area of segment 4 composed of 4–11 (8) ducts.

Affinities The yellow scale and California red scale for many years were treated as a single species. However, based on differences in natural enemies, appearance, and feeding sites, it was evident to biological control experts that two closely related species were involved. To establish that the two species were distinct, Nel (1933) performed crossbreeding experiments and found that yellow and California red scales were completely incompatible. With this in mind, McKenzie (1937) found that the scleroses anterolaterad of the vulva were well developed in California red scale and were absent in yellow scale. Unfortunately, this character has not been effective in all cases. For a detailed treatment of the differences between these species see the affinities section of *A. aurantii*. This species also is very similar to Asiatic red scale. For a comparison see the affinities section of *A. taxus*.

Hosts Polyphagous but *Citrus* is by far the most common host. Borchsenius (1966) lists hosts in 18 families and 28 genera; Dekle (1977) reports it from 19 genera in Florida. We have examined specimens from *Aegle, Arbutus, Areca, Aucuba, Cananga, Carissa, Citrus, Cordia, Cymbidium, Garcinia, Gardenia, Grevillea, Jasminum, Lansium, Litsea, Loranthus, Mangifera, Manilkara, Melaleuca, Murraya, Musa, Narcissus, Nerium, Olea, Persea, Piper, Plumeria, Psidium, Robinia, Rosa, Ruscus,* and *Vitis.* The most common quarantine host is *Citrus.*

Distribution Occurs in CA, FL, TX. This species is not as widely distributed as *A. aurantii.* We have examined material taken in quarantine from Australia, China, Fiji, India, Jamaica, Mexico, Pakistan, Philippine Islands, South Africa, Taiwan, and Thailand. Nakahara (1982) records it from the New World (Chile, Jamaica, Mexico, Nicaragua); Africa (Guinea, South Africa, Zaire); former Soviet Union (Armenia, Azerbaijan, Crimea, Georgia, Krasnodar); Asia (Afghanistan, Bangladesh, China, India, Indonesia, Iran, Japan, Korea, Malaysia, Nepal, Okinawa, Pakistan, Philippines, South Yemen, Taiwan, Thailand, Turkey); others (Australia, Bonin Islands, Fiji, New Guinea, Western Samoa). Longo et al. (1994) report it from Italy. A distribution map of this species was published by CAB International (1997).

Biology McKenzie (1956) states that yellow scale infestations are usually on the leaves and fruit of *Citrus* spp. Only in heavy infestations will this species inhabit twigs and larger branches. The yellow scale seems to prefer inland areas although it may be found in small numbers in coastal regions. The life history of this species is apparently similar to that of California red scale. Nel (1933) found that development from newborn crawlers to reproducing adult females took about 64 days in a lath house in Riverside, California. At about 28 °C the reproductive period of the female lasts about 60 days. Crawlers are deposited directly onto the host and males are necessary for reproduction. Two or 3 generations are reported in the Georgian SSR (Hadzibejli 1983) and Azerbaijan (Borchsenius 1950); 3 generations occur in Turkey and the Aegean region, and the overwintering stage primarily is the second instar (Onder 1982); 3 generations also are reported in Pakistan (Alam and Sattar 1965). A pheromone has been identified for this species (Gieselmann et al. 1979a).

Economic Importance The presence of yellow scale on citrus fruit can cause rejection of the fruit at packing houses. Heavy infestations may cause defoliation, dieback of branches, and fruit drop. Generally, yellow scale is less of a problem than California red scale. According to Talhouk (1975) it is an important pest in the former Soviet Union, Pakistan, Japan, California, Florida, Mexico, and Australia. It is interesting that he does not consider it to be a 'very important pest' any place in the world; his data suggest that it is a pest primarily in drier climates. The literature gives various reports of the severity of this pest. Examples include: may cause heavy leaf drop but is not as important as California red scale (Quayle 1932); a secondary pest of camellias (Morrison 1946); an important problem in central California (LaFollette 1949); a pest of *Laurus nobilis* and *Osmanthus fragrans* on the Black Sea coast of western Georgia (Vashadze 1955); a pest of tea, citrus, *Laurus*, camphor, and camellia in the Georgian SSR (Kobakhidze 1954); causes great harm to budwood of lemon, mandarin, orange, poncirus, persimmon, laurel, tung, and eucalyptus in Georgia (former Soviet

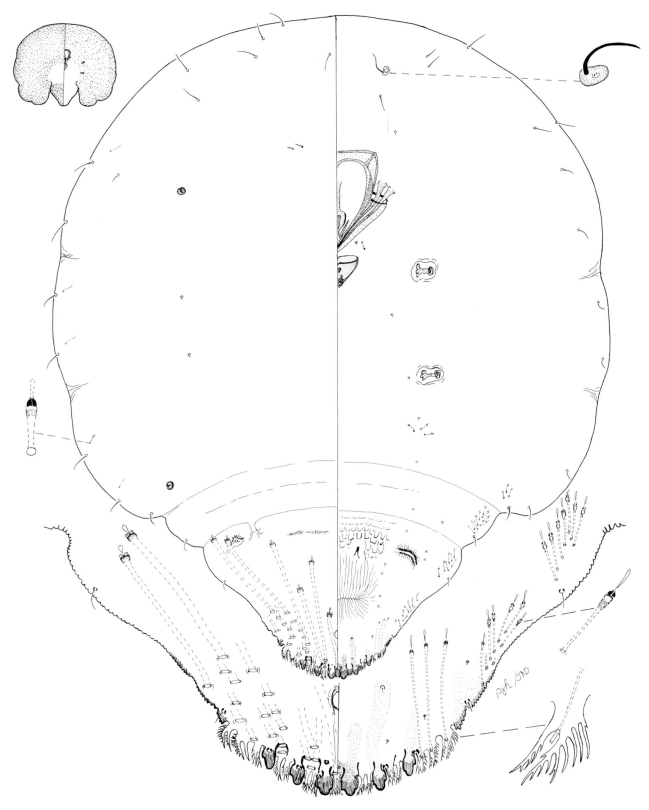

Figure 11. *Aonidiella citrina* (Coquillett), yellow scale, Mexico, intercepted at Brownsville, TX, on *Jasminum* sp., V-3-1947.

Union) (Sikharulidze 1958); sometimes an economic species in Florida (Dekle 1977); a serious pest of citrus in North America (Swan and Papp 1972); causes severe damage in citrus areas in temperate climates in western Africa (Villardebo 1974); a serious enough pest on tangerines in Turkey to warrant chemical control (Tuncyürek and Erkin 1981); a greenhouse pest of citrus (Ter-Grigorian 1954); a minor pest of citrus in California (Gill 1982); became one of the most serious pests of citrus in central and southern California in the late 1940s (Flanders 1948), but in recent years is of minor importance in California (Rosen and DeBach 1978). The California red scale has displaced the yellow scale in southern California south of Ventura County (DeBach and Sundby 1963), but it remains a species of some consequence in the San Joaquin Valley (Clausen 1958). The yellow scale is listed as one of the 43 principal armored scale pests of the world by Beardsley and González (1975) and is considered an important pest by Miller and Davidson (1990). *Comperiella bifasciata* (yellow scale strain) and *Aphytis melinus* have controlled this pest in most areas of California (Fisher and DeBach 1976) but seem to be less effective in the San Joaquin Valley. *Comperiella bifasciata* is especially effective where generations overlap broadly, providing a ready supply of susceptible stages of the host at all times of the year (Rosen and DeBach 1978).

Selected References Longo et al. (1994); McKenzie (1937); McKenzie (1956).

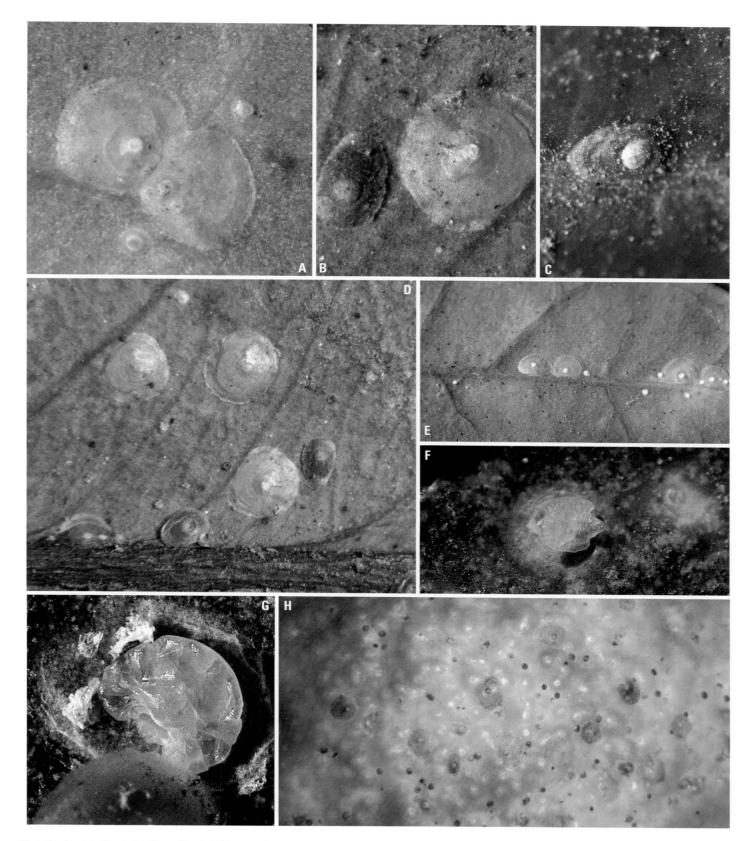

Plate 9. *Aonidiella citrina* (Coquillett), **Yellow scale**

A. Close-up of nearly transparent adult female cover on English laurel with yellowish adult female body beneath (R. J. Gill).
B. Close-up of newly formed adult female cover and old male cover on citrus (R. J. Gill).
C. Close-up of newly formed male cover on citrus (D. R. Miller).
D. Distance view of male and female covers on citrus (R. J. Gill).
E. Female covers and several white-capped crawlers on citrus (D. R. Miller).

F. Body of newly formed adult female before pygidial retraction on citrus (J. A. Davidson).
G. Body of mature adult female showing pygidial retraction on citrus (D. R. Miller).
H. Damage to orange fruit (J. A. Davidson).

Aonidiella orientalis (Newstead)

Suggested Common Name Oriental armored scale (also called Oriental scale, Oriental yellow scale, orientalis yellow scale, Oriental yellow scale, Oriental red scale).

Common Synonyms and Combinations *Aspidiotus osbeckiae* Green, *Aspidiotus (Diaspidiotus) osbeckiae* Green, *Aspidiotus (Evaspidiotus) osbeckiae* Green, *Aspidiotus (Evaspidiotus) orientalis* Newstead, *Aspidiotus taprobanus* Green, *Aspidiotus (Chrysomphalus) pedronis* Cockerell, *Chrysomphalus pedronis* (Cockerell), *Aspidiotus cocotiphagus* Marlatt, *Aspidiotus (Chrysomphalus) taprobanus* Green, *Chrysomphalus pedroniformis* Cockerell and Robinson, *Chrysomphalus orientalis* (Newstead).

Field Characters (Plate 10) Adult female cover flat, circular to oval; opaque, yellow to yellowish brown, margin light; shed skins subcentral, transparent or brown. Male cover oval, same color and texture as female; shed skin submarginal. Body of adult female yellow; eggs and crawlers yellow. Body of mature adult female with thorax slightly produced into lateral lobes, but not as reniform as many other species of genus, sclerotized submarginally on thorax and head. Commonly on leaves and fruit of host; also common on trunks of palms at base of fronds.

Slide-mounted Characters Adult female (Fig. 12) with 3 pairs of well-developed lobes, fourth lobe represented by series of sclerotized points; paraphyses usually weakly developed, difficult to discern, when present, paraphysis formula of 3-3-1, 3-3-0, or 3-2-2. Median lobes separated by space 0.3–0.5 (0.4) times width of median lobe, with noticeable paraphysis-like sclerotization on medial and lateral margins of lobe, without yoke, usually with medial and lateral margins parallel, with 1 lateral notch, with 1 medial notch; second lobes simple, slightly smaller than median lobes, with 1–2 (2) lateral notches, 1 medial notch; third lobes simple, about same size or slightly smaller than second lobe, with 1–2 (1) lateral notches, 0–1 (0) medial notches; fourth lobe represented by series of sclerotized points. Plates fimbriate, those in third space simple, elongate, sickle shaped, with microduct emptying on mesal side of plate near base, rarely 1 plate in third space bifurcate; plate formula 2-3-3; with 2 broad, apically fringed plates between median lobes, equal to or longer than median lobes. Macroducts of 1 variable size, all about same width, 1 macroduct between median lobes, extending 1.6–2.3 (2.0) times distance between posterior apex of anal opening and base of median lobes, 51–78 (60) μ long, longest macroduct in first space 51–69 (59) μ long, with 15–25 (21) on each side of pygidium on segments 5 to 8, ducts in submarginal and marginal areas, total of 26–94 (52) macroducts on each side of body, some macroduct orifices anterior of anal opening; prepygidial macroducts of 2 sizes, larger size about same size as on pygidium, in distinct submarginal clusters on segments 2 to 4, smaller size in marginal and sometimes submarginal areas of metathorax to segment 3 or 4. Pygidial microducts on venter in submarginal areas of segment 5, rarely segment 6, with 5–7 (6) ducts; prepygidial ducts of 1 size, submarginal ducts on each of segments 1 or 2 to 4, submedial ducts near mouthparts, near posterior and anterior spiracles, often present on metathorax and segments 1 to 3; pygidial microducts absent from dorsum, of 1 size on prepygidial areas, in submarginal areas of prothorax to segment 1 or 2, often in submedial areas of mesothorax or metathorax to segments 1 or 2. Perivulvar pores usually in 4 groups, rarely 5, 6–14 (9) pores on each side of body. Perispiracular pores absent. Anal opening located 2.1–2.9 (2.4) times length of anal opening from base of median lobes, anal opening 10–15 (13) μ long. Dorsal seta laterad of median lobes 1.0–1.7 (1.4) times length of median lobe. Eyes absent. Antennae each with 1 seta. Cicatrices usually on prothorax and segment 1.

Body with lateral areas of thorax and segment 1 slightly swollen in mature specimens but not surrounding pygidium, marginal and submarginal areas of head and thorax sclerotized. Apophyses anterolaterad of vulva, when present, without scleroses. Ventral cluster of microducts on submarginal area of segment 4 composed of 4–7 (5) ducts.

Affinities The Oriental armored scale resembles *Aonidiella aurantii* and *A. citrina*, the California red scale and the yellow scale, but is different in having perivulvar pores and marginal clusters of prepygidial macroducts, lacking scleroses and/or apophyses anterior of the vulva, lacking sclerotization marginally on segments 2 to 4, and normally not having reniform shape in fully mature females. The California red scale and yellow scale lack perivulvar pores and clusters of prepygidial macroducts, have apophyses and scleroses near the vulva, have sclerotization marginally on segments 2 to 4, and have a reniform body shape in fully mature females.

Hosts Polyphagous, with palms the most common host in the United States. Borchsenius (1966) lists hosts in 26 families and 36 genera; Dekle (1977) reports it from 68 genera in Florida. We have examined specimens from *Agave*, *Annona*, *Areca*, *Artocarpus*, *Arundo*, *Asparagus*, *Atalantia*, *Atalaya*, *Attalea*, *Azadirachta*, *Bauhinia*, *Bischofia*, *Cajanus*, *Carica*, *Cassia*, *Cattleya*, *Cherimoya*, *Chrysalidocarpus*, *Citrus*, *Cocos*, *Corypha*, *Cycas*, *Cymbidium*, *Delonix*, *Diospyros*, *Dysoxylum*, *Erythrina*, *Eucalyptus*, *Euonymus*, *Ficus*, *Flacourtia*, *Garcinia*, *Hibiscus*, *Ilex*, *Iris*, *Jasminum*, *Lagerstroemia*, *Laguncularia*, *Ligustrum*, *Litchi*, *Loranthus*, *Mangifera*, *Melia*, *Moringa*, *Murraya*, *Musa*, *Nerium*, *Olea*, *Pachira*, *Pedilanthus*, *Persea*, *Philodendron*, *Phoenix*, *Plumeria*, *Podocarpus*, *Prosopis*, *Psidium*, *Rhizophora*, *Ricinus*, *Rosa*, *Ruta*, *Sclerocarya*, *Sida*, *Solanum*, *Spondias*, *Strelitzia*, *Tamarindus*, *Vanda*, *Vicia*, *Vitis*, *Zanthoxylum*, and *Zizyphus*. The most common quarantine hosts are palms, especially *Cocos*, and *Mangifera* and *Annona*.

Distribution Found out-of-doors in FL. We have examined a single slide from cycad from Mission, TX, but this record must be confirmed. It has been collected in greenhouses in IL and DC. We have examined material taken in quarantine from Australia, Bahamas, Burma, Cambodia, Celebes, China, Colombia, Cuba, Dominican Republic, Guam, Guatemala, Haiti, India, Iran, Iraq, Jamaica, Japan, Laos, Malaysia, Mauritius, Mexico, Pakistan, Panama, Philippine Islands, Puerto Rico, Rota Island, Saipan, Saudi Arabia, Singapore, Sri Lanka, Sudan, Taiwan, and Thailand. Nakahara (1982) reported it from Afghanistan, Brazil, Hong Kong, Indonesia, Kenya, Madagascar, Mariana Islands, Nepal, Nicobar Island, Somalia, South Africa, Tanzania. It was first reported in Israel in 1980 (Ofek et al. 1997) and is a pest in Egypt (Mohammad et al. 2001). A distribution map of this species was published by CAB International (1978) but it appears that the species has spread fairly rapidly in recent years.

Biology According to Glover (1933) *A. orientalis* occurs on the leaves, stems, twigs, and trunks of *Zizyphus jujuba* and *Schleichera trijuga* with the former becoming relatively immune when more than 2 to 4 years old. He indicates there are three generations per year with crawlers appearing October–November, February–March, and May–June. Infestations on the leaves and stems of roses are noted in the same report from India. In Saudi Arabia, Badawi and Al-Ahmed (1990) found continuous reproduction but suggest that population peaks on December 22, January 20, March 16, and April 13 might represent different generations. In Saudi Arabia, Moussa (1986) suggests there were 4 generations per year. At temperatures of about 20 °C, eggs hatched in 2 to 3 days; at temperatures of 30 °C, they

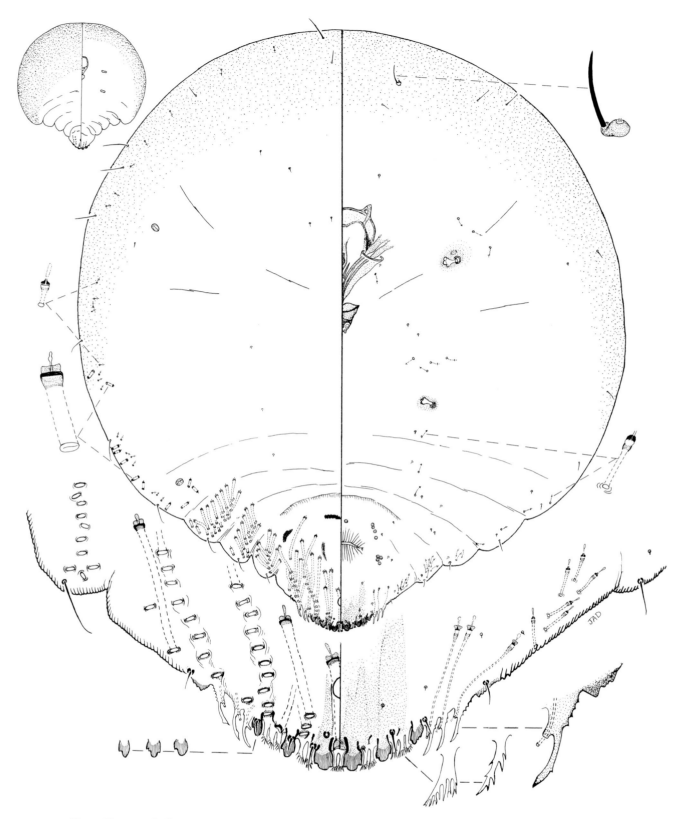

Figure 12. *Aonidiella orientalis* (Newstead), Oriental armored scale, Hofut, Arabia, on *Phoenix dactylifera*, VII-30-1953.

hatched in 7 to 19 hours. At about 22°C, the duration of the first instar was 11 to 19 days and the second instar took 21 to 29 days to complete. The total life cycle took 101 days in the winter generation, 78 days in the spring generation, and 68 days in the summer generation; the fourth generation was not measured (Moussa 1986). Populations were held to low densities from May to early December, probably because of high temperatures. In Egypt, in 1996, four population peaks were recorded, but in 1997 only three were found (Mohammad et al. 2001).

Economic Importance According to Dekle (1965) this species is a serious pest of coconut palms and ivy in Florida. Glover (1933) reports it as a pest of *Zizyphus jujuba*, *Schleichera trijuga*, and *Rosa* (roses) in India, whereas Dutta and Baghel (1991) consider it to be a pest of many trees and plants, with *Dalbergia sissoo* being the most common in the same country. Rajagopal and Krishnamoorthy (1996) list it as a pest of arecanut, banana, citrus, coconut, guava, mulberry, peach, pomegranate, and tamarind, and treat it as a major pest of arecanut and coconut in south India. Sen (1937) notes a severe attack on host trees of *Kerria lacca* (Kerr) causing problems in culturing lac insects for the production of shellac. In Saudi Arabia it is a serious pest on citrus, shade trees, and ornamentals (Moussa 1986). A series of papers has been published on the biological control of this pest in Australia (Elder and Bell 1998; Elder and Smith 1995; Elder et al. 1997; Elder et al. 1998). In recent years this species has become a troublesome pest of neem (*Azadirachta indica*) (Lale 1998). In Egypt the species is a pest on *Ficus nitida* (Mohammad et al. 2001), and in Israel it causes damage to the leaves, twigs, and fruit of mango (Ofek et al. 1997). Miller and Davidson (1990) consider this species to be a serious pest in a small area of the world.

Selected References Glover (1933); McKenzie (1937).

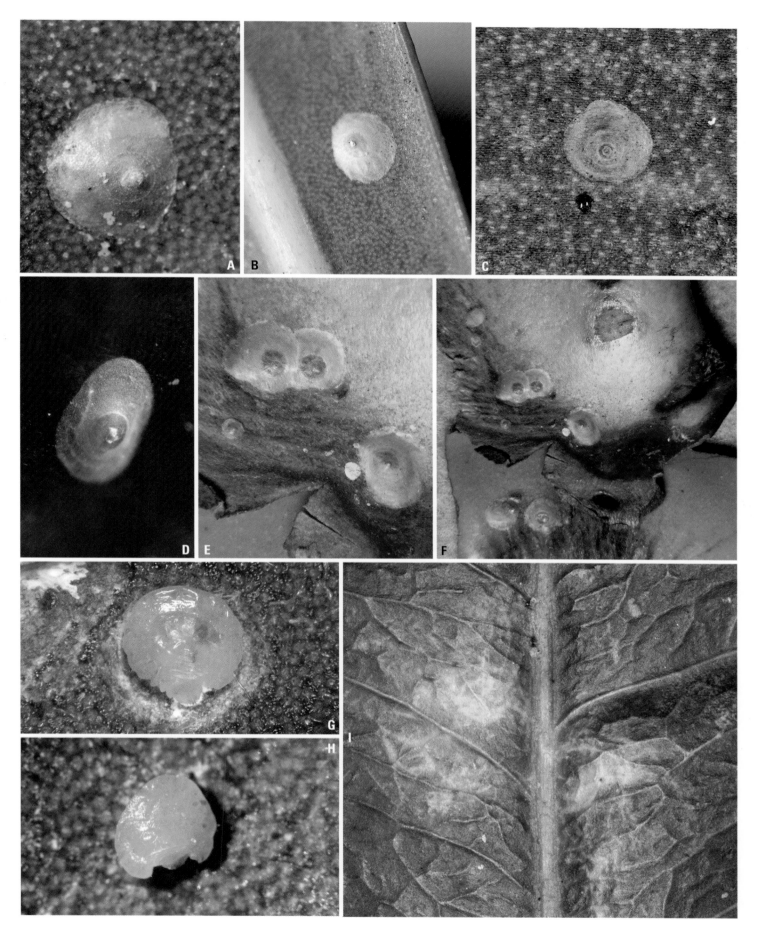

Plate 10. *Aonidiella orientalis* (Newstead), Oriental armored scale

A. Old adult female cover on cycad (J. A. Davidson).
B. Newly formed adult female cover on cycad (J. A. Davidson).
C. Adult female cover on succulent (J. A. Davidson).
D. Adult male cover on cycad (D. R. Miller).
E. Scale covers on coconut (R. J. Gill).
F. Scale covers on coconut (R. J. Gill).

G. Body of young adult female with pygidium protruding on cycad (J. A. Davidson).
H. Body of mature adult female with pygidium retracted on cycad (J. A. Davidson).
I. Damage on *Murraya* (J. A. Davidson).

61

Aonidiella taxus Leonardi

Suggested Common Name Asiatic red scale.

Common Synonyms and Combinations *Chrysomphalus taxus* (Leonardi), *Aspidiotus taxus* (Leonardi).

Field Characters (Plate 11) Adult female cover flat, circular, translucent; shed skins central to subcentral, translucent. Male cover elongate oval, similar in color and texture to female cover; shed skin subcentral, translucent. Body of mature adult female kidney shaped, changing from yellow to reddish brown with age, giving older scale cover reddish-brown appearance; eggs yellow; crawlers brownish yellow. Mainly on undersides of leaves causing yellow spots on top. Males normally occurring on upper leaf surfaces. Crawler cap white with darkened central area. Dorsal cover of mature female with broad, light, marginal band beyond body perimeter.

Slide-mounted Characters Adult female (Fig. 13) with 3 pairs of well-developed lobes, fourth lobe represented by series of sclerotized points; paraphyses usually obvious, paraphysis formula of 2-3-1, 2-3-0, or 2-3-2. Median lobes separated by space 0.5–0.7 (0.5) times width of median lobe, with noticeable paraphysis-like sclerotization on medial and lateral margin of lobe, without yoke, usually with medial and lateral margins parallel, with 1 lateral notch, with 1 medial notch; second lobes simple, slightly smaller or equal to median lobes, with 1–2 (1) lateral notches, 0–1 (1) medial notch; third lobes simple, usually slightly smaller than second lobe, occasionally about equal, with 1–2 (1) lateral notches, 0–1 (0) medial notches; fourth lobe represented by series of sclerotized points. Plates fimbriate, those in third space deeply bifurcate, with microduct emptying between main branches; plate formula 2-2-3, with plate in second space divided giving appearance of 3 plates; with 2 apically fringed plates between median lobes, equal to or longer than median lobes. Macroducts of 1 variable size, all about same width, 1 macroduct between median lobes, extending 1.6–2.2 (2.0) times distance between posterior apex of anal opening and base of median lobes, 67–81 (74) μ long, longest macroduct in first space 83–103 (91) μ long, with 24–38 (32) on each side of pygidium on segments 5 to 8, ducts in submarginal and marginal areas, total macroduct on each side of body 29–41 (33), some macroduct orifices anterior of anal opening; prepygidial macroducts of 1 size, of small size only, in marginal areas of metathorax and segment 1. Pygidial microducts on venter in submarginal areas of segment 5, rarely segment 6, with 5–7 (7) ducts; prepygidial ducts of 1 size, submarginal ducts on each of segments 1 or 2 to 4, submedial ducts near mouthparts, near posterior and anterior spiracles, often present on metathorax and segments 1 to 3; pygidial and prepygidial microducts absent from dorsum. Perivulvar pores absent. Perispiracular pores absent. Anal opening located 1.6–2.9 (2.2) times length of anal opening from base of median lobes, anal opening 15–20 (17) μ long. Dorsal seta laterad of median lobes 0.7–0.9 (0.8) times length of median lobe. Eyes represented by sclerotized projection with 1 to several points, on body margin laterad of clypeolabral shield. Antennae each with 1 seta. Cicatrices usually on prothorax and segment 1. Body with lateral areas of thorax and segment 1 swollen in mature specimens, partially surrounding pygidium, marginal and submarginal areas of head and thorax sclerotized. Apophyses anterolaterad of vulva without scleroses. Ventral cluster of microducts on submarginal area of segment 4 composed of 6–9 (8) ducts. Marginal setae on segment 3 originating on tubercle.

Affinities Of the economic species in the United States, Asiatic red scale is most similar to *Aonidiella citrina*, yellow scale, but differs by having marginal tubercles on segment 3, by lacking sclerotized apophyses anterolaterad of vulva, and by having 24–38 (32) macroducts on each side of body. Yellow scale lacks marginal tubercles on segment 3, has apophyses, and 20–26 (23) macroducts. Outside of the United States, Asiatic red scale is most similar to *A. inornata* Leonardi but differs by having the second and third lobes about the same width as the median lobes. *Aonidiella inornata* has the second and third lobes noticeably narrower than the median lobes. In the field the Asiatic red scale and *A. aurantii* (California red scale) are similar. The former has a wide, light band beyond the body perimeter; this band is narrow in the California red scale.

Hosts Very limited host range. Borchsenius (1966) records hosts in 1 family and 2 genera; Dekle (1977) reports it from *Podocarpus* only. We have examined material from *Podocarpus* and *Taxus*. Nakahara (1982) records it from *Cephalotaxus*. There is also a series of specimens from Japan recorded from *Quercus*, but it is possible that they are mislabeled. Other hosts include *Bladhia sieboldii* and *Cephalotaxus drupacea* (Murakami 1970).

Distribution This is a subtropical species that probably is native to the Orient. It occurs out-of-doors in AL, CA (eradicated), FL, LA, and in greenhouses in the North. We have examined specimens from CA, FL, LA, MD, PA in the United States, and from Argentina, Italy, and Japan. It is found widely in Argentina, Brazil, and Italy and often is taken in quarantine from Japan. It is reported in southern areas of the former Soviet Union (Kuznetsov 1971) and China (Qin et al. 1997).

Biology Stoetzel (1975) indicates that this species feeds primarily on the leaves of the host. In the greenhouse, she found many overlapping generations with eggs laid continuously at a rate of 3–5 each day until death. In the Crimea area of the former Soviet Union, Kuznetsov (1971) reports 2 generations per year. Qin et al. (1997) concludes that *Aonidiella taxus* has 3 or 4 generations per year in Shanghai, China. Fertilized adult females of the third generation and second-instar nymphs of the fourth generation overwinter on leaves of *Podocarpus* and *Taxus*. Immatures of each generation are present from mid-May to late July, late July to the beginning of October, mid-September to late November, and mid-October to late November. Uematsu (1978) discusses the bionomics of this species.

Economic Importance Johnson and Lyon (1976) report Asiatic red scale as a severe pest of *Podocarpus macrophylla* and *Taxus baccata*. Dekle (1965) considers this species to be an economic pest in Florida nurseries. It feeds primarily on the undersides of leaves causing yellow spots on the upper surfaces of mature leaves. It may cause notches on the margins of developing leaves. Miller and Davidson (1990) consider this species to be an occasional pest.

Selected References Johnson and Lyon (1976); McKenzie (1956); Stoetzel (1975).

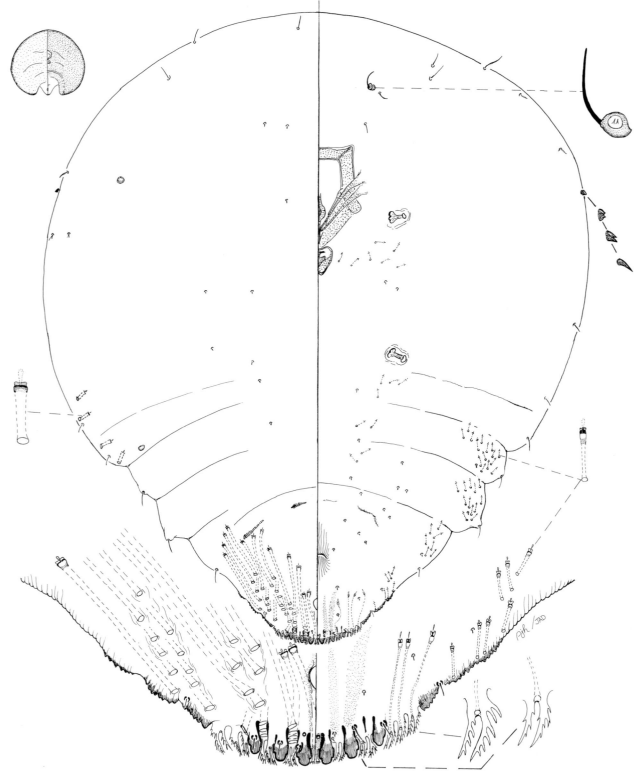

Figure 13. *Aonidiella taxus* Leonardi, Asiatic red scale, College Park, MD, on *Podocarpus* sp., VII-19-1971.

Plate 11. *Aonidiella taxus* Leonardi, Asiatic red scale

A. Infestation on taxus (J. A. Davidson).
B. Adult female (left) and male (right) covers on taxus (J. A. Davidson).
C. Distance view of scale covers on taxus (J. A. Davidson).
D. Body of adult female with characteristic retracted pygidium (D. R. Miller).

E. Bodies of young and mature adult females on podocarpus (D. R. Miller).
F. Heavily infested podocarpus (J. A. Davidson).
G. Damage to podocarpus leaves (J. A. Davidson).

64

Plate 12. *Aspidiella sacchari* (Cockerell), Sugarcane scale

A. Mature adult female covers (A. Hamon).
B. Rubbed adult female cover showing large second-instar shed skin (J. A. Davidson).

C. Female and male covers (J. A. Davidson).
D. Cluster of covers at base of leaf sheath (J. A. Davidson).
E. Body of newly matured adult female (J. A. Davidson).

Aspidiella sacchari (Cockerell)

ESA Approved Common Name Sugarcane scale.

Common Synonyms and Combinations *Aspidiotus sacchari* Cockerell, *Targionia sacchari* (Cockerell).

Field Characters (Plate 12) Adult female cover convex, circular, grayish brown to slightly purplish brown, opaque; shed skins central to subcentral, light yellow when not covered by wax. Male cover elongate oval, same color and texture as female cover; shed skin marginal, yellow. Body of female pale white; eggs probably white. In sheaths of leaves, on stems, and on crowns; often found just below soil surface.

Slide-mounted Characters Adult female (Fig. 14) with 2 pairs of well-developed lobes, third lobe represented by swelling only slightly larger than protrusions along remainder of pygidium; paraphyses absent, but sometimes marginal macroducts with sclerotized orifice similar in appearance to paraphyses. Median lobes separated by space 0.0–0.1 (0.1) times width of median lobe, with large, conspicuous basal sclerosis, without yoke, with medial margins parallel, lateral margins converging apically, with 1–2 (1) lateral notches, with 0–1 (1) medial notch; second lobes simple, conspicuously smaller than median lobes, approximately half size, 1–4 (2) lateral notches, 0–1 (0) medial notches; third lobes normally represented by low protrusion, sometimes difficult to distinguish from other marginal crenulations, less than half size of second lobes, usually without notches rarely with 1 or 2 laterally. Plates usually bifurcate, with protruding microduct at apex; plate formula 2-3-0, rarely 2-2-0; with 2 slender, bifurcate plates between median lobes, about same length as lobes, usually difficult to see because hidden by lobes. Macroducts of 1 size, unusually slender, 2–3 (2) macroducts between median lobes, posterior duct extending 0.7–0.8 (0.7) times distance between posterior apex of anal opening and base of median lobes, 58–66 (62) µ long, longest macroduct in first space 49–59 (55) µ long, with 32–56 (42) on each side of pygidium on segments 5 to 8, ducts in submarginal and marginal areas, total dorsal macroducts on each side of body 86–108 (97), some macroduct orifices anterior of anal opening; prepygidial macroducts present from metathorax to segment 4. Pygidial microducts similar in appearance to macroducts on venter in submarginal areas of segments 5, 6, and usually 7 with 23–42 (31) ducts, submedial ducts present on segments 5 and 6; prepygidial ducts of 2 sizes, longest ducts similar in appearance to macroducts, on margin and submargin of segment 4, shorter ducts in submarginal areas of head, thorax, and segments 1 to 4, submedial ducts in clusters near mouthparts and spiracles and on head and segments 1 to 4, medial ducts sometimes present on mesothorax, metathorax, and segments 1 to 2; pygidial microducts on dorsum usually present submedially on segments 5 and 6, of 2 sizes on prepygidial areas, short ducts present along body margin from head or prothorax to segment 2 or 3, long ducts present in same area from head to segment 2 or 3, short ducts in submedial areas of head, thorax, and segments 1 to 4. Perivulvar pores in 4 groups, 6–15 (12) pores on each side of body. Perispiracular pores absent. Anal opening located 5.8–8.6 (7.0) times length of anal opening from base of median lobes, anal opening 9–15 (12) µ long. Dorsal seta laterad of median lobes 1.9–3.0 (2.4) times length of median lobe. Eyes represented by weakly sclerotized area on body margin at level of anterior part of mouthparts. Antennae each with 1 seta. Cicatrices usually on prothorax, segment 1 and 3. Body pear shaped to oval. Spiracles each with area of sclerotization on surrounding derm. Abdominal margin sclerotized, strongly crenulate to segment 1.

Affinities In the United States the sugarcane scale is unique by having 3 pairs of lobes, numerous thin macroducts, no plates anterior of the third lobes, and a crenulate, abdominal margin. Outside of the United States the sugarcane scale is most similar to *Aspidiella hartii* but differs by having plates anterior of the third lobes and lacking the crenulate abdominal body margin.

Hosts Occurs on many genera of grasses and is commonly collected on sugarcane. Borchsenius (1966) lists hosts in 2 families and 6 genera; Dekle (1977) reports it from 21 genera in Florida including several dicotyledonous hosts. We have examined specimens from *Ananas*, *Andropogon*, *Annona*, *Axonopus*, *Canna*, *Chloris*, *Cocos*, *Costus*, *Cymbopogon*, *Cynodon*, *Digitaria*, *Dioscorea*, *Dracaena*, *Eremochloa*, *Ficus*, *Ligustrum*, *Mangifera*, *Neyraudia*, *Opizia*, *Panicum*, *Paspalum*, *Pennisetum*, *Saccharum*, *Sporobolus*, *Stenotaphrum*, *Tripsacum*, and *Vetiveria*.

Distribution Found in the United States in FL, HI, TX. We have examined material taken in quarantine from Antigua, Bahamas, Barbados, Cuba, Fiji, Guam, Guatemala, Haiti, Honduras, Jamaica, Java, Liberia, Malaysia, Mariana Islands, Marshall Islands, Mexico, New Guinea, Nicaragua, Nigeria, Pakistan, Panama, Puerto Rico, Sri Lanka, Surinam, Tahiti, Trinidad, Virgin Islands, Yap Island. Nakahara (1982) reports it from Colombia, Guyana, Indonesia, Kenya, Madagascar, Mauritius, Laos, Reunion, Rodrigues, Sierra Leone.

Biology Little is known about the biology of this species. It occurs beneath the leaf sheaths of its host and may be found above or below the soil surface.

Economic Importance Dekle (1965) reports the sugarcane scale as a pest of sugarcane and lawn grasses in Florida. Miskimen and Bond (1970) consider it to be the most serious armored scale pest of sugarcane in the Virgin Islands, but it is repeatedly listed as a minor pest of this host in most parts of the world (Wolcott 1948). Miller and Davidson (1990) consider this species to be an occasional pest.

Selected References Balachowsky (1958); Rao and Sankaran (1969).

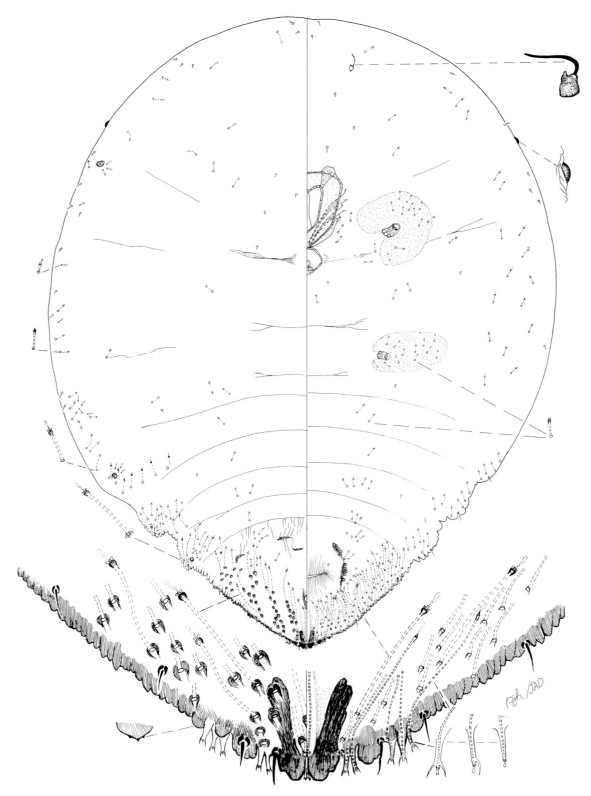

Figure 14. *Aspidiella sacchari* (Cockerell), sugarcane scale, McAllen, TX, on *Axonopus affinis*, III-19-1975.

Aspidiotus cryptomeriae Kuwana

Suggested Common Name Cryptomeria scale.
Common Synonyms and Combinations None.
Field Characters (Plate 13) Adult female cover flat, elongate oval, grayish brown, slightly transparent; shed skins central, yellow. Male cover elongate, same color and texture as female; shed skin subcentral, yellow. Body of female oval, yellow; eggs and crawlers yellow. On undersides of needles. Covers often under thin layer of host cuticle.

Slide-mounted Characters Adult female (Fig. 15) with 3 pairs of well-developed lobes; paraphyses absent. Median lobes separated by space 0.5–1.0 (0.6) times width of median lobe, with noticeable sclerotized area anterior of lobe, without yoke, usually with medial and lateral margins parallel or diverging slightly apically, with 1 lateral notch, with 0–1 (1) medial notch; second lobes simple, about same size as median lobes, sometimes slightly shorter and wider, with 1 lateral notch, without a medial notch; third lobes simple, about same size as second lobe, with 1 lateral notch, without medial notch. Plates fimbriate, with microduct emptying on mesal side of most plates; plate formula 2-3-3 or 2-3-4, with 4 or 5 plates anterior of seta marking segment 5; with 2 broad, apically fringed plates between median lobes, equal to or longer than median lobes. Macroducts of 1 size, 1 macroduct between median lobes, extending 0.8–1.2 (1.0) times distance between posterior apex of anal opening and base of median lobes, 60–75 (61) µ long, longest macroduct in first space 73–91 (83) µ long, with 11–22 (14) on each side of pygidium on segments 5 to 8, ducts in submarginal and marginal areas, total macroducts on each side of body 11–25 (14), usually without macroduct orifices anterior of anal opening, rarely with 1; prepygidial macroducts rarely present on segment 4, of same size and shape as on pygidium. Pygidial microducts on venter in submarginal areas of segment 5, rarely segments 6 and 7, with 1–5 (2) ducts; prepygidial ducts of 1 size, submarginal ducts on each of segments 1 or 2 to 4, occasionally on head and prothorax, submedial ducts near mouthparts, near posterior and anterior spiracles, often present on metathorax and segments 1 to 3, sometimes on mesothorax; pygidial microducts absent from dorsum, of 1 size on prepygidial areas, in submarginal areas of head to metathorax, usually in submedial areas of thorax. Perivulvar pores usually in 4 groups, rarely 5, 10–23 (15) pores on each side of body. Perispiracular pores absent. Anal opening located 2.5–4.3 (3.6) times length of anal opening from base of median lobes, anal opening 17–23 (20) µ long. Dorsal seta laterad of median lobes 1.0–1.9 (1.4) times length of median lobe. Eyes represented by circular sclerotized area on body margin at level of anterior margin of mouthparts. Antennae each with 1 seta. Cicatrices usually on prothorax and segment 1, occasionally with small cicatrix on segment 3. Body oval or pyriform. Median lobes not appearing recessed into pygidium.

Affinities The cryptomeria scale is surprisingly similar to *Aspidiotus destructor*, the coconut scale, but differs by occurring on coniferous hosts, by having the anal opening located 2.5–4.3 (3.6) times the length of the anal opening from the base of the median lobes, the anal opening 17–23 (20) µ long, the dorsal seta laterad of the median lobes 1.0–1.9 (1.4) times the length of the median lobe, the longest macroduct in the first space 73–91 (83) µ long, the plates between the median lobes about the same length as the lobes, by lacking apical serrations on the median lobes, and by having median lobes that protrude

as far from the pygidium as the second lobes. The coconut scale normally does not occur on conifers, the anal opening is located 1.3–2.2 (1.5) times length of anal opening from the base of the median lobes, the anal opening is 23–30 (26) µ long, the dorsal seta laterad of the median lobes is 2.0–3.1 (2.3) times the length of the median lobe, the longest macroduct in the first space is 51–81 (65) µ long, the plates between the median lobes extend beyond the apex of the lobes, the median lobes have many apical serrations, and the median lobes often are retracted and do not protrude from the pygidium as far as the second lobes.

Hosts On many coniferous hosts. Borchsenius (1966) records hosts in 2 families and 7 genera. We have examined specimens from the United States on *Abies*, *Cedrus*, *Cryptomeria*, *Picea*, and *Tsuga*. Nakahara (1982) reports it on *Cephalotaxus*, *Chamaecyparis*, *Keteleeria*, *Pinus*, *Pseudotsuga*, *Taxus*, and *Torreya*. It has been reported from Japan by Kuwana (1933) on *Chamaecyparis* and *Pinus*, and by Takagi (1969) on *Cephalotaxus*, *Pseudotsuga*, *Taxus*, and *Torreya*.

Distribution Cryptomeria scale is known in the United States from CT, DE, IN, MD, NY, PA. It also occurs in China, Japan Korea, Sakhalin, and Taiwan and often has been taken in quarantine from Japan.

Biology This species has 2 generations per year and overwinters as second instars in Maryland (Stoetzel and Davidson 1974). Adults of the overwintering generation appear in March and April, and 30–40 eggs are laid in June. Adults of the summer generation are present in July, and eggs and crawlers appear in late August and September. Males are winged. Although Stimmel (1986) indicates that this species probably has a single generation per year in Pennsylvania, more recent research by Pennsylvania Department of Agriculture technician Sandra Gardosik has shown that most populations pass through two annual generations (Stimmel 2002, personal communication). Crawler emergence peaks for these bivoltine populations occur in early- to mid-June, and again in mid-August; *A. cryptomeriae* consistently overwinters as second instars in Pennsylvania. Local habitat conditions (e.g., pockets of lower temperature) appear to affect the number of annual generations and the timing of crawler emergence. Murakami (1970) indicates there are 2 to 3 generations per year in Japan, and that the overwintering stage is the nymph. Danzig (1980) states that the species has only a single generation each year in Sakhalin. Research in Japan has shown the existence of host races in this species with subtle morphological and molecular differences (Miyanoshita et al. 1991; Miyanoshita et al. 1993; Don et al. 1995; Miyanoshita and Tatsuki 1995).

Economic Importance Stimmel (1986) indicates that heavy populations in Pennsylvania cause noticeable chlorosis of the leaves, stunting and distortion of new growth, premature leaf drop, and probably death of individual trees. He considers it to be a serious and severe pest in Pennsylvania (Stimmel 2002 personal communication). The cryptomeria scale often is encountered on hemlocks in nurseries in Maryland (Gimpel 1998, personal communication). It is an important pest of conifers in Japan especially in nurseries where young trees are planted densely or are shaded by shelters (Murakami 1970). This species is listed as an occasional pest by Miller and Davidson (1990).

Selected References Don et al. (1995); Murakami (1970); Stoetzel and Davidson (1974).

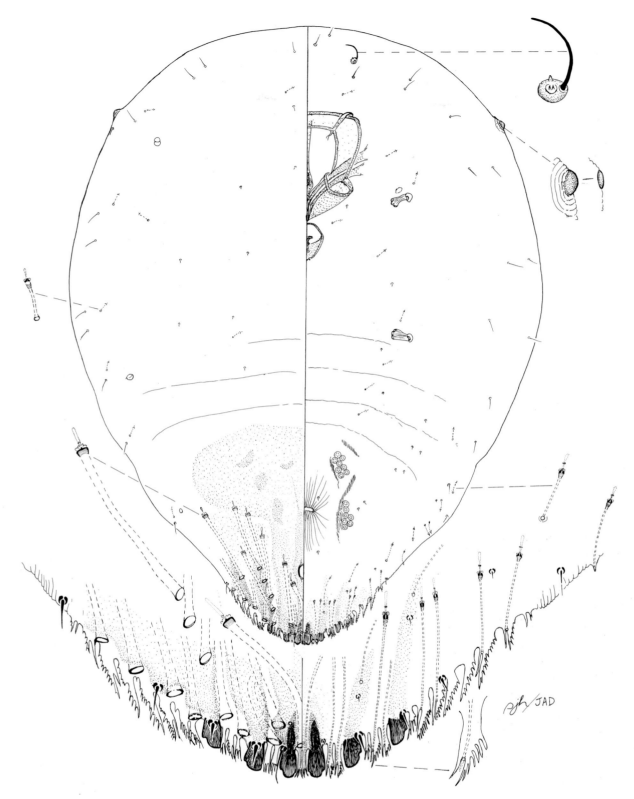

Figure 15. *Aspidiotus cryptomeriae* Kuwana, cryptomeria scale, Long Island, NY, on *Abies Nordmanniana*, VIII-30-1933.

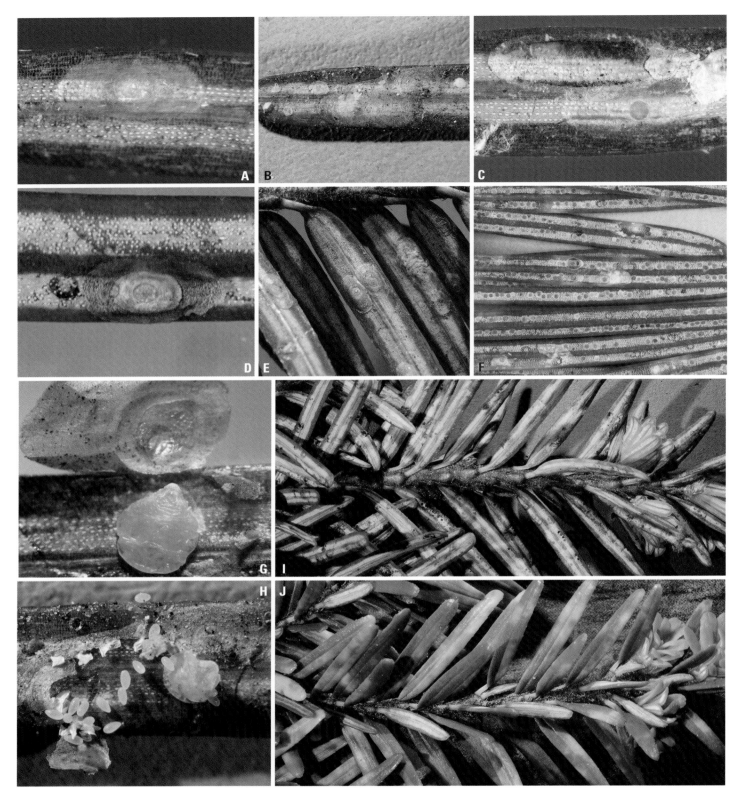

Plate 13. *Aspidiotus cryptomeriae* Kuwana, Cryptomeria scale

A. Young adult female cover on underside of hemlock needle (J. A. Davidson).

B. Covers of several female instars on hemlock (J. A. Davidson).

C. Female cryptomeria scale (bottom), female elongate hemlock scale on hemlock (J. A. Davidson).

D. Old cover of female cryptomeria scale on hemlock (J. A. Davidson).

E. Distance view of female and male covers (J. A. Davidson).

F. Heavy infestation composed mostly of capped crawlers on fir (J. A. Davidson).

G. Body of young adult female on hemlock (J. A. Davidson).

H. Body of mature adult female with eggs on hemlock (J. A. Davidson).

I. Scale covers on lower surface of hemlock needles (J. A. Davidson).

J. Chlorotic or yellow areas on upper surface of hemlock needles caused by feeding scales on undersurface (J. A. Davidson).

Plate 14. *Aspidiotus destructor* Signoret, Coconut scale

A. Yellow adult females beginning to lay eggs under transparent covers on palm (J. A. Davidson).
B. Clear covers showing eggs arranged in concentric rings on palm (R. J. Gill).
C. Transparent covers with predominantly white egg shells beneath on palm (J. A. Davidson).
D. Old scale covers showing dead brown females and hatched eggs on palm (D. R. Miller).

E. Heavy infestation on palm (J. A. Davidson).
F. Bodies of adult females (J. A. Davidson).
G. Undersurfaces of infested and uninfested leaves of *Persea* (J. A. Davidson).
H. Upper surfaces of uninfested and infested leaves of *Persea* (J. A. Davidson).
I. Infestation on *Alpinia speciosa* (R. J. Gill).

Aspidiotus destructor Signoret

ESA Approved Common Name Coconut scale (also called transparent scale and bourbon scale).

Common Synonyms and Combinations *Aspidiotus transparens* Green, *Aspidiotus fallax* Cockerell, *Aspidiotus cocotis* Newstead, *Aspidiotus simillimus translucens* Cockerell, *Aspidiotus destructor transparens* Green, *Aspidiotus translucens* Cockerell, *Aspidiotus vastatrix* Leroy, *Aspidiotus oppugnatus* Silvestri.

Field Characters (Plate 14) Adult female cover translucent, flat, circular; shed skins central or subcentral, yellow or yellowish brown. Male cover elongate oval, similar color and texture as female cover; shed skin submarginal. Body of adult female yellow orange, clearly visible through scale cover; eggs white when first laid, turning yellow after few days. Normally in dense infestations on undersides of leaves, occasionally on bark and fruit.

Slide-mounted Characters Adult female (Fig. 16) with 3 pairs of well-developed lobes; paraphyses absent. Median lobes separated by space 0.4–0.7 (0.6) times width of median lobe, with noticeable sclerotized area anterior of lobe, without yoke, occasionally with weak sclerotization between lobes, but not typical of yoke, usually with medial and lateral margins parallel, sometimes with lobe margins converging apically, with 1 lateral notch, 1 medial notch, lobe apex with small serrations on specimens that have unworn lobes; second lobes simple, usually noticeably larger than median lobes, sometimes about same size or smaller, with 1–2 (1) lateral notches, without a medial notch; third lobes simple, about same size as second lobe or slightly smaller, with 1–2 (1) lateral notch, without medial notch. Plates fimbriate, with microduct often emptying on mesal side of plate; plate formula 2-3-3 or 2-3-4, with 4 or 5 plates anterior of seta marking segment 5; with 2 broad, apically fringed plates between median lobes, usually conspicuously longer than median lobes. Macroducts of 1 size, 1 macroduct between median lobes, extending 1.0–1.7 (1.3) times distance between posterior apex of anal opening and base of median lobes, 49–60 (54) μ long, longest macroduct in first space 51–81 (65) μ long, with 15–27 (18) on each side of pygidium on segments 5 to 8, ducts in submarginal and marginal areas, total macroducts on each side of body 15–29 (19), without macroduct orifices anterior of anal opening; prepygidial macroducts rarely present on segment 4, of same size and shape as on pygidium. Pygidial microducts on venter in submarginal areas of segment 5, rarely segments 6 and 7, with 1–5 (2) ducts; prepygidial ducts of 1 size, submarginal ducts usually on each of meso- and metathoracic segments and on segments 1 to 4, occasionally on head and prothorax, submedial ducts near mouthparts, sometimes near posterior and anterior spiracles, often present on metathorax and segments 1 to 3; pygidial microducts absent from dorsum, of 2 sizes on prepygidial areas, shorter ducts in submarginal areas of head to metathorax or segments 1 or 2, longer ducts usually in submedial areas of thorax and segments 1 and 3. Perivulvar pores usually in 4 groups, rarely 5, and 13–18 (15) pores on each side of body. Perispiracular pores absent. Anal opening located 1.3–2.2 (1.5) times length of anal opening from base of median lobes, anal opening 23–30 (26) μ long. Dorsal seta laterad of median lobes 2.0–3.1 (2.3) times length of median lobe. Eyes represented by circular sclerotized area on body margin at level of anterior margin of mouthparts, rarely absent. Antennae each with 1 seta. Cicatrices usually on prothorax and segment 1, occasionally with small cicatrix on segment 3. Body oval or pyriform. Median lobes extremely variable, often appearing recessed into pygidium, sometimes protruding beyond apex of lobe 2.

Affinities The coconut scale is similar to *Aspidiotus spinosus*, spinose scale, but differs by having a larger anal opening (23–30 (26) μ long), median lobes separated by space 0.4–0.7 (0.6) times width of median lobe, third lobes usually about same size as second lobes, without noticeable sclerotized rim at dermal orifice of all macroducts, with several plates anterior of seta marking segment 5, macroduct between median lobes 49–60 (54) μ long, without macroduct orifices anterior of anal opening, 13–18 (15) perivulvar pores on each side of body. *Aspidiotus spinosus* has a small anal opening (7–15 (10) μ long), median lobes separated by space 0.2–0.3 (0.3) times width of median lobe, third lobes noticeably smaller than second lobes, with noticeable sclerotized rim at dermal orifice of macroducts, without plates anterior of seta marking segment 5, macroduct between median lobes 29–34 (32) μ long, with macroduct orifices anterior of anal opening, and 7–14 (10) perivulvar pores on each side of body. The coconut scale also is similar to the cryptomeria scale (*Aspidiotus cryptomeriae*) and the aglaonema scale (*Aspidiotus excisus*); for a comparison see affinities of the latter species. Of species occurring outside of the United States, *Aspidiotus rigidus* closely resembles *A. destructor* and the two cannot be distinguished morphologically. Reyne (1948) reported the following differences on coconut palms in Sangi (North Celebes): (1) In *A. rigidus* only a semilunar area under the scale cover is filled with eggs while in *A. destructor* eggs are deposited in a complete circle around the female, (2) In *A. rigidus* the average duration of the life cycle at sea level is 46 days while only 32 days in *A. destructor*, (3) *Garcinia mangostana* is the common host for *A. rigidus*, while specimens of *A. destructor* placed on this host died before reaching maturity, and (4) there is a difference in the diversity and species of predators and parasites attacking these pests.

Hosts Polyphagous, Borchsenius (1966) records it from 76 genera in 46 families of host plants; Dekle (1977) reports it from 132 host genera. We have examined U.S. specimens from the following host genera: *Acer, Acorus, Aegle, Agave, Aglaonema, Alectryon, Anacardium, Ananas, Annona, Areca, Artocarpus, Aspidistra, Aucuba, Barringtonia, Callistemon, Calocarpum, Calophyllum, Camellia, Carica, Carludovica, Cassia, Ceiba, Chamaedorea, Chrysalidocarpus, Cinnamomum, Citrus, Coccothrinax, Cocos, Colocasia, Congea, Cryosophila, Cryptocarya, Cycas, Cymbidium, Cyrtopodium, Dalbergia, Dendrobium, Dictyosperma, Dieffenbachia, Dillenia, Dioscorea, Dipteryx, Elaeis, Eugenia, Euphorbia, Eupritchardia, Fatsia, Ficus, Garcinia, Gnetum, Grevillea, Hernandia, Heterospathe, Hevea, Hylocereus, Jasminum, Kentia, Lagerstroemia, Laurus, Licuala, Ligustrum, Litsea, Livistona, Lonicera, Loranthus, Maesa, Magnolia, Mammea, Mangifera, Manihot, Maranta, Michelia, Moringa, Murraya, Musa, Myristica, Nepenthes, Nephelium, Nerium, Nipa, Normanbya, Olea, Osmanthus, Palaquium, Pandanus, Persea, Peperomia, Phoenix, Plumeria, Podocarpus, Poinciana, Poncirus, Prunus, Psidium, Ptychosperma, Punica, Rhododendron, Rosa, Sabal, Saccharum, Sonneratia, Spondias, Terminalia, Thea, Vitis, Washingtonia,* and *Zingiber.* Common hosts are avocado, coconut, screw pine (*Pandanus*), and banana.

Distribution In the United States coconut scale has been reported out-of-doors in CA, FL, GA, HI; it apparently is eradicated from CA (Ryan 1946; Gill 1997). It occasionally is taken in other states in greenhouses and on foliage plants; Nakahara (1982) mentioned CT and PA, but it probably is encountered in other states as well. This species often is taken in quarantine from tropical and subtropical areas in Asia, Australia, the Caribbean islands, Central and South Africa, Central and South

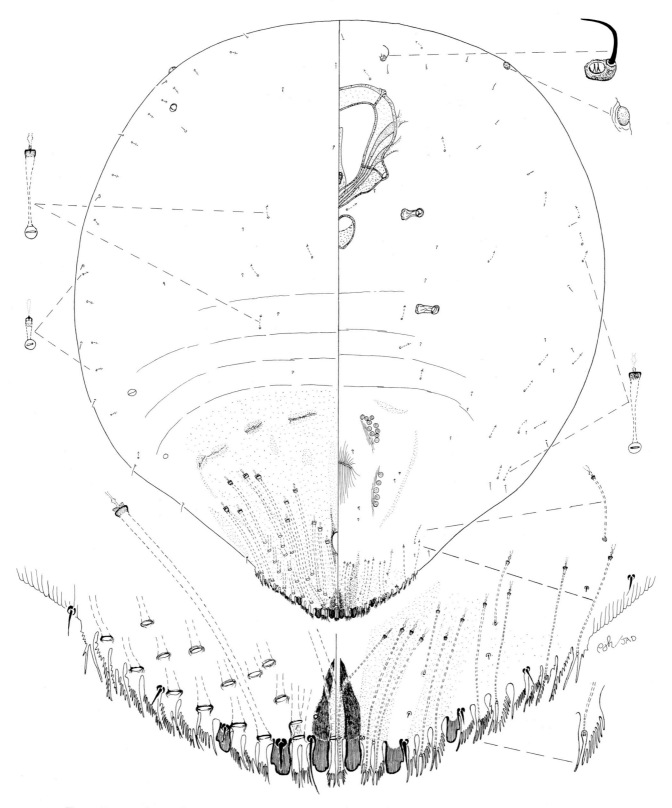

Figure 16. *Aspidiotus destructor* Signoret, coconut scale, Honduras, intercepted at Miami, FL, on *Mourea* sp., VIII-29-1975.

America, Mexico, and the Pacific Islands. Nakahara (1982) reports it from the New World (widespread, not reported in Argentina, Canada, Chile, Paraguay, or Uruguay); Europe (Germany greenhouse); former Soviet Union (Azerbaijan, Georgia); Africa (many countries south of the Sahara on the continent, no record from North Africa except Egypt, Canary Islands, Cape Verde, Madagascar, Madeira Island, Reunion, Seychelles); Asia (Iran eastward to Indonesia, north to Japan and Korea); others (Australia, Bonin Islands, Caroline Islands, Fiji, Hawaii, Mariana Islands, Marshall Islands, New Britain, New Caledonia, New Guinea, New Hebrides, Samoa, Society Islands, Solomon Islands, Tuamotu Island, Wallis Island). A distribution map of this species was published by CAB International (1966).

Biology Simmonds (1921) reports that *A. destructor* females in Fiji laid about 80 eggs in 10 days. These were positioned in 3 or 4 concentric rings around the body of each female. The eggs began to hatch in about 7 days, and 38 days were required for a complete life cycle. Taylor (1935) reports that the time required to develop from the egg to the adult male was 32 days and from the egg to the adult female 34.5 days. If development were relatively even throughout the year, more than 9 generations could be produced annually. Goberdhan (1962) found that in the coconut areas of Trinidad this scale breeds continuously year-round with some slow down during wet periods. In Japan coconut scale is reported to have 1 generation annually with the nymphs appearing in May (Murakami 1970). Borchsenius (1950) reports 2 to 3 generations per year in Adzharia, and Imamkuliev (1966) states there are 3 per year in Azerbaijan in the former Soviet Union. In studies in Hunan, China, Zhou et al. (1993) found that *Aspidiotus destructor* produced 3 generations annually, with the fertilized female overwintering on the stem of *Actinidia* trees. The sex ratio (females to males) was 1:1.04. Oviposition began in early to mid-April. The average eggs laid by one female was 32–42. The effective accumulated temperature was 755.89 day-degrees for females and 727.6 day-degrees for males.

Economic Importance Coconut scale is primarily a pest of coconut palm, but it occasionally can be destructive on other palms, bananas, mango, guava, ornamentals, and tea. Taylor (1935) reports *A. destructor* as causing extensive damage in Fiji to coconuts, bananas, 'yaqona,' and avocado. In heavy infestations leaves and fruit can be completely covered with the scale. Feeding sites often turn yellow, and leaves usually fall prematurely. In Mauritius at one time it was so destructive that coconut palms were no longer planted (Moutia and Mamet 1946). Until successful biological control agents were introduced, the copra industry was virtually eliminated in areas that were infested with this pest. As an example of the seriousness of the coconut scale, Reyne (1948) noted an outbreak in Sangi (North Celebres) in which 400,000 palms were infested, 30,000 were killed, and no fruit was produced anywhere in the infested region. Recent publications have mentioned coconut scale as a pest of black pepper (Devasahayam 1992), mango (Handa and Dahiya 1999), mangrove (Kathiresan 1993), and papaya (Maheswari and Purushotham 1999). With effective methods of biological control, this species is rarely a serious pest. Occasionally, plants become heavily infested when coconut scale occurs with honeydew producing species that are tended by ants. The coconut scale is listed as one of the 43 principal armored scale pests of the world by Beardsley and González (1975) and Miller and Davidson (1990). This pest is under effective biological control in most areas where it once was a serious problem. Some of the more spectacular successes have occurred in Fiji with *Cryptognatha nodiceps* (Taylor 1935), in Mauritius with *Chilocorus politus* and *C. nigritus* (Moutia and Mamet 1946), in the New Hebrides with *Rhizobius pulchellus* (Cochereau 1969), in Guam with *Telsimia nitida* (Fullaway 1946), in the Ivory Coast with *Chilocorus schioedtei* and *C. dohrni* (Mariau and Julia 1977), and in Puerto Rico with *C. cacti* (Wolcott 1960). Rosen and DeBach (1978) give a detailed summary of many of the biological control efforts implemented against this scale. In most areas where the pest was a serious problem, it now is difficult to find any scales at all.

Selected References Beardsley (1970); Goberdhan (1962); Reyne (1948); Simmonds (1921); Taylor (1935).

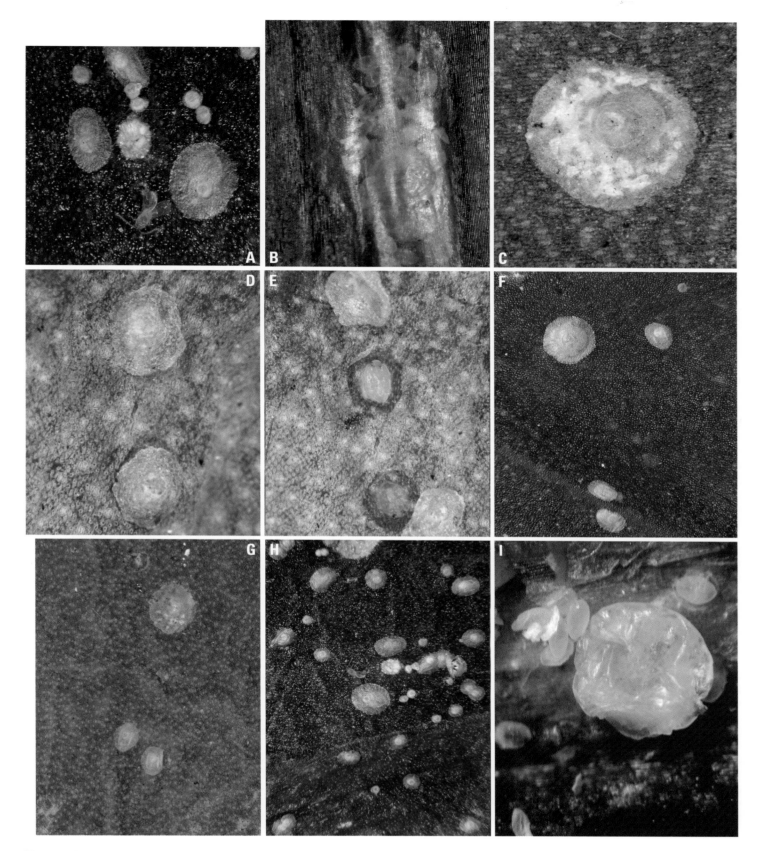

Plate 15. *Aspidiotus excisus* Green, Aglaonema scale

A. Close-up of adult female cover (right), two narrower male covers (left), and adult male between, on aglaonema (R. J. Gill).
B. Adult female cover with eggs and egg shells on dracaena (R. J. Gill).
C. Adult female cover with egg shells on agave (J. A. Davidson).
D. Male covers on agave (J. A. Davidson).

E. Bodies of second-instar males (J. A. Davidson).
F. Adult female cover and three male covers on aglaonema (R. J. Gill).
G. Female and male covers on agave (J. A. Davidson).
H. Covers of most instars on aglaonema (R. J. Gill).
I. Body of adult female and eggs on dracaena (R. J. Gill).

Aspidiotus excisus Green

Suggested Common Name Aglaonema scale.

Common Synonyms and Combinations *Evaspidiotus excisus* (Green), *Temnaspidiotus excisus* (Green).

Field Characters (Plate 15) Adult female cover translucent white, slightly convex, circular or irregular; shed skins central or subcentral, yellow. Male cover elongate oval, similar in color and texture to female cover. Body of adult female, visible through scale cover, yellow; eggs yellow. Normally on leaves and green stems.

Slide-mounted Characters Adult female (Fig. 17) with 3 pairs of well-developed lobes; paraphyses absent. Median lobes separated by space 0.3–0.5 (0.4) times width of median lobe, with noticeable sclerotized area anterior of lobe, without yoke, usually with medial and lateral margins parallel or diverging apically, with 1 lateral notch and 1 medial notch; second lobes simple, often longer and wider than median lobes, sometimes about same size or rarely smaller, with 1–2 (1) lateral notches, with 1–2 (1) medial notches; third lobes simple, usually smaller than second lobes, sometimes about same size, with 0–1 (1) lateral notch, 0–1 (0) medial notch. Plates fimbriate, with microduct emptying on mesal side of many plates; plate formula 2-3-3 or 2-3-4, with 3 to 5 plates anterior of seta marking segment 5; with 2 broad, bifurcate plates between median lobes, longer than median lobes. Macroducts of 1 size, slightly smaller anteriorly, 1 macroduct between median lobes, extending 0.4–0.7 (0.6) times distance between posterior apex of anal opening and base of median lobes, 26–33 (29) μ long, longest macroduct in first space 27–36 (31) μ long, with 13–20 (16) on each side of pygidium on segments 5 to 8, ducts in submarginal and marginal areas, total macroducts on each side of body 22–38 (30), with macroduct orifices anterior of anal opening; prepygidial macroducts of 2 sizes, larger size of same size and shape as on pygidium, present on segment 4, smaller size on segments 2 and 3. Pygidial microducts on venter usually in submarginal areas of segment 5, rarely segments 6 and 7, with 0–4 (2) ducts; prepygidial ducts of 1 size, submarginal ducts on each of segments 1 or 2 to 4, also on head and thorax, submedial ducts near mouthparts and anterior spiracle, usually present near posterior spiracle, often present on head to metathorax, sometimes on segments 1 to 3; pygidial microducts absent from dorsum, of 2 sizes on prepygidial areas, smaller size in submarginal areas from head to segments 2 or 3, larger size in submarginal areas of head to metathorax, usually in submedial areas from head to segment 2, rarely on segment 3. Perivulvar pores usually in 4 groups, rarely 5, 12–24 (16) pores on each side of body. Perispiracular pores absent. Anal opening located 2.6–4.3 (3.1) times length of anal opening from base of median lobes, anal opening 12–20 (18) μ long. Dorsal seta laterad of median lobes 1.4–2.2 (1.8) times length of median lobe. Eyes represented by circular sclerotized area on body margin at level of anterior margin of mouthparts. Antennae each with 1 seta. Cicatrices usually on prothorax, segment 1, and 3. Body oval or pyriform, apical apex truncate. Median lobes usually recessed into pygidium, usually with series of small serrations apically.

Affinities The aglaonema scale is similar to *Aspidiotus destructor*, the coconut scale, but differs by having a notch in the medial margin on the second lobe, the macroduct between the median lobes extending 0.4–0.7 (0.6) times the distance between the posterior apex of the anal opening and the base of the median lobes, the macroduct between the median lobes 26–33 (29) μ long, some macroduct orifices anterior of the anal opening, the anal opening located 2.6–4.3 (3.1) times the length of the anal opening from the base of the median lobes, the anal opening 12–20 (18) μ long, the dorsal seta laterad of the median lobes is 1.4–2.2 (1.8) times as long as the median lobes, and the median lobes usually not protruding beyond the apices of the second lobes. The coconut scale has no notch on the medial margin on the second lobe, the macroduct between the median lobes extends 1.0–1.7 (1.3) times the distance between the posterior apex of the anal opening and the base of the median lobes, the macroduct between the median lobes is 49–60 (54) μ long, there usually are no macroduct orifices anterior of the anal opening, the anal opening is located 1.3–2.1 (1.5) times the length of the anal opening from the base of the median lobes, the anal opening is 23–30 (26) μ long, the dorsal seta laterad of the median lobes is 2.0–3.1 (2.3) times as long as the median lobes, and the median lobes sometimes protrude beyond the apices of the second lobes. For a comparison of this species with oleander scale (*A. nerii*) see the affinities section of the latter.

Hosts Found on several unrelated hosts. Borchsenius (1966) records it from 10 genera in 9 families of host plants; Dekle (1977) reports it from 13 host genera. We have examined specimens from the following genera: *Adenium, Aechmea, Aglaonema, Alocasia, Carica, Caryota, Chlorophytum, Citrus, Cocos, Coffea, Cyanotis, Cycas, Euphorbia, Gardenia, Ipomoea, Jacobinia, Lonicera, Mangifera, Melicoccus, Murraya, Musa, Pentas, Pritchardia, Psidium,* and *Thespesia.* The most common host in the United States is *Aglaonema.* Based on material in the USNM Collection aglaonema scale often is intercepted in quarantine on *Aglaonema, Citrus, Lonicera.*

Distribution In the United States this species is reported from FL only. It often is taken in U.S. quarantine from Asia, the Caribbean islands, South and Central America, and the Pacific Islands. Nakahara (1982) reports it from the New World (Antigua, Colombia, Costa Rica, Dominican Republic, Ecuador, El Salvador, Grenada, Guatemala, Guyana, Honduras, Jamaica, Martinique, Mexico, Panama, Puerto Rico, St. Croix, Surinam, Trinidad, Venezuela); Asia (China, India, Indonesia, Japan, Pakistan, Philippines, Singapore, Sri Lanka, Taiwan, Thailand); others (Fiji, New Guinea, Palau Islands, Saipan, Tinian, Yap Island).

Biology This species is commonly encountered on the leaves of its host where it may build to heavy populations. Takahashi (1936) reports that this species caused gall-like folds on the leaves of *Clerodendron inerme.*

Economic Importance Dekle (1966) considers this scale to be a serious pest in Florida, especially on *Aglaonema* and *Hoya carnosa.* Jepson (1915) states that the species is an injurious pest on banana in Fiji. Talhouk (1975) considers it to be of some economic importance in India and China on citrus and of little importance in Pakistan. Miller and Davidson (1990) consider this species to be a serious pest in a small area of the world.

Selected References Dekle (1966); Ferris (1938a).

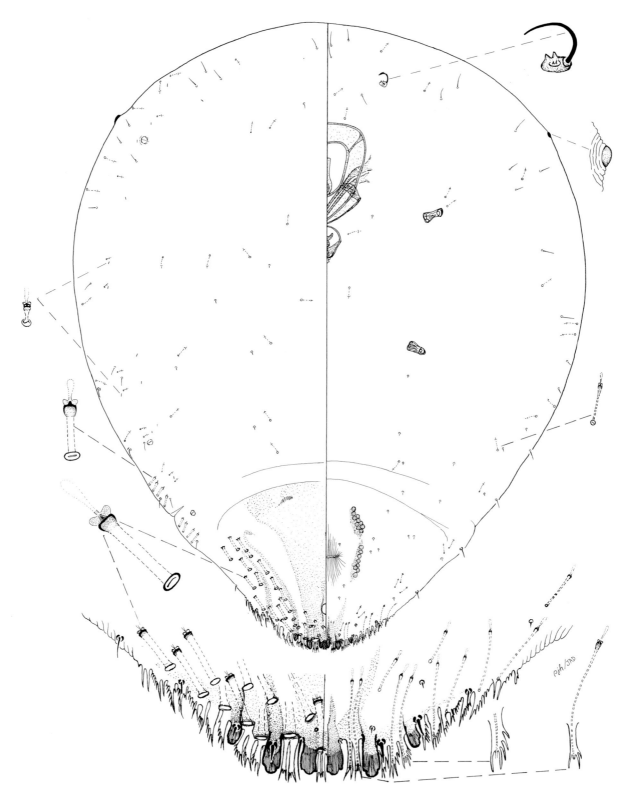

Figure 17. *Aspidiotus excisus* Green, aglaonema scale, Dominican Republic, intercepted at San Juan, Puerto Rico, on *Aglaonema* sp., II-27-1947.

Aspidiotus nerii Bouché

ESA Approved Common Name Oleander scale (also called ivy scale, lemon peel scale, and white scale).

There is strong evidence suggesting that this species is a complex of species, particularly biparental and uniparental taxa (Gerson and Hazan 1979). Papers by DeBach and Fisher (1956), Gerson and Hazan (1979), and Schmutterer (1952a) demonstrate important differences among these siblings in characteristics of life history, host preference, environmental tolerance, natural enemies, and scale cover color. *Aspidiotus nerii* generally is regarded as the bisexual form; Schmutterer describes the unisexual form he studied in Germany as *A. hederae* (= *nerii*) *unisexualis* Schmutterer, and Gerson and Hazan treat the unisexual taxon they studied in Israel as *A. paranerii* Gerson. It has not been possible to discriminate these species morphologically other than by subtle differences in the color of the scale cover. *Aspidiotus nerii unisexualis* has a yellowish cover, whereas the nominal subspecies has a white cover. In the case of *A. paranerii* the scale cover is yellowish red, compared to *A. nerii* which is described as yellow. Although we concur with these authors that biological species should be recognized and treated as separate entities, it is virtually impossible for us to sort out the literature pertinent to each species, or for that matter even understand how many species exist within the confines of what traditionally is considered to be *A. nerii*. Therefore, we give a single treatment for the several entities that represent this species complex. It is important to realize there is more than a single species involved. Caution should be exercised when applying the general information given in the following treatment.

Common Synonyms and Combinations There are more than 40 synonyms of this species. We have listed most of those encountered in American literature. *Aspidiotus genistae* Westwood, *Chermes aloes* Boisduval, *Chermes ericae* Boisduval, *Chermes cycadicola* Boisduval, *Aspidiotus cycadicola* (Boisduval), *Aspidiotus Bouchéi* Targioni-Tozzetti, *Aspidiotus caldesii* Targioni-Tozzetti, *Aspidiotus denticulatus* Targioni-Tozzetti, *Aspidiotus villosus* Targioni-Tozzetti, *Aspidiotus aloes* (Boisduval), *Aspidiotus budleiae* Signoret, *Aspidiotus epidendri* Signoret, *Aspidiotus ericae* (Boisduval), *Aspidiotus ilicis* Signoret, *Aspidiotus oleastri* Colvee, *Aspidiotus affinis* Targioni-Tozzetti, *Aspidiotus hederae* Signoret, *Aspidiotus ceratoniae* Signoret.

Field Characters (Plate 16) Adult female cover flat or convex, circular, white to opaque tan, transparent; shed skins central, yellow, light brown, or reddish brown. Male cover similar in color, smaller, oval with submarginal shed skin. Body of female orange yellow; eggs pale yellow. Normally on leaves, occasionally on stems, bark, or fruit.

Slide-mounted Characters Adult female (Fig. 18) with 3 pairs of well-developed lobes; paraphyses absent. Median lobes separated by space 0.3–0.7 (0.5) times width of median lobe, with noticeable sclerotized area anterior of lobe, without yoke, usually with medial and lateral margins parallel, with 1–2 (1) lateral notches, 0–1 (1) medial notch; second lobes simple, unusually variable in size, normally slightly smaller than median lobes and of same shape, occasionally nearly equal in size or conspicuously smaller and with pointed apex, with 0–1 (1) lateral notches, 0–1 (0) medial notches; third lobes simple, conspicuously smaller than second lobes and with pointed apex, with 0–1 (1) lateral notches, without medial notch. Plates fimbriate, with microduct emptying near apex of most plates; plate formula 2-3-4, with 2 to 4 plates anterior of seta marking segment 5; with 2 broad, apically fringed plates between median lobes, usually protruding slightly beyond apex of median lobes. Macroducts of 1 size, 1 macroduct between median lobes, extending 0.3–0.6 (0.4) times distance between posterior apex of anal opening and base of median lobes, 19–26 (23) μ long, longest macroduct in first space 19–26 (22) μ long, with 11–16 (13) on each side of pygidium on segments 5 to 8, ducts in submarginal and marginal areas, total macroducts 17–52 (28) on each side of body, with macroduct orifices anterior of anal opening; prepygidial macroducts of 2 sizes, larger size of same size and shape as on pygidium, present on segments 1 or 2 to segment 4, smaller size on ventral margin of segments 1 or 2 to segments 3 or 4. Pygidial microducts on venter in submarginal areas of segment 5, occasionally segments 6 and 7, with 1–7 (4) ducts; prepygidial ducts of 1 size, submarginal ducts usually on head, thorax, and segments 1 to 4, submedial ducts near mouthparts, posterior and anterior spiracles, prothorax to metathorax, and on 1 or more of segments 1 to 3; pygidial microducts absent from dorsum, of 2 sizes on prepygidial areas, shorter ducts in submarginal areas of head to metathorax or segments 1 or 2, longer ducts usually in submedial areas of thorax and segments 1 to 3. Perivulvar pores usually in 4 groups, rarely 5, 10–17 (15) pores on each side of body. Perispiracular pores absent. Anal opening located 2.0–3.0 (2.5) times length of anal opening from base of median lobes, anal opening 17–25 (21) μ long. Dorsal seta laterad of median lobes 1.2–2.3 (1.6) times length of median lobe. Eyes represented by circular sclerotized area on body margin at level of anterior margin of mouthparts. Antennae each with 1 seta. Cicatrices usually on prothorax and segment 1. Body oval or pyriform. Specimens collected primarily on palm hosts tend to be different in the structure of their plates. The plate between lobes 2 and 3 that is adjacent to lobe 2 is usually heavily fringed, and the plates anterior of lobe 3 tend to be longer than the other plates. Specimens from other hosts usually have the plate between lobes 2 and 3 that is adjacent to lobe 2 simple or with 1 or 2 tines, and the plates anterior of lobe 3 tend to be of equal length or shorter than the other plates.

Affinities The oleander scale is similar to *Aspidiotus excisus*, the aglaonema scale, but differs by having the median lobes protruding from the pygidium as far as the second lobes, the second lobes usually without a medial notch, the third lobes apically pointed, the macroduct between the median lobes 19–26 (23) μ long, and small-sized macroducts on prepygidial margin. The aglaonema scale has the median lobes less protuberant from the pygidium than the second lobes, the second lobes usually with 1–2 (1) medial notches, the third lobes apically rounded, the macroduct between the median lobes 26–33 (29) μ long, and small-sized macroducts absent from prepygidial margin. For a comparison of this species with *A. spinosus* see the affinities section of that species.

Hosts Polyphagous. Borchsenius (1966) records it from 235 genera in 86 families; Dekle (1977) reports it from 61 host genera. We have examined specimens from the following host genera: *Acacia, Aechmea, Agave, Aleurites, Aloe, Alyxia, Ananas, Andromeda, Annona, Anthurium, Arabis, Araucaria, Arbutus, Arctostaphylos, Ardisia, Areca, Argania, Asparagus, Aspidistra, Atriplex, Aucuba, Banksia, Berberis, Bupleurum, Calycotome, Camellia, Cassia, Catalpa, Ceanothus, Cedrus, Ceratonia, Cereus, Chamaerops, Choisya, Citrus, Cneoridium, Cocos, Codiaeum, Conium, Cordyline, Coronilla, Corylus, Crinodendron, Cycas, Cydonia, Cymbidium, Cyperus, Cypripedium, Cytisus, Daphne, Dendrobium, Desfontainea, Dianthus, Dioon, Diospyros, Dodonaea, Dracaena, Echeveria, Echinocactus, Erythea, Eucryphia, Euonymus, Euphorbia, Fatsia, Ficus, Francoa, Fraxinus, Furcraea, Gardenia, Garrya, Gleditsia, Hakea, Haworthia, Hedera, Heliotropium, Hemerocallis, Hepatica, Hibiscus, Huernia, Ilex, Iris, Jasminum,*

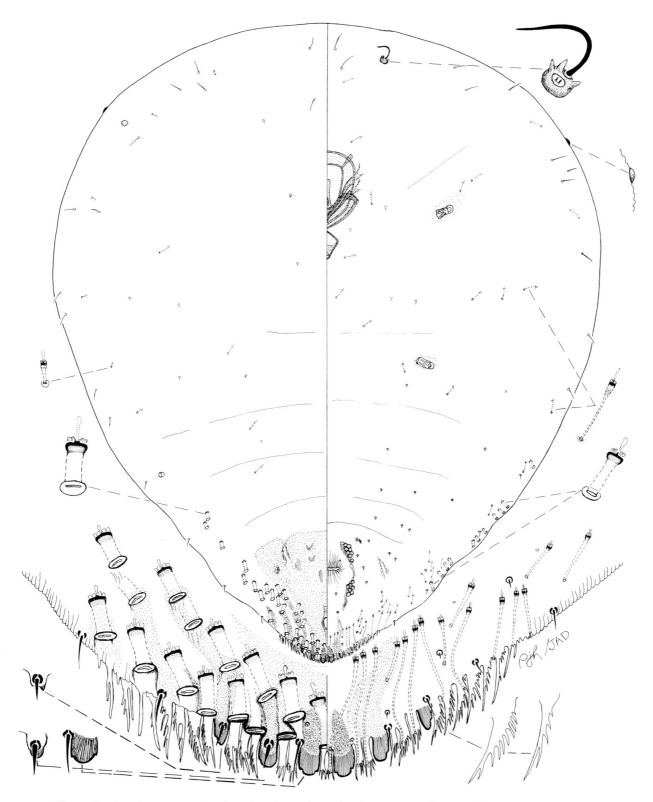

Figure 18. *Aspidiotus nerii* Bouché, oleander scale, Colombia, on *Passiflora* sp., V-17-1972.

Juglans, Kentia, Laelia, Lapageria, Laurelia, Laurus, Lavatera, Ligustrum, Liriope, Livistona, Lonicera, Macadamia, Macrozamia, Magnolia, Mangifera, Medicago, Melia, Meryta, Miltonia, Mimosa, Morus, Musa, Myoporum, Nerium, Ocimum, Odontoglossum, Olea, Osmanthus, Osmarea, Passiflora, Peperomia, Persea, Peumus, Phoenix, Phoradendron, Phormium, Phyllostachys, Phytelephas, Phytolacca, Pitcairnia, Pittosporum, Platanus, Protea, Prunus, Psidium, Pterogyne, Pueraria, Punica, Pyrus, Rhododendron, Rhopalostylis, Ribes, Robinia, Rosa, Rosmarinus, Rubia, Ruta, Sabal, Salvia, Sapindus, Sarracenia, Sedum, Silene, Smilax, Solanum, Sophora, Sorbus, Stapelia, Strelitzia, Streptosolen, Tamarix, Teucrium, Thuja, Tillandsia, Trochocarpa, Trochodendron, Umbellularia, Vaccinium, Vinca, Vitex, Vitis, and *Yucca.* Common hosts are avocado, bay, citrus, ivy, oleander, olive, and palms.

Distribution We have examined U.S. specimens from AL, AR, AZ, CA, CO, CT, DC, FL, GA, HI, IL, IN, LA, MA, MD, ME, MI, MO, MS, MT, NC, NH, NJ, NM, NY, OH, OR, PA, SC, SD, TN, TX, UT, VA, VT, WA, WI, WY. The oleander scale occurs out-of-doors in the southern states only. It commonly is taken in quarantine from most tropical or subtropical areas of the world. Nakahara (1982) reports it from the New World (Antigua, Argentina, Bermuda, Bolivia, Brazil, Canada, Chile, Colombia, Costa Rica, Cuba, Ecuador, Dominican Republic, Grenada, Guatemala, Guadeloupe, Haiti, Jamaica, Mexico, Panama, Peru, Puerto Rico, Uruguay); Europe (widespread); former Soviet Union (Armenia, Azerbaijan, Crimea, Georgia, Krasnodar); Africa (North Africa, Canary Islands, Ethiopia, Kenya, Madagascar, Madeira Island, Malawi, Mozambique, Nigeria, Rhodesia, South Africa, Southwestern Africa, Tanzania, Uganda, Zaire); Asia (China, Cyprus, Iran, Israel, Jordan, Korea, Lebanon, Japan, Saudi Arabia, Syria, Turkey); others (Australia, Hawaii, New Caledonia, New Zealand). A distribution map of this species was published by CAB International (1970a).

Biology There is a great deal of biological variation reported for this species, a fact that supports the hypothesis that this scale actually is a complex of several sibling species. The oleander scale develops almost continuously throughout the year, but the predominant form in the winter is the newly mature adult female. Schmutterer (1952a) indicates that a generation of both the biparental and uniparental taxa required approximately 90 days from egg to egg at about 77°F. DeBach and Fisher (1956) found that each of the biparental and uniparental taxa they studied required 49 days at 75°F. The results of Gerson and Hazan (1979) also are different; they found that the uniparental species required about 45 days to mature at 75°F, but the biparental species required more than 60 days at the same temperature. Generations tend to overlap broadly because the oviposition period can be very long. Gerson and Hazan (1979) found oviposition periods of about 37 days in the biparental species and about 24 days in the uniparental species at 75°F. Most authors report 3 generations per year, for example, in Azerbaijan (Imamkuliev 1966), Egypt (Salama and Hamdy 1974a), Crete (Neuenschwander et al. 1977), Israel (Berlinger et al. 1999), in greenhouse (Saakyan-Baranova 1954); others indicate 3 to 4 generations, for example, in Greece (Argyriou and Kourmadas 1981), Israel (Avidov and Harpaz 1969), Middle East (Talhouk 1969); and some indicate 4 generations, for example, Greece (Argyriou 1976), Romania (Savescu 1961). There is one

report of a yearly cycle of 2 and a partial third generation in Georgia (former Soviet Union) (Hadzibejli 1983). Crawlers usually begin to appear in May in the Northern Hemisphere (Avidov and Harpaz 1969). Gerson and Hazan (1979) counted an average of about 100 crawlers produced by the biparental species at about 78°F on potatoes and slightly more than 40 crawlers by the uniparental species under the same conditions. Bartra (1978) reported that the average number of eggs per female in a bisexual taxon was about 200 in Spain. Males (when present) generally settle on the upper sides of leaves and the females are located on the under surface (Avidov and Harpaz 1969). Life table information was presented by Karaca and Uygun (1993) on different host plants in Turkey. The sex pheromone produced by the female is quite unusual and has been synthesized (Boyer and Ducrot 1999a, 1999b; Einhorn et al. 1998; Petschen et al. 1999, 2000).

Economic Importance This species is sometimes considered a pest of oleander, olive, citrus, palms, kiwi fruit, and many woody ornamentals. Talhouk (1975) considers the oleander scale to be a 'very important pest' of citrus in Italy only. In other citrus-growing areas of the world he considered it of minor importance. A sample of the literature that discusses the economic importance of this scale is as follows: Steiner and Elliott (1983) consider it to be a pest on indoor plants; Gill (1982) indicates that it is a minor pest of olive in California; Andres (1979) regards it as a major pest of olive in Spain; Alexandrakis et al. (1977) demonstrate that it reduced fruit weight and oil content in olives grown in Crete; Smee (1936) reports it as a serious pest of tung (*Aleurites montana*) in Africa; Batiashvili (1954) considers it to be a serious pest of tung in the Black Sea area of Georgia; McKenzie (1935a) regards it as a pest of avocado in California; Szeremlei et al. (1979) consider it to be a pest of poinsettias in Hungary; Morris (1927) states that it is a serious pest on carob and wattle in Cyprus. On olive, when this pest settles on the fruit, the area surrounding the scale remains green and must be culled since the infested fruit will not pickle properly. Fruit that is attacked early in its development is pitted at the feeding site, and the oil content of the olive may be reduced by as much as 25% (Ebeling 1959). On fruit in New Zealand (Blank et al. 1999) and Chile (González 1989), oleander scale is considered an important pest. Evidence is presented that the highest infestations are nearest areas of alternative host plants, supporting the hypothesis that primary sources of infestation are from other host plants (Blank et al. 1999). The oleander scale is listed as one of the 43 principal armored scale pests of the world by Beardsley and González (1975) and is considered an important pest by Miller and Davidson (1990). Successful biological control of oleander scale has been achieved in many areas using *Aphytis melinus* and *A. chilensis* (DeBach and Rosen 1976). Alexandrakis and Neuenschwander (1980) have shown that the latter parasite is capable of high levels of parasitization over a wide range of pest densities in Crete on olives. Oleander scale can be very abundant on jojoba in Israel and the United States, but it never seems to reach pest status (Berlinger et al. 1999). In Chile it is abundant, but in this situation it is considered a pest by reducing the rate of photosynthesis and by damaging the fruit (Quiroga et al. 1991).

Selected References DeBach and Fisher (1956); Gerson and Hazan (1979); Gill (1997); Stoetzel and Davidson (1974).

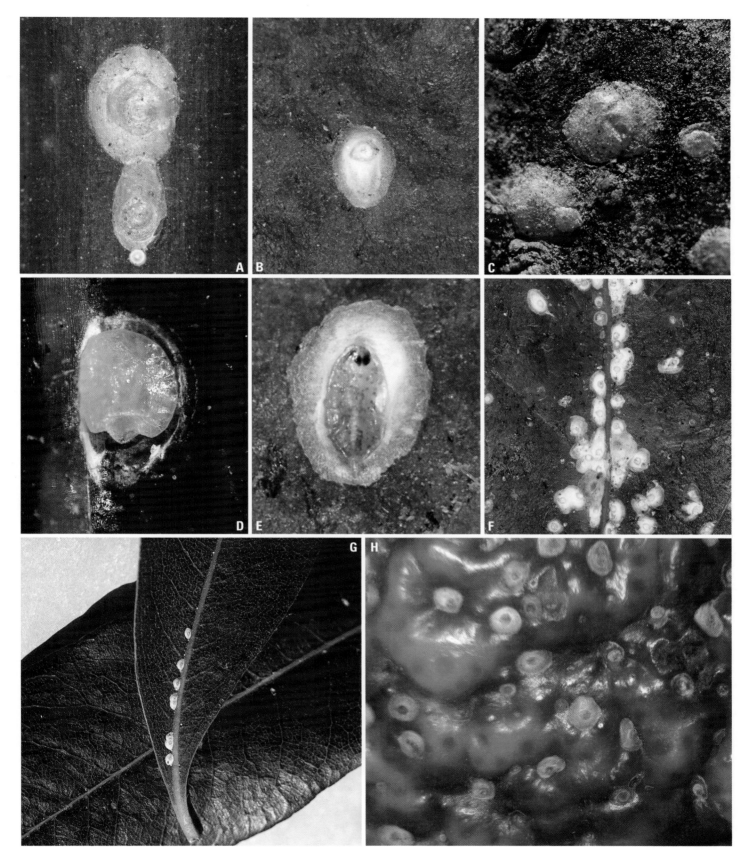

Plate 16. *Aspidiotus nerii* Bouché, Oleander scale

A. Adult female cover, male cover, and capped crawler on ivy (J. A. Davidson).
B. Adult male cover on ivy (J. A. Davidson).
C. Two adult female covers on olea (J. A. Davidson).
D. Body of adult female on pittosporum (J. A. Davidson).
E. Male pupa with cover removed on ivy (J. A. Davidson).
F. Infestation on ivy (J. A. Davidson).
G. Infestation on pittosporum (J. A. Davidson).
H. Damage caused to orange (J. A. Davidson).

Aspidiotus spinosus Comstock

Suggested Common Name Spinose scale (also called spined scale and avocado scale).

Common Synonyms and Combinations *Aspidiotus persearum* Cockerell.

Field Characters (Plate 17) Adult female cover light brown to white, opaque, slightly convex, circular; shed skins central. Male cover oval, similar color and texture as female cover; shed skin submarginal. Body of adult female yellow. On leaves or bark.

Slide-mounted Characters Adult female (Fig. 19) with 3 pairs of well-developed lobes, rarely with a small fourth lobe; paraphyses absent. Median lobes separated by space 0.2–0.3 (0.3) times width of median lobe, with noticeable sclerotized area anterior of lobe, without yoke, usually with medial margins parallel and lateral margins converging slightly, apically with or without irregular serrations, with 1 lateral notch, 1 medial notch; second lobes simple, usually noticeably smaller than median lobes, rarely about same size, usually expanded near lobe apex, sometimes of same shape as median lobes, with 1 lateral notch, 0–1 (1) medial notch; third lobes simple, noticeably smaller than second lobe, represented by small point or narrow, finger-like projection, with 0–1 (1) lateral notch, 0–1 (0) medial notch, fourth lobe rarely present, represented by unnotched finger-like projection similar in shape to third lobe. Plates fimbriate, often with microduct emptying in middle of plates; plate formula 2-3-5, rarely 2-3-4, without plates anterior of seta marking segment 5; with 2 broad, apically fringed or bifurcate plates between median lobes, usually reaching apex of median lobes. Macroducts of 1 variable size, smaller at pygidial apex, increasing anteriorly, 1 macroduct between median lobes, extending 1.4–2.0 (1.8) times distance between posterior apex of anal opening and base of median lobes, 29–34 (32) μ long, longest macroduct in first space 29–41 (35) μ long, with 10–18 (14) on each side of pygidium on segments 5 to 8, ducts in submarginal and marginal areas, total macroduct on each side of body 18–25 (22), with macroduct orifices anterior of anal opening; prepygidial macroducts located near body margin on segments 1 or 2 to 3, smaller than on pygidium. Pygidial microducts on venter in submarginal areas of segments 5 and 6, rarely absent or present on segment 7, with 2–4 (3) ducts; prepygidial ducts of 1 size, submarginal ducts usually on head and thoracic and abdominal segments, submedial ducts usually near mouthparts and anterior spiracles, normally absent from posterior spiracular area, usually on head, prothorax, metathorax, and segments 1 to 3; pygidial microducts absent from dorsum, of 2 sizes on prepygidial areas, shorter ducts in submarginal areas of head to segment 3, longer ducts usually in submedial areas of head, thorax, and segments 1 and 3. Perivulvar pores in 4 groups, 7–14 (10) pores on each side of body. Perispiracular pores absent. Anal opening located 1.6–2.2 (1.8) times length of anal opening from base of median lobes, anal opening 7–15 (10) μ long. Dorsal seta laterad of median lobes 0.8–1.6 (1.4) times length of median lobe, not extending beyond lobe apex. Eyes represented by circular sclerotized area on body margin at level slightly anterior of juncture between clypeolabral shield and labium, rarely absent. Antennae each with 1 seta. Cicatrices on prothorax, segment 1, and 3. Body pyriform. Median lobes sometimes variable, often appearing recessed into pygidium, usually protruding well beyond apex of lobe 2. Pygidium with definite sclerotized areas. Setae associated with lobes 2 and 3 greatly enlarged basally, nearly as wide as third lobe. Macroduct orifices with sclerotized rim at dermal apex.

Affinities Spinose scale is the most distinctive species of *Aspidiotus* in the United States because of the small size and posterior placement of the anal opening. It is similar to *Aspidiotus nerii*, oleander scale, but lacks plates anterior of the seta marking segment 5, has the macroduct between the median lobes 29–34 (32) μ long, has large macroducts restricted to the area posterior of segment 5, has 7–14 (10) perivulvar pores on each side of the body, and has the anal opening 7–15 (10) μ long. Oleander scale has plates anterior of the seta marking segment 5, has the macroduct between the median lobes 19–26 (23) μ long, has large macroducts present anterior of segment 5, has 10–17 (15) perivulvar pores on each side of body, and has the anal opening 17–25 (21) μ long. For a comparison of this species with the coconut scale see the affinities section of *Aspidiotus destructor*.

Hosts Polyphagous. Borchsenius (1966) records it from 40 genera in 19 families; Dekle (1977) reports from 62 host genera in Florida. We have examined specimens from the following genera: *Aleurites, Arenga, Argania, Bactris, Cajanus, Camellia, Caryota, Chamaedorea, Chrysalidocarpus, Citrus, Cocos, Dioon, Diospyros, Eriobotrya, Euonymus, Euphorbia, Fatsia, Ficus, Haworthia, Laelia, Laurus, Lycaste, Mangifera, Monstera, Persea, Pouteria, Pritchardia, Roystonea, Trachycarpus,* and *Vitis.* Common hosts are avocado, camellia, palms, mango, orchids, and persimmon.

Distribution In the United States the spinose scale is reported from AL, CA, FL, GA, HI, LA, MS, TX. It has been taken indoors in DC, NY, VA. It commonly is taken in quarantine from Japan, Mexico, and South America. Nakahara (1982) reports it from the New World (Bahamas, Bermuda, Brazil, Central America, Colombia, Cuba, Dominican Republic, Mexico, Peru, Puerto Rico, Uruguay); Europe (Azores, Italy, Spain, England, Germany); former Soviet Union (Georgia, Leningrad); Africa (Algeria, Canary Islands, Madagascar, Madeira Island, Mozambique, South Africa, Tanzania); Asia (China, Israel, Japan, Syria, Turkey).

Biology The life history of this species is unknown. Biparental (Schmutterer 1952b) and uniparental (Gerson and Zor 1973) populations are known. In Israel the species tends to infest the upper surface of the leaves of avocado near the midrib. It is not abundant on avocados in Israel, accounting for only about 4% of the total number of armored scales on this host (Gerson and Zor 1973).

Economic Importance Spinose scale is not often regarded as an economic problem, although it has been listed as a pest of mango in Israel by Wysoki (1997) and a problem on ornamentals in Italy by Lozzia (1985).

Selected References Ferris (1938a); Gerson and Zor (1973); Schmutterer (1952b).

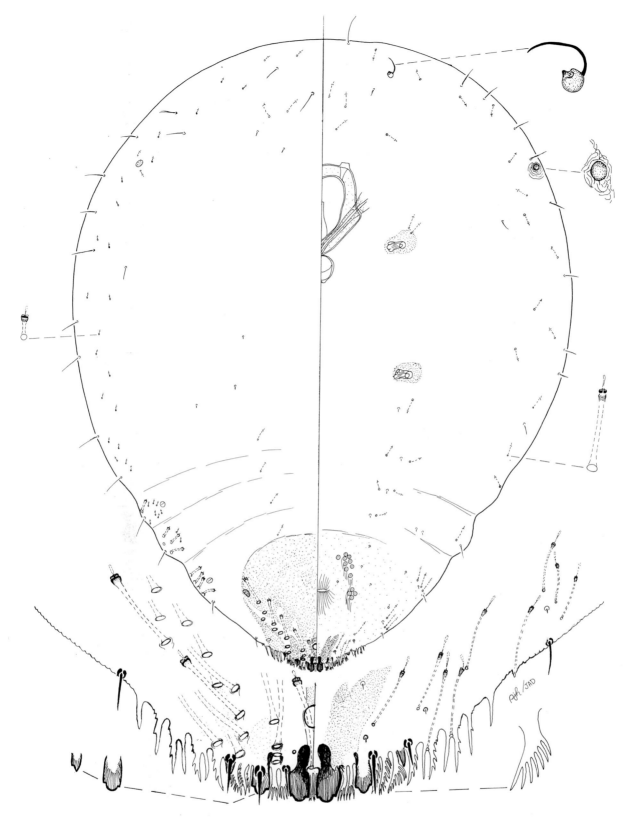

Figure 19. *Aspidiotus spinosus* Comstock, spinose scale, Brownsville, TX, on *Euonymus* sp., V-21-1978.

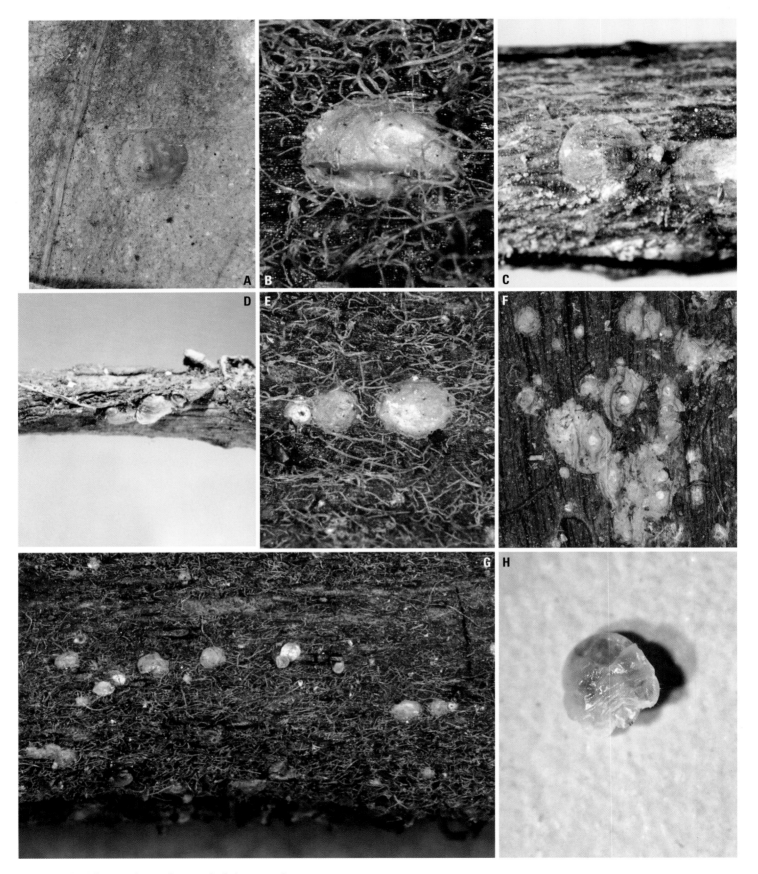

Plate 17. *Aspidiotus spinosus* Comstock, Spinose scale

A. Old adult female cover on *Sterculia* sp. (J. A. Davidson).
B. Adult female cover on rose (J. A. Davidson).
C. Mature adult female cover (J. A. Davidson).
D. Adult female covers (J. A. Davidson).

E. Adult female covers on rose (J. A. Davidson).
F. Colony on coconut husk (J. A. Davidson).
G. Distance view on rose (J. A. Davidson).
H. Body of adult female (J. A. Davidson).

Plate 18. *Aulacaspis rosae* (Bouché), Rose scale

A. Adult female cover on *Rubus* (R. J. Gill).
B. Adult female and male covers on *Rubus* (J. A. Davidson).
C. Female and male covers on *Rubus* (J. A. Davidson).
D. Female and male covers on *Rubus* (J. A. Davidson).
E. Close-up of adult male cover on *Rubus* (R. J. Gill).
F. Body of adult female with eggs on *Rubus* (R. J. Gill).
G. Distance view on *Rubus* (J. A. Davidson).
H. Distance view of heavy infestation on *Rubus* (R. J. Gill).

Aulacaspis rosae (Bouché)

ESA Approved Common Name Rose scale (also called rosa scale).

Common Synonyms and Combinations *Aspidiotus rosae* Bouché, *Chermes rosae* (Bouché), *Diaspis rosae* (Bouché).

Field Characters (Plate 18) Adult female cover flat, circular or oval, dirty white; shed skins marginal or submarginal, yellow to brown. Male cover elongate, parallel sided, felted, with 3 longitudinal ridges, white; shed skin marginal, yellow or brown. Body of adult female red or reddish brown; eggs red or reddish brown. On stems or bark above or below soil surface.

Slide-mounted Characters Adult female (Fig. 20) with 3 pairs of well-developed lobes, fourth lobes represented by sclerotized raised area; paraphyses absent. Median lobes separated by space 0.2–0.3 (0.2) times width of median lobe, without sclerotized area anterior of lobe, with yoke, medial margin parallel sided one-third to one-half length of lobe, then strongly divergent to apices, lateral margins converging slightly, with 0–3 (0) lateral notches, medial margins serrate with 4–9 (7) notches; second lobes bilobed, noticeably smaller than median lobes, medial lobule largest with 0–1 (0) lateral notches, 0–1 (0) medial notches, lateral lobule with 0–1 (0) lateral notches, 0–1 (0) medial notches; third lobes bilobed, about same size as second lobes, medial lobule largest with 0–2 (0) lateral notches, 0–1 (0) medial notches, lateral lobule with 0–6 (2) lateral notches, without medial notches; fourth lobes simple, represented by flat projection with 4–8 (6) serrations, lobe rarely absent. Gland-spine formula 1-1-1, with 16–24 (20) gland spines near body margin from segment 2 to 4; without gland spines between median lobes. Macroducts of 1 size, without macroduct between median lobes, longest macroduct in first space 24–32 (28) μ long, with 12–19 (16) on each side of pygidium on segments 5 to 8, ducts in submedial areas on segments 6 and 5, submarginal on segment 5, and marginal on segments 5 to 7, with total of 36–65 (51) macroducts on each side of body, with macroduct orifices anterior of anal opening; prepygidial macroducts of 2 sizes, larger size approximately same size and shape as on pygidium located in submedial areas on segments 3 and 4, submarginal on segments 3 and 4, and marginal areas of segments 4 and sometimes 3, smaller ducts located near body margin on segments 1 or 2 to segment 3. Pygidial microducts on venter in submarginal areas of segments 5 to 7, with 4–9 (7) ducts, in submedial area of segment 5; prepygidial ducts of 1 size, submarginal ducts rarely present on mesothorax and metathorax, present on segments 2 and 3, submedial ducts on metathorax and segments 2 and 3; pygidial microducts absent from dorsum, of same size as on venter on prepygidial areas, in submarginal areas of head to segment 3, in submedial areas of head. Perivulvar pores in 5 groups, 51–68 (64) pores on each side of body. Perispiracular pores with 3 loculi, anterior spiracle with 17–35 (23) pores, posterior spiracle with 4–13 (8) pores. Anal opening located 6.9–14.8 (9.8) times length of anal opening from base of median lobes, anal opening 8–15 (12) μ long. Dorsal seta laterad of median lobes 0.5–0.9 (0.6) times length of median lobe, extending beyond lobe apex. Eyes represented by circular or irregularly sclerotized area on body margin at level slightly posterior of anterior margin of clypeolabral shield, rarely absent. Antennae each with 1 seta. Cicatrices on prothorax and segment 1. Body of characteristic shape, with head, prothorax, and mesothorax swollen and rectangular, this area usually sclerotized. Submedian ducts on segment 6 variable, 1–5 (3) ducts in each cluster. Pygidium with definite sclerotized areas. Setae associated with lobes 2 and 3 greatly enlarged basally, nearly as wide as third lobe. Macroduct orifice with sclerotized rim at dermal apex.

Affinities The rose scale is one of three *Aulacaspis* species that occur in the continental United States and is easily recognized by the characteristic body shape which has the head and thoracic area noticeably wider than the abdomen. It is most similar to *A. rosarum* Borchsenius, which differs by having a submedial group of macroducts on segment 2. *Aulacaspis rosarum* occurs in Hawaii in the United States. For a comparison of the rose scale with the cycad scale, *A. yasumatsui*, see the affinities section of the latter species.

Hosts Oligophagous. Borchsenius (1966) records it from 3 genera in the family Rosaceae; Dekle (1977) reports it from 6 host genera in 3 families (Rosaceae, Oleaceae, Geraniaceae) in Florida. We have examined specimens from *Rosa* and *Rubus*. This species is occasionally reported from strawberries, but we have not examined specimens from this host.

Distribution We have examined U.S. specimens from AR, CA, CT, DC, FL, GA, IA, IL, IN, KS, KY, LA, MA, MD, MI, MO, MS, NC, NH, NJ, NM, NY, OH, OR, PA, SC, TX, UT, VA, WA, WY, WV. The literature records this species in the following states: AL (Montgomery 1921), AZ (Glick 1922), ME, RI (Montgomery 1921), TN (Bentley and Bartlett 1931), WI (Severin and Severin 1909). Records of *Aulacaspis rosae* from Hawaii are misidentifications of *A. rosarum*. Nakahara (1982) reports it from the New World (Argentina, Brazil, Canada, Chile, Colombia, Costa Rica, Dominican Republic, El Salvador, Guatemala, Jamaica, Peru, Puerto Rico, Venezuela); Europe (widespread); former Soviet Union (many countries); Africa (North Africa, Canary Islands, Cape Verde, Madeira Island, South Africa, Southwestern Africa, Tanzania, Zaire); Asia (China, India, Iran, Iraq, Israel, Japan, Korea, Philippines, Taiwan, Turkey); others (Australia, Bonin Islands, New Caledonia, New Zealand).

Biology In Oregon (Schuh and Mote 1948) the rose scale has one generation per year and crawlers appear 'during the spring.' Two generations per year apparently occur in Ohio (Kosztarab 1963) and second-generation crawlers are observed in late July. Two generations per year have been reported in New York (Johnson and Lyon 1976) with crawlers noted from late May to early June and again in August. One generation has been reported in England (Boratynski 1953), Germany (Schmutterer 1959), Romania (Savescu 1961), and Hungary (Balás and Sáringer 1982); 2 in Japan (Murakami 1970); and 3 in France (Bénassy 1959). The overwintering stages are reported as follows: egg—in Iowa (Hibbs 1956), Oregon (Schuh and Mote 1948), and Hungary (Balás and Sáringer 1982); adult female—in Japan (Murakami 1970) and Germany (Schmutterer 1959); in England, Boratynski (1953) indicated that the species did not have a regular life cycle and survived only in sheltered areas; in France, Bénassy (1959) indicated that the most common overwintering stage was the first instar but that some of the second instars go into a diapause; in Russia, Danzig (1959) found all stages present in the winter and indicated there is a very high level of winter mortality. The biology reported for the rose scale is unusually variable. It may overwinter in most stages, that is, egg, first and second instar, and adult female. Apparently this species undergoes some development in cold months although winter diapause has been reported in France. It should be noted that *A. rosarum* has been confused with the rose scale, and that some of the data given on *A. rosae* actually may pertain to *A. rosarum*.

Economic Importance This species is sometimes considered a pest of roses, raspberries, blackberries, boysenberries, dewberries, loganberries, and strawberries. Although the species is relatively easy to control with pruning and occasional chemical

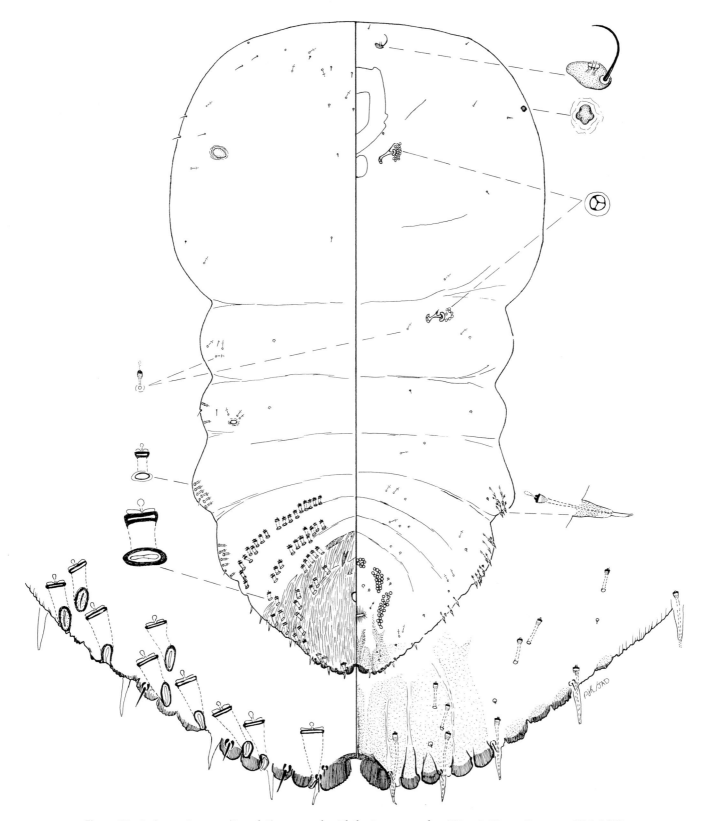

Figure 20. *Aulacaspis rosae* (Bouché), rose scale, Chile, intercepted at Miami, FL, on *Rosa* sp., III-1-1972.

control, neglected plantings can be severely damaged. A sample of the literature that discusses the economic importance is as follows: Seabra (1918) considers the rose scale to be a very destructive pest of roses in Portugal; Merrill and Chaffin (1923) list it as a serious pest of blackberry and rose in Florida but indicated that it was easily controlled; in British Columbia, Glendenning (1923) considers it to be a serious pest of roses and cane fruits when neglected; in Iowa infested roses were destroyed and were not allowed to be sold (Drake 1935); on the Maltese Islands it was reported as a common and injurious pest on roses, particularly climbing roses (Borg 1932); Szeremlei et al. (1979) treat it as an important greenhouse pest on roses in Hungary; in Utah it was considered to be a serious pest of raspberries and strawberries (Knowlton and Smith 1936); it was listed as an occasional pest of roses in greenhouses in England (Wilson 1938); this species was a very injurious pest in the Transcarpathian area of the former Soviet Union on *Rubus* (Tereznikova 1963); it is a serious pest of raspberries in some commercial plantings in Maryland (Gimpel 1986, personal communication). The rose scale is listed as one of the 43 principal armored scale pests of the world by Beardsley and González (1975) and is considered a serious pest in a small area of the world by Miller and Davidson (1990). Williams (1991) discusses the impact of this pest in cultivated *Rubus* fields in the United States, and Guilleminot and Apablaza (1986) consider it to be an occasional pest on raspberries in Chile.

Selected References Bénassy (1959); Ferris (1937).

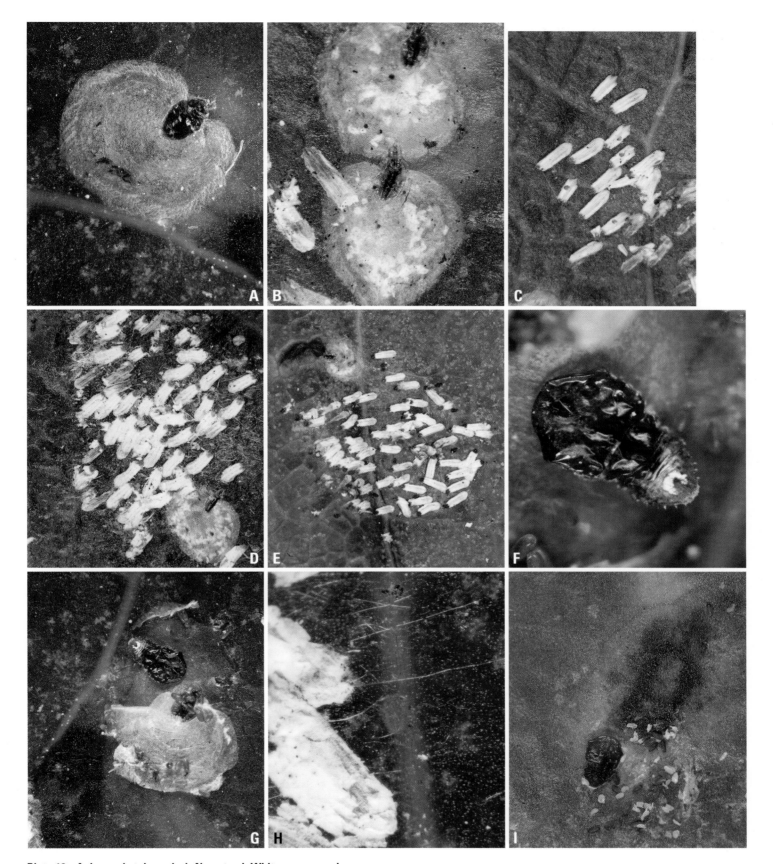

Plate 19. *Aulacaspis tubercularis* Newstead, White mango scale

A. Adult female cover with eggs faintly visible through cover on mango (J. A. Davidson).

B. Adult female cover with egg shells visible through cover on mango (J. A. Davidson).

C. Adult male covers on mango (J. A. Davidson).

D. Adult female cover surrounded by male covers on mango (J. A. Davidson).

E. Distance view of female and male covers on mango (J. A. Davidson).

F. Old adult female body on mango (J. A. Davidson).

G. Adult female body with eggs on mango (J. A. Davidson).

H. Close-up of crawler on mango (J. A. Davidson).

I. Feeding damage caused by female (J. A. Davidson).

Aulacaspis tubercularis Newstead

Suggested Common Name White mango scale (also called cinnamomum scale, cinnamon scale, escama blanca del mango, escama del mango, and mango scale).

Common Synonyms and Combinations *Aulacaspis (Diaspis) tubercularis* Newstead, *Aulacaspis cinnamomi* Newstead, *Diaspis (Aulacaspis) cinnamomi mangiferae* Newstead, *Diaspis mangiferae* (Newstead), *Diaspis cinnamomi-mangiferae* (Newstead), *Diaspis (Aulacaspis) cinnamomi* (Newstead), *Diaspis (Aulacaspis) tubercularis* Newstead.

Field Characters (Plate 19) Adult female cover flat, circular, transparent or solid white, sometimes wrinkled; shed skins marginal, yellow to brown, with dark brown median line. Male cover elongate, parallel sided, felted, with 3 longitudinal ridges, white; shed skin marginal, yellow or brown. Body of adult female red; eggs and crawlers red or pink. On leaves, new stems, and fruit.

Slide-mounted Characters Adult female (Fig. 21) with 3 pairs of well-developed lobes, fourth and sometimes fifth lobes represented by sclerotized raised areas; small paraphyses usually attached to medial and lateral margins of medial lobule of second lobes and to medial margin of medial lobule of third lobes, occasionally attached to all lobules of second and third lobes. Median lobes separated by space 0.2–0.4 (0.3) times width of median lobe, without sclerotized area anterior of lobe, with yoke, medial margin parallel sided one-quarter to one-half length of lobe then strongly divergent to apices, lateral margins diverging slightly, with 0–2 (0) lateral notches, medial margins serrate with 6–12 (9) notches; second lobes bilobed, noticeably smaller than median lobes, medial lobule largest with 0–4 (2) lateral notches, 0–2 (0) medial notches, lateral lobule with 0–1 (0) lateral notches, without medial notches; third lobes bilobed, usually slightly smaller than second lobes, medial lobule largest with 0–3 (2) lateral notches, without medial notches, lateral lobule with 0–5 (3) lateral notches, without medial notches; fourth lobes simple, represented by flat projection with 4–8 (6) serrations. Gland-spine formula 1-1-1, with 11–22 (15) gland spines near body margin from segment 2 to 4; without gland spines between median lobes. Macroducts of 1 size, without macroduct between median lobes, longest macroduct in first space 22–30 (26) μ long, with 7–13 (10) on each side of pygidium on segments 5 to 8, ducts in submedial areas on segments 6 and 5, rarely absent from 6, submarginal on segment 5, and marginal on segments 5 to 7, with total of 27–43 (38) macroducts on each side of body, with macroduct orifices anterior of anal opening; prepygidial macroducts of 2 sizes, larger size approximately same size and shape as on pygidium, located in submedial areas on segments 3 and 4, submarginal on segments 3 and 4, and marginal areas of segments 3 and 4, smaller ducts located near body margin on segments 2 to 3. Pygidial microducts on venter in submarginal areas of segments 5 to 6, with 3–4 (4) ducts, in submedial area of segment 5; prepygidial ducts of 1 size, submarginal ducts present on segment 3 and 4, submedial ducts on prothorax and metathorax near spiracles and segment 1; pygidial microducts absent from dorsum, shorter than on venter on prepygidial areas, in submarginal areas of head to segment 1. Perivulvar pores in 5 groups, 37–57 (48) pores on each side of body. Perispiracular pores with 3 loculi, anterior spiracle with 6–17 (11) pores, posterior spiracle with 3–10 (6) pores (often oriented sideways). Anal opening located 8.9–13.8 (10.4) times length of anal opening from base of median lobes, anal opening 8–11 (10) μ long. Dorsal seta laterad of median lobes 0.4–0.6 (0.5) times length of median lobe, extending beyond lobe apex. Eyes represented by small sclerotized area on body margin at apex of tubercle, sometimes absent. Antennae each with 1 seta. Cicatrices on prothorax, segments 1 and 3, prothoracic pair unusually large, nearly as big as anterior spiracle. Body of characteristic shape, with head, prothorax, and mesothorax swollen and rectangular. Submedian ducts on segment 6 variable, 0–2 (1) ducts in each cluster. Lateral margin of prothorax on older females with 1 swollen tubercle on each side of body, often with inconspicuous eyes at apex. Labium set in groove, with sclerotized areas on each side.

Affinities The white mango scale is distinct among species of *Aulacaspis* that occur in the United States by having a tubercle on each side of prothorax on older females and by having the labium set in a groove with sclerotized areas on the side of the groove. Neither *A. yasumatsui* or *A. rosae* have these structures.

Hosts Polyphagous. Borchsenius (1966) records it from 7 genera in 4 families; Miller and Gimpel (2002) report it from 12 genera in 8 families. We have examined specimens from the following host genera: *Aglaia*, *Cinnamomum*, *Citrus*, *Cucumis*, *Cucurbita*, *Desmos*, *Gaiadendron*, *Laurus*, *Litchi*, *Litsea*, *Luffa*, *Mangifera*, *Nephelium*, *Persea*, *Prunus*, and *Psidium*. Hamon (2002) reports it in Florida from the following additional hosts: *Acer*, *Cocos*, *Dietes*, *Dimocarpus*, *Machilus*, *Phoebe*, and *Pittosporum*. It most commonly is collected on mango.

Distribution This species recently was discovered in the continental United States in Florida (Hamon 2002). Miller and Gimpel (2002) report it from Afrotropical (Ghana, Kenya, Madagascar, Mauritius, Mozambique, Reunion, Rodrigues Islands, South Africa, Tanzania, Uganda, Zanzibar, Zimbabwe); Australasian (New Caledonia, Vanuatu); Neotropical (Aruba, Bermuda, Brazil, British Virgin Islands, Colombia, Dominican Republic, Grenada, Guadeloupe, Martinique, Puerto Rico, St. Croix, Trinidad and Tobago, U.S. Virgin Islands, Venezuela); Oriental (China, India, Indonesia, Malaysia, Pakistan, Philippines, Ryukyu Islands, Sri Lanka, Taiwan, Thailand); Palearctic (Egypt, Iraq, Israel, Italy, Japan). It most commonly is taken in quarantine from the Caribbean and southeast Asia. A distribution map of this species was published by CAB International (1993).

Biology The life history of this species was studied intensively in South Africa (Labuschagne et al. 1995) in preparation for implementing an integrated pest management system. The white mango scale does not have discrete generations in South Africa primarily because the ovipositional period is about 45 days long (82 days for one particularly long-lived female). Since the duration of a generation (from egg to egg) was about 52 days under a regime of 26 °C daytime temperatures and 13 °C nighttime temperatures, it is possible that the offspring of an adult female could begin laying eggs before the mother had finished oviposition. Thus, over a period of a year, the generations became completely overlapping and all life stages were present at all times of the year. Based on these observations, a generation takes about two months to complete; thus, it is feasible for there to be 5 or 6 generations per year, depending on temperature. Duration of the life stages was about 13 days for the crawler, 15 days for the second instar, and 12 days before the adult female began laying eggs. Labuschagne et al. (1995) provided evidence that temperatures of above 30 °C caused high scale mortality and that populations tended to build up on the cooler sides of the mango tree. Halteren (1970) investigated the life history of this species on mango in Ghana and suggested that a generation required 35 to 40 days from egg laying to the adult female stage. If a 12-day preoviposition period is added, as suggested in Labuschagne et al. (1995), then generation times reported in South Africa and Ghana are fairly consistent. Male development was much quicker than in the female, taking 23

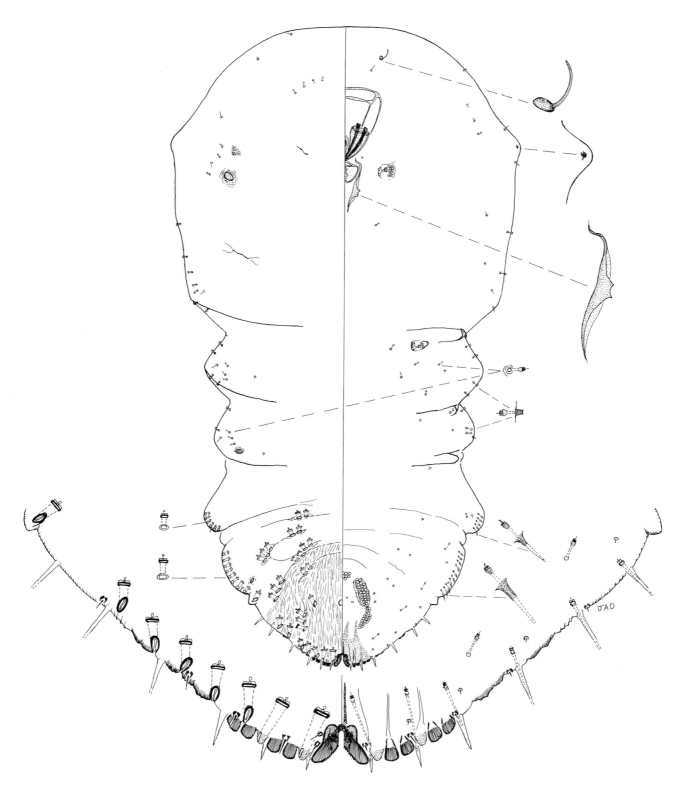

Figure 21. *Aulacaspis tubercularis* Newstead, white mango scale, Venezuela, intercepted at San Juan, PR, on lauraceous leaf, V-22-1976.

to 28 days (Halteren 1970). He did not observe the long oviposition period mentioned by Labuschagne et al. (1995) but indicated that adult females laid eggs for 8 to 12 days and deposited a total of 80 to 200 eggs. Eggs hatched in 7 or 8 days.

Economic Importance This species is a serious pest of mangos in many parts of the world. According to Labuschagne et al. (1995) damage is primarily cosmetic, causing blemishes in the fruit that make it unsightly but does not detract from the flavor or texture. Halteren (1970) suggests that feeding causes significant damage to the leaves forming chlorotic areas around the body of the insect. Literature that discusses this species as a pest of mango is as follows: Brazil (Wolff and Corseuil 1993), Colombia (Kondo and Kawai 1995), Japan (Kinjo et al. 1996), Mauritius (Moutia and Mamet 1947), Mozambique (Almeida 1972), Pakistan (Mahmood and Mohyuddin 1986), Philippines (Webster 1920), Portugal (Fernandes 1989), Puerto Rico (Gallardo-Covas 1983), Rhodesia (Chorley 1939), South Africa (Labuschagne et al. 1995), and Venezuela (Penella 1942). The parasitic wasp, *Aphytis chionaspis* Ren (given as *Aphytis* sp. in many papers), and the predaceous nitidulid beetle, *Cybocephalus binotatus* Grouvelle, have been used as biological control agents against this pest in IPM programs in mango orchards in South Africa (Labuschagne et al. 1996).

Selected References Hamon (2002); Labuschagne et al. (1995); Takagi (1970).

Plate 20. *Aulacaspis yasumatsui* Takagi, Cycad aulacaspis scale

A. Adult female cover on cycad (J. A. Davidson).
B. Adult and second-instar female covers (J. A. Davidson).
C. Adult male and two female covers (J. A. Davidson).
D. Distance view of male and female covers (J. A. Davidson).
E. Distance view of covers on cycad frond (J. A. Davidson).

F. Heavy infestation at base of cycad frond (J. A. Davidson).
G. Body of adult female with cover removed (J. A. Davidson).
H. Eggs and single yellow crawler (J. A. Davidson).
I. Feeding damage to cycad frond (J. A. Davidson).

Aulacaspis yasumatsui Takagi

Suggested Common Name Cycad aulacaspis scale (also called cycad scale and sago palm scale).

Common Synonyms and Combinations None.

Field Characters (Plate 20) Adult female cover flat, circular to pear shaped, but often distorted due to crowding among female covers or veins on cycad host, adult female cover white, sometimes translucent enough to see body of adult through cover; shed skins marginal, light yellow or white. Male cover elongate, parallel sided, felted, with 3 longitudinal ridges, white; shed skin marginal, light yellow or white. Body of adult female orange; eggs orange. In light infestations, on undersides of fronds; in heavy infestations, on both top and bottom surfaces; also as deep as 60 cm in soil. Main roots covered in heavily infested plants.

Slide-mounted Characters Adult female (Fig. 22) with 3 pairs of well-developed lobes, fourth and fifth lobes represented by sclerotized raised areas, fifth lobes sometimes absent; paraphyses absent. Median lobes separated by space 0.2–0.5 (0.4) times width of median lobe, without sclerotized area anterior of lobe, with yoke, medial margin parallel sided one-third to one-half length of lobe then strongly divergent to apices, lateral margins diverging slightly, without lateral notches, medial margins serrate with 6–12 (8) notches; second lobes bilobed, noticeably smaller than median lobes, medial lobule largest with 0–4 (3) lateral notches, 0–2 (0) medial notches, lateral lobule with 0–3 (1) lateral notches, 0–2 (1) medial notches; third lobes bilobed, equal to or slightly larger than second lobes, medial lobule largest with 1–5 (3) lateral notches, without medial notches, lateral lobule with 0–6 (3) lateral notches, without medial notches; fourth lobes bilobed, represented by flat projection with 0–6 (3) serrations on medial lobe, 0–6 (3) serrations on lateral lobe; fifth lobes rarely absent, when present bilobed, with 0–5 (1) serrations on medial lobe, 0–7 (2) serrations on lateral lobe. Gland-spine formula 1-1-1, with 12–22 (16) gland spines near body margin from segment 1 or 2 to 4; without gland spines between median lobes. Macroducts of 1 size, without macroduct between median lobes, longest macroduct in first space 23–43 (30) µ long, with 11–17 (14) on each side of pygidium on segments 5 to 8, ducts in submedial areas on segments 5 and 6, submarginal on segment 5, and marginal on segments 5 to 7, with 28–52 (40) macroducts on each side of body, with macroduct orifices anterior of anal opening; prepygidial macroducts of 2 sizes, larger size approximately same size and shape as on pygidium, located in submedial areas on segments 3 and 4, submarginal areas on segments 3 and 4, and marginal areas of segments 4 and sometimes 3, smaller ducts located near body margin on segments 2 and 3. Pygidial microducts on venter in submarginal areas of segments 5 to 6 or 7, with 2–11 (6) ducts, rarely present in submedial area of segment 5; prepygidial ducts of 1 size, submarginal ducts present on head or prothorax to segments 3 and 4, submedial ducts on any or all of mesothorax, metathorax and segments 1 to 4; pygidial microducts absent from dorsum, of same size as on venter on prepygidial areas, in submarginal areas of head to segment 2 or 3, in submedial areas of mesothorax to segments 2 and 3. Perivulvar pores in 5, rarely 4, groups, 37–57 (45) pores on each side of body. Perispiracular pores usually with 3 loculi, anterior spiracle with 10–29 (18) pores, posterior spiracle with 6–30 (15) pores. Anal opening located 5.7–8.8 (7.4) times length of anal opening from base of median lobes, anal opening 10–15 (13) µ long. Dorsal seta laterad of median lobes 0.4–0.7 (0.5) times length of median lobe, often extending beyond lobe apex. Eyes represented by circular or irregularly sclerotized area on body margin at level slightly posterior of anterior margin of clypeolabral shield, frequently absent. Antennae each with 1 seta. Cicatrices often absent, when present, on prothorax, segment 1 and 3. Body of characteristic shape, with head, prothorax, and mesothorax slightly swollen but with rounded margins, not rectangular, this area not usually sclerotized. Submedian ducts on segment 6 relatively consistent, 1–3 (2) ducts in each cluster. Pygidium with definite sclerotized areas.

Affinities The cycad aulacaspis scale is very similar to *Aulacaspis rosae*, the rose scale differing primarily in the shape of the swollen prosoma. In *A. yasumatsui* the prosoma has rounded lateral margins and does not form a distinct rectangle whereas in *A. rosae* the prosoma is distinctly rectangular. There are minor differences in the number of perivulvar pores and the distribution of microducts but these characters overlap and are somewhat variable.

Hosts This species has a restricted host range occurring on *Cycas, Dioon, Encephalartos, Microcycas, Stangeria* (Howard et al. 1999), and *Macrozamia* (Halbert 1998) in Florida. It seems to prefer *Cycas* spp.

Distribution This species is known only from Florida in the continental United States, but specimens have been identified from Hawaii also (Heu and Chun 2000). The species also is known from the Cayman Islands, China, Hong Kong, Puerto Rico, Thailand, Singapore, and the U.S. Virgin Islands. We have examined material from Thailand and Florida. It commonly is taken in quarantine on *Cycas* spp. from Thailand. A distribution map of this species was published by CAB International (2000).

Biology The cycad aulacaspis scale is a tropical or subtropical species and apparently continues to grow throughout the year. Howard et al. (1999) reports the time required from egg hatch to adult females was about 28 days at 25 °C. A series of clean plants was placed next to heavily infested plants and developmental times were recorded. In April the time required from first infestation to second instars was about 16 days and from second instars to adult females was about 12 days. A generation from infestation of experimental plants to the appearance of second-generation crawlers required slightly more than 41 days. Howard et al. (1999) infested a second set of clean plants in August and found that after 15 days of exposure to infested plants some individuals had reached second instars; adults comprised about 75% of the population after 21 days, and after 35 days most of the population consisted of mature females. Females can lay more than 100 eggs, which hatch in 8 to 12 days.

Economic Importance This species is considered to be a serious pest of cycads in Florida, and Hawaii. Since its original discovery in the southern part of Miami, Florida, in 1996 (Halbert 1996) it has spread to more than 20 counties in the state (1996 to 2000) and its continued spread seems inevitable. Its effect can be quite devastating. When Howard et al. (1999) conducted life-history experiments, they placed clean plants in close proximity to heavily infested plants; the clean plants were quickly infested with crawlers within 2 weeks. After a month the fronds of the previously clean plants showed a few chlorotic spots around the feeding scales, and within 270 days the plants were heavily desiccated and brown, and in a year they were dead. When a plant is heavily infested, both sides of the fronds are white because of the thick layers of scale bodies covering all parts of the host. Even after the scales die they adhere tightly to the host and are nearly impossible to remove from the frond surfaces. Another troublesome characteristic of this pest is that it infests subterranean portions of the plant and apparently uses these nearly 'invisible' infestations as reservoirs for rein-

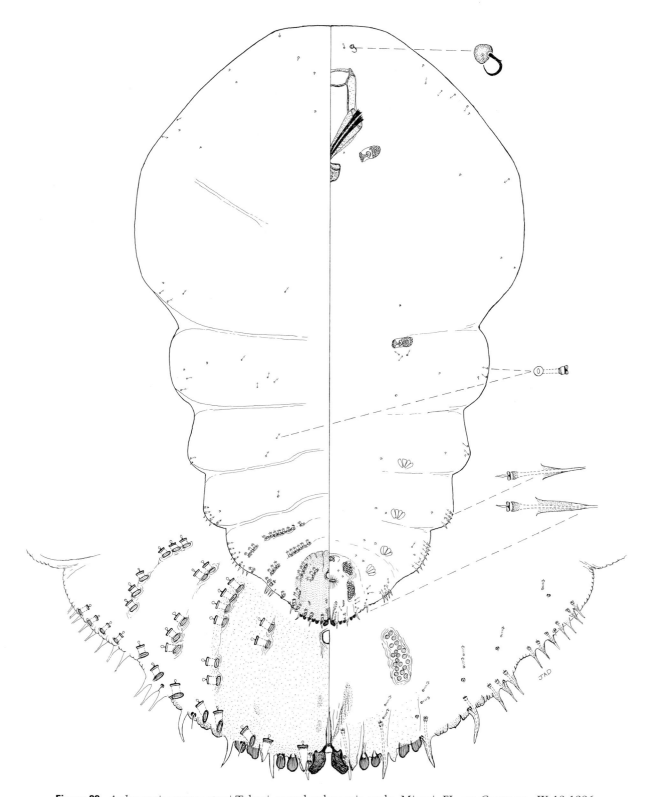

Figure 22. *Aulacaspis yasumatsui* Takagi, cycad aulacaspis scale, Miami, FL, on *Cycas* sp., IX-19-1996.

festation when aboveground populations are destroyed (Howard et al. 1999). In 1998, R. M. Baranowski and H. B. Glenn, University of Florida, Tropical Research & Education Center, Homestead, FL, collected a nitidulid predator, *Cybocephalus binotatus* Grouvelle, and an aphelinid parasite, *Coccobius fulvus* (Compere and Annecke), in Thailand and reared them in quarantine for about a year. Eventually these two natural enemies were released in Florida as biological control agents of the cycad aulacaspis scale. In several instances, they have given excellent control of the pest, but in other situations their impact has been less noticeable (Howard 2000, personal communication). They apparently are well established in Florida and have been released at more than 40 different sites (Baranowski and Glenn 1999).

Selected References Howard et al. (1999); Takagi (1977).

Carulaspis juniperi (Bouché)

ESA Approved Common Name Juniper scale.

Common Synonyms and Combinations *Aspidiotus juniperi* Bouché, *Diaspis juniperi* (Bouché). This species often has been misidentified as *Carulaspis visci* (Schrank) and *C. minima* (Targioni-Tozzetti).

Field Characters (Plate 21) Adult female cover moderately convex, circular, white; shed skins central, yellow. Male cover elongate, parallel sided, felted, with 1 or 3 longitudinal ridges, white; shed skin marginal, yellow. Body of adult female yellow with greenish mottling when young, reddish brown when older; eggs yellow; body of adult male orange, with wings. On foliage.

Slide-mounted Characters Adult female (Fig. 23) with 2 pairs of well-developed lobes, third and fourth lobes represented by sclerotized raised area; paraphyses absent. Median lobes separated by space 0.4–0.8 (0.6) times width of median lobe, without sclerotized area anterior of lobe, without yoke, medial margin parallel or slightly divergent apically, lateral margins parallel or diverging slightly, without notches; second lobes bilobed, about same size or slightly larger than median lobes, without notches; third lobes bilobed or simple, broader and differently shaped than second lobes, medial lobule largest, apically pointed, with 0–5 (3) lateral notches, without medial notches, lateral lobule without notches; fourth lobes usually simple, sometimes bilobed, medial lobule represented by flat projection with 0–8 (4) serrations, lateral lobule when present with 0–2 (0) notches. Gland-spine formula 2-2-1, with 3–7 (5) gland spines near body margin from segments 2 or 3 to 4; with 2 gland spines between median lobes, about same length as median lobes. Macroducts of 2 sizes, larger size present in marginal areas of segments 5 to 8, smaller size present in submarginal and submedial area of segment 5 and rarely 6, with macroduct between median lobes 15–20 (19) μ long, macroduct in first space 17–20 (18) μ long, with 11–15 (13) on each side of pygidium on segments 5 to 8, ducts in submarginal and submedial areas, with total of 56–90 (72) macroducts on each side of body, with macroduct orifices anterior of anal opening; prepygidial macroducts of 2 sizes on dorsum, larger size approximately same size and shape as on pygidium located in marginal area of segment 4, smaller ducts located near body margin in submedial areas of segments 2 or 3 to 4; macroducts on venter of small size, located on submargin of any of segments 1, 2, or 3. Pygidial microducts on venter in submarginal areas of segments 5 and usually 6, with 2–4 (3) ducts, in submedial area of segment 5; prepygidial ducts of 1 size, submarginal ducts rarely present on segment 1, present on segments 2 and/or 3 and 4, submedial ducts usually present on thorax and segments 1 or 2 to 4; pygidial microducts absent from dorsum, of same size as on venter on prepygidial areas, present from meso- or metathorax to segment 1 or 2 in submarginal areas, present on any or all of prothorax, mesothorax, metathorax, segment 1, and segment 2 in submedial areas. Perivulvar pores in 5 groups (rarely with an extra group), 19–32 (27) pores on each side of body. Perispiracular pores with 3 loculi, anterior spiracle with 1 or 2 (1) pores, posterior spiracle without pores. Anal opening located 3.7–5.8 (4.5) times length of anal opening from base of median lobes, anal opening 12–16 (14) μ long. Dorsal seta laterad of median lobes 0.9–1.4 (1.2) times length of median lobe, extending beyond lobe apex. Eyes absent or represented by irregularly sclerotized spot on body margin at level slightly posterior of anterior margin of clypeolabral shield. Antennae each with 1 seta. Cicatrices absent. Body broadly pyriform.

Affinities The juniper scale is very similar to *Carulaspis minima*, the minute cypress scale, and until recently these species have been confused. The juniper scale has 1 macroduct between the median lobes, usually has more than 60 macroducts on each side of body, has crawlers with 7-segmented antennae, and has egg-filled adult females that are chestnut brown. The minute cypress scale lacks a macroduct between the median lobes, usually has less than 60 macroducts on each side of the body, has crawlers with 6-segmented antennae, and has egg-filled adult females that are red.

Hosts Restricted to certain groups of conifers; Borchsenius (1966) records it from 11 genera in 1 family of host plant. We have examined U.S. specimens from the following host genera: *Chamaecyparis*, *Cupressus*, *Juniperus*, *Libocedrus*, *Pinus*, *Sequoia*, and *Thuja*. Other hosts reported are: *Biota*, *Callitris*, *Cryptomeria*, *Cupressocyparis*, *Libocedrus*, *Taxodium* (Nakahara 1982). The juniper scale is most commonly collected on *Juniperus* spp.

Distribution In the United States we have examined specimens from AR, CA, CT, DC, DE, FL, GA, ID, IN, KS, KY, MA, MD, MI, MO, NC, NJ, NY, OH, OK, OR, PA, RI, SC, TX, UT, VA, WA, WI. The species also has been reported from TN. The species may be taken in quarantine from Argentina, Australia, Canada, Chile, Europe, Iran, North Africa, and Turkey. Nakahara (1982) reports it from the New World (Argentina, Brazil, Canada, Chile); Europe (many countries including the Azores); former Soviet Union (Crimea, Caucasus, Turkmenistan, Uzbekistan); Africa (Algeria, Canary Islands, Egypt ?, Madeira Island, Morocco); Asia (Iran, Turkey); Australia; New Zealand.

Biology The juniper scale apparently has only 1 generation per year, although Brown and Eads (1967) state, 'There is at least one generation annually and very likely several are produced.' Mated adult females overwinter and begin laying eggs in March in warm areas to May in cooler regions (Johnson and Lyon 1976). Crawlers appear in late May in Delaware (Bray 1974); in late May or early June in Connecticut (Schread 1955), New Jersey (Allen and Weiss 1953), and Ohio (Neiswander 1951); and in June in Maryland (McComb and Davidson 1969), and Illinois (English 1970). Crawlers first appear in March or April in southern California (Brown and Eads 1967), but Gill (1997) suggests that this paper most likely refers to *C. minima*. Egg laying and crawler emergence continues over a 30 to 45 day period. Adult males may be present as early as mid-July and as late as August in Ohio (Neiswander 1951, 1966). Males are winged. Winter kill is variable; Neiswander (1951) found 85% mortality in 1948 and 13% in 1950. Mussey and Potter (1997) studied the timing of egg hatch over a three-year period in Kentucky and reported that the day-degree (DDF) requirement with a base of 45 °F was 987 in 1992, 799 in 1993, and 791 in 1994. They found that flowering plant activity was a much better predictor of egg hatch than day degrees or absolute date. Egg hatch of juniper scale is closely correlated with 95% bloom of *Ilex opaca* and *Cladrastris kentuckea*. Garcia (1998) studied the interaction of this species on the fruit of *Juniperus communis* and the viability of the fruit.

Economic Importance Heavy juniper scale infestations cause foliage to turn yellow. Heavy populations cause needle drop and dieback. Entire plants have been killed. Johnson and Lyon (1976) report this scale to be more serious in the eastern part of the United States. In California it is not a pest and seems to prefer the cooler parts of the state (Gill 1997). Beardsley and González (1975) consider this scale to be one of 43 serious armored scale pests, and Miller and Davidson (1990) consider it to be a serious world pest.

Selected References Boratynski (1957); Stimmel (1979).

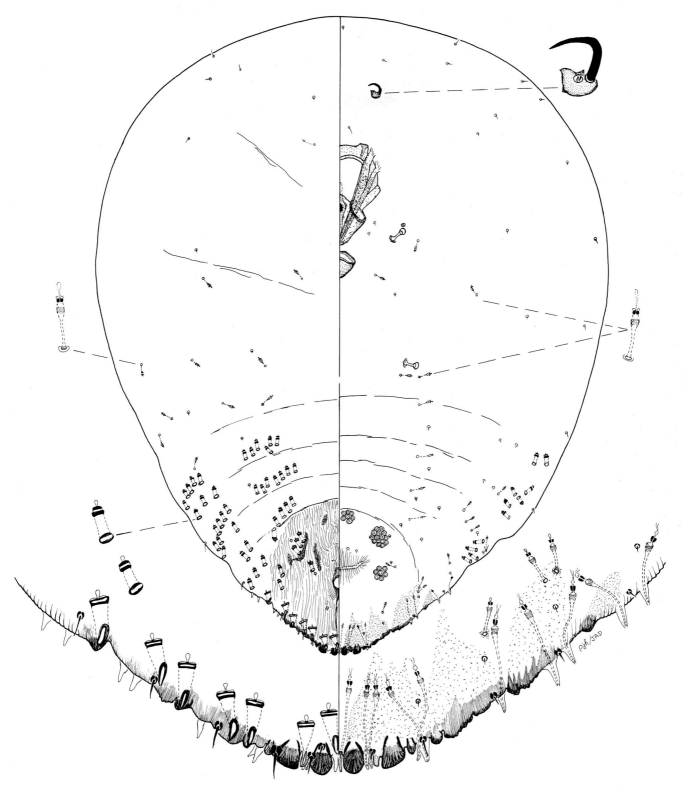

Figure 23. *Carulaspis juniperi* (Bouché), juniper scale, Oswego, KS, on *Juniperus* sp., VI-18-1958.

Plate 21. *Carulaspis juniperi* (Bouché), Juniper scale

A. Adult female cover on juniper (J. A. Davidson).
B. Adult female covers on juniper (J. A. Davidson).
C. Many white-ridged male covers on juniper (J. A. Davidson).
D. Distance view of female covers on juniper (J. A. Davidson).
E. Distance view of male covers on juniper (J. A. Davidson).

F. Female and male covers on juniper (J. A. Davidson).
G. Body of young adult female on juniper (J. A. Davidson).
H. Damage caused by heavy infestation on juniper (J. A. Davidson).
I. Damaged scale covers caused by parasites and predators (J. A. Davidson).

98

Plate 22. *Carulaspis minima* (Targioni Tozzetti), Minute cypress scale

A. Adult female cover on *Juniperus virginiana* (J. A. Davidson).
B. Adult female cover on redwood (J. A. Davidson).
C. Adult male covers on redwood (J. A. Davidson).
D. Female and male covers on Leyland cypress (J. A. Davidson).
E. Body of adult female and scale cover in April (J. A. Davidson).

F. Egg-laying female on juniper in May (J. A. Davidson).
G. Adult female scale covers and settled crawlers on juniper (J. A. Davidson).
H. Heavily damaged rocket juniper (J. A. Davidson).

Carulaspis minima (Targioni Tozzetti)

Suggested Common Name Minute cypress scale (also called Bermuda cedar scale).

Common Synonyms and Combinations *Diaspis carueli* Targioni Tozzetti. There are some unresolved nomenclatural issues that make it difficult to determine whether the correct species epithet for this scale is *minima* or *carueli*.

Field Characters (Plate 22) Adult female cover slightly convex, circular to oval, white; shed skins central or subcentral, yellow. Male cover elongate oval, parallel sided, felted, with 1 or 3 faint ridges, white; shed skin marginal, light yellow. Body of newly formed adult yellow with mottling of green, body of egg-filled adult female red; eggs reddish yellow. On foliage, sometimes on fruit.

Slide-mounted Characters Adult female (Fig. 24) with 2 pairs of well-developed lobes, third and fourth lobes represented by sclerotized raised area; paraphyses absent. Median lobes separated by space 0.4–1.5 (0.9) times width of median lobe, without sclerotized area anterior of lobe, without yoke, medial margin parallel or slightly divergent apically, lateral margins parallel or diverging slightly, without notches; second lobes bilobed, about same size or slightly smaller than median lobes, without notches; third lobes bilobed, smaller and differently shaped than second lobes, medial lobule largest, apically pointed, with 1–3 (2) lateral notches, without medial notches, lateral lobule without notches; fourth lobes simple, represented by flat projection with 0–2 (1) serrations. Gland-spine formula 2-2-1, with 5–8 (6) gland spines near body margin from segments 1, 2, or 3 to 4; with 2 gland spines between median lobes, about same length as median lobes. Macroducts of 2 sizes, larger size present in marginal areas of segments 5 to 7, smaller size present in submarginal and submedial area of segment 5, without macroduct between median lobes, macroduct in first space 17–20 (19) μ long, with 8–10 (9) on each side of pygidium on segments 5 to 8, ducts in submarginal and submedial areas, with total of 38–53 (43) macroducts on each side of body, with macroduct orifices anterior of anal opening; prepygidial macroducts of 2 sizes on dorsum, larger size approximately same size and shape as on pygidium located in marginal area of segment 4, smaller ducts located near body margin in submedial areas of segments 1 or 2 to 4; macroducts on venter of small size, located on submargin of any of segments 1, 2, or 3. Pygidial microducts on venter in submarginal areas of segments 5 and 6, with 2–5 (4) ducts, in submedial area of segment 5; prepygidial ducts of 1 size, submarginal ducts rarely present on segment 1, present on segments 2 and/or 3 and 4, submedial ducts usually present on thorax and segments 1 or 2 to 4; pygidial microducts absent from dorsum, of same size as on venter on prepygidial areas, usually absent from submarginal areas, occasionally present on mesothorax, metathorax, segment 1, and segment 2, in submedial areas of prothorax to segment 1 or 2. Perivulvar pores in 5 groups, 20–24 (23) pores on each side of body. Perispiracular pores with 3 loculi, anterior spiracle with 1 or 2 (1) pores, posterior spiracle without pores. Anal opening located 2.7–4.2 (3.5) times length of anal opening from base of median lobes, anal opening 13–20 (16) μ long. Dorsal seta laterad of median lobes 0.7–1.5 (1.2) times length of median lobe, extending beyond lobe apex. Eyes absent or represented by irregularly clear area on body margin at level slightly posterior of anterior margin of clypeolabral shield. Antennae each with 1 seta. Cicatrices usually absent, sometimes on prothorax and segment 1. Body broadly pyriform.

Affinities For many years the minute cypress scale and juniper scale were treated as a single species. Balachowsky (1954) and Boratynski (1957) distinguished the adult females. Baccetti (1960) studied the species in Italy and pointed out species differences among each instar. He found a remarkable difference in the antennal segmentation of the crawler, which we have verified. In the field this species tends to be slightly smaller than the juniper scale (Boratynski 1957). For a comparison, see the affinities section of juniper scale (*C. juniperi*).

Hosts Restricted to certain conifers. Borchsenius (1966) records it from 7 genera of conifers; Dekle (1977) reports it from 6 genera in Florida. We have examined specimens from the following: *Cedrus*, *Chamaecyparis*, *Cryptomeria*, *Cupressus*, *Juniperus*, *Picea*, *Sequoia*, and *Thuja*. Other hosts reported are *Callitris* and *Cupressocyparis* (Nakahara 1982).

Distribution We have examined U.S. specimens from CA, FL, HI, KS, MO, NC, NM, NY, OK, TX, UT, VA. The species has been reported from GA, LA, PA, TN, VT, WA (Nakahara 1982). Nakahara also reported it from the New World (Argentina, Bermuda, Chile, Colombia, Cuba, Mexico, Uruguay); Europe (Azores, Austria, England, France, Greece, Italy, Portugal, Romania, Spain, Sweden); former Soviet Union (Armenia, Crimea, Krasnodar, Transcaucasia, Ukraine); Africa (Algeria, Ethiopia, Guinea, Madeira, Morocco, South Africa); Asia (Iran, Israel, Jordan); others (Hawaii).

Biology The minute cypress scale occurs mainly in the southern United States as it prefers a warm climate. It has 1 generation per year in the northern part of its distribution and possibly 2 in the south. Baccetti (1960) reports 2 generations per year in Italy and Boratynski (1957) found 1 generation in England. It is interesting that in the same locality in Italy, the minute cypress scale life cycle is about 2 weeks ahead of the juniper scale's life cycle, while in England, the opposite occurs. In Pennsylvania they are in synchrony (Stimmel 1979). In Pennsylvania, the overwintering stage is the fertilized adult female; oviposition is in late April to mid-May; eggs hatch in about 1 week; peak crawler emergence is late May; second instars appear in mid-June; adults are present beginning in late June; and adult males were predominant in early July.

Economic Importance This scale may be a serious pest of juniper. In Bermuda, in combination with *Lepidosaphes pallida* (Green), it nearly decimated the endemic Bermuda cedar, *Juniperus bermudiana* (Bennett and Hughes 1959). Damage to the host may include dieback of the branches, leaf drop, and even death. Trees weakened by this scale are susceptible to other pests. Beardsley and González (1975) consider it to be one of the 43 principal armored scale pests in the world, and Miller and Davidson (1990) include it in their list of armored scale pests. Dekle (1977) indicates that this species is an important pest of ornamental hosts in commercial nurseries and control sprays with pesticides may be required. Gill (1982) considers it to be a minor pest of ornamental cypress and juniper in California. Gill (1997) states that it was not a pest except in some wholesale nurseries.

Selected References Baccetti (1960); Balachowsky (1954); Boratynski (1957); Ferris (1938a); Stimmel (1979).

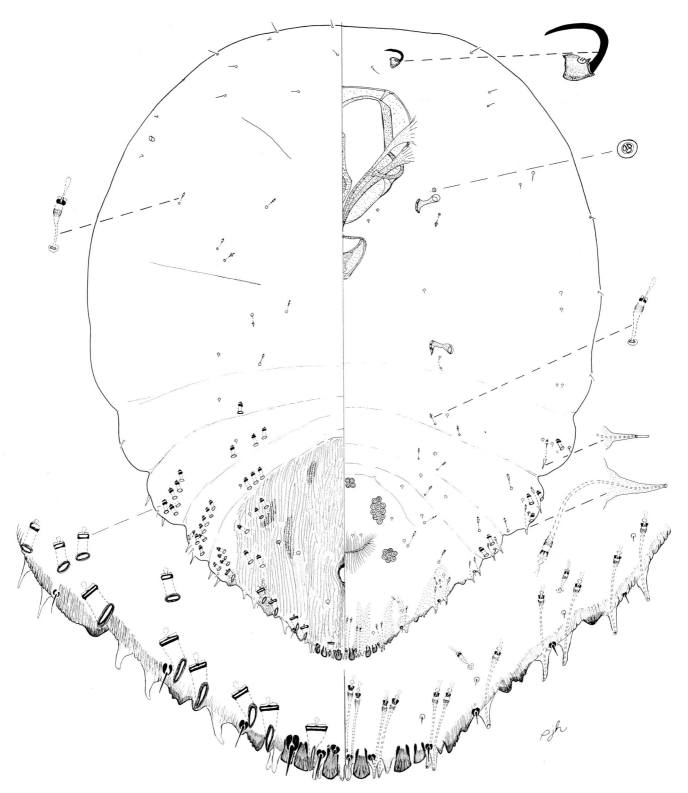

Figure 24. *Carulaspis minima* (Targioni Tozzetti), minute cypress scale, Lincoln County, NC, on *Juniperus chinensis*, IX-12-1970.

Chionaspis americana Johnson

ESA Approved Common Name Elm scurfy scale.

Common Synonyms and Combinations *Chionaspis furfurus* var. *ulmi* Cockerell, *Fundaspis americana* (Johnson), *Jaapia americana* (Johnson), *Chionaspis ulmi* Cockerell.

Field Characters (Plate 23) Adult female cover flat, grayish to white, oyster-shell shaped; shed skins marginal, brown to yellow. Male cover elongate oval, white, with 3 ridges; shed skin marginal, beige to light yellow. Body of adult female without eggs orange, with eggs red or purple; eggs yellowish red; crawlers red. Females normally on stems and branches. Males predominately on undersides of leaves. Adult males both apterous and brachypterous (Bullington et al. 1989).

Slide-mounted Characters Adult female (Figs. 25 and 26) with 3 or 4 pairs of definite lobes; without paraphyses. Median lobes closely appressed from base to apex or from base to half length of lobes, with conspicuous long yoke, medial margins parallel, lateral margins diverging or parallel, with 1–4 (2) lateral notches, 0–1 (0) medial notch; second lobes bilobed, conspicuously smaller than median lobes, medial lobule largest with 1–3 (2) lateral notches, without medial notches, lateral lobule with 0–1 (0) lateral notches, without medial notches; third lobes bilobed, smaller than second lobes, represented by low series of notches, medial lobule largest with 0–3 (2) lateral notches, lateral lobule with 1–3 (2) notches; fourth lobes when present simple, represented by low series of notches, with 0–7 (4) notches. Gland-spine formula usually 1-2-2, sometimes 1-1-2, 1-2-3, 1-2-2, or 1-1-2, with 15–23 (18) gland spines near each body margin from prothorax, mesothorax, or metathorax to segment 4; without gland spines between median lobes. Macroducts of 2 sizes, largest in submedial areas of segments 3 to 5 and marginal areas of segments 3 to 7, one specimen from lilac with 1 such duct on mediolateral area of segment 6, 2 specimens from *Celtis* without mediolateral cluster on segment 3, small size ducts in marginal areas from meso- or metathorax to segment 3, without duct between median lobes, macroduct in first space 16–25 (20) μ long, with 10–17 (13) macroducts on each side of body on segments 5 to 7, total macrotubular ducts on each side of body 42–119 (79), some macroduct orifices anterior of anal opening. Pygidial microducts on venter in submarginal areas of segments 5 to 7, with 4–8 (6) ducts; prepygidial ducts of 1 size, highly variable, extremes with many ducts present on all segments laterally from head to segment 4, and many ducts near anterior and posterior spiracles and near medial setae on segments 1 to 4, other extreme with only 1 or 2 such ducts in lateral areas, with only 1 or 2 near spiracles, and with or without ducts near medial setae; pygidial microducts usually absent from dorsum, rarely with 1 or 2 interspersed with mediolateral macroducts on segment 5; prepygidial microducts highly variable, when numerous present in mediolateral clusters on segments 1 to 4. Perivulvar pores in 5 groups, with 33–71 (48) pores on each side of body. Perispiracular pores with 3 loculi, anterior spiracles with 8–15 (13) pores, posterior spiracles with 2–4 (3) pores. Anal opening located 8.0–13.3 (9.8) times length of anal opening from base of median lobes, anal opening 10–18 (14) μ long. Dorsal seta laterad of median lobes 0.7–1.2 (0.9) times length of median lobe. Eyes represented by small sclerotized area, sometimes on

small tubercle, located on body margin at level of middle of clypeolabral shield. Antennae each with 1 long seta. Cicatrices usually present on segment 1, sometimes visible on prothorax and segment 3. Body elongate. Median lobes generally protruding beyond second lobes.

Affinities Elm scurfy scale is distinctive among the economic species of *Chionaspis* in the United States because of the large and closely appressed median lobes. It is similar to *Chionaspis floridensis* Takagi but differs in lacking or having only 1 submedian macroduct on segment 4. *Chionaspis floridensis* has 3 or more submedian macroducts on segment 4. *Chionaspis americana* has an unusually large amount of variation in the distribution of the prepygidial microducts (Willoughby and Kosztarab 1974). We include illustrations of two extreme forms.

Hosts Borchsenius (1966) reports this species from 4 plant genera in 2 families; Dekle (1977) reports it in Florida from *Fraxinus* and *Ulmus*. Based on the information provided by Takagi (1969) we believe that the *Fraxinus* record pertains to *Chionaspis floridensis*. This species primarily occurs on *Ulmus* and secondarily on *Celtis*. We have examined specimens from *Ligustrum*, *Prunus*, *Quercus*, *Salix*, and *Syringa*. Records from the literature include *Carpinus*, *Crataegus*, *Morus*, *Platanus*, and *Tilia*.

Distribution Elm scurfy scale is native to the United States and apparently does not occur elsewhere. The USNM Collection contains material from AL, CA, CO, CT, DC, DE, FL, GA, IA, IL, IN, KS, LA, MA, MD, MI, MN, MO, MS, NB, NC, NJ, NY, OH, OK, PA, RI, SC, SD, TN, TX, VA, WI, WV.

Biology This species has 2 generations per year and overwinters in the egg stage. Crawlers have been reported in May in Delaware, Illinois, and Ohio, and late April to May in Virginia. Willoughby and Kosztarab (1974) report that in Virginia first instars molt to second instars in late May and early June. Adults appear in June and July. Females lay eggs in July. Crawlers of the second generation hatch in July and develop into adults in late August and early September. Overwintering eggs are laid in October and November. Females occur mainly on the bark while males prefer the undersides of leaves. Adult males are predominantly apterous in the fall generation and apterous and brachypterous in the spring. Hollinger (1923) indicates that the species may have 3 'broods' in Missouri.

Economic Importance The elm scurfy scale has been treated as a pest in Connecticut (Britton and Friend 1935), Iowa (Drake 1934), Michigan (Pettit 1928), Minnesota (Lugger 1900), Ohio (Houser 1908), and Pennsylvania (Dodge and Rickett 1943; Trimble 1929). In Missouri it has been reported to kill elm twigs, branches, and sometimes small trees. Large trees are weakened by heavy infestations. The feeding of second-instar males on leaves causes small chlorotic spots (Hollinger 1923). Miller and Davidson (1990) consider this species to be an occasional economic pest. Note that most records of damage are older than the 1950s. The species has not been reported as a pest in recent years.

Selected References Baker (1972); Bullington et al. (1989); Liu et al. (1989b); Willoughby and Kosztarab (1974).

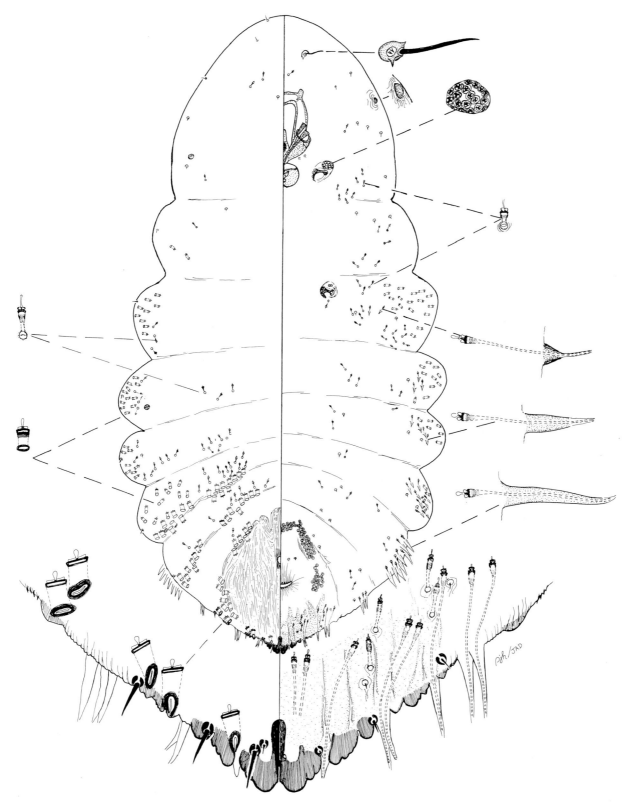

Figure 25. *Chionaspis americana* Johnson, elm scurfy scale, Ames, IA, on *Ulmus americana*, X-8-1954 (with many dorsal macroducts).

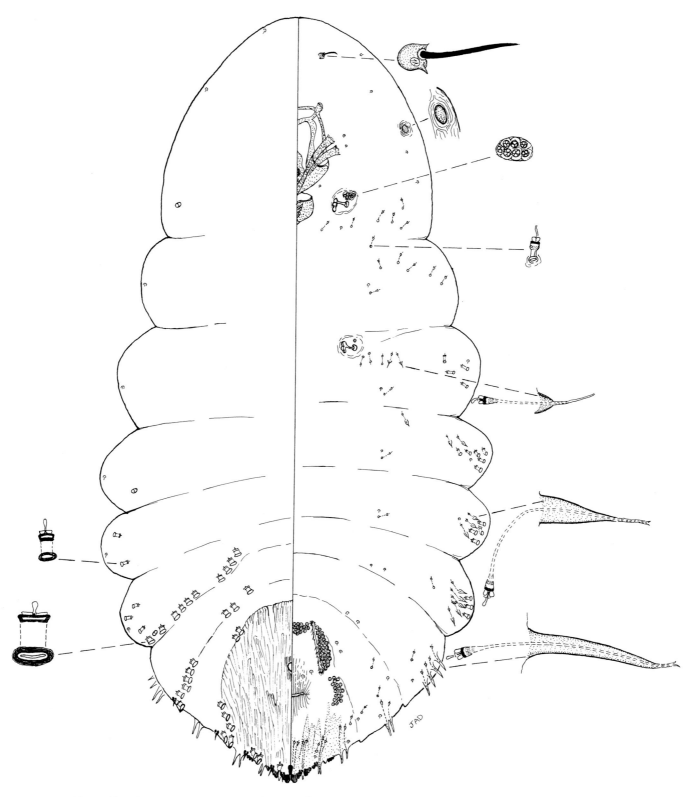

Figure 26. *Chionaspis americana* Johnson, elm scurfy scale, Brownsville, TX, on *Celtis* sp., V-25-1978 (with few dorsomedial macroducts).

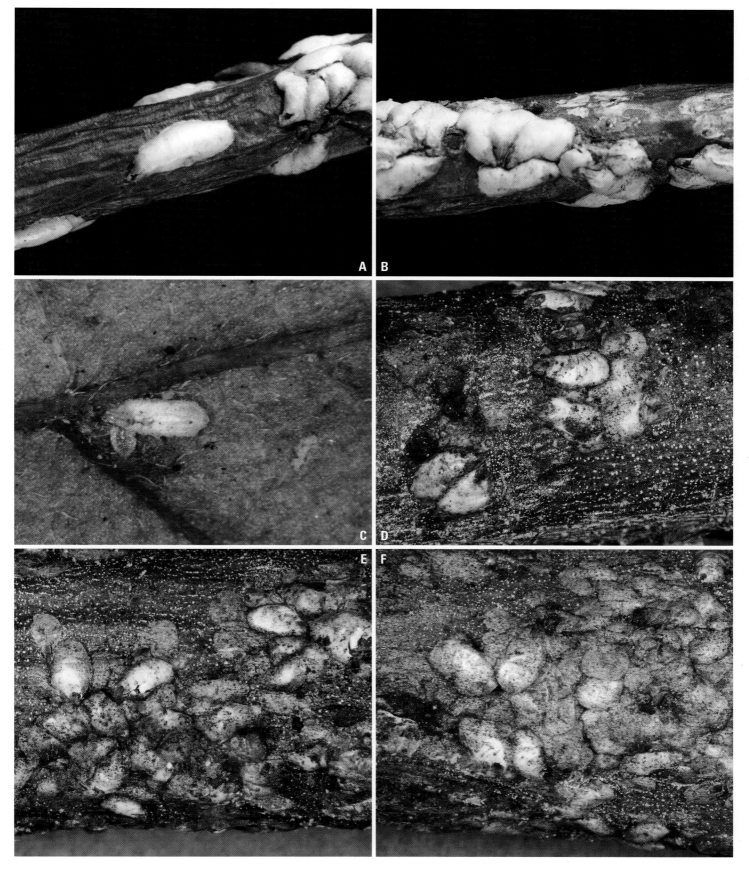

Plate 23. *Chionaspis americana* Johnson, Elm scurfy scale

A. Adult female cover on elm (J. A. Davidson).
B. Dense infestation of adult female covers on elm (J. A. Davidson).
C. Male cover and crawler on leaf of elm (J. A. Davidson).

D. Adult female covers on elm (J. A. Davidson).
E. Heavy infestation on elm (J. A. Davidson).
F. Heavy infestation on elm (J. A. Davidson).

Chionaspis corni Cooley

ESA Approved Common Name Dogwood scale.
Common Synonyms and Combinations None.
Field Characters (Plate 24) Adult female cover unusually thin, flat white, oyster-shell shaped, expanded posteriorly; shed skins marginal, orangish yellow. Male cover elongate, parallel sides, white, with 3 ridges; shed skin marginal, yellow or light brown. Body of adult female most likely purple or red; eggs purple. On bark.

Slide-mounted Characters Adult female (Fig. 27) usually with 3 pairs of definite lobes, occasionally with fourth lobes represented by low, weakly sclerotized series of points; without paraphyses or with small paraphysis attached to medial margin of medial lobule of second lobes. Median lobes closely appressed from base to one-third length of lobe, strongly diverging to apex, with conspicuous, short yoke, medial margins parallel basally, diverging apically, lateral margins diverging or parallel, usually without notches, newly molted females with 1–2 (2) lateral notches, 3–6 (4) medial notches; second lobes bilobed, smaller than median lobes, medial lobule largest with 0–2 (1) lateral notches, without medial notches, lateral lobule without notches; third lobes bilobed, smaller than second lobes, represented by low series of notches, medial lobule largest with 0–3 (1) lateral notches, lateral lobule with 0–4 (2) notches; fourth lobes usually absent, when present, bilobed, represented by low series of notches, medial lobule with 4–6 (5) notches, lateral lobule with 3–5 (4) notches. Gland-spine formula usually 2-2-2, sometimes 1-2-2, 1-1-1, or 1-1-2, with 13–25 (19) gland spines near each body margin from prothorax, mesothorax, or metathorax to segment 4; without gland spines between median lobes. Macroducts of 2 sizes, largest in submedial areas of segments 2 or 3 to 6, with 4–7 (5) on each side of 6, and marginal areas of segments 2 to 7, small size ducts in marginal areas from prothorax to segments 2 or 3, without duct between median lobes, macroduct in first space 20–24 (22) μ long, with 19–24 (21) macroducts on each side of body on segments 5 to 7, total macroducts on each side of body 86–148 (120), some macroduct orifices anterior of anal opening. Pygidial microducts on venter in submarginal areas of segments 5 to 7, with 7–11 (9) ducts; prepygidial ducts of 1 size, variable, most specimens with many ducts present on all segments laterally from head to segment 4, and many ducts anterior of mouthparts, near anterior and posterior spiracles, and near medial setae on segments 1 to 4, a few specimens with only 1 or 2 such ducts in lateral areas, with only 1 or 2 near spiracles, and with or without ducts near medial setae; pygidial microducts absent from dorsum; prepygidial microducts variable, when numerous, present in mediolateral clusters from mesothorax to segment 4, sometimes entirely absent. Perivulvar pores with 34–66 (50) pores on each side of body. Perispiracular pores with 3 loculi, anterior spiracles with 7–21 (15) pores, posterior spiracles with 3–9 (6) pores. Anal opening located 8–13 (10) times length of anal opening from base of median lobes, anal opening 10–15 (13) μ long. Dorsal seta laterad of median lobes 1.0–1.2 (1.1) times length of median lobe. Eyes absent or represented by inconspicuous clear area on derm, located on body margin at level of middle of clypeolabral shield. Antennae each with 1 long seta. Cicatrices usually present on segment 1, sometimes visible on prothorax and segment 3. Body elongate. Median lobes generally protruding beyond second lobes. As in *Chionaspis americana*, this species has forms with and without the dorsal macroducts on dorsum of thorax and anterior abdominal segments.

Affinities *Chionaspis corni* resembles *C. ortholobis* Comstock in arrangement of the dorsal macroducts but differs in having median lobes that are angulate and strongly divergent; whereas, in *C. ortholobis* the median lobes are evenly rounded and are not strongly divergent. The dogwood scale also is similar to *C. lintneri* Comstock but differs by having the medial margin of the medial lobes conspicuously longer than the lateral margin; whereas, in *C. lintneri* the medial and lateral margins of the median lobes are about equal in size.

Hosts Borchsenius (1966) reports this species from 2 plant genera in 2 families; Liu et al. (1989b) report it from the same 2 hosts: *Cornus* and *Ribes*. This species is collected most commonly on *Cornus*. We have examined specimens from *Cornus*, *Lindera*, and *Viburnum*.

Distribution The dogwood scale is native to the United States. We have examined U.S. specimens from CA, ID, IL, IN, LA, MA, MS, NY, OH, and PA. It is reported in the literature from CT, KS, MD, MI, UT, WI, and Ontario, Canada.

Biology The dogwood scale overwinters as eggs on the bark (Houser 1918). Kosztarab (1963) found adult females in mid-September in Ohio. To our knowledge there is no detailed life history of this species.

Economic Importance According to Baker (1972) this scale often damages ornamental dogwoods in the Midwest. Weigel and Baumhofer (1948) report that heavily infested branches of dogwoods were killed. This species is considered to be an occasional pest by Miller and Davidson (1990).

Selected References Bullington et al. (1989); Liu et al. (1989b).

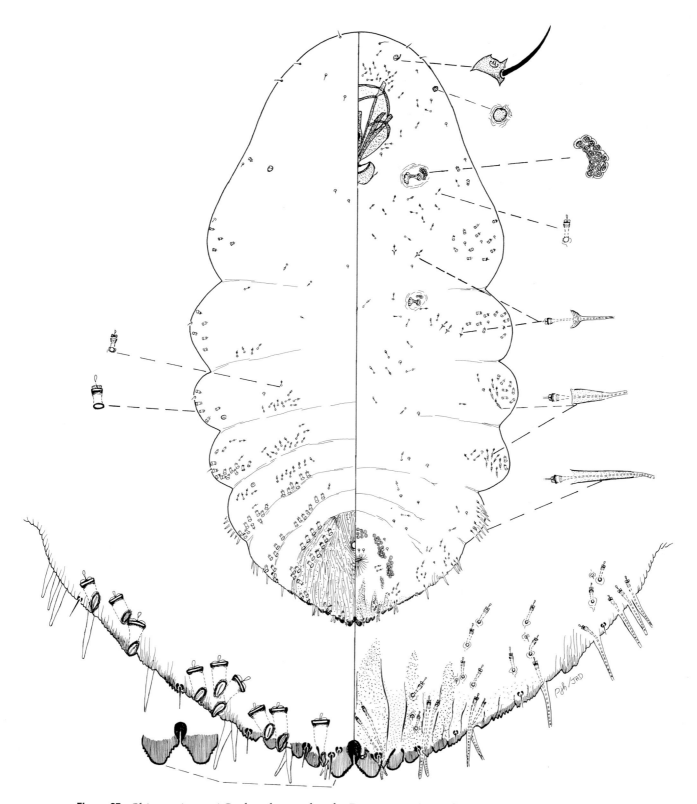

Figure 27. *Chionaspis corni* Cooley, dogwood scale, Davenport, IA, on *Cornus* sp., VI-25-1976.

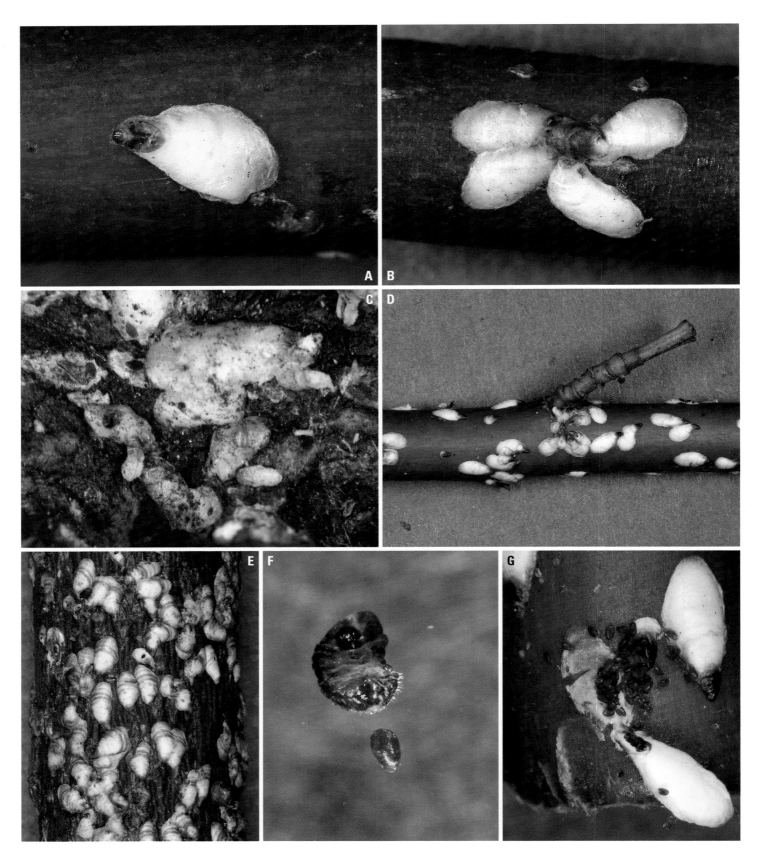

Plate 24. *Chionaspis corni* Cooley, Dogwood scale

A. Adult female cover on dogwood (J. A. Davidson).
B. Adult female covers on dogwood (J. A. Davidson).
C. Small white male covers among larger female covers on dogwood (J. A. Davidson).
D. Distance view of adult female covers on dogwood (J. A. Davidson).

E. Distance view of heavy infestation on dogwood (J. A. Davidson).
F. Female body with egg (J. A. Davidson).
G. Cover removed to show eggs (J. A. Davidson).

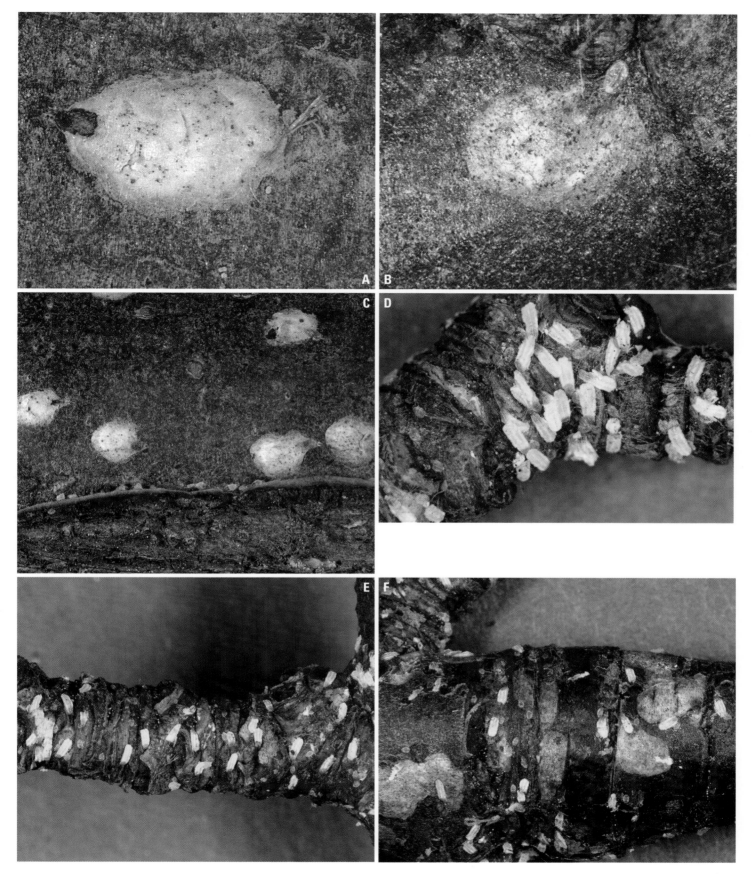

Plate 25. *Chionaspis furfura* (Fitch), Scurfy scale

A. Adult female cover on apple (J. A. Davidson).
B. Adult female covers on apple (J. A. Davidson).
C. Adult female covers on apple (J. A. Davidson).
D. Adult male covers on apple (J. A. Davidson).

E. Distance view of male covers on apple (J. A. Davidson).
F. Distance view of male and female covers on apple (J. A. Davidson).

Chionaspis furfura (Fitch)

ESA Approved Common Name Scurfy scale.

Common Synonyms and Combinations *Aspidiotus furfurus* Fitch, *Aspidiotus cerasi* Fitch, *Coccus harrisi* Walsh, *Aspidiotus harrisii* (Walsh), *Diaspis harrisii* (Walsh), *Chionaspis furfurus* (Fitch), *Chionaspis furfurus* var. *fulvus* King, *Chionaspis furfurea* Lindinger.

Field Characters (Plate 25) Adult female cover white, flattened, broadly oyster-shell shaped; shed skins marginal, yellowish brown. Male cover elongate oval, white, with 3 longitudinal ridges; shed skin marginal, yellow. Body of adult female before egg formation yellowish brown, after egg formation red; eggs red. Normally on bark.

Slide-mounted Characters Adult female (Fig. 28) with 3 or 4 pairs of definite lobes; transverse paraphysis present from lateral margin of each median lobe, small paraphysis on medial margin of first lobule of lobes 2 and 3. Median lobes closely appressed from base to about half length of lobe, usually with conspicuous, clubbed-shape yoke, yoke rarely inconspicuous, medial margins of median lobes parallel in basal half, rounded to apex, lateral margins diverging or parallel, with 0–6 (3) inconspicuous, lateral notches, 0–4 (2) inconspicuous, medial notches; second lobes bilobed, smaller than median lobes, medial lobule largest with 0–4 (2) lateral notches, without medial notches, lateral lobule with 0–1 (0) lateral notches, without medial notches; third lobes bilobed, smaller than second lobes, represented by low series of notches, medial lobule largest with 3–5 (4) lateral notches, lateral lobule with 0–7 (4) notches; fourth lobes, when present, bilobed, represented by low series of notches, medial lobule with 4–9 (5) notches, lateral lobule with 3–9 (5) notches. Gland-spine formula usually 1-1-1, with 10–27 (19) gland spines near each body margin from metathorax or segment 1 to segment 4; without gland spines between median lobes. Macroducts of 2 sizes, largest in submedial areas of segments 4 to 5, rarely on segments 3 and 6, and marginal areas of segments 3 to 7, small size ducts in marginal areas from metathorax or segment 1 to segment 3, without duct between median lobes, macroduct in first space 23–27 (25) μ long, with 8–13 (10) macroducts on each side of body on segments 5 to 7, total macroducts on each side of body 37–71 (55), some macroduct orifices anterior of anal opening. Pygidial microducts on venter in submarginal areas of any or all of segments 5 to 7, with 3–4 (3) ducts; prepygidial ducts of 1 size, highly variable in number, usually scattered in lateral and submedial areas, without clusters near mouthparts or spiracles; pygidial microducts usually absent from dorsum, rarely with 1 or 2 interspersed with mediolateral macroducts on segment 5; prepygidial microducts highly variable, when numerous, present in mediolateral clusters on segments 1 to 4. Perivulvar pores in 5 groups, with 24–83 (66) pores on each side of body. Perispiracular pores with 3 loculi, anterior spiracles with 5–9 (7) pores, posterior spiracles with 0–2 (1) pores. Anal opening located 8–14 (11) times length of anal opening from base of median lobes, anal opening 10–15 (13) μ long. Dorsal seta laterad of median lobes 0.9–1.1 (1.0) times length of median lobe. Eyes represented by distinct sclerotized area sometimes on small protrusion, located on body margin at level of middle of clypeolabral shield. Antennae each with 1 long seta. Cicatrices usually present on segment 1, sometimes visible on prothorax and segment 3. Body elongate. Median lobes generally protruding beyond second lobes.

Affinities *Chionaspis furfura*, *C. ortholobis* Comstock, and *C. salicis* (Linnaeus) are similar by possessing short, broad, rounded median lobes; scurfy scale differs by having fewer than 15 macroducts on each side of the body on segments 5 to 7 and

a gland-spine formula that normally is 1-1-1. *Chionaspis ortholobis* and *C. salicis* each have more than 20 macroducts on segments 5 to 7 and a gland-spine formula that normally is 2-2-2.

Hosts Borchsenius (1966) reports this species from 9 plant genera in 2 families; Liu et al. (1989b) report it on 16 host genera. Scurfy scale is commonly found on rosaceous plants and is collected most often on apple and pear. We have examined specimens from *Amelanchier*, *Cotoneaster*, *Cydonia*, *Juglans*, *Malus*, *Prunus*, *Pyracantha*, *Pyrus*, *Ribes*, *Sorbus*, and *Zanthoxylum*. Many additional hosts are recorded in the literature. Liu et al. (1989b) mention *Annona*, *Betula*, *Chaenomeles*, *Crataegus*, and *Rhamnus*. Literature citations of hosts such as *Cornus*, *Salix*, and *Ulmus* probably are misidentifications of other *Chionaspis* species.

Distribution Scurfy scale is native to the United States and apparently does not occur elsewhere except Canada. The USNM Collection contains material from AL, CA, CT, DC, DE, GA, IL, IN, MD, MI, MO, MS, NC, NJ, NY, OH, PA, RI, SC, TN, VA, WI, WV. Additional records from the literature include CO, FL, IA, KS, KY, MA, ME, MN, MT, NB, NM, TX, UT.

Biology This species has 1 generation per year in northern areas (e.g., CT, northern IL, MT, OH) and 2 in southern areas (e.g., DC, NC, NY, southern IL, and VA). Crawlers have been reported from April to early May and again in mid-July in VA (Hill 1952) and NC (Turnipseed and Smith 1953); in late April in DE (Bray 1974); in late May or early June in OH (Houser 1918), CT (Britton 1903), MT (Cooley 1900), and IL (English 1970); and from May to early June and again in mid-July in NY (Hammer 1938). Eggs are the overwintering stage. Hill (1952) has correlated crawler hatch from overwintering eggs with the bloom period of apples; hatch is complete by petal fall. Turnipseed and Smith (1953) report the following life history in NC: Crawlers that hatch in early April molt to second instars by May 10, and molt to adult females by June 7. Eggs are laid by June 28, which hatch about mid-July. Second instars are present in mid-August and adult females are collected in early September. Egg laying begins in late September. Adult males are present from late May to mid-June and again in mid-August to early September. Preferred feeding sites seem to be on old wood, but heavy infestations that have two generations each year may infest new growth, fruit, and even leaves. There seems to be a preference for the woody undersides of lower branches. Winged morphs of males are prevalent, but Bullington et al. (1989) indicate that apterous morphs also exist based on their examination of a pupa.

Economic Importance Scurfy scale is usually not a serious pest. Damaging populations have been reported, especially from apple and pear orchards, where spray schedules were not maintained. High populations have been reported to kill young trees and cause loss of vigor and dieback on old trees. This species prefers lower branches of the host but in high populations it spreads to the new growth and fruit. On apples the scales cause small, red depressions, which reduce the value of the fruit. It has been reported as a pest in Canada (Swaine and Hutchings 1926), Connecticut (Schread 1970), Illinois (Flint and Farrar 1940), Maryland (Langford and Cory 1939), New York (Blackman 1916; Hammer 1938), Ohio (Houser 1908), and Pennsylvania (Trimble 1929). Miller and Davidson (1990) consider this species to be a serious pest in a small area of the world.

Selected References Bullington et al. (1989); Hill (1952); Liu et al. (1989b).

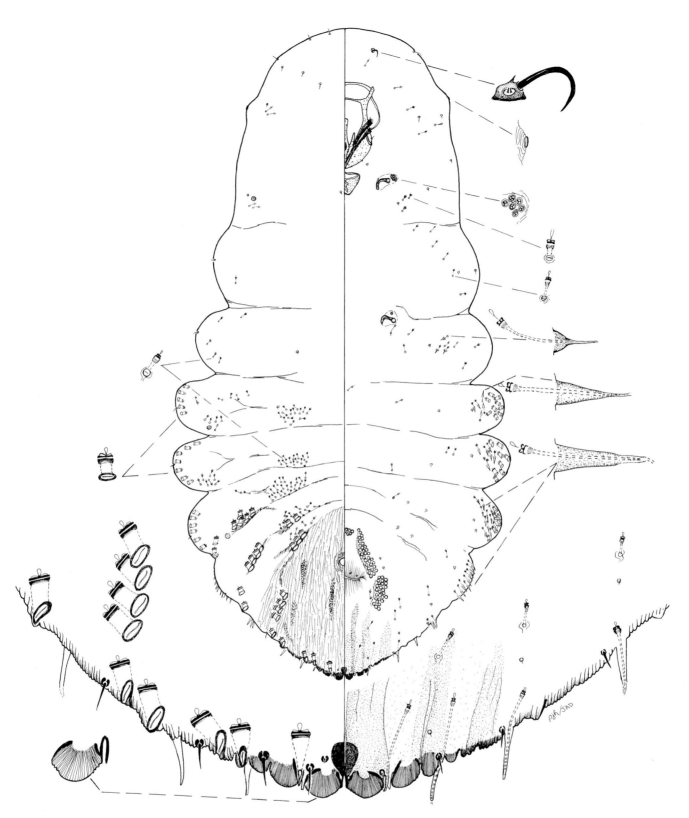

Figure 28. *Chionaspis furfura* (Fitch), scurfy scale, Madison County, IN, on mountain ash, VIII-28-1975.

Chionaspis heterophyllae Cooley

Suggested Common Name Pine scale.

Common Synonyms and Combinations *Chionaspis pinifoliae heterophyllae* Cooley, *Phenacaspis heterophyllae* (Cooley). This species has been commonly misidentified as *Chionaspis pinifoliae* (Fitch).

Field Characters (Plate 26) Adult female cover slightly convex, white, oyster-shell shaped, amount of posterior expansion dependent on diameter of host needles; shed skins marginal, yellowish brown or translucent. Male cover elongate, parallel sided, white, with 3 longitudinal ridges; shed skin marginal, light yellow or translucent. Body of adult female elongate, with obvious lateral protrusion on anterior abdominal segments, light yellow in newly mature forms, brownish purple in specimens containing eggs, dark spot in center of body near juncture of abdomen and thorax; eggs and newly hatched crawlers brownish purple; settled crawlers light yellow with dark, longitudinal line in central part of body. On needles.

Slide-mounted Characters Adult female (Fig. 29) with 3 pairs of definite lobes, 4th pair represented by low series of sclerotized points; without paraphyses. Median lobes separated by space 0.4–1.0 (0.8) times width of median lobe, with slightly protruding yoke, lateral margins of lobes slightly divergent, medial margins strongly divergent, with 0–4 (2) lateral notches and 0–6 (3) medial notches; second lobes bilobed, smaller than median lobes, medial lobule largest with 0–4 (1) lateral notches, 0–4 (1) medial notches, lateral lobule with 0–2 (1) lateral notches, 0–2 (0) medial notches; third lobes bilobed, with medial lobule similar in appearance to medial lobule of second lobe but slightly smaller, lateral lobule composed of flattened sclerotized area, medial lobule with 0–5 (2) lateral notches, 0–1 (1) medial notches, lateral lobule with 1–6 (5) notches; fourth lobes bilobed, represented by low series of notches, medial lobule with 4–9 (6) notches, lateral lobule with 2–9 (5) notches. Gland-spine formula 1-1-1, with 8–13 (11) gland spines near each body margin from prothorax, mesothorax, or metathorax to segment 4; without gland spines between median lobes. Macroducts of 2 sizes, largest in submedial areas of segments 3 to 6, with 2–4 (3) on segment 6, and marginal areas of segments 3 to 7, small size ducts in submedial areas of any or all of segments 2 to 4, in marginal areas from pro- or mesothorax to segment 3, without duct between median lobes, macroduct in first space 19–22 (20) µ long, with 14–23 (18) macroducts on each side of body on segments 5 to 7, total macroducts on each side of body 75–148 (115), some macroduct orifices anterior of anal opening. Pygidial microducts on venter in submarginal areas of segments 5 to 7, with 2–4 (3) ducts; prepygidial microducts of 1 size, on venter from head to segment 4, in marginal or submarginal areas from prothorax to segment 4, in submedial areas from head to segment 2 or 3, with conspicuous cluster on head, with or without clusters laterad of spiracle; pygidial microducts usually absent from dorsum, prepygidial microducts on dorsum from head near mouthparts to segment 3. Perivulvar pores in 5 groups, 33–51 (40) pores on each side of body. Perispiracular pores primarily with 3 loculi, anterior spiracles with 4–8 (6) pores, posterior spiracles with 2–3 (3) pores. Anal opening located 7–11 (9) times length of anal opening from base of median lobes, anal opening 13–18 (14) µ long. Dorsal seta laterad of median lobes 1.5–3.5 (2.4) times length of median lobe. Eyes represented by small sclerotized area, located on body margin at level near anterior end of clypeolabral shield. Antennae each with 1 long seta. Cicatrices usually present on segment 1, sometimes visible on prothorax and segment 3. Body elongate. Second lobes generally not protruding beyond median lobes.

Affinities *Chionaspis heterophyllae* resembles *C. pinifoliae* (Fitch), pine needle scale, but has narrow, divergent median lobes with apices that are widely separated and a yoke that does not strongly protrude anteriorly. *Chionaspis pinifoliae* has wide, round, median lobes with apices that are not widely separated and a yoke that strongly protrudes anteriorly. Other minor differences include: Pine scale with lateral lobule of third lobe with 1–6 (5) notches, large size submedial macroducts on segments 3 to 6, ventral pygidial microducts on segments 5 to 7, with 2–4 (3) ducts on each side of body. Pine needle scale has the lateral lobule of third lobe with 0–3 (1) notches, large size submedial macroducts on segments 4 or 5 to 6, ventral pygidial microducts on segment 7, rarely 6, with 1–3 (1) ducts on each side of body.

Hosts We have only examined specimens from *Pinus*, but it also is reported from *Abies* and *Picea* (Liu et al. 1989b).

Distribution Pine scale is native to the United States and North America. We have examined U.S. specimens from AL, AR, CT, DC, FL, GA, LA, MA, MD, MO, MS, NC, NH, NY, OH, PA, RI, SC, VA, Canada, and Mexico. It is reported in the literature from CA, DE, IN, KY, MI, NJ, TN, TX, WA, WV.

Biology Kosztarab (1963) reports that the species overwinters as eggs, which begin hatching in mid-April. Egg-laying females were noted in mid-May. Crawlers occur in early June in Maryland, in mid-April in Ohio (Kosztarab 1963). Nielsen and Johnson (1973) misidentified this species as *Phenacaspis pinifoliae* (Shour 1986) but undertook a fairly detailed study of the life history and population dynamics of this pest. Their research was done in central New York on mugo pine, red pine, and Scots pine. They observed 2 generations per year. The overwintering stage was the egg, which began hatching in early to mid-May. About 90% of the eggs hatched within a 3-day time span. Within a week of settling the crawlers changed from bright red to pale amber. Nearly all crawlers molted to the second instar by June 1, and adult females began to appear in mid-June. Adult males were evident in the third week of June, and egg laying began in early July. The summer generation crawlers were present from mid-July to early September. Adults first appeared in mid-August and egg laying began in early September. The pine scale occurs on the needles of its host. Only alate males have been collected (Bullington et al. 1989). Shour (1986) examined feeding preferences based on needle surface (convex vs. concave) and needle age and also examined differences in settling site for males and females. There were slight differences compared with *Chionaspis pinifoliae*, pine needle scale.

Economic Importance We have observed the pine scale causing extensive dieback to mature specimen plantings of Mugo pine in College Park, Maryland. Until a few years ago economic entomologists generally were unaware of the existence of the pine scale. Some of the economic literature dealing with the closely related pine needle scale (*C. pinifoliae*) undoubtedly refers to pine scale or mixed infestations of both species. Negron and Clarke (1995) indicate that this species became a serious pest in loblolly pine seed orchards when the trees were treated with pesticides. Miller and Davidson (1990) consider this species to be an occasional pest.

Selected References Bullington et al. (1989); Liu et al. (1989b); Nielsen and Johnson (1973); Shour and Schuder (1987).

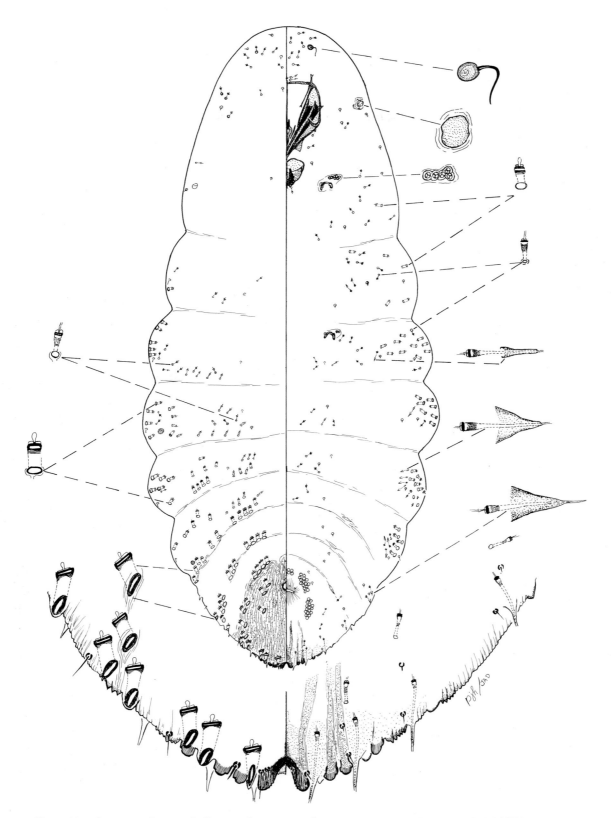

Figure 29. *Chionaspis heterophyllae* Cooley, pine scale, Vienna, MD, on *Pinus* sp., V-16-1976.

Plate 26. *Chionaspis heterophyllae* Cooley, Pine scale

A. Small male cover and larger female covers on mugo pine (J. A. Davidson).
B. Male cover on mugo pine (J. A. Davidson).
C. Distance view of male and female covers on mugo pine (J. A. Davidson).
D. Body of female with eggs on mugo pine (J. A. Davidson).
E. Adult female cover pulled back to show eggs and egg shells on mugo pine (J. A. Davidson).

F. Reddish newly settled crawler (top); yellow feeding crawler (bottom) on mugo pine (J. A. Davidson).
G. Heavy infestation on mugo pine (J. A. Davidson).
H. Damage to needles on mugo pine (J. A. Davidson).
I. Parasite emergence holes on mugo pine (J. A. Davidson).

114

Plate 27. *Chionaspis pinifoliae* (Fitch), Pine needle scale

A. Wide adult female cover on mugo pine (J. A. Davidson).
B. Narrow adult female cover on mugo pine (J. A. Davidson).
C. Small adult covers on dwarf Alberta spruce (J. A. Davidson).
D. Small faintly ridged male covers in biparental population from Jeffrey pine (R. J. Gill).
E. Body of adult female in fall before laying eggs on mugo pine (J. A. Davidson).

F. Dead adult female body and overwintering eggs in early spring on mugo pine (J. A. Davidson).
G. Red, active, non-feeding crawler on mugo pine (J. A. Davidson).
H. Yellow, settled, feeding crawler on mugo pine (J. A. Davidson).
I. Damage caused by feeding of scale on mugo pine (J. A. Davidson).
J. Heavily infested, dying mugo pine (J. A. Davidson).

Chionaspis pinifoliae (Fitch)

ESA Approved Common Name Pine needle scale.

Common Synonyms and Combinations *Aspidiotus pinifoliae* Fitch, *Mytilaspis pinifoliae* (Fitch), *Chionaspis pinifoliae semiaureus* Cockerell, *Chionaspis pinifolii* (Fitch), *Trichomytilus pinifolii* (Fitch), *Polyaspis pinifolii* (Fitch), *Phenacaspis pinifoliae* (Fitch).

Field Characters (Plate 27) Adult female cover slightly convex, white, oyster-shell shaped; shed skins marginal, normally yellow, occasionally reported as transparent or red. Male cover elongate, parallel sided, white, with 3 longitudinal ridges; shed skin marginal, light yellow or translucent. Body of female elongate, with obvious lateral protrusion on anterior abdominal segments, yellow, with dark spot in center of body near juncture of abdomen and thorax, body red or purple when gravid; eggs and newly hatched crawlers red, settled crawlers turning light yellow, nearly transparent, with dark, longitudinal line in central part of body. On needles.

Slide-mounted Characters Adult female (Fig. 30) with 3 pairs of definite lobes, 4th pair represented by series of low, sclerotized points; without paraphyses. Median lobes separated by space 0.2–0.6 (0.4) times width of median lobe, with strongly protruding yoke, lateral margins of lobes slightly divergent, medial margins parallel or convergent from base to about two-thirds of distance to apex, apical one-third rounded, without lateral notches, with 0–4 (2) medial notches; second lobes bilobed, smaller than median lobes, medial lobule largest usually without notches, occasionally with 1, lateral lobule with 0–1 (0) lateral notches, without medial notches; third lobes bilobed, with medial lobule similar in appearance to medial lobule of second lobe but slightly smaller, lateral lobule composed of flattened sclerotized area, medial lobule with 0–3 (0) lateral notches, 0–1 (0) medial notches, lateral lobule with 0–3 (1) notches; fourth lobes bilobed, represented by low series of notches, medial lobule with 2–8 (3) notches, lateral lobule with 3–4 (3). Gland-spine formula 1-1-1, with 10–16 (14) gland spines near each body margin from mesothorax or metathorax to segment 4; without gland spines between median lobes. Macroducts of 2 sizes, largest in submedial areas of segments 4 or 5 to 6, with 2–6 (3) on segment 6, and marginal areas of segments 3 to 7, small size ducts in submedial areas of any or all of segments 3 and 4, in marginal areas from meso- or metathorax to segment 3, without duct between median lobes, macroduct in first space 20–27 (22) μ long, with 14–23 (17) macroducts on each side of body on segments 5 to 7, with total of 61–127 (88) macroducts on each side of body, some macroduct orifices anterior of anal opening. Pygidial microducts usually on venter in submarginal areas of segment 7, rarely on segment 6, with 1–3 (1) ducts; prepygidial microducts of 1 size, on venter from head to segment 4, in marginal or submarginal areas from prothorax to segment 4, in submedial areas from head to segment 2 or 3, with conspicuous cluster on head, with or without clusters laterad of spiracle; pygidial microducts absent from dorsum, prepygidial microducts on dorsum from head near mouthparts to segment 4, often in conspicuous clusters submedially on segments 1 to 4. Perivulvar pores in 5, rarely 4, groups, 25–48 (36) pores on each side of body. Perispiracular pores primarily with 3 loculi, anterior spiracles with 3–8 (5) pores, posterior spiracles with 1–3 (2) pores. Anal opening located 6–11 (9) times length of anal opening from base of median lobes, anal opening 10–15 (13) μ long. Dorsal seta laterad of median lobes 1.9–3.6 (2.6) times length of median lobe. Eyes represented by small sclerotized area, located on body margin at level near anterior end of clypeolabral shield. Antennae each with 1 long seta. Cicatrices usually present on segment 1, sometimes visible on pro-

thorax and segment 3. Body elongate. Second lobes generally protruding beyond median lobes.

Affinities The pine needle scale is very similar to *Chionaspis heterophyllae*, the pine scale. For a comparison of these species see affinities section of *C. heterophyllae*. Some evidence presented by Luck and Dahlsten (1974) and Stimmann (1969) suggests that the pine needle scale may be a complex of at least 2 sibling species.

Hosts Borchsenius (1966) reports this scale from 7 plant genera in 2 families. We have examined specimens from *Abies*, *Picea*, *Pinus*, *Pseudotsuga*, *Torreya*, and *Tsuga*. It has been reported on *Cedrus*, *Juniperus*, and *Taxus*.

Distribution Pine needle scale is native to the United States. We have examined U.S. specimens from AZ, CA, CO, CT, DC, GA, IA, ID, IL, IN, KY, MA, MD, ME, MI, MN, NC, ND, NE, NH, NJ, NM, NY, OH, OR, PA, SD, TN, TX, UT, VA, VT, WV, WY. We also have seen material from Canada, Central America, England, and Germany. The species has been reported by reliable sources (Liu et al. 1989b; Nakahara 1982) from FL, MT, WA, WI, Cuba, and Mexico. Because of the frequent confusion between the pine needle scale and the pine scale, it is unwise to rely on literature records of this species.

Biology In Canada, the western United States, and the northern-most eastern United States. 1 generation per year is reported. From the southern United States, Rhode Island, Connecticut, Massachusetts, Michigan, Ohio, Pennsylvania, and parts of New York there are 2 generations per year. This species usually overwinters in the egg stage, but Luck and Dahlsten (1974) discovered that gravid adult females overwintered on Jeffrey pine at South Lake Tahoe, California, while those on lodgepole pine in the same area overwintered in the normal manner as eggs. The lodgepole population is parthenogenetic, while the Jeffrey pine population has about a 50:50 mix of males and females. A generalized life history for the Jeffrey population is as follows: Egg laying began in late May, although a relatively small number of eggs were laid on warm days in the winter. Crawlers appeared in mid-June and persisted until August, second instars were present from late July through September, adult males were present in September, and adult females appeared in early September. Stimmann (1969) found that adult females laid eggs throughout the winter in Oregon on lodgepole pine. Burden and Hart (1989) developed a degree-day model for egg eclosion of this species. Eliason and McCullough (1997) demonstrated different survival rates and fecundity on four varieties of Scotch pine suggesting the possibility of host-plant resistance. Stimmel (2002, personal communication) indicated that this species overwinters as both eggs and gravid females in Pennsylvania. Shour (1986) examined feeding preferences based on needle surface (convex vs. concave) and needle age and also examined differences in settling site preferences between males and females. There were slight differences compared with *Chionaspis heterophyllae*, pine scale.

Economic Importance The pine needle scale can build to large populations that cause the needles of infested trees to appear gray from a distance. In heavy infestations needles turn yellow and eventually drop. The lower branches usually die first and entire trees may be killed. Cumming (1953) considers this species to be a serious pest of spruce and pine in ornamental plantings in the Prairie Provinces of Canada. Miller and Davidson (1990) consider it to be a serious pest in a small area of the world.

Selected References Bullington et al. (1989); Liu et al. (1989b); Luck and Dahlsten (1974).

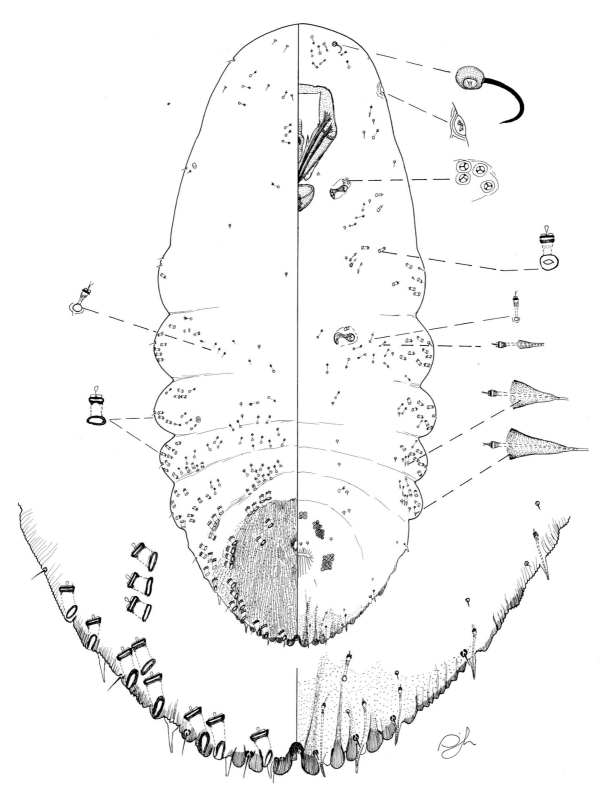

Figure 30. *Chionaspis pinifoliae* (Fitch), pine needle scale, Indianapolis, IN, on *Abies concolor*, VIII-12-1950.

Chionaspis salicis (Linnaeus)

Suggested Common Name Willow scale (also called black willow scale).

Common Synonyms and Combinations *Coccus salicis* Linnaeus, *Chionaspis cryptogamus* (Dalman), *Aspidiotus minimus* Baerensprung, *Aspidiotus salicis-nigrae* Walsh, *Chionaspis salicisnigrae* (Walsh), *Aspidiotus vaccinii* Bouché, *Aspidiotus populi* Baerensprung, *Mytilaspis maquarti* Targioni Tozzetti, *Chionaspis aceris* Henschel, *Mytilaspis salicis* Le Baron, *Chionaspis salicis* (Le Baron), *Chionaspis alni* Signoret, *Chionaspis fraxini* Signoret, *Lecanium myrtilli* Kaltenbach, *Chionaspis ortholobis bruneri* Cockerell, *Chionaspis micropori* Marlatt, *Chionaspis montana* Borchsenius, *Chionaspis polypora* Borchsenius, *Chionaspis salicisnigrae* was synonymized with *C. salicis* by Danzig (1970) due to the large amount of overlapping variation found in populations of each 'species.' Liu et al. (1989b) treated the species as separate based on the absence of dorsal microducts in the submedian areas of segments 5 and 6 in *C. salicis* and their presence in *C. salicisnigrae*. We have examined several specimens of *C. salicisnigrae* that lack microducts in this area and therefore conclude that Danzig (1970) is correct in synonymizing the species.

Field Characters (Plate 28) Adult female cover white, oyster-shell shaped; shed skins marginal, yellowish brown. Male cover white, elongate oval, with 3 longitudinal ridges weakly indicated; shed skin marginal, yellowish brown. Body of adult female red; eggs and newly hatched crawlers red, purple, or black. On twigs, branches, and trunk; males also on upper surfaces of leaves near primary veins.

Slide-mounted Characters Adult female (Fig. 31) usually with 3 pairs of definite lobes, fourth lobes represented by low, weakly sclerotized series of points; second and third lobes usually with paraphysis on mesal margin of medial lobule, occasionally with small paraphysis on lateral margin of medial lobule. Median lobes variable, either closely appressed medially from base to one-half of length of lobe, with rounded apex, or with lobes rounded and divergent from base, with short, broad yoke, often with 2 small, sclerotized rods between lobes, medial margins parallel basally, diverging apically, or divergent from base, lateral margins diverging or parallel, with 0–5 (2) very small lateral notches, with 0–4 (2) medial notches; second lobes bilobed, smaller than median lobes, medial lobule largest with 0–2 (2) lateral notches, without medial notches, lateral lobule with 0–3 (1) lateral notches, without medial notches; third lobes bilobed, slightly smaller or equal in size to second lobes, medial lobule largest with 2–5 (3) lateral notches, without medial notches, lateral lobule with 2–4 (3) notches, without medial notches; fourth lobes bilobed, represented by low series of notches, medial lobule with 3–7 (5) notches, lateral lobule with 5–7 (6) notches. Gland-spine formula highly variable, including 2-2-2, 1-2-2, or 2-3-3, with 24–35 (30) gland spines near each body margin from prothorax, mesothorax, or metathorax to segment 4; without gland spines between median lobes. Duct distribution unusually variable, sometimes large macroducts replaced by small macroducts or by microducts. Macroducts of 2 sizes, largest in submedial areas of segments 1 or 2 to 5 or 6, when present on 6 with 3–7 (5) large macroducts on each side of segment, and in marginal areas of segments 1 or 2 to 7, small size ducts in submedial areas of none, any, or all of segments 1, 2, 5, or 6, marginal areas from pro-, meso-, or metathorax to segments 3 or 4, without duct between median lobes, macroduct in first space 19–23 (21) μ long, with 17–27 (24) macroducts on each side of body on segments 5 to 7, with total of 151–269 (173) macroducts on each side of body, some macroduct orifices anterior of anal opening. Pygidial microducts on venter in sub-

marginal areas of segments 5 to 7, with 4–8 (6) ducts; prepygidial ducts of 1 size, variable, most specimens with many ducts present on all segments laterally from head to segment 4, and many ducts anterior of mouthparts, near anterior and posterior spiracles, and near medial setae on segments 1 to 4, a few specimens with only 1 or 2 such ducts in lateral areas, with only 1 or 2 near spiracles, and with or without ducts near medial setae; pygidial microducts often replace macroduct clusters in submedial area of segment 6, often present on segment 5 also; prepygidial microducts variable, when numerous, present in lateral and mediolateral areas of head and thorax, and in mediolateral clusters on segments 1 to 4. Perivulvar pores arranged in 5 groups with 49–98 (70) pores on each side of body. Perispiracular pores with 3 loculi, anterior spiracles with 7–15 (11) pores, posterior spiracles with 3–5 (4) pores. Anal opening located 8–13 (10) times length of anal opening from base of median lobes, anal opening 13–20 (16) μ long. Dorsal seta laterad of median lobes 0.9–1.2 (1.0) times length of median lobe. Eyes absent or represented by inconspicuous clear area on derm, located on body margin at level of anterior margin of clypeolabral shield. Antennae each with 1 long seta. Cicatrices usually not visible, occasionally on prothorax. Body elongate. Median lobes generally protruding beyond second lobes.

Affinities *Chionaspis salicis* resembles *C. corni* and *C. ortholobis*. In *C. corni* the median lobes are strongly divergent with a straight medial margin, and small submedian microducts are absent from the dorsum of segments 5 and 6. *Chionapsis salicis* has median lobes that are only slightly divergent with a rounded medial margin, and small submedian microducts often are present on the dorsum of segments 5 and 6. In *C. ortholobis* there are more than 10 ventral microducts on segments 5 to 7, there are no dorsal microducts on segments 5 and 6, and there are fewer than 5 dorsal microducts on each side of segment 4, whereas, in North American specimens of *C. salicis* there are fewer than 10 ventral microducts on segments 5 to 7, there usually are many dorsal microducts on segments 5 and 6, and there are more than 8 dorsal microducts on each side of segment 4.

Hosts Borchsenius (1966) records this species from 38 plant genera in 13 families. In the United States, the willow scale is usually found on *Populus* or *Salix*, but in other parts of the world it has been recorded on several other hosts including *Ribes*. We have examined material from *Cornus* and *Fraxinus*, and have assigned these specimens to other species of *Chionaspis*. The species also is recorded from *Amelanchier* and *Liriodendron*, but we suspect these may be misidentifications. Host genera given by Borchsenius are *Acer*, *Alnus*, *Amelanchier*, *Andromeda*, *Arctostaphylos*, *Betula*, *Corylus*, *Cornus*, *Cotoneaster*, *Cytisus*, *Erica*, *Euonymus*, *Fraxinus*, *Genista*, *Grossularia*, *Helianthemum*, *Jasminum*, *Ligustrum*, *Ledum*, *Liriodendron*, *Lyonia*, *Myrtus*, *Populus*, *Pyrus*, *Quercus*, *Rhamnus*, *Rhododendron*, *Ribes*, *Rosa*, *Salix*, *Sarothamnus*, *Sorbus*, *Syringa*, *Ulmus*, *Vaccinium*, *Viburnum*, and *Vitis*.

Distribution We have examined U.S. specimens from AR, AZ, CO, IA, ID, IL, IN, KS, LA, MD, MI, MN, MO, MS, NC, ND, NE, NJ, NM, NY, OH, OK, PA, SD, TX, WI, WY. It has been reported from AL, CA, DC, FL, MA, MT, TN, VA, Canada, and Mexico. In the Old World it occurs in China, Europe, Korea, and the former Soviet Union.

Biology According to Langford (1926) this species has 2 generations per year in Colorado and overwinters in the egg stage. First-generation crawlers appear in late April and molt in mid-May. Adults are present from early to mid-June and eggs begin

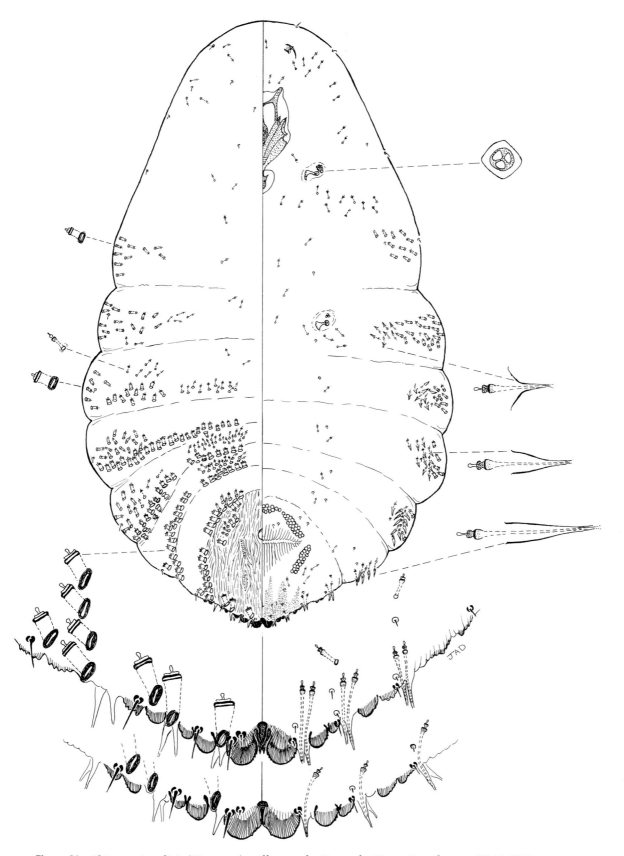

Figure 31. *Chionaspis salicis* (Linnaeus), willow scale, Decorah, IA, on *Populus* sp., IX-13-1938.

to appear in late June. Crawlers of the second generation are present in July and adults appear in late August. Overwintering eggs are laid in September. Crawlers have been reported in Missouri in mid-June (Hollinger 1923) and in Ohio in late May to early June (Kosztarab 1963). In Tennessee this species has 3 generations per year and overwinters as eggs. Crawlers of each generation begin emerging during the third week of April, the second week of June, and the fourth week of August; second instars appear the first week of May, second week of July, and first week of September. Adult males begin emerging the fourth week of May, first week of August, and the second week of September; adult females are present the third week of May, fourth week in July, and third week of September. Adult females laid from 18–265 eggs with an average of 152; this is very different from that found in Colorado where 11–54 eggs were laid, with an average of 33 (Lambdin 1990).

Economic Importance The willow scale can build to large populations that completely cover twigs and branches. It has been reported to kill branches, and Langford (1926) states it may 'become so severe as to cause the dying of branches, and in some cases the entire tree.' Young trees are reported to be most affected. According to Danzig (1980) the willow scale is a pest of currants, aspen, and willow. This species was apparently a more serious pest in the early part of the last century, but Lambdin (1990) reports that it can be a pest on ornamental willows in Tennessee where it causes loss of plant vigor, dieback, stunting, and eventual death of the affected plant. Miller and Davidson (1990) consider this species to be an occasional pest.

Selected References Danzig (1980); Ferris (1937); Lambdin (1990); Langford (1926).

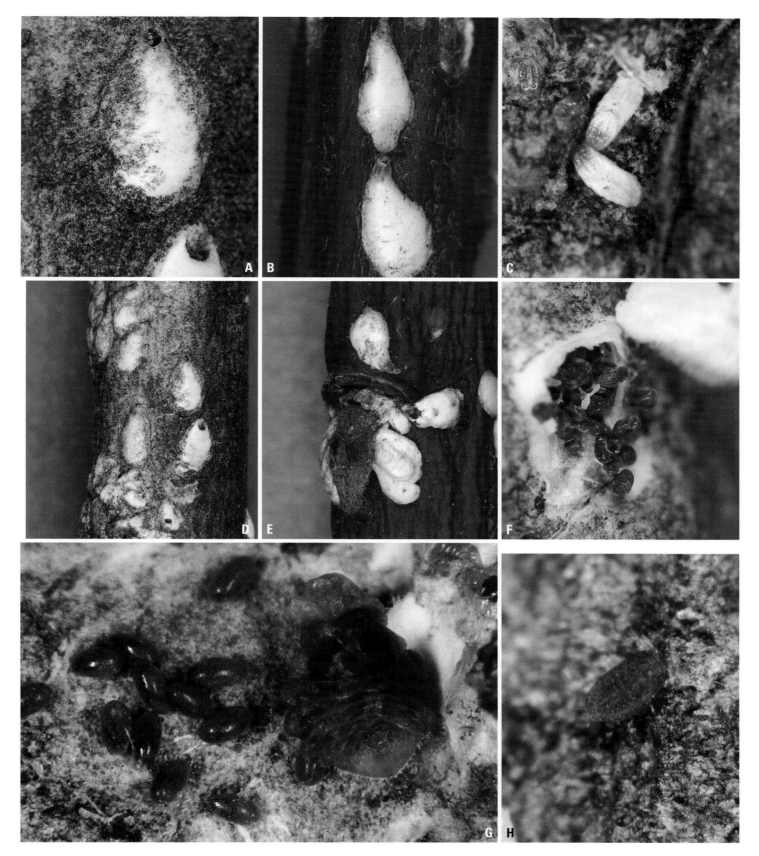

Plate 28. *Chionaspis salicis* (Linnaeus), Willow scale

A. Adult female cover on black willow (J. A. Davidson).
B. Adult female covers on dogwood (J. A. Davidson).
C. Adult male covers on black willow (J. A. Davidson).
D. Male and female covers on black willow (J. A. Davidson).
E. Distance view of female covers around dogwood bud (J. A. Davidson).

F. Eggs on black willow in December (J. A. Davidson).
G. Eggs on black willow in July with female (J. A. Davidson).
H. Crawler on black willow in August (J. A. Davidson).

121

Chrysomphalus aonidum (Linnaeus)

ESA Approved Common Name Florida red scale (also called black scale, circular black scale, circular scale, fig scale, circular purple scale, orange brown scale; Egyptian black scale).

Common Synonyms and Combinations *Coccus aonidum* Linnaeus, *Chrysomphalus ficus* (Ashmead), *Aspidiotus ficus* Ashmead.

Field Characters (Plate 29) Adult female cover slightly convex, circular, dark brown, black, or reddish brown; shed skins central, lighter than scale produced by adult, reddish yellow. Male cover same color as female cover, elongate; exuviae submarginal. Body of female pale yellow when first mature, dark yellow later; eggs yellow; male orange. Normally on leaves and fruit.

Slide-mounted Characters Adult female (Fig. 32) with 3 pairs of definite lobes, fourth lobes represented by low series of sclerotized points; paraphysis formula usually 2-2-0, rarely 2-3-1 or 2-2-1, when all paraphyses present, with large paraphyses attached to lateral margin of median lobe, to medial margin of lobes 2 and 3, and in space between lobes 2 and 3, small paraphyses attached to lateral margins of lobes 2 and 3 and in space between lobes 3 and 4. Median lobes separated by space 0.5–1.0 (0.7) times width of median lobe, without basal sclerotization, with paraphysis attached to medial margin, without yoke, medial margins and lateral margins converging slightly, with 1 lateral notch, 0–1 (0) medial notch; second lobes simple, about same size as median lobes, with rounded apex, with 1–2 (1) lateral notches, without medial notch; third lobes simple, about same size and shape as second lobes, with 2–3 (3) lateral notches, without medial notches; fourth lobes represented by series of 2 or 3 low swellings, with 12–35 (28) notches. Plates between lobes fimbriate, plates in third space each with microduct and protruding orifice near center, without conspicuous, clavate process on plate between lobes 3 and 4; plate formula 2-3-3, rarely 2-3-4; plates between median lobes 0.8–1.0 (0.9) times as long as median lobes. Macroducts of 2 sizes, larger size marginal between median lobes and in spaces between median lobe and lobe 2 and between lobe 2 and 3, smaller size elsewhere on pygidium, duct between median lobes extending 1.9–3.1 (2.3) times distance between posterior apex of anal opening and base of median lobes, 109–151 (120) μ long, large macroduct in first space 112–155 (133) μ long, with 40–56 (49) macroducts on each side of pygidium on segments 5 to 8, ducts in submarginal and marginal areas, with total of 49–75 (65) macroducts on each side of body, some macroduct orifices anterior of anal opening; prepygidial macroducts in cluster on margin of segment 2 composed of 9–27 (16) large ducts, small ducts scattered along body margin from meso- or metathorax to segments 2, 3, or 4. Pygidial microducts on venter in submarginal and marginal areas of segment 5, with 2–4 (3) ducts; prepygidial ducts of 1 size, in submedial area of abdomen, in submarginal and marginal areas of segment 4 rarely 3; dorsal pygidial ducts absent; prepygidial microducts of 2 sizes, larger size in medial areas of any or all of pro-, meso-, or metathorax to segment 3, smaller size in submarginal areas of mesothorax to segment 1, 2, or 3. Perivulvar pores arranged in 4 or 5 groups, 9–13 (11) pores on each side of body. Pores absent near spiracles. Anal opening located 1.8–3.1 (2.3) times length of anal opening from base of median lobes, anal opening 12–18 (15) μ long. Dorsal seta laterad of median lobes 0.6–0.8 (0.6) times length of median lobe. Eyes spurlike, on mesothorax. Antennae each with 1 long seta. Cicatrices absent. Body pear shaped.

Affinities This species can be distinguished from all other U.S. species of *Chrysomphalus* by having a single marginal cluster of prepygidial macroducts on each side of body. *Chrysomphalus*

bifasciculatus has 2 such clusters, and *C. dictyospermi* lacks these clusters entirely.

Hosts Polyphagous. Borchsenius (1966) records the species from 200 genera in 74 families; Dekle (1977) records it from 259 genera in Florida. It is commonly found on citrus and various kinds of palms. We have examined specimens from the following host genera: *Acrocomia, Agave, Aglaonema, Allamanda, Aloe, Annona, Anthurium, Aralia, Areca, Arenga, Arundina, Asparagus, Aspidistra, Bambusa, Barringtonia, Bauhinia, Broussaisia, Bruguiera, Buxus, Caladium, Calanthe, Callistemon, Calophyllum, Camellia, Canna, Carissa, Cassia, Castilla, Celtis, Chamaedorea, Chrysalidocarpus, Cinnamomum, Citrus, Cocos, Coelia, Coelogyne, Coffea, Cordia, Cycas, Cyclamen, Cymbidium, Cynodon, Dendrobium, Dianthus, Diospyros, Dracaena, Epidendrum, Epipremnum, Eucalyptus, Eugenia, Euphorbia, Euonymus, Eurya, Ficus, Garcinia, Gardenia, Gnetum, Grevillea, Haitia, Hedera, Heliconia, Hibiscus, Howea, Ilex, Jasminum, Kentia, Lantana, Laurus, Ligustrum, Livistona, Malus, Mammea, Mangifera, Melicoccus, Monstera, Musa, Nipa, Nerium, Odontoglossum, Olea, Pandanus, Papyrus, Persea, Philodendron, Phoenix, Pimenta, Pinus, Plumeria, Podocarpus, Poinsettia, Polyalthia, Pothos, Psidium, Ptychosperma, Punica, Rheedia, Rosa, Roystonea, Ruta, Salix, Scindapsus, Sideroxylon, Solanum, Stokesia, Strelitzia, Syzygium, Tamarindus, Terminalia, Trichopilia, Trigonidium, Ulmus, Vanda, Veitchia, Vitis, Yucca, Zamia,* and *Zingiber.*

Distribution This species occurs out-of-doors in most tropical and subtropical areas of the world and is found in greenhouses elsewhere. It is apparently indigenous to Southeast Asia. We have examined U.S. material from: out-of-doors—CA (eradicated), FL, GA, HI, LA, MS, TX; greenhouse—DC, IL, IN, KS, MA, MD, MI, MO, NC, NH, NJ, NY, OH, OK, OR, PA, RI, SC, WV. It frequently is taken in quarantine at U.S. ports-of-entry. A distribution map of this species was published by CAB International (1988).

Biology The Florida red scale may have 3 to 6 generations per year depending on the species of host, the position on the host, and environmental conditions. Development is continuous throughout the year. Avidov and Harpaz (1969) give a good summary of studies carried out in Israel. Egg production varies considerably. On lemon fruit an average of 224 offspring are produced per female in the spring, 108 in the summer, and 38 in the fall. On citrus leaves there were 19 offspring in spring, 80 in summer, 44 in fall. The fruit of citrus seems to be the preferred feeding area, although individuals are commonly encountered on leaves and rarely on branches. During warm parts of the year 3 times as many scales occur on the fruit of citrus as on the leaves. The adult female lays eggs that hatch from within a few hours up to 10 days depending on weather conditions. The time from egg hatch to maturity is 40–55 days in summer, and up to 180 days in winter. The production of a generation requires about 875 day-degrees. Males are present and are apparently necessary for reproduction. In Hunan, China, this species is the main pest of oranges; it completes 3 to 4 generations per year, and larvae hatch from May to July in the first generation, mid-June to early September in the second, September to November in the third, and a partial fourth generation is present after November (Gan et al. 1993). A model was developed for forecasting the occurrence of the pest in Hunan (Gan et al. 1993). Lu (1989) demonstrated that there is virtually no competition between the Florida red scale and the California red scale on citrus when they co-occur on the same trees.

Economic Importance On citrus the Florida red scale can cause considerable damage. Feeding on leaves causes yellowing and

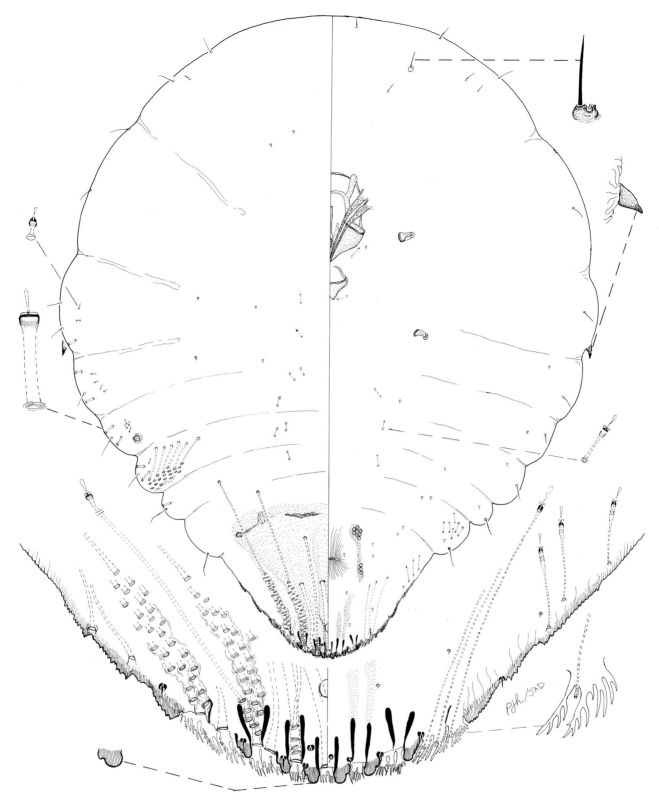

Figure 32. *Chrysomphalus aonidum* (Linnaeus), Florida red scale, East Meadow, Long Island, NY, on *Pandanus* sp., I-27-1949.

eventual leaf drop. Fruit that is infested while it is small will drop from the tree or will be dwarfed. Contaminated fruit must be culled. Trees that are infested for several years become weakened. This is an especially difficult pest to control with insecticides because it may occur on a wide range of weedy hosts including conifers and grasses. Once an infestation is under control, reinfestation may be very rapid. Biological control agents, especially *Aphytis holoxanthus* DeBach, have successfully controlled this pest in most citrus-growing areas of the world. The Florida red scale also is reported as an occasional pest of palms (Dhileepan 1992), mango (Wysoki 1997), bananas (Rosen and DeBach 1978), avocados (Ebeling 1959), and grape (Gomez-Menor Ortega 1939). A great deal of recent research has been done on the biological control of this species and the effects of parasites on the scale and its natural enemies (Cohen et al. 1987; Cohen et al. 1994; Havron and Rosen 1994; Malipatil et al. 2000; Rehman et al. 1999; Steinberg et al. 1987). Beardsley and González (1975) consider this scale to be one of 43 serious armored scale pests. Miller and Davidson (1990) consider it to be a serious world pest.

Selected References Avidov and Harpaz (1969); Bodenheimer (1951); McKenzie (1939).

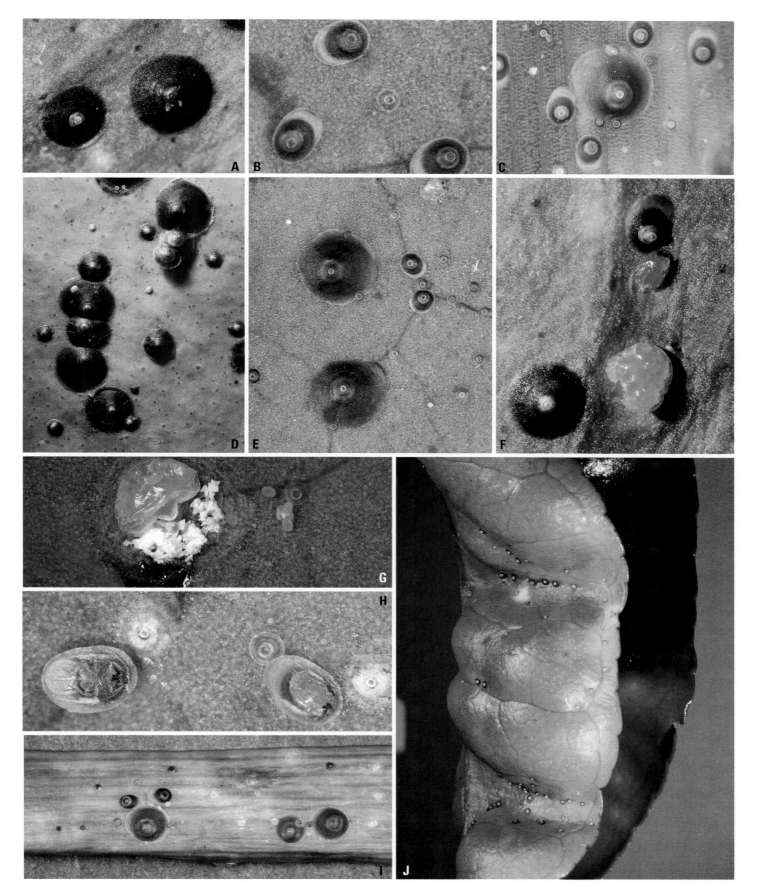

Plate 29. *Chrysomphalus aonidum* (Linnaeus), Florida red scale

A. Adult female covers on citrus (J. A. Davidson).
B. Male covers on shefflera (J. A. Davidson).
C. Adult female, male, and immature covers on palm (M. B. Stoetzel).
D. Adult female and immature covers on citrus (J. A. Davidson).
E. Adult female and male covers with many capped crawler covers on shefflera (J. A. Davidson).
F. Yellow adult female body before egg laying and orange male pupa on citrus (J. A. Davidson).

G. Orange adult female with eggs, crawlers, and egg shells on shefflera (J. A. Davidson).
H. Covers turned over to show adult male (top) and pupa (bottom) on shefflera (J. A. Davidson).
I. Chlorotic discoloration on liriope (J. A. Davidson).
J. Scale colonies associated with leaf distortion on citrus (J. A. Davidson).

Chrysomphalus bifasciculatus (Ferris)

Suggested Common Name Bifasciculate scale (also called false Florida red scale).

Common Synonyms and Combinations None. Often misidentified as *Chrysomphalus aonidum* Linnaeus (= *C. ficus* Ashmead).

Field Characters (Plate 30) Adult female cover slightly convex, circular, brown, dark gray, or black; shed skins central, lighter than scale produced by adult, reddish yellow to reddish brown. Male cover same color as female cover, elongate; shed skin submarginal. Body of adult female pale yellow when first mature, dark yellow later; eggs yellow. Normally on leaves. In bifasciculate scale, cover of the adult female lighter in color than Florida red scale and therefore does not contrast as strongly with shed skins.

Slide Mounted Characters Adult female (Fig. 33) with 3 pairs of definite lobes, fourth lobes represented by low series of sclerotized points; paraphysis formula variable, 2-1-0 is most common, also 2-3-0, 2-2-2, 2-2-0, 2-2-1, when all paraphyses present with large paraphyses attached to lateral margin of median lobe, to medial margin of lobes 2 and 3, and in space between lobes 2 and 3, small paraphyses attached to lateral margins of lobes 2 and 3 and in space between lobes 3 and 4. Median lobes separated by space 0.5–0.7 (0.7) times width of median lobe, without basal sclerotization, with paraphysis attached to medial margin, without yoke, medial margins and lateral margins parallel or converging or diverging slightly, with 1 lateral notch, 0–1 (1) medial notch; second lobes simple, usually slightly smaller than median lobes, with rounded apex, with 0–2 (1) lateral notches, with 0–2 (1) medial notches; third lobes simple, slightly smaller than second lobes, with 2–4 (3) lateral notches, with 0–1 (0) medial notches; fourth lobes represented by series of 2 or 3 low swellings, with 16–37 (23) notches. Plates between lobes fimbriate, plates in third space each with microduct and protruding orifice near center, with slightly clavate process on plates between lobes 3 and 4; plate formula 2-3-3, rarely 2-2-3; plates between median lobes 0.6–1.0 (0.8) times as long as median lobes. Macroducts of 2 sizes, larger size marginal between median lobes and in spaces between median lobe and lobe 2 and between lobe 2 and 3, smaller size elsewhere on pygidium, duct between median lobes extending 2.3–3.5 (2.8) times distance between posterior apex of anal opening and base of median lobes, 77–99 (88) μ long, large duct in first space 88–112 (102) μ long with 23–33 (28) macroducts on each side of pygidium on segments 5 to 8, ducts in submarginal and marginal areas, with total of 43–59 (47) macroducts on each side of body, some macroduct orifices anterior of anal opening; prepygidial macroducts in cluster on margin of segments 2 and 3, composed of 8–13 (10) ducts on segment 2 and 5–12 (8) on segment 3, small ducts present along body margin from meso- or metathorax to segment 3. Pygidial microducts on venter in submarginal and marginal areas of segment 5, with 3–4 (3) ducts; prepygidial ducts of 1 size, in submedial area of abdomen, in submarginal and marginal areas of segment 4; dorsal pygidial ducts absent; prepygidial microducts of 1 size, in submedial areas of any or all of pro-, meso-, or metathorax to segment 3, smaller size ducts absent from submarginal areas.

Perivulvar pores arranged in 4 or 5 groups, 4–8 (6) pores on each side of body. Pores absent near spiracles. Anal opening located 2.0–3.5 (2.6) times length of anal opening from base of median lobes, anal opening 10–15 (12) μ long. Dorsal seta laterad of median lobes 0.5–0.8 (0.5) times length of median lobe. Eyes variable, flat sclerotized area, dome, or small spur, on mesothorax. Antennae each with 1 seta. Cicatrices often present on prothorax and segment 1. Body pear shaped.

Affinities This species can be distinguished from all other U.S. species of *Chrysomphalus* by having 2 marginal clusters of prepygidial macroducts on each side of the body. *Chrysomphalus aonidum* has 1 pair of clusters, and *C. dictyospermi* (Morgan) lacks these clusters. Bifasciculate scale has about 28 macroducts and about 6 perivulvar pores on each side of body. Florida red scale has about 49 macroducts and 11 perivulvar pores on each side of body.

Hosts Polyphagous. Borchsenius (1966) records it from 40 genera in 26 families. It is commonly encountered on holly and citrus. We have examined specimens from the following host genera: *Aucuba, Annona, Aspidistra, Buxus, Camellia, Castanopsis, Citrus, Cordyline, Daphne, Elaeagnus, Eugenia, Euonymus, Hedera, Ilex, Lantana, Ligustrum, Lonicera, Osmanthus, Pittosporum*, and *Rhapis*.

Distribution We have seen U.S. specimens from AL, CA, GA, LA, MD, NC, SC, TX, VA. It has been reported from NJ and OK. It occurs out-of-doors north to VA. This species has been collected in quarantine inspections from China, Hawaii, Japan, Korea, Mexico, Taiwan, Venezuela, and the former Soviet Union (greenhouses). The species apparently occurs out-of-doors much farther north than Florida red scale.

Biology Murakami (1970) reports that this species has 2 generations per year in Japan. Eggs are laid in May and hatch in late May and early June. Adults of the first generation appear in July and early August. Eggs of the second generation are laid and hatch in late July and August. Mature, fertilized adult females overwinter. Azim (1961) presents somewhat different data from Japan. He states that the species develops continuously but is very slow in the winter. The time from egg hatch to molting to the second instar requires about 15 to 20 days in April and June; the second instar takes 12 to 18 days; development from the egg to the adult female takes about 63 days in the field. At room temperature on squash plants the life cycle takes about 45 days with the first stage taking 14 days, the second about 11, and the adult to being gravid 20 days. Males were abundant. McComb and Davidson (1969) report crawlers in Maryland in May on *Ilex*. The species usually occurs on the leaves of the host.

Economic Importance In Japan, bifasciculate scale can build to such large populations on ornamentals that their leaves become brown and dried up; they are a particular problem on *Euonymus japonica* and *Aspidistra elatior* (Azim 1961). This species occasionally is a pest of selected ornamentals (Morrison 1946). Miller and Davidson (1990) consider this species to be an occasional pest.

Selected References Azim (1961); Ebeling (1959); McKenzie (1939); Murakami (1970).

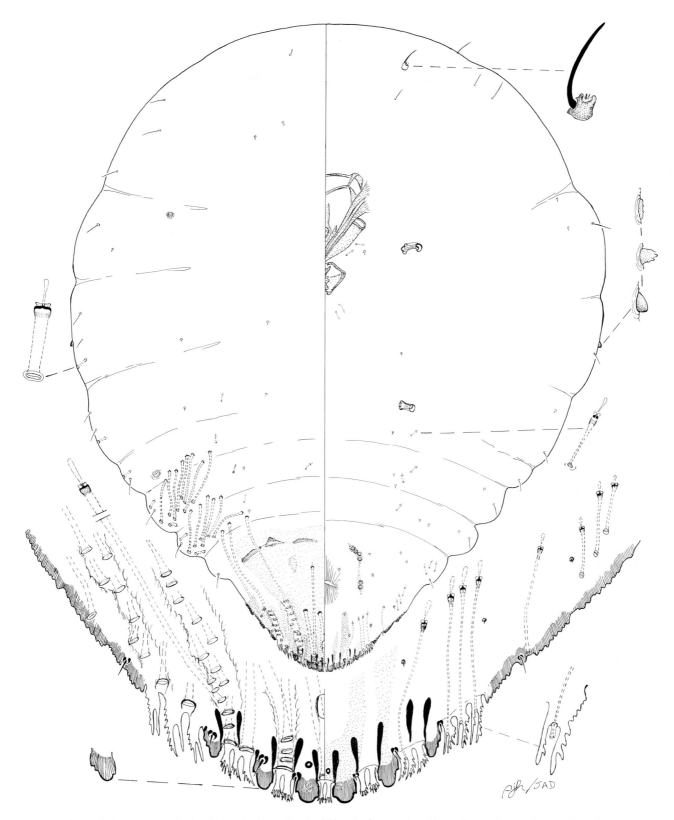

Figure 33. *Chrysomphalus bifasciculatus* Ferris, bifasciculate scale, Alexandria, VA, on *Ilex* sp., VIII-13-1954.

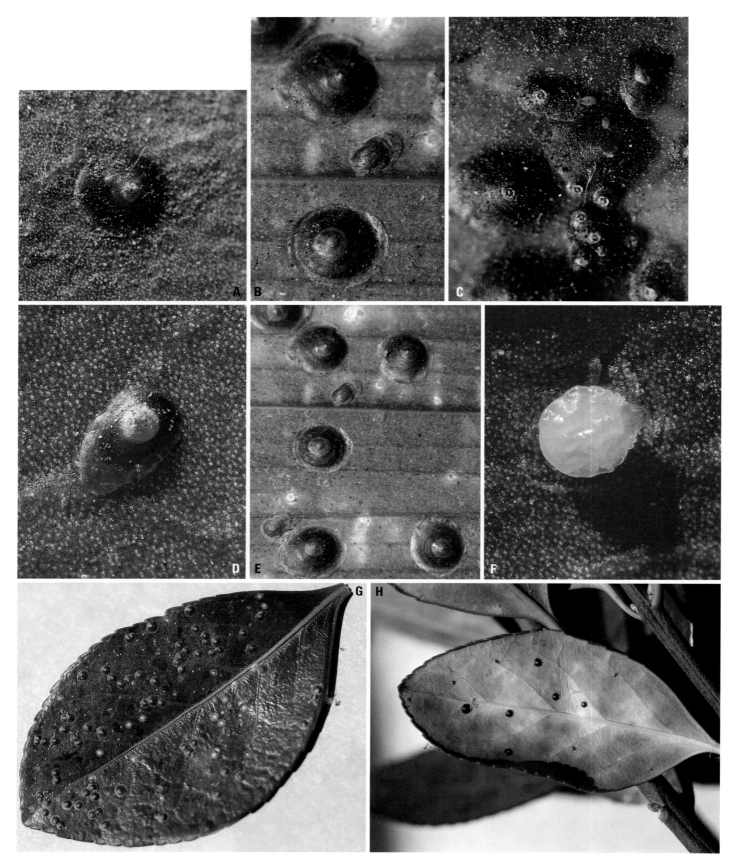

Plate 30. *Chrysomphalus bifasciculatus* Ferris, Bifasciculate scale

A. Young adult female cover on *Euonymus* (J. A. Davidson).
B. Adult female and male covers on lily (R. J. Gill).
C. Male and female covers and two crawlers on *Euonymus* (J. A. Davidson).
D. Male cover on *Euonymus* (J. A. Davidson).
E. Adult female and male covers on lily (R. J. Gill).

F. Young adult female body on *Euonymus* (J. A. Davidson).
G. Heavy infestation on top surface of *Euonymus* leaf (J. A. Davidson).
H. Chlorotic damage visible on bottom surface of heavily infested *Euonymus* leaf (J. A. Davidson).

128

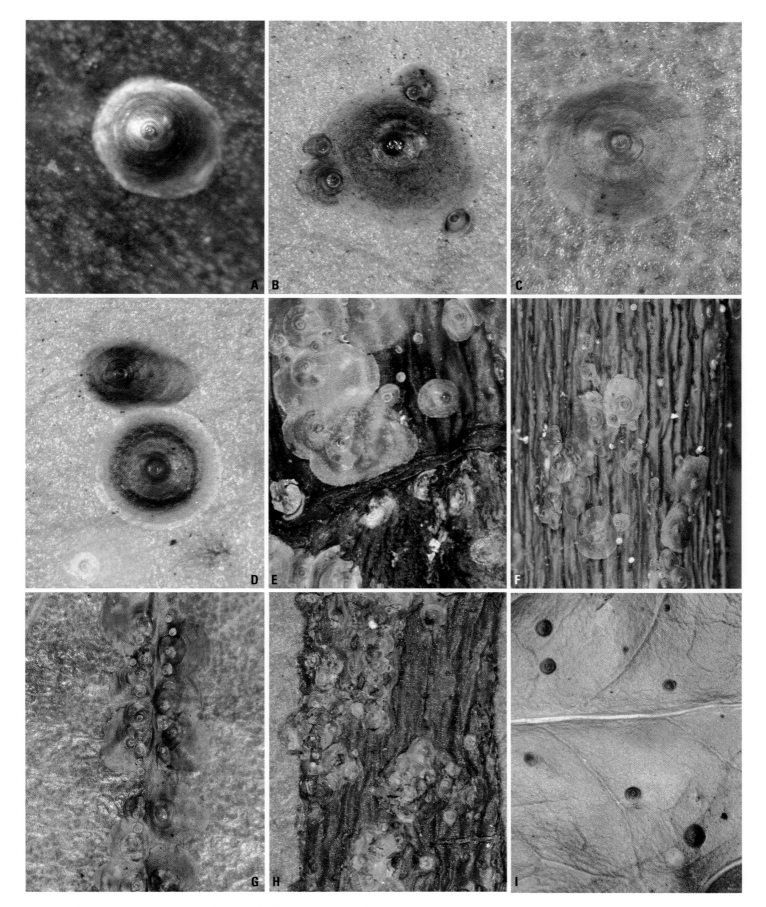

Plate 31. *Chrysomphalus dictyospermi* **(Morgan), Dictyospermum scale**

A. Adult female cover on dracaena (R. J. Gill).
B. Adult female cover with immature covers on holly leaf (J. A. Davidson).
C. Adult female cover on cinnamon leaf (J. A. Davidson).
D. Adult female (bottom) and male cover on holly leaf (J. A. Davidson).

E. Colony on rose (J. A. Davidson).
F. Colony on rose (J. A. Davidson).
G. Infestation on cinnamon (J. A. Davidson).
H. Distance view on *Persea* bark (J. A. Davidson).
I. Distance view of covers on holly (J. A. Davidson).

Chrysomphalus dictyospermi (Morgan)

ESA Approved Common Name Dictyospermum scale (also called Morgan's scale, Spanish red scale, red cochineal, la cochenille pou rouge).

Common Synonyms and Combinations *Aspidiotus dictyospermi* Morgan, *Chrysomphalus minor* Berlese, *Aspidiotus mangiferae* Cockerell, *Chrysomphalus mangiferae* (Cockerell), *Aspidiotus agrumincola* De Gregorio, *Aspidiotus dictyospermi* var. *jamaicensis* Cockerell, *Chrysomphalus jamaicensis* (Cockerell), *Aspidiotus dictyospermi* var. *arecae* Newstead, *Chrysomphalus arecae* (Newstead); *Chrysomphalus castigatus* Mamet, *Aspidiotus jamaicensis* (Cockerell).

Field Characters (Plate 31) Adult female cover circular, slightly convex, thin, reddish brown, copper, or gray; second shed skins central, yellow, first instar cover white when unrubbed. Male cover same color as female, elongate; shed skin submarginal. Body of female yellow; eggs yellow. Males may be present or absent. Normally on upper surface of leaves.

Slide-mounted Characters Adult female (Fig. 34) with 3 pairs of definite lobes, fourth lobes represented by low series of sclerotized points; paraphysis formula usually 2-2-0, rarely 2-2-1 or 2-3-0, when all paraphyses present with large paraphyses attached to lateral margin of median lobe, to medial margin of lobes 2 and 3, and in space between lobes 2 and 3, small paraphyses in space between lobes 3 and 4. Median lobes separated by space 0.4–0.8 (0.6) times width of median lobe, without basal sclerotization, with paraphysis attached to medial margin, without yoke, medial margins and lateral margins parallel or converging slightly, with 1–2 (1) lateral notch, 0–1 (0) medial notch; second lobes simple, equal to or slightly smaller than median lobes, with rounded apex, with 1 lateral notch, with 0–1 (0) medial notches; third lobes simple, slightly smaller than second lobes, with 1–4 (2) lateral notches, without medial notches; fourth lobes represented by series of 3 low swellings, with 17–25 (21) notches. Plates between lobes fimbriate, plates in third space each with microduct and protruding orifice near center, with conspicuous clavate process on plates between lobes 3 and 4; plate formula 2-3-3, rarely 2-2-3; plates between median lobes 0.5–0.8 (0.7) times as long as median lobes. Macroducts of 2 sizes, larger size marginal between median lobes and in spaces between median lobe and lobe 2 and between lobe 2 and 3, smaller size elsewhere on pygidium, duct between median lobes extending 2.5–3.4 (3.0) times distance between posterior apex of anal opening and base of median lobes, 75–109 (91) μ long, large macroduct in first space 88–133 (103) μ long, with 17–26 (20) macroducts on each side of pygidium on segments 5 to 8, ducts in submarginal and marginal areas, total of 22–34 (27) macroducts on each side of body, some macroduct orifices anterior of anal opening; prepygidial macroducts not forming cluster on margin of segments 2 or 3, small ducts present along body margin from mesothorax to segment 3. Pygidial microducts on venter in submarginal and marginal areas of segment 5, with 2–3 (3) ducts; prepygidial ducts of 1 size, in submedial areas of abdomen and thorax, in submarginal and marginal areas of segment 4 and sometimes segment 3; dorsal pygidial ducts absent; prepygidial microducts of 2 sizes, larger size in submedial areas of any or all of pro-, meso-, or metathorax to segment 3, smaller size in submarginal areas of head to segment 1, 2, or 3. Perivulvar pores arranged in 4 groups, 5–6 (6) pores on each side of body. Pores absent near spiracles. Anal opening located 1.8–3.9 (2.6) times length of anal opening from base of median lobes, anal opening 10–15 (12) μ long. Dorsal seta laterad of median lobes 0.6–0.8 (0.7) times length of median lobe. Eyes variable, flat sclerotized area, dome, or small spur, on mesotho-

rax. Antennae each with 1 seta. Cicatrices often present on prothorax, segments 1 and 3. Body pear shaped.

Affinities This species is easily recognizable among the economic species of *Chrysomphalus* because it lacks clusters of macroducts on the prepygidium, has conspicuously clavate processes on the plates between lobes 3 and 4, and has small dorsal microducts present on the submargin from the head to the anterior abdominal segments. *Chrysomphalus aonidum* has only 1 cluster of prepygidial macroducts on segment 2, lacks clavate processes on the plates between lobes 3 and 4, and has small dorsal microducts on the submargin from the mesothorax to the anterior abdominal segments. *Chrysomphalus bifasciculatus* has 2 clusters of prepygidial macroducts on segments 2 and 3, has weakly clavate processes on the plates between lobes 3 and 4, and lacks small dorsal microducts completely.

Hosts Polyphagous. Borchsenius (1966) reports it from 187 genera in 71 families; Dekle (1977) reports it from 150 plant genera in Florida. It often is found on avocado, citrus, orchids, and palms. We have examined specimens from the following host genera: *Acacia, Acorus, Agave, Aleurites, Aloe, Alpinia, Alyxia, Amherstia, Ananas, Anthurium, Araucaria, Area, Artocarpus, Asparagus, Aspidistra, Attalea, Aucuba, Baccharis, Barringtonia, Bassia, Bauhinia, Bougainvillea, Broughtonia, Buxus, Calathea, Camellia, Carissa, Caryota, Cassia, Castanea, Castanopsis, Cattleya, Cephalotaxus, Ceratonia, Chaetachme, Chamaedorea, Chlorophytum, Chrysalidocarpus, Cinnamomum, Citrus, Cocos, Coelogyne, Cordia, Croton, Cryptocoryne, Cryptomeria, Cycas, Cymbidium, Cypripedium, Davidsonia, Dendrobium, Dictyosperma, Dillenia, Diospyros, Dracaena, Epidendrum, Eriobotrya, Eugenia, Euphorbia, Euphoria, Euonymus, Fatsia, Ficus, Fortunella, Garcinia, Gardenia, Genista, Hedera, Heliconia, Hevea, Ilex, Illigera, Isochilus, Ixora, Jasminum, Jessenia, Juniperus, Kentia, Laelia, Lagerstroemia, Lantana, Laurus, Lecythis, Ligustrum, Litchi, Livistona, Lycaste, Macadamia, Magnolia, Malus, Mangifera, Maranta, Melaleuca, Microcitrus, Mimosa, Mimusops, Monodora, Monstera, Morus, Musa, Nephelium, Nerium, Odontoglossum, Olea, Oncidium, Ophiopogon, Pachycereus, Pandanus, Passiflora, Peristeria, Persea, Philodendron, Phoenix, Pimenta, Pinus, Piper, Platanus, Podocarpus, Polyandrocococos, Pothos, Pritchardia, Prunus, Psidium, Pyracantha, Pyrus, Quercus, Retinospora, Rhynchostylis, Rosa, Roystonea, Ruscus, Sabal, Salacca, Sansevieria, Sarcococca, Sarothamnus, Scindapsus, Serenoa, Smilax, Sonneratia, Spondias, Strelitzia, Swietenia, Synechanthus, Taxus, Tecoma, Thuja, Tillandsia, Trichopilia, Vanda, Vandopsis, Vellozia, Viburnum, Viola, Vitis,* and *Yucca.*

Distribution We have examined U.S. material from: Out-of-doors—AL, AZ, CA, FL, GA, HI, LA, MS, NM, SC, TX; Greenhouse—CO, CT, DC, IL, IN, KS, KY, MA, MD, MI, MN, MO, NC, NH, NJ, OH, PA, RI, TN, UT, VA, WV. It has been reported from greenhouses in NY. This species occurs out-of-doors in most tropical areas of the world. It most often is intercepted in quarantine from Mexico, Central and South America. A distribution map of this species was published by CAB International (1951).

Biology The dictyospermum scale may have 5 or 6 generations per year in the greenhouse (Quayle 1938), 3 or 4 generations (Quayle 1938), or 5 or 6 generations (Johnson and Lyon 1976) per year out-of-doors in California. In Louisiana the generations are overlapping and continuous (Cressman 1933). At 72 °F development of the crawler takes about 16 days, the second instar 18 days, and development from the second molt

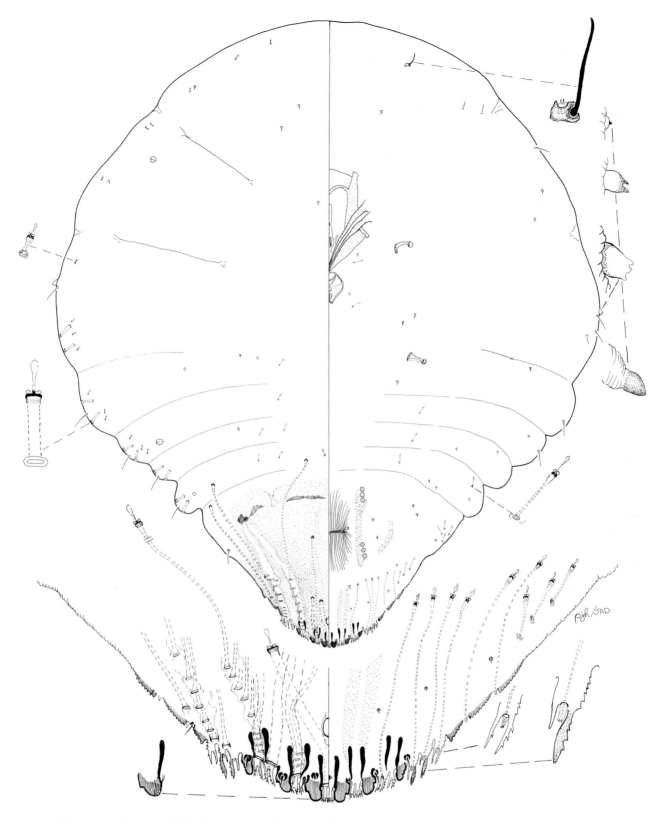

Figure 34. *Chrysomphalus dictyospermi* (Morgan), dictyospermum scale, Lutz, FL, on *Strelitzia reginae*, II-5-1970.

to the production of the first egg requires about 28 days. At 80 °F development of the same stages is 12 days, 13 days, and 22 days. Development time seems to be little affected by host differences (Cressman 1933). In the spring in California a generation may be completed in 7 weeks (Johnson and Lyon 1976). The upper surface of the leaves seems to be the preferred feeding area; although undersides of leaves, branches, and fruit may be infested also. The species has parthenogenetic forms and bisexual forms. Parthenogenetic populations are reported out-of-doors in Ventura County, California, southern Florida, and Mexico; and in greenhouses in Connecticut and Germany. Obligate sexual populations are known out-of-doors in New Orleans, Louisiana. Eggs may hatch from 1–24 hours after being laid or crawlers may be laid directly. In Egypt Salama (1970) found 2 generations per year, the first population peak was in April and the second was in October. In Italy Viggiani and Iannaccone (1973) indicated there were 3 generations per year; the first from mid-April to mid-July, the second from mid-July to mid-September, and the third from mid-September to mid-April.

Economic Importance This species is a serious pest of citrus, avocados, and an array of subtropical and tropical plants. Cressman (1933) considers it to be one of the most important scale pests in New Orleans. In Florida it is an important pest on avocado and may be serious on tangerines, oranges, and grapefruit (Thompson 1940). It is considered the most important citrus pest in the western Mediterranean basin (Ebeling 1959). Examples of papers implicating this species as a pest are as follows: citrus in Greece (Argyriou 1977); citrus in Morocco (Bénassy and Euverte 1967); citrus in Chile (González 1969); citrus in Tunisia (Sigwalt 1971); mango in Brazil (Fonseca 1963); palms and date palms in Iraq (Hussain 1974); coconut palms in the Seychelles (Vesey-FitzGerald 1953); avocados in Florida (Wolfenbarger 1951); mangos in Florida (Wolfenbarger 1955); lychee in Florida (Dekle 1954). The dictyospermum scale has been successfully controlled using biological control agents (Inserra 1970) in California and elsewhere, and *Aphytis melinus* is at least partially responsible for the success (DeBach 1969). In Greece this scale is effectively controlled by *Aphtis melinus* (Rosen and DeBach 1979). Beardsley and González (1975) consider this scale to be one of 43 serious armored scale pests. Miller and Davidson (1990) consider it to be a serious world pest.

Selected References Cressman (1933); Ebeling (1959); McKenzie (1939); Viggiani and Iannaccone (1973).

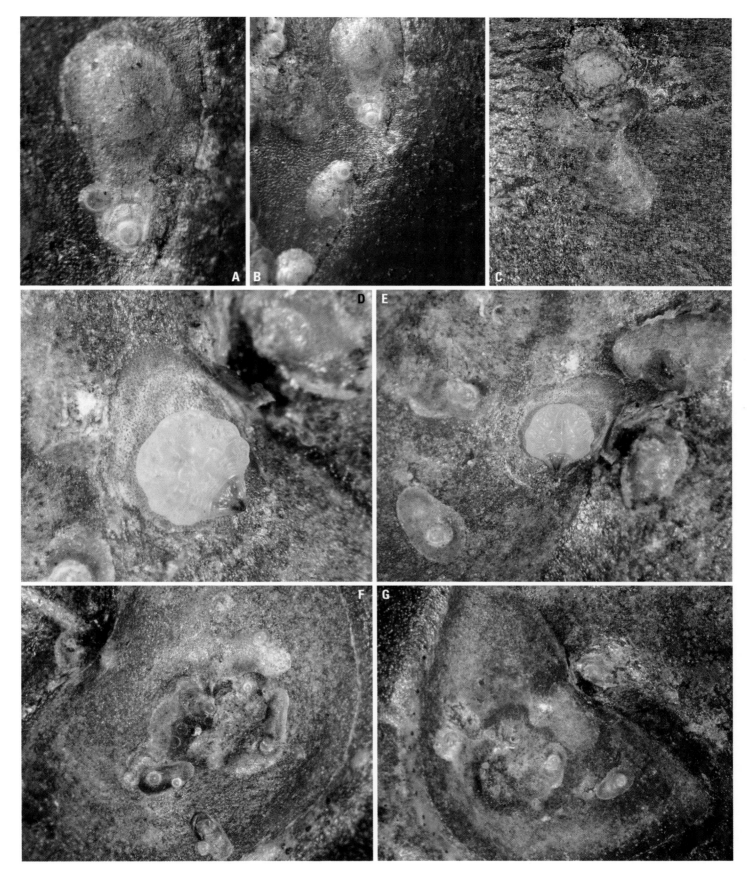

Plate 32. *Clavaspis herculeana* (Cockerell and Hadden), Herculeana scale

A. Adult female cover and immature covers on plumeria (R. J. Gill).
B. Female covers on plumeria (R. J. Gill).
C. Female cover on plumeria (J. A. Davidson).
D. Body of young adult female on plumeria (J. A. Davidson).
E. Body of adult female and male cover on plumeria (J. A. Davidson).

F. Distance view of male and cryptic female covers around plumeria leaf scars (J. A. Davidson).
G. Distance view of male and female covers on plumeria (J. A. Davidson).

133

Clavaspis herculeana

(Cockerell and Hadden)

Suggested Common Name Herculeana scale (also called herculean scale, escama del terminal).

Common Synonyms and Combinations *Aspidiotus herculeana* Cockerell and Hadden in Doane and Hadden, *Aspidiotus subsimilis v. anonae* Houser, *Clavaspis anonae* (Houser), *Aspidiotus symbioticus* Hempel, *Chrysomphalus alluaudi* Mamet, *Clavaspis alluadi* (Mamet), *Clavaspis symbioticus* (Hempel).

Field Characters (Plate 32) Adult female cover circular; thin, slightly convex; light brown; shed skins light to dark brown, subcentral, lighter than adult female cover. Male cover more elongate than female, slightly darker; shed skin marginal, caramel in color. Body of adult female clear yellow; eggs probably yellow (Mosquera 1976). Under bark of host, often inconspicuous. Also on fruit and leaves (Merrill 1953).

Slide-mounted Characters Adult female (Fig. 35) with 1 pair of definite lobes, occasionally with slight swellings in position of second and third lobes; paraphysis formula 2-1-0, 3-0-0 or 2-0-0, 3-2-0, or 2-2-0, when all paraphyses present with large paraphyses attached to medial margin of position of lobe 2 and space between lobe 2 and median lobe (this paraphysis is unusually large and mushroom shaped), small paraphyses attached to lateral margin of median lobes and position of lobe 3, attached to lateral margin of position of lobe 2, and in space between positions of lobes 2 and 3. Median lobes separated by space 0.1–0.2 (0.1) times width of median lobe, without basal sclerotization, with paraphysis attached to medial margin, without yoke, medial margins parallel or converging, lateral margins converging, with 1 lateral notch, 0–1 (0) medial notch. Plates between median lobes absent, plates either simple or with 1 or 2 small tines; plate formula 2-2-2, rarely 2-2-3, or 2-2-1. Macroducts of 2 sizes, larger size on pygidium between median lobes and on segments 5 to 7 in marginal and submarginal areas, duct between median lobes extending 2.2–3.4 (2.9) times distance between posterior apex of anal opening and base of median lobes, 59–82 (75) µ long, marginal macroduct in first space 70–88 (82) µ long, with 13–25 (17) macroducts on each side of pygidium on segments 5 to 8, ducts in submarginal and marginal areas, total of 22–36 (28) macroducts on each side of body, some macroduct orifices anterior of anal opening; smaller macroducts present along body margin from metathorax to segment 3. Pygidial microducts on venter in submarginal and marginal areas of segment 5, with 4–9 (7) ducts; prepygidial ducts of 1 size in submedial areas of abdomen and thorax, in submarginal and marginal areas of segment 3 and 4, rarely on segment 2; on dorsum pygidial ducts absent; prepygidial microducts of 2 sizes,

larger size in submedial areas of any or all of pro-, meso-, or metathorax to 4, smaller size in submarginal areas of pro- or mesothorax to segments 2 or 3. Perivulvar pores absent. Pores absent near spiracles. Anal opening located 1.8–2.7 (2.2) times length of anal opening from base of median lobes, anal opening 8–16 (12) µ long. Dorsal seta laterad of median lobes 0.8–1.4 (1.1) times length of median lobe. Eyes absent. Antennae each with 1 seta. Cicatrices rarely present on prothorax, segments 1 and 3. Body pear shaped.

Affinities The unusually large club at the apex of the longest paraphysis easily distinguishes herculeana scale from all other economic species of armored scales.

Hosts Borchsenius (1966) reports it from 14 families and 24 genera; Dekle (1977) reports it from 16 plant genera in Florida. We have examined specimens from the following hosts: *Afzelia, Annona, Artocarpus, Avicennia, Citrus, Brosimum, Caesalpinia, Calocarpum, Cassia, Clerodendron, Coccoloba, Cordia, Delonix, Elaphrium, Eriobotrya, Erythrina, Euphorbia, Genipa, Hedera, Hymenaea, Loranthus, Mammea, Mangifera, Manilkara, Morus, Phoradendron, Pithecellobium, Plumeria, Prosopis, Prunus, Psidium, Punica, Rosa, Spondias, Sterculia, Tabebuia,* and *Vitis.* It is most commonly intercepted on *Spondias* at U.S. ports-of-entry.

Distribution We have examined U.S. material from FL, HI, TX. We also have seen material from Caribbean islands, Central America, Japan, Mexico, Pacific Islands including the Philippines, and South America. This species has been reported from southern France, Tadzhikistan, and Malaya. According to Hiltabrand (1961) it was collected in a nursery in Los Angeles County, CA. Nakahara (1982) reports it from Mexico, the Caribbean islands, Central and South America, several countries in the Afrotropical Area, the Philippines and Singapore in Asia, Australia, Fiji, New Caledonia, and Tahiti.

Biology Little is known about the biology of this scale. It usually occurs under the bark of its host and often is associated with *Septobasidium* fungi.

Economic Importance The herculeana scale occurs on many important tropical and subtropical hosts and has considerable potential as a pest. It has been reported as a sporadic pest in Mauritius on *Spondias* and *Pyrus* (Moutia and Mamet 1947). Although it is not considered a pest, it is very abundant on pear in Colombia (Mosquera 1976). Brimblecombe (1961) indicates that it occasionally builds to noticeable infestations in isolated pockets on papaya in Australia. It is reported by Arnett (1985) and Miller and Davidson (1990) to be an occasional pest.

Selected References Balachowsky (1956); Merrill (1953).

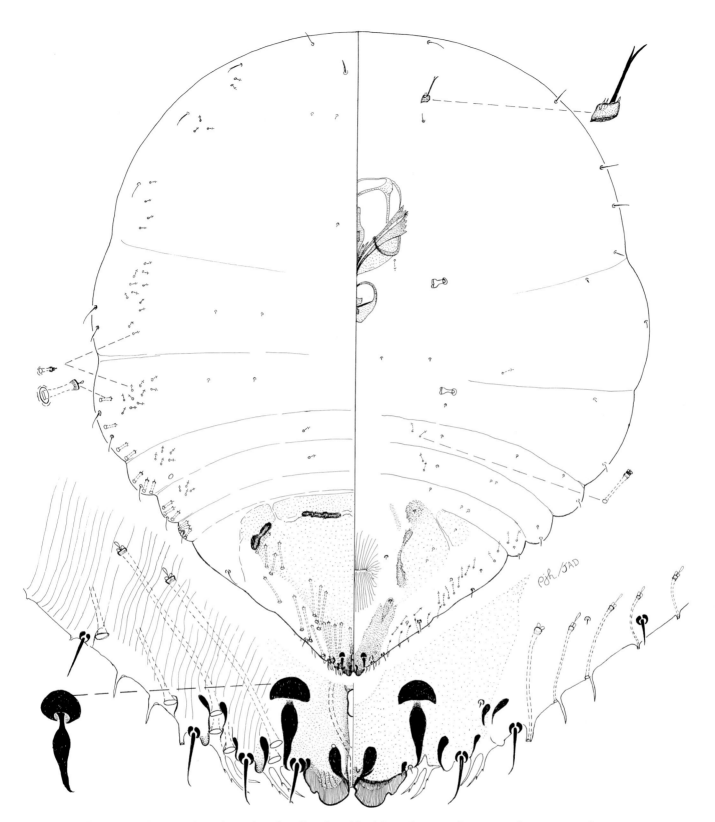

Figure 35. *Clavaspis herculeana* (Cockerell and Hadden), herculeana scale, Venezuela, intercepted at San Juan, PR, on tree bark, VI-2-1971.

Clavaspis ulmi (Johnson)

Suggested Common Name Elm armored scale (also called elm clavaspis scale; corky bark aspidiotus).

Common Synonyms and Combinations *Aspidiotus ulmi* Johnson, *Aonidiella ulmi* (Johnson), *Hendaspidiotus ulmi* (Johnson).

Field Characters (Plate 33) Adult female cover convex, circular, dirty white or tan, inside of cover white; shed skins central or subcentral, bright orange yellow. Male cover same as female except more elongate. Body of female and eggs yellow; adult male yellow. Normally deep in cracks on trunks.

Slide-mounted Characters Adult female (Fig. 36) with 1 pair of definite lobes, second lobes usually present but small, represented by sclerotized protrusion, third lobe normally absent, rarely similar in appearance to second lobes; paraphysis formula 2-1-0, or 2-2-0, when all paraphyses present with large paraphyses in space between lobe 2 and median lobe (this paraphysis is unusually large and clubbed shaped), small paraphyses attached to lateral margin of median lobes, medial margin of lobe 2, medial margin of lobe 3, and in space between lobe 2 and 3. Median lobes separated by space 0.2–0.3 (0.2) times width of median lobe, often with basal sclerotization, without yoke, medial margins parallel, lateral margins usually converging slightly, with 1 lateral notch and 1 medial notch. Plates between median lobes absent, plates either simple or with 1 or 2 small tines; plate formula 2-2-5, rarely 2-3-5, 2-3-3, or 2-3-4; without plates anterior of seta marking segment 5. Macroducts of 2 sizes, larger size on pygidium between median lobes and on segments 5 to 7, in marginal and submarginal areas, duct between median lobes extending 0.9–1.4 (1.1) times distance between posterior apex of anal opening and base of median lobes, 54–67 (75) μ long, of 10 specimens examined, 1 specimen lacked a duct between median lobes, marginal macroduct in first space 60–92 (70) μ long, with 25–34 (30) macroducts on each side of pygidium on segments 5 to 8, ducts in submarginal and marginal areas, total of 30–39 (35) macroducts on each side of body, some macroduct orifices anterior of anal opening; smaller macroducts present along body margin from mesothorax, metathorax, or segment 1 to segment 2 or 3. Pygidial microducts on venter in submarginal and marginal areas of segments 5 and 6, with 9–18 (13) ducts; prepygidial ducts of 1 size in submedial areas of abdomen and thorax, in submarginal and marginal areas of segments 3 and 4, sometimes on segment 2; on dorsum pygidial ducts absent; prepygidial microducts of 2 sizes, larger size in submedial areas of any or all of head, pro-, meso-, or metathorax to segment 3, smaller size in submarginal areas of head, pro-, or mesothorax to segment 2. Perivulvar pores absent from 7 of 10 specimens examined, when present in 2 clusters each composed of 1–3 (2) pores. Pores absent near spiracles. Anal opening located 3.0–4.7 (4.0) times length of anal opening from base of median lobes, anal opening 12–17 (15) μ long. Dorsal seta laterad of median lobes 0.8–1.2 (1.0) times length of median lobe. Eyes absent. Antennae each with 1 seta. Cicatrices rarely present on segment 1. Body pear shaped.

Affinities This species is similar to *Clavaspis texana* Ferris but differs by having 3 distinct pairs of macroduct furrows; these are absent on *C. texana*.

Hosts Borchsenius (1966) reports it from 9 host families and 10 genera. We have examined specimens from the following: *Acer, Aesculus, Catalpa, Cornus, Digitalis, Forestiera, Ginkgo, Lonicera, Magnolia, Malus, Ribes, Tilia,* and *Ulmus. Juglans, Robinia,* and *Euonymus* also are cited as hosts in the literature. Elm appears to be the preferred host.

Distribution This species probably is native to North America. We have examined U.S. material from CA, CO, CT, DC, GA, IL, KS, MD, MO, MS, NJ, NY, OH, SC, VA, WI, WV. We also have seen specimens from Canada. It is reported in the literature from IN, LA, PA, and TX.

Biology Little is known about the biology of this scale. In Ohio it overwinters as second instars, and crawlers are present in early to mid-May (Kosztarab 1963). Infestations normally occur on the trunk of the host.

Economic Importance This species is found on many ornamental and forest plants and is therefore considered to be a potentially economic species. Miller and Davidson (1990) consider it to be an occasional pest.

Selected References Ferris (1938a); Kosztarab (1963).

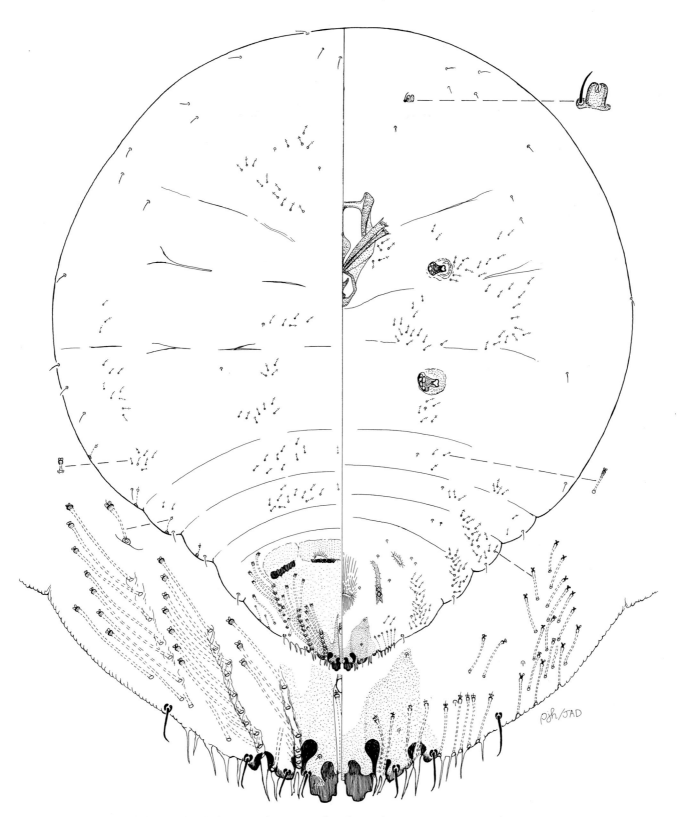

Figure 36. *Clavaspis ulmi* (Johnson), elm armored scale, Burlingame, CA, on *Catalpa* sp., X-19-1967.

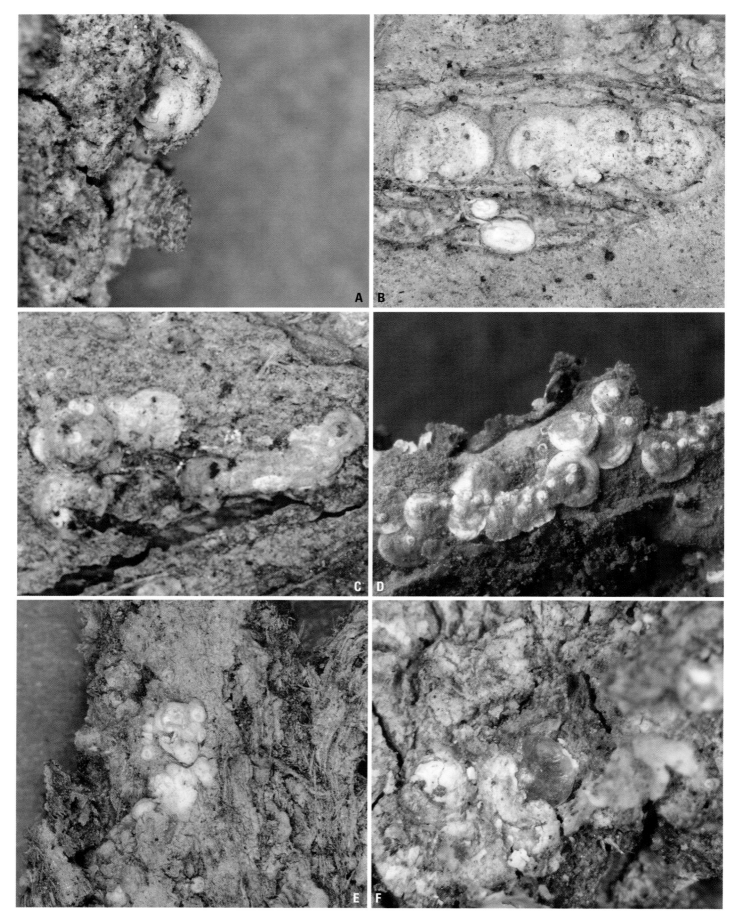

Plate 33. *Clavaspis ulmi* (Johnson), Elm armored scale

A. Side view of adult female cover on elm (J. A. Davidson).
B. Female covers under elm bark (J. A. Davidson).
C. Female covers on elm bark (J. A. Davidson).

D. Covers under catalpa bark (R. J. Gill).
E. Distance view of covers in cracks of elm bark (J. A. Davidson).
F. Body of old adult female on elm (J. A. Davidson).

Plate 34. *Comstockiella sabalis* (Comstock), Palmetto scale

A. Adult female cover on palm (D. R. Miller).
B. Adult female and two male covers on palm (R. J. Gill).
C. Covers removed to show adult females and eggs and four parasitized females on palm (J. A. Davidson).
D. Body of second-instar male on palm (J. A. Davidson).
E. Wingless adult male on palm (J. A. Davidson).
F. Wingless adult male mounted on microscope slide (J. A. Davidson).

G. Distance view of infestation composed mainly of small male covers on palm (J. A. Davidson).
H. Heavy infestation on palm (J. A. Davidson).
I. Damage viewed on top side of palm frond resulting from infestation on underside (D. R. Miller).

Comstockiella sabalis (Comstock)

Suggested Common Name Palmetto scale.

Common Synonyms and Combinations *Aspidiotus sabalis* Comstock, *Aspidiotus sabalis* v. *mexicana* Cockerell.

Field Characters (Plate 34) Adult female cover irregularly circular; white; slightly convex; without shed skins, sometimes incorporated in cover, but covered by wax, yellow. Male cover same color as female cover; smaller, elongate; shed skin incorporated in wax. Body of adult female light pink when newly molted, brownish red when more mature; eggs light pink or cream; adult male light pink or cream, wingless. Usually on leaves, occasionally on bark and fruit.

Slide-mounted Characters Adult female (Fig. 37) with 2 pair of lobes, series of sclerotized protrusions surrounding seta marking segment 6; paraphyses absent. Median lobes separated by space 0.1–0.2 (0.2) times width of median lobe, without basal sclerotization or yoke, medial margins and lateral margins usually slightly converging apically or parallel, composed of 3–4 (3) distinct protrusions and 2–3 (2) lateral notches, without medial notches; second lobes composed of 2–4 (3) protrusions, conspicuously smaller than median lobes, with 1–3 (2) lateral notches, without medial notches. Plates absent, possible plates between median lobes. Tubular ducts of 1 size over body, long and narrow, in position of macroducts and microducts in other species, with single bar and 8-shaped sclerotized apex; without duct between median lobes, present on pygidium in 2 grooves from median lobes forward to segment 5, in marginal, submarginal, and submedial areas, with 38–149 (93) ducts on each side of pygidium on segments 5 to 8, some duct orifices anterior of anal opening. Other pygidial ducts on venter in submarginal and marginal areas of segment 5 or 6 to 7 or 8, with 6–21 (11) ducts; prepygidial ducts in submarginal area of abdomen and near spiracles, also present near posterior end of mouthparts, ducts near anterior mouthparts sometimes shorter than on other parts of body; on dorsum, prepygidial ducts in submedial clusters on metathorax or segment 1 to segment 4. Perivulvar pores in 6–8 (6) clusters composed of 12–37 (24) pores on each side of body. Perispiracular pores primarily with 5 loculi, anterior spiracles with 1–10 (4) pores, posterior spiracles without pores. Anal opening located 4.1–7.0 (6.0) times length of anal opening from base of median lobes, anal opening 11–21 (14) μ long. Dorsal seta laterad of median lobes 0.7–1.3 (0.9) times length of median lobe. Eyes present on posterior margin of head, represented by flat sclerotized area, often obscured by sclerotization of anterior margin of head. Antennae each with 3–4 (4) clubbed seta. Cicatrices absent. Body pear shaped. Marginal setae unusually large. Spiracular plate unusually large.

Affinities The palmetto scale is unique because it has unusually shaped lobes, lacks plates, and has 6 or more groups of perivulvar pores. This is a very unusual armored scale in that it does not have ducts differentiated into macro- and microducts, the ducts are 8-shaped, the eyes are on the head, the lobes are crude expansions of the posterior part of the abdomen, there is more than 1 seta on the antenna, there are at least 6 clusters of perivulvar pores, the shed skins are not incorporated into the scale cover in a regular manner, and the perispiracular pores have 5 loculi.

Hosts Borchsenius (1966) reports it on 1 plant family and 2 genera; Dekle (1977) reports it from 17 plant genera in Florida. We have examined specimens from the following hosts: *Brahea*, *Citrus*, *Erythea*, *Sabal*, *Phoenix*, and *Washingtonia*. Other hosts reported in the literature include *Acoelorrhaphe*, *Agave*, *Arecastrum*, *Butia*, *Chamaerops*, *Cocos*, *Hyphaene*, *Lantana*, *Livistona*, *Rhapidophyllum*, *Roystonea*, *Serenoa*, and *Thrinax*. This species is most commonly collected on cabbage palm and saw palmetto.

Distribution The palmetto scale is native to the New World. It is commonly found in the Caribbean and Gulf of Mexico area. We have examined U.S. material from out-of-doors in CA, FL, GA, LA, NC, SC, TX, and from indoors in NJ and OH. It has been reported from MS.

Biology Little is known about the biology of this scale. It usually occurs on the leaves of its hosts but is sometimes found on the trunk or fruit. It often is associated with a fungus. It does not incorporate the shed skins into the scale cover in the same way as other armored scales; in fact, many specimens do not include it into the cover at all, others seem to incorporate them accidentally. The adult male is apterous (Howell 1979).

Economic Importance This species often causes yellow leaf splotches where it feeds on the palm host. In Bermuda it was a serious pest of *Sabal bermudiana*, the only native palm on Bermuda. After the introduction of several natural enemies, the palmetto scale has been kept under satisfactory control (Bennett and Hughes 1959). Evans and Pedata (1997) suggest that the most important natural enemy in Bermuda very well could be *Coccobius donatellae*, a parasitic wasp. Miller and Davidson (1990) consider this species to be an occasional pest.

Selected References Bennett and Hughes (1959); Ferris (1938a); Howell (1979); Howell and Tippins (1977).

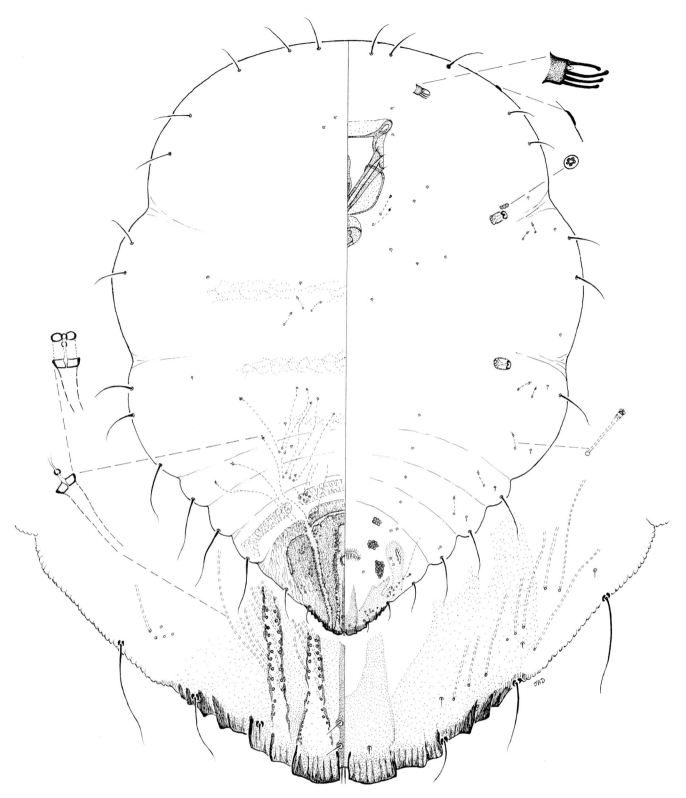

Figure 37. *Comstockiella sabalis* (Comstock), palmetto scale, Wilmington, NC, on palm, XI-25-1943.

Cupressaspis shastae (Coleman)

Suggested Common Name Redwood scale (also called red cedar scale, Utah cedar scale).

Common Synonyms and Combinations *Aspidiotus coniferarum* var. *shastae* Coleman, *Aonidia juniperi* Marlatt, *Aspidiotus shastae* Coleman, *Gonaspidiotus shastae* (Coleman), *Targionia juniperi* (Marlatt), *Cryptaspidiotus shastae* (Coleman), *Aonidia shastae* (Coleman).

Field Characters (Plate 35) Adult female cover convex, circular, gray or brown; shed skins central, yellow. Male cover oval, similar in color and texture to female cover except darker; shed skin submarginal. Females pupillarial, that is, enclosed in shed skin of second instar and not producing additional wax on cover. On foliage, appearing as hardened pitch on leaves.

Slide-mounted Characters Adult female (Fig. 38) with 2 pairs of well-developed lobes, third pair sometimes represented by small hyaline projection; paraphyses absent. Median lobes separated by space 0.4–1.2 (0.7) times width of median lobe, without paraphyses or associated sclerotizations, without notches; second lobes simple, one-half to one-quarter size of median lobes, usually pointed and sclerotized, sometimes represented by hyaline point, without notches; third lobes usually absent. Plates difficult to see, variable, in some specimens small, apical fimbriations present on broad plates; in others, plates represented by cone-shape structures without fimbriations, without associated microducts; plate formula 2-1-0 or rarely 2-2-0; with 2 broad or thin, apically fringed or acute plates between median lobes, equal to or shorter than median lobes. Macroducts of 1 size, 1 (rarely 2) macroduct between median lobes, extending 0.6–1.2 (0.9) times distance between posterior apex of anal opening and base of median lobes, 25–37 (30) µ long, longest macroduct in first space 28–39 (30) µ long, with 7–13 (11) on each side of pygidium on segments 5 to 8, ducts in submarginal and marginal areas, total macroducts 7–18 (11) on each side of body, some macroduct orifices anterior of anal opening; prepygidial macroducts present on 2 of 10 specimens examined, restricted to submarginal area of segment 4 on 1 specimen, on submarginal area of segments 3 to 4 on other specimen. Pygidial microducts on venter in submarginal areas of segment 5, rarely 6, with 1–3 (2) ducts; prepygidial ducts of 1 size, submarginal ducts absent, submedial ducts usually near anterior spiracle, sometimes also present near posterior spiracle and on any or all of segments 1 to 4; pygidial microducts absent from dorsum, usually of 1 size on prepygidial areas, rarely in submarginal areas of prothorax to segment 1 or 2, usually present in submedial areas of 1 or more of segments 1 to 4. Perivulvar and perispiracular pores absent. Anal opening located 2.2–3.7 (2.9) times length of anal opening from base of median lobes, anal opening 10–15 (12) µ long. Dorsal seta laterad of median lobes 0.7–1.6 (1.0) times length of median lobe. Eyes usually represented by small spur or dome, sometimes absent. Antennae each with 1 seta. Cicatrices normally absent but sometimes present on prothorax and segment 1. Body pear shaped, thorax and head without sclerotization in fully mature adult females.

Affinities The redwood scale is most similar to *Cupressaspis atlantica* (Ferris) because both are pupillarial, occur on juniper, and have the same general body appearance. The redwood scale differs by having more than 4 macrotubular ducts on each side of the pygidium and by having the anal opening about half the distance from the vulva to the posterior apex of the pygidium. *Cupressaspis atlantica* has 4 macroducts on each side of the pygidium and the anal opening is located directly above the vulva.

Hosts This species is known from *Cupressus, Juniperus, Libocedrus, Sequoia,* and *Torreya* (Gill 1997). It is most commonly found on juniper. It is a native species and has not been taken in quarantine.

Biology Biological information is unavailable for this species except that it is ovoviviparous and is at least partially pupillarial (Gill 1997).

Distribution We have examined U.S. specimens from AZ, CA, KS, MO, NJ, OK, UT. It has been reported in TX (Nakahara 1982).

Economic Importance Michener et al. (1957) consider this species to be a serious pest of red cedar in certain areas of Kansas in the 1950s and state that it had been a pest in Kansas previously in the 1930s. Others indicate that although the species builds to high numbers on junipers and redwoods, its economic impact is inconsequential (Gill 1997; Furniss and Carolin 1977).

Selected References Gill (1997); McKenzie (1956); Michener et al. (1957).

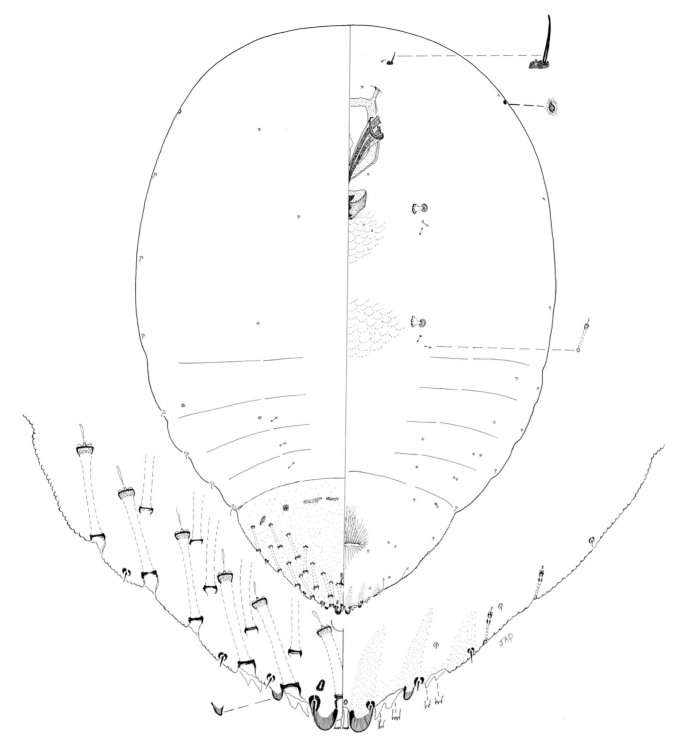

Figure 38. *Cupressaspis shastae* (Coleman), redwood scale, Kansas City, MO, on *Juniperus virginiana*, IV-4-1938.

Plate 35. *Cupressaspis shastae* (Coleman), Redwood scale

A. Adult female cover (upper left) on incense cedar (R. J. Gill).
B. Adult female cover (top) on nutmeg (R. J. Gill).
C. Adult male cover on nutmeg (R. J. Gill).
D. Distance view of female covers on incense cedar (R. J. Gill).
E. Distance view of covers on nutmeg (R. J. Gill).

F. Infestation on juniper berries (R. J. Gill).
G. Body of pupillarial female on nutmeg (R. J. Gill).
H. Distance view of exposed pupillarial females among scale covers on nutmeg (R. J. Gill).
I. Body of young female on juniper berry (J. A. Davidson).

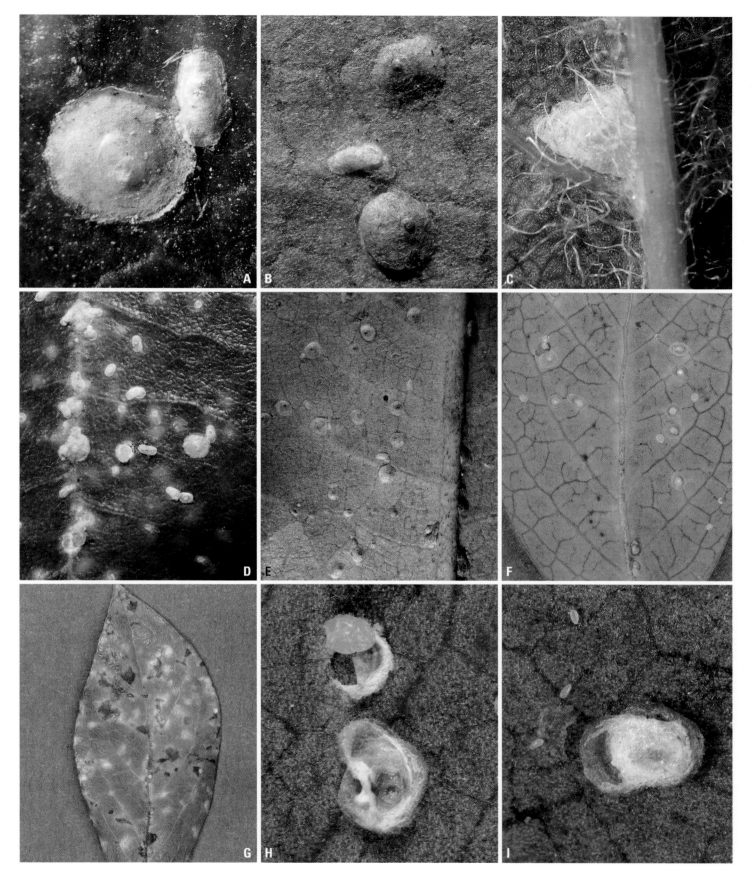

Plate 36. *Diaspidiotus ancylus* (Putnam), Putnam scale (leaf form)

A. Adult female cover and male cover on upper surface of rhododendron leaf (J. A. Davidson).
B. Two adult female and one male cover on lower surface of rhododendron leaf (J. A. Davidson).
C. Female cover on lower surface of maple leaf (J. A. Davidson).
D. Distance view of covers on upper surface of rhododendron leaf (J. A. Davidson).
E. Distance view of covers on lower surface of rhododendron leaf (J. A. Davidson).

F. Distance view of covers on lower surface of blueberry leaf (J. A. Davidson).
G. Distance view of damage on upper surface of rhododendron leaf in late summer (J. A. Davidson).
H. Body of adult female and necrotic feeding damage on rhododendron (J. A. Davidson).
I. Ventral cover of adult female and three eggs (J. A. Davidson).

Diaspidiotus ancylus (Putnam)

Suggested Common Name Putnam scale (also called maple leaf aspidiotus, Howard scale, rhododendron scale).

Common Synonyms and Combinations *Diaspis ancylus* Putnam, *Aspidiotus ancylus* (Putnam), *Aspidiotus convexus* Comstock, *Aspidiotus circularis* Fitch, *Aspidiotus ancylus* var. *serratus* Newell and Cockerell, *Aspidiotus* (*Evaspidiotus*) *convexus* Comstock, *Aspidiotus aesculi solus* Hunter, *Aspidiotus ancylus* var. *latilobis* Newell, *Aspidiotus ancylus* var. *ornatus* Leonardi, *Aspidiotus ohioensis* York, *Aspidiotus oxycrataegi* Hollinger, *Aspidiotus howardi* Cockerell, *Aspidiotus comstocki* Johnson, *Aspidiotus townsendi* Cockerell.

Field Characters (Plates 36 and 37) Adult female cover on bark dark gray (Plate 37), on fruit and leaves (Plate 36) white, convex, circular; shed skins subcentral, yellow or red. Considerable variation in scale cover depending on host, for example, on maple leaves cover thin and delicate, on rhododendron thick and robust. Male cover similar to female cover except oval with submarginal shed skin. Body of adult female yellow, anterior portion of old females brown; eggs and crawlers yellow. On bark, leaves, and fruit.

Slide-mounted Characters Putnam scale is extremely variable and probably is a complex of several species. We are here treating two commonly collected forms that predominate on stems and leaves respectively.

Ancylus (bark) form—Adult female (Fig. 39) with 1 pair of definite lobes, second lobes when present represented by small sclerotized protrusion; paraphysis formula 2-2-0, with large paraphyses in space between lobe 2 and median lobe, attached to medial margin of position of lobe 2, medial margin of position of lobe 3, and in space between positions of lobe 2 and 3. Median lobes separated by space 0.1–0.3 (0.2) times width of median lobe, with paraphysis attached to medial margin, without basal sclerotization or yoke, medial margins usually converging, sometimes parallel, lateral margins usually converging slightly, with 1–2 (1) lateral notches and without medial notch. Plates simple or with inconspicuous lateral tines; plate formula 2-2-2, 2-2-3, or 2-2-4; without plates anterior of seta marking segment 5; plates between median lobes inconspicuous about one-half length of median lobes. Macroducts of 2 sizes, larger size on pygidium between median lobes and on segments 5 to 7, in marginal and submarginal areas, duct between median lobes extending 1.2–1.5 (1.3) times distance between posterior apex of anal opening and base of median lobes, 44–49 (47) μ long, marginal duct in first space 38–60 (46) μ long, with 18–47 (31) macroducts on each side of pygidium on segments 5 to 8, ducts in submarginal and marginal areas, total of 27–58 (35) macroducts on each side of body, some macroduct orifices anterior of anal opening; smaller macroducts present along body margin from metathorax to segment 1, 2, or 3. Pygidial microducts on venter in submarginal and marginal areas of segment 5, with 2–4 (3) ducts; prepygidial ducts of 1 size in submedial areas of abdomen and thorax, in submarginal and marginal areas of segments 3 and 4; on dorsum pygidial ducts absent; prepygidial microducts of 2 sizes, larger size in submedial areas of any or all of metathorax to segment 3, smaller size in submarginal areas of head, pro-, or mesothorax to segments 2 or 3. Perivulvar pores in 4–5 (4) indefinite clusters, composed of 9–22 (15) pores on each side of body. Pores absent near spiracles. Anal opening located 1.8–2.5 (2.2) times length of anal opening from base of median lobes, anal opening 12–20 (15) μ long. Dorsal seta laterad of median lobes 1.1–1.2 (1.1) times length of median lobe. Eyes usually represented by 1 or more sclerotized points. Antennae each with 1 seta. Cica-

trices usually present on segment 1. Body pear shaped, thorax and head sometimes sclerotized.

Comstocki-howardi (leaf) form—This form is most easily recognized by having conspicuous second lobes. Adult female (Fig. 40) with 2 pairs of definite lobes, second lobes vary from smaller than median lobes to much larger, third lobes when present represented by small sclerotized protrusion; paraphysis formula usually 2-2-0, rarely 2-1-0, with large paraphyses in space between lobe 2 and median lobe, attached to medial margin of position of lobe 2, medial margin of position of lobe 3, and in space between positions of lobe 2 and 3. Median lobes separated by space 0.2–0.3 (0.2) times width of median lobe, with paraphysis attached to medial margin, without basal sclerotization or yoke, medial and lateral margins parallel or converging, with 1 lateral notch and 0–1 (0) medial notch; second lobes simple, equal to or conspicuously larger than median lobes, with rounded apex, with 0–1 (1) lateral notch, with 0–1 (0) medial notches; third lobes, when present, represented by small weakly sclerotized point, without notches. Plates conspicuously fimbriate, especially between median lobe and lobe 2; plate formula 2-2-3, 2-2-4, 2-3-4, or 2-3-5; plates between median lobes 0.5–0.8 (0.6) times as long as median lobes. Macroducts of 2 sizes, larger size on pygidium between median lobes and on segments 5 to 7, in marginal and submarginal areas, duct between median lobes extending 1.2–1.5 (1.4) times distance between posterior apex of anal opening and base of median lobes, 30–44 (37) μ long, marginal duct in first space 38–50 (41) μ long, with 12–22 (17) macroducts on each side of pygidium on segments 5 to 8, ducts in submarginal and marginal areas, total of 15–32 (20) macroducts on each side of body, some macroduct orifices anterior of anal opening; smaller macroducts present along body margin from metathorax to segment 2 or 3. Pygidial microducts on venter in submarginal and marginal areas of segment 5, with 1–4 (2) ducts; prepygidial ducts of 1 size in submedial areas of abdomen and thorax, in submarginal and marginal areas of segments 3 and 4, usually with only 1 or 2 ducts in this area; on dorsum pygidial ducts absent; prepygidial microducts of 2 sizes, in very small numbers over surface, larger size in submedial areas of any or all of metathorax to segment 3, smaller size in submarginal areas of head, pro-, or mesothorax to segments 2 or 3. Perivulvar pores in 4 indefinite clusters, composed of 8–10 (9) pores on each side of body. Pores absent near spiracles. Anal opening located 2.0–2.2 (2.1) times length of anal opening from base of median lobes, anal opening 10–19 (13) μ long. Dorsal seta laterad of median lobes 0.9–1.3 (1.1) times length of median lobe. Eyes usually represented by 1 or more sclerotized points. Antennae each with 1 seta. Cicatrices usually present on segment 1. Body pear shaped, thorax and head sometimes sclerotized.

The 2 forms are not always distinct because many specimens have intermediate characteristics. In general the stem and bark form (ancylus) has: 1 pair of lobes, numerous microducts, especially on the dorsal submargin of the thorax, plates with only a few short tines, and more macrotubular ducts and perivulvar pores. The leaf form (comstocki-howardi) has: 2 pair of lobes (the second lobes may be larger and protrude further than the median lobes), few microducts, plates fimbriate, with many long tines, few macrotubular ducts and perivulvar pores.

Affinities The Putnam scale is similar to *Diaspidiotus osborni*, Osborn scale. For a comparison of these species see affinities section of the Osborn scale. The grape scale, *D. uvae*, also is similar to Putnam scale but differs by lacking the macroduct

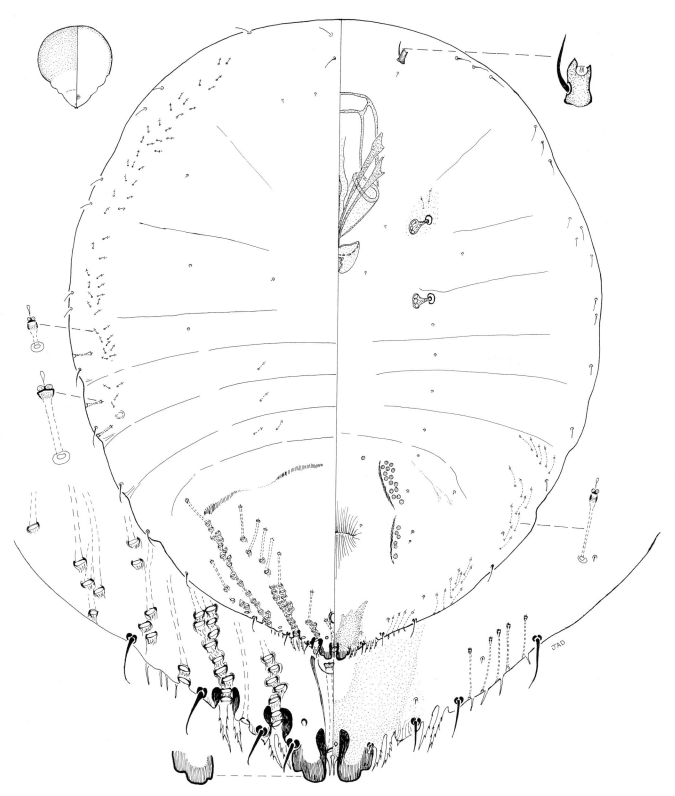

Figure 39. *Diaspidiotus ancylus* (Putnam), Putnam scale, Davenport, IA, on *Acer* sp., IV-30-1908 (bark form).

between the median lobes. The median lobes and plates also are slightly different, but a comparison of a series of specimens of each species is necessary to clearly see these differences. *Diaspidiotus piceus* (Sanders) resembles Putnam scale but the former has a continuous band of macroducts to segment 3 or 4. While the latter never has macroducts on segment 3, it rarely has 1 or 2 on segment 4.

Hosts Borchsenius (1966) reports it on 17 plant families and 30 genera; Dekle (1977) reports it from 16 plant genera in Florida. We have examined specimens from the following hosts: *Acer, Actinidia, Ailanthus, Asimina, Betula, Caragana, Carpinus, Carya, Castanea, Ceanothus, Celtis, Cladrastis, Cornus, Corylus, Cotoneaster, Diospyros, Euonymus, Fagus, Fraxinus, Gleditsia, Hedera, Hydrangea, Ilex, Maclura, Magnolia, Malus, Platanus, Populus, Prunus, Ptelea, Quercus, Rhamnus, Rhus, Rhododendron, Ribes, Rosa, Salix, Sorbus, Staphylea, Tamarix, Tilia, Tsuga, Ulmus, Vaccinium,* and *Ziziphus.* Additional hosts are recorded in the literature.

Distribution This species is native to North America. We have examined U.S. material from AL, AZ, CA, CO, CT, DC, DE, FL, GA, IA, ID, IL, IN, KS, KY, LA, MA, MD, ME, MI, MN, MO, MS, NB, NC, ND, NH, NJ, NM, NY, OH, OK, PA, RI, SC, SD, TN, TX, UT, VA, VT, WI, WV, WY. It is reported from Australia, Brazil, Canada, Chile, France, Germany, Mexico, Portugal, South Africa, and Spain (Danzig 1972; Nakahara 1982).

Biology This species has 1 generation per year in some areas (e.g., Iowa, parts of New Jersey, Ohio, and Pennsylvania) and 2 generations in others (e.g., southern Illinois, parts of New Jersey, and possibly Delaware). Crawlers are reported in late spring or early summer in Iowa (Putnam 1880), in May and July in Delaware (Bray 1974), before midsummer in Ohio (Houser 1918), in May or June and midsummer in Illinois (Stannard 1965), and peak in early June and late August in New Jersey (Polavarapu et al. 2000). Stimmel (1976) states that crawlers are present in Pennsylvania for 4 to 5 weeks and are active through late July. In Illinois, Tinker (1957) reports crawler peaks in the third week of June and the second week of August. In New Jersey, crawlers of the first generation were present in May, June, and July, and those of the second generation in August, September, and October. The species overwinters as second instars on the bark of twigs in both single-generation (Stimmel 1976) and double-generation areas (Tinker 1957; Polavarapu et al. 2000). Kosztarab (1963) indicates that adult females overwinter in Ohio. In Illinois, Tinker (1957) reports that females lay an average of 49 eggs at a rate of 2–3 eggs each day; eggs hatch in about 16 hours. Adults appear in May and July in Illinois (Tinker 1957), April in Pennsylvania (Stimmel 1976), and May and July in New Jersey (Polavarapu et al. 2000). Arancibia et al. (1990) found 2 generations per year and overwintering as second instars on *Robinia pseudoacacia.* When fruit of peaches, pears, and apples were infested artificially, development did not progress beyond the second instar. Stannard (1965) reports two distinct morphological forms in Illinois. The 'ancylus form' is most abundant on the bark and is predominant in single generation areas; the 'comstocki-howardi form' is predominant on the undersurface of leaves and is the common summer form in two-generation areas. Polavarapu et al. (2000) found the same two forms on blueberries in New Jersey with the ancylus form predominantly under the bark and the comstocki-howardi form on the leaves and fruit. It appears that most of the population remains on the main branches of the host and only a small portion infests the leaves, green stems, and fruit. They noted significant differences in the appearance of the scale cover and the microscopic characteristics of the adult female. Based on the scant biological information available on this species, it is likely that 'ancylus' is a complex of species that may each have several forms. Proper identity of the constituents of the complex awaits detailed biological studies.

Economic Importance Putnam scale is occasionally an economic pest. It can kill twigs and branches of heavily infested trees (Baker 1972). Outbreaks of the pest were found to follow foliar applications of pesticides, which presumably reduced the parasite populations. On blueberries Putnam scale can reduce plant vigor (Antonelli et al. 1992) and contaminate and cause deformities on the fruit (Polavarapu et al. 2000). Feeding on the leaves and green stems causes red areas around the feeding sites. Regular pruning to remove older canes appears to keep this pest from becoming a serious problem in blueberries (Marucci 1966). This species also is reported as a pest of walnuts, elms, other ornamentals, and peaches (Gill 1997). Six species of eulophid wasps were reared from this scale by Tinker (1957) in Illinois. For this species Polavarapu et al. (2000) reported 9 hymenopterous parasitoids, 2 lady beetles, and a mite associated with this pest in blueberries. Kosztarab (1996) also mentions natural enemies of Putnam scale. Miller and Davidson (1990) consider this species to be an occasional pest.

Selected References McKenzie (1956); Polavarapu et al (2000); Stannard (1965); Tinker (1957).

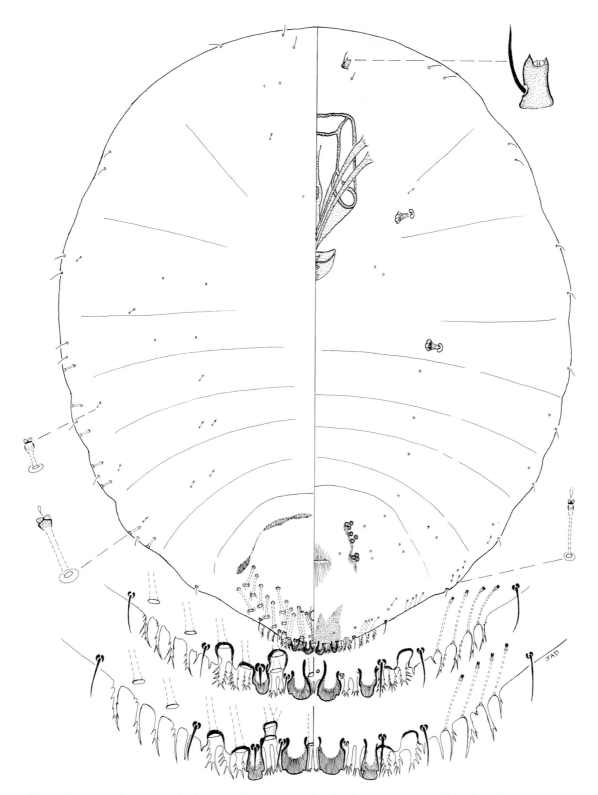

Figure 40. *Diaspidiotus ancylus* (Putnam), Putnam scale, Charleston, MO, on soft leaf maple, VIII-8-1977 (leaf form with moderately sized second lobes); Atlanta, GA, on *Acer* sp., VI-29-1944 (pygidium on lower right with large-sized second lobes).

Plate 37. *Diaspidiotus ancylus* (Putnam), Putnam scale (stem and bark form)

A. Circular adult female cover and elongate male cover on blueberry bark (J. A. Davidson).

B. Adult female cover on blueberry fruit (J. A. Davidson).

C. Male covers on blueberry fruit (J. A. Davidson).

D. Male covers on maple (J. A. Davidson).

E. Body of overwintering second-instar females on blueberry (J. A. Davidson).

F. Body of overwintering second-instar female on maple (J. A. Davidson).

G. Feeding damage on blueberry fruit (J. A. Davidson).

Plate 38. *Diaspidiotus forbesi* (Johnson), Forbes scale

A. Adult female covers on cherry (J. A. Davidson).
B. Adult female covers on cherry (J. A. Davidson).
C. Adult male cover on cherry (J. A. Davidson).
D. Male and female covers on cherry (J. A. Davidson).
E. Colony on cherry bark (J. A. Davidson).

F. Colony on apple bark (J. A. Davidson).
G. Distance view on cherry (J. A. Davidson).
H. Distance view on pear (J. A. Davidson).
I. Body of adult female on tree peony (J. A. Davidson).

151

Diaspidiotus forbesi (Johnson)

ESA Approved Common Name Forbes scale (also called cherry scale).

Common Synonyms and Combinations *Aspidiotus forbesi* Johnson, *Aspidiotus (Diaspidiotus) forbesi* Johnson, *Aspidiotus (Aspidiella) forbesi* Johnson, *Aspidiotus (Diaspidiotus) fernaldi hesperius* Cockerell, *Aspidiotus fernaldi hesperius* Cockerell, *Forbesaspis (Aspidiotus) forbesi* Johnson, *Aspidiotus hesperius* Cockerell, *Aspidiotus (Quadraspidiotus) forbesi* Johnson, *Quadraspidiotus forbesi* (Johnson).

Field Characters (Plate 38) Adult female cover convex to almost flat, oval to circular, thin, dirty gray; shed skins subcentral to central, orange when rubbed. Male cover smaller, elongate oval, dark gray, margin slightly lighter; shed skin submarginal, orange. Body of adult female light red when young, turning yellowish red when egg laying begins; eggs yellow red; crawlers yellow. On bark.

Slide-mounted Characters Adult female (Fig. 41) with 2 pairs of lobes; paraphysis formula usually 2-2-0, rarely 2-1-0. Median lobes separated by space 0.01–0.15 (0.06) times width of median lobe, with small basal sclerotization near base of dorsal seta, with conspicuous paraphysis-like sclerotization attached to medial margin, without yoke, medial and lateral margins converging apically, with 1 lateral notch, 0–1 (0) medial notch; second lobes simple, about one-half size of medial lobes, same shape, with 1–2 (1) lateral notches, 0–1 (0) medial notch. Plates simple, apparently without associated microduct; plate formula usually 1-2-0, rarely 0-2-0 or 1-3-0, without plates anterior of second space; without plates between medial lobes. Macroducts on pygidium of 1 size, ducts of some specimens tend to collapse but not in any consistent pattern, usually with macroduct between medial lobes 35–55 (48) μ long, extending 1.5–2.5 (2.0) times distance to posterior apex of anal opening, longest macroduct in first space 38–50 (43) μ long, with 13–23 (18) macroducts on each side of pygidium on segments 5 to 8, ducts in submarginal and marginal areas, some macroduct orifices anterior of anal opening; prepygidial macroducts usually absent, sometimes submarginally on segment 4. Pygidial microducts on venter in submarginal and marginal areas of segments 5 and 6, with 4–11 (8) ducts, marginal ducts sometimes with protruding truncate orifice; prepygidial ducts of 2 sizes, smaller size in medial area of metathorax or segment 1 to segment 2 or 3, occasionally in submedial areas near mouthparts, in marginal or submarginal areas of segments 2 or 3 to segment 4, larger size ventral microducts on marginal or submarginal area of metathorax or segment 1 to segment 2 or 3; pygidial microducts absent from dorsum; prepygidial microducts of small size only, in submedial areas of segments 1 to 3 or 4, in marginal or submarginal areas of head or prothorax to metathorax or segment 1. Perivulvar pores usually in 5 groups, 8–12 (10) pores on each side of body. Perispiracular pores absent. Anal opening located 1.3–2.2 (1.8) times length of anal opening from base of median lobes, anal opening 10–18 (13) μ long. Dorsal seta laterad of median lobes 0.9–1.7 (1.1) times length of median lobes. Eyes usually absent, when present located near body margin at level of anterior margin of clypeolabral shield. Antennae each with 1 long seta. Cicatrices usually absent. Body broadly oval.

Affinities Of the economic species in the United States, Forbes scale is most similar to *Diaspidiotus juglansregiae*, walnut scale. The former differs by lacking third lobes, by lacking a distinct constriction between the meso- and metathorax, and by lacking a dorsal sclerotization in old adult females. The walnut scale has third lobes, a marginal constriction between the meso- and metathorax, and the dorsal derm sclerotized in old adult females.

Hosts Polyphagous. Borchsenius (1966) records it from 20 genera in 11 families; Dekle (1977) notes it from 10 genera in Florida with *Carya*, *Cornus*, *Prunus*, and *Pyrus* the most frequently reported hosts. We have examined specimens from *Acer*, *Amorpha*, *Celtis*, *Ceanothus*, *Cotoneaster*, *Cephalanthus*, *Cornus*, *Corylus*, *Crataegus*, *Cydonia*, *Forsythia*, *Fraxinus*, *Ilex*, *Ligustrum*, *Lonicera*, *Malus*, *Paeonia*, *Prunus*, *Ribes*, *Robinia*, *Viburnum*, and *Vaccinium*.

Distribution This species probably is native to North America. We have seen U.S. specimens from AL, AR, CA, CO, CT, DC, DE, GA, IA, IL, IN, KS, KY, MD, MS, MO, NB, NC, NJ, NM, NY, OH, OK, PA, SC, TN, TX, UT, VA, WV. We also have seen material from Canada, Mexico, and South Africa.

Biology Forbes scale has 2 generations per year in Illinois (Felt 1901) with adult males present from July 10 to August 1. Crawlers were noted in August and September and 'partly grown' forms overwintered. In North Carolina (Turnipseed and Smith 1953), crawlers appeared from mid-June to mid-July and again in late August indicating 2 generations per year. Mated adult females overwintered. Kosztarab (1963) noted wingless males in mid-April and recorded egg hatching in early July. Two generations were seen in Ohio (Johnson 1896) where wingless males were observed in April. Westcott (1973) reported one to three generations per year.

Economic Importance Beardsley and González (1975) consider this species to be one of 43 serious armored scale pests. Miller and Davidson (1990) consider it to be a serious pest in a small area of the world. According to Hamilton and Summerland (1953), Forbes scale is primarily a pest of apple, plum, and cherry in the Midwest. Although it was first mentioned in the literature many years ago (Johnson 1896), it did not become numerous enough to cause damage until 1946. The 3 factors believed responsible for this population increase were (1) the use of DDT, which reduced predator numbers, (2) the decline of a formerly abundant competitor (San Jose scale), possibly also due to the use of DDT, and (3) the winters were mild at this time, with temperatures rarely falling below the thermal threshold for Forbes scale (−15 °F). A survey completed in 1951 shows that 72% of the apple orchards in Indiana, Illinois, Kentucky, and Tennessee had Forbes scale infestations, and 22% of the infestations were considered to be severe. At this point Forbes scale was the predominate species in apple, plum, and cherry orchards. It also has been recorded as an orchard pest in Georgia (Smith 1905), Michigan (Hutson 1933), Missouri (Hollinger 1923), Ohio (Kosztarab 1963), and Ontario, Canada (Bethune 1908).

Selected References Ferris (1938a); Turnipseed and Smith (1953).

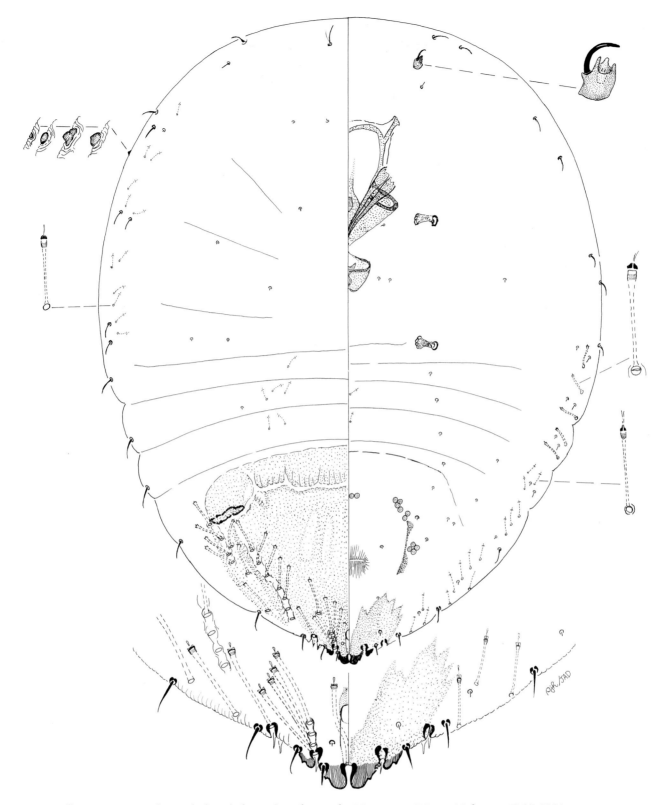

Figure 41. *Diaspidiotus forbesi* (Johnson), Forbes scale, Vincennes, IN, on *Malus* sp., II-27-1951.

Diaspidiotus gigas (Thiem and Gerneck)

Suggested Common Name Poplar scale (also called willow scale).

Common Synonyms and Combinations *Aspidiotus (Euraspidiotus) gigas* Thiem and Gerneck, *Aspidiotus multiglandulatus* Borchsenius, *Aspidiotus gigas* Thiem and Gerneck, *Quadraspidiotus gigas* (Thiem and Gerneck).

Field Characters (Plate 39) Adult female cover moderately convex, circular, gray; shed skins central to subcentral, orange yellow. Male cover smaller, elongate oval, gray; shed skin submarginal, orange yellow. Body of adult female yellow; eggs and crawlers yellow. Mainly on bark.

Slide-mounted Characters Adult female (Fig. 42) with 3 pairs of lobes, rarely with weakly indicated fourth lobe; paraphysis formula 2-2-0. Median lobes separated by space 0.1–0.3 (0.2) times width of median lobe, without basal sclerotization, usually without paraphysis-like sclerotization, rarely with small paraphysis-like sclerotization structure attached to medial margin, without yoke, medial and lateral margins converging apically, in unworn specimens each medial lobe with 1 lateral notch, without medial notch; second lobes each simple, about one-half size of median lobes, more acute medially than median lobes but with broadly rounded apex, with 0–2 (1) lateral notches, without medial notch; third lobes simple, smaller than second lobes, same shape, without notches; fourth lobes usually absent, rarely represented by small sclerotized protrusion. Plates fringed or simple, medial plate in first space simple, others usually at least bifurcate, occasionally medial plate in second space simple, without associated microducts; plate formula usually 2-2-0, rarely 2-3-0, without plates anterior of second space; plates between median lobes slender 0.4–0.8 (0.6) times as long as median lobes. Macroducts of 1 size, gradually decreasing anteriorly, ducts of some collapsed but not in any pattern, usually with 1 or rarely 2 ducts between median lobes 35–48 (41) μ long, extending 0.7–1.2 (1.0) times distance to posterior apex of anal opening, longest macroduct in first space 37–58 (45) μ long, with 33–60 (48) macroducts on each side of pygidium of segments 5 to 7 or 8, ducts in submarginal and marginal areas, total of 50–109 (83) macroducts on each side of body, some macroduct orifices anterior of anal opening; prepygidial macroducts present submarginally on segments 1, 2, or 3 to 4, segment 1 usually with none, rarely with 1, segment 2 with 0–6 (2) ducts, segment 3 with 1–16 (6) ducts, segment 4 with 10–25 (13) ducts. Pygidial microducts on venter in submarginal and marginal areas of segments 5 and 6, prepygidial ducts of 2 sizes, small size in submedial areas anterior and laterad of mouthparts, on any or all of metathorax to segment 2, large microducts on submargin of segments 3 and 4; on dorsum pygidial microducts absent; prepygidial ducts of 2 sizes, larger size in medial and submedial areas of segments 1 to 3 or 4, smaller size in submarginal areas of prothorax to segment 1, occasionally absent from 1 or both of prothorax and mesothorax. Perivulvar pores in 5 groups, 16–45 (32) pores on each side of body. Perispiracular pores absent. Anal opening located 2.4–3.3 (2.7) times length of anal opening from base of median lobes, anal opening 13–20 (16) μ long. Dorsal seta laterad of median lobes 1.0–1.5 (1.1) times length of median lobe. Eyes absent or small and inconspicuous, located near body margin at level of antenna. Antennae each with 1 long seta, occasionally bifurcate apically. Cicatrices absent or inconspicuous, when present located on prothorax, segment 1 and 3. Body pear shaped. Older specimens with head and thorax partially sclerotized. With 7–10 (8) macroducts in first space.

Affinities Poplar scale is very similar to *Diaspidiotus ostreaeformis*, European fruit scale. The former differs by having: 50–109 (83) macroducts on each side of the pygidium; 7–10 (8) macroducts in the first space; segment 4 with 10–25 (13) macroducts on each side; 16–45 (32) perivulvar pores on each side of the body; the perivulvar pores usually arranged in 5 groups; *D. ostreaeformis* has: 25–40 (35) total macroducts; 5–6 (6) macroducts in the first space; segment 4 with 3–9 (6) macroducts on each side; 8–24 (16) perivulvars arranged in 4 or 5 groups.

Hosts Poplar scale usually is restricted to *Populus* and *Salix*; in the United States, we have examined specimens from only these hosts.

Distribution This species probably is native to Europe. We have seen U.S. specimens from ID, MT, NY, OH, OR, PA, RI, UT, WA, WI, WY. We have seen material also from Canada, Germany, and Russia. Nakahara (1982) reports this species from 10 European countries, Algeria, and Turkey. It occasionally has been intercepted by U.S. quarantine officers on *Salix* and *Populus* from Europe and once on *Tilia* from Bulgaria.

Biology Lelláková-Dusková (1963) published a detailed study of poplar scale biology. Her study in Czechoslovakia encompassed four years, including weekly sampling during the growing season. She found that it preferred the smooth bark areas on trunks and thick branches rather than thin twigs. Occasionally it settled on petioles but not farther out on leaves. There was 1 generation per year with second-instar males and females overwintering. The male prepupal stage occurred from late March to late April. The male pupal stage was present from mid-April to early May. She also succeeded in transferring this species to *Tilia cordata* Mill., *Ulmus scabra* Mill., and *Fraxinus excelsior* L. Poplar scale completed several generations on these hosts. Adult males and females appeared in May. Mating was needed for reproduction. Eggs were laid from late June to early September. Females produced an average of 145 eggs each. Eggs hatched shortly after being laid, thus crawlers were present from late June to mid-September. The overwintering second instars appeared in mid-September. This species was reported to have one generation per year in Hungary (Kosztarab and Kozár 1978), China (Hu et al. 1982), and Germany (Schmutterer 1959). In China in Heilongjing Province the scale overwinters in the second instar, adults appear in mid-May, an average of 93 eggs are deposited per female, and crawlers are most abundant from late June to early July.

Economic Importance According to Lelláková-Dusková (1963) poplar scale is a serious pest of *Populus* spp. in Czechoslovakia. Large populations were noted in protected areas, such as parks and alleys. She noted that this species has the ability to inhibit tissue growth on branches, which makes the wood worthless for certain commercial purposes. In some areas she reported forest stands of several acres of *Populus* destroyed by this pest. Baker (1972) noted that in the United States, willow appears to be the favorite host of this scale, and that heavy infestations are especially serious to young trees. Poplar scale is a serious pest in northeastern China on *Populus berolinensis* and *P. simonii* (Hu et al. 1982; Xie et al. 1995). Considerable work has been done to understand the dynamics of outbreaks (Liu et al. 1997a) and to develop control strategies including the use of insect growth regulators (Chi et al. 1997a), kairomones and oviposition deterring pheromones (Chi et al. 1997b), and natural enemies (Li 1996; Ma et al. 1997). Xie et al. (1995) suggest that outbreaks may be caused by urban pollution. Miller and Davidson (1990) consider this species to be a serious pest in a small area of the world.

Selected References Kosztarab (1963); Lelláková-Dusková (1963).

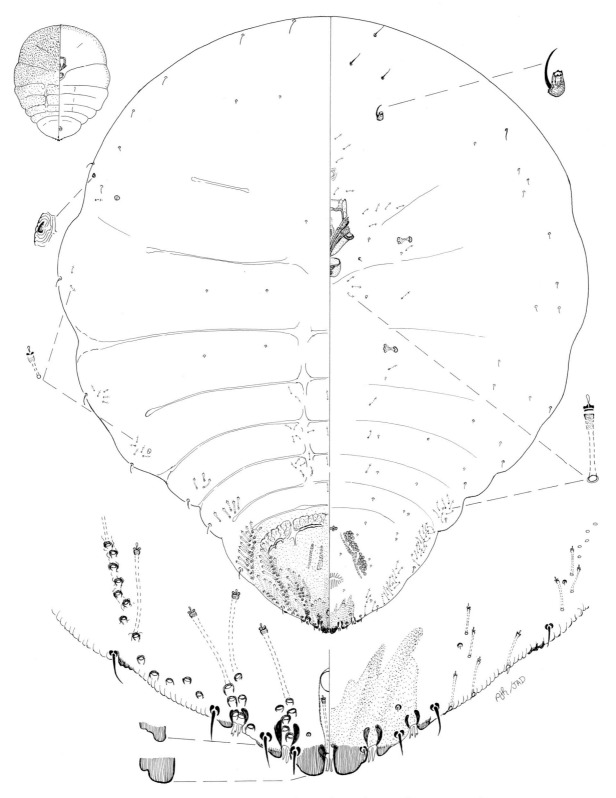

Figure 42. *Diaspidiotus gigas* (Thiem and Gerneck), poplar scale, Seattle, WA, on *Salix* sp., VI-20-1965.

Plate 39. *Diaspidiotus gigas* (Thiem and Gerneck), Poplar scale

A. Young adult female covers on basket willow (J. A. Davidson).
B. Adult male cover on basket willow (J. A. Davidson).
C. Dense infestation on bark (W. T. Johnson).
D. Distance shot on basket willow (J. A. Davidson).

E. Distance shot on basket willow (J. A. Davidson).
F. Cluster of female covers in basket willow bark depression (J. A. Davidson).
G. Distance view of infestation on bark (W. T. Johnson).

Plate 40. *Diaspidiotus juglansregiae* (Comstock), Walnut scale

A. Adult female cover on hemlock (J. A. Davidson).
B. Adult females on Japanese holly (J. A. Davidson).
C. Young adult female cover on red maple (J. A. Davidson).
D. Adult male covers (left) on red maple (J. A. Davidson).

E. Adult female body on plum (R. J. Gill).
F. Body of parasitized adult female on red maple (J. A. Davidson).
G. Distance view of male cover on Foster holly (J. A. Davidson).
H. Ventral view of parasitized adult female (J. A. Davidson).

Diaspidiotus juglansregiae (Comstock)

ESA Approved Common Name Walnut scale (also called English walnut scale, gopher scale).

Common Synonyms and Combinations *Aspidiotus juglans-regiae* Comstock, *Aspidiotus juglans-regiae* var. *pruni* Cockerell, *Aspidiotus juglans-regiae* var. *albus* Cockerell, *Aspidiotus (Diaspidiotus) juglans-regiae* Comstock, *Aspidiotus (Evaspidiotus) juglans-regiae* Comstock, *Aspidiotus fernaldi* Cockerell, *Aspidiotus fernaldi* var. *albiventer* Hunter, *Aspidiotus fernaldi* var. *cockerelli* Parrott, *Quadraspidiotus fernaldi* (Cockerell), *Aspidiotus (Diaspidiotus) glanduliferus* Cockerell, *Aspidiotus juglandis-regiae* Lindinger, *Furcaspis juglans-regiae* (Comstock), *Quadraspidiotus juglans-regiae* (Comstock), *Quadraspidiotus juglansregiae* (Comstock).

Field Characters (Plate 40) Adult female cover variable depending on age and host, relatively flat, circular, gray to reddish brown; shed skins central to subcentral, yellow to orange when rubbed. Male cover smaller, elongate oval, gray to brown; shed skin submarginal, yellow to orange. Body of adult female yellow; eggs and crawlers in Maryland orange to reddish orange; in California eggs yellow. On bark and stems of host, often under host epidermis. Body shape of adult female distinctive with 2 notches on each side of thorax. Males tend to settle under edge of female cover with apex of each male cover barely protruding outside female cover, giving effect of petals on a flower (Gill 1997).

Slide-mounted Characters We believe that walnut scale may encompass 2 species. This is suggested by the unusually large amount of variation in host diversity, life history, field appearance, and morphology. Unfortunately, we have been unable to discover sufficient, discrete characters to differentiate distinct taxa; further research is required.

Adult female (Fig. 43) with 3 pairs of lobes, fourth lobes rarely represented by small sclerotized point; paraphysis formula 2-2-0. Median lobes separated by space 0.02–0.20 (0.08) times width of median lobe, without basal sclerotization, usually with paraphysis-like sclerotization attached to medial margin, without yoke, medial margins parallel, lateral margins converging apically, with 1 lateral notch, 0–1 (0) medial notch; second lobes simple, about one-half size of median lobes, usually with rounded, medial apex, with 1–3 (2) lateral notches, without medial notch; third lobes simple, about one-quarter size of second lobe, apically acute, without notches; fourth lobes usually absent. Plates simple, occasionally with 1 or 2 additional, small tines; plate formula usually 2-3-0, rarely 1-3-0 or 2-2-0, second and third plates in first and second spaces, respectively, often obscured by larger structures, without plates anterior of second space; plate between median lobes 0.6–1.0 (0.8) times as long as median lobes. Macroducts of 1 size, duct between median lobes 45–54 (46) µ long, extending 1.2–2.5 (1.8) times distance between posterior apex of anal opening and base of median lobe, longest macroduct in first space 40–52 (46) µ long, with 30–75 (47) macroducts on each side of pygidium on segments 5 to 8, ducts in submarginal and marginal areas, total of 38–85 (55) macroducts on each side of body, some macroducts present submarginally on segment 4. Pygidial microducts on venter in submarginal and marginal areas of segments 5 and 6, with 6–14 (9) ducts; prepygidial ducts of 1 size, in submedial area near mouthparts and anterior spiracle, in submarginal areas on segments 2 or 3 to 4; on dorsum pygidial microducts absent; prepygidial microducts of 2 sizes, larger size in submedial areas of any or all of metathorax to segment 4, smaller size submarginally on any or all of head to segment 3. Perivulvar pores usually in 5 groups, rarely 4, 16–27 (21) pores on each side of body. Perispiracular pores absent. Anal opening located 1.6–4.3 (2.6) times length of anal opening from base of median lobes, anal opening 8–13 (11) µ long. Dorsal seta laterad of median lobes 0.7–1.3 (1.0) times length of median lobe. Eyes usually located near apex of protrusion formed by constriction between prothorax and mesothorax, usually in form of small sclerotized spur. Antennae each with 1 long seta. Cicatrices on any or all of prothorax, segment 1, or segment 3. Body broadly oval, with definite constrictions between prothorax and mesothorax, mesothorax and metathorax, segments 1 and 2, segments 2 and 3, and segments 3 and 4. Older specimens with entire dorsum sclerotized.

Affinities Walnut scale is distinctive among the economic scales of the United States by having definite, marginal constrictions between the prothorax and mesothorax and between the mesothorax and metathorax, and by having the dorsal surface sclerotized on more mature adult females. The species is most similar to *Diaspidiotus ostreaeformis*, European fruit scale, but differs by having the characters above and by having: 1–3 (2) lateral notches on the second lobe; simple plates in second space; no dorsomedial microducts on the abdomen; the macroduct between the median lobes extending beyond the anterior end of the anal opening; 30–75 (47) macroducts on each side of the pygidium; and usually lacking macroducts on segments 2 and 3. European fruit scale has: 0–1 (0) lateral notches on the second lobe; at least some plates bifurcate in second space; a line of dorsomedial microducts on some of the prepygidial segments of the abdomen; the macroduct between the median lobes not reaching the anterior apex of the anal opening; 15–28 (21) macroducts on each side of the pygidium; and usually by having macroducts on segments 2 and/or 3.

Hosts Polyphagous. Borchsenius (1966) records it from 14 genera in 10 families; Dekle (1977) records it from 40 genera in Florida with *Ilex* and *Prunus* the most frequently reported genera. We have examined specimens from the following: *Acer, Aesculus, Alnus, Arbutus, Betula, Carya, Cercis, Cornus, Crataegus, Fraxinus, Gleditsia, Ilex, Juglans, Liquidambar, Liriodendron, Mahonia, Malus, Picea, Pinus, Populus, Prunus, Ptelea, Pyrus, Salix, Syringa, Tilia, Tsuga, Ulmus, Viburnum,* and *Zanthoxylum.* It has been reported from *Celtis, Gymnocladus, Maclura, Robinia,* and *Sorbus* (Kosztarab 1963). Although this pest has been found on a number of species in *Juglans,* it most often is found on English walnut (*Juglans regia*). We have seen 2 specimens from Japanese walnut (*Juglans ailanthifolia*) collected in Wrightsville, Arkansas.

Distribution This species probably is native to North America. We have examined U.S. specimens from AL, AR, CA, CT, DC, DE, FL, GA, IA, IL, IN, KS, KY, LA, MA, MD, MI, MO, MS, NC, NJ, NM, NY, OH, PA, SC, TN, TX, VA, WV. It is known only from the Canada, Mexico, and United States.

Biology In a carefully monitored study in central Maryland, Stoetzel (1975) found that walnut scale had 1 generation per year with second-instar males and females overwintering on the twigs and branches of *Ilex opaca.* Adult males were alate and occurred from late April to mid-May. Eggs were laid in 1 clutch of 20–40 beginning in late June and then were laid individually through September. Crawlers appeared in late June, and although most had settled by mid-July, crawlers were present through September. Second instars appeared in early September. Gordon and Potter (1988) found a very different life history in Kentucky on red maple (*Acer rubrum*), more similar to the findings of Kosztarab (1963) in Ohio than Stoetzel in Maryland. Gordon and Potter found 2 generations per year with overwintering in second-instar males and females. Prepupae and

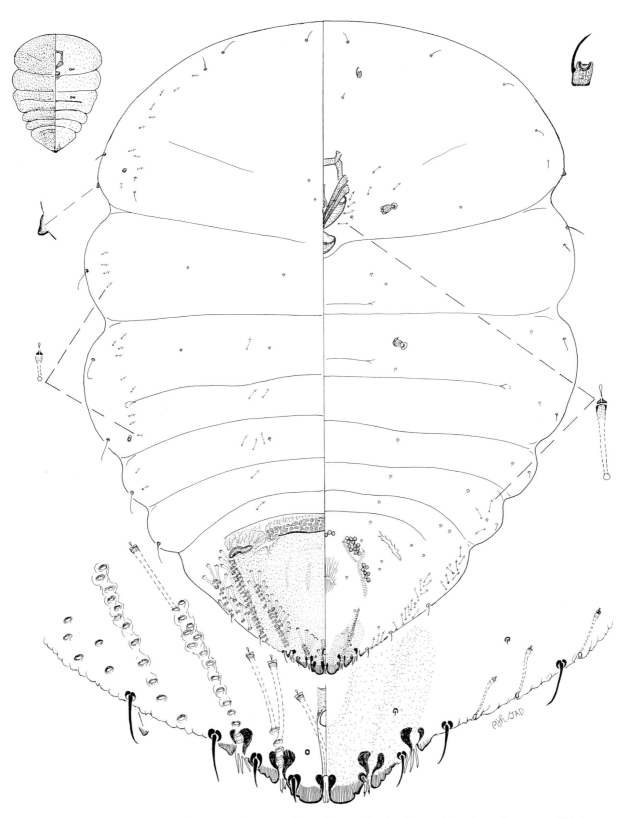

Figure 43. *Diaspidiotus juglansregiae* (Comstock), walnut scale, Scarborough, NY, on *Tsuga* sp., VI-14-1934.

pupae were present from late March to mid-April, and adult males were observed in the first three weeks of April. Adult females were present from April to late June, and eggs and crawlers occurred from mid-May to mid-June with a few stragglers in early July. Second instars were present from early July to early August, and prepupal and pupal males were observed in mid- to late July. Adult males were seen for about two weeks in late July coinciding with the first appearance of the adult females. The latter stage was present until early September. Eggs and crawlers were present from early or mid-August until mid-September, and second instars occurred from September through the winter. By mid-October nearly all of the population was in the second instar. Differences in the appearance of second-instar males and females was first observed in January. Lambdin et al. (1993) investigated the life history on dogwood trees in Tennessee. They essentially obtained the same results as Gordon and Potter, but the life cycle was about 2 weeks earlier. Hollinger (1923) stated that 'there are probably three full generations a year' in Missouri. In Ohio, Houser (1918) reported this species as overwintering in the adult stage. Females laid eggs in early spring, and the first generation was completed in June, at which time eggs were laid to begin the second generation. He suggested that 2 or more generations were present. Kosztarab (1963) also noted that adults overwintered and mated in early spring. Adult males were found as early as mid-March in Columbus, Ohio, and again in mid-May. Egg laying was observed from June to mid-July, and crawlers were seen in September. Johnson and Lyon (1976) also recorded 2 or more generations in Ohio with adults overwintering and mating in early spring. Egg laying occurred in June and July and crawlers appeared in September. We believe that the biologies mentioned above for Ohio may be inaccurate because adult males were reported to overwinter. As discussed in the biology section of the introduction, male prepupae, pupae, and adults each live for only a few hours or days, and therefore males apparently do not overwinter except as settled crawlers or second instars. The latter situation is the most common. Some species do overwinter as mated adult females. Gordon and Potter (1988) discuss discrepancies in life history information on this species. They suggest the possibility that the short second generation may have been missed by Stoetzel (1975), but also explore the possibility that there may be more than a single species involved. Walnut scale is widely distributed in the United States, has been reported from widely divergent hosts (including conifers and deciduous trees), and shows much variation in taxonomic characters of slide-mounted adult females. These factors provide evidence that more than 1 species may be involved.

Economic Importance This species is treated as a serious pest by Beardsley and González (1975) and Miller and Davidson (1990). Baker (1972) comments that infestations heavy enough to kill twigs and branches have been observed in the South. Riedl et al. (1979) mention walnut scale as the most common and important scale insect pest of walnut trees in the Midwest. Gill (1982, 1997) notes that in California, walnut scale is a pest of walnuts throughout the state as well as a serious pest of ash and birch trees in ornamental plantings. The pest is usually not a problem in California walnut orchards where heavy pesticide usage is absent (Ebeling 1959). It is held under control primarily by *Aphycus californicus* Howard (Ebeling 1959).

Selected References Ferris (1938a); Gordon and Potter (1988); Stoetzel (1975).

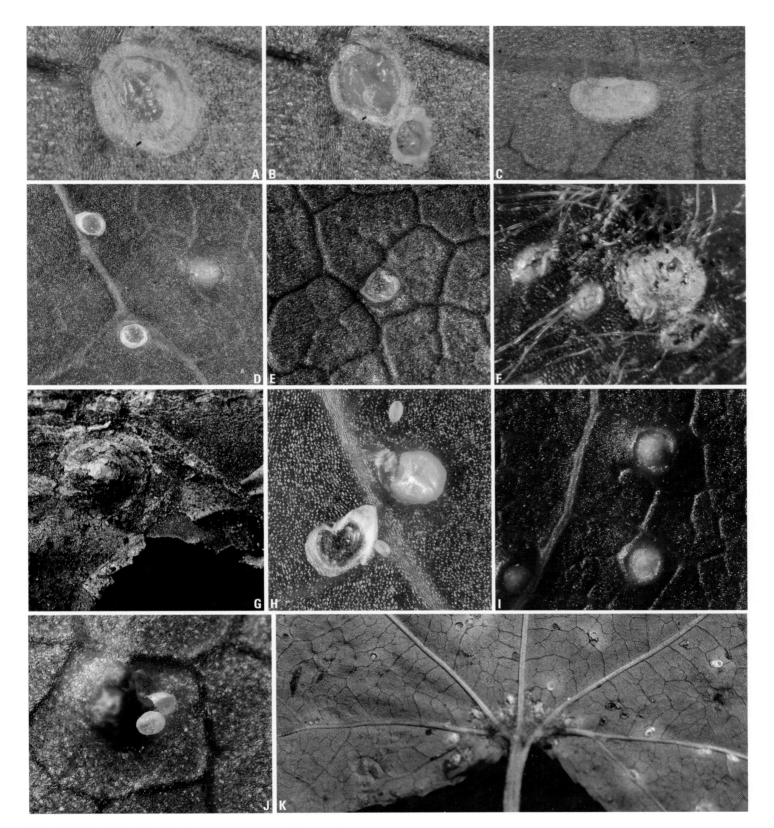

Plate 41. *Diaspidiotus liquidambaris* (Kotinsky), Sweetgum scale

A. Adult female cover on leaf of sweetgum (J. A. Davidson).
B. Adult female cover with crawler shed skin rubbed off on sweetgum (J. A. Davidson).
C. Male cover on sweetgum (J. A. Davidson).
D. Distance view of adult female covers on upper leaf surface of sweetgum (J. A. Davidson).
E. Distance view of adult female cover on lower leaf surface of sweetgum (J. A. Davidson).
F. Adult female cover on green shoot of sweetgum (J. A. Davidson).

G. Overwintering adult female cover on bark of sweetgum (J. A. Davidson).
H. Adult female cover removed to expose female body and two crawlers on sweetgum leaf (J. A. Davidson).
I. Dorsally produced galls on sweetgum (J. A. Davidson).
J. Female removed to show gall depression on lower leaf surface; also note crawler without cap (J. A. Davidson).
K. Ventral view of leaf showing scale covers concentrated at base of leaf veins (J. A. Davidson).

161

Diaspidiotus liquidambaris (Kotinsky)

Suggested Common Name Sweetgum scale (also called sweet gum scale).

Common Synonyms and Combinations *Cryptophyllaspis liquidambaris* Kotinsky, *Chemnaspidiotus liquidambaris* (Kotinsky), *Aspidiotus liquidambaris* (Kotinsky).

Field Characters (Plate 41) Adult female cover on leaves lacks cover and shed skin of crawler, composed of large, central shed skin of second instar with narrow margin of white wax. On stems adult female cover white or gray, circular, with shed skins of crawler and second instar located subcentral. Male cover oval, white, with yellow, submarginal shed skin. Body of adult female and eggs yellow. On stems and leaves. Small gall-like depressions formed on underside of leaves.

Slide-mounted Characters Adult female (Figs. 44 and 45) with 1 pair of definite lobes, second and third pairs absent or represented by unsclerotized points that may be as long as median lobes; paraphysis formula usually 2-2-0, rarely 2-2-2 or 2-1-0, when 2-2-0 with large paraphyses in space between lobe 2 and median lobe, attached to medial margin of lobe 2, medial margin of lobe 3, and in space between lobes 2 and 3. Median lobes separated by space 0.2–0.4 (0.3) times width of median lobe, without basal sclerotization, paraphyses, or yoke, medial margins diverging apically or parallel, lateral margins converging, with 1 lateral notch and 1 medial notch; second lobes simple, represented by unsclerotized point similar in appearance to plate, without notches; third lobes, when present, represented by unsclerotized point similar to second lobe, without notches. Plates between median lobes and second lobe and between second lobes and third lobes usually with conspicuous tines, those on segment 5 usually simple; plate formula 2-3-3 or 2-3-4; plates between median lobes 0.3–0.8 (0.6) times as long as median lobes. Macroducts of 2 sizes, larger size on pygidium between median lobes and on segments 5 to 7 in marginal and submarginal areas, also present on prepygidium in submarginal areas of segments 3 and 4, occasionally absent from 3, duct between median lobes extending 1.3–2.2 (1.7) times distance between posterior apex of anal opening and base of median lobes, 32–52 (43) μ long, marginal macroduct in first space 32–58 (45) μ long, with 18–30 (24) macroducts on each side of pygidium on segments 5 to 8, ducts in submarginal and marginal areas, total of 7–38 (20) macroducts on each side of body, some macroduct orifices anterior of anal opening; smaller macroducts present along body margin from meso- or metathorax to segment 3. Pygidial microducts on venter in submarginal and marginal areas of segments 5 and 6, with 6–10 (8) ducts; prepygidial ducts of 1 size in submarginal and marginal areas of segments 2 or 3 and 4, absent submedially; on dorsum pygidial ducts absent; prepygidial microducts of 2 sizes, larger size in submedial areas of any or all of metathorax to segment 4, smaller size in submarginal areas of meso- or metathorax to segments 1 or 2. Perivulvar pores absent. Pores absent near spiracles. Anal opening located 1.3–2.9 (2.0) times length of anal opening from base of median lobes, anal opening 9–17 (12) μ long. Dorsal seta laterad of median lobes 1.0–1.6 (1.2) times length of median lobe. Eyes either absent or in small dermal indentation on prothorax. Antennae each with 1 seta. Cicatrices usually present on prothorax and segment 1. Body pear shaped. Head and thorax often sclerotized in older adult females.

The leaf and stem forms of this species are similar in appearance, differing primarily in the number of macroducts on each side of the pygidium. The leaf form has 7–14 (10) large-sized macroducts on segments 5 to 8; the bark form has 18–38 (29) macroducts on segments 3 to 8. Intermediate forms were found.

Affinities The sweetgum scale is distinct from other *Diaspidiotus* species because it causes leaf galls, lacks perivulvar pores, and has relatively fimbriate plates.

Hosts Limited host range; we have examined specimens from *Liquidambar* and *Magnolia*. It is reported on *Acer* (Borchsenius 1966; Dekle 1977).

Distribution This species is native to North America. We have examined U.S. material from CT, DC, DE, FL, GA, IL, IN, LA, MD, MO, MS, NC, NJ, NY, OH, PA, SC, TN, VA. It has been reported from AL, CA, OK, TX.

Biology Stoetzel and Davidson (1974) found 2 generations per year in Maryland. Fertilized adult females overwinter on the bark. Eggs and crawlers are present from mid-May through June. Most crawlers settle on the leaves. Adults first appear in mid-June. Second-generation crawlers occur in early July to early October; most settle on the bark. Second-generation adults first appear in early September. Adult males are winged in the first generation and wingless in the second. In Ohio, the overwintering stages are unmated adult females and pupal males located at the base of twig buds; crawlers appearing in early June (Kosztarab 1963).

Economic Importance Galls formed on sweetgum leaves by the summer generation of the sweetgum scale are unsightly. Dekle (1965) states this scale is occasionally economic in Florida. Baker (1972) reports that it causes serious damage in nurseries in Missouri. Miller and Davidson (1990) consider this species to be an occasional pest.

Selected References Ferris (1938a); Stoetzel and Davidson (1974).

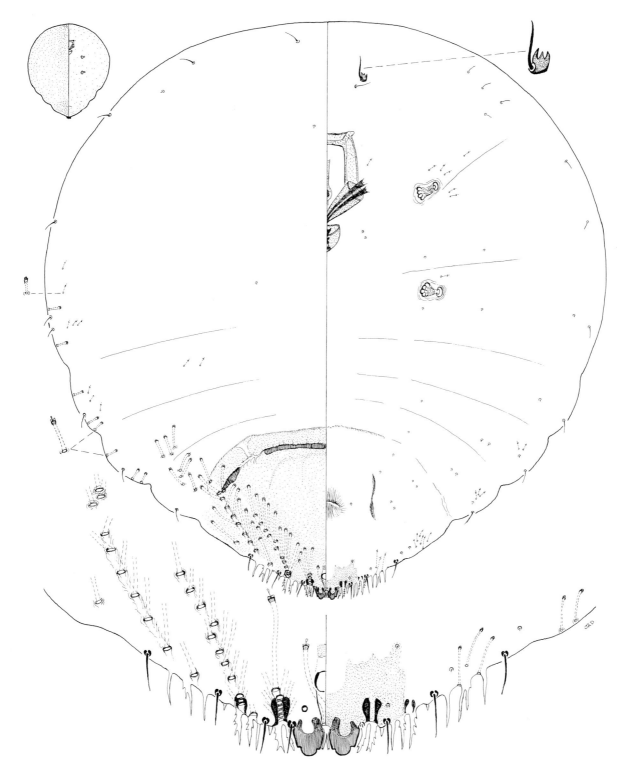

Figure 44. *Diaspidiotus liquidambaris* (Kotinsky), sweetgum scale, St. Louis, MO, on *Liquidambar* sp., II-5-1940 (bark form).

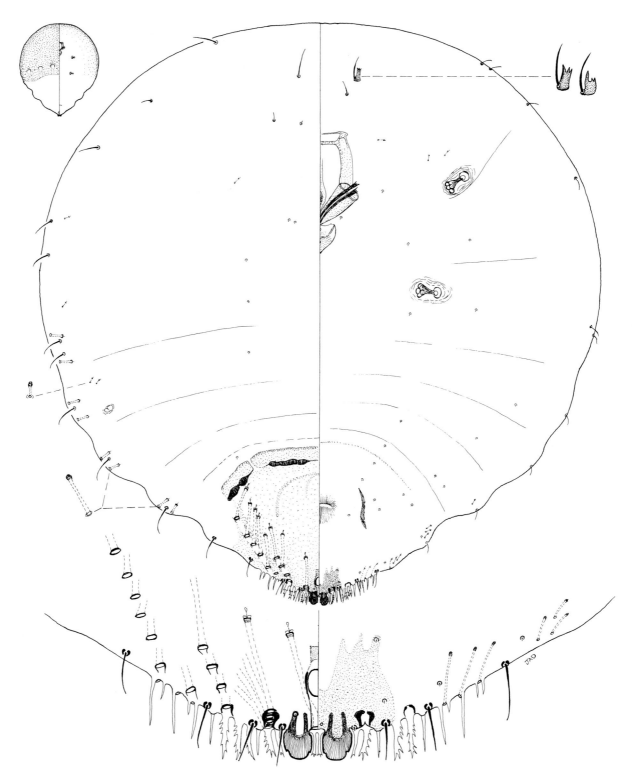

Figure 45. *Diaspidiotus liquidambaris* (Kotinsky), sweetgum scale, TN, on *Liquidambar* sp., V-30-1953 (leaf form).

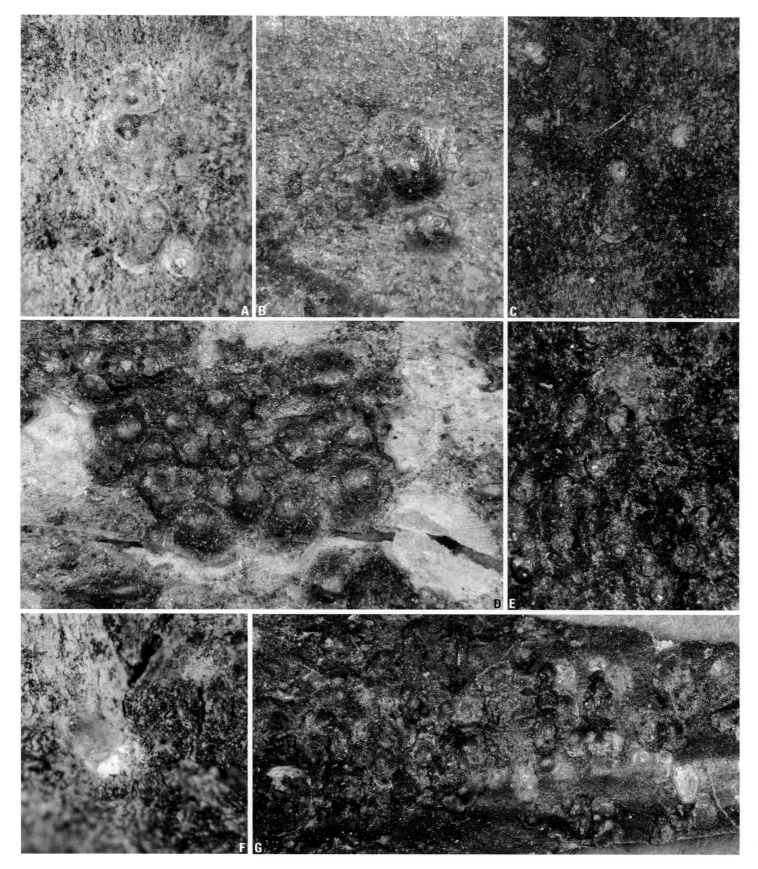

Plate 42. *Diaspidiotus osborni* (Newell and Cockerell), Osborn scale

A. Circular adult female cover and elongate male cover on oak (J. A. Davidson).
B. Young adult female covers on oak (J. A. Davidson).
C. Central male cover surrounded by sooty mold on oak (J. A. Davidson).

D. Cluster of mostly adult female covers on oak (J. A. Davidson).
E. Heavily infested oak bark (J. A. Davidson).
F. Body of young adult female on oak (J. A. Davidson).
G. Distance view of infested oak bark (J. A. Davidson).

Diaspidiotus osborni (Newell and Cockerell)

Suggested Common Name Osborn scale.

Common Synonyms and Combinations *Aspidiotus osborni* Newell and Cockerell, *Diaspis snowii* Hunter, *Aspidiotus (Diaspidiotus) osborni* Newell and Cockerell, *Aspidiotus yulupae* Bremner.

Field Characters (Plate 42) Adult female cover flat, circular, gray; shed skins subcentral, orange to yellow. Male cover same as female except smaller, oval, with submarginal shed skin. Body of adult female and eggs yellow. On stems and bark; usually blends with bark.

Slide-mounted Characters Adult female (Fig. 46) with 1 pair of definite lobes, second and third pairs absent or rarely represented by small, unsclerotized points; paraphysis formula usually 2-2-0, rarely 2-2-1, when 2-2-0 with paraphyses in space between position of lobe 2 and median lobe, attached to medial margin of position of lobe 2, medial margin of position of lobe 3, and in space between positions of lobes 2 and 3. Median lobes separated by space 0.1–0.3 (0.2) times width of median lobe, without basal sclerotization, usually with paraphysis attached to medial margin of median lobe, without yoke, medial margins parallel, converging, or slightly diverging apically, lateral margins converging, with 1–3 (1) lateral notches and without medial notch; second lobes normally absent, when present, simple, represented by unsclerotized point; third lobes usually absent, when present, similar to second lobe, without notches. Plates between median lobes and second lobe and between second lobes and third lobes usually without tines or with tines very small and infrequent; plate formula usually 2-2-0, occasionally 2-3-0, 1-2-0, or 2-1-0; plates between median lobes absent or very inconspicuous, when visible 0.1–0.4 (0.2) times as long as median lobes. Macroducts of 2 sizes, larger size on pygidium between median lobes and on segments 5 to 7 in marginal and submarginal areas, of 10 specimens examined 2 also had macroducts present on prepygidium in submarginal area of segment 4, duct between median lobes extending 1.3–1.9 (1.6) times distance between posterior apex of anal opening and base of median lobes, 40–57 (48) μ long, marginal macroduct in first space 34–58 (50) μ long, with 15–39 (22) macroducts on each side of pygidium on segments 5 to 8, ducts in submarginal and marginal areas, total of 17–41 (26) macroducts on each side of body, some macroduct orifices anterior of anal opening; smaller macroducts present along body margin from metathorax or segment 1 to segment 3 (absent on 1 specimen). Pygidial microducts on venter in submarginal and marginal areas of segment 5, sometimes 6, with 2–6 (4) ducts; prepygidial ducts of 1 size in submarginal and marginal areas of any or all of metathorax, segment 1 to segments 3 and 4, present submedially near spiracles on segments 1 to 3 or 4; on dorsum pygidial ducts absent; prepygidial microducts of 2 sizes, larger size in submedial areas of any or all of metathorax to 4, smaller size in submarginal areas of head or prothorax to segments 1, 2, or 3. Perivulvar pores in 4–5 (4) indefinite clusters, composed of 10–28 (17) pores on each side of body. Pores absent near spiracles. Anal opening located 1.8–3.6 (3.0) times length of anal opening from base of median lobes, anal opening 10–13 (10) μ long. Dorsal seta laterad of median lobes 0.6–1.2 (0.9) times length of median lobe. Eyes usually represented by small sclerotized spur or dome, rarely absent, on prothorax near intersegmental line with mesothorax. Antennae each with 1 seta. Cicatrices usually absent, sometimes present on prothorax and segment 1. Body pear shaped. Thorax and head sclerotized on old adult females.

Affinities The Osborn scale is similar to *Diaspidiotus ancylus*, Putnam scale, but differs by usually having the median lobes with short medial margins compared to the lateral margin, no conspicuous plates between median lobes, the plates simple or slightly fringed, and the plates absent anterior of segment 6. The Putnam scale usually has the median lobes with the medial margin only slightly shorter than the lateral margin, plates between the median lobes, conspicuously fimbriated plates, the plates usually present anterior of segment 6.

Hosts We have examined specimens from the following: *Carya, Castanea, Cornus, Ilex, Platanus, Populus, Quercus, Tilia,* and *Ulmus*; oak seems to be the most common host. We have been unable to verify the many hosts recorded in the literature and note that this species often has been misidentified. Gill (1997) indicated that it is common on walnuts and oaks in California.

Distribution This species probably is native to North America. We have examined U.S. material from AL, CT, DC, GA, IA, KS, LA, MD, MI, MO, MS, NC, NJ, NY, OH, OK, PA, SC, TN, TX, VA, VT, WI, WV. Nakahara (1982) reports it from CA, DE, FL, IL, KY, MA, NH, NM. The species is known also from Europe (Kozár et al. 1979; Pellizzari-Scaltriti and Camporese 1991).

Biology The Osborn scale has 2 generations per year and overwinters as mated adult females on oak stems in Maryland (Stoetzel and Davidson 1974). Eggs are laid at a rate of about 4–7 each day. Crawlers are present in late May and June and again in August. In Ohio crawlers are reported in early July (Kosztarab 1963). In Maryland adult males of the first generation are winged while those of the second are wingless.

Economic Importance This species is found on several ornamental and forest trees and is therefore considered to be a potentially economic species. Miller and Davidson (1990) consider this species to be an occasional pest.

Selected References Couch (1931); Ferris (1938a); Stoetzel and Davidson (1974).

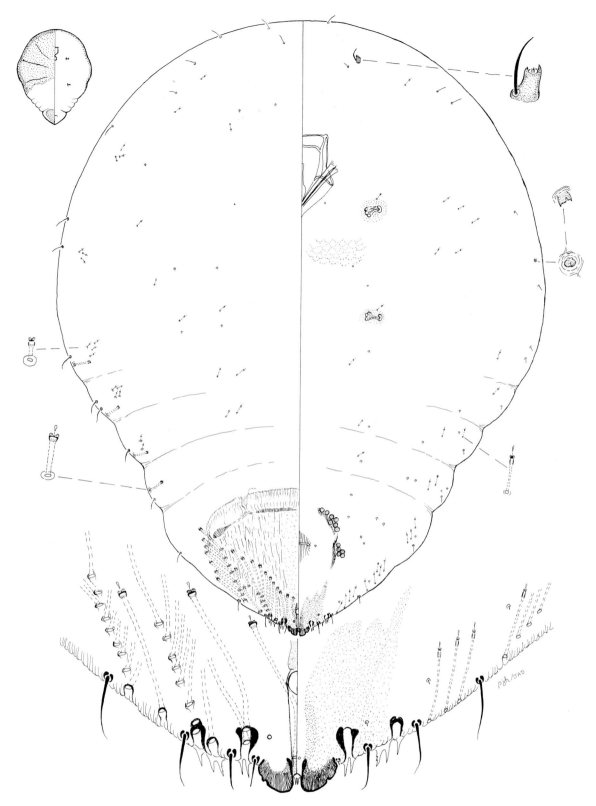

Figure 46. *Diaspidiotus osborni* (Newell and Cockerell), Osborn scale, Norfolk, VA, on *Quercus palustris*, I-25-1949.

Diaspidiotus ostreaeformis (Curtis)

ESA Approved Common Name European fruit scale (also called pear oyster scale, false San Jose scale, pear tree oyster scale, yellow apple scale, oystershell scale, oyster-shell scale, yellow oyster scale, green oyster scale, Curtis scale).

Common Synonyms and Combinations *Aspidiotus ostreaeformis* Curtis, *Aspidiotus betulae* Baerensprung, *Aspidiotus hippocastani* Signoret, *Aspidiotus oxyacanthae* Signoret, *Aspidiotus ostreaeformis oblongus* Goethe, *Aspidiotus ostreaeformis magnus* Goethe, *Aspidiotus (Euraspidiotus) ostreaeformis* Curtis, *Aspidiotus (Quadraspidiotus) ostreaeformis* Curtis, *Aspidiotus hunteri* Newell, *Quadraspidiotus gigas* (Thiem and Gerneck), *Quadraspidiotus ostreaeformis* (Curtis).

Field Characters (Plate 43) Adult female cover moderately convex, circular or oval, gray; shed skins subcentral, orange or yellow. Male cover smaller, elongate oval; shed skin submarginal, yellow. Body of adult female, eggs, and crawler light yellow. On bark and fruit.

Slide-mounted Characters Adult female (Fig. 47) with 3 pairs of lobes; paraphysis formula 2-2-0. Median lobes separated by space 0.1–0.3 (0.2) times width of median lobe, without basal sclerotization, usually with weakly indicated paraphysis-like sclerotization attached to medial margin, without yoke, medial margins parallel or converging apically, lateral margins converging apically, with 0–1 (1) lateral notch, 0–1 (0) medial notch; second lobes simple, about one-quarter size of median lobes, usually with acute, medial apex, with 0–1 (0) lateral and medial notch; third lobes simple, about one-half size of second lobe, usually apically acute, without notches; fourth lobes absent. Plates fringed or simple, medial plate in first space usually simple, others usually at least bifurcate, without associated microducts; plate formula usually 2-2-0, rarely 1-2-0, 2-3-0, or 2-4-0, without plates anterior of second space; plates between median lobes 0.4–1.7 (0.7) times as long as median lobes. Macroducts of 1 size, gradually decreasing in size anteriorly, duct between median lobes 31–41 (38) μ long, extending 0.3–0.5 (0.4) times distance to posterior apex of anal opening, longest duct in first space 35–52 (41) μ long, with 15–28 (21) macroducts on each side of pygidium on segments 5 to 8, ducts in submarginal and marginal areas, total of 25–40 (35) macroducts on each side of body, some macroduct orifices anterior of anal opening; prepygidial macroducts present submarginally on segments 2 or 3 to 4, segment 2 with 0–3 (1) ducts, segment 3 with 2–8 (4) ducts, segment 4 with 3–9 (6) ducts. Pygidial microducts on venter in submarginal and marginal areas of segments 5 and 6, with 7–15 (11) ducts; prepygidial ducts of 2 sizes, small size in submedial area on head, near mouthparts, anterior of each spiracle, and on any or all of metathorax to segment 3, submarginal small microducts, on any or all of metathorax to segment 4, large microducts on submargin of any or all of segments 1 to 4; on dorsum pygidial microducts absent; prepygidial microducts of 2 sizes, larger size in medial areas of metathorax, segment 1, or segment 2 to segment 3, in submedial areas of segments 1 or 2 to 3, in submarginal areas in position above anterior spiracle, rarely in marginal area near eye, smaller size submarginally on any or all of head to segment 2. Perivulvar pores in 4 or 5 groups, 8–24 (18) pores on each side of body. Perispiracular pores absent. Anal opening located 1.8–4.5 (2.9) times length of anal opening from base of median lobes, anal opening 10–17 (14) μ long. Dorsal seta laterad of median lobes 1.1–1.7 (1.3) times length of median lobe. Eyes represented by small dome or acute sclerotized area, located near body margin at level of antenna. Antennae each with 1 long seta. Cicatrices usually absent, rarely on prothorax and/or segment 1. Body pear shaped. Older specimens with head and thorax partially sclerotized. With 4–5 (5) macroducts in first space.

Affinities European fruit scale is similar to *Diaspidiotus gigas*, the poplar scale, for a comparison of these species see the affinities section of the latter.

Hosts Polyphagous. Borchsenius (1966) records it from 41 genera in 18 families of host plants. We have examined specimens from the following: *Caragana*, *Cornus*, *Betula*, *Malus*, *Pinus*, *Platanus*, *Populus*, *Prunus*, *Pyrus*, *Ribes*, *Schisandra*, *Sorbus*, *Syringa*, and *Tilia*. Many hosts that are reported in the literature may be in error.

Distribution This species probably is native to the colder parts of Europe. We have examined U.S. specimens from CO, CT, ID, MA, MI, MN, MT, NH, NY, OH, PA, RI, SD, UT, VT, WA, WI. It has been reported from IA, KS, ME, OR, and WY (Nakahara 1982). Outside of the United States it occurs in Argentina, the temperate portions of Asia, Australia, Canada, Europe, New Zealand, and Russia.

Biology The European fruit scale has 1 generation per year in all areas where it has been studied. In most situations it overwinters as second instars, but in England it is reported to overwinter as eggs (Anonymous 1977), and in New Zealand the predominant overwintering form is the first instar although second instars also occur (Richards 1962). In a different location in New Zealand the second instar is the predominant overwintering stage (McLaren 1989). In Germany only second instars survive the winter; overwintering first instars are dead by early spring (Schmutterer 1952a). Adult males and females are present in June, and crawlers first appear during the latter part of the month (Felt 1901). In British Columbia crawlers are present from June to September (Madsen and Arrand 1971). The European fruit scale seems to prefer a colder climate than the San Jose scale. This phenomenon is best demonstrated in New Zealand where San Jose scale does not occur south of Nelson, and its place as the dominant apple pest is occupied by the European fruit scale (Richards 1962). In British Columbia the two species rarely occur together but there is no clear-cut correlation with temperature (Morgan and Angle 1968). The European fruit scale is on the stems and branches of the host and prefers scaffold limbs with rough bark. As its name implies, it also is found on the fruit of its host, particularly at the calyx and stem ends. Crawlers and second instars are the only stages found on fruit.

Economic Importance Beardsley and González (1975) list this species as one of the principal armored scale pests in the world. Miller and Davidson (1990) also consider it to be an important pest. It is of economic importance in much of Europe (Anonymous 1977; Balachowsky 1950; Schmutterer 1959; Hippe et al. 1995), New Zealand (Richards 1962; McLaren 1989), Canada (Madsen and Arrand 1971; Morgan and Angle 1968), Australia (Williams 1970), and the northern United States (Felt 1901; Hutson 1936). It is most significant on apple but may be serious on pear, plum, apricot, and many other rosaceous hosts. The European fruit scale does not kill small twigs and branches like San Jose scale but causes damage primarily by contamination of the fruit where there is pitting and discoloration at feeding locations (Morgan and Angle 1968; Richards 1962). Certain varieties of apples seem to be more resistant to the scale than others (Morgan and Angle 1968). European fruit scale is able to survive on fruit in cold storage whereas San Jose scale is killed under such conditions (Morgan 1967).

Selected References Balachowsky (1950); McLaren (1989); Richards (1962).

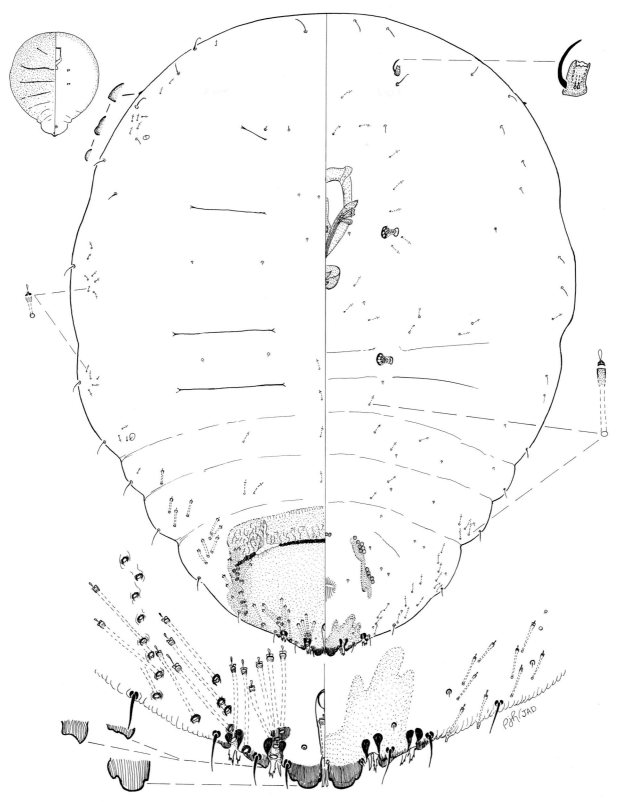

Figure 47. *Diaspidiotus ostreaeformis* (Curtis), European fruit scale, Castle Dale, UT, on *Malus* sp., V-14-1958.

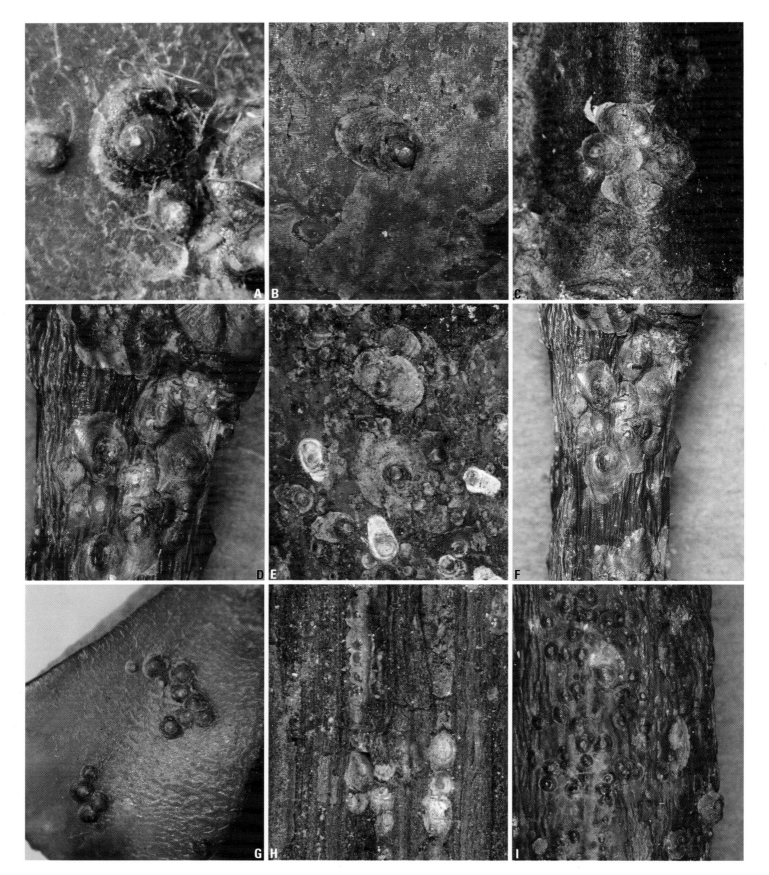

Plate 43. *Diaspidiotus ostreaeformis* (Curtis), European fruit scale

A. Adult female cover in center on apple (R. J. Gill).
B. Adult male cover on mountain ash (J. A. Davidson).
C. Male covers on currant (J. A. Davidson).
D. Female and male covers on willow (J. A. Davidson).
E. Infestation on willow (J. A. Davidson).

F. Infestation on willow (J. A. Davidson).
G. Distance view of female covers on apple (R. J. Gill).
H. Distance view on grape bark (J. A. Davidson).
I. Distance view (J. A. Davidson).

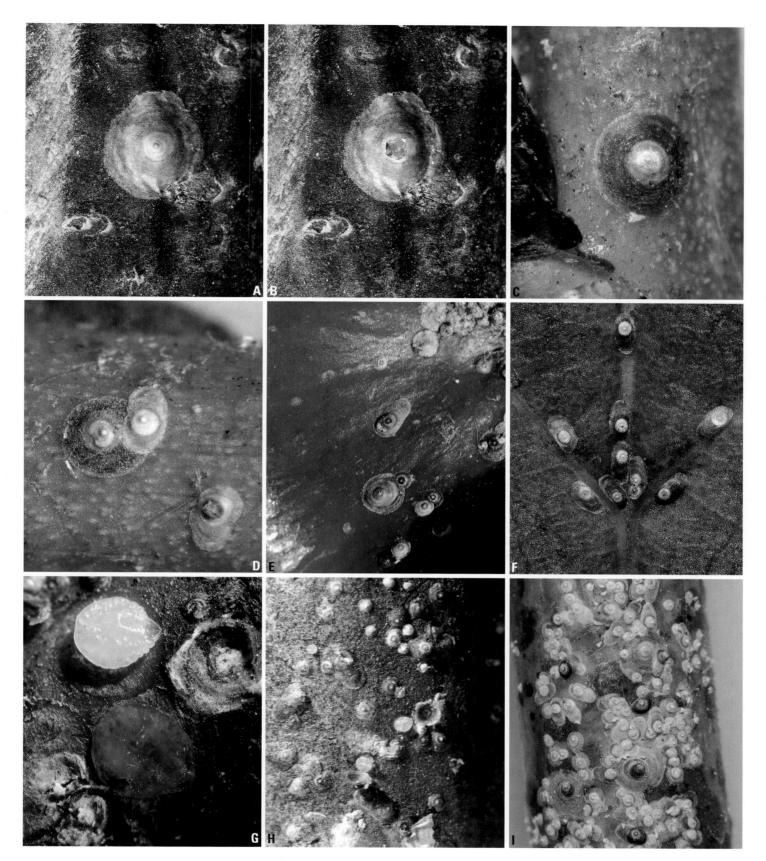

Plate 44. *Diaspidiotus perniciosus* (Comstock), San Jose scale

A. Unrubbed adult female cover on quince (J. A. Davidson).
B. Rubbed adult female cover showing yellow crawler skin (J. A. Davidson).
C. Dark adult female cover on almond (R. J. Gill).
D. Adult female cover and two male covers on almond (R. J. Gill).
E. Female, male, and immature covers on apple (J. A. Davidson).
F. Adult male covers on Japanese flowering cherry leaf (J. A. Davidson).

G. Body of unparasitized yellow adult female and body of parasitized brown adult female on quince (J. A. Davidson).
H. Covers removed to reveal overwintering crawlers on quince (J. A. Davidson).
I. Distance view of heavy infestation and discoloration damage on almond twig (R. J. Gill).

Diaspidiotus perniciosus (Comstock)

ESA Approved Common Name San Jose scale (also called Chinese scale and pernicious scale).

Common Synonyms and Combinations *Aspidiotus perniciosus* Comstock; *Aspidiotus* (*Comstockaspis*) *perniciosus* Comstock, *Aonidia fusca* Maskell, *Aonidiella perniciosa* (Comstock), *Comstockaspis perniciosus* (Comstock), *Aspidiotus* (*Hemiberlesia*) *perniciosus* (Comstock), *Aspidiotus* (*Quadraspidiotus*) *perniciosus* Comstock, *Hemiberlesiana perniciosa* (Comstock), *Quadraspidiotus perniciosus* (Comstock).

Field Characters (Plate 44) Adult female cover relatively flat, circular, gray to light brown; shed skins central to subcentral, yellow when rubbed. Male cover smaller, elongate oval, similar in color and texture to that of female; shed skin submarginal, yellow. Second-instar cover diagnostic—convex, circular, gray, with subcentral white ring. Body of adult female yellow; eggs yellow with orange; crawlers yellow orange. Mainly on bark but in high populations, leaves and fruit heavily infested. Characteristic red reaction appears on infested leaves, green stems, and cambium layer on some hosts. Males more abundant on leaves than females.

Slide-mounted Characters Adult female (Fig. 48) with 2 pairs of definite lobes, third lobes represented by narrow, plate-like structure; paraphysis formula 2-2-0, rarely 2-1-0. Median lobes separated by space 0.1–0.4 (0.2) times width of median lobe, without basal sclerotization, usually with paraphysis-like sclerotization attached to medial margin, without yoke, medial margins parallel or converging slightly, lateral margins converging apically, with 0–1 (1) lateral notch, 0–1 (0) medial notch; second lobes simple, about one-half size of median lobes, usually with rounded apex, with 0–2 (1) lateral notches, without medial notch; third lobes simple, unsclerotized, platelike, usually with acute apex; fourth lobes absent. Plates between lobes usually with 1 or more small tines, occasionally simple, plates between third lobe and seta marking segment 5 usually 3 in number with 1 or 2 conspicuous tines, marginal microducts anterior of segment 5 seta often with protruding orifice; plate formula 2-3-3, rarely 2-3-2 or 2-3-4; plates between median lobes 0.4–0.8 (0.6) times as long as median lobes. Macroducts of 1 size, duct between median lobes 45–59 (51) μ long, extending 1.4–2.3 (1.7) times distance between posterior apex of anal opening and base of median lobes, longest duct in first space 45–57 (49) μ long, with 13–19 (15) macroducts on each side of pygidium on segments 5 to 8, ducts in submarginal and marginal areas, some macroduct orifices anterior of anal opening; prepygidial macroducts absent. Pygidial microducts on venter in submarginal and marginal areas of segment 5, with 2–4 (3) ducts; prepygidial ducts of 2 sizes, small size in submedial area near mouthparts and from prothorax to metathorax or segment 1, in submarginal and marginal areas of segment 2 or 3 to 4, rarely absent from these areas, large microducts on submargin of mesothorax, metathorax, segment 1, or segment 2 to segment 3 or 4; on dorsum pygidial ducts occasionally on submargin of segment 5; prepygidial microducts of 2 sizes, larger size in medial areas of any or all of metathorax to segment 4, in submarginal areas of segments 1 or 2 to 4, smaller size in marginal areas of head or prothorax to metathorax, segment 1, or 2. Perivulvar and perispiracular pores absent. Anal opening located 1.6–2.6 (2.1) times length of anal opening from base of median lobes, anal opening 12–20 (14) μ long. Dorsal seta laterad of median lobes 0.7–1.0 (0.9) times length of median lobe. Eyes absent or small and shaped like spur or dome, located anterior and laterad of mouthparts. Antennae each with 1 long seta. Cicatrices usually absent, occasionally on prothorax and/or

segment 1. Body pear shaped. Area posterior and mesad of anterior spiracle with small dermal pocket.

Affinities The San Jose scale is distinctive by lacking perivulvar pores and by having 3 uniquely shaped plates anterior of seta marking segment 6.

Hosts Polyphagous. Borchsenius (1966) records it from 240 genera in 81 families of host plants; Dekle (1977) reports it from 35 genera in Florida with *Pyracantha*, *Pyrus*, *Prunus*, and *Rosa* the most frequently reported host genera. The national (USNM) scale insect collection contains nearly 2000 slides of this pest but the great majority were collected from fruit trees, particularly: *Citrus*, *Malus*, *Prunus*, and *Pyrus*. We have examined specimens from *Actinidia*, *Aleurites*, *Aronia*, *Asclepias*, *Betula*, *Buxus*, *Camellia*, *Capsicum*, *Carya*, *Ceanothus*, *Citrus*, *Codiaeum*, *Cornus*, *Cotoneaster*, *Crataegus*, *Cydonia*, *Diospyros*, *Eriobotrya*, *Eucalyptus*, *Euonymus*, *Fagus*, *Hypericum*, *Juglans*, *Magnolia*, *Mahonia*, *Malus*, *Morus*, *Olea*, *Paeonia*, *Photinia*, *Pinus*, *Prunus*, *Pyracantha*, *Pyrus*, *Quercus*, *Rhus*, *Ribes*, *Robinia*, *Rosa*, *Sorbaria*, *Sorbus*, *Spiraea*, *Symphoricarpos*, *Syringa*, *Tilia*, *Ulmus*, *Veronica*, *Viburnum*, *Vitis*, *Zelkova*, and *Ziziphus*.

Distribution This species is believed to have originated in the area of northern China–Soviet Far East–North Korea (Rosen and Debach 1978). It was discovered in the United States about 1870 in San Jose, California (hence its name), in the orchard of a private citizen who had imported much plant material from the Orient. In about 1886 a couple of nurseries in New Jersey bought infested 'curculio-proof' plum varieties from a California nursery and the scale became established on the East coast. From these foci it rapidly spread throughout the United States, and by 1897 was recorded from 33 states and Canada (Chambliss 1898). We have examined U.S. material from AL, AR, AZ, CA, CO, CT, DC, DE, FL, HI, IA, ID, IL, IN, KS, KY, LA, MA, MD, MI, MO, MS, NC, NH, NJ, NM, NY, OH, OK, OR, PA, RI, SC, TN, TX, UT, VA, VT, WA, WI, WV, WY. San Jose scale is widely distributed throughout the temperate and subtropical areas of the world and often is intercepted in quarantine on fruit, especially oranges and tangerines. A distribution map of this species was published by CAB International (1986).

Biology The San Jose scale is the most studied armored scale insect in the world. According to Rosen and DeBach (1978) this pest may undergo 1 to 5 generations per year with climate being the major limiting factor. In the colder, maritime territories of Russia 1 generation per year was reported, but in warmer areas 2 (Tereznikova 1969) and 2 to 4 (Smol'yannikov 1980) generations were found. In central Europe 2 generations per year prevailed, for instance, in northern France (Bénassy 1969), Bulgaria (Kr'steva 1977), Hungary (Balás and Sáringer 1982), and Czechoslovakia (Lelláková-Dusková 1969). In the Mediterranean region 3 generations were found, for instance, in France (Loucif and Bonafonte 1977), Italy (Zocchi 1960), Portugal (Freitas 1966), Greece (Navrozidis et al. 1999), and Turkey (Duzgunes 1969). Similarly, 3 generations were seen in New Zealand (Wearing 1976), Japan (Murakami 1970), India (Dinabandhoo and Bhalla 1980), Chile (González 1981), and Uruguay (Carbonell Bruhn and Briozzo Beltrame 1975). In North America Johnson and Lyon (1976) reported that up to 5 generations per year were possible. Generation times recorded for the United States include 2 in southern Oregon (Westigard et al. 1979), 3 to 4 in California (Ferris 1981), 2 or more in Ohio (Kosztarab 1963), and 3 in Maryland (Stoetzel 1975). The species has been reported to be oviparous (Rosen and DeBach 1978, Stoetzel 1975) and ovovi-

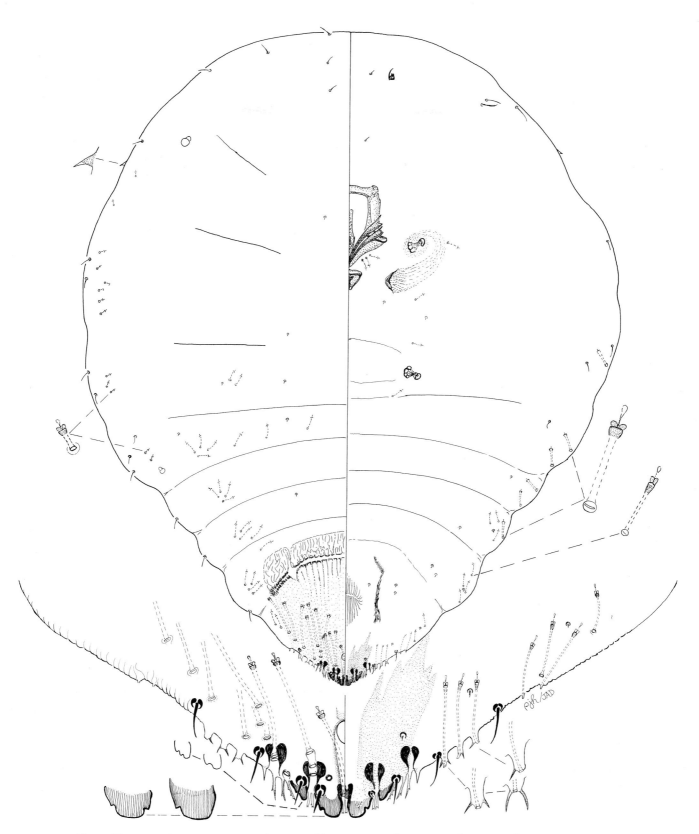

Figure 48. *Diaspidiotus perniciosus* (Comstock), San Jose scale, MD, on flowering cherry, X-30-1953.

viparous (many early authors, e.g., Chambliss 1898; Fernald 1899). There is a possibility that both conditions exist, but more than likely the delicate 'egg shell' (chorion or amnion) is broken prior to 'birth,' or is lost soon after oviposition. The existence of this delicate membrane around the developing embryo was noted by the observant Pergande (in Marlatt 1906), but he chose to consider the species oviparous. The overwintering stage is the settled crawler in most areas: Russia (Danzig 1964; Smol'yannikov 1980), Germany (Schmutterer 1959), Czechoslovakia (Huba 1962), Hungary (Balás and Sáringer 1982; Bognár and Vinis 1979), Bulgaria (Kr'steva 1977), Italy (Zocchi 1960), Portugal (Freitas 1966), Chile (González 1981), and New Zealand (Wearing 1976). In North America crawlers were the overwintering stage reported in Canada (Anonymous 1981), Oregon (Westigard 1979), California (Ferris 1981), and Maryland (Stoetzel 1975). A few workers have reported second instars as the overwintering stage: Loucif and Bonafonte (1977) in France and Italy and Wearing (1976) in New Zealand. Gentile and Summers (1958) found 80% of the pest population in Sacramento, California, in January were settled crawlers (black-cap phase). The term black-cap refers to the dark gray cover of the settled crawler (first instar). After 1 or 2 generations of development in warm areas there is little synchrony in crawler emergence, and most stages are present at the same time but in varying numbers. Kawecki (1956) reported that in Poland part of a generation of crawlers may go into diapause in the summer. Chumakova (1965) recorded that the onset of diapause is determined by temperatures below 20 °C or by a daily photoperiod shorter than 16 to 18 hours. Rock and McClain (1990) reported that this species hibernates in a nondiapausing state of dormancy and this dormancy is temperature dependent. According to studies by Marlatt (1906) in Washington, D.C., 4 complete generations develop annually. Males greatly outnumbered females in the overwintering generation, but females outnumbered males in late summer generations. Developmental times for male instars were as follows: hatching to prepupa about 18 days, to pupa an additional 2 days, to adult male an additional 4 to 6 days, or about 26 days from hatching. Developmental times for female instars were as follows: Hatching to second instar about 20 days; to third instar (adult) about 10 additional days. A new generation of crawlers appeared in 33 to 40 days. Females each produced about 400 crawlers. Gentile and Summers (1958) worked out the life history with colonies on potted peach trees in Sacramento, California. They found 3 major generations per year with first generation crawlers active in April, May, and June; second-generation crawlers in July and August; and third-generation crawlers in September, October, and November. Unimpregnated females did not produce fertile eggs. The threshold of development has been calculated at 7.3 °C, with a thermal constant of 770 effective day-degrees required for the completion of one generation (Huba 1962). Freitas reported that high summer temperature and low humidity are especially injurious to early instars of San Jose scale, and that these factors are more important than parasites in regulating populations on apple in Portugal. Mathys and Guignard (1967) found that crawler emergence did not occur below 50% RH. Rice (1974) documented the presence of a female-produced pheromone for this species by using virgin females to attract males in the field. Rice and Jones (1977) used gourds infested with virgin females to study male flight patterns in Fresno, California. The 4 main male flight periods were March–April, June, July–August, and September–October. Gieselmann et al. (1979b) identified and synthesized the San Jose scale sex pheromone. Bennett and Brown (1958) stated that in 1954, Linden reported a haplodiploid sex differential in this scale, which has 8 chromosomes in the female and 4 in the male. He assumed males developed from unfertilized eggs.

Economic Importance A review of the economic literature is in many ways a review of the evolution of economic entomology in the United States, and it partly explains why it developed so quickly. The impact of this foreign pest on newly infested orchards was devastating. For example, 1000 acres of mature apple trees were killed in Illinois alone in 1962 (Metcalf et al. 1962). Pear, peach, plum, prune, apricot, nectarine, and sweet cherry growers also suffered serious losses throughout the country. Marlatt (1906) chronicled the effects of this pest on fruit trees, noting that a halo developed around scales feeding on fruit and tender twigs. Heavily infested fruit became distorted, pitted, cracked, and often fell prematurely, or at least were unmarketable. The cambium layer of young, heavily infested twigs usually was stained deep red or purplish. Gentile and Summers (1958) noted that this insect infests all surface parts of young host plants from ground level to the tips of the shoots. Branches of mature trees were favored infestation sites. Leaves and fruits were colonized only in severe infestations. Heavily infested peach twigs were observed to die back 4 months after arrival of the first crawlers. Unchecked infestations killed young trees within 3 years. Large infestations on mature trees permanently affected their structure, vigor, and productivity. The realization that other serious pests could be imported accidentally in the same manner led the federal government to develop and finally institute in 1912 a series of quarantine laws governing the importation of plant materials. Most states had laws preventing the shipment of infested nursery stock by the end of the 1800s (Metcalf et al. 1962). Attempts to control this scale resulted in the development of a lime-sulphur spray, which Felt (1907) reported gave 'satisfactory results' if properly prepared and applied. This treatment was the recommended control by the experiment stations of 23 states (Surface 1907). This material was the first general use of an insecticide spray in the United States and was widely employed for control of San Jose scale until about 1922. The first cases of insect resistance were reported by Melander (1915) who found differences among populations of San Jose scale to lime-sulphur sprays in the fruit orchards of Washington. Later, it was discovered accidentally that lime-sulphur also controlled peach leaf curl disease and this led to experiments against other orchard diseases (Davis 1933). Although petroleum oil fractions were used as insecticides as early as 1871 (Riley 1892), the sudden popularity of lime-sulphur slowed the development of oil sprays for scale control. Eventually Jones (1910) reported that lubricating and crude oil emulsions were much more effective than lime-sulphur against the European fruit lecanium. Later Ackerman (1923) reported that oil emulsions were more effective than lime-sulphur for San Jose scale control and that oil sprays had superior handling qualities. This recommendation was rapidly adopted throughout the country and oils became the dormant spray of choice for San Jose and other scales. In modern times refined oils have become important components of control strategies for this pest including the use of soybean oils for dormant sprays (Pless et al. 1995; Hix et al. 1999). Continued refinements led to the addition of modern insecticides to oil sprays for increased efficacy against this pest (Anthon 1960). Early attempts to eradicate San Jose scale from shipments of nursery stock led to the use of gases for fumigation. Hydrocyanic acid gas was the first widely used fumigant utilized for this purpose (Johnson 1898). Tents were placed over orchard trees for field fumigation. Britton (1908) summarized the early work with carbon disulphide, carbon tetrachloride, and hydrocyanic acid gas fumigation tests against this scale. This adaptable pest eventually exhibited resistance to hydrocyanic acid gas (Beran 1943). Following World War II, DDT, the harbinger of the great synthetic organic insecticide industry, came into wide use for controlling scales and other insect pests. This residual insecticide was useful because its effectiveness covered the long crawler emergence periods. Barnes (1959) found that a foliar spray of DDT controlled San Jose scale on fruit trees. Occasionally San Jose scale outbreaks occurred in orchards repeatedly treated with synthetic organic insecticides (English 1955; Duda 1959), presumably due to the development of resistance, or due to a disturbance in the parasite–predator control complex. González (1981) conducted a detailed study on the ecology, biology, and pest status of the species in Chile. The earlier noted discovery that virgin females produced a male attracting pheromone, and the subsequent identification and synthesis of the compound, led applied entomologists to use this knowledge in the development of novel San Jose scale control programs. Three years of male trapping studies showed that small, open, sticky tent traps baited with fresh 300-μg pheromone loads, and hung high in the north or east quadrants of trees caught the most males (Hoyt et al. 1983). Rice et al. (1982) demonstrated that male traps could be used in conjunction with day-degree data to develop a crawler-spray control model. They found that May crawler sprays should be timed in California based on specified day-degrees following the first time males are trapped. Downing and Logan (1977) showed in laboratory tests that adult male San Jose scales are killed easily by insecticides. They then monitored field populations with the benefit of pheromone traps and timed sprays to emergence of the first males. They found that 1 spray of Penncap E or Diazinon at male emergence gave as good control of the scale as 3 sprays directed against the crawler stage. Because males don't feed after the second instar, the mode of action must be by contact. Estimates of scale infestation rates were based on the number of live scales found on twigs and apples among sprayed and control trees. Simulation models continue to be important tools for the control of this scale (McClain et al. 1990a; McClain et al. 1990b; Schaub et al. 1999). Mague and Reissig (1983) summarized the recent work on airborne scale crawler dispersal and studied San Jose scale crawler airborne dispersal in apple orchards. They found airborne transport was important in dispersing to new hosts. Within tree dispersal throughout the season varied from 1.6 to 20 times greater than between trees. Significantly more crawlers were recorded in the upper canopy than on lower limbs. It could not be determined if this resulted from better spray coverage of the lower canopy, or a behavioral component of the scale insect. Russian workers have studied fruit tree resistance to San Jose scale. Prints (1971) reported an inverse correlation between the thickness of secondary bark and the amount of necrosis this pest was able to cause on plum and apple. Busuiok (1972) found some apple varieties, such as Lithuanian pipping, mleev reinette, virgin pink, and sary-sinap, resistant to San Jose scale over a large area of the former Soviet Union. Gatina (1973) noted that in the flood plain of the Dnester, clyman, tragedy resistant, agen prune, and common prune only suffer slight damage from this pest. Kalabekov (1974) studied 240 varieties of fruit trees and found a group of apple varieties resistant enough to San Jose scale to be grown without the use of specific insecticide sprays for the scale. Bichina and Gatina (1976) studied the use of resistant apple varieties in an integrated pest control program. According to Rosen and DeBach (1978), in the years immediately after its introduction into the United States the San Jose scale population increased rapidly as the pest spread throughout the country leaving severely damaged fruit orchards in its wake. Eventually the scale infestations became much reduced in severity. This generally is attributed to the impact of predators and parasites, especially the parasite *Encarsia perniciosi* (Tower). Beardsley and González (1975) consider this scale to be one of 43 serious armored scale pests. Miller and Davidson (1990) consider it to be a serious world pest.

Selected References Ferris (1938a); Rosen and DeBach (1978).

Diaspidiotus uvae (Comstock)

Suggested Common Name Grape scale.

Common Synonyms and Combinations *Aspidiotus uvae* Comstock, *Aspidiotus (Diaspidiotus) uvae* Comstock, *Diaspidiotus uviae* (Comstock), *Aspidiotus uvaspis* Lindinger.

Field Characters (Plate 45) Adult female cover white or grayish white, flat, circular, or slightly elongate; shed skins central or subcentral, yellow or orange. Male cover similar to female cover except oval with submarginal shed skin. Body of adult female pale yellow; eggs and crawlers yellow. On stems and branches, often found under bark.

Slide-mounted Characters Adult female (Fig. 49) with 1 pair of definite lobes, second and third pairs absent or rarely represented by small, unsclerotized points; paraphysis formula usually 2-2-0, with paraphyses in space between position of lobe 2 and median lobe, attached to medial margin of position of lobe 2, medial margin of position of lobe 3, and in space between positions of lobes 2 and 3. Median lobes separated by space 0.1–0.3 (0.2) times width of median lobe, without basal sclerotization or yoke, with small paraphysis on medial margin, medial margin usually slightly diverging apically, lateral margins converging, with 1 lateral notch and 1 medial notch; second lobes normally absent, when present simple, represented by unsclerotized point; third lobes usually absent, when present, similar to second lobe, without notches. Plates between median lobes and second lobe and between second lobes and third lobes usually with noticeable tines; plate formula 2-2-3, 2-3-2, 2-3-3, or 2-3-4; plates between median lobes absent. Macroducts of 2 sizes, larger size on pygidium between median lobes and on segments 5 to 7 in marginal and submarginal areas, of 10 specimens examined 6 also had macroducts present on prepygidium in submarginal area of segment 4, duct between median lobes absent, marginal duct in first space 42–55 (49) µ long, with 20–31 (24) macroducts on each side of pygidium on segments 5 to 8, ducts in submarginal and marginal areas, total of 24–38 (29) macroducts on each side of body; some macroduct orifices anterior of anal opening; smaller macroducts present along body margin from metathorax or segment 1 to segment 3. Pygidial microducts on venter in submarginal and marginal areas of segment 5, sometimes on 6, with 5–9 (4) ducts; prepygidial ducts of 1 size in submarginal and marginal areas of segments 3 and 4, present submedially near mouthparts and spiracles; on dorsum pygidial ducts absent; prepygidial microducts of 2 sizes, larger size in submedial areas of any or all of metathorax to 4, smaller size in submarginal areas of head or prothorax to segments 2 or 3 (absent from 1 specimen). Perivulvar pores in 5 indefinite clusters (median 'cluster' composed of 1 or 2 pores), with 9–13 (11) pores on each side of body. Pores absent near spiracles. Anal opening located 2.1–3.7 (2.8) times length of anal opening from base of median lobes, anal opening 10–16 (14) µ long. Dorsal seta laterad of median lobes 0.9–1.2 (1.1) times length of median lobe. Eyes usually represented by small sclerotized spur or dome, rarely absent, on prothorax near intersegmental line with mesothorax. Antennae each with 1 seta. Cicatrices usually absent, sometimes present on prothorax and segment 1. Body pear shaped. Head and thorax often sclerotized in older adult females.

Affinities This species is similar to *Diaspidiotus ancylus*, Putnam scale. For a comparison of these species see affinities section of the Putnam scale. Grape scale can be distinguished from all economic species of *Diaspidiotus* in the United States by the absence of a macroduct between the median lobes.

Hosts We have seen grape scale only from grape (*Vitis*). We have examined specimens of *D. ancylus* misidentified as *D.*

uvae from the following hosts: *Acer, Carya, Catalpa, Fraxinus, Liquidambar, Maclura, Platanus, Populus, Prunus, Quercus, Syringa,* and *Vaccinium*. We believe that many records of this species from hosts other than grape are erroneous.

Distribution We have examined U.S. specimens from AR, DC, FL, GA, IL, IN, KY, MD, MO, MS, NC, OH, PA, VA. Literature records in the United States include AL, CA, CT, DE, KS, NJ, NY, TN, TX, WV. Foreign records include Brazil, Canary Islands, Spain, and the West Indies, but these may be based on misidentifications.

Biology The grape scale has 2 generations per year in Arkansas, with peak emergence of crawlers in the first generation around May 20, adult male peak emergence about June 26, with peak emergence of crawlers in the second generation around August 12, adult male peak emergence about September 18 (Johnson et al. 1999). They found winged males in the spring and wingless ones in the fall and were able to predict crawler emergence based on day-degree accumulations. This species has 1 generation per year in the District of Columbia (Zimmer 1912) and Missouri (Hollinger 1923). Crawlers are reported in May and June in D.C. (Zimmer 1912). Adult females lay 35–50 crawlers rather than eggs; the species overwinters in a 'nearly full grown' condition (Zimmer 1912). Whitehead (1963) found that the species overwinters as mated adult females in Arkansas. Eggs or crawlers were laid by the adult female; some females laid only crawlers and others laid a combination of crawlers and eggs, but none laid eggs alone. Crawlers were first observed around May 15. Crawler emergence occurred on May 26 and second instars of the first generation appeared about 12 days later; adult females first appeared around June 15. Adult males were present around June 25. Eggs and crawlers of the second generation were present in the field on July 25 one year and July 16 the next; second-generation second-instar females were present about 9 days later, and adult females were present beginning mid-September. Adult males, which were wingless, also were present in mid-September. Eggs didn't appear in the overwintering females until late April. Taylor (1908) states that the species overwinters as eggs in Missouri; however, Hollinger (1923) reports the overwintering stage as adult females in Missouri, and Kosztarab (1963) reports similar observations in Ohio. The largest populations of grape scale occur on 2- or 3-year-old canes.

Economic Importance Zimmer (1912) notes grape scale 'injuries have become important, requiring treatment for the preservation of the vines.' Hollinger (1923) reports this scale doing 'considerable damage' to improperly maintained vineyards in Missouri. Ebeling (1959) notes in heavy infestations 'the growth of vines may be retarded.' Johnson et al. (1999) indicate that in Arkansas before 1950 the grape scale was of little economic importance, but since that time this pest has caused many vineyards to be abandoned or removed because the scale killed vines. Whitehead (1963) suggests that the problems in Arkansas were the result of insecticide use for the control of other vineyard pests. He indicates that natural enemies that normally keep the scale under natural control are severely impacted by the pesticides. Kimouilo and Costa (1987) suggest the strong possibility that grape scale produces a toxin that causes 'yellow net' symptoms on grapes in Brazil. According to Gobbato (1936) grape scale is one of the principal pests of grape in Brazil. Miller and Davidson (1990) consider this species to be an occasional pest.

Selected References Ferris (1938a); Johnson et al. (1999); Ruiz Castro (1944); Zimmer (1912).

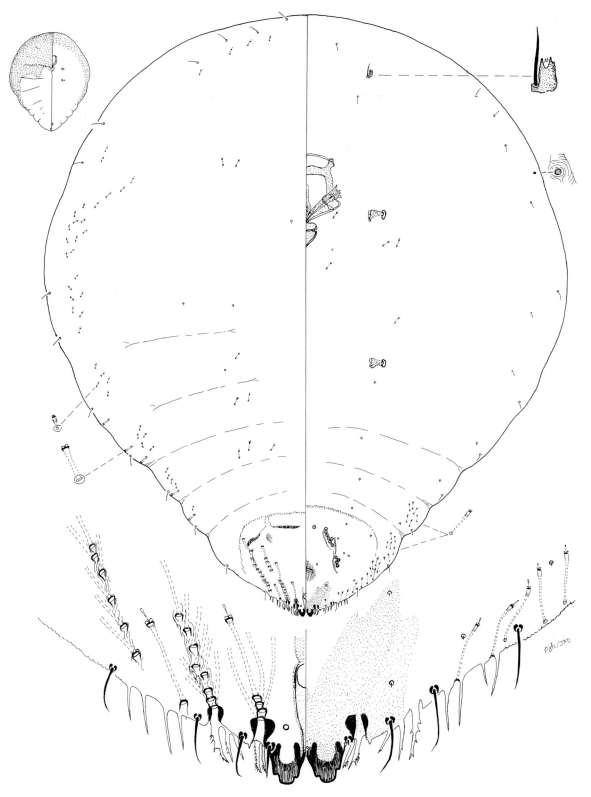

Figure 49. *Diaspidiotus uvae* (Comstock), grape scale, Batavia, OH, on *Vitis* sp., II-26-1974.

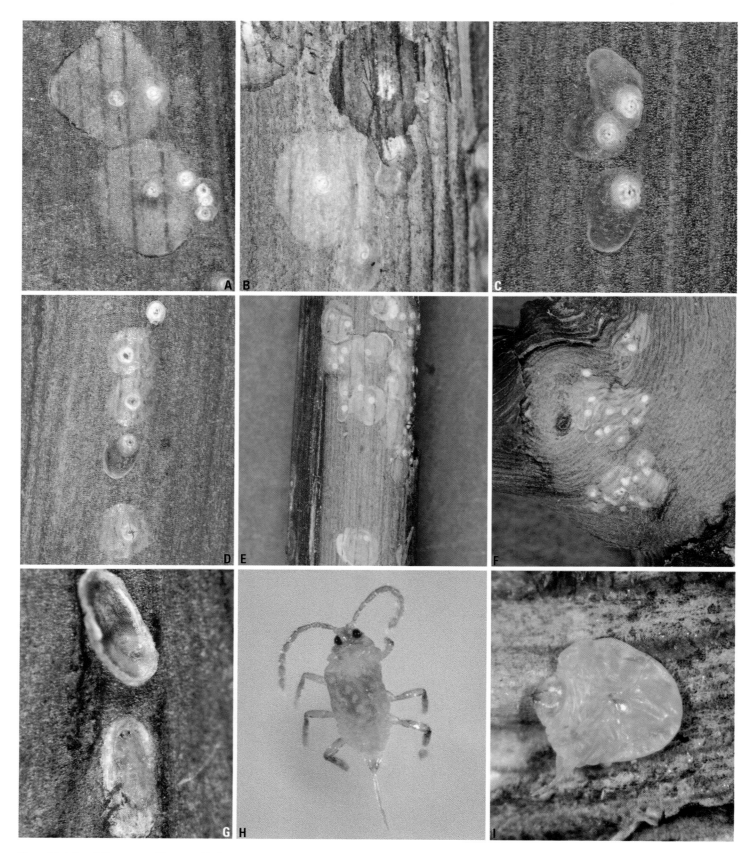

Plate 45. *Diaspidiotus uvae* (Comstock), Grape scale

A. Adult female covers on grape (J. A. Davidson).
B. Large adult female cover and small male cover under epidermis of bark of grape. Note scars where other scales were removed (J. A. Davidson).
C. Male covers on grape (J. A. Davidson).
D. Dark male and young adult female covers on grape (J. A. Davidson).

E. Cluster of covers on grape (J. A. Davidson).
F. Distance view of covers around twig base of grape (J. A. Davidson).
G. Pupa of wingless male (J. A. Davidson).
H. Wingless adult male (J. A. Davidson).
I. Body of adult female on grape (J. A. Davidson).

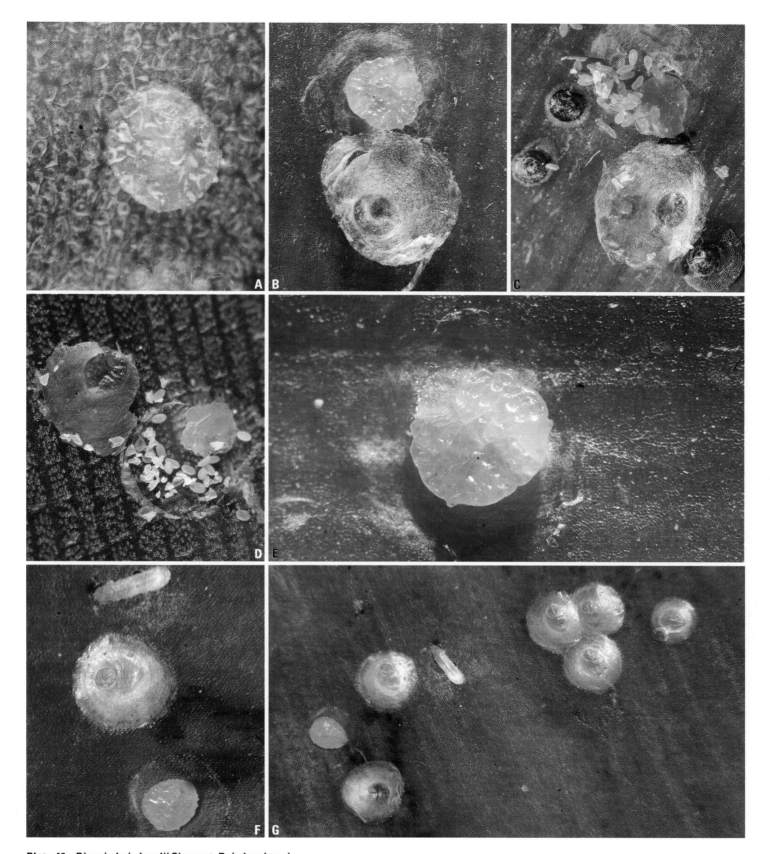

Plate 46. *Diaspis boisduvalii* Signoret, Boisduval scale

A. Translucent adult female cover on bromeliad (J. A. Davidson).
B. Young adult female with cover removed on bromeliad (J. A. Davidson).
C. Adult female with eggs on bromeliad. Black covers are *Gymnaspis aechmeae* (flyspeck scale) (J. A. Davidson).
D. Adult female with egg shells, eggs, and crawlers on bird-of-paradise (J. A. Davidson).

E. Body of young adult female showing protuberances bearing eye spots on bromeliad (J. A. Davidson).
F. Body of adult female, cover, and male cover on orchid (R. J. Gill).
G. Distance view of male and female covers on orchid (R. J. Gill).

179

Diaspis boisduvalii Signoret

Suggested Common Name Boisduval scale (also called cocoa-nut snow scale, cocos scale).

Common Synonyms and Combinations *Aulacaspis boisduvalii* (Signoret), *Aulacaspis cattleya* Cockerell, *Diaspis cattleya* (Cockerell), *Diaspis trinacis* Colvee.

Field Characters (Plate 46) Adult female cover white, transparent, flat, circular shed skins subcentral, light yellow. Male cover white, elongate, tricarinate, with shed skin marginal. Body of adult female bright yellow; eggs clear when laid, changing from yellow to orange before hatching. Primarily on leaves and pseudobulbs. When scale cover removed, adult female with single hornlike projection on either side of body near junction of head and thorax (Gill 1997).

Slide-mounted Characters Adult female (Fig. 50) with 4 pairs of definite lobes; with small paraphyses associated with first 3 pairs of lobes, most apparent on medial margin of medial lobule of each lobe. Median lobes separated by space 0.8–1.4 (1.0) times width of median lobe, with lateral and medial paraphyses, small basal sclerosis, without yoke, lateral margins of lobes usually slightly divergent, occasionally parallel, medial margins strongly divergent apically, with 0–1 (0) lateral notches, with 5–10 (8) medial notches; second lobes bilobed, smaller than median lobes, medial lobule largest without notches, lateral lobule without notches; third lobes bilobed, similar in appearance to second lobes, about equal in size to second lobes, without notches; fourth lobes simple, represented by flat projection containing series of 2–6 (5) notches. Gland-spine formula 1-1-1, with 10–12 (11) gland spines near each body margin from segments 2 to 4; without gland spines between median lobes. Macroducts of 2 sizes, largest in submarginal areas of segments 6 and 7 and in marginal areas of segments 4 to 8, small size ducts usually in submedial areas of segments 6 and 7, occasionally also on segment 5 or absent completely, in marginal and submarginal areas from segments 2 to 6, with large duct between median lobes 18–22 (20) μ long, macroduct in first space 20–24 (22) μ long, with 15–21 (20) macroducts on each side of body on segments 5 to 7, total of 29–52 (40) macroducts on each side of body, some macroduct orifices anterior of anal opening. Pygidial microducts usually on venter in submarginal areas of segments 5 to 7, rarely absent on segment 7, with 3–8 (5) ducts; prepygidial ducts of 1 size in submarginal and marginal areas of segments 3 and/or 4, present submedially near any or all of mouthparts, posterior spiracle, and segments 1 to 4; on dorsum pygidial ducts absent; prepygidial microducts of 2 sizes, larger size sometimes in submedial areas of metathorax to segment 2, also forming a band from body margin to submedial area of metathorax, smaller size in submarginal areas of mesothorax to segments 1. Perivulvar pores in 5 groups, 28–46 (38) pores on each side of body. Perispiracular pores with 3 loculi, anterior spiracles with 1–7 (3) pores, posterior spiracles without pores. Anal opening located 2.7–4.3 (3.4) times length of anal opening from base of median lobes, anal opening 15–22 (19) μ long. Dorsal seta laterad of median lobes 1.3–2.1 (1.6) times length of median lobe. Eyes apparently represented by small raised areas on prosomal lobes, located on body margin on prothorax. Antennae each with 1 long seta. Cicatrices usually present on prothorax, segment 1, and segment 3, sometimes absent from prothorax. Body turbinate. Prosomal lobes usually present on margin of prothorax, rarely absent or inconspicuous. Conspicuous, sclerotized spur on body margin near seta marking segment 4, represented by narrowly pointed process.

Affinities Of the economic species in the United States, Boisduval scale is most similar to *Diaspis bromeliae*, pineapple

scale. The former has: 2 large, submarginal macroducts on each side of the pygidium; many small macroducts; 15–21 (20) macroducts on each side of body on segments 5 to 7; anal opening located 2.7–4.3 (3.4) times length of anal opening from base of median lobes; and anterior spiracles with 1–7 (3) associated perispiracular pores. Pineapple scale has: more than 2 large, submarginal macroducts on each side of the pygidium; few small macroducts; 10–13 (11) macroducts on each side of body on segments 5 to 7; anal opening located 3.8–6.5 (5.6) times length of anal opening from base of median lobes; and anterior spiracles with 12–16 (14) associated perispiracular pores.

Hosts Polyphagous. Borchsenius (1966) records the species from 44 genera in 15 families; Dekle (1977) records it from 65 genera in Florida. In the United States, Boisduval scale is found most often on orchids and palms in greenhouses. The orchid genera *Cattleya* and *Cymbidium* seem to be common hosts. In the tropics this pest is most often found on orchids, palms, bromeliads, and bananas. We have examined specimens from the following host genera: *Achras, Acineta, Aechmea, Ananas, Angraecum, Areca, Astrocaryum, Attalea, Bactris, Billbergia, Bletia, Brassavola, Brassia, Broughtonia, Calathea, Catopsis, Cattleya, Caularthron, Chamaedorea, Chamaerops, Chrysalidocarpus, Cocos, Codiaeum, Corozo, Cycnoches, Cymbidium, Dendrobium, Dracaena, Elaeis, Epidendrum, Epidendrum, Epipremnum, Ferocactus, Ficus, Guilielma, Heliconia, Hylocereus, Kentia, Laelia, Latania, Livistona, Lucaena, Lycaste, Maranta, Maxillaria, Miltonia, Musa, Neofinetia, Neoglaziovia, Nidularium, Nypa, Odontoglossum, Oncidium, Opuntia, Pandanus, Peristeria, Persea, Phoenix, Pitcairnia, Pleurothallis, Puya, Ravenala, Renanthera, Rhipsalis, Ronnbergia, Roystonea, Schomburgkia, Sophronitis, Stanhopea, Strelitzia, Symplocos, Tillandsia, Trachycarpus, Trichopilia, Vanda, Vitis, Vriesea, Washingtonia,* and *Xylobium.*

Distribution We have examined U.S. specimens from AK, AL, CA, CT, DC, GA, HI, IL, LA, MA, MD, MO, NH, NJ, OH, PA, VA. This species also has been reported from CO, DE, FL, IA, KY, MS, NC, OK, TN, TX, WA, WI. It occurs throughout the tropics of the world and in greenhouses in the North. It most often is intercepted in quarantine from Central and South America and Mexico.

Biology The following information is taken from Bohart (1942). In the greenhouse, on orchids, the period from egg to egg-laying females requires about 50 days; adult males require about 33 days. Eggs hatch in 5 to 7 days after being laid. The crawler stage lasts about 9 days, the second-instar female about 8 days. Adult males leave the scale cover about 15 days after the first molt. Adult females may live as long as 7 months, and each produces about 200 eggs. The ratio of males to females is about 1 : 1. Settling occurs on all aerial parts of the plant, but there appears to be a preference for the midrib and the part of the petiole that is covered by the sheath.

Economic Importance Boisduval scale is the most important insect pest of orchids in Florida (Dekle 1977). Bohart (1942) reports that only a few scales are necessary to cause damage. A few days after settling, a large chlorotic spot develops surrounding the body of the scale. Heavy infestations cause leaves to turn yellow and fall from the plant; entire plants are sometimes killed. Steinweden (1945) indicates that in 5 months 7 adult females on *Cattleya* produced 10,000 scales. The species also is reported as a minor pest of bananas, pineapple (Chua and Wood 1990), coffee (Balachowsky 1929), and coconuts (Munoz Ginarte 1937). Examples of countries or states where this species is considered to be a pest are as follows: California (Koehler 1964; Gill 1997), Colorado (Cockerell 1922), Hawaii

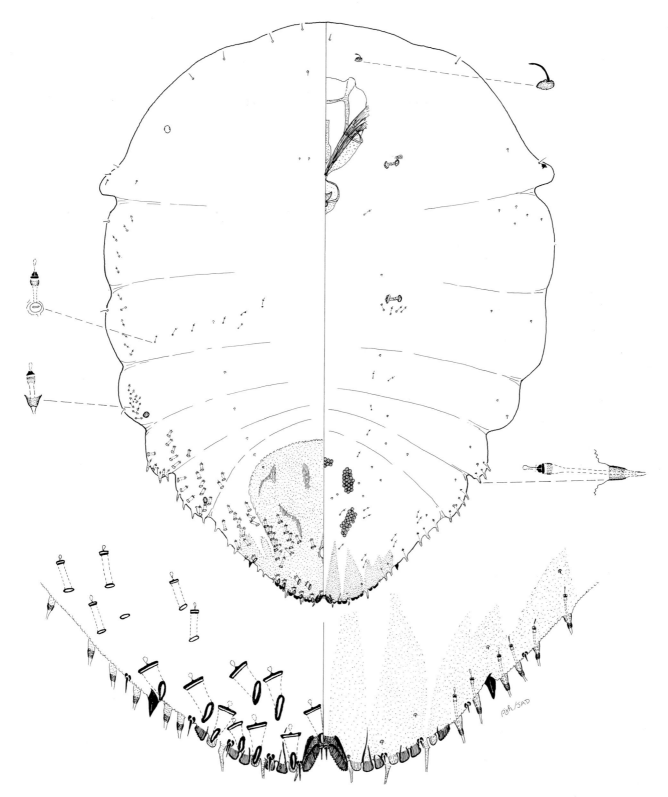

Figure 50. *Diaspis boisduvalii* Signoret, Boisduval scale, Alexandria, VA, on *Cattleya* sp., V-7-1961.

(Kotinsky 1909), North Carolina (Baker 1994), Ohio (Steiner 1987), Zaire (Balachowsky 1929), West Indies (Ballou 1922), Argentina (Chiesa Molinari 1948), Spain and Portugal (Efimoff 1937), French West Africa and Togo (Mallamaire 1954), England (Miles and Miles 1935), Mauritius (Moutia and Mamet 1947), Bulgaria (Tzalev 1964), Trinidad (Urich 1893), Cuba (Munoz Ginarte 1937), France (Panis and Pinet 1998), Czechoslovakia (Zahradnik 1990a), Canary Island (Gomez-Menor Ortega 1958),

and Sierra Leone (Hargreaves 1937). Panis and Pinet (1998) discuss the biological control of this pest in greenhouses using the parasite *Coccidencyrtus malloi* Blanchard. Beardsley and González (1975) consider this scale to be one of 43 serious armored scale pests. Miller and Davidson (1990) consider it to be a serious world pest.

Selected References Bohart (1942); Howell (1975); McKenzie (1956).

Diaspis bromeliae (Kerner)

ESA Approved Common Name Pineapple scale.

Common Synonyms and Combinations *Coccus bromeliae* Kerner, *Aspidiotus bromeliae* (Kerner), *Chermes bromeliae* (Kerner), *Diaspis tillandsiae* Del Guercio, *Aulacaspis bromeliae* (Kerner).

Field Characters (Plate 47) Adult female cover white, semi translucent, flat, circular; shed skins light brown or yellow, subcentral. Male cover elongate oval, white, with 3 longitudinal ridges; shed skin marginal, yellow. Body of adult female, eggs, and newly hatched crawlers yellow. Principally on leaves, normally under outer layer of leaf surface, occasionally on fruit.

Slide-mounted Characters Adult female (Fig. 51) with 4 pairs of definite lobes; with small paraphyses associated with first 3 pairs of lobes, most apparent on medial margin of medial lobule of each lobe. Median lobes separated by space 0.8–1.7 (1.3) times width of median lobe, with lateral and medial paraphyses, without basal sclerosis or yoke, lateral margins of lobes slightly divergent, medial margins strongly divergent apically, with 0–1 (0) lateral notches, with 3–5 (4) medial notches; second lobes bilobed, smaller than median lobes, medial lobule largest, without notches, lateral lobule without notches; third lobes bilobed, similar in size and appearance to second lobes, without notches; fourth lobes simple, represented by flat projection containing series of 3–5 (4) notches. Gland-spine formula 1-1-1, with 9–10 (10) gland spines near each body margin from segments 2 to 4; without gland spines between median lobes. Macroducts of 2 sizes, largest in submarginal areas of segments 3, 4, or 5 to 7 and in marginal areas of segments 4 to 8, small size ducts usually absent submedially, sometimes present on segment 5 and/or 4, in marginal and submarginal areas from segments 2 to 4, with large duct between median lobes 22–28 (26) μ long, macroduct in first space 22–27 (24) μ long, with 10–13 (11) macroducts on each side of body on segments 5 to 7, total of 24–38 (30) macroducts on each side of body, some macroduct orifices anterior of anal opening. Pygidial microducts usually on venter in submarginal areas of segment 5 to 7, rarely absent on segment 7, with 3–5 (4) ducts; prepygidial ducts of 1 size in submarginal and marginal areas of segments 3 and 4, present submedially near mouthparts, near each spiracle, and on segments 1 to 3 or 4; on dorsum pygidial ducts absent; prepygidial microducts of 2 sizes, larger size in submedial areas of metathorax to segment 2, also forming a band from body margin to submedial area of meso- and metathorax, smaller size in submarginal areas of mesothorax to segment 1, usually with small projecting orifice. Perivulvar pores in 5 groups, 27–49 (39) pores on each side of body. Perispiracular pores with 3 loculi, anterior spiracles with 12–16 (14) pores, posterior spiracles without pores. Anal opening located 3.8–6.5 (5.6) times length of anal opening from base of median lobes, anal opening 14–21 (16) μ long. Dorsal seta laterad of median lobes 1.1–1.6 (1.3) times length of median lobe. Eyes represented by small raised areas on prosomal lobes (when present), located on body margin on prothorax. Antennae each with 1 long seta. Cicatrices usually present on prothorax, segment 1, and segment 3, sometimes absent from prothorax. Body turbinate. Prosomal lobes usually present on margin of prothorax, absent on about half of specimens. Conspicuous sclerotized spur on body margin near seta marking segment 4, represented by narrowly pointed process.

Affinities This species is most similar to *Diaspis boisduvalii*, Boisduval scale. For a comparison of these species see the description of *D. boisduvalii*.

Hosts The USNM Collection contains about 200 slides from pineapple and other bromeliads. The Boisduval scale often has been recorded on other hosts, particularly species of Palmae and Orchidaceae. We have seen only two collections from hosts other than Bromeliaceae; 11 specimens from *Cattelya* and 15 from 'orchids,' 1920, Columbus, Ohio. We believe that most records from other hosts are misidentifications. We have examined specimens from the following host genera: *Aechmea, Ananas, Billbergia, Bromelia, Cattleya, Chevalieria, Guzmania, Neoregelia, Nidularium,* and *Tillandsia*.

Distribution We have examined U.S. specimens from CA, DC, FL, HI, IL, MO, NJ, OH, PA, VA, WY. It has been reported from NY and TX. The pineapple scale occurs throughout the tropics of the world and in greenhouses in the northern latitudes where bromeliads are grown. It most often is intercepted in quarantine from the Caribbean islands. A distribution map of this species was published by CAB International (1973).

Biology Little information is available on this species. Brimblecombe (1956) states that development from the egg to adult takes about two months in the summer in Australia; several generations occur each year. Males are common. The preferred feeding site seems to be at the base of the leaves. Heavy infestations also occur on the fruit and the leaf blade. Dziedzicka (1989) indicates that the 'species produces few generations per year.' According to Murray (1980, 1982), in Australia the pineapple scale lays eggs that hatch in about 7 days, and development from the egg to the adult takes about 2 months. Reproduction is continuous with eggs present at all times of the year although there are several periods when adult females are most abundant. Adult males are common.

Economic Importance The pineapple scale occasionally builds to severe infestations in pineapple plantations in Australia, but the populations usually are very localized and apparently are brought under control relatively rapidly by natural enemies. According to Brimblecombe (1956) dense colonies give the leaves a gray appearance and cause a lack of vigor in the plant. Heavily infested plants produce small fruit and numerous suckers. On many hosts the area surrounding the scale cover turns chlorotic. Ornamental bromeliads subjected to heavy infestations become unsightly with abundant yellow spotting. Murray (1980) states that heavily infested plants become weak and stunted, show conspicuous dying back of the foliage, and produce undersized, pinched-looking fruit in pineapple fields in Australia. Williams and Watson (1988) mention this species as a pest of pineapple. Beardsley and González (1975) consider this scale to be one of 43 serious armored scale pests. Miller and Davidson (1990) consider it to be a serious world pest.

Selected References Brimblecombe (1956); Ferris (1938a); Howell 1975.

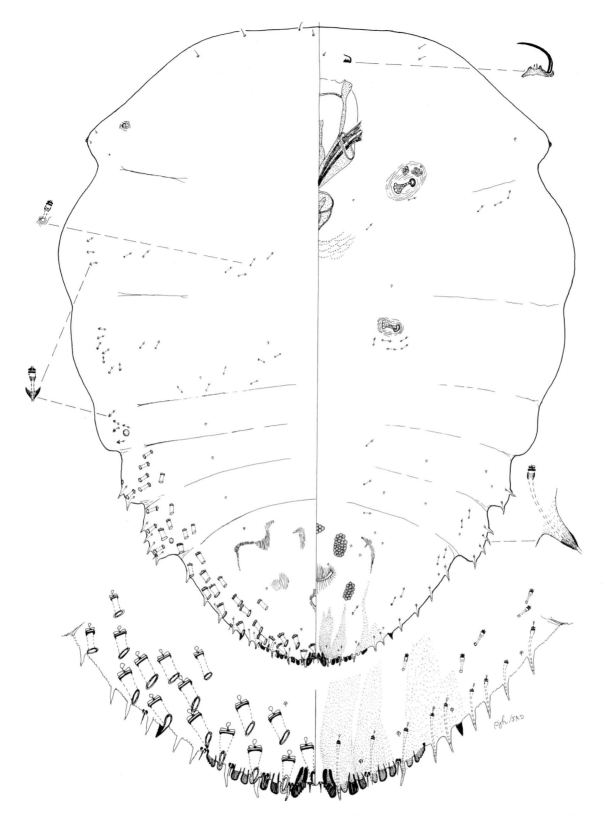

Figure 51. *Diaspis bromeliae* (Kerner), pineapple scale, Holland, intercepted at Miami, FL, on bromeliad, VI-20-1973.

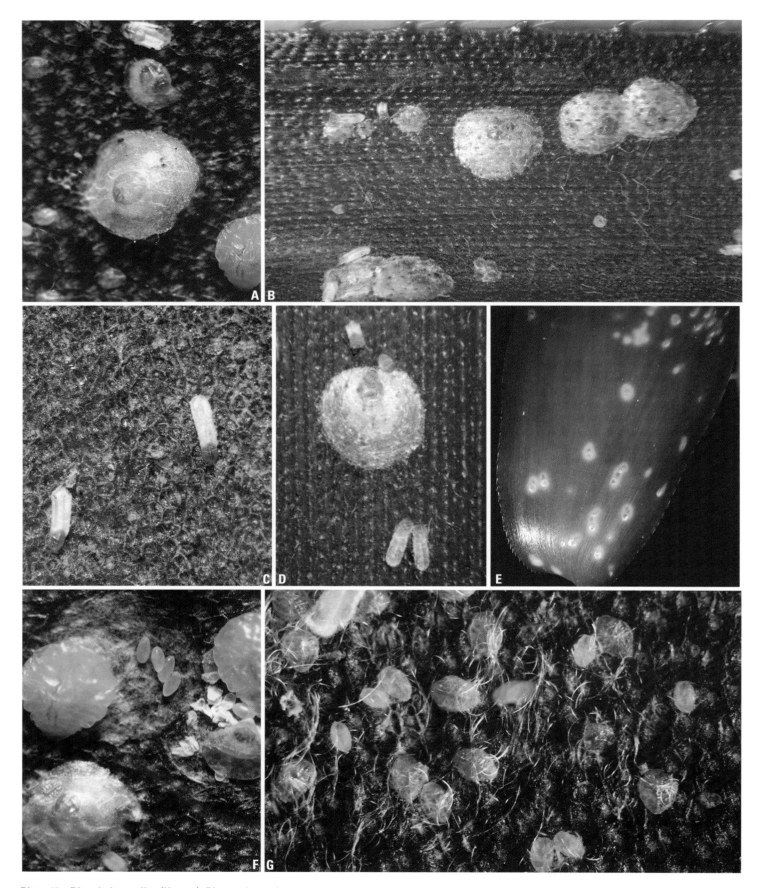

Plate 47. *Diaspis bromeliae* (Kerner), Pineapple scale

A. Adult female cover and body on bromeliad (J. A. Davidson).
B. Mostly adult female covers on pineapple (R. J. Gill).
C. Adult male covers on bromeliad (J. A. Davidson).
D. Adult female and male covers on pineapple (R. J. Gill).

E. Distance view of covers and white damaged areas on bromeliad (J. A. Davidson).
F. Body of adult female with eggs on bromeliad (J. A. Davidson).
G. Small, newly settled crawlers on bromeliad (J. A. Davidson).

Plate 48. *Diaspis echinocacti* (Bouché), **Cactus scale**

A. Adult female cover on cactus (J. A. Davidson).
B. Adult female and male covers on cactus (J. A. Davidson).
C. Second-instar female body, adult female cover, and male covers on cactus (J. A. Davidson).
D. Distance view of scale covers on cactus (J. A. Davidson).
E. Distance view of heavy infestation on opuntia cactus (D. R. Miller).

F. Body of young adult female on opuntia cactus (J. A. Davidson).
G. Bodies of mature adult females on opuntia cactus (J. A. Davidson).
H. Body of adult female with white eggs on cactus (J. A. Davidson).

185

Diaspis echinocacti (Bouché)

ESA Approved Common Name Cactus scale (also called prickly pear scale).

Common Synonyms and Combinations *Aspidiotus echinocacti* Bouché, *Diaspis calyptroides* Costa, *Chermes echinocacti* (Bouché), *Diaspis cacti* Comstock, *Diaspis opuntiae* Newstead, *Diaspis opunticola* Newstead, *Diaspis cacti* var. *opuntiae* Cockerell, *Diaspis cacti* var. *opunticola* Newstead, *Diaspis calyptroides* var. *opuntiae* Cockerell; *Diaspis calyptroides* var. *cacti* Comstock, *Diaspis echinocacti cacti* Comstock, *Diaspis echinocacti opuntiae* Cockerell, *Diaspis dactylproides* Bodenheimer, *Diplacaspis echinocacti* (Bouché) (see Mann 1969); *Chionaspis cacti* Comstock; *Carulaspis calyptroides* Costa.

Field Characters (Plate 48) Adult female cover white, flat, circular; shed skins dark brown when rubbed, often covered with a grayish white film, central to subcentral. Male cover elongate oval, white, with 3 longitudinal ridges; shed skin marginal, white. Body of female yellowish white; eggs and newly hatched crawlers white. On stems, fruit, and pads of cacti.

Slide-mounted Characters Adult female (Fig. 52) with 4 pairs of definite lobes; with small paraphyses associated with first 3 pairs of lobes, most apparent on medial margin of medial lobule of each lobe. Median lobes separated by space 0.3–1.1 (0.7) times width of median lobe, with lateral and medial paraphyses, with or without small basal sclerosis, without yoke, lateral margins of lobes usually slightly divergent, occasionally parallel, medial margins variable from rounded to parallel basally and strongly divergent on apical two-thirds of margin, without notches; second lobes bilobed, usually about same size as median lobes, medial lobule largest without notches, lateral lobule without notches; third lobes bilobed, similar in size and appearance to second lobes, without notches; fourth lobes simple, represented by flat projection containing series of 0–8 (5) notches. Gland-spine formula normally 1-1-1, rarely 2-2-1 or 1-2-1, 7–13 (10) gland spines near each body margin from segments 2 to 4; without gland spines between median lobes. Macroducts of 2 sizes, largest in submarginal areas of segments 6 and 7 and in marginal areas of segments 4 to 8, smaller ducts in submedial areas of segments 2 or 3 to 7, in marginal and submarginal areas from segments 1 or 2 to 6, with large duct between median lobes 20–25 (23) μ long, macroduct in first space 22–26 (24) μ long, with 23–38 (30) macroducts on each side of body on segments 5 to 7, total of 92–139 (111) macroducts on each side of body, some macroduct orifices anterior of anal opening. Pygidial microducts usually on venter in submarginal areas of segments 5 to 7, rarely absent on segment 7, with 5–9 (7) ducts; prepygidial ducts of 1 size in submarginal and marginal areas of segments 3 and 4, present submedially near mouthparts and near each spiracle; on dorsum pygidial ducts absent; prepygidial microducts of 2 sizes, larger size in submedial areas of any or all of mesothorax, metathorax, or segment 1, smaller size in submarginal areas of mesothorax to segments 1 (most with small protruding orifice). Perivulvar pores in 5 groups, 22–44 (35) pores on each side of body. Perispiracular pores with 3 loculi, anterior spiracles with 2–4 (3) pores, posterior spiracles without pores. Anal opening located 2.8–5.3 (4.3) times length of anal opening from base of median lobes, anal opening 14–20 (16) μ long. Dorsal seta laterad of median lobes 0.7–2.0 (1.2) times length of median lobe. Eyes absent. Antennae each with 1 long seta. Cicatrices usually present on segment 1 and segment 3, usually absent from prothorax. Body turbinate. Prosomal lobes usually absent, rarely represented by small protrusions on margin of prothorax. Inconspicuous sclerotized spur on body margin near seta marking segment 4, represented by broadly pointed process.

Affinities Of the economic species in the United States, cactus scale is most similar to *Diaspis boisduvalii*, Boisduval scale, but differs by having median lobes that lack notches, by having submedial macroducts on abdominal segments 2 or 3 to 7, and by having 23–38 (30) macroducts on each side of body on segments 5 to 7. Boisduval scale has median lobes with notches on medial margin, lacks submedial macroducts on abdominal segments 2 to 4, and has 15–21 (20) macroducts on each side of body on segments 5 to 7.

Hosts The USNM Collection contains about 500 slides with specimens collected from many genera of the Cactaceae. We have examined specimens from the following cactus genera: *Acanthocereus, Ancistrocactus, Ariocarpus, Astrophytum, Cephalocereus, Cereus, Copiapoa, Coryphantha, Dendrocereus, Echinocactus, Echinocereus, Echinofossulocactus, Echinopsis, Epiphyllum, Ferocactus, Harrisia, Heliocereus, Hylocereus, Lemaireocereus, Leuchtenbergia, Lobivia, Lophocereus, Mammillaria, Melocactus, Myrtillocactus, Nopalxochia, Opuntia, Pachycereus, Parodia, Pelecyphora, Peniocereus, Pereskia, Rebutia, Thelocactus, Trichocereus, Turbinocactus, Weberocereus, Wilcoxia,* and *Zygocactus.* We have examined specimens collected on two different succulents as well, *Dudleya* and *Cotyledon.* We suspect that literature reports of this species on hosts other than cactus or succulents probably are misidentifications. Oetting (1984) tested 19 different cacti and 21 non-cacti as hosts and found that of the 19 cacti, 16 were accepted as hosts; whereas, of the 21 non-cacti, only 1 (*Portulaca*) was acceptable.

Distribution We have examined U.S. specimens from AL, AZ, CA, DC, FL, HI, KS, KY, LA, MD, MI, MO, MS, NC, NJ, NM, NY, OH, OK, OR, PA, TX, VA, WY. The cactus scale occurs out-of-doors in the south and greenhouses in the north. This species occurs worldwide wherever cacti are grown. Most quarantine interceptions are from Mexico. Additional common records include Central and South America and the Caribbean islands.

Biology The only definitive study on the life history of this species is by Oetting (1984). He found 'that development from the egg to adult required from 23 to 26 days under constant 27 °C or greenhouse conditions with males requiring 1 to 2 days longer than females. Adult females had a preoviposition period of 12 to 23 days followed by 1 to 2 months of oviposition.' An individual female laid about 100 to 200 eggs. The species is reported to have 2 generations per year in greenhouses in Moscow, USSR (Saakyan-Baranova 1954). It frequently is found on the pads and fruit of its cactus host; it is occasionally reported on the subterranean crown and roots.

Economic Importance Damage appears to be variable. Many publications state that the cactus scale has little or no effect on its host, while others record it as an important pest. This variation probably reflects the presence or absence of natural enemies and/or the susceptibility of the various host plants. According to Russo and Siscaro (1994) this species recently has become a problem on prickly pear (*Opuntia ficus-indica*) on some farms in Sicily, requiring specific measures of chemical control. Feeding resulted in the presence of chlorotic areas on the fruit. The encyrtid *Plagiomerus diaspidis* was the most effective natural enemy in the study area, giving rates of parasitism greater than 40%. This species is considered to be an important world pest by Miller and Davidson (1990).

Selected References Ferris (1937); Oetting (1984).

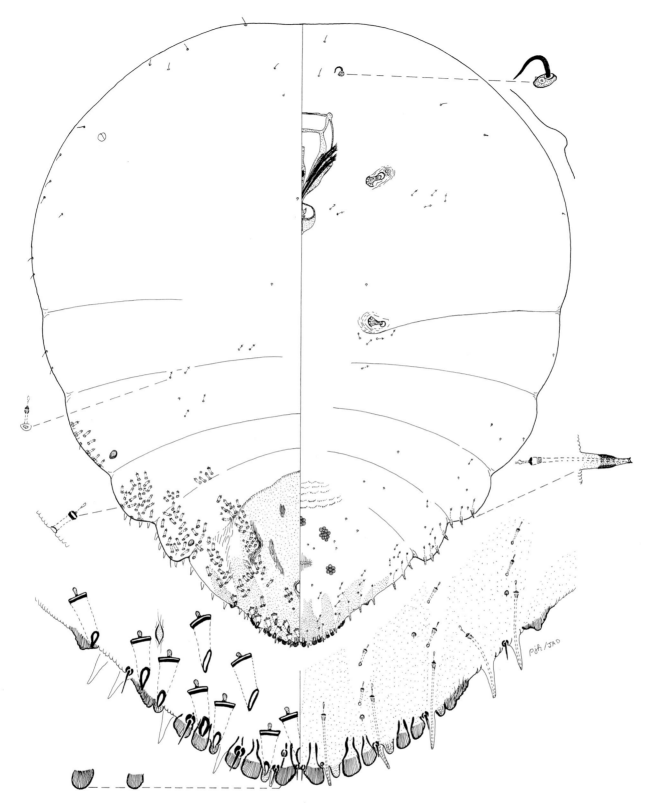

Figure 52. *Diaspis echinocacti* (Bouché), cactus scale, Sonora, Mexico, intercepted at Nogales, AZ, host unknown, III-15-1944.

Duplaspidiotus claviger (Cockerell)

Suggested Common Name Camellia mining scale.

Common Synonyms and Combinations *Pseudaonidia clavigera* Cockerell, *Pseudaonidia iota* Green and Laing.

Field Characters (Plate 49) Adult female cover grayish, covered by epidermis of host, circular, convex; shed skins central to subcentral, yellow orange. Male unknown. Body of adult female, eggs, and crawlers light purple. Found under the bark of twigs and branches. Detected by swelling of bark and first instar shed skin on bark surface; scraping of bark will reveal white ventral cover. Thorax and head of adult female heavily sclerotized on old adult females, with clear areas on head and prothorax.

Slide-mounted Characters Adult female (Fig. 53) with 3 pairs of definite lobes, pygidial margin irregularly serrate from third lobes to segment 3; paraphysis formula 1-1-0, paraphyses in first and second spaces each composed of elongate sclerotization and detached, oval sclerotization; each of 3 pairs of lobes sometimes with small paraphysis like sclerotization on medial and lateral margins, but usually so small as to be unrecognizable except in perfect specimens. Median lobes separated by space 0.1–0.3 (0.2) times width of median lobe, with narrow sclerotization between median lobes apparently homologous to yoke but normally not attached to lobe, with small paraphyses on each lobe margin in perfect specimens, without basal sclerotization, medial margin usually slightly diverging apically, lateral margins parallel or converging, with 1 lateral notch and 0–1 (0) medial notch; second lobes smaller than median lobes, simple, with 1 lateral notch and without medial notch; third lobes same as second except smaller. Plates simple, without tines, normally with truncate or slightly rounded apex; plate formula usually 2-3-2, rarely 2-3-1 or 1-3-2; plates between median lobes absent. Macroducts of 1 size, unusually narrow, similar in appearance to microducts, present in marginal or submarginal areas of segments 1 to 8, with 4–8 ducts between median lobes, these ducts of 2 kinds, long ducts similar to microducts arising from plates, shorter ducts similar in appearance to dorsal macroducts, longest macroduct between median lobes extending 0.9–1.2 (1.1) times distance between posterior apex of anal opening and base of median lobes, 94–106 (99) µ long, marginal macroduct in first space 70–98 (84) µ long, with 63–86 (74) macroducts on each side of pygidium on segments 5 to 8, ducts in submarginal and marginal areas, total of 179–235 (213) macroducts on each side of body, some macroduct orifices anterior of anal opening. Pygidial microducts on venter in submarginal and marginal areas of segments 5 and 6, with 15–21 (17) ducts; prepygidial ducts of 1 size in submarginal and marginal areas of segments 1 to 3, present in submarginal cluster on mesothorax, present submedially anterior of mouthparts, lateral of mouthparts, near each spiracle, and near submedial setae on segments 1 to 4, ducts near spiracles sometimes absent; on dorsum pygidial ducts absent; prepygidial microducts of 2 sizes, larger size in submedial areas of any or all of metathorax to 4, smaller size in submarginal areas of prothorax or mesothorax to metathorax or segment 1, in 1 specimen larger ducts were completely absent. Perivulvar pores usually in 3 clusters, occasionally divided into 4 or 5 clusters, with 29–39 (34) pores on each side of body. Perispiracular pores normally with 5 loculi, anterior spiracles with 6–13 (19) pores, posterior spiracles usually without pores, 1 specimen with 1 pore on 1 side of body. Anal opening located 5.3–7.3 (6.2) times length of anal opening from base of median lobes, anal opening 14–17 (15) µ long. Dorsal seta laterad of median lobes 0.5–0.8 (0.6) times length of median lobe. Eyes absent. Antennae each with 1 seta, often with little or no basal structure. Cicatrices absent. Body pear shaped, with conspicuous constrictions between prothorax and mesothorax and between metathorax and segment 1. Prosoma sclerotized in old females, with 2 pairs of clear areas on dorsum on each side of head and prothorax. Dorsum of pygidium with series of conspicuous areolations. Perivulvar pores usually oriented in vertical plane in relation to slide cover slip.

Affinities The combination of the presence and peculiar arrangement of perivulvar pores and the unusually shaped paraphyses distinguishes this species from all others.

Hosts Polyphagous. Borchsenius (1966) reports it from 8 genera in 7 families; Dekle (1977) records it on 40 host genera in Florida and states that *Camellia*, *Ligustrum*, and *Viburnum* are the most frequently reported hosts. We have examined material from *Calophyllum*, *Camellia*, *Citrus*, *Eugenia*, *Ficus*, *Gardenia*, *Hibiscus*, *Inga*, *Lantana*, *Macadamia*, *Malus*, *Metrosideros*, *Myricaria*, *Nipa*, *Polyalthia*, *Santalum*, *Styphelia*, and *Vitis*.

Distribution In the United States camellia mining scale occurs in FL and HI. It has been taken in quarantine from Guam, Panama, and Tahiti. It is reported from China, Guam, Indonesia, Japan, Java, Seychelles, South Africa, Sri Lanka, and Taiwan.

Biology We have been unable to find a detailed account of the biology of this species. This pest often is referred to as a mining scale. It can bury itself beneath the host epidermis with only the first shed skin visible. Usually there is a tell-tale swelling indicating its presence. Camellia mining scale commonly occurs on twigs and branches.

Economic Importance Dekle and Kuitert (1975) and Dekle (1977) note that heavy infestations of this scale on *Camellia* and *Ligustrum* cause dieback of twigs and branches and limit new growth. Fullaway (1928) reports this species killing grape twigs in Hawaii. Miller and Davidson (1990) consider this species to be an occasional pest.

Selected References Balachowsky (1958); Dekle (1977); Dekle and Kuitert (1975).

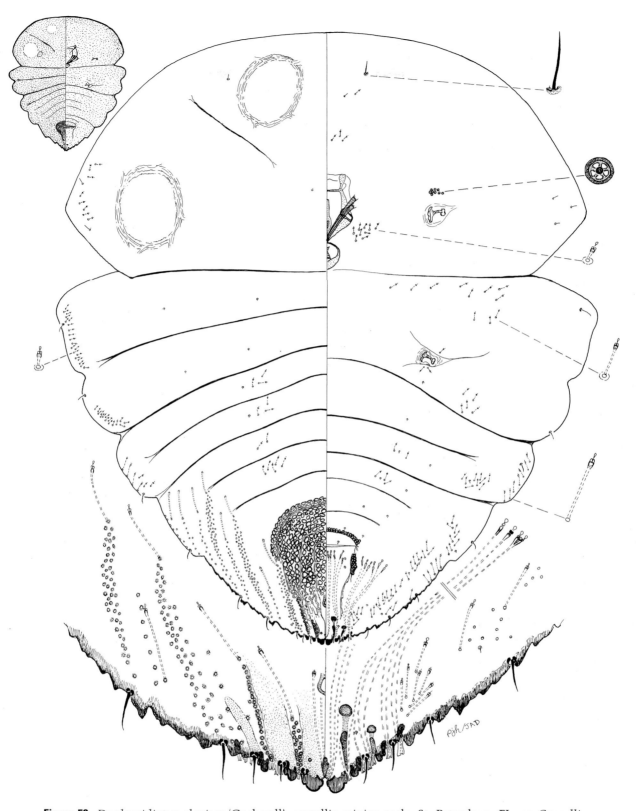

Figure 53. *Duplaspidiotus claviger* (Cockerell), camellia mining scale, St. Petersburg, FL, on *Camellia* sp., I-18-1962.

Plate 49. *Duplaspidiotus claviger* (Cockerell), Camellia mining scale

A. Adult female cover partially covered by lichens and bark on unknown host (J. A. Davidson).
B. Rubbed adult female cover on apple (J. A. Davidson).
C. Adult female cover covered by lichens and bark on unknown host (J. A. Davidson).
D. Infestation on apple (J. A. Davidson).

E. Body of second instar on unknown host (J. A. Davidson).
F. Body of parasitized adult female (J. A. Davidson).
G. Covers on apple bark (J. A. Davidson).
H. Covers on apple bark (J. A. Davidson).
I. Distance view of heavy infestation on apple (J. A. Davidson).

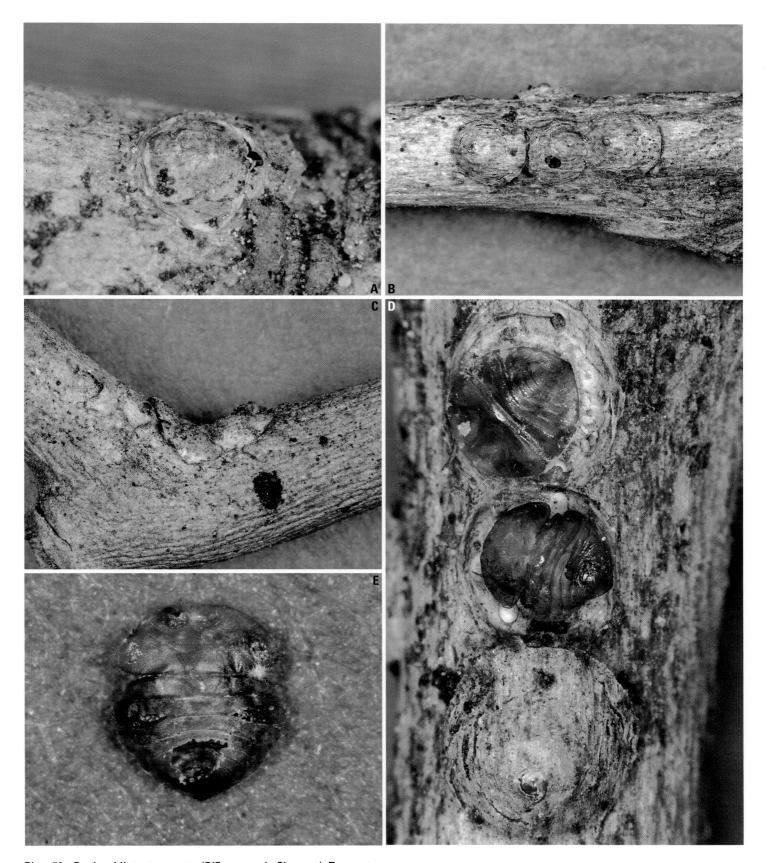

Plate 50. *Duplaspidiotus tesseratus* (D'Emmerez de Charmoy), **Tesserate scale**

A. Adult female cover on camellia (J. A. Davidson).
B. Adult female covers; one with a parasite emergence hole (J. A. Davidson).
C. Inconspicuous cluster of female covers in twig fork (J. A. Davidson).
D. Bodies of two adult females and one scale cover (J. A. Davidson).
E. Dead female body (J. A. Davidson).

191

Duplaspidiotus tesseratus
(D'Emmerez de Charmoy)

Suggested Common Name Tesserate scale.

Common Synonyms and Combinations *Aspidiotus* (*Diaspidiotus*) *tesseratus* D'Emmerez de Charmoy, *Pseudaonidia tesseratus* (D'Emmerez de Charmoy), *Lattaspidiotus tesseratus* (D'Emmerez de Charmoy), *Lattaspidiotus oreodoxae* Rutherford.

Apparently the correct last name of the author of this species is as given above, not 'Charmoy' as it has been used in most cases of scale insect literature. Also the correct author of the species is D'Emmerez de Charmoy not Grandpre and Charmoy (Mamet 1953).

Field Characters (Plate 50) Adult female cover gray, circular, convex, usually thick, often masked by a layer of the host plant; shed skins subcentral, reddish brown. Male cover smaller and more elongate than female cover; shed skin submarginal. Body of mature adult female light to dark brown, probably light purple when newly mature. On bark of host, usually under outer layer of bark. Thorax and head of adult female heavily sclerotized on old females, clear areas on head and prothorax.

Slide-mounted Characters Adult female (Fig. 54) with 3 pairs of definite lobes, pygidial margin with 2 or 3 spurs from third lobes to segment 3; paraphysis formula 1-1-1, paraphyses in first and second spaces each composed of elongate sclerotization and detached, oval sclerotization, paraphysis in third space represented by oval sclerotization only, third sclerotization rarely absent. Median lobes separated by space 0.1–0.3 (0.2) times width of median lobe, with narrow sclerotization between median lobes apparently homologous to yoke but normally not attached to lobe, with small paraphyses on medial margin, without basal sclerotization, medial margin usually parallel for first third of lobe then diverging apically, lateral margins usually parallel for first third of lobe then converging apically, with 1 lateral notch and usually without medial notch 0–1 (0); second lobes smaller than median lobes, simple, with 1 lateral notch and without medial notch; third lobes same as second except smaller. Plates simple, without tines, normally with truncate or slightly rounded apex; plate formula usually 2-3-0, rarely 2-2-0; plates between median lobes 0.3–0.4 (0.4) times as long as median lobes. Macroducts of 1 size, unusually narrow, similar in appearance to microducts, present in marginal or submarginal areas of segments 3 to 8, with 4–8 ducts between median lobes, longest macroduct between median lobes extending 1.0–1.8 (1.4) times distance between posterior apex of anal opening and base of median lobes, 79–119 (92) µ long, marginal macroduct in first space 62–108 (86) µ long, with 102–235 (185) macroducts on each side of pygidium on segments 5 to 8, ducts in submarginal and marginal areas, total of 205–337 (272) macroducts on each dorsal side of body, some macroduct orifices anterior of anal opening. Pygidial microducts on venter in submarginal and marginal areas of segment 5 and 6, with 10–33 (19) ducts; prepygidial ducts of 1 size in submarginal and marginal areas of metathorax or segment 1 to segment 4, present in large cluster on mesothorax; on dorsum pygidial ducts absent; prepygidial microducts of 2 sizes, larger size in submedial areas from metathorax or segment 1 to segment 3 or 4, smaller size in conspicuous submarginal clusters on pro- and metathorax. Perivulvar pores absent, vulvar flap conspicuous. Perispiracular pores normally with 5 loculi, anterior spiracles with 6–27 (12) pores, posterior spiracles without pores. Anal opening located 3.9–6.7 (5.6) times length of anal opening from base of median lobes, anal opening 10–15 (12) µ long. Dorsal seta laterad of median lobes 0.6–0.8 (0.7) times length of median lobe. Eyes absent. Antennae each with 1 seta, often with little or no basal structure. Cicatrices absent. Body pear shaped, with conspicuous constriction between prothorax and mesothorax. Prosoma sclerotized in old females, with 2 pairs of clear areas on dorsum on each side of head, 1 clear area mesad of cluster of small microducts on dorsum of prothorax. Dorsum of pygidium with series of conspicuous areolations.

Affinities The combination of the unusually shaped paraphyses and the absence of perivulvar pores distinguishes this species from all others.

Hosts Polyphagous. Borchsenius (1966) reports the species on 27 genera in 17 families; Dekle (1977) reports it on 17 host genera in Florida and states that *Camellia* and *Hibiscus* are the most frequently reported hosts. We have examined material from *Adansonia, Aegle, Bignonia, Bougainvillea, Bryanthus, Calocarpum, Casuarina, Catasetum, Cattleya, Coryanthes, Ficus, Guaiacum, Haematoxylum, Hibiscus, Kokia, Malvaviscus, Manilkara, Melicoccus, Melochia, Parkia, Petrea, Prosopis, Pyrus, Reinhardtia, Rosa, Smilax,* and *Vanda.*

Distribution In the United States tesserate scale occurs in FL and HI. It has been taken in quarantine from South and Central America and from the Caribbean islands.

Biology We have been unable to find a detailed account of the biology of this species. Dekle (1977) and Balachowsky (1958) each describe the tesserate scale as usually covered by the outer layers of bark. Other authors do not mention this habit and illustrate the scale cover without an overlaying bark cover.

Economic Importance Tesserate scale has been reported by Wilson (1923) to cause severe injury to grapes in the Virgin Islands and by Foldi and Soria (1989) to damage the same host in Brazil. Moutia and Mamet (1947) consider it to be a minor pest of grapes in Mauritius. Miller and Davidson (1990) consider this species to be an occasional pest.

Selected References Balachowsky (1958); Dekle (1977).

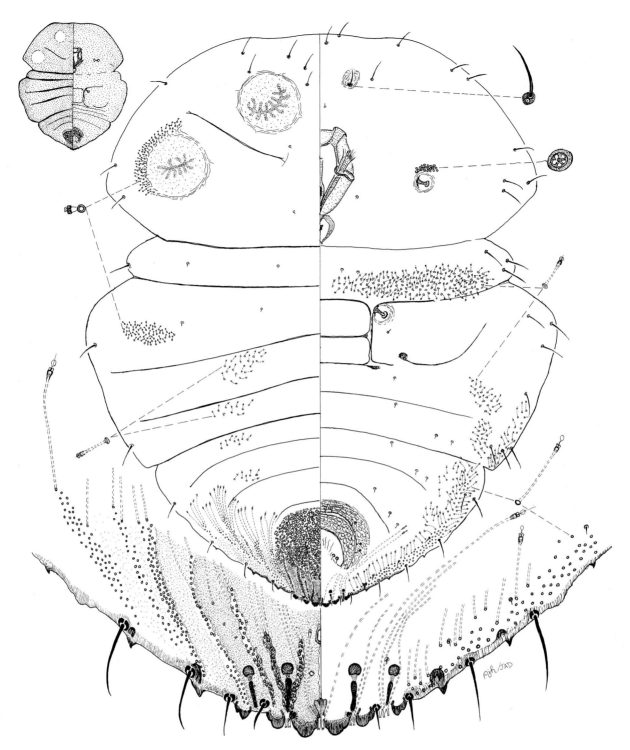

Figure 54. *Duplaspidiotus tesseratus* (D'Emmerez de Charmoy), tesserate scale, Colombia, intercepted at Brownsville, TX, on orchid, VIII-8-1944.

Dynaspidiotus britannicus (Newstead)

ESA Approved Common Name Holly scale (also called laurel scale).

Common Synonyms and Combinations *Aspidiotus britannicus* Newstead, *Aspidiotus (Dynaspidiotus) britannicus* Newstead, *Aspidiotus (Evaspidiotus) britannicus* Newstead, *Aspidiotus latastei* Hall.

Field Characters (Plate 51) Adult female cover gray or light brown, circular, flat; shed skins subcentral, yellow. Male cover small, elongate oval, similar in color and texture to that of the female; shed skin submarginal. Body of adult female probably yellow; eggs and crawlers yellow. On leaves, twigs, and berries.

Slide-mounted Characters Adult female (Fig. 55) with 3 pairs of definite lobes, fourth lobes absent or represented by low series of sclerotized points; paraphysis formula usually 2-2-0, rarely 3-2-1 or 2-3-0, when all paraphyses present with small paraphyses attached to lateral and medial margins of median lobe and lobe 2, to medial margin of lobe 3, and in first and second spaces. Median lobes separated by space 0.5–1.0 (0.7) times width of median lobe, without basal sclerotization, with paraphysis attached to medial margin, without yoke, medial margins parallel or converging slightly, lateral margins parallel or diverging slightly, with 1 lateral notch and 1 medial notch; second lobes simple, slightly smaller than median lobes, with rounded apex, with 1 lateral notch and 1 medial notch; third lobes simple, slightly smaller than second lobes, with 1–2 (2) lateral notches, 0–1 (1) medial notches; fourth lobes absent or represented by series of 1 or 2 low swellings, with 4–9 (7) notches. Plates between lobes fimbriate, microducts opening on some of plates in third space, anterior plate sometimes with 1 or 2 clavate processes; plate formula 2-3-4, rarely 2-3-3 or 2-3-5; plates between median lobes 0.8–1.0 (0.9) times as long as median lobes. Macroducts on pygidium of 1 size (marginal ducts in first and second space slightly wider and shorter than others on pygidium), present submarginally and marginally on segments 5 to 8, duct between median lobes extending 0.8–1.2 (1.0) times distance between posterior apex of anal opening and base of median lobes, 35–51 (44) μ long, marginal duct in first space 45–68 (53) μ long, with 16–24 (20) macroducts on each side of pygidium on segments 5 to 8, ducts in submarginal and marginal areas, total of 25–36 (30) macroducts on each side of body, some macroduct orifices anterior of anal opening; prepygidial macroducts small, present along body margin from metathorax to segment 3. Pygidial microducts on venter in submarginal and marginal areas of segment 5, with 0–3 (2) ducts; prepygidial ducts of 1 size, in submedial areas near mouthparts and spiracles, and on segments 1 to 3 or 4, in submarginal and marginal areas of metathorax or segment 1 to segments 3 or 4; dorsal pygidial ducts absent; prepygidial microducts of 1 size, in submedial areas of any or all of pro-, meso-, or metathorax to segment 3. Perivulvar pores arranged in 4 or 5 groups, 9–17 (10) pores on each side of body. Pores absent near spiracles. Anal opening located 0.8–1.5 (1.2) times length of anal opening from base of median lobes, anal opening 15–25 (19) μ long. Dorsal seta laterad of median lobes 0.8–1.5 (1.2) times length of median lobe. Eyes variable, often with 1 to 3 spurs, sometimes absent. Antennae each with 1 seta. Cicatrices absent or weakly indicated on segments 1 and 3. Body pear shaped.

Affinities The holly scale pygidium is similar to some species of *Aonidiella* and *Chrysomphalus* but *Aonidiella* has prepygidial lobes and *Chrysomphalus* has definite pygidial furrows; these structures are absent on *Dynaspidiotus*.

Hosts Polyphagous according to Borchsenius (1966) who records it from 23 host genera in 18 families. We have examined material from *Buxus, Hedera, Ilex, Laurus, Lonicera, Rhamnus, Ruscus, Sterculia,* and palm. Most material in the USNM Collection at Beltsville is from *Laurus* or *Ilex*.

Distribution This scale probably is European in origin. In the United States, occurs in CA, IL, MI, NJ, OR, PA, WA. It is reported by Nakahara (1982) from IN and MA. It has been taken in quarantine from Bermuda, Brazil, Canada, many European countries including most Mediterranean areas, Russia, Africa, and Asia (Palestine and Turkey). Many specimens in the USNM Collection are second instars. This species overwinters as immatures and frequently is collected from holly cuttings used to make wreaths. It is possible that some records are from such collections and do not represent established populations.

Biology Roaf and Mote (1935) found that holly scale preferred *Ilex aquifolium* in Oregon. Male and female scales occur on both surfaces of leaves, but females prefer upper leaf surfaces while males prefer lower leaf surfaces. During July and August females lay 1–3 eggs per day until about 100 eggs are produced. Eggs hatch in 4 to 6 days. Second-instar females overwinter. Adult females are present in March–July. Males are reported to overwinter as prepupae and pupae, but because these instars are not known as overwintering stages in other armored scales this information is questionable. Adult males occur in March; there is one generation per year. Del Bene (1984) studied the biology of this species in Tuscany, Italy, on *Ruscus racemosus*. The species had 2 generations per year, overwintered as second instars, and eggs were laid in June to July for the first generation and September to October in the second.

Economic Importance The holly scale was considered the most important insect pest of English holly in Oregon by Roaf and Mote (1935). The scale weakens infested plants and may be so abundant on twigs, leaves, and berries as to make holly plants and holly Christmas decorations unsaleable. In recent years this species has been considered a pest of olives (Argyriou 1990), ornamentals (Zahradník 1990b), *Ruscus racemosus* (Del Bene 1984), and lemons (Swirski 1985). Tranfaglia and Viggiani (1986) indicate that it is controlled by natural enemies in many situations in Italy. Miller and Davidson (1990) consider this species to be an occasional pest.

Selected References Del Bene (1984); Ferris (1938a); Roaf and Mote (1935).

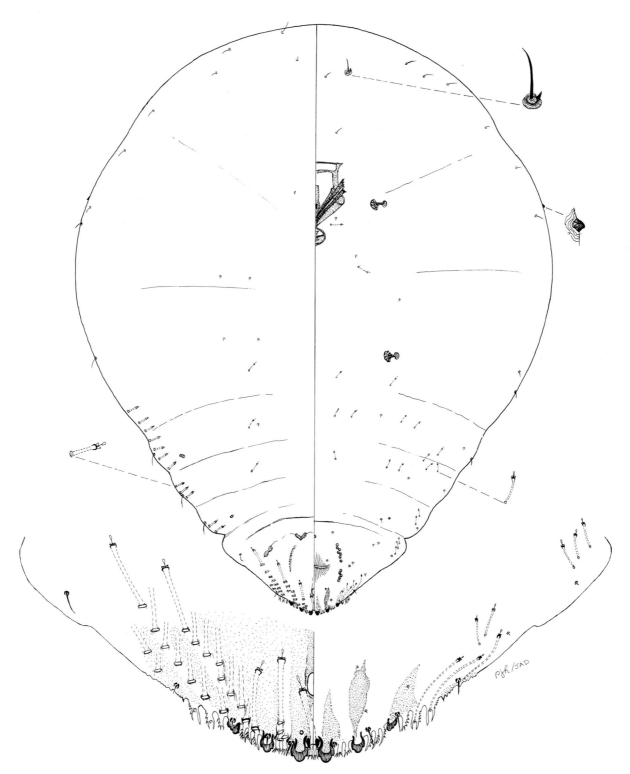

Figure 55. *Dynaspidiotus britannicus* (Newstead), holly scale, Italy, intercepted at HI, on *Ruscus* sp., V-21-1952.

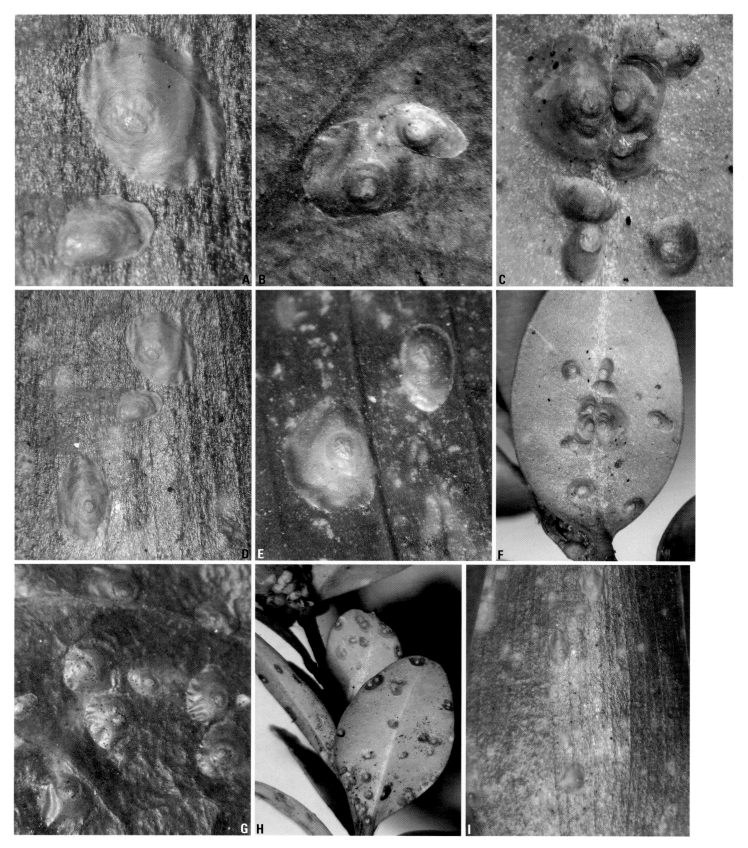

Plate 51. *Dynaspidiotus britannicus* **(Newstead), Holly scale**

A. Adult female and male covers on *Ruscus*; note female body through cover (R. J. Gill).

B. Adult female and male covers on ivy (R. J. Gill).

C. Female and male covers on boxwood (R. J. Gill).

D. Female and male covers on *Ruscus* (R. J. Gill).

E. Female and male covers on *Ruscus* (R. J. Gill).

F. Scale covers on underside of boxwood leaf (R. J. Gill).

G. Mostly adult female covers on ivy (R. J. Gill).

H. Distance view of covers on underside of boxwood leaves (R. J. Gill).

I. Damaged leaf apparently caused by feeding of scale (R. J. Gill).

Plate 52. *Epidiaspis leperii* (Signoret), Italian pear scale

A. Cluster of adult female covers on walnut (J. A. Davidson).
B. Adult female and male covers exposed from beneath associated lichens on walnut (R. J. Gill).
C. Bodies of young adult females on walnut (J. A. Davidson).
D. Body of old adult female on walnut (R. J. Gill).
E. Body of second-instar female on walnut (R. J. Gill).

F. Distance view of scale covers engulfed in lichens on walnut (R. J. Gill).
G. Walnut twigs showing lichen encrustation (J. A. Davidson).
H. Branch dieback on walnut (J. A. Davidson).
I. Dieback on entire walnut tree (J. A. Davidson).

Epidiaspis leperii (Signoret)

ESA Approved Common Name Italian pear scale (also called red pear scale, European pear scale, grey pear scale, pear tree oyster scale).

Common Synonyms and Combinations *Diaspis leperii* Signoret, *Diaspis ostreaeformis* Signoret (not Curtis), *Aspidiotus piricola* Del Guercio, *Diaspis fallax* Horvath, *Diaspis piricola* (Del Guercio), *Aspidiotus pyricola* Del Guercio, *Diaspis pyri* Colvee, *Epidiaspis betulae* Lindinger (not Bearensprung).

Field Characters (Plate 52) Adult female cover white or gray, convex, circular; shed skins covered with white secretion, when rubbed, yellow or brown, subcentral. Male cover elongate, white, without ridges; shed skin marginal, yellow; male without wings (Kosztarab 1963), body orange or yellow. Body of adult female pink, becoming darker red as egg laying progresses; eggs white when first laid but red over time; first instars wine red (Geoffrion 1976). On bark.

Slide-mounted Characters Adult female (Fig. 56) with 1 pair of definite lobes and 4 pairs of smaller lobes, last 1 or 2 in shape of spur; paraphysis formula usually 1-1-2, rarely 1-1-1, when all paraphyses present with conspicuous, comma-shaped paraphyses attached to lateral margins of lobes 2, 3, and 4, attached to medial margins of lobes 4 and 5, with normal paraphysis attached to lateral margin of median lobes, paraphyses unusual, comma shaped, associated with segmental setal bases except on median lobe. Median lobes separated by space 0.1–0.2 (0.2) times width of median lobe, with lateral and medial paraphyses, without basal sclerosis and yoke, lateral margins of lobes usually parallel for first third of lobe, converging apically on apical two-thirds of lobe, medial margins usually converging apically, occasionally diverging, with 0–2 (1) lateral notches, without medial notches; second lobes bilobed, conspicuously smaller than median lobes, medial lobule largest without notches, lateral lobule with 0–1 (0) lateral notches; third lobes simple, similar in appearance to second lobes, slightly larger than second lobes, usually without notches; fourth lobes simple or bilobed, about same size as third lobes, without notches; fifth lobes simple, spurlike, without notches. Gland-spine formula normally 1-1-1, rarely 0-1-1, with 8–14 (12) gland spines near each body margin from segment 2 to 4; with minute gland spines between median lobes. Macroducts of 2 sizes, largest in submarginal areas of segments 4 or 5 to 7 and in marginal areas of segments 4 to 8, small size ducts in submedial areas of segments 4 or 5 to 7, in marginal and submarginal areas from segments 2 or 3 to 4, with large duct between median lobes, 1 specimen out of 10 examined without duct between lobes, duct between lobes 20–29 (24) µ long, macroduct in first space 19–25 (22) µ long, with 16–25 (30) macroducts on each side of body on segments 5 to 7, total of 23–31 (27) macroducts on each side of body, some macroduct orifices anterior of anal opening. Pygidial microducts usually on venter in submarginal areas of segments 5 to 6, rarely absent on segment 6, with 1–6 (2) ducts; prepygidial ducts of 1 size in submarginal and marginal areas of segments 3 and 4, present submedially near mouthparts, near each spiracle, and any or all of segments 1 to 3, often present in submarginal cluster on prothorax; on dorsum pygidial ducts absent; prepygidial microducts of 2 sizes, larger size in submedial areas of meso- or metathorax to segment 1, 2, or 3, smaller size in submarginal areas of meso- or metathorax to segments 1 or 2. Perivulvar pores in 5 groups (2 specimen out of 10 with sixth supernumerary cluster anterior of normal clusters), 24–31 (27) pores on each side of body. Perispiracular pores with 3 loculi, anterior spiracles with 3–10 (5) pores, posterior spiracles with 1–2 (1) pores. Anal opening located 5.0–8.9 (6.6) times length of anal opening from base of median lobes, anal opening 10–17 (13) µ long. Dorsal seta laterad of median lobes 0.7–1.0 (0.8) times length of median lobe. Eyes absent. Antennae each with 1 long seta. Cicatrices absent. Body pear shaped, unsclerotized. Gland spine usually curved.

Affinities The combination of large, protruding median lobes, small second to fifth lobes, and one paraphysis in each of spaces 2 to 4, distinguishes this species from all others in the United States. *Epidiaspis leperii* resembles *E. peragrata* Ferris from Mexico. The latter has an acute pygidium, centrally placed anus, and is known only from *Struthanthus* and *Hyperbaena*. *Epidiaspis leperii* has an obtuse pygidium, posteriorly placed anus, and is known from many hosts, but mainly rosaceous plants.

Hosts Polyphagous. Borchsenius (1966) records it from one genus in each of 7 families and from 11 genera of Rosaceae. We have seen Italian pear scale from *Crataegus*, *Ficus*, *Heteromeles*, *Juglans*, *Malus*, *Olea*, *Persea*, *Prunus*, *Pyrus*, and *Ribes*. Gill (1997) indicates that it is common on walnuts and stone and pome fruits in California. He also indicates that it often is found on *Heteromeles* in natural habitats.

Distribution We have examined U.S. specimens from CA, CT, MD, MO, MT, NY, OK, PA, RI. Records outside of the United States include Algeria, Argentina, Azores, Chile, many European countries, Madeira Island, Iran, Mexico, Morocco, Tunisia, Turkey, and Uruguay. It has been taken most commonly in quarantine from Europe on fruit tree cuttings, such as apple, pear, nectarines, and cherry.

Biology A detailed study of this species was carried out in Switzerland by Geier (1949). There is one generation per year and the species overwinters as mated, adult females. Eggs are present in late May, June, and early July. Crawlers predominate in June but also are present in May, July, and early August. Adult males are apterous and are found in July and August; adult females are present at all times of the year. The preferred feeding site seems to be on two-year branches in protected areas. In Hungary there is one generation per year, there is an obligate diapause over the winter, mated adult females are the overwintering stage, egg laying and hatching can last up to two months, egg laying begins at the end of May when pears are in blossom, most first instars emerge between June 5 and June 15, the first instar lasts about one month, the second instar requires about 20 to 30 days, and adults are first observed at the end of July or early August (Kosztarab and Kozár 1988). Bodenheimer (1953) observed 1 generation per year in Turkey with the 'immature females' overwintering and oviposition taking place in April and May. Crawlers hatched in June. A pheromone is known for this species (El-Kareim and Kozár 1988b). A detailed study of the biology, economic damage, and natural enemies was carried out by Geoffrion (1976) in France.

Economic Importance The Italian pear scale is an economic pest on many rosaceous plants including pears, apples, prunes, and peaches. It causes distortion of branches, gummosis, and early fruit drop. It is an important pest in some parts of Europe. In Hungary it has caused serious problems on pears (Kosztarab and Kozár 1988); it can cause serious problems on olives in the Mediterranean (Argyriou 1990). El-Kareim and Kozár (1988a) report that egg production, sex ratio, and level of infestation differ significantly among cultivars of plum, apple, and pear; in fact, the apple varieties Jonathan and Starling were free of infestation in one experiment. Gill (1997) reported damage to walnuts in California; large populations caused weakened trees and smaller nut size. Examples of countries where this species is considered of economic importance are: Republic of Georgia (Aleksidze 1995; Yasnosh 1995), France (Bianchi and Bénassy

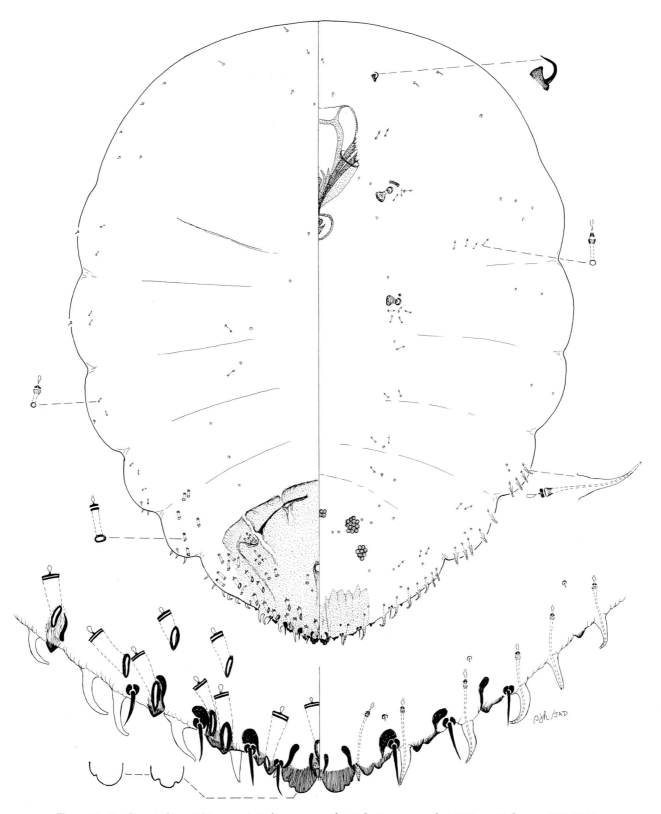

Figure 56. *Epidiaspis leperii* (Signoret), Italian pear scale, Italy, intercepted at NY, on *Malus* sp., II-6-1954.

1979; Blaisinger 1979; Geoffrion 1976), Argentina (Chiesa Molinari 1948), Hungary (El-Kareim and Kozár 1988a; Kosztarab and Kozár 1988; El-Kareim et al. 1988), Norway (Fjelddalen 1996), England and Wales (Fryer 1936), Switzerland (Geier 1949), California (Gill 1997), Italy (Melis 1949), Uruguay (Trujillo Peluffo 1942), Turkey (Bodenheimer 1953). Beardsley and González (1975) consider this scale to be one of 43 serious armored scale pests. Miller and Davidson (1990) consider it to be a serious world pest.

Selected References Ferris (1937); Geier (1949); Geoffrion (1976); Gill (1997).

Fiorinia externa Ferris

ESA Approved Common Name Elongate hemlock scale (also called fiorinia hemlock scale).

Common Synonyms and Combinations None.

Field Characters (Plate 53) Adult female pupillarial, completely enclosed in second shed skins, which is elongate, reddish brown, parallel sides, rounded at each end, and overlaid with a thin, translucent, waxy layer, without medial carina; shed skin marginal. Male cover smaller, elongate, oval, white, felted, first shed skin marginal. Body of adult female yellowish brown; eggs and crawlers yellow. Heavily infested leaves coated with white filamentous secretion. On undersurface of needles.

Slide-mounted Characters Adult female (Fig. 57) with 2 definite pairs of lobes, third pair represented by series of small, sclerotized points, usually divided into 2 lobules; paraphyses attached to medial and lateral margins of median lobes and medial and lateral margins of medial lobule of second lobes. Median lobes contiguous basally, with lateral and medial paraphyses and conspicuous yoke, without basal sclerosis, lateral and medial margins of lobes divergent, with 1–3 (2) lateral notches, with 1–4 (3) medial notches; second lobes bilobed, conspicuously broader than median lobes, medial lobule largest with 0–3 (1) lateral notches, 1–3 (1) medial notches, lateral lobule with 0–2 (0) lateral notches, 0–2 (0) medial notches; third lobes bilobed or simple, represented by series of sclerotized points, medial lobule with 3–5 (4) notches, lateral lobule with 2–6 (3) notches. Gland-spine formula normally 1-1-0, with 10–21 (15) gland spines or gland tubercles near each body margin from metathorax or segment 1 to segment 3 or 4; without gland spines between median lobes. Macroducts of 1 size, present along body margin from segment 4 to 7, without duct between median lobes, macroduct in first space 18–22 (21) μ long, with 4–5 (5) macroducts on each side of body on segments 5 to 7, total of 5–6 (6) macroducts on each side of body, without macroducts anterior of anal opening. Pygidial microducts usually on venter in submarginal areas of segments 5 to 7, rarely absent from segment 5, with 3–4 (3) ducts; prepygidial ducts of 1 size, present submedially near mouthparts and near each spiracle, with 1 or 2 sometimes present in submedial areas of segments 1 to 4; on dorsum pygidial ducts absent; prepygidial microducts of 1 size in submedial areas of segments 1 to 4. Perivulvar pores in 5 indistinct groups, 38–49 (42) pores on each side of body. Perispiracular pores with 3 loculi, anterior spiracles with 2–6 (4) pores, posterior spiracles without pores. Anal opening located 11.2–18.5 (14.9) times length of anal opening from base of median lobes, anal opening 10–17 (13) μ long. Dorsal seta laterad of median lobes 0.9–2.0 (1.3) times length of median lobe. Eyes absent. Antennae each with 1 long seta. Cicatrices normally present on segments 1 and 3. Body elongate, unsclerotized. Antennae submarginal, close together, narrower than width of clypeolabral shield, each with 2–4 (3) points.

Affinities In the United States this species is most similar to *Fiorinia fioriniae*, palm fiorinia scale, but differs by having 5 or 6 macroducts on each side of body, antennae located on submargin, 38–49 (42) perivulvar pores on each side of body, anal opening located 11.2–18.5 (14.9) times length of anal opening from base of median lobes, by lacking a median longitudinal carina on the second shed skin, and by infesting coniferous hosts. Palm fiorinia scale has 3 or 4 macroducts on each side of body, antennae located on margin, 21–36 (26) perivulvar pores on each side of body, anal opening located 7.7–10.6 (9.0) times length of anal opening from base of median lobes, median carina, and occurs on many hosts.

Hosts We have seen specimens from *Abies, Pseudotsuga*, and *Tsuga*. It is recorded from *Cedrus, Picea, Pinus*, and *Taxus*

(McClure and Fergione 1977). *Tsuga* appears to be the preferred host (McClure and Fergione 1977).

Distribution This scale probably was introduced into the United States from Japan; it was recently discovered in England (Williams 1988) and probably was introduced there from the United States. We have examined U.S. specimens from CT, DC, MD, NJ, NY, OH, PA. It is recorded from MA, RI, VA.

Biology The biology and ecology of this species have been studied extensively by Mark McClure of the Connecticut Agricultural Experiment Station. He found that the species has 1 generation and a partial second per year in Connecticut (McClure 1978); most individuals in the second generation perish during the winter months. Overwintering takes place as eggs within the pupillarial female or as fully mature adult females. Seasonal development of the species varies considerably depending on the climate. At two localities in Connecticut, McClure found a difference of 3 to 4 weeks in crawler activity. At Westport, Connecticut, crawlers appeared in May, second instars were present in June, adult males and females were present in early July. Species of *Tsuga* are the preferred hosts for this pest. According to Davidson and McComb (1958) the elongate hemlock scale has 2 generations per year in Maryland with crawler production peaks occurring in spring and fall. Heller (1977) reports 2 generations per year in Pennsylvania. Stimmel (1980) found multiple overlapping generations in Pennsylvania with eggs, second-instar males and females, and adult females present during the winter. Wallner (1964) reports one complete and one partial generation per year on Long Island. Monthly observations have shown adult females overwinter and active crawlers are present throughout the warm months (May–October) indicating overlapping generations. Unlike most armored scales, members of this genus live as adult females within the enlarged shed skin of the second instar. As the adult female shrinks inside the second shed skin, eggs are laid in two rows with their ends meeting in the median longitudinal axis of the shed skin. The average number of eggs found behind one female at one time is 6; as these hatch more are laid. Females lay up to 20 eggs over a period of 1 to 1.5 weeks. Eggs hatch in 3 to 4 weeks. Crawlers molt into second instars in 3 to 4 weeks. In 4 more weeks adult females appear. These mate with males and begin producing second-generation eggs 6 to 8 weeks later. Crawlers from these eggs produce the overwintering generation. McClure (1980b) reports a direct positive correlation between the concentration of nitrogen in young foliage and the rate of mortality and number of progeny produced by female elongate hemlock scales.

Economic Importance In Connecticut, the elongate hemlock scale is reported to cause death of trees of various sizes both in natural and ornamental situations (McClure 1977c). The insects occur on the undersides of leaves and remove fluids from the internal mesophyll cells. Feeding eventually causes chlorosis, early leaf drop, decreased growth, and death in heavy infestations. Economic populations have been reported in CT, MD, and PA. This species is considered to be a serious pest in Pennsylvania, particularly on *Abies* Christmas trees (Stimmel 2002, personal communication). McClure (1977c) reports heavy parasitism in several instances. Because crawlers occur throughout much of the summer, control with contact insecticides is relatively ineffective. Use of foliar systemic insecticides has proved to be quite effective, as have 'superior' dormant oils (Wallner 1964). In Japan this species is apparently not an economic problem (Murakami 1970). Miller and Davidson (1990) consider this species to be an occasional pest.

Selected References Davidson and McComb (1958); Ferris (1942); McClure (1977a, b, c, d).

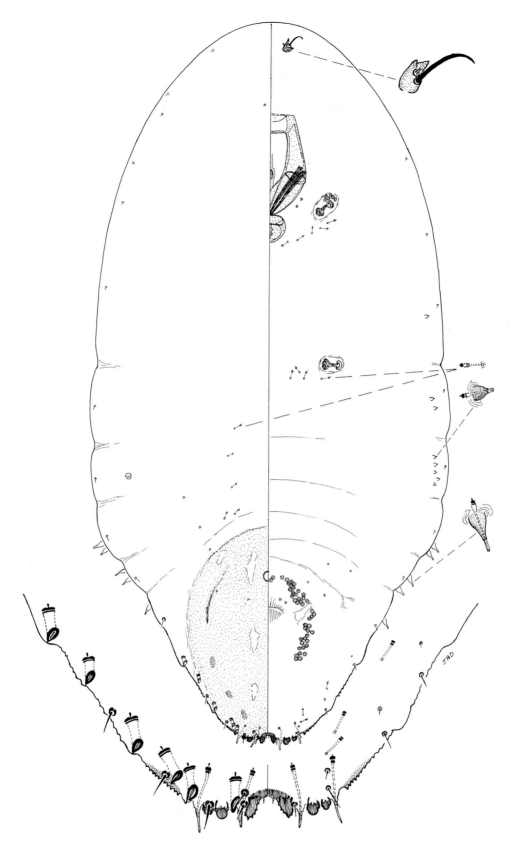

Figure 57. *Fiorinia externa* Ferris, elongate hemlock scale, Trenton, NJ, on *Tsuga* sp., VI-21-1974.

Plate 53. *Fiorinia externa* Ferris, Elongate hemlock scale

A. Infestation on hemlock (J. A. Davidson).
B. Rubbed female covers. One specimen turned reveals two rows of white egg shells within second-instar shed skin (J. A. Davidson).
C. Adult female and male covers on hemlock (J. A. Davidson).
D. Distance view of covers on hemlock (J. A. Davidson).
E. Distance view of covers on hemlock (J. A. Davidson).

F. Distance view on hemlock needles with wax threads produced by crawlers (J. A. Davidson).
G. Distance view with yellow feeding crawlers, adult female covers, and male covers on hemlock (J. A. Davidson).
H. Winged adult male on hemlock (J. A. Davidson).
I. Distance view of underside of hemlock showing white washed appearance of infested needles (J. A. Davidson).

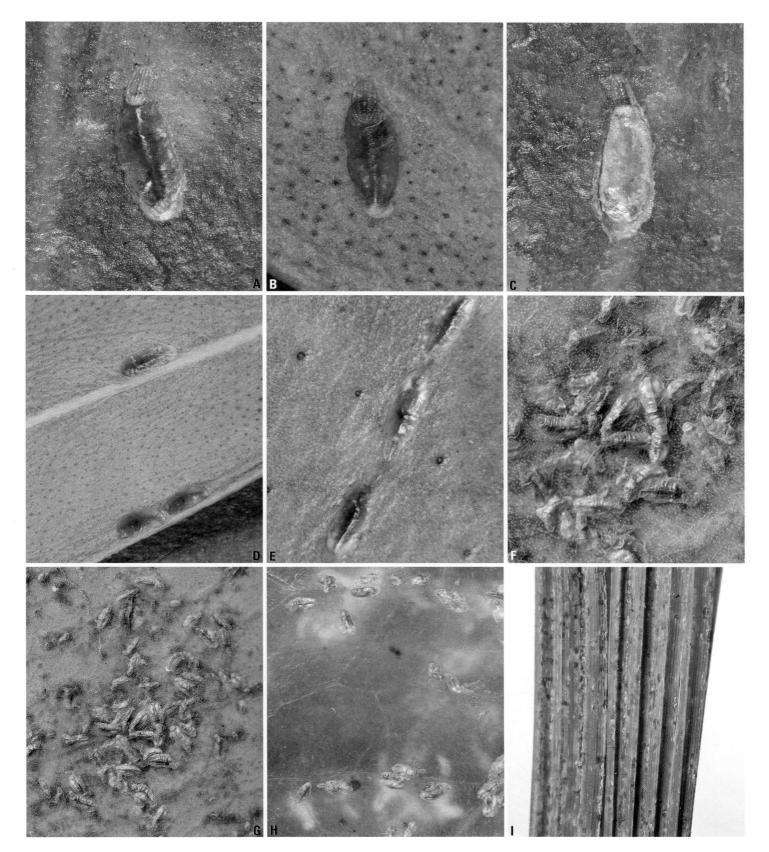

Plate 54. *Fiorinia fioriniae* (Targioni Tozzetti), Palm fiorinia scale

A. Pupillarial female on camellia (J. A. Davidson).
B. Adult female with eggs faintly visible through second-instar shed skin on bottle brush (J. A. Davidson).
C. Ventral view of adult female on camellia (J. A. Davidson).
D. Adult female covers on underside of bottle brush leaf (J. A. Davidson).
E. Adult female covers on camellia leaf (R. J. Gill).

F. Heavy infestation on *Persea* leaf (J. A. Davidson).
G. Distance view of infestation on *Persea* leaf (J. A. Davidson).
H. Distance view showing stylet track damage on camellia (J. A. Davidson).
I. Distance view of heavy infestation and damage on palm (D. R. Miller).

203

Fiorinia fioriniae (Targioni Tozzetti)

Suggested Common Name Palm fiorinia scale (also called avocado scale, European fiorinia scale, fiorinia scale, ridged scale, camellia scale).

Common Synonyms and Combinations *Diaspis fioriniae* Targioni Tozzetti, *Fiorinia pellucida* Targioni Tozzetti, *Fiorinia camelliae* Comstock, *Uhleria fioriniae* (Comstock), *Uhleria camelliae* (Comstock), *Fiorinia palmae* Green, *Chermes arecae* Boisduval.

Field Characters (Plate 54) Adult female pupillarial, completely enclosed in second shed skin, which is elongate, brownish yellow to brownish orange, elliptical, with faint mediolongitudinal carina; shed skin marginal. Males rare, but reported Danzig (1964), Ferris (1938a), Gill (1997), and Kuwana (1925). Previous descriptions of male covers vary. Male cover white and felted like other *Fiorina* (Ferris 1938a); dull white, nearly transparent, scarcely noticeable (Gill 1997). Body of adult female yellow to brown; eggs and crawlers probably yellow. On upper and lower leaf surfaces, prefers lower surface.

Slide-mounted Characters Adult female (Fig. 58) with 2 definite pairs of lobes, third pair represented by series of small, sclerotized points, without clear division into 2 lobules; paraphyses attached to medial and lateral margins of median lobes, medial and lateral margins of medial lobule of second lobes, and medial margin of third lobes. Median lobes contiguous basally, with lateral and medial paraphyses, with conspicuous yoke, without basal sclerosis, lateral and medial margins of lobes divergent, with 0–1 (0) lateral notches, with 2–4 (3) medial notches; second lobes bilobed, conspicuously broader than median lobes, medial lobule largest with 0–2 (1) lateral notches, 0–1 (0) medial notches, lateral lobule with 1–2 (1) lateral notches, without medial notches; third lobes simple, represented by series of sclerotized points, with 3–6 (4) notches. Gland-spine formula normally 1-1-0, with 11–19 (15) gland spines or gland tubercles near each body margin from mesothorax to segment 3 or 4; without gland spines between median lobes. Macroducts usually of 1 size, occasionally with 1 on segment 4 smaller, present along body margin from segment 4 or 5 to 7, without duct between median lobes, macroduct in first space 13–20 (17) μ long, with 2–4 (3) macroducts on each side of body on segments 5 to 7, without prepygidial macroducts, without macroducts anterior of anal opening. Pygidial microducts usually on venter in submarginal areas of segments 5 to 7, rarely absent from segment 5, with 2–4 (3) ducts; prepygidial ducts of 1 size, present submedially near mouthparts and near each spiracle, with 1 or 2 sometimes present in submedial areas of segments 1 to 4; on dorsum pygidial ducts sometimes present on submedial area of segment 5; prepygidial microducts of 1 size in submedial areas of segments 1 to 4. Perivulvar pores in 5 indistinct groups, 21–33 (26) pores on each side of body. Perispiracular pores with 3 loculi, anterior spiracles with 2–5 (3) pores, posterior spiracles without pores. Anal opening located 7.7–10.6 (9.0) times length of anal opening from base of median lobes, anal opening 10–15 (12) μ long. Dorsal seta laterad of median lobes 0.9–1.7 (1.2) times length of median lobe. Eyes absent. Antennae each with 1 long seta. Cicatrices normally present on segments 1. Body elongate, unsclerotized. Antennae marginal, close together, nearly touching one another, each with 0–3 (2) points.

Affinities In the United States this species is most similar to *Fiorinia externa*, elongate hemlock scale. For a comparison of this species see the treatment of elongate hemlock scale.

Hosts Polyphagous. Borchsenius (1966) records 45 host genera in 23 families; Dekle (1977) reports the palm fiorinia scale on 26 host genera in Florida and states that palms and *Laurus* are the most frequently reported hosts. We have examined material from *Agave, Aleurites, Anacardium, Artocarpus, Buxus, Callistemon, Camellia, Capsicum, Cocos, Cordyline, Cyathodes, Cycas, Diospyros, Eugenia, Euonymus, Garcinia, Gardenia, Hedera, Kentia, Latania, Laurus, Licuala, Ligustrum, Mangifera, Manilkara, Morus, Musa, Myrtus, Persea, Phoenix, Phormium, Polyalthia, Psidium, Ptychosperma, Quercus, Rosa, Ruscus,* and *Ruta*. It is taken most commonly in quarantine on camellia, mango, and avocado.

Distribution This scale probably was introduced into the United States from the Orient. We have examined U.S. specimens from AL, DC, FL, GA, NY, OH, TX. It has been recorded from CA (in San Diego County only [Gill 1997]), CT, IL, MA, MS, OK, PA, SC. This pest commonly is intercepted in quarantine from most tropical areas of the world and occasionally from greenhouses in northern countries.

Biology There are few reports dealing with the life cycle of palm fiorinia scale. Johnson and Lyon (1976) indicate overlapping generations in the southern United States. Murakami (1970) states that it has 3 generations per year in Japan with eggs laid in May, July, and August.

Economic Importance The palm fiorinia scale is regarded as a pest of avocado (Williams and Watson 1988; Perez Guerra 1986; Cohic 1958), palms (Hodgson and Hilburn 1991; Dekle 1977), tea (Nagarkatti and Sankaran 1990), and ornamentals (Dekle 1977). Beardsley and González (1975) consider this scale to be one of 43 serious armored scale pests. Miller and Davidson (1990) consider it to be a serious world pest.

Selected References Ferris (1938a); Johnson and Lyon (1976); Murakami (1970).

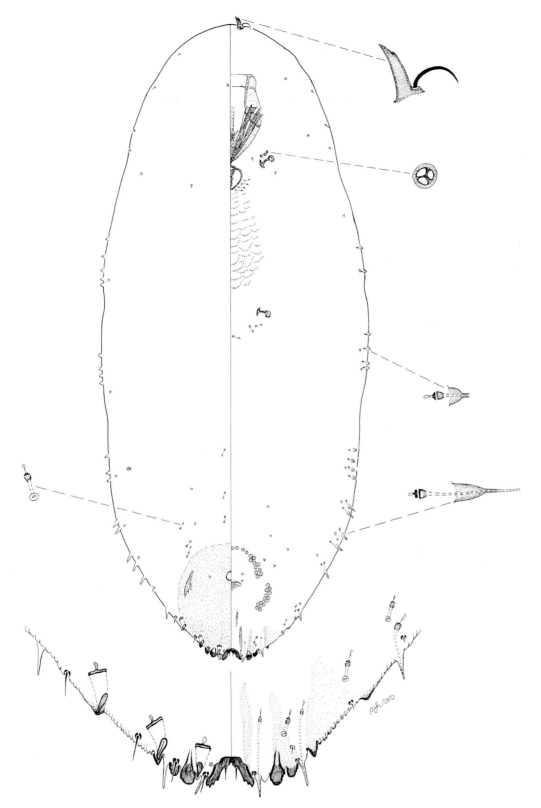

Figure 58. *Fiorinia fioriniae* (Targioni Tozzetti), palm fiorinia scale, Lima, Peru, on *Persea americana*, I-24-1975.

Fiorinia japonica Kuwana

Suggested Common Name Coniferous fiorinia scale (also called juniper fiorinia scale and Japanese scale).

Common Synonyms and Combinations *Fiorinia fioriniae* var. *japonica* Kuwana, *Fiorinia juniperi* Leonardi.

Field Characters (Plate 55) Adult female pupillarial, completely enclosed in second shed skin, elongate, brownish yellow, usually with dark median area, elliptical, with mediolongitudinal carina, sometimes with white powdery substance; shed skin yellow, marginal. Male cover elongate oval, white, felted, first shed skin yellow, marginal. Body of adult female yellow; eggs and crawlers probably yellow. On upper and lower leaf surfaces.

Slide-mounted Characters Adult female (Fig. 59) with 2 definite pairs of lobes, third pair represented by series of small, sclerotized points, sometimes with division into 2 lobules; paraphyses sometimes represented by small sclerotizations attached to medial and lateral margins of medial lobule of second lobes, and medial margin of lateral lobule of second lobes. Median lobes contiguous basally, without lateral and medial paraphyses, with conspicuous yoke, without basal sclerosis, lateral and medial margins of lobes divergent, with 0–1 (0) lateral notches, with 2–4 (3) medial notches; second lobes bilobed, broader than median lobes, medial lobule largest with 1–3 (2) lateral notches, 0–1 (0) medial notches, lateral lobule with 0–2 (1) lateral notches, with 0–1 (0) medial notches; third lobes simple or weakly bilobed, represented by series of sclerotized points, with 4–15 (9) notches. Gland-spine formula normally 1-1-0, with 9–19 (16) gland spines or gland tubercles near each body margin from mesothorax to segment 3 or 4; without gland spines between median lobes. Macroducts usually of 2 sizes, usually with 1 or 2 on segment 5 and/or 4, smaller nearly indistinguishable from microducts except with sclerotized dermal orifice, present along body margin from segment 4 or 5 to 7, without duct between median lobes, macroduct in first space 15–22 (19) μ long, with 4–6 (5) macroducts on each side of body on segments 5 to 7, total of 5–9 (7) on each side of body, without macroducts anterior of anal opening. Pygidial microducts usually on venter in submarginal areas of segments 5 to 7, rarely absent from segment 5, with 4–7 (5) ducts; prepygidial ducts of 1 size, present submedially near mouthparts and near each spiracle, often with a medial cluster on segment 1 and with 1 or 2 sometimes present in submedial areas of segments 2; on dorsum pygidial ducts sometimes present on submedial area of segment 5; prepygidial microducts of 1 size in submedial and sublateral areas of segments 3 and 4, sometimes present on segments 1 and 2. Perivulvar pores in 5 groups, 30–54 (42) pores on each side of body. Perispiracular pores with 3 loculi, anterior spiracles with 3–12 (7) pores, posterior spiracles without pores. Anal opening located 6.2–14.7 (10.4) times length of anal opening from base of median lobes, anal opening 9–18 (11) μ long. Dorsal seta laterad of median lobes 0.6–1.1 (0.8) times length of median lobe. Eyes usually represented by sclerotized spot, located on body margin near anterior edge of clypeolabral shield. Antennae each with 1 long seta, sometimes with basal portion of antenna developed into pointed process, but this is not always present. Cicatrices normally present on segment 1, rarely present on prothorax. Body elongate, unsclerotized. Antennae marginal, close together, nearly touching one another, with 0–2 (1) points.

Affinities In the United States this species is most similar to *Fiorinia fioriniae*, palm fiorinia scale, but differs by having 4 large marginal macroducts on each side of pygidium on segments 5 to 7 and 30–54 (42) perivulvar pores on each side of body; whereas, palm fiorinia scale has 3 large marginal macroducts on each side of pygidium on segments 5 to 7 and 21–33 (26) perivulvars on each side of body.

Hosts This species is restricted to conifers. Borchsenius (1966) records 8 host genera in 2 families; Nakahara (1982) records it from 9 coniferous genera. We have examined material from *Abies, Cephalotaxus, Cupressus, Juniperus, Picea, Pinus, Podocarpus, Sciadopitys, Taxus, Torreya*, and *Tsuga*. It is taken most commonly in quarantine on pine and *Taxus*.

Distribution This scale probably was introduced into the United States from the Orient. We have examined U.S. specimens from DC, MD, and VA. It has been recorded from CA (now reported eradicated). This pest commonly is intercepted in quarantine from Japan and Taiwan.

Biology As far as we can determine, the biology of this species has never been studied.

Economic Importance The coniferous fiorinia scale is a pest of conifers in the Washington, D.C. area; it causes chlorosis of the leaves, leaf drop, and an unsightly appearance. Tang (1984) considers it to be a serious pest of pine trees in Beijing, China. Miller and Davidson (1990) consider this species to be an occasional pest.

Selected Reference Takagi (1970).

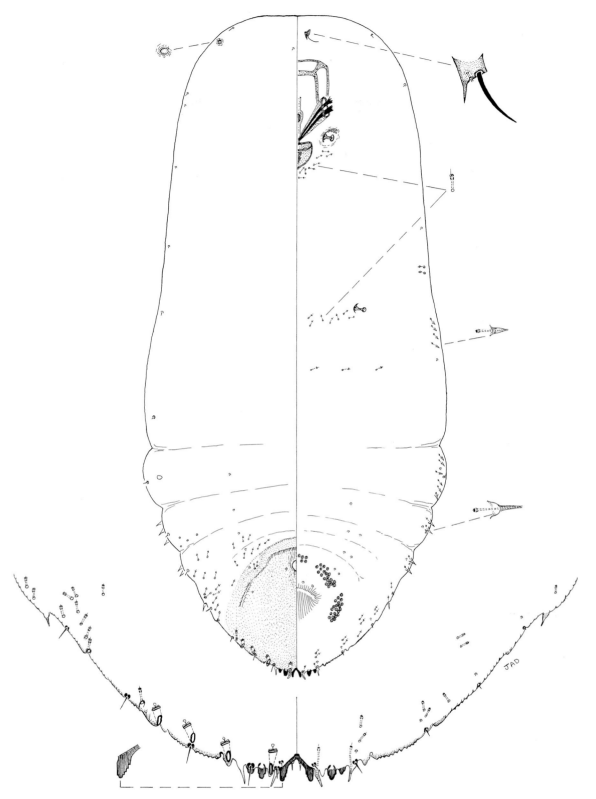

Figure 59. *Fiorinia japonica* Kuwana, coniferous fiorinia scale, Bethesda, MD, on *Tsuga canadensis*, IV-14-1987.

Plate 55. *Fiorinia japonica* Kuwana, **Coniferous fiorinia scale**

A. Adult female showing distinct dorsal black spot on blue spruce (J. A. Davidson).
B. Adult females showing less distinct dorsal black spot on blue spruce (J. A. Davidson).
C. Fuzzy white male cover with associated settled crawlers producing white threads on Atlas cedar (J. A. Davidson).

D. Adult female and male covers on spruce (J. A. Davidson).
E. Adult female covers on blue spruce (J. A. Davidson).
F. Crawler on blue spruce (J. A. Davidson).
G. Moderate infestation on spruce (J. A. Davidson).
H. Heavy infestation on spruce (J. A. Davidson).
I. Moderate infestation on Atlas cedar (J. A. Davidson).

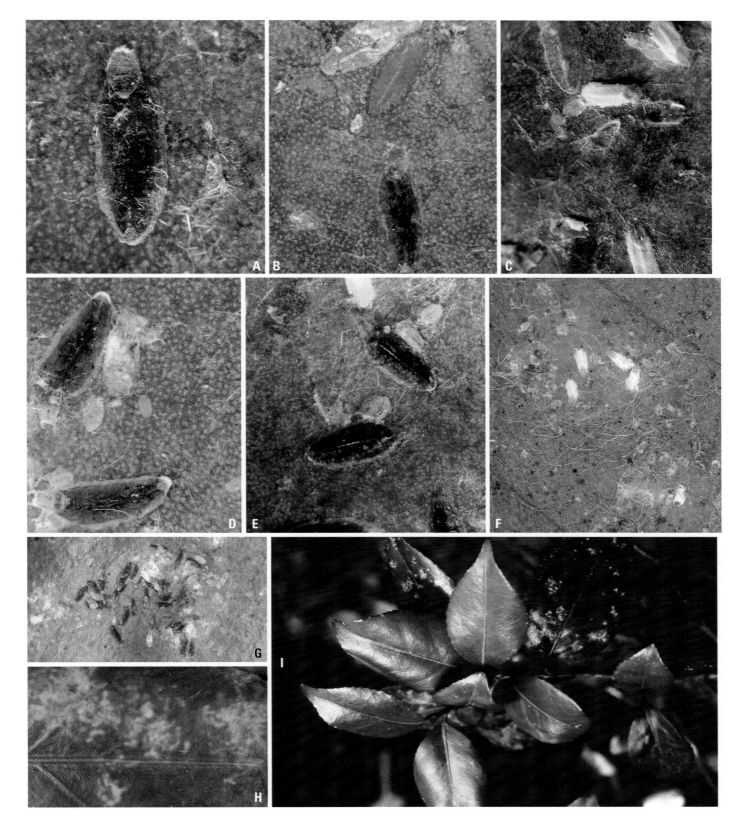

Plate 56. *Fiorinia theae* Green, Tea scale

A. Old pupillarial adult female cover on camellia (J. A. Davidson).
B. Mature yellowish second-instar female (top) with thin, transparent, waxy cover. Dark adult female cover (bottom) on camellia (J. A. Davidson).
C. Tricarinate, ridged male covers on holly (J. A. Davidson).
D. Adult female covers, active yellow crawlers, and whitish feeding crawlers on camellia (J. A. Davidson).
E. Adult female covers, male covers, and feeding crawlers on holly (J. A. Davidson).

F. Distance view showing long white waxy threads produced by crawlers from wax pores on head (J. A. Davidson).
G. Distance view of covers on camellia (J. A. Davidson).
H. Visible damage on top of leaf caused by feeding scales on undersurface (J. A. Davidson).
I. Distance view of damage caused by feeding on camellia leaves (D. R. Miller).

209

Fiorinia theae Green

ESA Approved Common Name Tea scale.
Common Synonyms and Combinations None.

Field Characters (Plate 56) Adult female pupillarial, enclosed in second shed skin, elongate, with conspicuous longitudinal ridge, cover parallel sided, pointed at posterior apex, light gray to nearly black; shed skin marginal, clear to yellow. Male cover smaller, white, felted, faintly tricarinate, first shed skin marginal. Body of adult female, eggs, and crawlers yellow. Heavily infested leaves with white, filamentous secretion. Primarily on underside of leaves, rarely on upper surfaces.

Slide-mounted Characters Adult female (Fig. 60) usually with 2 definite pairs of lobes, occasionally with second lobes represented by series of low points, third lobes either present or represented by series of small, sclerotized points, without clear division into 2 lobules; paraphyses attached to medial and lateral margins of median lobes. Median lobes contiguous basally, with lateral and medial paraphyses, with conspicuous yoke, without basal sclerosis, lateral and medial margins of lobes parallel or divergent, notching continuous on lobe not distinctly lateral or medial, with 2–6 (4) notches; second lobes usually bilobed, smaller than median lobes, medial lobule largest with 2–5 (3) notches, lateral lobule with 1–4 (2) notches; third lobes present in 4 of 10 specimens, when present, simple, represented by series of sclerotized points, with 3–6 (4) notches; fourth lobes present in 1 of 10 specimens, with 2 notches. Gland-spine formula normally 1-1-1, sometimes 1-1-0, with 17–36 (15) gland spines or gland tubercles near each body margin from pro- or mesothorax to segments 3 or 4; without gland spines between median lobes. Macroducts usually of 1 size, about same size as microtubular ducts, present along body margin from segments 4 to 7, without duct between median lobes, macroduct in first space 12–19 (15) μ long, with 6–8 (7) macroducts on each side of body on segments 5 to 7, total of 6–12 (9) macroducts on each side of body, without macroducts anterior of anal opening. Pygidial microducts absent from venter of pygidium; prepygidial ducts of 1 size, present submedially near mouthparts and near each spiracle; on dorsum pygidial ducts submedially on segment 5; prepygidial microducts of 1 size in submedial areas of segments 2 or 3 to 4. Perivulvar pores in 5 indistinct groups, 24–37 (30) pores on each side of body. Perispiracular pores with 3 loculi, anterior spiracles with 2–4 (3) pores, posterior spiracles without pores. Anal opening located 6.7–9.4 (8.1) times length of anal opening from base of median lobes, anal opening 9–13 (11) μ long. Dorsal seta laterad of median lobes 0.5–0.8 (0.6) times length of median lobe. Eyes normally represented by series of 1 or more small domes on lateral margin anterior of mouthparts. Antennae each with 1 long seta. Cicatrices normally present on segment 1. Body elongate, unsclerotized. Distance between antennae less than width of clypeolabral shield apart, not close enough to touch one another, with characteristic protrusion or tubercle between them, located submarginally, each with 0–3 (2) points.

Affinities This species is unusual by having a tubercle or unusually shaped tubercle between the antennae and by having the antennae set close together, that is, less than the width of the clypeolabral shield.

Hosts Polyphagous. Borchsenius (1966) records 7 host genera in 5 families; Dekle (1977) reports it on 18 host genera in Florida. We have seen specimens collected on *Camellia, Citrus, Euonymus, Eurya, Gardenia, Ilex, Ostodes, Rhododendron,* and *Spondias.* The tea scale is most commonly collected on *Camellia* and *Ilex.*

Distribution This species probably is native to the Oriental Region. We have examined U.S. material from AL, AR, DC, FL, GA, LA, MA, MD, MO, MS, NC, NY, OK, PA, SC, TX, VA. Tea scale has been reported from CA (eradicated), IN, TN. It is known to overwinter as far north as Maryland. Records from northern states represent indoor infestations. This pest has been intercepted in quarantine inspections from Bahamas, China, Honduras, India, Indonesia, Japan, Mexico, Nepal, Sri Lanka, and Taiwan.

Biology Tea scale is most commonly found on camellias in the United States. English and Turnipseed (1940) reported a 60 to 70 day life cycle in warm weather in Alabama with nymphs hatching throughout the year. They noted overlapping generations with continuous crawler production from March to November. Females are reported to lay up to 4 eggs each day with an average of 32 eggs per female (Das and Das 1962). Egg production totals reported by English and Turnipseed (1940) are 10–16 per female. Eggs are laid in 2 longitudinal rows within the second shed skin, which hatch in 7 to 21 days in Alabama (English and Turnipseed 1940) or 4 to 6 days in India (Das and Das 1962). Total developmental time from egg hatching to emergence of adult females is 24 to 27 days; for adult males it is 22 to 24 days in May in Tocklai, India. Winged males often are abundant and apparently are required for the production of viable crawlers. Das and Das (1962) report that this scale appears to prefer shaded areas in tea plantations in India. In Florida the biology of this species was studied in the laboratory at 25 °C and 70% RH on butternut squash (Munir and Sailer 1985). Females lay 17–43 (28) eggs that hatch in about 10 days. Crawlers move over the host for 1 to 4 days, settle, and insert their mouthparts into the host. Molting to the second instar occurs about 10 days after settling. The duration of the male life stages is as follows: second instar 11 days, third-instar prepupa 5 days, fourth-instar pupa 4 days, adult male 1 day; the male life cycle is completed in about 34 days. The duration of the female life stages is: second instar 6 days; adult female 17 days; the female cycle is completed in about 65 days. Munir and Sailer (1985) indicate there is a strong male-biased sex ratio of about 2 : 1. In the field they observed that the tea scale is multivoltine with overlapping generations, and development continues throughout the year but is slowed in the winter. They indicated that camellia and holly seemed to be preferred hosts.

Economic Importance The tea scale is a serious pest of camellias in the southeastern United Sates (English 1990). Heavily infested leaves are covered with a cottony mass on the under surfaces, have chlorotic spots around the cover of the female, and eventually become completely yellow and fall from the plant (English 1990). Plants become stunted, have an unsightly appearance, and have a decrease in bloom productivity. This scale has been considered the most important pest of camellia in the south (English and Turnipseed 1940) and one of the ten most important scale pests in Florida (Dekle 1977). The tea scale apparently is not a serious problem on tea in certain parts of India (Das and Das 1962) because it is kept in check by a complex of natural enemies including several parasites in the genera *Aphytis* and *Encarsia*, a complex of predators, and fungi in the genera *Aschersonia* and *Fusarium* (Nagarkatti and Sankaran 1990). The tea scale is commonly found on several species of holly in the eastern United States (McComb 1986). Beardsley and González (1975) consider this scale to be one of 43 serious armored scale pests. Miller and Davidson (1990) consider it to be a serious world pest.

Selected References Das and Das (1962); English and Turnipseed (1940); Gill (1997); Munir and Sailer (1985).

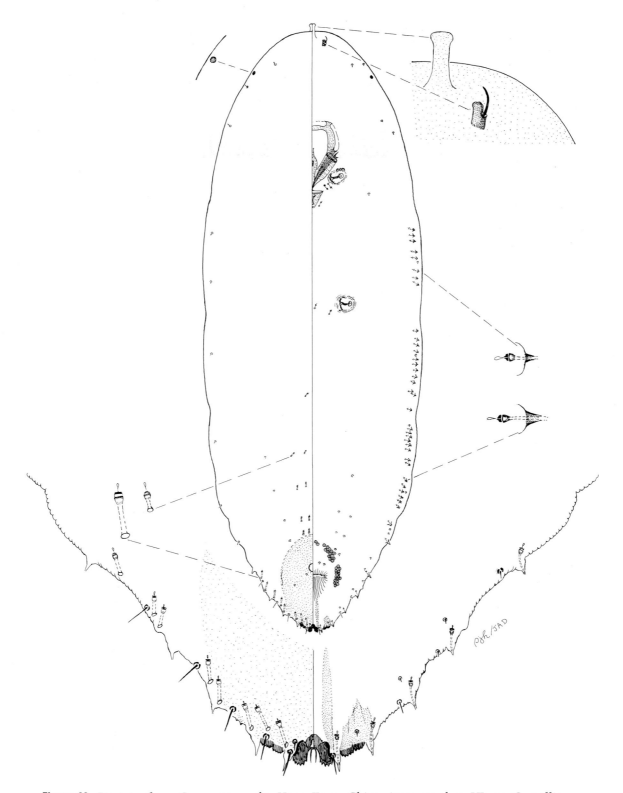

Figure 60. *Fiorinia theae* Green, tea scale, Hong Kong, China, intercepted at HI, on *Camellia* sp., I-13-1949.

Froggattiella penicillata (Green)

Suggested Common Name Penicillate scale.

Common Synonyms and Combinations *Odonaspis penicillata* Green, *Anoplaspis penicillata* (Green), *Dycryptaspis penicillata* (Green).

Field Characters (Plate 57) Adult female cover moderately convex, elongate oval, light brown or tan, ventral cover well developed; shed skins marginal, shiny yellow or yellow brown. Male cover elongate oval, similar in color and texture to female cover; shed skin marginal, yellow. Body of adult female white; eggs and crawlers presumed white also. On stems or nodes under leaf sheaths.

Slide-mounted Characters Adult female (Fig. 61) with lobes indefinite, median lobes apparently represented by projection laterad of large medial seta, lobe 2 apparently represented by projection with paraphysis attached basally, lobe 3 possibly represented by lobe of same shape as lobe 2 slightly anterior on segment 6, lobes 4 and 5 possibly represented by largest projections mesad of seta on segments 4 and 5; with 2 conspicuous paraphyses on each side of pygidium, 1 attached to medial margin of median lobe and 1 attached to remnant of lobe 2. Gland spines present in clump of 4–6 (5) conspicuous spines between median lobe remnants. Macroducts of 1 size over entire body, ducts in positions normally occupied by microducts of same size and shape as macroducts, on dorsum in segmental rows on segments 5 to 8, in submarginal and marginal clusters on metathorax or segment 1 to segment 8, without macroducts between median lobes, macroduct in first space about 5 μ long, with approximately 250–400 (315) dorsal macroducts on each side of body on segments 5 to 8, total of about 375–580 (450) dorsal macroducts on each side of body, some orifices occurring anterior of anal opening, ventral macroducts in marginal and submarginal clusters on posterior margin of head to segment 8. Pygidial microducts on venter present only in association with terminal gland spines, 100–155 (125) μ long, extending 0.5–0.9 (0.7) times distance from posterior apex of anal ring to base of median lobes; prepygidial microducts in medial or submedial clusters near mouthparts, spiracles, and segments 1 to 8, in lateral or sublateral areas from head to segment 8, with rows of ducts from body margin to near antennae and to spiracles. Perivulvar pores absent. Perispiracular pores usually with 5 loculi, anterior spiracles each with 4–10 (7) pores, posterior spiracles without pores. Anal opening located 13–23 (19) times length of anal opening from base of median lobes, anal opening 8–12 (10) μ long. Dorsal seta laterad of median lobes 20–34 (29) μ long. Eyes absent. Antennae each with 1 large seta. Cicatrices absent. Setae on margin of pygidium with base set in dermal pocket. Antennae placed more posteriorly than normal, nearly laterad of anterior edge of clypeolabral shield. Body nearly oval in newly matured females but elongate in older females.

Affinities Of the economic species, penicillate scale is most similar to *Odonaspis ruthae*, Bermuda grass scale, but is easily separated by the lack of perivulvar pores, the lack of distinct microducts on the venter, and the presence of the cluster of conspicuous gland spines at the apex of the pygidium. Bermuda grass scale has perivulvar pores, has distinct microducts on the venter, and lacks gland spines entirely.

Hosts Limited in its host range to bamboo species. Borchsenius (1966) records it from 4 genera; Dekle (1977) reports it in Florida on *Bambusa* only; and Ben-Dov (1988) records it from 4 bamboo genera worldwide. We have examined material from the following genera of bamboo: *Arundinaria*, *Bambusa*, *Dendrocalamus*, *Gigantochloa*, and *Phyllostachys*.

Distributions We have examined U.S. specimens from AL, CA, DC, FL, GA, HI, LA, MS, TX. The record from DC is from a greenhouse. This species is taken most frequently in quarantine at U.S. ports-of-entry on bamboo from the Caribbean islands, China, Japan, South and Central America, and Vietnam.

Biology We have been unable to find anything about the biology of this species.

Economic Importance Penicillate scale generally is not considered to be a problem, but Gill (1982, 1997) mentions it as a common pest in most suburban areas of California and recommends treatment of infested plants, especially in nurseries.

Selected References Ben-Dov (1988); Gill (1997); Takagi (1969).

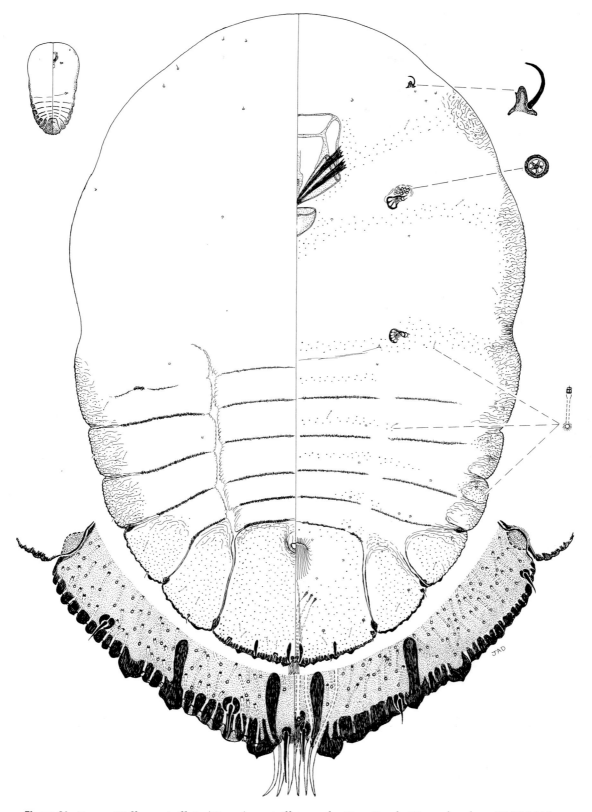

Figure 61. *Froggattiella penicillata* (Green), penicillate scale, Vero Beach, FL, on bamboo, IV-25-1946.

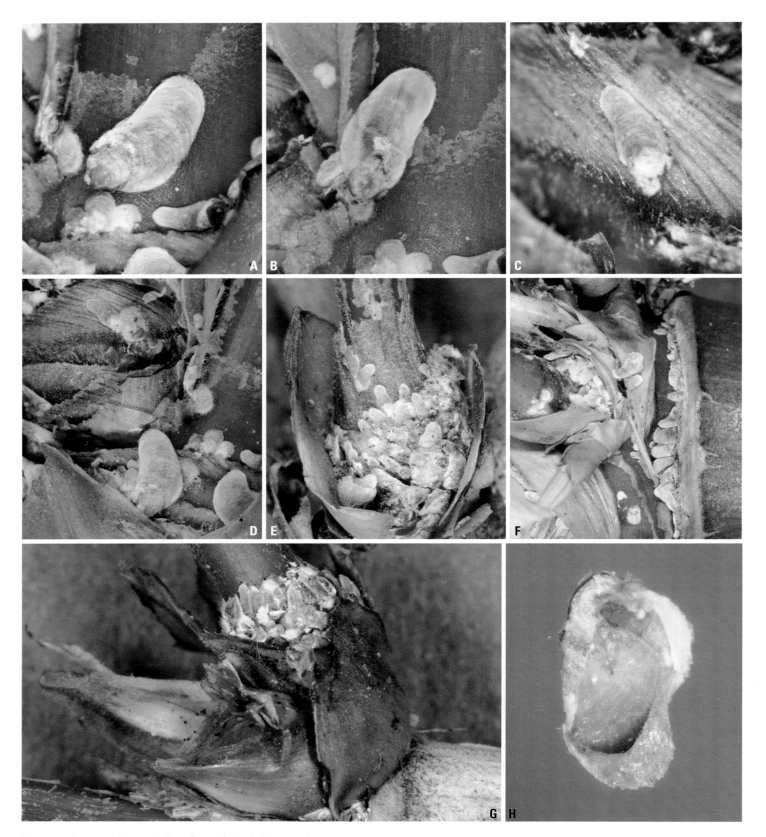

Plate 57. *Froggattiella penicillata* (Green), Penicillate scale

A. Adult female cover and smaller adult male cover on bamboo (J. A. Davidson).

B. Same adult female with cover turned over to show complete ventral cover (J. A. Davidson).

C. Adult male cover on bamboo (J. A. Davidson).

D. Infestation at base of bamboo leaves (J. A. Davidson).

E. Female covers concentrated at base of bamboo stem with male covers (J. A. Davidson).

F. Infestation surrounding bamboo leaf sheath base (J. A. Davidson).

G. Distance view of concentrated covers with leaf sheath base removed (J. A. Davidson).

H. Cover partially removed showing adult female body (J. A. Davidson).

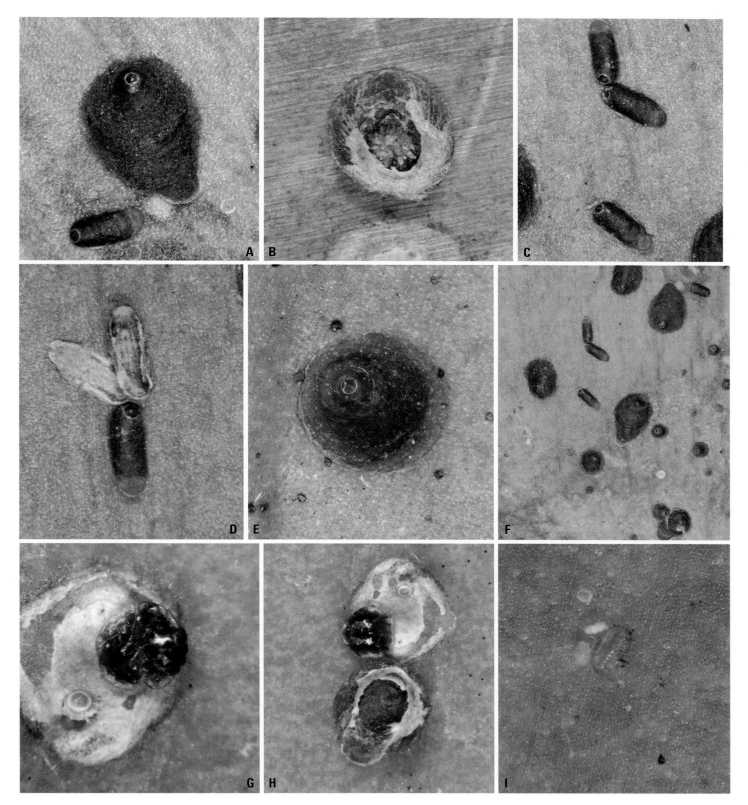

Plate 58. *Furcaspis biformis* (Cockerell), Orchid scale

A. Adult female cover and smaller male cover on orchid (J. A. Davidson).

B. Ventral view of adult female cover showing partial ventral cover and dead body of adult female on orchid (J. A. Davidson).

C. Male covers with light, posterior exit flap on orchid (J. A. Davidson).

D. Adult male cover turned over to expose white, 'hinged' exit flap on orchid (J. A. Davidson).

E. Adult female cover before production of crawler flap on cattleya orchid leaf (J. A. Davidson).

F. Distance view of infestation on cattleya orchid leaf (J. A. Davidson).

G. Body of adult female on cattleya orchid leaf (J. A. Davidson).

H. Body of adult female with crawlers on cattleya orchid leaf (J. A. Davidson).

I. Settled crawler on cattleya orchid leaf (J. A. Davidson).

215

Furcaspis biformis (Cockerell)

Suggested Common Name Orchid scale (also called red orchid scale).

Common Synonyms and Combinations *Aspidiotus biformis* Cockerell, *Aspidiotus biformis* (Cockerell), *Aspidiotus biformis* var. *cattleyae* Cockerell, *Aspidiotus* (*Chrysomphalus*) *biformis* Cockerell, *Aspidiotus* (*Chrysomphalus*) *biformis* var. *cattleyae* Cockerell, *Aspidiotus* (*Evaspidiotus*) *biformis* Cockerell, *Chrysomphalus biformis* (Cockerell), *Chrysomphalus biformis cattleyae* (Cockerell), *Aspidiotus biprominens* Kuwana and Muramatsu, *Targionia biformis* (Cockerell).

Field Characters (Plate 58) Adult female cover moderately convex, circular or slightly elliptical, dark reddish brown, scale margin sometimes lighter; shed skins central or subcentral, about same color as rest of cover. Male cover same color as female cover, except posterior end lighter in color, more elongate; shed skin submarginal. Body of newly matured adult female clear, turning violet or purple at maturity. Normally on leaves or pseudobulbs.

Slide-mounted Characters Adult female (Fig. 62) with 3 definite pairs of lobes, third pair represented by normal lobule similar to median and second lobes but also with row of 3 additional lobules represented by series of sclerotized points, fourth lobe represented by 3, 4, or 5 indistinct lobules each with series of points, fifth lobe represented by 1 or 2 projections with 2–8 points (some specimens with lateral lobule of second and third lobes indistinct, unsclerotized, or absent); paraphysis formula variable from 2-2 to 4-3, usually with several small paraphyses in space between third and fourth lobes also. Median lobes separated by space 0.6–1.0 (0.8) times width of median lobe, with small paraphysis attached to medial margin of each median lobe, without basal sclerotization, lateral margin diverging apically, medial margins converging, apically rounded and without medial or lateral notches; second lobes slightly smaller than median lobes, same shape; third lobes same shape but smaller and with series of apically notched lobules; fourth lobes composed of series of apically notched lobules; fifth lobes usually composed of 1 or 2 projections. Plates with bifurcate or clubbed apex; plate formula 2-3-1; plates between median lobes conspicuous, 0.9–1.3 (1.0) time length of lobe. Macroducts usually of 1 size, quite narrow and similar in appearance to microducts, occasionally with a few between median lobes and on segments 3 and 4 shorter, present along body margin from segments 3 or 4 to 8, with 3–7 (5) ducts between median lobes, longest duct 138–153 (146) μ long, marginal macroduct in first space 88–130 (105) μ long, with 28–48 (37) macroducts on each side of body on segments 5 to 8, total of 38–56 (44) macroducts on each side of body, with several macroducts anterior of anal opening. Pygidial microducts usually on venter in submarginal areas of segments 5 and 6, rarely absent from segment 5, with 2–6 (4) ducts; prepygidial ducts of 1 size, present in submedial areas of meso- or metathorax and segments 1 to 4, in lateral clusters from prothorax or mesothorax to segment 4; on dorsum pygidial ducts absent; prepygidial microducts of 1 or 2 sizes in submedial areas of metathorax and segments 1 to 4, in lateral areas from mesothorax or metathorax to segment 4. Gland tubercles present in lateral areas of meso- and metathoracic segments. Perivulvar pores in 5 half-circle bands around vulva, not separated into distinct groups, 6–11 (8) pores on each side of body. Perispiracular pores with 3 loculi, anterior spiracles with 3–14 (9) pores, posterior spiracles without pores. Anal opening located 4.3–8.4 (6.8) times length of anal opening from base of median lobes, anal opening 10–18 (14) μ long. Dorsal seta laterad of median lobes about 0.6 times length of median lobe. Eyes often absent, when present represented by weakly sclerotized dome. Antennae each with 2 very long seta and 2 slightly shorter. Cicatrices normally present on prothorax and segment 1, rarely on segment 3, cicatrix on prothorax often heavily sclerotized and divided into 2 distinct cicatrices. Body pear shaped, dorsal surface with a series of sclerotized areas similar to species of *Melanaspis*.

Affinities In the United States this species is unique by having furcate or clubbed plates between the median lobes and in spaces 1, 2, and 3.

Hosts Usually restricted to orchids. Borchsenius (1966) records 8 host genera in 1 family; Dekle (1977) reports the orchid scale on 4 host genera in Florida including *Philodendron*. We have examined material from *Agave*, *Aloe*, *Brassavola*, *Brassia*, *Bromelia*, *Cattleya*, *Cymbidium*, *Dendrobium*, *Epidendrum*, *Gongora*, *Ionopsis*, *Laelia*, *Masdevallia*, *Oncidium*, *Pedilanthus*, *Polyrrhiza*, *Renanthera*, *Rodriguezia*, *Schomburgkia*, *Vanda*, and *Yucca*. It is taken most commonly in quarantine on cattleya orchids.

Distribution This scale probably was introduced into the United States from the tropics. We have examined U.S. specimens of the orchid scale from FL and HI. It has been recorded from CO, DC, NJ, WA in greenhouses. This pest is commonly intercepted in quarantine from most tropical areas of the world.

Biology There is no information on the life history of this species other than it occurs on the leaves and pseudobulbs of its host.

Economic Importance The orchid scale is regarded as a minor pest of orchids (Schmutterer et al. 1957; Zimmerman 1948). Miller and Davidson (1990) consider this species to be an occasional pest.

Selected References Ferris (1938a); Williams and Watson (1988).

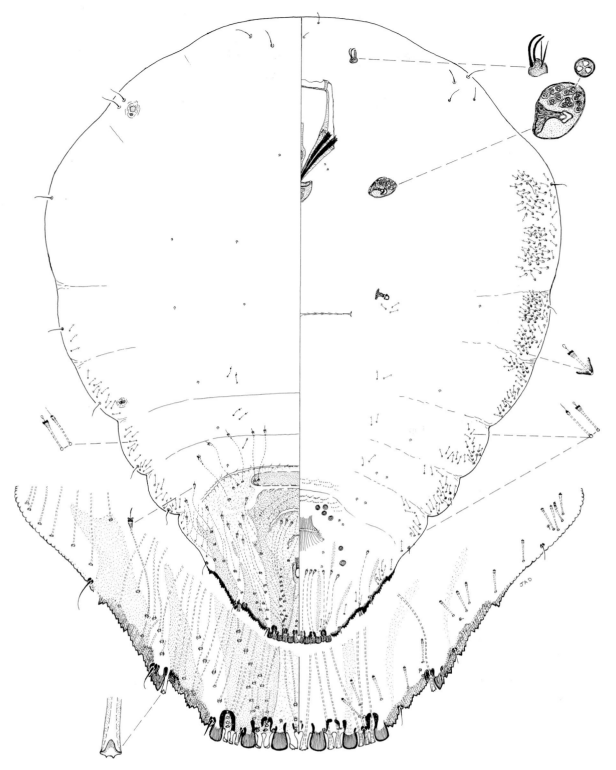

Figure 62. *Furcaspis biformis* (Cockerell), orchid scale, Colombia, intercepted at San Juan, PR, on orchid, XII-7-1948.

Furchadaspis zamiae (Morgan)

Suggested Common Name Zamia scale (also called cycad scale).

Common Synonyms and Combinations *Diaspis zamiae* Morgan, *Howardia elegans* Leonardi in Berlese and Leonardi, *Furchadiaspis elegans* Leonardi, *Howardia ramiae* (Morgan), *Diaspis rhusae* Brain, *Diaspis rhois* Lindinger, *Aulacaspis zamiae* (Morgan), *Diaspis jamiae* Morgan, *Megalodiaspis zamiae* (Morgan).

Field Characters (Plate 59) Adult female cover white, nearly round, moderately convex; shed skins marginal, light yellow or white, often covered with woolly threads of white wax. Males apparently absent in California, mentioned as present by Brain (1915). Eggs yellow. Body of adult female yellow. Found primarily on undersides of leaves among hairs of host, occasionally on stems (Gill 1997).

Slide-mounted Characters Adult female (Fig. 63) with lobes variable, normally with 3 pairs of definite lobes, lobes 4 and 5 sometimes present; with small paraphyses usually associated with medial lobule of lobes 2 and 3. Median lobes separated by space 0.8–1.7 (1.2) times width of median lobe, with lateral and medial paraphyses, without sclerosis or yoke, lateral margins of lobes usually slightly divergent, occasionally parallel, medial margins strongly divergent apically, without lateral notches, with 6–9 (8) medial notches; second lobes bilobed, usually slightly smaller than median lobes, occasionally larger or conspicuously smaller than median lobes, medial lobule largest, usually rounded apically, rarely pointed, with 0–2 (1) lateral notches, 0–1 (0) medial notches, lateral lobule often less heavily sclerotized, usually pointed, rarely rounded apically, with 0–1 (0) notches; third lobes usually bilobed, similar in appearance to second lobes, slightly smaller than second lobes, medial lobule with 0–2 (1) lateral notches, without medial notches, lateral lobule with 0–1 (0) lateral notches, without medial notches; fourth lobes simple or bilobed, usually represented by small sclerotized point, without notches; fifth lobe present on 3 of 10 specimens examined represented by small sclerotized point. Gland spines usually apically divided between median lobes and between lobes 4 and 3, 3 and 2, and 2 and 1, gland-spine formula 1-1-1, with 15–46 (30) gland spines near each body margin from head or prothorax to segment 4; with conspicuous gland spines between median lobes. Macroducts of 2 sizes, largest in submarginal areas of segments 5 to 7 and in marginal areas of segments 4 to 8, smaller size usually in medial areas in any or all of mesothorax to segment 6, in mediolateral areas from meso- or metathorax to segment 5 or 6, in marginal and submarginal areas from head, pro- or mesothorax to segment 8, with 1 or 2 ducts between median lobes extending 0.2 times distance between posterior apex of anal opening to base of median lobes, with duct between median lobes 12–18 (15) μ long, macroduct in first space 12–16 (14) μ long, with 18–33 (24) macroducts on each side of body on segments 5 to 7, total of 156–284 (196) macroducts on each side of body, some macroduct orifices anterior of anal opening. Pygidial microducts usually on venter in submarginal and marginal areas of segment 5 to 7, rarely in submedial area, with 3–27 (10) ducts; prepygidial ducts of 1 size, variable in distribution, usually in marginal areas of segments 3 and 4 and on head and thorax, rarely present from segment 4 to head, present submedially near mouthparts, near anterior and posterior spiracles and on segments 1 to 4, present medially on any or all of mesothorax to segment 6; on dorsum microtubular ducts absent. Perivulvar pores absent. Perispiracular pores with 3 loculi, anterior spiracles with 3–6 (4) pores, posterior spiracles with 1–3 (2) pores. Anal opening located 2.2–3.9 (3.3) times length of anal opening from base of median lobes, anal opening 20–28 (23) μ long. Dorsal seta laterad of median lobes 1.6–2.5 (2.0) times length of median lobe. Eyes usually absent, sometimes represented by small raised pigmented spot, located on body margin laterad or anterior edge of clypeolabral shield. Antennae usually each with 1 long basally divided seta, rarely with 2 setae. Cicatrices absent. Body oval or turbinate. Macroducts ventral on head, prothorax, or mesothorax to segments 2 or 3. Mature females with marginal sclerotization on head and thorax. Projection usually present on body margin adjacent to marginal macroduct between median lobe and lobes 2, occasionally present between lobes 2 and 3, and 3 and 4.

Affinities Of the economic species in the United States, zamia scale is most similar to species of *Diaspis*, such as Boisduval scale, pineapple scale, and cactus scale, by having median lobes that are strongly divergent medially, 2 sizes of macrotubular ducts, and a turbinate body. Zamia scale differs from all of these in lacking perivulvar pores and a pygidial spur and by having macroducts on the thorax and bifurcate gland spines on the pygidium. Species of *Diaspis* have perivulvar pores and a pygidial spur and lack macrotubular ducts on the thorax and bifurcate gland spines on the pygidium.

Hosts Relatively restricted host range and seems to prefer cycads such as *Zamia* and *Cycas*. Borchsenius (1966) records the species from 9 genera in 3 families; Gill (1997) records it from 10 genera. We have examined specimens from the following host genera: *Cussonia, Cycas, Elaeodendron, Encephalartos, Macrozamia, Metroxylon, Rhus, Strelitzia,* and *Zamia.* It has been recorded from *Aralia, Maytenus, Musa, Thevetia, Trachycarpus* (Gill 1997); *Ceratozamia, Dioon, Stangeria* (Borchsenius 1966). It is taken most commonly in quarantine on *Cycas revoluta.*

Distribution We have examined U.S. specimens from CA, HI, MA, MD, MO, NY, PA, WI. It is found in greenhouses and conservatories in all but CA and HI. It occurs throughout the warm areas of the world and in greenhouses in the North. It most often is intercepted in quarantine from South Africa.

Biology According to Saakyan-Baranova (1954) this species has 2 generations per year under greenhouse conditions. Nothing else is known about the biology of zamia scale except that it is ovoviviparous and is parthenogenetic at least in California (Gill 1997), Turkey (Bodenheimer 1953), and Russia (Danzig 1964).

Economic Importance Zamia scale is considered to be a serious pest of cycads in California causing chlorosis of the leaves, reducing the aesthetic characteristics of ornamental plants (Brown and Eads 1967). The species is an occasional pest in greenhouses and ornamental gardens in the following: California (Dreistadt et al. 1994), Egypt (Alfieri 1929), Poland (Komosinska 1968), Germany (Lengerken 1932), Russia (Saakyan-Baranova 1954), and Czechoslovakia (Zahradník 1990a).

Selected References Brown and Eads (1967); Gill (1997); Saakyan-Baranova (1954).

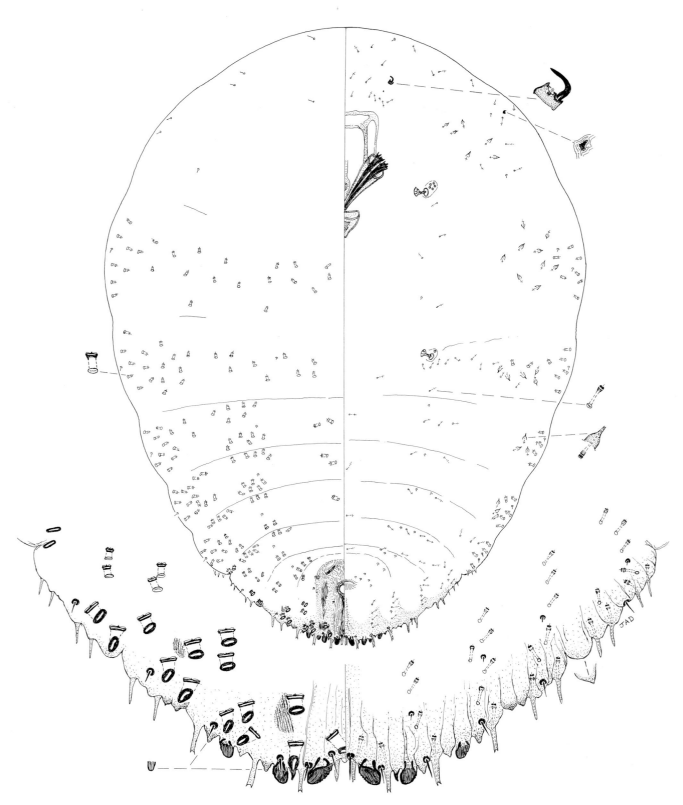

Figure 63. *Furchadaspis zamiae* (Morgan), zamia scale, Sicily, intercepted at Washington, DC, *Elaeodendron ilicifolia*, VI-10-1957.

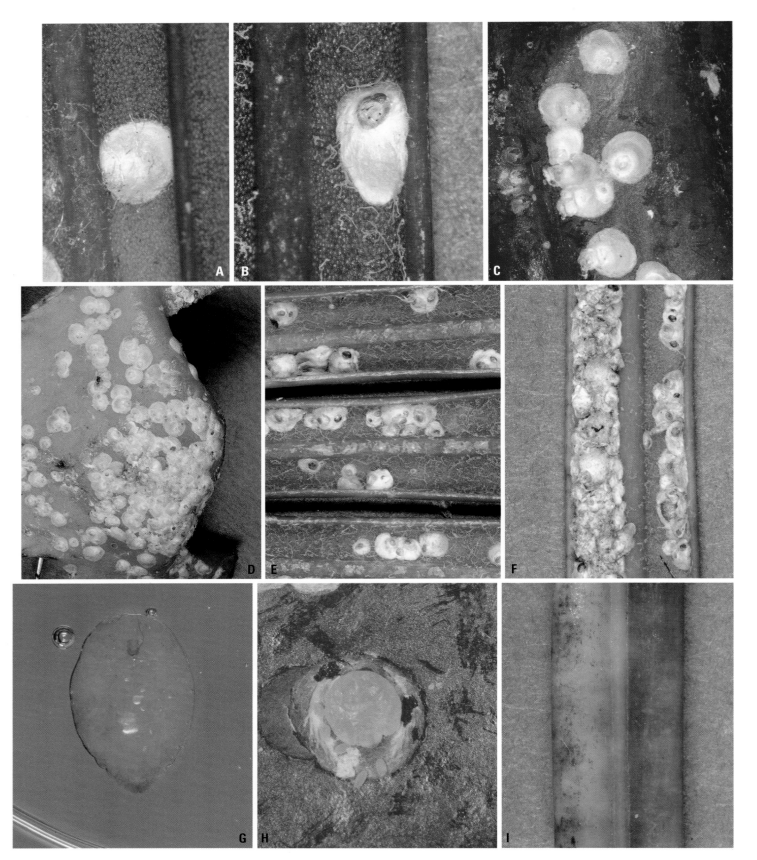

Plate 59. *Furchadaspis zamiae* (Morgan), Zamia scale

A. Adult female scale cover on cycad (J. A. Davidson).
B. Male cover on cycad (J. A. Davidson).
C. Mostly adult female covers on cycad cone (J. A. Davidson).
D. Heavy infestation at base of cycad cone (J. A. Davidson).
E. Infestation on underside of cycad leaves showing parasite emergence holes (J. A. Davidson).

F. Damaging infestation on cycad leaf (J. A. Davidson).
G. Slide-mounted adult female distended with eggs (J. A. Davidson).
H. Adult female with crawlers (J. A. Davidson).
I. Upper surface of leaf showing yellowing caused by infestation on undersurface of cycad (J. A. Davidson).

220

Plate 60. *Gymnaspis aechmeae* Newstead, Flyspeck scale

A. Pupillarial adult female cover (top) and typical male cover on bromeliad (J. A. Davidson).

B. Rubbed cover showing shiny black second-instar shed skin that contains adult female on bromeliad (J. A. Davidson).

C. Second-instar shed skin containing adult female on bromeliad (J. A. Davidson).

D. Male cover (top) and three small female covers on bromeliad (J. A. Davidson).

E. Male cover turned to show male pupa (J. A. Davidson).

F. Yellow spots on bromeliad leaves where scales have fed (J. A. Davidson).

G. Distance view of flyspeck scale on bottom side of bromeliad leaf (J. A. Davidson).

H. Distance view of bromeliad leaf showing yellowing from feeding and white spots where scales fell off of leaf (J. A. Davidson).

221

Gymnaspis aechmeae Newstead

Suggested Common Name Flyspeck scale (also called achmeae scale).

Common Synonyms and Combinations *Aonidia picea* Leonardi.

Field Characters (Plate 60) Adult female cover black, very convex, with only 1 shed skin because adult female remains inside rotund, shiny black second-instar shed skin, giving appearance of fly defecation. Male cover with white to brown wax around body margin, large submarginal shed skin of first instar covering most of cover. Body of adult female reddish violet; crawlers probably reddish violet also. Usually on upper surfaces of leaves.

Slide-mounted Characters Adult female (Fig. 64) pupillarial, median lobes absent, second lobes rarely represented by small sclerotized nob, third lobes usually present, represented by slightly larger inconspicuous knob, sometimes absent; paraphyses absent. Pygidium with fringe of conspicuous gland spines each containing 1 microduct, present from segments 5 to 8, with 8–13 (10) on each side of pygidium. Plates absent. Without macroducts, replaced by gland spine microducts. Pygidial microducts absent from venter except those in gland spines; prepygidial ducts of 2 sizes: shorter ducts in marginal areas from head to segment 4, most abundant on thorax and anterior abdominal segments and in cluster near clypeolabral shield; longer ducts most abundant near mouthparts, scattered over thorax and abdomen; on dorsum prepygidial and pygidial ducts absent. Perivulvar pores absent. Pores absent near spiracles. Anal opening located 2.8–3.7 (3.2) times length of anal opening from base of median lobes, anal opening 12–18 (15) μ long. Longest pygidial seta on segment 8 ranging from 2–7 (5) μ long. Eyes represented by small sclerotized area on slight swelling on body margin near anterior spiracles. Antennae each with 2 or 3 conspicuous setae. Cicatrices often absent, sometimes present on prothorax and segments 1 and 3. Body oval, with pygidium strongly protruding. Pygidium with distinct sclerotized pattern.

Affinities Flyspeck scale is distinct from all other U.S. armored scales by lacking lobes, having a pygidial fringe of gland spines, by lacking perivulvar and spiracular setae, and by being pupillarial. The second-instar female is very scalelike with 3 or 4 pairs of lobes, plates that are trifurcate each with a conspicuous microduct, macrotubular ducts that are oriented like those of *Parlatoria* species, and remnants of legs.

Hosts Found primarily on bromeliads. Borchsenius (1966) records 15 host genera in 3 families; Dekle (1977) reports the flyspeck scale on 17 host genera in Florida. We have examined specimens of *G. aechmeae* from the following hosts: *Aechmea, Ananas, Aregelia, Billbergia, Bromelia, Karatas, Monstera, Neoregelia, Nidularium, Quesnelia, Tillandsia,* and *Vriesea.* It is taken most often in quarantine on bromeliads.

Distribution We have examined confirmed infestations of this species in FL only. It has been reported in greenhouses in DC, MD, MO, NJ, NY, PA, WV. Literature records out-of-doors include AL. Although this species is commonly encountered in quarantine shipments in California, it is not established there (Gill 2002, personal communication). We have examined specimens from 2 separate collections in Oahu but they may be quarantine records. The species has not been confirmed to occur in either CA or HI. It commonly is taken at U.S. ports-of-entry from Europe, South and Central America, and the former Soviet Union.

Biology This species has 2 generations per year in the warm moist parts of the greenhouse (Saakyan-Baranova 1954). Adult males are present (Schmutterer 1959), and first instars, rather than eggs, are laid by the adult female (Zahradník 1968). According to Danzig (1964) the scale cover of the 'young female' (probably second instar) is brown with a black crawler shed skin. The adult female loses the scale and remains inside of the second-instar shed skin.

Economic Importance This species is considered to be an important pest of bromeliads in Florida (Dekle 1977) and in greenhouses in Europe (Schmutterer et al. 1957) and the former Soviet Union (Borchsenius 1963). Miller and Davidson (1990) consider this species to be an occasional pest.

Selected References Ferris (1937); Saakyan-Baranova (1954).

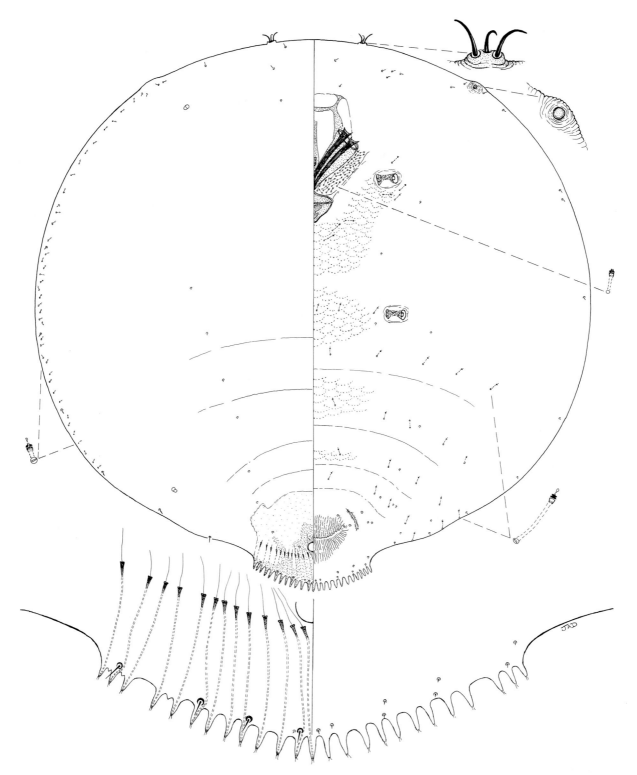

Figure 64. *Gymnaspis aechmeae* Newstead, flyspeck scale, Germany, intercepted at Hoboken, NJ, on *Nidularium* sp., XI-2-1978.

Hemiberlesia lataniae (Signoret)

ESA Approved Common Name Latania scale (also called quince scale).

Common Synonyms and Combinations *Aspidiotus lataniae* Signoret, *Aspidiotus cydoniae* Comstock, *Aspidiotus punicae* Cockerell, *Aspidiotus diffinis* v. *lateralis* (Cockerell), *Aspidiotus (Hemiberlesia) cydoniae* Comstock, *Aspidiotus (Hemiberlesia) crawii* Cockerell, *Aspidiotus (Diaspidiotus) punicae* Cockerell, *Aspidiotus (Diaspidiotus) greeni* Cockerell, *Aspidiotus cydoniae* v. *tecta* Maskell, *Aspidiotus greeni* Cockerell, *Aspidiotus (Evaspidiotus) punicae* Cockerell, *Aspidiotus (Evaspidiotus) cydoniae* Comstock, *Aspidiotus lateralis* Cockerell, *Aspidiotus crawii* Cockerell, *Aspidiotus (Evaspidiotus) crawii* Cockerell, *Aspidiotus (Evaspidiotus) greeni* Cockerell, *Aspidiotus (Hemiberlesia) lataniae* Signoret, *Aspidiotus cydoniae punicae* Cockerell, *Aspidiotus cydoniae crawii* (Cockerell), *Diaspidiotus lataniae* (Signoret), *Diaspidiotus (Aspidiotus) lataniae* (Signoret), *Aspidiotus aspleniae* Sasaki, *Aspidiotus tectus* Maskell.

Field Characters (Plate 61) Adult female cover variable, apparently depending on position on host. On leaves and smooth stems: cover of female convex, circular, light gray to white. On older stems and branches: cover of female circular, not as convex as on leaves, gray to brown; shed skins central to subcentral, gray, dark brown or yellow when rubbed. Male cover elongate oval, same color and texture as female cover; shed skin submarginal. Body of adult female yellow; eggs and crawlers yellow. On stems and leaves.

Slide-mounted Characters Adult female (Fig. 65) with 3 pairs of lobes, second and third lobes represented by hyaline or sclerotized points, fourth lobes usually absent, sometimes with small sclerotized area in position of lobe; paraphysis formula 2-2-0, with paraphyses in space between lobe 2 and median lobe, attached to medial margin of lobe 2, medial margin of lobe 3, and in space between lobes 2 and 3. Median lobes separated by space 0.1–0.3 (0.2) times width of median lobe, usually with small paraphysis attached to medial margin, without basal sclerosis or yoke, medial margins of lobe parallel or slightly converging, lateral margins rarely parallel, usually convergent apically, with 1 lateral notch and 1 medial notch; second lobes hyaline or sclerotized, varying from one-quarter length to as long as median lobes, pointed, without notches; third lobes structurally same as second lobes, normally smaller; fourth lobe usually absent, when present represented by small sclerotized swelling. Plates between median lobes and second lobe, between second lobes and third lobes, and between third lobes and fourth lobes with increasingly smaller tines, without plates anterior of position of fourth lobes, plates apparently without microducts; plates in third space simple, without tines and microduct; plate formula 2-3-0 to 2-3-5, with 2 protruding microducts at base of plates in second space, with several protruding microducts anterior of setae marking segments 5 and 6; median lobes usually with 2 slender plates between them about 0.8–1.3 (1.0) times as long as median lobes, sometimes plates absent between median lobes. Macroducts of 1 size, on segments 5 to 7 or 8 in marginal and submarginal areas, duct between median lobes present or absent, when present, extending 1.2–3.6 (2.5) times distance between posterior apex of anal opening and base of median lobes, 54–74 (66) µ long, macroduct in first space 50–68 (58) µ long, with 11–20 (16) macroducts on each side of segments 5 to 7 or 8, without macroducts anterior of pygidium, some orifices occurring anterior of anal opening. Pygidial microducts on venter in submarginal clusters on segments 5 and 6, with 10–16 (14) ducts on each side of body; prepygidial microducts of 1 size in submarginal and marginal areas of mesothorax, metathorax, or segment 1 to segments 3 or 4, also present submedially near spiracles and on 1 or more of segments 1 to 4; on dorsum pygidial ducts absent; prepygidial microducts of 2 sizes, larger size in submedial areas of any or all of thorax and anterior abdominal segments, smaller size in submarginal areas of head or prothorax to segments 2 or 3. Perivulvar pores in 4 indistinct groups, 7–29 (18) pores. Perispiracular pores absent. Anal opening located 0.6–1.6 (1.1) times length of anal opening from base of median lobes, anal opening 20–32 (26) µ long. Dorsal seta laterad of median lobe 1.1–1.4 (1.2) times length of lobe. Eyes absent or represented by small spot or sclerotized spur. Antennae usually each with 1 large seta and 3 spurs. Prosoma not sclerotized. Cicatrices variable, sometimes on prothorax, segments 1 and 3.

As presently understood latania scale is an unusually variable species and may be a complex of species. Life history information, host relationships, geographical data, and diversity of intergrading morphologies support this hypothesis. Unfortunately, we have been unable to find distinct morphological segregates within the limits of *Hemiberlesia lataniae*. Morphological variation of note is the range in numbers of perivulvar pores, the converging or parallel median lobes, the presence or absence of a macroduct between the median lobes, and cover shape and color.

Affinities Latania scale is similar to *Hemiberlesia rapax*, greedy scale, except the former has perivulvar pores and a relatively smaller, more anteriorly placed anal opening. Greedy scale lacks perivulvar pores and has a larger, more posteriorly placed anal opening.

Hosts Polyphagous. Borchsenius (1966) records it from 224 genera in 78 families; Dekle (1977) reports it from 276 genera in Florida and indicats that it is most frequently found on *Casuarina*, *Eriobotrya*, *Rosa*, and various palms. McKenzie (1956) conjectures that it probably would feed on any woody plant except oaks and certain conifers. Dziedzicka (1989) indicates that this scale insect occurs on more than 600 host plants. We have examined specimens from the following host genera: *Abies, Acacia, Acer, Acokanthera, Acrocomia, Actinidia, Adenium, Agave, Albizia, Aleurites, Aloe, Ambrosia, Amherstia, Amygdalus, Anthurium, Antigonon, Aralia, Araucaria, Ardisia, Areca, Argania, Aristolochia, Arundina, Asparagus, Astianthus, Atalaya, Attalea, Aucuba, Averrhoa, Baccharis, Bactris, Barringtonia, Batis, Bauhinia, Beaucarnea, Berberis, Bertholletia, Beta, Bignonia, Blighia, Borrichia, Bougainvillea, Bromelia, Bryophyllum, Bursera, Buxus, Caesalpinia, Callistemon, Callitris, Calocarpum, Calodendrum, Calycophyllum, Calyptrogyne, Camellia, Canavalia, Canna, Carica, Carya, Casimiroa, Cassia, Castanea, Casuarina, Cattleya, Ceanothus, Cedrela, Cedrus, Ceiba, Celtis, Centaurium, Ceratonia, Cereus, Chamaedorea, Chamaerops, Chlorophytum, Chrysanthemum, Chrysophyllum, Chysis, Cinnamomum, Citharexylum, Citrus, Clethra, Cleyera, Clinostigma, Coccoloba, Cocos, Codiaeum, Coffea, Conocarpus, Cordia, Cordyline, Corokia, Corylus, Crataegus, Crateva, Crotalaria, Croton, Cupressus, Curcurbita, Cussonia, Cyathodes, Cycas, Cydonia, Cymbidium, Cypripedium, Dahlia, Dalbergia, Datura, Dendrobium, Dianthus, Diaphananthe, Dioon, Diospyros, Dracaena, Durio, Echeveria, Elaeagnus, Elaeis, Encephalartos, Epidendrum, Erigeron, Eriobotrya, Eucalyptus, Euclea, Eugenia, Euonymus, Euphorbia, Euterpe, Ficus, Firmiana, Flacourtia, Furcraea, Garcinia, Gardenia, Gelonium, Gladiolus, Gliricidia, Gnaphalium, Grevillea, Gronophyllum, Haworthia, Hedera, Heliconia, Hevea, Hibiscus, Hyophorbe, Hypericum, Ilex, Indigofera, Ipomoea, Iris, Iva, Jacaranda, Jasminum, Jubaea, Juglans,*

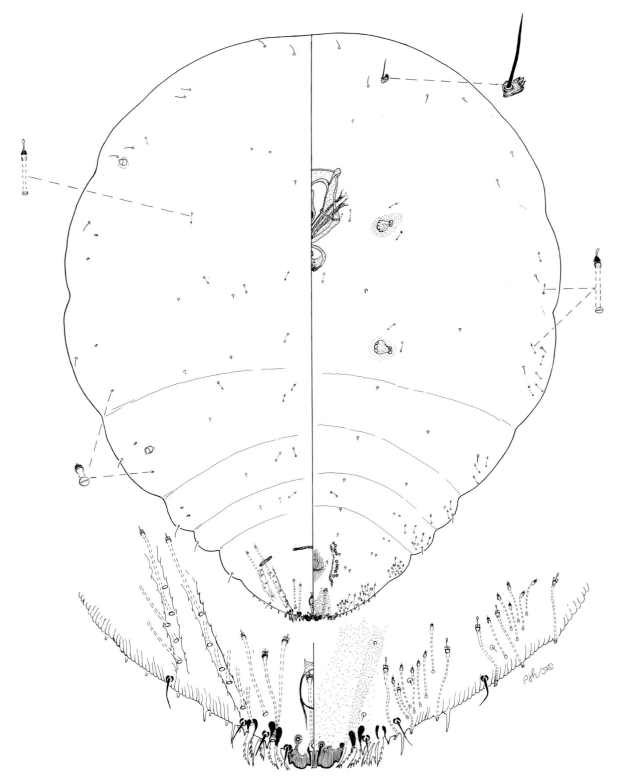

Figure 65. *Hemiberlesia lataniae* (Signoret), latania scale, Dominica, on *Terminalia* sp., III-12-1964.

Kalmia, Kydia, Lagerstroemia, Larrea, Lantana, Laurus, Lespedeza, Leucaena, Leucophyllum, Leucothoe, Licuala, Litsea, Livistona, Lonicera, Loranthus, Lucuma, Macadamia, Maesa, Magnolia, Malus, Malvaviscus, Mammea, Mangifera, Manilkara, Maranta, Melia, Melicoccus, Metrosideros, Miconia, Microcos, Mimosa, Morinda, Morus, Muehlenbeckia, Musa, Neomoorea, Nepenthes, Nephthytis, Nerium, Nolina, Odontoglossum, Olea, Oncosperma, Oncidium, Ophiopogon, Opuntia, Orbignya, Oreopanax, Ormosia, Osteomeles, Pachystroma, Pandanus, Parkinsonia, Passiflora, Pelargonium, Peltophorum, Persea, Petrea, Philodendron, Phoenix, Phoradendron, Physalis, Pimenta, Pinanga, Pinus, Piper, Pithecellobium, Platanus, Plocosperma, Plumeria, Poinciana, Polyalthia, Populus, Pouteria, Prinsepia, Pritchardia, Prosopis, Protea, Prunus, Psidium, Pterocarya, Ptychosperma, Punica, Pyracantha, Pyrus, Quisqualis, Rapanea, Retama, Rheedia, Rhododendron, Rhodomyrtus, Rhynchostylis, Robinia, Rosa, Rosmarinus, Roystonea, Rubus, Salix, Salvia, Samanea, Sapium, Saraca, Schefflera, Schinopsis, Sedum, Serrisa, Solanum, Spathoglottis, Spondias, Stachyphrynium, Stahlia, Stenocarpus, Strelitzia, Strongylodon, Stylophyllum, Symplocos, Syngonium, Syringa, Tecoma, Tecomaria, Terminalia, Tetracera, Theobroma, Thevetia, Thrinax, Thuja, Tillandsia, Tipuana, Trachelospermum, Ulmus, Vaccinium, Vanda, Vateria, Veitchia, Verbesina, Viburnum, Vigna, Viola, Viscum, Vitis, Wikstroemia, Wisteria, Yucca, and *Ziziphus.* This species is taken in quarantine most commonly on avocado and other tropical fruit.

Distribution The indigenous area of latania scale is unknown, but it is probably an Old World species. We have examined U.S. material from AL, AR, AZ, CA, DC, FL, GA, HI, IN, KS, LA, MD, MO, MS, NC, NJ, NY, OH, OK, TX, VA. It has been recorded in the literature from CO, DE, IL, MA, PA, SC, TN. In the United States latania scale occurs out-of-doors as far north as Maryland. It is taken in quarantine from all tropical and subtropical areas of the world and also occasionally occurs in temperate regions. A distribution map of this species was published by CAB International (1976).

Biology The latania scale occurs on the leaves and/or bark of its host. Stoetzel and Davidson (1974) reported 2 generations per year in Maryland on *Ilex* 'Foster' with second-instar males and females overwintering on twigs and leaves. Alate males were found throughout May. Each female of the first generation laid eggs in a batch of 15–20 and then 3–5 a day until the female died. Crawlers were present from mid-June to mid-July. Alate males of the second generation were found during the middle 2 weeks of August. Eggs were laid in 1 batch of 15–25 in September. The eggs hatched immediately after being laid. Ebeling (1959) described the life cycle of the female latania scale in southern California as requiring 2 weeks for the first molt and 16 to 19 days for the second. In 26 to 30 days after the second molt eggs were laid. He stated 'males have not been found in California.' McKenzie (1935a) found the life cycle required 56 to 65 days in coastal San Diego County. Gerson and Zor (1973)

reported 4 generations per year in Israel. El-Minshawy et al. (1971) found 3 overlapping generations per year in Egypt, and Salama and Hamdy (1974b) found only 2 generations in a different area of Egypt. It should be noted that this scale has been reported to be unisexual in California and Israel while in Maryland males are readily apparent. Wang and Su (1989) demonstrated that the length of time to complete a life stage, fecundity rate, net reproductive rate, and growth rate are all correlated with temperature.

Economic Importance Dekle (1977) reports the latania scale to be one of the most common scale insects in Florida where it was a serious pest of palms, Australian pine, and other ornamental plants. He also reports that under proper conditions several wasp parasites were capable of controlling this pest. McKenzie (1935) considers it to be the most important scale insect pest of avocado in California. Wolfenbarger (1963) states that this scale was not a serious avocado pest in Florida where it usually was found on the undersides of avocado leaves in late fall and winter. Gerson and Zor (1973) report it as an avocado pest in Israel. Heavy infestations killed small twigs, while feeding on thin-skinned varieties caused protuberant growth from the rind, and infested seedlings developed a 'bottle-neck' like growth. It was recorded as a pest of *Annona muricata* in the West Indies by Wolcott (1933). Bitancourt et al. (1933) consider it to be a minor citrus pest in Brazil. Wilcox and Holt (1913) found latania scale causing occasional injury to hibiscus in Hawaii. In Greece, Koronéos (1934) notes it as a frequent greenhouse pest. Smee (1936) describes a serious outbreak of this pest on *Aleurites montana* in Nyasaland. Jepson (1915) records it as an injurious pest of *Musa* and *Cocos* in Fiji. In recent years, it has become a very serious pest of kiwifruit in New Zealand in combination with *Hemiberlesia rapax* (Blank et al. 1992), and several pest management procedures and biological control agents have been introduced (Hill et al. 1993). It is reported as a pest of kiwifruit in Chile (González 1989) and New Zealand (Blank et al. 1999) also; avocado in New Zealand (Blank et al. 1993); loquats in China (Liu 1994); macadamia in Israel (Wysoki 1977); olives in Israel, Egypt, Turkey, Chile, and California (Argyriou 1990); oil palm in India (Dhileepan 1992); date palm in India (Murlidharan 1993); tea in India, Japan, Malawi, Sri Lanka, and the former Soviet Union (Nagarkatti and Sankaran 1990); and conifers in several areas of the world (Zahradník 1990b). The species often is controlled by natural enemies, such as fungi, mites, parasitic wasps, and beetles. Blank et al. (1999) demonstrate that the highest infestations in kiwifruit blocks were in areas adjacent to alternative host plants, supporting the hypothesis that scale distributions were caused by aerial dispersal of crawlers from alternative hosts. Beardsley and González (1975) consider this scale to be one of 43 serious armored scale pests. Miller and Davidson (1990) consider it to be a serious world pest.

Selected References Ebeling (1959); Ferris (1938a); Gerson and Zor (1973); McKenzie (1935); Stoetzel and Davidson (1975).

Plate 61. *Hemiberlesia lataniae* (Signoret), Latania scale

A. Adult female cover on euonymus (J. A. Davidson).
B. Adult female covers on shefflera (J. A. Davidson).
C. Adult female cover and body on euonymus (J. A. Davidson).
D. Adult female and male covers in biparental population on *Dioon spinulosum* (R. J. Gill).
E. Immature and mature female covers on acuba (R. J. Gill).
F. Distance view of covers on euonymus (J. A. Davidson).
G. Bodies of adult females on euonymus (J. A. Davidson).
H. Body of adult female, egg, and crawler on cotoneaster (J. A. Davidson).
I. Dieback caused by heavy infestation on pyracantha (J. A. Davidson).

Hemiberlesia neodiffinis Miller and Davidson

Suggested Common Name False diffinis scale.

Common Synonyms and Combinations This species previously was misidentified as *Aspidiotus diffinis* and *Hemiberlesia diffinis* in the following publications:

Aspidiotus diffinis Newstead; Marlatt 1899:75, 1900:425–427; Couch 1935:16, 1938:107; Ferris 1921:125; Bibby 1931:591; Wescott 1973:411.

Hemiberlesia diffinis (Newstead); MacGillivray 1921:437; Ferris 1938a:238, 1942:446; Schmidt 1940:193; Kosztarab 1963:34; Dekle 1965:69, 1977:71; McDaniel 1969:107; Tippins and Beshear 1970:9; Beshear et al. 1973:6; Stoetzel and Davidson 1974:501; Stoetzel 1976:323; Lambdin and Watson 1980:80; Miller and Howard 1981:166; Mead 1982:4; Nakahara 1982:41; MacGowan 1983:7; Mead 1984:2; MacGowan 1987:9; Miller and Davidson 1990:302.

Field Characters (Plate 62) Adult female cover oval, highly convex, gray; shed skins submarginal, brownish yellow. Male cover elongate oval, same color and texture as female cover; shed skin submarginal. Body of adult female yellow; eggs and crawlers probably yellow. On twigs and branches.

Slide-mounted Characters Adult female (Fig. 66) with 3 pairs of definite lobes, fourth lobes, when present, represented by small sclerotized swelling; paraphysis formula usually 2-2-0, with paraphyses in space between lobe 2 and median lobe, attached to medial margin of lobe 2, medial margin of lobe 3, and in space between lobes 2 and 3. Median lobes separated by space 0.1–0.2 (0.2) times width of median lobe, with small paraphysis attached to medial margin, without basal sclerotization or yoke, medial margin usually slightly converging apically, lateral margins converging, with 1 lateral notch and 0–1 (0) medial notch; second lobes sclerotized, pointed, usually with lateral notch, about one-third to one-quarter size of median lobes; third lobes sclerotized, pointed, without notches, equal to or smaller than second lobes. Plates between median lobes and second lobe, between second lobes and third lobes, and between third lobes and fourth lobes with increasingly larger tines, sometimes with 2 or more simple plates anterior of fourth lobes, plates in first and second spaces apparently without microducts; plates in third space distinctly shaped, each with 1 microduct, plates anterior of seta marking segment 5 with single microduct; plate formula 2-3-3; median lobes each with 2 slender plates between them about 0.8–1.1 (1.1) times as long as median lobes. Macroducts of 1 size, on segments 5 to 7 in marginal and submarginal areas, duct between median lobes absent, marginal macroduct in first space 64–75 (69) μ long, with 10–15 (12) macroducts on each side of pygidium on segments 5 to 7, without prepygidial macroducts, some macroduct orifices anterior of anal opening.

Pygidial microducts on venter in submarginal and marginal areas of segments 5 and 6, with 6–13 (9) ducts; prepygidial ducts of 2 sizes, longer size in submarginal and marginal areas of segments 1 to 3 or 4, also present submedially near spiracles, shorter size present along body margin from head to segments 3 or 4; on dorsum pygidial ducts absent; prepygidial microducts of 2 sizes, larger size in submedial areas of any or all of mesothorax to 4, smaller size in submarginal areas of head or prothorax to segments 2 or 3. Perivulvar pores absent. Pores absent near spiracles. Anal opening located 1.1–1.6 (1.3) times length of anal opening from base of median lobes, anal opening 17–30 (23) μ long. Dorsal seta laterad of median lobes 1.0–1.4 (1.2) times length of median lobe. Eyes usually represented by small sclerotized spur or dome on mesothorax. Antennae each with 1 seta. Cicatrices usually present on prothorax and segment 1. Body pear shaped.

Affinities The false diffinis scale is similar to *Hemiberlesia diffinis*, which differs by having a macroduct between the median lobes and large plates in the second space that each possess 2 or 3 associated microducts. *Hemiberlesia neodiffinis* has no macroduct between the median lobes and small plates in the second space that each possess 1 associated microduct.

Hosts Polyphagous. Dekle (1977) reports it (misidentified as *H. diffinis*) on 8 genera of host plants in Florida. We have examined specimens from the following: *Carya, Celtis, Cephalanthus, Cinnamomum, Ficus, Fraxinus, Juglans, Liriodendron, Magnolia, Nerium, Persea, Phoradendron, Pterocarya, Quercus, Salix, Syringa, Tilia,* and *Ulmus*.

Distribution This species apparently is of New World origin and is known only from this region. We have examined U.S. material from AL, AR, DC, FL, GA, IL, LA, MD, MO, MS, NC, NJ, NY, OH, SC, TN, TX. It is reported from KS and VA. It has been taken in quarantine from Canada.

Biology Stoetzel and Davidson (1974) report 1 generation per year in Maryland on tuliptree, with second-instar males and females overwintering on twigs and branches. Alate males appeared in late May. A microscope slide is available containing settled crawlers collected 27 July on tulip tree in Maryland.

Economic Importance This species has been reported as a pest of lilac and may be a problem on linden. Injury is described as considerable with shoots being weakened or killed (Marlatt 1900). An incident of severe damage to tuliptree is known in Maryland. Miller and Davidson (1990) consider this species to be an occasional pest.

Selected References Kosztarab (1963); Miller and Davidson (1998); Stoetzel and Davidson (1974).

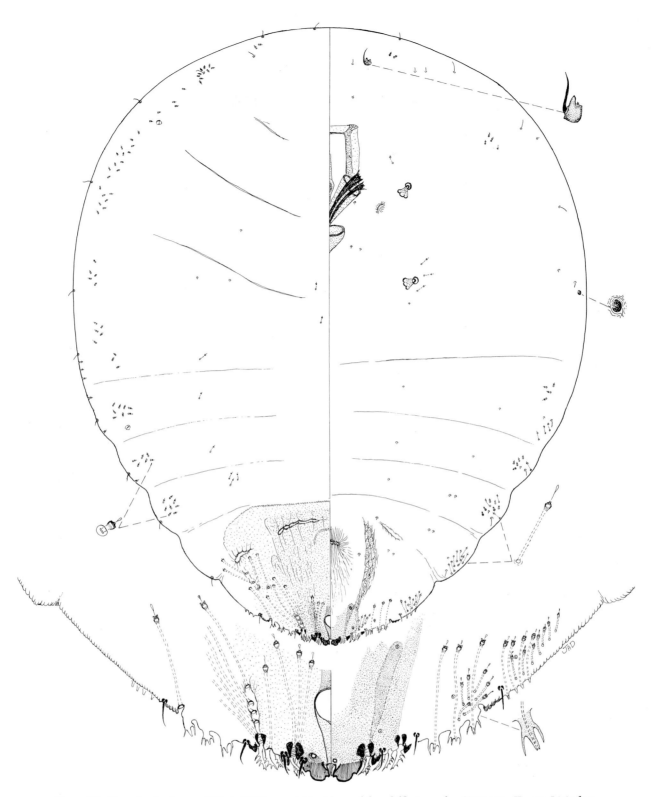

Figure 66. *Hemiberlesia neodiffinis* Miller and Davidson, false diffinis scale, Simpson, IL, on *Liriodendron tulipfera*, VIII-7-1969.

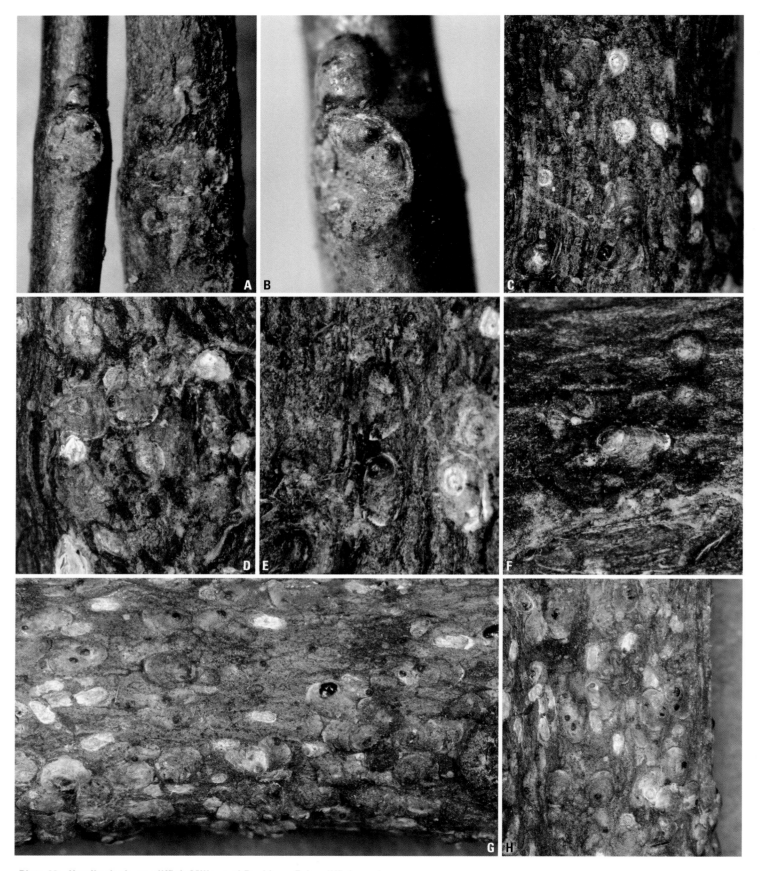

Plate 62. *Hemiberlesia neodiffinis* Miller and Davidson, False diffinis scale

A. Female covers on tulip poplar (R. J. Gill).
B. Female covers on tulip poplar (R. J. Gill).
C. Cryptic infestation on lilac (J. A. Davidson).
D. Infestation on lilac (J. A. Davidson).
E. Male cover (bottom left) and female cover (top left) on lilac (J. A. Davidson).

F. Covers concealed by sooty mold on lilac (J. A. Davidson).
G. Distance view of infestation on lilac (J. A. Davidson).
H. Distance view of covers on lilac (J. A. Davidson).

230

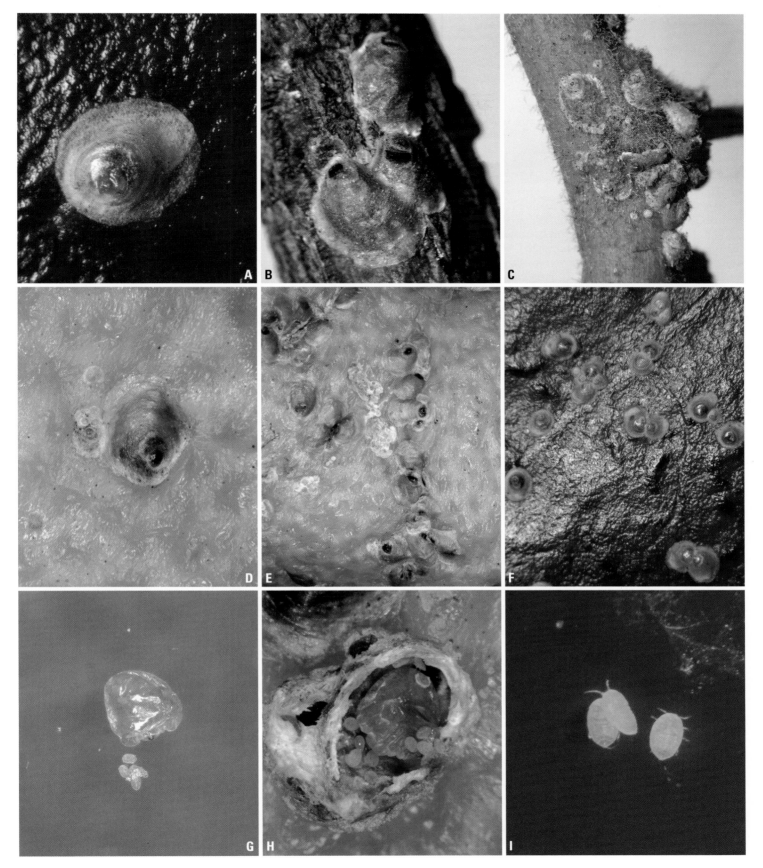

Plate 63. *Hemiberlesia rapax* (Comstock), **Greedy scale**

A. Rubbed adult female cover on camellia (R. J. Gill).
B. Adult female covers on *Euonymus myrtifolia* (R. J. Gill).
C. Adult female covers on passion vine (R. J. Gill).
D. Female covers on citrus (J. A. Davidson).
E. Distance view of female covers on citrus (J. A. Davidson).

F. Distance view of female covers on camellia (R. J. Gill).
G. Young adult female, crawlers, and egg shells on citrus (J. A. Davidson).
H. Old adult female, eggs, and crawlers (J. A. Davidson).
I. Crawlers from citrus (J. A. Davidson).

231

Hemiberlesia rapax (Comstock)

ESA Approved Common Name Greedy scale.

Common Synonyms and Combinations *Aspidiotus camelliae* Signoret, *Aspidiotus rapax* Comstock, *Aspidiotus flavescens* Green, *Aspidiotus (Hemiberlesia) rapax* Comstock, *Hemiberlesia camelliae* (Signoret), *Aspidiotus (Hemiberlesia) tricolor* Cockerell, *Aspidiotus lucumae* Cockerell, *Aspidiotus (Hemiberlesia) camelliae* Signoret, *Hemiberlesia tricolor* (Cockerell), *Hemiberlesia argentina* Leonardi, *Aspidiotus argentina* (Leonardi), *Hendaspidiotus tricolor* (Cockerell), *Hemiberlesiana camelliae* (Signoret), *Diaspidiotus camelliae* (Signoret).

Field Characters (Plate 63) Adult female cover circular, strongly convex, gray to whitish; shed skins subcentral, yellowish brown. Male cover (when present) oval, same color and texture as female cover; shed skin submarginal. Body of adult female yellow; eggs and crawlers yellow. On twigs, leaves, and fruits.

Slide-mounted Characters Adult female (Fig. 67) with 3 definite pairs of lobes, second and third lobes represented by small sclerotized points, fourth lobes often absent, sometimes with small sclerotized swelling or hyaline point; paraphysis formula 2-2-0, with paraphyses in space between lobe 2 and median lobe, attached to medial margin of lobe 2, medial margin of lobe 3, and in space between lobes 2 and 3. Median lobes separated by space 0.2–0.3 (0.2) times width of median lobe, usually with small paraphysis attached to medial margin, without basal sclerosis or yoke, medial margins of lobe parallel or slightly converging, lateral margins convergent apically, with 1 lateral notch and 1 medial notch; second lobes sclerotized, pointed, without notches, one-eighth to one-quarter size of median lobes; third lobes similar to second lobes; fourth lobes usually present represented by small sclerotized swelling or rarely hyaline point. Plate formula 2-3-2 or 2-3-3, with 1–3 protruding microduct openings at base of plates in third space, absent elsewhere; plates absent from area anterior of seta marking segment 5; plates in first and second spaces fimbriate except for occasional simple plate in second space, without microducts; plates anterior of second space simple, without microducts; median lobes with 2 slender plates between them, about 1.1–1.4 (1.2) times as long as median lobes. Macroduct of 1 size, on segments 5 to 7 in marginal and submarginal areas, without macroduct between median lobes, macroduct in first space 32–47 (41) μ long, with 6–24 (12) ducts on each side of pygidium on segments 5 to 7, without macroducts anterior of segment 7, some orifices occurring anterior of anal opening. Pygidial microducts on venter in submarginal clusters on segments 5 and 6, with 12–22 (16) on each side of body; prepygidial microducts of 1 size in submarginal and marginal areas of mesothorax, metathorax, or segment 1 to segments 3 or 4, also present submedially near spiracles and on 1 or more of segments 1 to 4; on dorsum pygidial ducts absent; prepygidial microducts of 2 sizes, larger size in submedial areas of any or all of thorax and anterior abdominal segments, smaller size in submarginal areas on head or prothorax to segments 2 or 3. Perivulvar and peristigmatic pores absent. Anal opening located 0.2–0.6 (0.4) times length of anal opening from base of median lobes, anal opening 25–35 (29) μ long. Dorsal seta laterad of median lobe 1.0–1.3 (1.2) times length of lobe. Eyes represented by small sclerotized area with 2 or 3 sclerotized projections or with single spur. Antennae usually each with 1 large seta and 3 spurs. Prosoma not sclerotized. Cicatrices absent or on prothorax and segment 1.

Affinities The greedy scale is most similar to *Hemiberlesia lataniae*, the latania scale, but the former lacks perivulvar pores and protruding microducts at the bases of the plates in the second space, and has a larger anal opening located near the median lobes, whereas, latania scale has perivulvar pores and protruding microducts at the bases of plates in the second space, and a smaller, more anteriorly placed anal opening.

Hosts Polyphagous. Borchsenius (1966) records it from hosts in 117 genera in 60 families; Dekle (1977) records it from 38 genera with *Carya illinoensis* the most frequently reported host in Florida. McKenzie (1956) states, 'There is a possibility this pest can be found on almost any woody plant.' Based on material in the USNM Collection, camellia appears to be one of the most common hosts in the United States. We have examined specimens from the following genera: *Acacia, Achras, Actinidia, Agathis, Agonis, Annona, Aralia, Asclepias, Baccharis, Banksia, Bauhinia, Callistemon, Camellia, Casimiroa, Ceanothus, Celtis, Cephalocereus, Ceratonia, Chamaecyparis, Cheirostemon, Cinchona, Cinnamomum, Citrus, Cobaea, Coccoloba, Cocos, Cornus, Crataegus, Cupressus, Cycas, Cymbidium, Cyperus, Daphne, Diospyros, Dracaena, Elaeagnus, Elaeodendron, Ephedra, Erica, Eucalyptus, Euonymus, Euphorbia, Fagara, Feijoa, Ficus, Fraxinus, Fuchsia, Gardenia, Gaultheria, Genista, Gnidia, Gomortega, Haworthia, Howea, Ilex, Impatiens, Jasminum, Juglans, Laurus, Leucodendron, Leucopogon, Liquidambar, Litchi, Lithraea, Livistona, Lonicera, Loranthus, Malus, Mangifera, Maranta, Marcgravia, Meryta, Metrosideros, Mimosa, Morus, Musa, Myrtus, Nerium, Olea, Osbeckia, Osmanthus, Pachypodium, Pedilanthus, Persea, Petrea, Phoenix, Phoradendron, Phormium, Phylica, Populus, Protea, Prunus, Psidium, Punica, Pyrus, Quercus, Rhododendron, Rosmarinus, Ruta, Salix, Schinus, Sophora, Staphylea, Styphelia, Thuja, Tristania, Ulmus, Umbellularia, Vaccinium, Veronica, Vinca, Viscum, Vitis,* and *Yucca.*

Distribution We have examined U.S. material from AL, CA, DC, FL, GA, LA, MD, MS, NM, NY, OH, OR, PA, SC, TN, TX, VT, WI, WV. This species has been reported from AZ, ID, IL, IN, MI, NJ, SD, VA, WA. A distribution map of this species was published by CAB International (1987). Greedy scale is commonly intercepted in quarantine from most tropical areas of the world and occasionally from greenhouses in northern countries.

Biology Schuh and Mote (1948) report greedy scale to have continuous generations in the greenhouse; bark is preferred but leaves and fruit are sometimes infested. According to Schmutterer (1952b) and Nur (1971) greedy scale reproduces without males. The species is reported to have two overlapping generations per year in New Zealand (Tomkins et al. 1992) and Italy (Bianchi et al. 1994). Rates of development were determined by Greaves et al. (1994) for the various instars of the scale under different temperature regimes. Blank et al. (1990) demonstrated that kiwifruit fields were reinfested from adjacent stands of *Beilschmiedia tarairi* and other alternate hosts in New Zealand.

Economic Importance Beardsley and González (1975) consider the greedy scale to be one of 43 serious armored scale pests. Miller and Davidson (1990) consider it to be a serious world pest. It is an important pest of kiwifruit in New Zealand (Blank et al. 1992), and pest management strategies and biological control agents have been introduced to reduce its impact (Stevens et al. 1994). Dekle (1977) indicates that it is sometimes a serious pest in Florida. According to McLean (1931) it is occasionally injurious to avocados in California. Das and Ganguli (1961) record it as destructive to young tea bushes. Ebeling (1959) commented that were it not for the practically universal use of pest-control measures on citrus for other insects, the greedy scale would be a serious pest on this crop. It has been reported as a pest of olives in California and Chile (Argyriou 1990); mango in the Philippines and Oceania (Chua and Wood

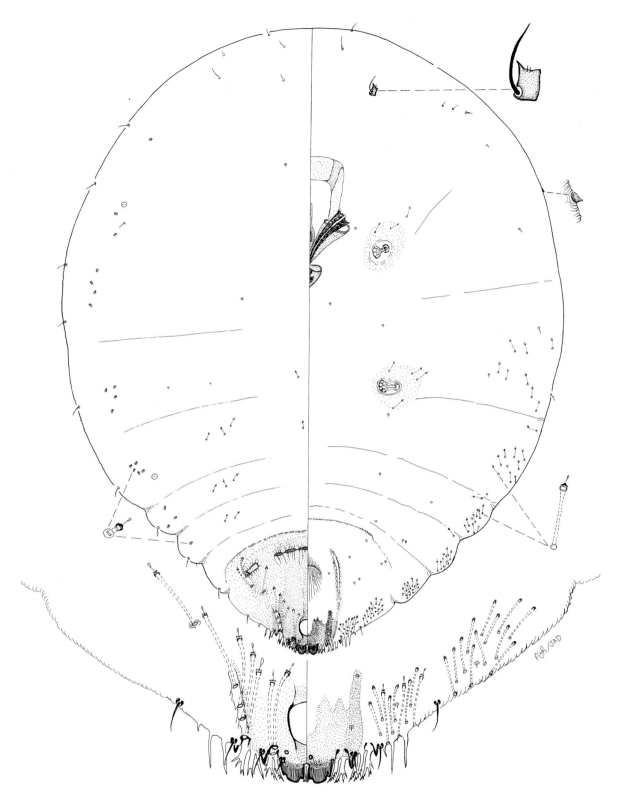

Figure 67. *Hemiberlesia rapax* (Comstock), greedy scale, England, intercepted at NY, on bay tree, XII-20-1949.

1990); kiwifruit in Chile (González 1989); ornamentals including roses, prunus, oaks, cotoneaster, camellia, magnolia, orchid, and aloes in many parts of the world (Myburgh 1990); tea in India, Japan, Malawi, Sri Lanka, and the former Soviet Union and can kill seedlings (Nagarkatti and Sankaran 1990); citrus in Iran and South America (Rose 1990b); and ornamentals in Italy (Tranfaglia and Viggiani 1986). The pest often is controlled by natural enemies, when they are present.

Selected References Ebeling (1959); Ferris (1938a); Schuh and Mote (1948); Tomkins et al. (1992).

Howardia biclavis (Comstock)

ESA Approved Common Name Mining scale (also called burrowing scale and biclavate scale).

Common Synonyms and Combinations *Chionaspis? biclavis* Comstock, *Aspidiotus theae* Green, *Chionaspis biclavis v. detecta* Maskell, *Howardia biclavis detecta* (Maskell), *Chionaspis (Howardia) biclavis* Comstock, *Howardia (Chionaspis) biclavis* (Comstock).

Field Characters (Plate 64) Adult female cover broadly oval to circular, moderately convex, whitish, grayish, or yellowish; shed skins subcentral, light brown. Males absent. Body of adult female white when newly mature, turning purplish brown; crawler light purple. Usually concealed by epidermis of host; color of cover obscured. On twigs and branches, rarely on leaves and fruit. Body of adult female with single notch between meso- and metathorax.

Slide-mounted Characters Adult female (Fig. 68) with 2 definite pairs of lobes, third and fourth pairs of lobes represented by 2 to 4 strongly sclerotized projections; without paraphyses except median lobes with conspicuous apically clubbed paraphysis touching medial margin of each median lobe and transverse paraphysis on each margin of each median lobe. Median lobes separated by space 0.1–0.3 (0.2) times width of median lobe, without basal sclerosis and yoke, medial margins of lobes diverging apically, lateral margins converging, with 5–12 (8) lateral notches, 2–5 (3) medial notches; second lobes simple, without notches, about one-eighth size of median lobes; third lobe area with 2 sclerotized projections, each projection normally with many serrations occasionally simple and similar to second lobe; fourth lobe area with 3 or 4 projections, posterior 2 projections with series of serrations, third projection when present simple or serrate, fourth projection spurlike, anterior of gland spines on segment 5. Gland spines each with microduct, gland-spine formula normally 2-3-4, sometimes 2-3-3, with 18–21 (20) gland spines near each body margin anterior of lobe 4 area; median lobes with 2 gland spines between them about 0.5–0.8 (0.7) times as long as median lobes. Gland tubercles on submargin of metathorax and segment 1. Macroducts of 1 size, short and slender, decreasing in length anteriorly, pygidial ducts present on segments 5 to 7, on submargin of prepygidium from mesothorax to segment 4, on submedian of segment 3 to 6, clusters of small ducts on submargin from segment 1 to prothorax, without a duct between median lobes, ducts in first space 16–19 (18) µ long, with 52–94 (74) ducts on pygidium, total of 148–194 (164) macroducts on dorsum on each side of body, orifices of pygidial ducts occurring anterior of anal opening. Pygidial microducts on venter in submarginal clusters on segments 5 to 7, with 14–16 (20) ducts on each side of body; prepygidial microducts of 2 sizes, shorter ducts in clusters on submargin from prothorax to segment 2, longer ducts present near body margin and in submedial areas of thorax and segments 1 to 3 or 4; on dorsum pygidial ducts absent; prepygidial microducts of 2 sizes, shorter ducts present in lateral areas of thorax, longer ducts in submedial areas of thorax and head. Perivulvar pores absent. Perispiracular pores predominately with 5 loculi, anterior spiracles each with 7–11 (9) pores, posterior pairs with 1–5 (3). Anal opening located 7–12 (9) times length of anal opening from base of median lobes, anal opening 20–35 (25) µ long. Dorsal seta laterad of median lobe 0.5–0.9 (0.8) times length of lobe. Eyes absent. Antennae each with 5 or 6 conspicuous setae. Prepygidium sclerotized from head to segment 2. Cicatrices absent. Anterior spur on each lateral margin of segments 2 to 4 or 5, smaller posterior spur on any or all of segments 2 to 4, spur on segment 5 rarely similar to lobelike processes with apical serrations. Space between median lobes with 2 large setae.

Affinities This species is unique in the United States by possessing the following combination of characters: third and fourth lobe areas with 3 or 4 sclerotized projections; a large, apically clubbed paraphysis touching the mesal margin of each median lobe; abdominal segments 2 to 4 each with a lateral sclerotized spur; all pygidial macroducts short, about equal in size.

Hosts Polyphagous. Borchsenius (1966) records it from 71 genera in 41 families; Dekle (1977) records it from 99 genera with *Ixora, Jasminum, Achras*, and *Tabernaemontana* as the most frequently reported hosts in Florida. We have examined specimens from the following host genera: *Acacia, Achras, Allamanda, Alternanthera, Annona, Anodendron, Antigonon, Bauhinia, Bixa, Bougainvillea, Brunfelsia, Caesalpinia, Calycanthus, Camellia, Cananga, Carica, Caryocar, Casearia, Casuarina, Cedrela, Centaurea, Centella, Chrysophyllum, Cinchona, Cinnamomum, Citrus, Clusia, Coffea, Cordia, Croton, Cupania, Delostoma, Derris, Dischidia, Dovyalis, Duranta, Elaeocarpus, Erythrina, Erythroxylum, Euphorbia, Feroniella, Ficus, Gardenia, Genipa, Grevillea, Hibiscus, Hymenaea, Inga, Ixora, Jasminum, Lantana, Leucaena, Licania, Ligustrum, Litchi, Lonicera, Loranthus, Lucuma, Macadamia, Malus, Mammea, Mangifera, Manilkara, Matisia, Mimusops, Monodora, Monstera, Montezuma, Mussaenda, Myrica, Nephelium, Pachira, Passiflora, Persea, Petrea, Phoebe, Photinia, Piper, Plumeria, Pongamia, Prunus, Psidium, Punica, Pyracantha, Pyrus, Randia, Sapium, Spathodea, Steriphoma, Tabebuia, Tabernaemontana, Terminalia, Theobroma, Trichilia, Waltheria*, and *Wisteria*.

Distribution We have examined U.S. material from FL and HI and from greenhouse infestations in the following northern states: DC, MD, MO, NY, OH, PA. It has been reported from KS and is considered to have been eradicated from CA. A distribution map of this species was published by CAB International (1957). Mining scale occurs in tropical areas throughout the world but most often is intercepted in quarantine from the Caribbean islands, Central and South America, and Mexico. Mining scale probably originated in the New World tropics (Balachowsky 1954).

Biology To our knowledge, a detailed study of the biology of this species has not been undertaken. According to Brown (1965) mining scale reproduces parthenogenetically without males. Kuwana (1911) and Murakami (1970) indicate that it has 1 generation per year in Japan. Of particular interest is the manner in which this pest becomes covered by a thin layer of the host's epidermis. There are many comments in the literature about how the scale mines or burrows into the host, but these statements almost certainly are inaccurate as the crawlers have no unusual adaptations for burrowing. It seems more likely that crawlers find cracks in the bark and settle in the bark tissue. As the scale cover enlarges, it slowly forces its way under the epidermal layer. This hypothesis is substantiated by Fennah (1947) who observed that crawlers settle in minute crevices of the bark.

Economic Importance Dekle (1977) reports the mining scale as an economic pest on many woody ornamentals in Florida despite the fact that it often is parasitized. Azevedo Marques (1923) considers it to be an important economic pest of coffee in Brazil. Plank and Winters (1949) describe it as causing heavy damage to *Cinchona ledgeriana* in Puerto Rico. It has been reported to cause damage to tea in Sri Lanka (Ramakrishna Ayyar 1930); India, Japan, and Malawi (Nagarkatti and Sankaran 1990); to citrus in Cuba (Grillo et al. 1983); to coffee in Cuba (Köhler et al. 1980); and to macadamia in India (Ebeling 1959). Beardsley and González (1975) consider this scale to be one of 43 serious armored scale pests. Miller and Davidson (1990) consider it to be a serious world pest.

Selected References Ferris (1937); Takagi (1992).

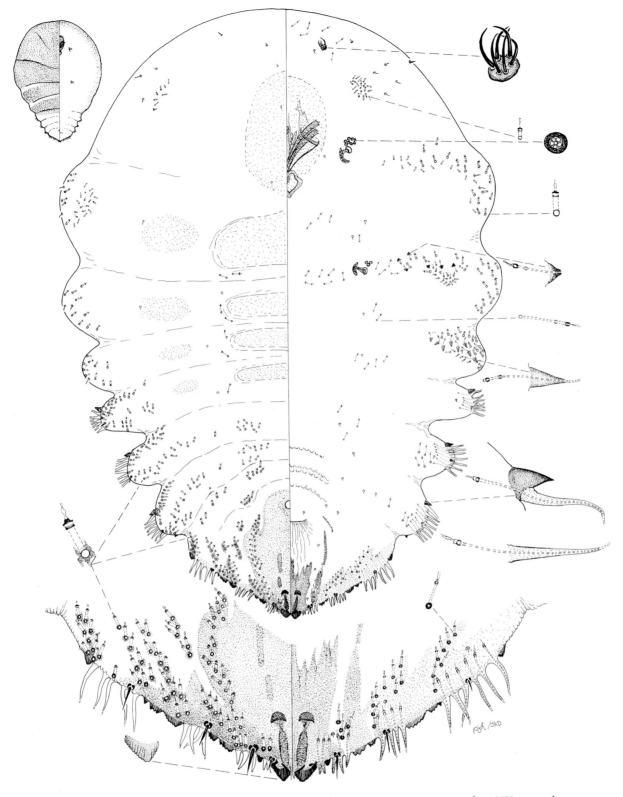

Figure 68. *Howardia biclavis* (Comstock), mining scale, Puerto Rico, intercepted at NY, on unknown host, III-24-1949.

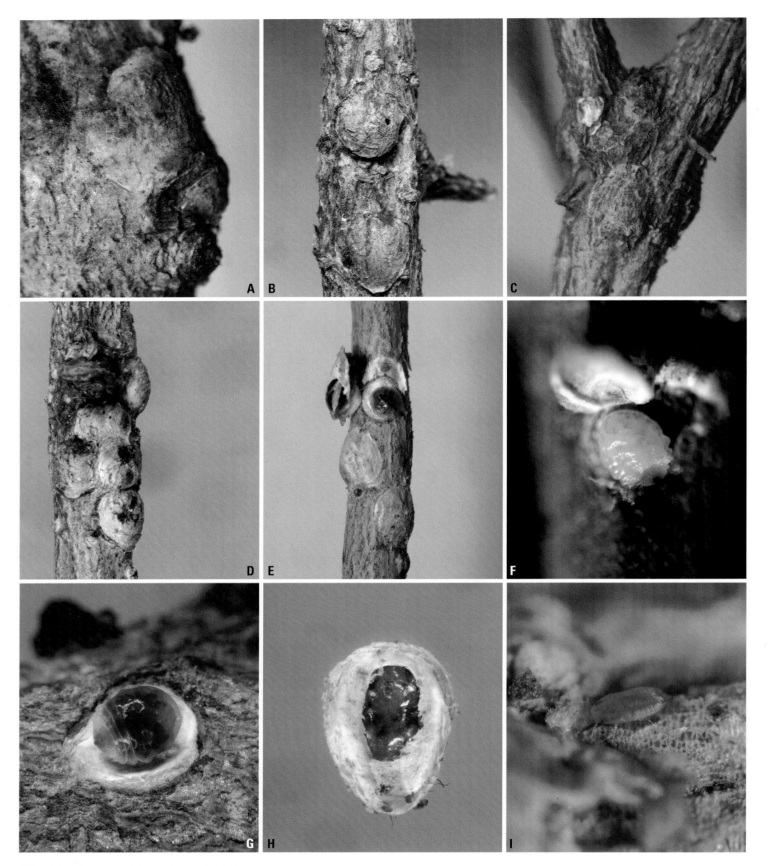

Plate 64. *Howardia biclavis* (Comstock), **Mining scale**

A. Cryptic female cover on *Ficus benjamina* (R. J. Gill).
B. Female covers on *Ficus* (J. A. Davidson).
C. Female covers under bark of *F. benjamina* (R. J. Gill).
D. Distance view of old covers on *Ficus* (J. A. Davidson).
E. Distance view of covers and body of adult female on *F. benjamina* (R. J. Gill).

F. Young adult female on hibiscus (D. R. Miller).
G. Body of mature adult female on *Ficus* (J. A. Davidson).
H. Underside of adult female showing extensive ventral cover (J. A. Davidson).
I. Pink crawler on *Ficus* (J. A. Davidson).

Plate 65. *Ischnaspis longirostris* (Signoret), Black thread scale

A. Adult female covers on *Ctenanthe oppenheimeriana* (R. J. Gill).
B. Female covers on *Chamaerops humilis* (D. R. Miller).
C. Distance view of covers on palm (J. A. Davidson).
D. Distance view of covers on palm (J. A. Davidson).
E. Ventral view of body of clear young female in cover on *C. oppen-heimeriana* (R. J. Gill).

F. Body of mature female on palm (J. A. Davidson).
G. Damage on *C. humilis* (D. R. Miller).
H. Infestation on palm (J. A. Davidson).
I. Distance view of heavy infestation on palm (J. A. Davidson).

Ischnaspis longirostris (Signoret)

ESA Approved Common Name Black thread scale (also called black lined scale, thread scale).

Common Synonyms and Combinations *Mytilaspis longirostris* Signoret, *Ischnaspis filiformis* Douglas, *Mytilaspis ritzemae basi* Leonardi, *Lepidosaphes ritsema-basi* (Leonardi).

Field Characters (Plate 65) Adult female cover elongate, approximately 8 times as long as broad, strongly convex, normally shiny black, sometimes covered with mealy bloom; shed skins marginal, pale yellow. Males not observed. Body of adult female yellow; eggs orange; crawlers yellow. On leaves, bark, and fruit.

Slide-mounted Characters Adult female (Fig. 69) with 2 definite pairs of lobes, third and fourth pairs represented by 1 to 3 sclerotized projections; paraphyses absent except on median lobe, medial lobule of lobe 2 with 2 large paraphyses, lateral lobule sometimes with 1 or 2 small paraphyses. Median lobes separated by space 0.6–0.8 (0.7) width of median lobe, with weak basal sclerosis, without yoke, medial margin of lobe diverging apically, lateral margin converging, with 8–11 (10) lateral notches, 8–11 (9) medial notches; second lobes bilobed, without notches, medial lobule largest; third lobe area with 2 or 3 projections, medial projection formed by swelling of marginal macroduct, middle projection about same size, either spurlike or serrate, lateral projection normally absent, occasionally represented by small, simple swelling; fourth lobe area usually with 3 spurlike projections, rarely with 1, 2, or 4. Gland spines each with microducts, gland-spine formula 1-1-1, with 11–21 (16) gland spines near each body margin anterior of lobe 4 area; median lobes without gland spines between them. Gland tubercles on submargin of metathorax, segment 1, and sometimes segment 2. Macroducts of 2 sizes, larger size on margin only, short and wide, duct in first space 13–20 (16) µ long, 1 in first space and 1 in second, small ducts thin, marginal and submedial on segments 5 and 6, also marginal on segment 4, 8–11 (10) macroducts on each side of pygidium on segments 5 to 7, prepygidial ducts restricted to marginal area except submedial clusters on segments 3 and 4, total of 35–47 (46) macroducts on dorsum of each side of body, orifices of pygidial ducts occurring anterior of anal opening. Pygidial microducts on venter in submarginal clusters on segments 5 and 6, with 6–11 (8) ducts on each side of body; prepygidial ducts scattered along body margin, in small submarginal clusters on abdominal segments; on dorsum pygidial ducts absent; prepygidial microducts marginal, represented by few ducts that are part of clusters from ventral areas. Perivulvar pores in 5 groups, with 6–9 (7) pores on each side of pygidium. Perispiracular pores with 3 loculi, anterior spiracles each with 1 pore, posterior spiracles without pores. Anal opening located 2.5–4.5 (4.0) times length of anal opening from base of median lobes, anal opening 10–13 (11) µ long. Dorsal seta laterad of median lobe 0.7–1.0 (0.8) times length of lobe. Eyes represented by small sclerotized area near body at level of anterior portion of mouthparts, rarely absent. Antennae each with 1 conspicuous seta with forked base. Prepygidium membranous. Cicatrix-like structures often on segments 1 and 3. Pygidial dorsum with conspicuous, coarse reticulations. Median lobes in concave depression. Body shape unusually elongate. Space between median lobes with 2 large setae. Space between median lobe and second lobe with 3 setae.

Affinities This species is distinct in the United States due to the following combination of characters: median lobes situated in a posterior depression, pygidial dorsum coarsely reticulate, and body elongate.

Hosts Polyphagous. Borchsenius (1966) records it from plants in 70 genera in 35 families; Dekle (1977) indicates that it occurs on plants in 133 genera in Florida and most frequently is found on palms, especially *Chamaedorea elegans* and *Monstera deliciosa*. We have examined specimens from the following host genera: *Acacia, Achras, Acrocomia, Amherstia, Annona, Anubias, Areca, Artocarpus, Arum, Asclepias, Asparagus, Bauhinia, Bromelia, Brownea, Camellia, Canna, Caryota, Cattleya, Caularthron, Cephaelis, Chamaedorea, Chrysalidocarpus, Citrus, Coccothrinax, Cocos, Coffea, Congea, Coronilla, Cycas, Dictyosperma, Didymosperma, Diospyros, Duranta, Elaeis, Eriobotrya, Eucalyptus, Eugenia, Ficus, Gardenia, Gliricidia, Gonocaryum, Gronophyllum, Heliconia, Heterospathe, Hibiscus, Howea, Inga, Iriartella, Ixora, Jasminum, Landolphia, Lantana, Litchi, Livistona, Lonicera, Mangifera, Maranta, Monstera, Moraea, Musa, Myristica, Napoleona, Nephelium, Norantea, Nypa, Odontoglossum, Oncidium, Pandanus, Philodendron, Phoenix, Pritchardia, Psychotria, Roystonea, Rubus, Sabal, Schomburgkia, Strelitzia, Swietenia, Tabernaemontana, Thunbergia, Vriesea, Wallichia,* and *Washingtonia*.

Distribution We have examined U.S. material from DC, FL, GA, HI, LA, MA, MD, MO, NJ, NY, OH, PA, VA. It also has been reported from CT, IL, OK, TN. Many of these records are from greenhouse infestations. It is recorded as eradicated from CA. A distribution map of this species was published by CAB International (1967). Black thread scale is commonly intercepted in quarantine from the Caribbean islands, Central and South America, and Mexico. This pest is found out-of-doors in tropical and subtropical areas throughout the world and in greenhouses in cold climates. It is believed to have originated from tropical Africa (Balachowsky 1954).

Biology Vesey-FitzGerald (1940) studied the life history of black thread scale in the Seychelles and found that females produced 20 to 30 eggs each. Eggs hatch soon after being laid, and crawlers settle to feed in about 24 hours. The second instar appears in about 3 days. Development proceeds throughout the year. Brown (1965) reported that this species is parthenogenetic.

Economic Importance According to Dekle (1977) the black thread scale is occasionally a serious pest of palms and greenhouse plants in Florida. Martorell (1940) found it causing heavy chlorosis and defoliation of Honduras mahogany (*Swietenia macrophylla*) in Puerto Rico. Bondar (1924) reports the black thread scale as a pest of coffee in Brazil. In Guam it was noted as a pest of mango by Swezey (1936). In the Seychelles, Vesey-FitzGerald (1940) records it as a serious pest of *Annona reticulata, A. cherimolia, Musa* (all varieties), cinnamon, *Coffea robusta*, litchi, mango, oil palm, *Raphia, Zizyphus jujuba*. He did not find any effective parasites or predators. DeBach and Rosen (1976) later report that the introduced coccinellids *Chilocorus nigritus* (Fabricius) and *C. distigma* (Klug) achieved complete control of the black thread scale on these islands. It is reported as a greenhouse pest in many parts of the world (Burger and Ulenberg 1990), and is a pest of coconut in Malaysia (Chua and Wood 1990), of oil palms in India (Dhileepan 1992), of mango and avocado in the Canary Islands (Perez Guerra 1986), and of coffee in Mexico (Ibarra-Nunez 1990). Beardsley and González (1975) consider this scale to be one of 43 serious armored scale pests. Miller and Davidson (1990) consider it to be a serious world pest.

Selected References Ben-Dov (1974); Ferris (1937); Vesey-FitzGerald (1940).

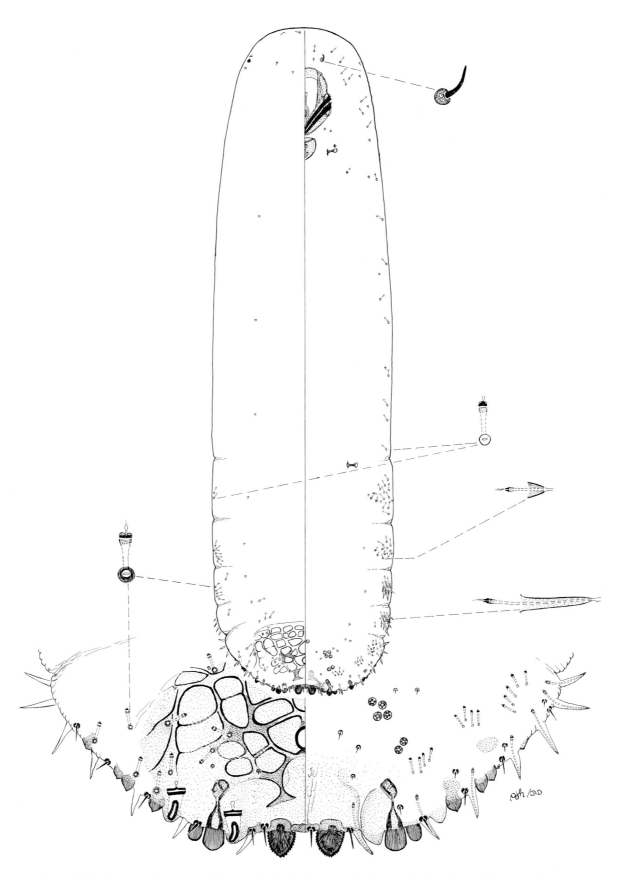

Figure 69. *Ischnaspis longirostris* (Signoret), black thread scale, Bangor, Indonesia, on *Myristica fragrans*, XI-?-1953.

Kuwanaspis pseudoleucaspis (Kuwana)

Suggested Common Name Bamboo diaspidid.

Common Synonyms and Combinations *Leucaspis bambusae* Kuwana, *Lepidosaphes bambusae* (Kuwana), *Lepidosaphoides bambusae* (Kuwana), *Chionaspis bambusae* (Kuwana), *Chionaspis pseudoleucaspis* Kuwana, *Mytilaspis bambusae* (Kuwana), *Tsukushiaspis pseudoleucaspis* (Kuwana), *Tsukushiaspis bambusae* (Kuwana).

Field Characters (Plate 66) Adult female cover white or tan, elongate, oyster-shell shaped, slightly convex; shed skins marginal, yellow or tan. Male cover white with 1 median ridge; shed skin marginal and yellow. Body of adult female elongate, yellow, with deep notch at junction of thorax and abdomen; eggs presumed yellow. Primarily on stems, particularly at nodes (Gill 1997).

Slide-mounted Characters Adult female (Fig. 70) normally with 2 pairs of definite lobes, lobes 3, 4, and sometimes 5 replaced by toothed process, rarely medial lobule of lobe 3 normal in appearance; paraphyses associated with medial and lateral lobules of lobe 2. Median lobes separated by space 0.8–1.6 (1.1) times width of median lobe, with lateral and medial paraphyses, without sclerosis or yoke, lateral margins of lobes usually parallel apically, occasionally slightly divergent or convergent, medial margins parallel apically, occasionally slightly divergent or convergent, with 0–2 (1) lateral notches, with 1–3 (2) medial notches; second lobes bilobed, equal to or slightly smaller than median lobes, medial lobule largest, rounded apically, with 0–2 (1) lateral notches, 0–2 (1) medial notches, lateral lobule rounded apically, with 0–2 (1) lateral notches, with 1–3 (1) medial notches; third lobe area usually with 3 or 4 dentate processes, rarely with medial lobule lobelike; fourth lobe area with 2–5 (4) dentate processes; fifth lobe area with 0–4 (2) dentate processes. Gland spines simple, gland-spine formula 2-2-2, 1-1-1, or 1-1-2, with 15–31 (22) gland spines near each body margin on segments 1 to 4; with dentate process between median lobes. Macroducts of 2 indistinct sizes, largest in marginal and submarginal areas of segments 2, 3, or 4 to 8, and in submedial areas of segments 2, 3, or 4 to 6, small size ducts usually in marginal and submarginal areas of metathorax or segment 1 to segments 2, 3, or 4, with 1 or 2 ducts between median lobes extending 0.1 times distance between posterior apex of anal opening to base of median lobes, with duct between median lobes 12–15 (14) μ long, macroduct in first space 11–15 (13) μ long, with 29–44 (36) macroducts on each side of body on segments 5 to 7, total of 73–133 (105) macroducts on each side of body; some macroduct orifices anterior of anal opening. Pygidial microducts usually on venter in medial, submarginal, and marginal areas of segments 5 to 7, with 8–16 (10) ducts; prepygidial ducts of 1 size, variable in distribution, usually with 1 or 2 in marginal areas of segment 4 and sometimes segment 3, usually with lateral cluster on posterior margin of metathorax and segment 1, submedial cluster near each spiracle and on any or all of segments 1 to 4, and 1 medial duct on any or all of metathorax to segment 4, absent from head; on dorsum microducts absent. Perivulvar pores in 5 indistinct groups, 10–25 (20) pores on each side of body. Perispiracular pores usually with 3 loculi, anterior spiracles with 5–12 (8) pores, posterior spiracles without pores. Anal opening located 8.8–13.5 (11.0) times length of anal opening from base of median lobes, anal opening 12–17 (14) μ long. Dorsal seta laterad of median lobes 1.0–1.8 (1.4) times length of median lobe. Eyes represented by dome-shaped area, located on body margin between anterior edge of clypeolabral shield and antenna. Antennae usually each with 2 setae, rarely 1. Cicatrices absent. Body elongate, broadest posteriorly. Macroducts ventral on metathorax to segments 2 or 3. Mature females without sclerotization except on pygidium.

Affinities Of the economic species in the United States, bamboo diaspidid is very distinctive by having: 2 pairs of lobes; dentate processes between the median lobes; dentate processes on body margin from segment 4 to segment 6; and by occurring on bamboo. It is most similar to other species of *Kuwanaspis* that occur in the United States, especially *K. howardi* (Cooley). They differ primarily in the position of a row of small-sized macroducts that extends across the venter of abdominal segment 1; it is present on *K. howardi*, whereas on *K. pseudoleucaspis* it is restricted to the marginal areas on the venter of segment 1.

Hosts Restricted to grasses, primarily bamboo. Borchsenius (1966) records it from 3 genera in 1 family. We have examined specimens from the following host genera: *Arundinaria*, *Bambusa*, *Fragesia*, and *Phyllostachys*. It has been recorded from *Cynodon*, *Paspalum* (Hodgson and Hilburn 1990), *Pleioblastus*, *Sasa*, *Sasamorpha* (Murakami 1970), *Semiarundinaria* (Nakahara 1982), and *Sinobambusa* (Murakami 1970).

Distribution We have examined U.S. specimens from CA, DC, FL, HI, MD, NY, VA. It has been reported from AL, GA, NJ, SC (Nakahara 1982), and PA (?) (Trimble 1929). It probably is native to the Oriental Region including China, Japan, Korea, and Taiwan but also has been introduced to Africa, Asia, and Europe. It rarely is taken at U.S. ports-of-entry.

Biology The bamboo diaspidid is reported to have 2 generations per year in China (Yan Aojin et al. 1985) and a single generation per year in Georgia of the former Soviet Union (Hadzibejli 1983). Although Brown (1965) suggests that the species is parthenogenetic, Kuwana (1928) and Gill (1997) provide a description of the male cover. Howell and Tippins (1973) and Danzig (1964) also note the absence of males.

Economic Importance In severe outbreaks, the culms of bamboo turn white as they become encrusted with scale covers giving the plants an unthrifty appearance. Heavily infested stands of bamboo sustain a general reduction in growth vigor (Wang et al. 1998b).

Selected References Gill (1997); Wang et al. (1998b); Yan Aojin et al. (1985).

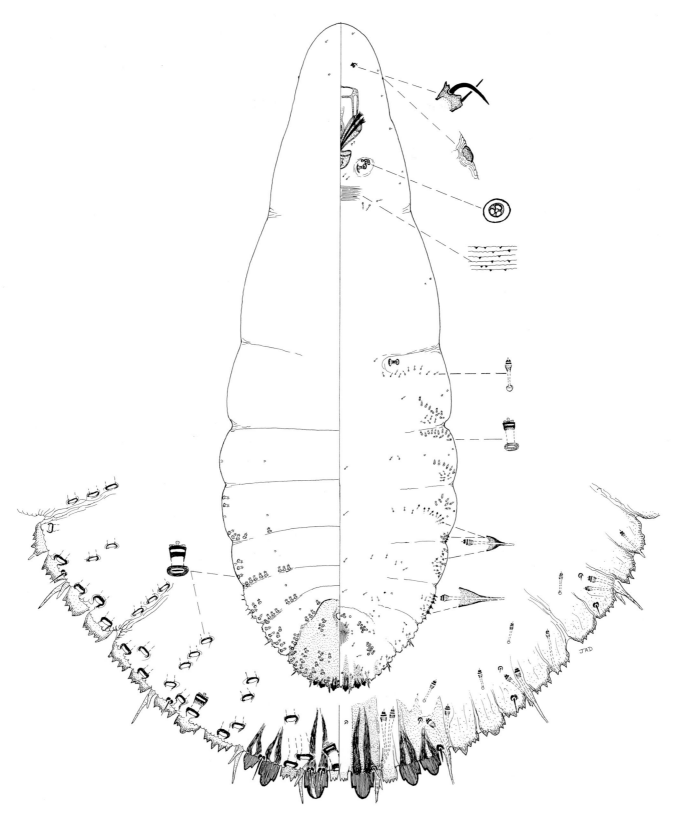

Figure 70. *Kuwanaspis pseudoleucaspis* (Kuwana), bamboo diaspidid, Kawaiha, HI, XI-29-1994.

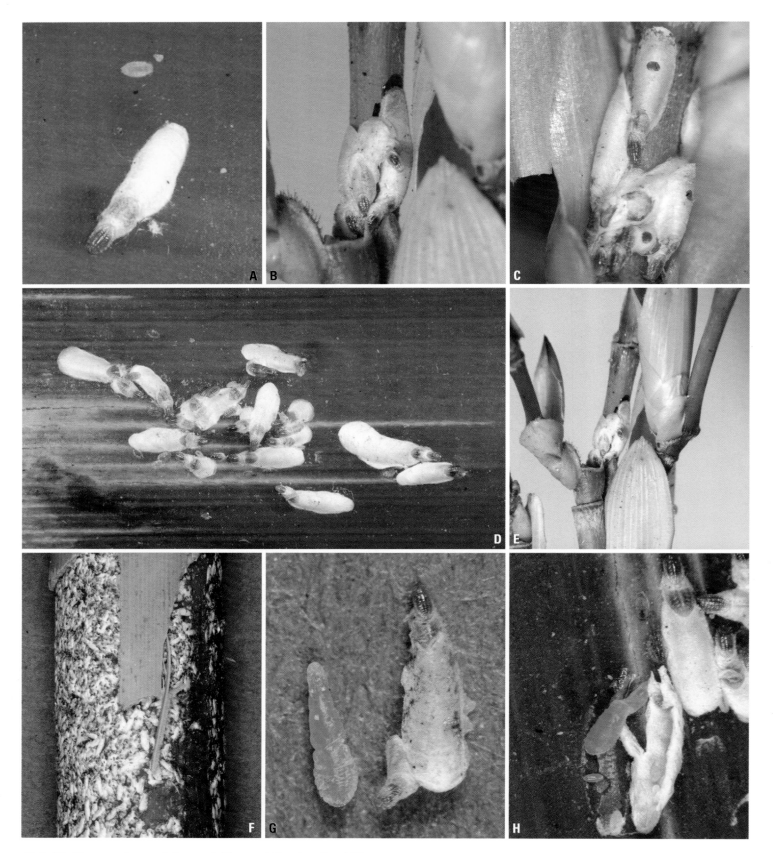

Plate 66. *Kuwanaspis pseudoleucaspis* (Kuwana), **Bamboo diaspidid**

A. Female cover and crawler on bamboo (J. A. Davidson).
B. Cluster of female covers on bamboo (R. J. Gill).
C. Female covers on bamboo showing parasite emergence holes (R. J. Gill).
D. Female covers on bamboo (J. A. Davidson).
E. Cluster of covers where bamboo sheath has been removed (R. J. Gill).

F. Distance view of heavy infestation on bamboo (J. A. Davidson).
G. Body of adult female removed from cover on bamboo (J. A. Davidson).
H. Cover removed from female and eggs (J. A. Davidson).

242

Plate 67. *Lepidosaphes beckii* (Newman), Purple scale

A. Adult female cover on citrus leaf (J. A. Davidson).
B. Adult female cover on citrus leaf (D. R. Miller).
C. Adult female cover on citrus fruit (R. J. Gill).
D. Adult female cover on citrus bark (J. A. Davidson).
E. Adult female cover on citrus and small back male cover (R. J. Gill).

F. Distance view of covers on orange (R. J. Gill).
G. White body of adult female in cover on citrus (J. A. Davidson).
H. Body of adult female with newly laid white egg (J. A. Davidson).
I. Branch dieback on citrus from heavy infestations of purple scale and Glover scale (J. A. Davidson).

243

Lepidosaphes beckii (Newman)

ESA Approved Common Name Purple scale (also called citrus mussel scale, orange scale, orange mussel scale, common and grass seed scale).

Common Synonyms and Combinations *Coccus beckii* Newman, *Aspidiotus citricola* Packard, *Coccus anguinus* Boisduval, *Mytilaspis flavescens* Targioni-Tozzetti, *Mytilaspis citricola* (Packard), *Mytilaspis fulva* Targioni-Tozzetti, *Mytilaspis citricola tasmaniae* Maskell, *Mytilaspis tasmaniae* Maskell, *Mytilaspis beckii* (Newman), *Lepidosaphes pinnaeformis* (Bouché), *Lepidosaphes citricola* (Packard), *Lepidosaphes (Mytilaspis) beckii* (Newman), *Mytilococcus beckii* (Newman), *Cornuaspis beckii* (Newman).

Field Characters (Plate 67) Adult female cover oyster-shell shaped, straight or curved, slightly to moderately convex, thick, purplish brown to pale yellow brown; with dorsal cover extending partially across venter especially in anterior region, white, ventral cover present. Male cover shorter, narrower, similar in color and texture to female cover; shed skin marginal, brown. Body of adult female white; eggs white; crawlers whitish yellow. On bark, leaves (both surfaces), and fruit of host.

Slide-mounted Characters Adult female (Fig. 71) with 2 pairs of definite lobes, third and fourth lobes sometimes represented by series of small points; without paraphyses except transverse sclerotizations on median lobes and on medial lobule of second lobes. Median lobes separated by space 0.4–0.6 (0.5) times width of median lobe, without basal sclerosis or yoke, lobe axes parallel, with 1–4 (2) lateral notches, 2–4 (3) medial notches; second lobes bilobed, about one-half size of median lobe, medial lobule largest, with 0–5 (1) lateral notches, absent medially, lateral lobule without notches; third and fourth lobe areas frequently with series of small points and swelling near marginal macroducts. Gland spines each with microduct, gland-spine formula 2-2-2, with 13–32 (23) gland spines near each body margin from metathorax to fourth lobe area, some having appearance of gland tubercles; median lobes with 2 gland spines between them. Macroducts of 2 sizes, larger size on margin only, short and wide, duct in first space 32–35 (33) μ long, 6 large ducts on each side of pygidium, present on segments 5 to 7; smaller size abundant, 1 submarginal duct on segment 7, submarginal clusters present from prothorax or mesothorax to segment 6, increasing in size posteriorly, submedial clusters from segments 2 or 3 to 6, ducts on thorax anteriorly becoming more abundant on venter than on dorsum, in ventral band posterior of hind spiracle, 36–62 (52) macroducts on pygidium on segments 5 to 7, total of 216–294 (248) macroducts on each side of body, orifices of macroducts occurring anterior of anal opening. Pygidial microducts on venter in submarginal clusters on segments 5 and 6, with 5–8 (6) ducts on each side of body; prepygidial ducts scattered along body margin, in small submarginal clusters on most abdominal segments; on dorsum pygidial ducts absent; prepygidial microducts marginal, represented by few ducts. Perivulvar pores in 5 groups, with 17–30 (23) pores on each side of pygidium. Perispiracular pores with 3 loculi, anterior spiracles each with 3–8 (5) pores, posterior spiracles without pores. Anal opening located 12–17 (14) times length of anal opening from base of median lobes, anal opening 10–13 (12) μ long. Dorsal seta laterad of median lobe 0.7–0.9 (0.8) times length of lobe. Eyes normally represented by small, flat, sclerotized area, at level of clypeolabral shield, rarely absent. Antennae each with 2 large setae, rarely 3. Prepygidium membranous. Pigmented cicatrices normally on segments 1, 2, and 4, rarely also on segments 3 and 5. Space between median lobes with 2 setae. Some specimens with small lateral spur on segment 4.

Affinities Of the economic species, purple scale is most similar to *Lepidosaphes ulmi*, oystershell scale, but differs by lacking lateral spurs or by having spurs restricted to segment 4; by having pigmented cicatrices on segments 1, 2, and 4, rarely on 3 and 5; and by having differently shaped median lobes. Purple scale also is similar to *Lepidosaphes chinesis*, Chinese lepidosaphes scale, but differs by having the above-mentioned characters except Chinese lepidosaphes scale normally has a series of pigmented cicatrices on segments 1 to 5. Purple scale also is similar to *Lepidosaphes cycadicola* Kuwana, a Taiwan species, but differs by having only 1 small submarginal macroduct on segment 7; *L. cycadicola* has 2 or 3 in this area.

Hosts Polyphagous. Borchsenius (1966) records it from 14 genera in 11 families; Dekle (1977) reports 45 genera in Florida, with the most frequently reported being citrus. We have examined specimens from the following genera: *Ananas*, *Camellia*, *Chalcas*, *Citrus*, *Coffea*, *Cymbidium*, *Elaeagnus*, *Erythrina*, *Fortunella*, *Ilex*, *Mangifera*, *Melia*, *Murraya*, *Passiflora*, *Phoradendron*, *Psidium*, *Salix*, and *Verbena*. It is reported on *Croton*, *Eucalyptus*, *Ficus*, and *Olea* by Fernandes (1987); and on *Balsamocitrus*, *Cassia*, *Cercidiphyllum*, *Codiaeum*, *Cycas*, *Malpighia*, *Persea*, *Psidium*, and *Quercus* by González and Hernandez (1988).

Distribution We have examined U.S. material from AL, CA, FL, GA, HI, LA, MO, NC. It has been reported from KS, NY, OK, PA, SC. A distribution map of this species was published by CAB International (1982). Northern records are from indoor collections. Of all the citrus scale insect pests, purple scale is the most widely distributed throughout the tropics of the world (Ebeling 1959). It most often is intercepted in quarantine on *Citrus* from the Caribbean islands, Central and South America, and Mexico.

Biology Ebeling (1959) points out that purple scale prefers humid areas. In California it does not occur in the interior districts, but it is an important pest in the coastal areas. Females produce 40–48 eggs over a period of about 3 weeks. Hatching begins 2 weeks after oviposition in summer and after 2 months in winter. The first molt occurs 2 weeks after hatching in summer. The second molt requires about 3 weeks in summer. Mating occurs soon after the second molt, and eggs are laid about 15 days later. Johnson and Lyon (1976) state that purple scale females lay 60–80 eggs. Quayle (1912) found that in Whittier, California, purple scale completed its life cycle in 3 months or less in the summer, with a fourth generation possible during warm winters. Thompson and Griffiths (1949) report a life cycle of about 3 months in the summer in Florida, with crawler peaks in March–April, June–July, and September–October. Outdoor insectary studies at Auburn, Alabama, by English and Turnipseed (1940), resulted in minimum and maximum generation times from birth to egg hatch of 42 and 198 days, respectively, with 77.3 days the average. In Chile, Zuniga (1971) reports 3 generations per year in the northern region, 2 to 3 generations per year in the central region, and 1 generation per year in the southern region. In Australia, Hely and Gellatley (1961) found that purple scale preferred *Citrus*, but heavy infestations also occurred on *Ilex*. They found that the first deposited eggs hatch when the parent female is about 90 days old. The average number of crawlers produced by each female is about 150 in spring and 70 in autumn. Generally only 2.5 generations occur each year between September and May. In Israel, Avidov and Harpaz (1969) report that summer development of purple scale required at least 50 days (44 days for males) while in winter about 110 days were needed.

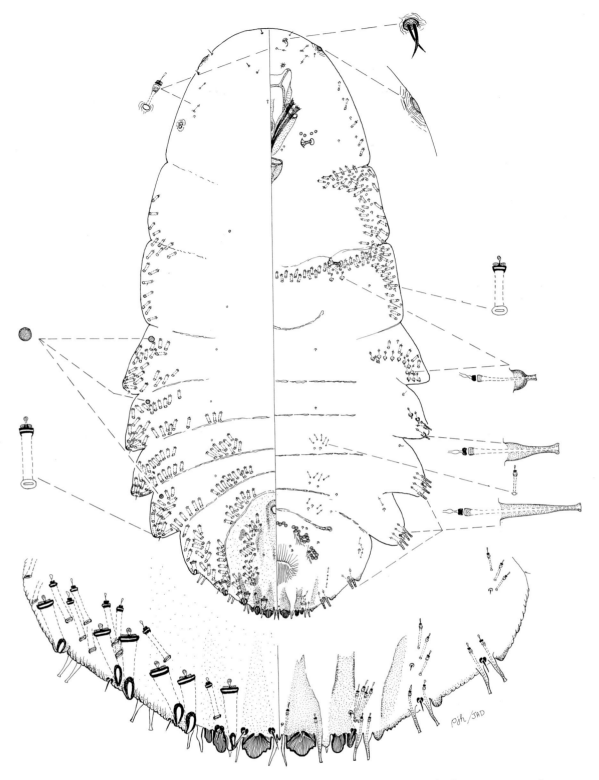

Figure 71. *Lepidosaphes beckii* (Newman), purple scale, Sanlo, New Hebrides, on orange leaves, XII-1-1950.

They concluded that the temperature threshold of development is 8 °C, and 1104 day-degrees are required for the production of 1 generation. They noted that this scale prefers trees with thick foliage and settles more on leaves and fruit than on young branches. They also found that females on fruit were more fecund than females on leaves. In the coastal plain, Bodenheimer and Steinitz (1937) found 4 generations per year: April, late June, late August to early September, and November to January. Bodenheimer (1951) reports the sex ratio from 373 individual matings to be 39.4% female and 60.6% male. He found facultative parthenogenesis common. Lindgren and Dickson (1941) never observed oviposition from isolated females in California. Bénassy et al. (1975) report that purple scale has 2 generations per year in France at Cote D-Azur, 4 in Italy at Naples, 4 in Egypt, and 3 in Tunisia. Smirnoff (1960) found 4 generations per year in Morocco. In South Africa, Hulley (1962) notes that egg hatch of purple scale is influenced by light and temperature. He found a positive phototaxis but no geotaxis. Most eggs hatched in the morning and crawlers settled more rapidly on dusty than clean leaves. In Nigeria, Eguagie (1972) found that purple scale fecundity was identical on orange, grapefruit, tangerine, and lime, with an incubation period of 10 to 25 days during June–July, at a mean temperature of 25.1 °C and 77.5% RH at 1600 hrs. Claps (1987) reports 5 generations in the insectary in Argentina and provided descriptions and illustrations of all stages (Claps 1991). Rodrigo and Garcia-Mari (1994) studied the seasonal variation of populations of the purple scale and its distribution in different parts of the tree. Foldi (1990) provides information on the formation of the scale cover of this species.

Economic Importance According to Dekle (1977) the purple scale has been reduced to a minor citrus pest in Florida since the appearance of the parasite *Aphytis lepidosaphes* Compere. Muma and Clancy (1961) state this specific parasite was discovered in Florida in 1958. It apparently spread into Florida from Mexico and Texas from California-reared material. It was imported into the latter from China in 1948–1949. It is most common in the southern United States on citrus. Simanton (1976) states that in Florida a 50% parasitism rate appears adequate to keep purple scale effectively controlled. Ebeling (1959) states that purple scale attacks all parts of the tree, and injury can be severe. In California defoliation and dieback usually occur on the lower north side of trees. Feeding scales cause yellow spots on leaves. Similar areas on the fruit do not attain their normal color and remain green. The purple scale was ranked the fourth most important pest of Texas citrus in 1950 (Ebeling 1950). He also notes that in parts of Mexico, the West Indies, Central America, Peru, Chile, Paraguay, Brazil, and certain areas of southeast Asia, purple scale is the most important citrus pest. The authors note that *Aphytis lepidosaphes* may have been introduced into some of these areas since Ebeling wrote the above, and this situation may have changed. Dean (1975) reports that purple scale, although once the fourth most important pest of Texas citrus, was now under complete biological control by *Aphytis lepidosaphes*. Avidov and Harpaz (1969) found that in Israel, populations of purple scale on citrus fruit are highest between September and January, and in certain cases about 30% of the yield was rejected for export because of contamination by the many adhering scales. They also found that heavily infested leaves are shed prematurely, and that when too many leaves are lost, trees are stunted and dieback occurs. Damage is more serious in mature groves where foliage is dense, and less serious in young groves where the foliage is less abundant and trees are not crowded together. This species is considered a major pest of citrus in South Africa (Bedford and Cilliers 1994) and in Spain (Rodrigo and Garcia-Mari 1990). Beardsley and González (1975) consider this scale to be one of 43 serious armored scale pests. Miller and Davidson (1990) consider it to be a serious world pest.

Selected References Dean (1975); Fabres (1980); Williams and Watson (1988).

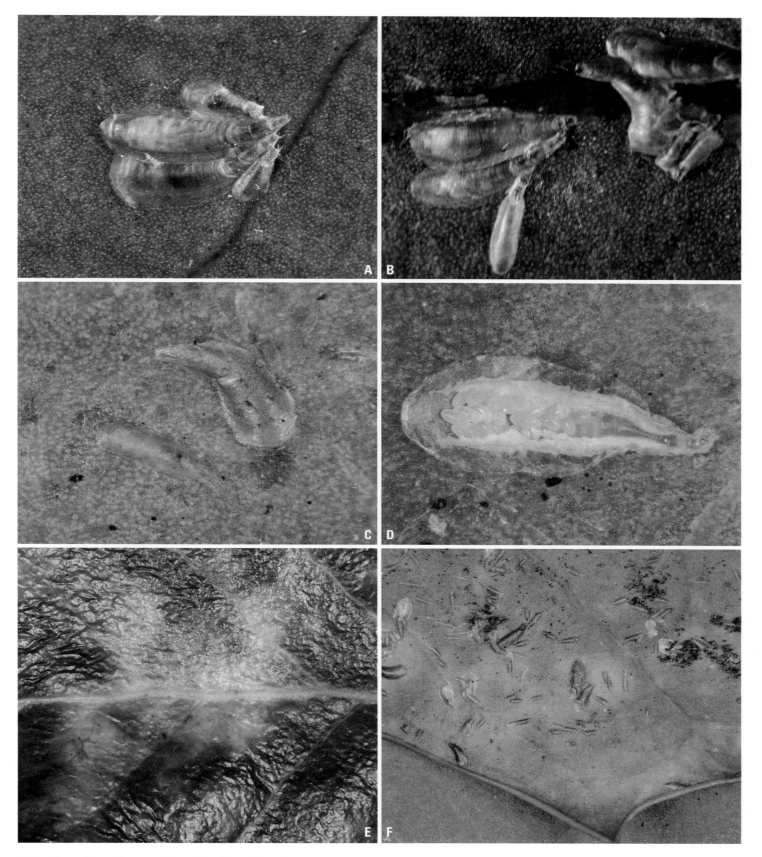

Plate 68. *Lepidosaphes camelliae* Hoke, Camellia scale

A. Translucent adult female scale covers and immature covers on camellia (J. A. Davidson).
B. Female and male covers on camellia (J. A. Davidson).
C. Opaque female and male covers on holly (J. A. Davidson).
D. Female cover turned to show eggs and female body (J. A. Davidson).

E. Damage to holly leaf viewed from top of leaf (J. A. Davidson).
F. Distance view of infestation and damage on underside of holly leaf (J. A. Davidson).

247

Lepidosaphes camelliae Hoke

ESA Approved Common Name Camellia scale.

Common Synonyms and Combinations *Insulaspis camelliae* (Hoke).

Field Characters (Plate 68) Adult female cover oyster-shell shaped, straight unless crowded, slightly convex, moderately thick, dark to light brown; shed skins marginal, yellowish brown. Male cover shorter, narrower, similar in texture, coppery brown; shed skin marginal, greenish gold. Body of adult female purple; eggs semitransparent with purple tint; crawlers pale cream, except head and pygidium reddish brown. On leaves of host.

Slide-mounted Characters Adult female (Fig. 72) with 2 definite pairs of lobes, third and fourth pairs of lobes represented by low, sclerotized areas with series of small projections; with thin paraphyses on lateral and medial margins of median lobe and medial lobule of second lobe. Median lobes separated by space 1.0–1.5 (1.3) times width of median lobe, without basal sclerosis or yoke, lobe axes parallel, with 1 lateral notch and 1 medial notch; second lobes bilobed, about one-half size of median lobe, medial lobule sometimes with lateral notch, medial lobule largest; third and fourth lobe areas represented by series of small projections. Gland spines each with microduct, gland spines between median lobes, first space, and sometimes second space with basal fringing, gland-spine formula 2-2-2 or 1-2-2, with 23–32 (29) gland spines near each body margin anterior of lobe 4 area, present to metathorax; median lobes with 2 conspicuous gland spines between them. Macroducts of 2 or 3 sizes, larger size on margin only, duct in first space 23–33 (28) μ long, 1 in first space, 2 in second, and 3 on segment 5, 1 smaller duct near lateral lobule of second lobe, small ducts in submarginal and submedial areas of segments 5 and 6, with 16–19 (17) macroducts on each side of segments 5 to 7; prepygidial ducts on marginal areas of mesothorax to segment 4, dorsal on all segments, ventral on mesothorax to segment 1 on submedial and submarginal areas of segments 1 or 2 to 4, total of 88–131 (111) macroducts on each side of body, orifices of pygidial ducts occurring anterior of anal opening. Pygidial microducts on venter in submarginal areas of segments 5 to 6, with 2–9 (5) ducts on each side of body; prepygidial ducts in submedial areas of segments 1 to 3 or 4, on submarginal areas of mesothorax, metathorax, and segment 4; dorsal microducts absent. Perivulvar pores in 5 groups, with 16–21 (18) pores on each side of pygidium. Perispiracular pores with 3 loculi, anterior spiracles each with 3–4 (4) pores, posterior spiracles without pores. Anal opening located 10–16 (13) times length of anal opening from base of median lobes, anal opening 10–14 (12) μ long. Seta lateral of median lobe 0.7–1.2 (1.0) times as long as length of median lobe. Eyes absent or represented by weakly sclerotized area slightly posterior of level of antennae. Antennae each with 2 or 3 conspicuous setae, 1 sometimes forked at base. Prepygidium membranous. Cicatrices absent. Body elongate. Space between median lobes with 2 setae. Lateral spur usually present on anterior margin of segment 4.

Affinities Camellia scale is similar to *Lepidosaphes conchiformis*, fig scale, except the former has an intermediate macroduct just anterior of the second lobule of the second lobe and has basal fringing of gland spines between the median lobes. Fig scale lacks the extra marginal macroduct anterior of lobe 2 and has simple gland spines between the median lobes.

Hosts Borchsenius (1966) records this species from *Camellia* and *Thea*; Dekle (1977) records it from 8 genera with *Camellia* and *Ilex* the most frequently reported hosts in Florida. Based on material in the USNM Collection, camellia is the most common host. We have seen material from *Camellia*, *Croton*, *Gardenia*, *Hedera*, *Ilex*, and *Symplocos*. It has been reported from *Cleyera*, *Ligustrum*, *Magnolia*, *Raphiolepis*, and *Ternstroemia*.

Distribution We have examined U.S. material from AL, AR, CT, FL, GA, LA, MA, MD, MO, MS, NC, NY, OK, PA, SC, TX, VA. It has been reported from DE and OR. This species has been eradicated from CA. The camellia scale occurs in China, Japan, Mexico, Okinawa. It occasionally is intercepted in quarantine from Japan. Northern records probably represent greenhouse infestations.

Biology In Georgia, Cooper and Oetting (1989) studied the life history of the camellia scale and estimated there were 4 or 5 overlapping generations. At 25 °C, development from egg hatching to the adult on *Camellia japonica* was about 29 days for males and 23 days for females. Females reared at this temperature laid about 96 eggs, which hatched in about 10 days. In Alabama, English and Turnipseed (1940) found that females deposit 25–55 eggs, which hatched in 11 to 24 days. The first molt occurred 12 to 27 days after hatching, and the second molt occurred 6 to 10 days later. Eggs were deposited 38 to 54 days after the eggs of the previous generation hatch. The life cycle was completed in 60 to 70 days. English and Turnipseed (1940) found the camellia scale to be less active than the tea scale on camellia in the winter. Few crawlers appeared from October to March on plants grown outdoors, but hatching continued in cold frames. Because of overlapping generations, all stages of the scale were present in the summer months.

Economic Importance Dekle (1977) states the camellia scale is occasionally a serious pest on camellia and holly in central and northern Florida. English and Turnipseed (1940) report that this pest may severely damage young plants and cuttings in nurseries. Even though it does not discolor leaves in heavy infestations, the foliage is devitalized and leaves drop prematurely. Cooper and Oetting (1989) consider it to be a pest in Georgia. Miller and Davidson (1990) consider this species to be a serious pest in a small area of the world.

Selected References Cooper and Oetting (1989); English and Turnipseed (1940); Ferris (1938a); Liu et al. (1989a).

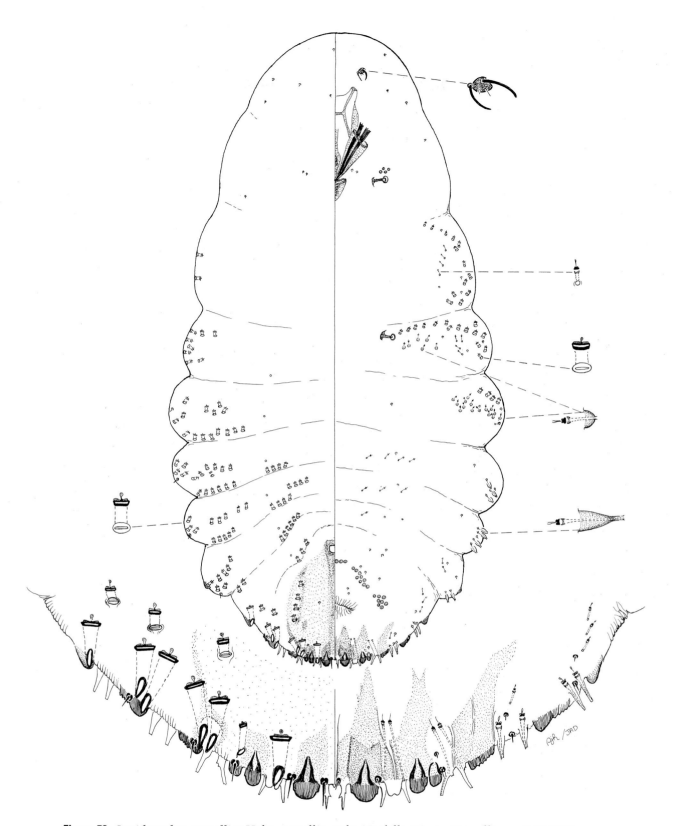

Figure 72. *Lepidosaphes camelliae* Hoke, camellia scale, Norfolk, VA, on *Camellia* sp., I-22-1952.

Lepidosaphes conchiformis (Gmelin)

ESA Approved Common Name Fig scale (also called Mediterranean fig scale, narrow fig scale).

Common Synonyms and Combinations *Coccus conchiformis* Gmelin, *Diaspis conchiformis* (Gmelin), *Mytilaspis linearis* Targioni Tozzetti, *Mytilaspis conchiformis* (Gmelin), *Mytilaspis ficus* Signoret, *Lepidosaphes ficus* (Signoret), *Lepidosaphes conchiformis-ulmi* Koroneos, *Mytilococcus conchiformis* (Gmelin), *Mytilococcus linearis* (Targioni Tozzetti), *Mytilaspis minima* Newstead, *Lepidosaphes minima* (Newstead), *Mytilaspis ficifolii* Berlese, *Lepidosaphes ficifolii* (Berlese), *Lepidosaphes ficifoliae ulmicola* Leonardi, *Mytilococcus ficifoliae ulmicola* Leonardi, *Mytilococcus minimus* (Newstead), *Lepidosaphes turkmenica* Borchsenius and Bostshik, *Insulaspis minima* (Newstead).

Field Characters (Plate 69) Adult female cover oyster-shell shaped, regularly curved on bark, irregular on leaf, slightly convex, relatively thick, dark brown on bark, thin and pale on leaf; leaf forms normally smaller than bark forms; shed skins marginal, yellow or pale brown. Male cover shorter, narrower than adult female, otherwise similar. Body of adult female, eggs, and crawler white, sometimes reported as pink to violet. On leaves and bark; males primarily on leaves.

Slide-mounted Characters This species has 2 distinctive forms, which are called the leaf and bark forms. We will describe the bark form first then give contrasting features of the leaf form.

Bark form: Adult female (Fig. 73) with 2 definite pairs of lobes, third and fourth pairs represented by low, sclerotized areas with or without series of small projections; without paraphyses except sometimes with transverse sclerotizations on median lobes and on medial lobule of second lobes. Median lobes separated by space about 0.5–0.9 (0.7) times width of medial lobe, without basal sclerosis or yoke, lobe axes parallel, with 2–4 (2) lateral notches and 1 medial notch; second lobes bilobed, conspicuously smaller than medial lobes, medial and lateral lobules without notches, lateral lobule pointed, medial lobule rounded or pointed, medial lobule largest; third lobe area with 2 swellings near marginal macroducts, posterior swelling with series of small points; fourth lobe area with 3 swellings near marginal macroducts, anterior and posterior swellings smooth, central swelling with or without series of small points. Gland spines normally simple with single microduct, occasionally bifurcate with 2 microducts, gland-spine formula 2-2-2, with 13–23 (17) gland spines near each body margin anterior of lobe 4 area, present to metathorax; median lobes with 2 gland spines between them. Macroducts of 2 sizes, larger size on margin only, duct in first space 20–28 (24) μ long, 1 marginal duct in first space, 2 in second space, and 3 on segment 5, without smaller duct near lateral lobule of second lobe, small ducts in submarginal and submedial areas of segments 5 and 6, with 16–36 (28) macroducts on each side of segments 5 to 7; prepygidial ducts on marginal areas of meso- or metathorax to segment 4, on submedial areas of segments 2 to 4, on medial areas of segments 2 or 3 to 4, total of 173–220 (204) macroducts on each side of body, orifices of pygidial ducts occurring anterior of anal opening. Pygidial microducts on venter in submarginal areas of segments 5 to 6 or 7, with 6–12 (8) ducts on each side of body; prepygidial ducts in submedial areas of segments 1 to 3 or 4, on submarginal areas of head, mesothorax, metathorax, and segment 4; dorsal microducts absent except 1 or 2 on head. Perivulvar pores in 5 groups, with 16–22 (19) pores. Perispiracular pores with 3 loculi, anterior spiracles each with 2–3 (3) pores, posterior spiracles without pores. Anal opening located 8–11 (10) times length of anal opening from bases of median lobes, anal opening 13–15 (13) μ long. Seta lateral of medial lobe 0.3–0.6 (0.4) times length

of lobe. Eyes usually represented by weakly sclerotized area lateral of antennae, sometimes absent. Antennae each with 2 or 3 setae. Posterior portion of body sclerotized in old females, younger specimens with sclerotization restricted to segments 2, 3, and 4, newly matured females unsclerotized. Cicatrices absent. Body shape elongate. Space between lobes with 2 setae. Spurs absent. With 14–29 (20) microducts on head near antennae.

Leaf form: Same as bark form, except median lobes each with 2 distinctive paraphyses, and 1 lateral and 1 medial notch, separated by space about 0.1–0.2 (0.2) times width of medial lobe; second lobes with medial lobule more rounded than on leaf form. Dorsal ducts absent from medial areas of segments 2 to 4. Each side of dorsum with 15–19 (17) macroducts on segments 5 to 7, total of 75–113 (97) macroducts on each side of body. Perivulvar pores in 3 or 5 indistinct clusters, with 9–11 (10) pores on each side of pygidium. Anal opening 9–13 (10) μ long. With 8–13 (11) microducts on head near antennae.

Affinities Fig scale closely resembles *Lepidosaphes granati* Koroneos of the Mediterranean region, but the former lacks posterior spiracular pores while the latter has at least 1 posterior spiracular pore on each side.

Hosts Borchsenius (1966) reports this scale from 11 host genera in 7 families; Komosinska (1975) records it from 23 genera. We have examined material from the following: *Ficus*, *Juglans*, *Malus*, *Pyrus*, *Tilia*, and *Ulmus*. Kosztarab and Kozár (1988) record it on *Acer*, *Betula*, *Carpinus*, *Celtis*, *Corylus*, *Diospyros*, *Fraxinus*, *Pistacia*, *Syringa*, and *Zelkova*.

Distribution The species is restricted to CA, and we have examined specimens from that state. Fig scale occurs in Argentina, Chile, Europe, Iran, Iraq, Israel, North Africa, Pakistan, former Soviet Union, Syria, Turkey. It most often is intercepted in quarantine from Italy.

Biology Lupo (1943) discovered that this species has a leaf form and stem form that are quite different morphologically. Stafford and Barnes (1948) studied the fig scale on Adriatic figs in the vicinity of Fresno, CA. They found that about 5% of the overwintering twig-inhabiting females were ovipositing in February when local almonds were in bloom. About 90% of the females were ovipositing when apricot blooms were falling. At this time Adriatic fig tree buds were about three-quarters of an inch long. Overwintering females on bark produced an average of 30.2 eggs each. Hatching began on April 5 and was nearly complete on June 24. Male crawlers settled mostly on the upper surfaces of leaves while female crawlers settled mainly on the lower leaf surfaces. More than 90% of the crawlers settle on leaves in this generation. Summer generation females on leaves produced an average of 12.4 eggs each, which hatch from July to October. These crawlers tend to settle on twigs. Nearly all first-generation males emerged by mid-July. Second-generation males were present from late September to November. These males fertilized females on bark, which became the overwintering generation. An overwintering, a first summer, and partial second summer generation are indicated on figs in Fresno, California. Bodenheimer (1924) reports 2 generations per year in Palestine. Komosinska (1975) reports only 1 generation per year in Poland. Fertilized adult females constitute the overwintering stage. Eggs are laid in May. Crawlers appear from mid-May to late June. Second instars occur from late June to late July. Male prepupae and pupae occur from late July to early August. Adult males appear around August and mate with adult females.

Economic Importance Stafford and Barnes (1948) report that although heavy populations of fig scale occur on the leaves and twigs, little immediate damage results. Such populations,

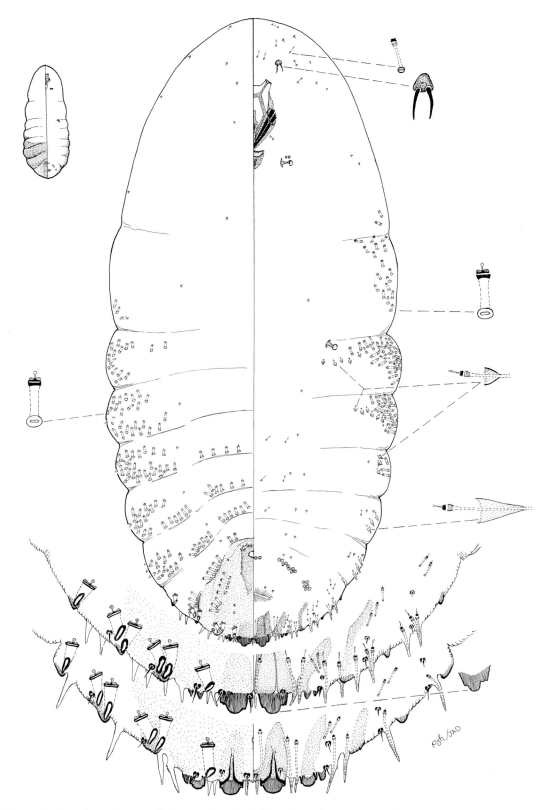

Figure 73. *Lepidosaphes conchiformis* (Gmelin), fig scale, Italy, intercepted at Detroit, MI, on *Ficus* sp., VI-21-1974.

however, will result in fig fruit infestation, and this will seriously affect the market value of figs because the scales cause dark green spots on fruits. In 1944, infested fruit grown for canning had to be used for jam stock, and hence were worth only about 70% of their value canned. Ahmad and Ghani (1971) note the fig scale seriously damaged *Citrus* and *Ricinus*. Komosinska (1969) found that heavy infestations caused bark 'ruptures' on *Tilia*. Chua and Wood (1990) suggest that this species is an important pest of figs. Beardsley and González (1975) consider this scale to be one of 43 serious armored scale pests. Miller and Davidson (1990) consider it to be a serious world pest.

Selected References Komosinska (1975); Stafford and Barnes (1948); Williams and Watson (1988).

Plate 69. *Lepidosaphes conchiformis* (Gmelin), Fig scale

A. Dark adult female cover on bark of fig (D. R. Miller).
B. Distance view of covers on bark of fig (D. R. Miller).
C. Covers with parasite emergence holes on fig (J. A. Davidson).
D. Light covers on green fig fruit (J. A. Davidson).
E. Covers on ripe fig fruit (J. A. Davidson).
F. Light adult female cover on leaf of fig (J. A. Davidson).
G. Male cover on leaf of fig (J. A. Davidson).

H. Heavy infestation on underside of fig leaf (J. A. Davidson).
I. Body of white adult female with eggs (J. A. Davidson).
J. Adult male on fig fruit (J. A. Davidson).
K. Yellow spots on top of fig leaf caused by fig scale feeding (J. A. Davidson).
L. Severe dieback damage to fig (J. A. Davidson).

252

Plate 70. *Lepidosaphes gloverii* (Packard), Glover scale

A. Adult female cover on citrus twig (J. A. Davidson).
B. Adult female covers on citrus leaf (J. A. Davidson).
C. Adult female cover on citrus leaf (J. A. Davidson).
D. Male cover on citrus (D. R. Miller).
E. Distance view of covers on citrus leaf (R. J. Gill).
F. Distance view of covers on twig of citrus (J. A. Davidson).
G. Egg and pink body of adult female inside turned cover (J. A. Davidson).

H. Two rows of tightly packed egg shells inside turned adult female cover (J. A. Davidson).
I. Body of second-instar male inside turned cover (J. A. Davidson).
J. Distance view of covers and feeding damage on citrus leaf (J. A. Davidson).

Lepidosaphes gloverii (Packard)

ESA Approved Common Name Glover scale (also called long scale, long mussel scale, citrus long scale).

Common Synonyms and Combinations *Aspidiotus gloverii* Packard, *Mytilaspis gloverii* (Packard), *Mytilaspis (Aspidiotus) gloverii* (Packard), *Mytiella sexspina* Hoke, *Mytilococcus gloverii* (Packard), *Insulaspis gloverii* (Packard).

Field Characters (Plate 70) Adult female cover light to dark brown, long, slender, sides nearly parallel, straight or curved; shed skins marginal, yellow. Male cover similar to that of female, shorter. Body of adult female pale pink to white; eggs white. On bark, leaves, and fruit. According to Ebeling (1959), Glover scale with ventral cover divided longitudinally in median area.

Slide-mounted Characters Adult female (Fig. 74) with 2 pairs of definite lobes, third and fourth lobes represented by series of points; with thin paraphyses on lateral and medial margins of median lobe and medial lobule of second lobe. Median lobes separated by space 0.5–1.0 (0.8) times width of median lobe, with thin paraphysis-like sclerotization attached to each margin of median lobes, lobe axes parallel, with 1–2 (2) lateral notches, 1–3 (2) medial notches; second lobes bilobed, medial lobule largest, with 1 lateral and 1 medial notch, lateral lobule with lateral margin with 1 notch, medial margin without notches; third lobes simple, represented by swelling surrounding marginal macroduct and series of conspicuous points; fourth lobes same as third lobe except with 2 macroduct swellings, 1 attached to series of points, and 1 anterior of seta marking lobe 5. Gland spines each with microduct, gland spines between medial lobes and sometimes in first space with weak basal fringing, gland-spine formula 2-2-2, with 16–23 (19) gland spines near each body margin anterior of lobe 4 area, present to metathorax; with 2 conspicuous gland spines between median lobes. Macroducts of 2 sizes, larger size on margin only, 1 duct in first space, 2 in second, and 3 on segment 5, 1 small duct near lateral lobule of second lobe; smaller size in submedial areas of segments 5 and 6, submarginal on segment 5, duct in first space 23–30 (27) μ long, with 21–27 (23) macroducts on each side of pygidium; prepygidial ducts on marginal areas of pro- or mesothorax to segment 5, these ducts dorsal from segments 1 to 4, ventral on prothorax to segment 2, submedial ducts on segments 2 or 3 to 4, total of 165–198 (178) macroducts on each side of body. Pygidial microducts on venter in submarginal areas of segments 5 and 6, with 3–7 (4) ducts on each side of body; prepygidial ducts in submedial areas of segments 1 to 4 and in cluster near posterior spiracles, on submarginal areas of segments 3 and 4; on dorsum pygidial and prepygidial ducts absent. Perivulvar pores in 4 or 5 indistinct groups, 10–15 (13) pores on each side of pygidium. Perispiracular pores with 3 loculi, anterior spiracles each with 2–4 (3) pores, posterior spiracles without pores. Anal opening located 12–17 (15) times length of anal opening from base of median lobes, anal opening 10–13 (12) μ long. Seta lateral of median lobe 0.5–1.0 (0.7) times length of lobe. Eyes represented by flat, weakly sclerotized area laterad of antenna. Antennae each with 2–4 (2) conspicuous setae. Prepygidium sclerotized in distinctive pattern (Fig. 74). Cicatrices absent. Body shape elongate. Space between median lobes with 2 setae. Lateral spurs on each of segments 2 to 4. Without microducts near antennae. With sclerotized pockets on venter of prothorax and mesothorax.

Affinities Glover scale is most similar to *Lepidosaphes yanagicola* Kuwana, fire bush scale, but can be separated by the distinctive prepygidial sclerotization pattern.

Hosts Borchsenius (1966) records this species from 9 genera in 8 families; Dekle (1977) records it from 19 genera with *Citrus*

the most frequently reported host in Florida. Based on material in the USNM Collection, *Citrus* is the preferred host. We have seen material from *Aglaia*, *Citropsis*, *Elaeagnus*, *Euonymus*, *Ficus*, *Maclura*, *Psidium*, and *Prunus*. González and Hernandez (1988) have recorded it on *Annona*, *Calophyllum*, *Fortunella*, and *Murraya* from Cuba; Williams and Watson (1988) reported it from *Alocasia*, *Cocos*, *Codiaeum*, and *Erythrina* from the South Pacific.

Distribution We have examined U.S. material from AL, CA, DC, FL, GA, LA, MS, NB, SC, TX. It has been reported from OK. Northern records indicate greenhouse infestations. Glover scale is widely distributed in the tropical and subtropical areas of the world and often is intercepted in quarantine on *Citrus*. A distribution map of this species was published by CAB International (1962a).

Biology This pest prefers humid climates and often is found associated with purple scale. According to Murakami (1970) Glover scale has 2 generations per year in Japan. Mated females overwinter and oviposit in March. The eggs are arranged in 2 rows beneath the scale cover. First-generation males appear in late July, and second-generation males occur in late October. Eguagie (1972) found that in Nigeria females produce an average of 46 eggs on orange, 38 on grapefruit, 31 on tangerine, and 27 on lime; there are 2 generations per year. The incubation period in June–July is 28 to 30 days at 25.1 °C mean temperature at 77.5% RH at 1600 hrs GMT. In November–December incubation takes 20 to 24 days at 26.5 °C mean temperature at 60.0% RH at 1600 hrs GMT. According to Simanton (1976) Glover scale is now the most widely distributed scale insect on Florida citrus. Before 1959, it occurred as light infestations with heavier purple scale infestations. It began to increase in 1959, following the 1958 appearance of *Aphytis lepidosaphes* Compere, which depressed purple scale populations. In about 1966, both species more or less stabilized with Glover scale populations highest. Under insectary conditions, Claps (1987) recorded 6 generations per year when reared on *Citrus* in Argentina. According to Bruwer and Schoeman (1990) the four factors influencing the population size of this scale in South Africa are: predation, primarily by *Chilocorus nigritus*; ectoparasitism by *Aphytis lepidosaphes*; endoparasitism by *Aspidiotiphagus citrinus*; and an unknown mortality factor.

Economic Importance According to Dekle (1977), Glover scale is a serious pest of Florida citrus, where it occurs more often on twigs and small branches than on leaves and fruit. Ebeling (1959) notes that this species was of little importance in California, generally being confined to the San Juan Capistrano Valley in southern Orange County. In South Africa it is not a problem when orchards are sprayed for California red scale, but it can be found in integrated control orchards around Nelspruit (Bedford 1978). Bruwer (1998) also discusses the economic importance of this pest in South Africa. Rose (1990b), in his summary of armored scale pests of citrus, reports Glover scale as a pest in Spain, Japan, Texas, Mexico, the Caribbean, South America, the Mediterranean, former Soviet Union, and South Africa. It has been reported as a citrus pest by Konar and Ghosh (1990) in India and also as a pest of mango in India (Chua and Wood 1990). Beardsley and González (1975) consider this scale to be one of 43 serious armored scale pests. Miller and Davidson (1990) consider it to be a serious world pest.

Selected References Bruwer (1998); Claps (1991); Eguagie (1972); Ferris (1938a).

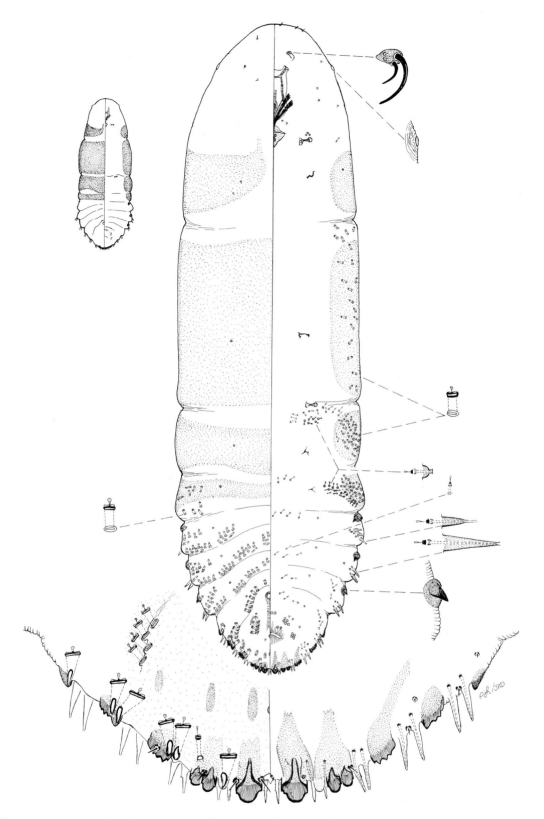

Figure 74. *Lepidosaphes gloverii* (Packard), Glover scale, Yunguilla, Ecuador, on orange leaves, XII-1-1954.

Lepidosaphes pallida (Maskell)

Suggested Common Name Maskell scale.

Common Synonyms and Combinations *Mytilaspis pallida* Maskell, *Mytilaspis pallida* var. *maskelli* Cockerell, *Lepidosaphes pallida maskelli* (Cockerell), *Insulaspis maskelli* (Cockerell), *Lepidosaphes maskelli* (Cockerell), *Insulaspis pallida* (Maskell).

Field Characters (Plate 71) Adult female cover oyster-shell shaped, straight, moderately convex, light brown, occasionally white posteriorly; shed skins marginal, yellow. Male cover shorter, narrower than female cover, same color and texture; shed skin marginal, yellow. Body of adult female white; eggs and crawlers white. On foliage.

Slide-mounted Characters Adult female (Fig. 75) with 2 definite pairs of lobes, third and fourth pairs of lobes represented by series of small points; with thin paraphyses on lateral and medial margins of median lobe, normally on medial lobule of second lobes, occasionally with one or two paraphyses on lateral lobule. Median lobes separated by space 0.7–1.0 (0.8) times width of median lobe, with conspicuous, curved paraphysis-like sclerotizations attached to each margin of median lobes, without yoke, lobe axes parallel, with 1 lateral and 1 medial notch; second lobes bilobed, medial lobule largest, with 0–1 (1) lateral notches, 0–1 (0) medial notches, lateral lobule with lateral margin with 0–1 (0) notches, medial margin without notches; third lobes single, represented by swelling surrounding marginal macroduct and series of small teeth; fourth lobes represented by 2 macroduct swellings, 1 swelling often with a series of teeth, anterior swelling in front of seta marking lobe 5. Gland spines each with microduct, gland spines between median lobes and sometimes in first space with basal fringing, gland-spine formula 2-2-2, with 17–36 (26) gland spines near each body margin anterior of lobe 4 area, present to metathorax or segment 1; median lobes with 2 conspicuous gland spines between lobes. Macroducts of 2 sizes, larger size on margin only, 1 duct in first space, 2 in second, and 3 on segment 5, 1 small duct anterior of second lobe, small ducts in submedial areas of segments 5 and 6, submarginal on segment 5, duct in first space 20–30 (26) μ long, with 11–17 (13) macroducts on each side of segments 5 to 7; prepygidial ducts on marginal areas of meso- or metathorax to segment 4, these ducts dorsal and ventral on meso- and metathorax and segment 1, dorsal only on segments 2 to 4, submedial ducts on segments 2 to 4, total of 65–97 (77) macroducts on each side of body. Pygidial microducts on venter in submarginal areas of segments 5 to 6 or 7, with 3–5 (4) ducts on each side of body; prepygidial ducts in submedial areas of segments 1 to 4, submarginal on segments 3 and sometimes 4; dorsal pygidial and prepygidial microducts absent. Perivulvar pores in 4 or 5 indistinct groups, 10–14 (12) pores on each side of pygidium. Perispiracular pores with 3 loculi, anterior spiracular pores each with 3–6 (4) pores, posterior spiracles without pores. Anal opening located 8–12 (10) times length of anal opening from base of medial lobes, anal opening 9–15 (11) μ long. Seta lateral of median lobes 0.5–1.0 (0.8) times length of lobe. Eyes difficult to see, usually represented by small membranous dome near body margin at level of antenna. Antennae each with 1 or 2 large setae, single seta occasionally with bifurcate base. Body unsclerotized. Cicatrices absent. Body shape elongate. Space between median lobes with 2 setae. Lateral spurs absent, though small swellings sometimes occur on lateral margins of segments 3 and 4. With 1–4 (3) microducts near antennae.

Affinities Maskell scale can be separated from other economic species of *Lepidosaphes* by having only a single lateral and medial notch on the median lobes, by having a thin dorsal macroduct (about the same width as a microduct) anterior of lobe 2, by lacking lateral spurs on the abdominal segments, and by occurring on coniferous hosts.

Hosts Borchsenius (1966) records this species from 9 genera in 2 families; Dekle (1977) reports it on 12 genera with *Juniperus* the most frequently reported host in Florida. We have examined specimens from *Cephalotaxus, Chamaecyparis, Cryptomeria, Cupressus, Juniperus, Picea, Podocarpus, Sciadopitys, Sequoia, Taxus,* and *Thuja.* It has been reported on *Araucaria, Citrus* (?), *Pinus, Taxodium,* and *Torreya* (Miller and Gimpel 2002). We have seen this scale only on the needles of its hosts.

Distribution We have examined U.S. material from AL, AR, CA, DC, DE, FL, HI, LA, MD, MS, NC, NJ, NY, PA, VA. This scale has been intercepted in quarantine inspections from Bermuda, China, Korea, Japan, Taiwan, and the former Soviet Union. It is reported in the literature from central and western Europe and from Turkey and Lebanon.

Biology The life history of this species is poorly known. According to Waterston (1949), in Bermuda the winter is spent as adult females. Eggs are laid beginning in late March. Crawlers are present between July and November and probably occur earlier in the year. Males are reported. In Europe, Schmutterer (1951) found that it had 1 generation per year and overwintered as fertilized females. Crawlers begin hatching in June. Adult males and females appeared in mid-August. In Maryland adult males are present in late September to early October along with all other stages, that is, eggs, crawlers, second instars, and adult females.

Economic Importance Dekle (1977) reports that Maskell scale is occasionally a serious pest of juniper in Florida. In concert with minute cypress scale, Maskell scale nearly eliminated the Bermuda cedar on Bermuda. The foliage of heavily infested trees turns pale yellow, and entire trees may be killed if remedial action is not taken. Hodgson and Hilburn (1991) indicate that Maskell scale is uncommon on Bermuda, being replaced by *Carulaspis minima.* The authors have seen dieback due to this pest on *Cryptomeria* in Maryland. Miller and Davidson (1990) consider this species to be a serious pest in a small area of the world.

Selected References McKenzie (1956); Schmutterer (1951); Stimmel (1999); Waterston (1949).

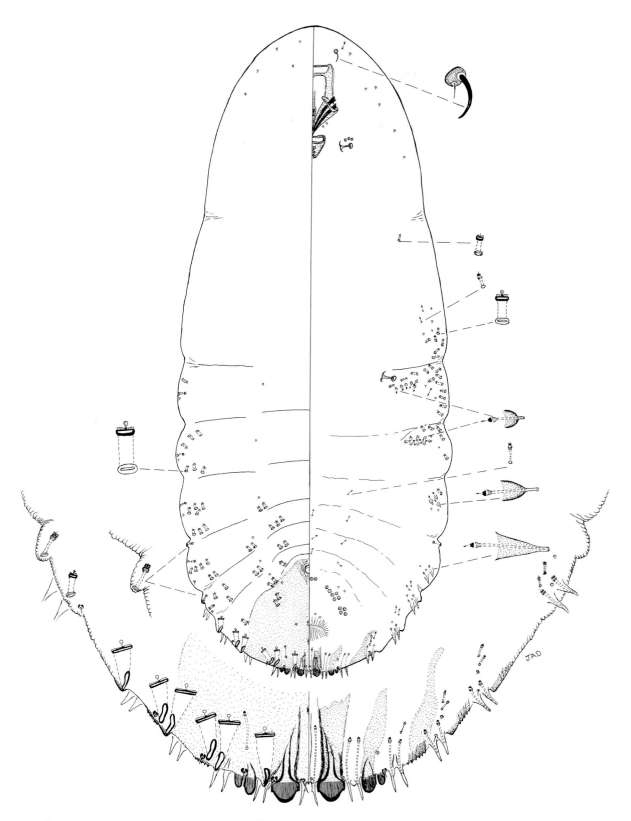

Figure 75. *Lepidosaphes pallida* (Maskell), Maskell scale, National Arboretum, Washington, DC, on *Cryptomeria* sp., VIII-17-1978.

Plate 71. *Lepidosaphes pallida* (Maskell), Maskell scale

A. Narrow form of adult female cover and male cover on cryptomeria needle (J. A. Davidson).
B. Narrow and wide forms of adult female covers on Leyland cypress (J. A. Davidson).
C. Young adult female beginning to lay eggs on cryptomeria (J. A. Davidson).
D. Male cover (right) with immature female cover (J. A. Davidson).

E. Heavy infestation on cryptomeria (J. A. Davidson).
F. Infestation on taxus (J. A. Davidson).
G. Distance view on cryptomeria (J. A. Davidson).
H. Body of white adult female (J. A. Davidson).
I. Body of adult female with eggs (J. A. Davidson).
J. Body of white crawler (J. A. Davidson).
K. Dieback on heavily infested cryptomeria (J. A. Davidson).

258

Plate 72. *Lepidosaphes pini* (Maskell), Pine oystershell scale

A. Rubbed adult female covers on pine (J. A. Davidson).
B. Irregularly crowded female covers on pine (J. A. Davidson).
C. Covers at base of needles on Thunbergia pine (J. A. Davidson).
D. Removal of needle shows depth of penetration of scale in needle sheath of Thunbergia pine (J. A. Davidson).
E. Distance view of heavy infestation causing yellowing of pine needles (J. A. Davidson).

F. Turned female cover shows extensive ventral cover on pine (J. A. Davidson).
G. Body of white adult female on pine (J. A. Davidson).
H. Body of adult female and eggs (J. A. Davidson).

Lepidosaphes pini (Maskell)

Suggested Common Name Pine oystershell scale (also called Oriental pine scale).

Common Synonyms and Combinations *Poliaspis pini* Maskell, *Chionaspis (Poliaspis) pini* (Maskell), *Mytilococcus pinorum* Lindinger, *Insulaspis pini* (Maskell).

Field Characters (Plate 72) Adult female cover oyster-shell shaped, straight or slightly curved, moderately convex, light brown; shed skins marginal, light yellow. Male cover shorter, narrower than female cover, same color and texture; shed skin marginal, yellow. Body of adult female white; eggs and crawlers white. On needles of host, often near needle sheath. Cryptic at base of pine needles, between needles, or in needle follicles (Stimmel 1994).

Slide-mounted Characters Adult female (Fig. 76) with 2 definite pairs of lobes, third lobes represented by series of small points, fourth lobes normally represented by sclerotized area only, occasionally with series of points; with paraphyses on lateral and medial margins of median lobe, small paraphyses normally on medial and lateral lobule of second lobes. Median lobes separated by space 1.0–1.7 (1.3) times width of median lobe, with conspicuous, curved paraphysis-like sclerotizations attached to each margin of median lobes, without yoke, lateral margin of lobe diverging slightly, medial margin parallel, with 1–2 (1) lateral notches and 1–3 (2) medial notches; second lobes bilobed, medial lobule largest, with 1–2 (1) lateral notches, 1 medial notch, lateral lobule with lateral margin with 0–1 (1) notch, medial margin simple; third lobes single, represented by swelling surrounding marginal macroduct and series of small teeth; fourth lobes represented by 2 macroduct swellings, 1 swelling sometimes attached to series of teeth, anterior swelling in front of seta marking lobe 5. Gland spines each with microduct, gland spines between medial lobes and sometimes in first space with basal fringing, gland-spine formula 2-2-2, anterior gland spines often with more than 1 associated microduct, with 17–36 (27) gland spines on each body margin anterior of lobe 4 area, present to metathorax; median lobes with 2 conspicuous gland spines between lobes. Macroducts of 2 sizes, larger size on margin only, 1 duct in first space, 2 in second, and 3 on segment 5, without small duct anterior of second lobe, smaller size in submedial areas of segments 5 and 6, submarginal on segment 5, marginal macroduct in first space 31–36 (23) μ long, 16–24 (19) macroducts on each side of segments 5 to 7; prepygidial ducts on marginal areas of meso- or metathorax to segment 4, these ducts dorsal and ventral on meso- and metathorax and segment 1, dorsal only on segments 2 to 4, submedial ducts on segments 2 to 4, ventral only on prothorax, total of 108–148 (130) macroducts on each side of body. Pygidial microducts on venter in submarginal areas of segments 5 to 7, with 3–9 (6) ducts on each side of body; prepygidial ducts in submedial areas of metathorax or segment 1 to segment 4 or 5, sub-marginal on head, prothorax, and metathorax, and on segment 3 and sometimes 4; dorsal pygidial and prepygidial microducts absent. Perivulvar pores in 7 or 8 distinct groups, 16–27 (23) pores on each side of pygidium. Perispiracular pores usually with 3 loculi, anterior spiracular pores each with 3–5 (4) pores, posterior spiracles without pores. Anal opening located 10–14 (12) times length of anal opening from base of medial lobes, anal opening 10–14 (12) μ long. Seta lateral of median lobes 0.6–1.0 (0.8) times length of lobe. Eyes represented by small sclerotized dome near body margin at level of anterior edge of mouthparts. Antennae each with 2 or rarely 3 large setae, single seta occasionally with bifurcate base. Body unsclerotized. Cicatrices absent. Body shape elongate. Space between median lobes with 2 setae. Lateral spurs absent, though small swellings sometimes occur on lateral margins of segments 3 and 4. With 0–4 (1) microducts near antennae.

Affinities Pine oystershell scale can be separated from other economic species of *Lepidosaphes* by having 7 or 8 groups of perivulvar pores; other species have 5 or less.

Hosts Borchsenius (1966) records this species from 4 genera in 2 families; Murakami (1970) found it on 2 host genera. We have examined specimens from *Pinus* only. It has been reported on: *Abies*, *Cycas*, *Podocarpus*, *Taxus*, and *Torreya*.

Distribution We have examined U.S. material from HI, MD, NJ, PA. It most commonly is taken in quarantine on pine from Japan but is reported from the Bonin Islands, China, Japan, Korea, and Taiwan.

Biology The life history of this species is poorly known. Smith-Fiola (1997, personal communication) indicated that in New Jersey the adult female is the overwintering stage; eggs are present in March and August; and crawlers occur in June and September. According to Murakami (1970), in Japan it has 2 generations per year and overwinters as fertilized adult females. In April, about 30 eggs are laid, and adult males are reported in August and mid-October. Xu (1981) studied this species in China and reports that the species has 2 generations per year and overwinters as adult females. Stimmel (1994) hypothesized that it has only 1 generation per year in Pennsylvania, with eggs serving as the overwintering stage.

Economic Importance The pine oystershell scale has been reported as a pest of pines by Schmutterer et al. (1957), Tang (1984), Zahradník (1990b), Chai et al. (1999), and Xu (1981). In recent years it has caused damage in the coastal areas of New Jersey and has required the use of chemical controls (Smith-Fiola 1994) (misidentified as *Lepidosaphes pallida*). This pest causes chlorosis of the base of the needles (Stimmel 1994). Miller and Davidson (1990) consider this species to be an occasional pest.

Selected References Kosztarab (1996); Murakami (1970); Xu (1981).

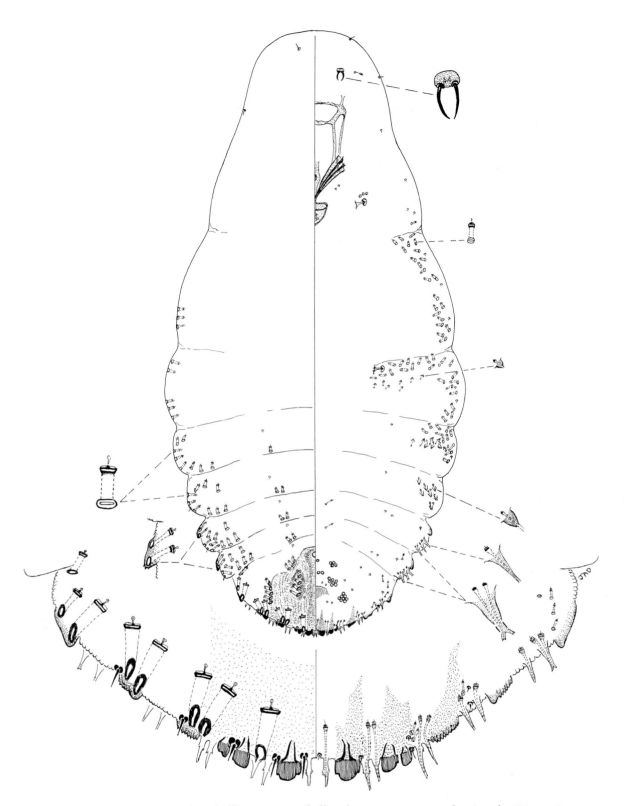

Figure 76. *Lepidosaphes pini* (Maskell), pine oystershell scale, Japan, intercepted at Seattle, WA, on *Pinus* sp., XII-30-1948.

Lepidosaphes pinnaeformis (Bouché)

Suggested Common Name Cymbidium scale (also called machilus oystershell).

Common Synonyms and Combinations *Aspidiotus pinnaeformis* Bouché, *Mytilaspis pinnaeformis* (Bouché), *Mytilaspis machili* Maskell, *Lepidosaphes machili* (Maskell), *Lepidosaphes tuberculata* Malenotti, *Lepidosaphes cymbidicola* Kuwana, *Eucornuaspis machili* (Maskell), *Scrupulaspis machili* (Maskell), *Lepidosaphes ezokihadae* Kuwana, *Lepidosaphes cinnamomi* Takahashi, *Lepidosaphes piniformis* (Bouché), *Mytilococcus piniformis* (Bouché), *Mytilococcus tuberculatus* (Malenotti). This species has been confused with *Lepidosaphes beckii*, and much of the literature pertaining to *L. pinnaeformis* on citrus refers to *L. beckii* and not *L. pinnaeformis*. There also is controversy about the distinctness of *L. pinnaeformis* and *L. machili*. We have accepted the synonymy presented by Borchsenius (1966) as there seem to be no obvious differences between the two. Both are elongate scales that are most common on orchids, particularly cymbidium orchids.

Field Characters (Plate 73) Adult female cover oyster-shell shaped, usually curved, moderately convex, brown, often with lighter periphery; shed skins marginal, orange or tan. Male cover shorter, narrower than female cover, same color and texture; shed skin marginal, orange or tan. Body of adult female white to light violet; eggs and crawlers probably white. On foliage.

Slide-mounted Characters Adult female (Fig. 77) with 2 definite pairs of lobes, third, fourth, and occasionally fifth pairs of lobes represented by series of small points; with thin paraphyses on lateral and medial margins of median lobe, normally on medial and lateral lobule of second lobes. Median lobes separated by space 1.0–1.8 (1.3) times width of median lobe, with conspicuous, curved paraphysis-like sclerotizations attached to each margin of median lobes, without yoke, lobe axes usually parallel sometimes slightly diverging, with 1–2 (1) lateral notches and 1–4 (2) medial notches; second lobes bilobed, medial lobule largest, lobules without notches; third lobes single, represented by series of small teeth, with 3–11 (5) notches; fourth lobes series of teeth with 6–11 (8) notches; fifth lobes present on 3 of 10 specimens examined, when present with 2–5 (3) notches. Gland spines each with microduct, without basal fringing, gland-spine formula 2-2-2, with 34–59 (26) gland spines near each body margin anterior of lobe 4 area, present to mesothorax; median lobes with 2 conspicuous gland spines between lobes. Macroducts of 2 sizes, larger size on margin only, 1 duct in first space, 2 in second, and 2 on segment 5; smaller size in marginal and submedial area of segments 5 and 6, duct in first space 22–30 (25) μ long, with 37–51 (46) macroducts on each side of segments 5 to 7; prepygidial ducts on marginal areas of pro- or mesothorax to segment 4, these ducts dorsal and ventral on pro- or mesothorax to segment 1 or 2, dorsal only on segments 2 or 3 to 4, submedial ducts on segments 2 or 3 to 4, medial ducts sometimes present on segments 3 or 4; with 179–297 (235) macroducts on each side of body. Pygidial microducts on venter in submarginal and sometimes submedial areas of segment 5 and usually 6, with 1–7 (4) ducts on each side of body; prepygidial ducts in submedial areas near antennae, near spiracles, and on segments 1 to 4, submarginal on head, thorax, and segments 2 or 3 to 4; dorsal pygidial microducts absent; prepygidial microducts in row on meso- or metathorax to segment 1 or rarely 2. Perivulvar pores in 5 distinct groups, 33–45 (40) pores on each side of pygidium. Perispiracular pores with 3 rarely 5 loculi, anterior spiracular pores each with 8–17 (11) pores, posterior spiracles without pores. Anal opening located 7–12 (9) times length of anal opening from base of medial lobes, anal opening 12–18 (16) μ long. Seta lateral of median lobes 1.5–1.9 (1.6) times length of lobe. Eyes in form of sclerotized anteriorly projecting spur near body margin between level of anterior part of clypeolabral shield and antenna. Antennae each with 3 or 4 large setae. Body unsclerotized. Cicatrices absent. Body shape elongate. Space between median lobes with 2 setae. Lateral spurs or sclerotized lobes conspicuous on segments 2 to 4, sometimes with small teeth. With 1–4 (3) microducts near antennae.

Affinities Cymbidium scale can be separated from all other species of armored scales in the United States by the uniquely shaped eyes that are spurlike. The species also has a unique combination of characters including the possession of numerous perivulvar pores, no cicatrices, macroducts in submedial areas of segments 2 or 3 to 6, large-sized macroducts restricted to margins of segments 5 to 7, dorsomedial microducts on metathorax and segment 1, and lateral spurs or sclerotized projections on segments 2 to 4.

Hosts Borchsenius (1966) records this species from 11 genera in 6 families. We have examined specimens from *Cinnamomum*, *Cymbidium*, *Dendrobium*, *Litsea*, *Machilus*, *Neolitsea*, *Ophiopogon*, and *Rhapis*. It has been reported on *Banksia*, *Cattleya*, *Cercidiphyllum*, *Citrus* (?), *Croton*, *Cycas*, *Daphniphyllum*, *Elaeagnus*, *Ficus*, *Illicium*, *Lindera*, *Magnolia*, *Michelia*, *Phellodendron*, *Phoebe*, *Pomaderris*, *Prunus*, *Pyrus*, *Quercus*, *Stauntonia*, *Taxus*, and *Tetradymia*. We suspect that some of these host records are erroneous because of confusion about the identity of the cymbidium scale. This species most frequently is taken at U.S. ports-of-entry on cymbidium orchids.

Distribution We have examined U.S. material from CA, FL, HI, MA, MD, NJ, NY. It is reported from DC, DE, PA (Nakahara 1982). In the 1940s and 1950s this scale was intercepted most frequently from England but recently is intercepted more commonly from the Oriental Region and Australia.

Biology The life history of this species is poorly known. According to Schmutterer (1959) it has about 4 generations per year in greenhouses. Females lay 31 to 132 eggs, which hatch in about 10 days. Males are present in all colonies.

Economic Importance The cymbidium scale was considered a serious pest of orchids grown in nurseries in California in the 1930s to 1950s (McKenzie 1956) but now is not of major importance because it is easily controlled (Gill 1997). It can build to heavy populations and cause significant damage when left unchecked. It is a pest in nearly any area of the world where cymbidium orchids are grown. It has been considered to be a pest in the following: Britain (Miles and Miles 1935), California (Steinweden 1948), Germany (Schmutterer 1959), and Russia (Saakyan-Baranova 1954), to name a few. Miller and Davidson (1990) consider this species to be an occasional pest.

Selected References McKenzie (1956); Schmutterer (1959); Takagi (1970).

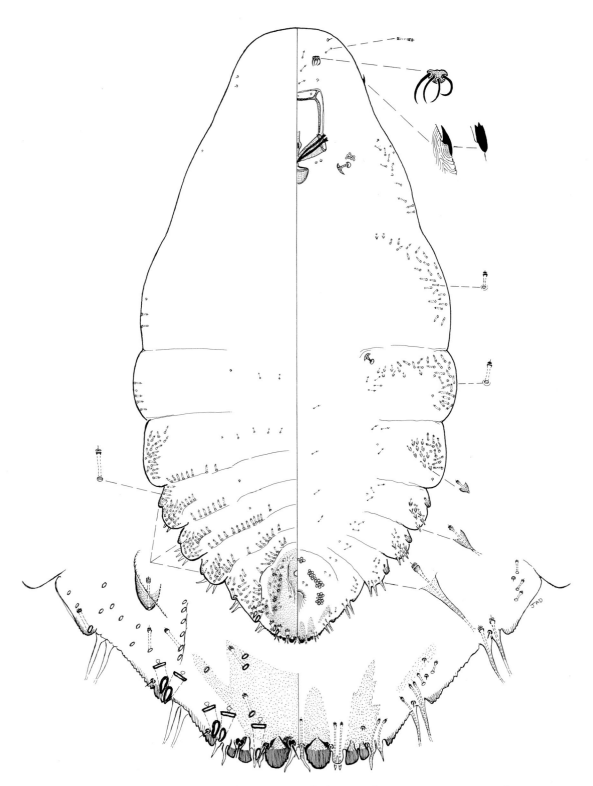

Figure 77. *Lepidosaphes pinnaeformis* (Bouché), cymbidium scale, Formosa, intercepted at HI, on *Cymbidium* sp., XI-21-1952.

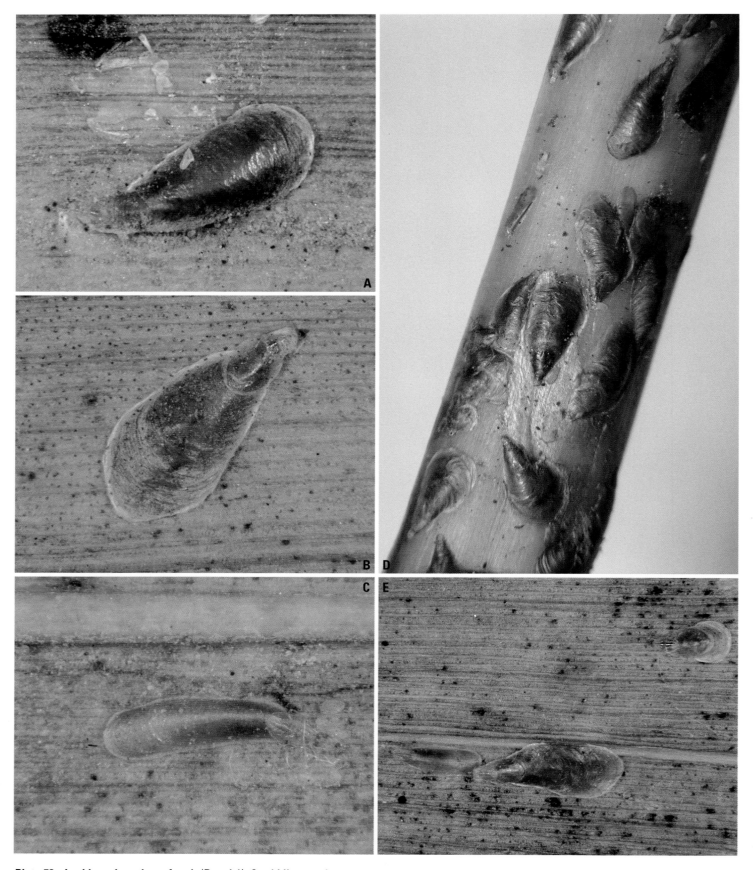

Plate 73. *Lepidosaphes pinnaeformis* (Bouché), Cymbidium scale

A. Rubbed adult female cover on cymbidium orchid (J. A. Davidson).

B. Unrubbed adult female cover on cymbidium orchid (J. A. Davidson).

C. Adult male cover on cymbidium orchid (J. A. Davidson).

D. Male and female covers with white crawler on orchid stem (R. J. Gill).

E. Distance view of adult male (left) and female center, with young adult female cover (upper right) on cymbidium orchid (J. A. Davidson).

Plate 74. *Lepidosaphes ulmi* (Linnaeus), Oystershell scale

A. Rubbed adult female cover on striped maple (J. A. Davidson).
B. Rubbed and unrubbed adult female covers on striped maple (J. A. Davidson).
C. Large female and smaller male covers on pachysandra (J. A. Davidson).
D. Adult female covers with eggs on sugar maple (J. A. Davidson).
E. Adult female covers on dogwood (J. A. Davidson).

F. Heavy infestation of gray cover form on ash (left) and brown cover form on maple (right) (J. A. Davidson).
G. Body of adult female on ash (J. A. Davidson).
H. Cover raised to show white eggs on pachysandra (J. A. Davidson).
I. Eggs dusted with white powdery wax produced by perivulvar pores (J. A. Davidson).

265

Lepidosaphes ulmi (Linnaeus)

ESA Approved Common Name Oystershell scale (also called apple mussel scale, apple comma scale, appletree bark louse, and mussel scale).

Common Synonyms and Combinations *Coccus ulmi* Linnaeus, *Coccus amygdali* Schrank, *Coccus berberidis* Schrank, *Diaspis linearis* Costa, *Aspidiotus falciformis* Baerensprung, *Aspidiotus pomorum* Bouché, *Aspidiotus juglandis* Fitch, *Mytilaspis juglandis* (Fitch), *Mytilaspis pomorum* (Bouché), *Mytilaspis pomicorticis* Riley, *Mytilaspis ulmicorticis* Riley, *Mytilaspis vitis* Goethe, *Mytilaspis ulicis* Douglas, *Mytilaspis ceratoniae* Gennadius, *Mytilaspis ulmi* (Linnaeus), *Mytilaspis pomorum* var. *candidus* Newstead, *Mytilaspis pomorum* var. *ulicis* Douglas, *Lepidosaphes pomorum* (Bouché), *Lepidosaphes ceratoniae* (Gennadius), *Lepidosaphes juglandis* (Fitch), *Aspidiotus pyrusmalus* Kennicott, *Lepidosaphes ulmi vitis* (Goethe), *Lepidosaphes ulmi candida* (Newstead), *Mytilaspis (Lepidosaphes) pomorum* (Bouché), *Lepidosaphes vulva* Nel, *Lepidosaphes ulmi-cotini* Koroneos, *Lepidosaphes ulmi-rosae* Koroneos, *Lepidosaphes oleae* Koroneos, *Lepidosaphes ulmi bisexualis* (Thiem), *Lepidosaphes mesasiatica* Borchsenius, *Lepidosaphes populi* Savescu, *Lepidosaphes tiliae* Savescu. The list of synonyms for this species is extensive; it is difficult to know if many of them are correct. We have synthesized the listings given by Borchsenius (1966) and Danzig and Pellizzari (1998).

Field Characters (Plate 74) Adult female cover oyster-shell shaped, straight unless crowded, convex, thick, dark to light brown or gray; shed skins marginal, brown or yellow. Certain populations with brown covers, others with gray covers. Male cover, when present, shorter, narrower, similar in texture and color; shed skin yellow to brown. Body of adult female white with brown pygidium; eggs and crawlers white. On bark.

Slide-mounted Characters Adult female (Fig. 78) with 2 definite pairs of lobes, third and fourth pairs of lobes represented by low, sclerotized series of points; with converging paraphyses on lateral and medial margins of median lobe and sometimes with small paraphysis on each margin of medial lobule of second lobes. Median lobes separated by space 0.4–0.6 (0.5) times width of median lobe, without basal sclerosis or yoke, lobe axes parallel or slightly divergent, with 1–3 (1) lateral notches, 1–2 (1) medial notches; second lobes bilobed, medial lobule largest, medial lobule often with 1 lateral notch, medial notch absent, lateral lobule simple; third and fourth lobe areas with series of small projections and swelling near marginal macroducts. Gland spines each with microduct, without basal fringing, gland-spine formula usually 2-2-2, rarely 2-2-1, with 20–42 (29) gland spines near each body margin anterior of lobe 4 area, present to metathorax or segment 1; median lobes with 2 gland spines between them varying from longer than lobes to slightly shorter. Macroducts of 2 sizes, larger size on margin only, 1 in first space, 2 in second, and 3 on segment 5, cluster of small ducts often present anterior of median lobe and second lobe, smaller size in segmental rows from marginal to medial areas on segments 5 and 6 in addition to submedial clusters on segments 7 and 8, marginal duct in first space 25–38 (29) μ long, 52–85 (67) macroducts on each side of segments 5 to 7; prepygidial ducts on marginal areas of dorsum of prothorax, mesothorax, or metathorax to segment 4, on venter from prothorax to segment 1 or 2, on dorsosubmedial areas of segments 1 to 4, orifices of pygidial ducts occurring anterior of anal opening, total of 234–337 (285) macroducts on each side of body. Pygidial microducts on venter in submarginal areas of segments 5 to 6, with 3–14 (8) on each side of body; prepygidial ducts on submedial areas of segments 1, 2, or 3 to 4, usually on submarginal segment areas of segments 2 or 3 to 4, rarely from prothorax to segment 5; dorsal pygidial and prepygidial microducts absent. Perivulvar pores in 5 groups, with 27–67 (45) pores on each side of pygidium. Perispiracular pores each with 3 loculi, anterior spiracles each with 3–8 (6) pores, posterior spiracles without pores. Anal opening located 11–18 (14) times length of anal opening from base of medial lobes, anal opening 13–28 (17) μ long. Seta lateral of median lobes 0.6–0.8 (0.7) times length of lobe. Eyes absent or represented by flat sclerotized area on body margin posterior of level of antennae. Antennae each with 2—3 (2) conspicuous setae. Prepygidial area membranous. Cicatrices variable, pigmented or membranous, on segments 1 to 6 when maximum number present, with 0–6 (3) on each side of body. Body elongate. Space between medial lobes with 2 setae. Lateral spurs variable in size, usually present on segments 2, 3, or 4. With 6–15 (10) microducts between antennae.

Affinities Oystershell scale is most similar to *Lepidosaphes beckii*, purple scale, but differs by having at least 2 pairs of lateral spurs, by usually lacking the cicatrix pattern of purple scale, by having differently shaped median lobes, by having 52–85 (67) macroducts on each side of pygidium, by having more than 1 macroduct on submargin anterior of lateral lobule of second lobe, and by having 27–67 (45) perivulvar pores on each side of pygidium. Purple scale has 0–1 pairs of lateral spurs, usually has cicatrices on segments 1, 2, and 4, has 36–62 (52) macroducts on each side of pygidium, has 1 macroduct on submargin anterior of second lobe, and has 17–30 (23) perivulvar pores on each side of pygidium.

Hosts Polyphagous. Borchsenius (1966) records it from 85 host genera in 33 families; Dekle (1977) records it from *Ficus* in Florida and points out that it has not been found in the state since 1921. Kosztarab and Kozár (1988) record 102 host genera in Central Europe. We have examined specimens from the following: *Abies, Acer, Aesculus, Amelanchier, Amorpha, Arabis, Artemisia, Berberis, Beta, Betula, Buxus, Calluna, Castanea, Catalpa, Ceanothus, Celastrus, Celtis, Centaurium, Ceratonia, Cercis, Chamaecyparis, Chionanthus, Citrus, Clematis, Cornus, Cotinus, Cotoneaster, Crataegus, Cytisus, Diospyros, Elaeagnus, Fagus, Ficus, Fitchia, Forsythia, Fraxinus, Geum, Ginkgo, Gleditsia, Holboellia, Hydrangea, Hypericum, Juglans, Koelreuteria, Ligustrum, Lonicera, Loranthus, Lupinus, Maclura, Magnolia, Malus, Mespilus, Michelia, Olea, Oxydendrum, Pachysandra, Paeonia, Panax, Persea, Phellodendron, Phlox, Picea, Pistacia, Planera, Populus, Prunus, Pyrus, Quercus, Rhododendron, Ribes, Robinia, Rosa, Rubus, Salix, Sarcococca, Sarothamnus, Sassafras, Sorbaria, Sorbus, Spartium, Spiraea, Spondias, Staphylea, Syringa, Taxus, Tilia, Tsuga, Ulmus, Vaccinium, Viburnum, Vitis,* and *Yucca.* Based on material in the USNM Collection, the Salicaceae, Rosaceae, and Oleaceae contain the most commonly infested host genera. Among the most commonly infested hosts in the United States are lilac, beech, birch, ash, maple, willow, elm, boxwood, apple, pear, and *Prunus* spp. This species is a shade tree and orchard pest and probably will attack any woody, deciduous plant as well as some conifers.

Distribution The oystershell scale is a northern species that occurs throughout most of the United States. We have seen material from all states except AK, AL, AZ, LA, MS, NM, NV, SC, TX. It has been reported in the literature from TN. This pest may be encountered in quarantine inspections from most areas of the world with temperate climates. A distribution map of this species was published by CAB International (1958).

Biology In North America, Griswold (1925), Samarasinghe and LeRoux (1966), and Garrett (1972) made detailed life-history studies of oystershell scale. Griswold (1925) noted an apple and

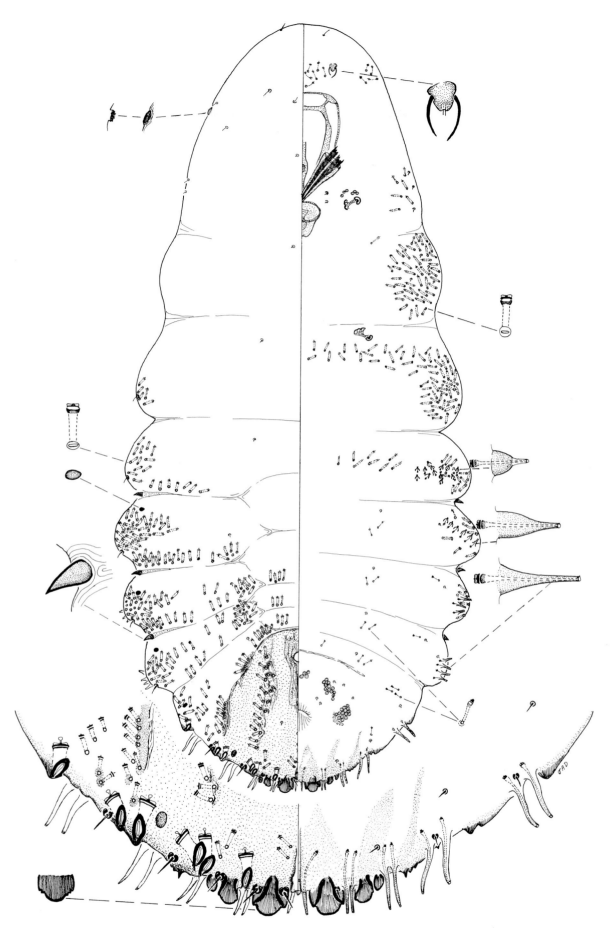

Figure 78. *Lepidosaphes ulmi* (Linnaeus), oystershell scale, Plummers Island, MD, on *Acer* sp., IX-18-1970.

a lilac form in New York. The former has longer, more narrow scale covers that are uniformly brown, has fewer pores, and develops about 2 weeks earlier than the lilac form. The latter has shorter, wider scale covers that are banded when new and pale gray when old, has more perivulvar pores, and develops about 2 weeks later than the apple form. She found the lilac or banded form on a wider variety of host plants, particularly shade trees, than the apple or brown form. She was able to induce apple forms to develop on lilac but the reverse host transfers failed. In Quebec, Canada, Samarasinghe and LeRoux (1966) studied a brown form on apple that was unisexual and univoltine. Females laid 20–110 eggs each from late August to early September. These hatched from early May to mid-June. Second instars occurred from mid-June to late July. Adults appeared from early August to late September. They found *Hemisarcoptes malus* and *Aphelinus mytilaspidus* to be the 2 most important mortality factors in the egg and adult stages of oystershell scale. In Maryland, Garrett (1972) worked with a yellow-banded, unisexual, univoltine form on poplar and willow that produced 20–101 eggs per female and hatched in late May. He also studied a gray, bisexual, bivoltine form on maple, lilac, and boxwood. Females of the latter produced 16–94 eggs, which hatched from late April to mid-May. In addition, the eggs and second instars were larger than in the univoltine form. There was no significant difference in perivulvar pore numbers between the uni- and bivoltine forms. Apple was an unacceptable host to both forms studied by Garrett. The following egg-hatching dates have been observed in North America: June and July, British Columbia (Madsen and Arrand 1971); May, Oregon (Schuh and Mote 1948); late June to early July, Wyoming (Spackman 1980); early May to early June, North Dakota (McBride 1975); one week after apple-petal fall, generally late May to early June, Minnesota (Hodson and Lofgren 1970); May and July, Indiana (Sadof 1992); about apple-blossom time, Wisconsin (McDowell 1960); late May, Illinois (English 1970); late May to early June, Pennsylvania (Heller 1977); mid-May to late May, Vermont (Nielsen 1970); late May to mid-June, Connecticut (Schread 1970); between silver tip and pre-pink of apple, first week of April to first week of May with second-generation crawlers in July, North Carolina (Turnipseed and Smith 1953). Kozár (1990a) found that eggs begin to hatch when there has been an accumulation of >130.8 °C day-degrees, and this event closely coincides with the appearance of apple blossoms. In Russia, Danzig (1959) found a unisexual, univoltine form on apple, barberry, pear, cotoneaster, and walnut and a bisexual, bivoltine form on poplar, willow, lilac, boxwood, and a few others. The latter form was more tolerant of low temperatures. Schmutterer (1951) noted a bisexual, bivoltine form in Germany (*L. ulmi bisexualis*). In Japan, Murakami (1970) recorded 1 generation per year with egg hatch in May. In Chile, it is reported to have a single generation per year and to overwinter in the egg stage (Carrillo et al. 1995).

Economic Importance Griswold (1925) summarizes numerous references in the early literature about the destructive nature of oystershell scale. The most serious of these outbreaks was noted by Sterrett (1915) in northern Ohio where entire stands of ash trees were killed. Garrett (1972) also discusses the economic importance of oystershell scale in the United States. Spackman (1980) describes oystershell scale as 'probably the insect pest most destructive to lilacs in Wyoming.' He noted that heavy infestations kill entire lilac bushes as well as ash and willow trees. Turnipseed and Smith (1953) state that oystershell scale is a serious pest of apples at higher elevations in western North Carolina. Recent literature discussing economic damage to specific hosts are as follows: Apples (Aleksidze 1995; Erol and Yasar 1996), olives (Argyriou 1990; Katsoyannos 1992), ornamentals (Davidson and Miller 1990), and walnut (Chua and Wood 1990). It appears this pest is not as abundant today as it was in the early 1900s. In many situations the large diversity of natural enemies prevents the oystershell scale from becoming a pest (Kozár 1990b). An interesting discovery by Mendel et al. (1992) is that if this species feeds on alkaloid-containing plants, the predators that feed on the scale have significantly diminished levels of survival. Beardsley and González (1975) consider this to be one of 43 major armored scale pests. Miller and Davidson (1990) consider it to be a world pest.

Selected References Ferris (1937); Garrett (1972); Griswold (1925); Samarasinghe and LeRoux (1966).

Plate 75. *Lepidosaphes yanagicola* Kuwana, Fire bush scale

A. Immature and mature female covers on unknown host (R. J. Gill).

B. Infestation on winged euonymus (J. A. Davidson).

C. Young light-colored and old dark-colored covers on winged euonymus (J. A. Davidson).

D. Distance view of heavy infestation on winged euonymus (J. A. Davidson).

E. Male covers on winged euonymus leaf (J. A. Davidson).

F. Distance view of covers on winged euonymus twig in fall color (J. A. Davidson).

G. Body of adult female on winged euonymus (J. A. Davidson).

H. White eggs on winged euonymus (J. A. Davidson).

I. Winged euonymus killed by severe infestations (J. A. Davidson).

Lepidosaphes yanagicola (Kuwana)

Suggested Common Name Fire bush scale (also called yanagicola oystershell scale and firebush scale).

Common Synonyms and Combinations *Insulaspis yanagicola* (Kuwana), *Lepidosaphes atunicola* Siraiwa.

Field Characters (Plate 75) Adult female cover narrow, elongate, oyster-shell shaped, straight unless crowded, brown; shed skins marginal, orange brown. Male cover shorter, similar in color and texture; shed skin marginal, orange brown. Body of adult female white; eggs white; crawlers white. On stems and bark; in heavy infestations near leaf midveins.

Slide-mounted Characters Adult female (Fig. 79) with 2 definite pairs of lobes, third-, fourth-, and occasionally fifth-lobe areas represented by series of sclerotized points; paraphysis formula 2-1-0 or 2-2-0, with slender paraphysis-like sclerotization attached to each medial margin of median lobes. Median lobes separated by space 0.5–0.9 (0.8) times width of median lobe, with distinctive sclerotization attached to anterior edge of median lobe (Fig. 79), lateral and medial margins of lobes parallel, with 1–3 (2) lateral notches, 1–3 (2) medial notches; second lobes bilobed, smaller than median lobes, medial lobule largest, with 1–2 (1) lateral notches and 0–2 (0) medial notches, lateral lobule without notches; third lobes represented by swelling surrounding marginal macroduct and series of small points; fourth and fifth lobes when present, simple, with series of small points. Gland spines each with 1 or 2 microducts, some with basal fringing, gland-spine formula normally 2-2-2, occasionally 2-2-3 or 2-3-3, with 22–50 (34) gland spines near each body margin anterior of lobe 4 area, present to metathorax; 2 conspicuous gland spines between median lobes. Macroducts of 2 sizes, larger size on margin from segments 5 to 7, smaller size present in submarginal areas from mesothorax to segment 5, in submedial areas from segments 2 to 6, submedial clusters on segment 6 with 1–4 (2) ducts, without duct between median lobes, macroduct in first space 22–30 (25) µ long, with 13–23 (17) macroducts on each side of body on segments 5 to 7, total of 97–128 (108) macroducts on each side of body, some orifices occurring anterior of anal opening. Pygidial microducts on venter of segments 5 to 7 with total of 4–11 (8) ducts; prepygidial ducts in submarginal clusters on meso- and metathorax also on segments 1 to 4, submedial ducts on segments 1 or 2 to 4; dorsal pygidial and prepygidial microducts absent. Perivulvar pores in 5 groups, 17–39 (29) pores on each side of pygidium. Perispiracular pores usually with 3 loculi, anterior spiracles each with 5–12 (8) pores, posterior spiracles without pores. Anal opening located 8–13 (10) times length of anal opening from base of median lobes, anal opening 12–17 (14) µ long. Seta lateral of median lobes 0.7–0.9 (0.8) times length of lobe. Eyes represented by flat, unsclerotized area laterad of anterior portion of clypeolabral shield. Antennae each with 1–2 (2) conspicuous setae. Cicatrices absent. Body shape elongate. Space between median lobes with 2 setae. Lateral spurs on each of segments 2 to 4. Without microducts near antennae.

Affinities Fire bush scale is surprisingly similar to *Lepidosaphes gloverii*, Glover scale, but differs by having the anal opening located 8–13 (10) times the length of the anal opening from the base of the median lobes and by lacking a sclerotized prepygidial pattern on older adult females. Glover scale has the anal opening located 12–17 (15) times the length of the anal opening from the base of the median lobes and has a distinctive sclerotized pattern on older adult females (Fig. 74).

Hosts Borchsenius (1966) records this species from *Acer*, *Albizia*, *Alnus*, *Fraxinus*, *Maackia*, *Morus*, *Salix*, and *Syringa*. Murakami (1970) lists it from *Albizia*, *Salix*, and *Tilia* in Japan. In the United States it occurs primarily on *Euonymus alatus*. We have examined material from *Celastrus*, *Diospyros*, *Euonymus*, *Pachysandra*, *Salix*, *Thea*, and *Tilia*.

Distribution We have examined U.S. specimens from GA, MA, MD, NJ, OH, PA, RI, TN, VA, WV. It has been reported from IN and OK. It is known from Japan, Korea, former Soviet Union (Sakhalin Island, Ussuri area).

Biology Porter et al. (1959) report that in Ohio females overwinter and produce eggs from early June into July. Crawlers appear by June 20 and all stages are present after late July. McComb and Davidson (1969) report crawlers in June in Maryland.

Economic Importance The authors have observed heavy infestations of fire bush scale causing premature leaf drop and dieback of the host in Beltsville, Maryland. According to Kosztarab (1963) several cases of heavy infestations and damage have been reported in Ohio nurseries. Kosztarab (1996) indicates that damage can be caused with premature leaf drop and twig dieback in nurseries in the northeastern United States on winged euonymus. Stimmel (2002, personal communication) indicates that although this scale builds to heavy populations in Pennsylvania, it rarely causes damage to firebush. Miller and Davidson (1990) consider this species to be an occasional pest.

Selected References Kosztarab (1963); Kosztarab (1996); Porter et al. (1959).

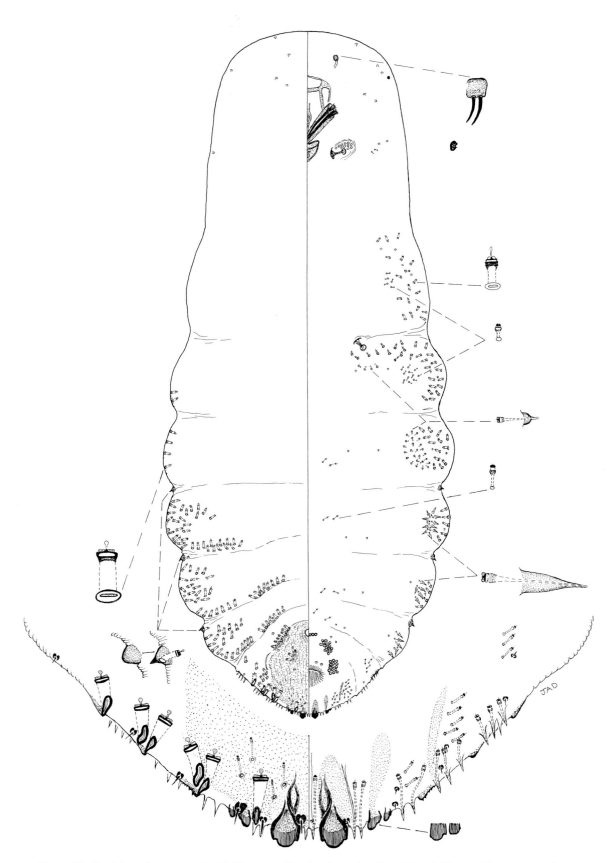

Figure 79. *Lepidosaphes yanagicola* Kuwana, fire bush scale, Republic, MO, on *Euonymus alatus*, IV-7-1976.

Lindingaspis rossi (Maskell)

Suggested Common Name Black araucaria scale (also called circular black scale, Ross' black scale, Ross scale, grey scale, rose scale).

Common Synonyms and Combinations *Aspidiotus rossi* Maskell, *Chrysomphalus rossi* (Maskell), *Aonidiella subrossi* Laing, *Melanaspis rossi* (Maskell), *Chrysomphalus niger* Lindinger, *Chrysomphalus subrossi* (Laing).

Field Characters (Plate 76) Adult female cover flat, circular, dark brown or black, often with concentric rings of lighter colored wax; shed skins subcentral, dark brown or black. Male cover same color as female cover, oval; shed skins submarginal. Body of female clear or light pink. Normally on leaves or needles, sometimes on fruit.

Slide-mounted Characters Adult female (Fig. 80) with 3 pairs of definite lobes, fourth lobes represented by low series of sclerotized points; paraphysis formula usually 2-3-4, rarely 2-3-5 or 2-3-7, paraphyses all approximately same length, becoming narrower anteriorly; numerous paraphyses associated with lobe 4. Median lobes separated by space 0.8–1.3 (1.1) times width of median lobe, without basal sclerotization, with paraphysis attached to medial margin, without yoke, medial margins and lateral margins parallel, with 1 lateral notch, 0–1 (0) medial notch; second lobes simple, slightly smaller or about same size as median lobes, with rounded apex, with 1–2 (1) lateral notches, without medial notch; third lobes simple, slightly smaller than second lobes, about same shape, with 1–3 (2) lateral notches, without medial notches; fourth lobes represented by series of 5–8 (6) points. Plates between lobes fimbriate, finger-like process on anterior 2 plates between lobes 3 and 4; plate formula 2-3-3, rarely 1-3-3 or 2-2-3; plates between median lobes slightly longer than median lobes. Macroducts of 2 sizes, larger size marginal between median lobes and in spaces between median lobe and lobe 2, between lobe 2 and 3, and between lobe 3 and lobe 4, also present at apex of sclerotized area anterior of lobe 3; smaller size similar in appearance to microduct but larger, duct between median lobes extending 0.7–0.9 (0.8) times distance between posterior apex of anal opening and base of median lobes, 112–148 (127) μ long, marginal macroduct in first space 118–150 (131) μ long, with 15–22 (18) macroducts on each side of pygidium on segments 5 to 8, ducts in submarginal and marginal areas, total of 45–84 (61) macroducts on each side of body, some macroduct orifices anterior of anal opening; prepygidial macroducts in submarginal areas of segments 3 and 4. Pygidial microducts on venter in submarginal and marginal areas of segments 5 or 5 and 6, with 3–7 (5) ducts; prepygidial ducts of 2 sizes, large size in marginal row along body margin from eye on mesothorax to segment 4, small number near body margin laterad of anterior spiracle, also present in submedial areas near mouthparts, spiracles, and any or all of segments 1 to 4; smaller ducts in submarginal clusters on metathorax and segment 1, occasionally with 1 or 2 laterad of anterior spiracle; dorsal pygidial ducts absent; prepygidial microducts of 1 size in submedial areas of pro-, meso-, or metathorax to segment 3.

Perivulvar pores arranged in 4 groups, 11–15 (13) pores on each side of body. Pores absent near spiracles. Anal opening located 4.4–6.5 (5.6) times length of anal opening from base of median lobes, anal opening 21–32 (27) μ long. Dorsal seta laterad of median lobes 0.7–1.2 (0.9) times length of median lobe. Eyes dome shaped, often projecting on body margin, located laterad of posterior spiracle. Antennae each with 1 long seta. Cicatrices represented by clear oval areas on derm on prothorax, segment 1, and 3. Body pear shaped.

Affinities *Lindingaspis rossi* is most similar to species of *Chrysomphalus* in the shape of the body, lobe arrangement, and organization of plates, but differs by having a row of marginal microducts along the body margin from the position of the eye to the pygidium and by having a series of paraphyses associated with lobe 4. Species of *Chrysomphalus* lack these structures.

Hosts Polyphagous. Borchsenius (1966) records it from 139 genera in 62 families; Munro and Fouche (1936) report it from 103 genera in South Africa. It is commonly found on conifers including *Araucaria* species. We have examined specimens from the following host genera: *Acacia, Araucaria, Banksia, Buxus, Citrus, Codiaeum, Dendrobium, Elaeagnus, Eucalyptus, Euonymus, Ficus, Geijera, Haworthia, Hedera, Laurus, Leptospermum, Leucospermum, Ligustrum, Macrozamia, Magnolia, Musa, Nerium, Nuytsia, Olea, Pinus, Pleiospilos, Protea, Prunus,* and *Styphelia.* In recent years it has been taken most commonly in quarantine at U.S. ports-of-entry on cut flowers such as *Protea.*

Distribution This species occurs out-of-doors in most tropical and subtropical areas of the world. We have examined U.S. material from CA, HI, NJ, NY. It frequently is taken in quarantine at U.S. ports-of-entry from Africa and Australia.

Biology The black araucaria scale is reported to have 1 generation per year on Monterey pines in New Zealand. Egg-laying adult females were present in September and October and crawlers were observed in early November (Timlin 1964a). Swailem et al. (1976) indicate there are 2 or 3 annual peaks of abundance of this scale on various hosts and at several different localities in Egypt. It is likely that peaks of abundance are correlated with generation times.

Economic Importance In California the black araucaria scale is considered to be a serious pest of California redwoods and araucaria trees. It feeds on the needles of the host and causes heavy chlorosis and sometimes leaf drop. Under certain circumstances it is necessary to use control strategies (Brown and Eads 1967). This pest is reported to cause red staining on the fruit of apples and can build to very heavy populations on Monterey pine in New Zealand (Timlin 1964a). Srinivasan et al. (1974) applied granular insecticides for the control of this pest on ornamentals in India. It is listed as a noxious pest of conifers by Zahradník (1990b) and is considered an occasional pest by Miller and Davidson (1990).

Selected References Gill (1997); Timlin (1964a).

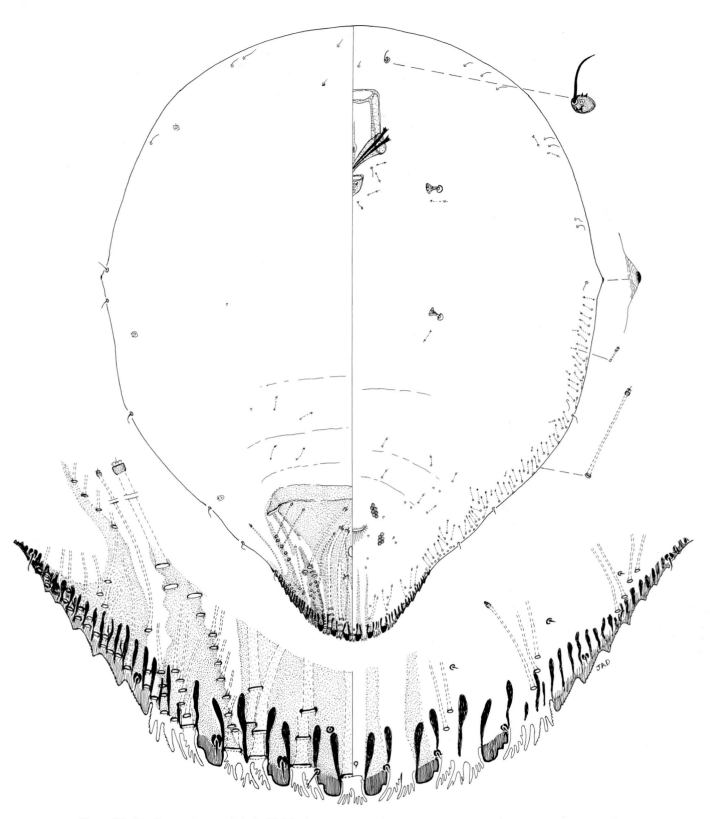

Figure 80. *Lindingaspis rossi* (Maskell), black araucaria scale, Jamaica, intercepted at Miami, FL, on *Codiaeum* sp., VII-13-1968.

Plate 76. *Lindingaspis rossi* (Maskell), Black araucaria scale

A. Mature circular female scale cover on *Araucaria heterophylla* (J. A. Davidson).
B. Mature elongate female scale cover on *A. heterophylla* (J. A. Davidson).
C. Second-instar female cover on *A. heterophylla* (J. A. Davidson).
D. Mature male covers on *A. heterophylla* (J. A. Davidson).
E. Various instar covers on *A. heterophylla* (J. A. Davidson).
F. Body of light-colored young female from *A. heterophylla* (J. A. Davidson).

G. Ventral cover partially torn to reveal darker body of older female and orange crawler on *A. heterophylla* (J. A. Davidson).
H. Male covers turned to reveal second instars on *A. heterophylla* (J. A. Davidson).
I. Distance view of yellowing damage on *A. heterophylla* (J. A. Davidson).

274

Plate 77. *Lopholeucaspis japonica* (Cockerell), Japanese maple scale

A. Adult female covers on pyracantha (J. A. Davidson).
B. Covers and settled crawlers on apple (J. A. Davidson).
C. Pupillarial covers sprayed with oil showing second-instar shed skins on privet (J. A. Davidson).
D. Male cover on privet (J. A. Davidson).

E. Distance view on apple (J. A. Davidson).
F. Distance view on pyracantha (J. A. Davidson).
G. Body of adult female just before egg laying (J. A. Davidson).
H. Adult female after laying eggs (J. A. Davidson).
I. Severe dieback on pyracantha (J. A. Davidson).

275

Lopholeucaspis japonica (Cockerell)

Suggested Common Name Japanese maple scale (also called Japanese scale and pear white scale).

Common Synonyms and Combinations *Leucaspis japonica* Cockerell, *Leucodiaspis hydrangae* Takahashi, *Leucodiaspis japonica* (Cockerell), *Leucaspis hydrangeae* (Takahashi).

Field Characters (Plate 77) Adult female pupillarial, covered dorsally by thickened second shed skin, open beneath, second shed skin elongate, reddish brown, irregularly oyster-shell shaped, and overlain by grayish white wax; shed skin marginal, brown. Male cover smaller but similar to female cover. Body of adult female white; eggs white with purple tinge; crawlers white. Normally on bark but in heavy infestations observed on leaves.

Slide-mounted Characters Adult female (Fig. 81) with 2 pairs of well-developed lobes; with converging paraphyses on lateral and medial margins of median lobes. Median lobes separated by space 1.1–2.2 (1.5) times width of median lobe, without basal sclerosis and yoke, medial margin axes parallel or diverging apically, lateral margins parallel or converging apically, with 0–2 (1) lateral notches, 0–1 (1) medial notches; second lobes simple, slightly smaller than median lobes, with 0–3 (1) lateral notches, 0–2 (1) medial notches. Plate formula normally 2-2-0, rarely 2-1-0, 2-3-0, or 3-3-0; segments 4 to 6 with lateral processes similar in appearance to plates but broader and shorter than plates, also with processes with rounded apices; gland spines present from area laterad of anterior part of mouthparts to segment 4, with 33–68 (60) gland spines; median lobes with 2 plates between them. Macroducts and microducts integrating in size, those on pygidium approximately equal in size, 1–2 (2) macroducts between median lobes, 12–18 (15) µ long, extending 0.1–0.3 (0.2) times length of distance between base of median lobes and posterior apex of anal opening, macroduct in first space 13–17 (15) µ long, with 32–46 (40) macroducts on each side of pygidium on segments 5 to 8, submarginal ducts on segments 5 to 7, submedial on segments 5 and 6; some macroduct orifices occurring anterior of anal opening; prepygidial macroducts decreasing in size anteriorly, on margin and submargin of segments 1 or 2 to 4, ventral on margin of prothorax to segments 1 or 2, total of 80–164 (122) macroducts on each side of body. Pygidial microducts on venter of segments 4 to 6 or 7, with 8–20 (13) microducts on pygidium, prepygidial microducts, other than those mentioned on dorsum (macroducts), submarginal from segments 2 or 3 to 4, submedial from metathorax or segment 1 to 4; dorsal microducts absent. Perivulvar pores normally in 7 groups, rarely as many as 12 groups, 38–58 (45) pores on each side of pygidium. Perispiracular pores primarily with 3 loculi; anterior spiracles with 8–13 (11) pores, posterior spiracles without pores. Anal opening located 4–7 (5) times length of anal opening from base of median lobes, anal opening 14–18 (16) µ long. Dorsal seta laterad of median lobes 0.2–0.4 (0.3) times length of lobes. Eyes represented by small clear area laterad of antennae. Antennae each with 3–6 (4) setae and 2 bisetose sensilla. Cicatrices absent. Small spines sometimes present near anterior margin of head. Macroducts with filamentous threads longer than length of pygidium. Dorsum of pygidium with 9 or 10 small sclerotized areas.

Affinities Japanese maple scale is most similar to *Lopholeucaspis cockerelli*, Cockerell scale, but can be separated by having 32–46 (40) macroducts on each side of pygidium, gland spines from the prothorax to segment 4, and by lacking a sclerotized spine lateral of each antenna. Cockerell scale has 11–18 (15) macroducts on each side of the pygidium, gland spines from the prothorax to segment 5 or 6, and a sclerotized spine lateral of each antenna.

Hosts Borchsenius (1966) reports this species from 16 genera in 13 families. We have examined specimens from *Acer, Castanea, Chaenomeles, Citrus, Cornus, Cotoneaster, Cydonia,* *Diospyros, Euonymus, Ilex, Ligustrum, Magnolia, Malus, Paeonia, Paulownia, Poncirus, Prunus, Pyracantha, Pyrus, Rosa, Styrax, Syringa, Ulmus, Wisteria, Zelkova,* and *Ziziphus.* It has been reported on *Alnus, Camellia, Celastrus, Cytisus Enkianthus, Ficus, Fraxinus, Laurus, Menyanthes, Populus, Salix,* and *Syringa.*

Distribution This scale probably was introduced into the United States from Asia. We have examined U.S. specimens from CT, DC, DE, GA, MD, NJ, NY, PA, RI, VA. This pest most commonly has been intercepted in quarantine inspections from Japan. We have examined material from China and Korea. It is reported in the literature from Afghanistan, Brazil, Burma, China, India, Iran, Korea, Okinawa, Pakistan, Taiwan, Turkey, and the former Soviet Union. A distribution map of this species was published by CAB International (1998).

Biology The life history of this species is poorly understood in the United States, although, Stimmel (1995) indicates that it has 1 generation per year and overwinters as adult females in Pennsylvania. According to Murakami (1970) the Japanese maple scale has 1 generation per year and overwinters as fertilized adult females in Japan. In Caucasus there are 2 generations per year, and overwintering takes place in the second-instar male and female (Kozarzhevskaya 1956). Although molting individuals can be observed as early as late March, adult females are not found until later in April. Prepupae and pupae appear at the end of March and adult males are present from late April to late May. In the spring, egg laying begins in late April and lasts until late June or early July. Crawlers appear in late May and are present until early August. Second instars are present from the last of June to late August and adult females can be found from July to October. Males in prepupal, pupal, and adult instars are present from the middle of July to the last of September. Eggs of the fall generation appear at the end of July and continue until October. Crawlers are present from August until October. Because egg-laying periods are so long, generations are overlapping and it is possible to detect two generations only by using comparative information on percentages of individual instars through time. McComb and Davidson (1969) report that crawlers were present in June in Maryland. Greenwood (1991, personal communication) studied the biology of this scale in Norfolk, Virginia. There are two overlapping generations per year, and overwintering occurs in the adult female. Adult males appear in the spring in early April. Oviposition begins in early May and continues for over a month, with an average of 25 eggs under each scale. Crawlers first appear in early June and are present until mid-July. Adults of the first generation appear in late July, and eggs and crawlers are present in August to October. The male was described by Bienkowski (1993). A model was developed by Zheng et al. (1993) to make control decisions in China on tea.

Economic Importance Japanese maple scale is a serious orchard pest, especially of citrus in southern areas of the former Soviet Union (Yasnosh 1986). In the United States it is known to damage maple and pyracantha (Davidson and Miller 1990), causing severe dieback. It has been reported as a pest of deciduous fruit trees (Kozár 1990b) and holly (McComb 1986). In Pennsylvania it is not a major pest, but it sometimes causes dieback, and limbs and smaller branches can be killed (Stimmel 1995). Recent literature portrays it as a serious pest of citrus and ornamentals in the former Soviet Union (Konstantinova 1992); of citrus, other fruit, tea, tung, and various ornamentals in the Republic of Georgia (Tabatadze and Yasnosh 1997; Tabatadze and Yasnosh 1998); and of tea in China (Zheng et al. 1993). Miller and Davidson (1990) consider it to be an important world pest.

Selected References Bienkowski (1993); Ferris (1938a); Kozarzhevskaya (1956).

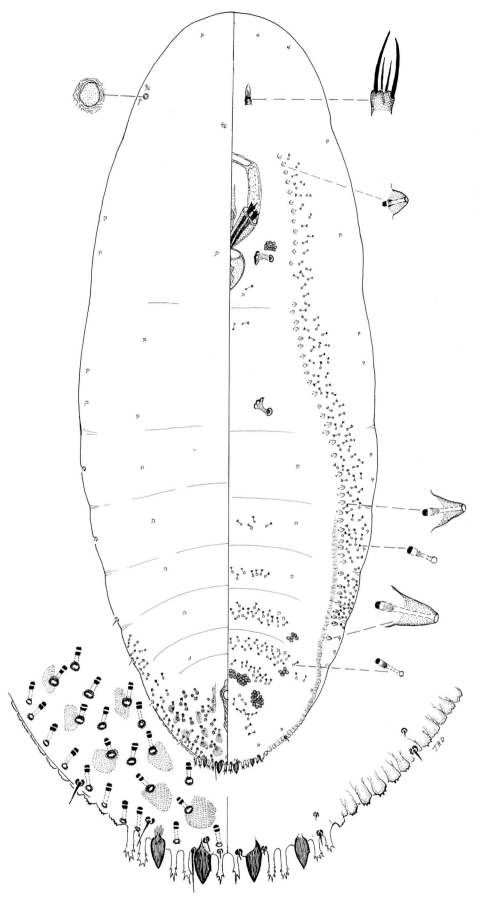

Figure 81. *Lopholeucaspis japonica* (Cockerell), Japanese maple scale, Japan, intercepted at Seattle, WA, on *Cydonia speciosa*, III-31-1948.

Melanaspis lilacina (Cockerell)

Suggested Common Name Dark oak scale.

Common Synonyms and Combinations *Aspidiotus (Chrysomphalus) lilacinus* Cockerell, *Aspidiotus lilacinus* Cockerell, *Chrysomphalus lilacinus* (Cockerell), *Pelomphala lilacinus* (Cockerell).

Field Characters (Plate 78) Adult female cover relatively flat, circular, black or dark gray; shed skins subcentral. Male cover similar in texture and color to female cover, elongate; shed skin submarginal. Body of adult female probably light pink to purple. On bark and leaves (Gill 1997).

Slide-mounted Characters Adult female (Fig. 82) with 4 pairs of well-developed lobes; paraphysis formula normally 2-2-3, occasionally 3-2-3 or 2-2-4. Median lobes separated by space 0.2–0.5 (0.3) times width of median lobe, without basal sclerosis and yoke, with conspicuous medial paraphysis, medial margin with axes parallel or diverging apically, lateral margins converging apically, with 0–2 (1) lateral notches, without medial notches; second lobes simple, about same size and shape as median lobes, with 1–2 (2) lateral notches, 0–1 (0) medial notches; third lobes simple, wider than second lobes, with 2–5 (3) lateral notches, without medial notches; fourth lobes usually with more than 1 lobule, with 2–5 (4) lateral notches on medial lobule, without medial notches on medial lobule, with 1–2 (1) lateral notches on lateral lobule, without medial notches on lateral lobule; fifth lobes absent. Gland-spine formula of 3-3-4; with 2–5 (4) gland-spine-like structures anterior of segment 5, absent elsewhere; with gland spines between median lobes that are three-quarters to equal to length of median lobes. Macroducts decreasing in size anteriorly, 2 macroducts between median lobes 138–235 (177) µ long, 0.6–1.4 (1.2) times as long as distance between base of median lobe and posterior apex of anal opening; macroduct in first space 123–198 (171) µ long, with 12–16 (14) macroducts on each side of pygidium on segments 5 to 8, marginal ducts on segments 5 to 8, submarginal on segments 5 to 7; macroduct orifices not occurring anterior of anal opening; prepygidial macroducts on margin and submargin of segment 4, total of 18–22 (20) macroducts on each side of body. Pygidial microducts on venter of segment 5 and sometimes 6, with 6–13 (8) microducts on each side of pygidium; prepygidial microducts on venter from mesothorax, metathorax, or segment 1 to segment 4 in submarginal areas, from anterior thorax to anterior abdominal segments in submedial areas; pygidial microducts absent from dorsum, prepygidial microducts on dorsum from head to segment 3 in submarginal areas. Perivulvar pores absent. Perispiracular pores absent. Anal opening located 6–15 (9) times length of anal opening from base of median lobes, anal opening 10–18 (15) µ long. Dorsal seta laterad of median lobes 0.3–0.8 (0.5) times length of median lobes. Eyes rarely present, represented by small sclerotized area. Antennae normally each with 1 long seta and 1 small seta. Cicatrices usually absent, rarely on body margin of prothorax or segment 1. Pygidium with sclerotized areas and with at least 1 interlobular paraphysis longer than other paraphyses. Macroducts restricted to medial margin of sclerotized area arising from base of third lobe.

Affinities Of the economic species, dark oak scale resembles *Melanaspis obscura*, obscure scale, but differs by lacking perivulvar pores that obscure scale possesses.

Hosts This species has been recorded only from the white oak and black oak groups of *Quercus*.

Distribution This species occurs in the western states. We have seen U.S. material from AZ, CA, NM. We also have seen specimens from Mexico.

Biology Nothing is known about the life history of this species except that it occurs on the leaves and bark of its host (Gill 1997).

Economic Importance Dark oak scale was reported as injurious to oaks in Arizona and New Mexico (Gill 1997).

Selected References Deitz and Davidson (1986); Gill (1997).

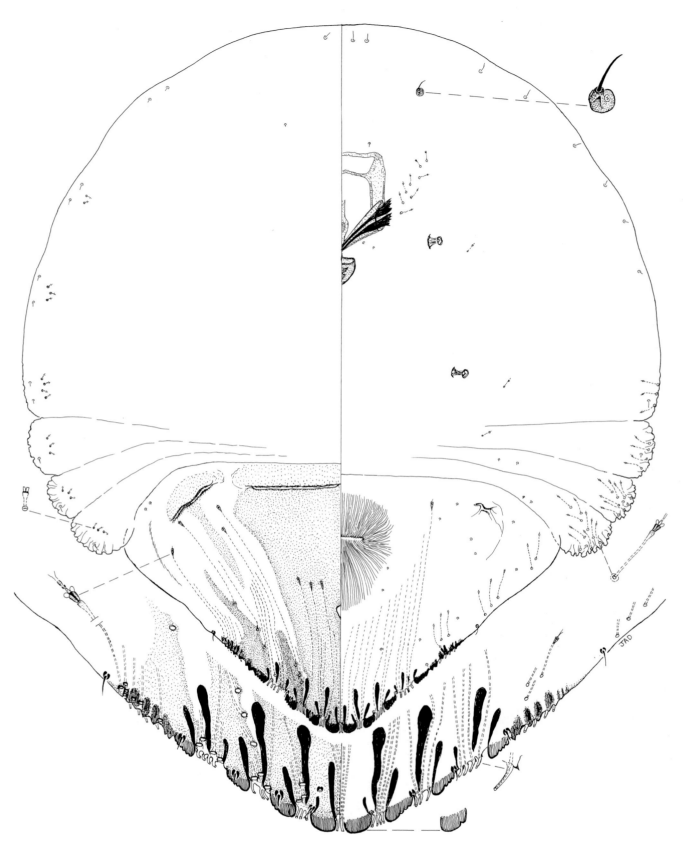

Figure 82. *Melanaspis lilacina* (Cockerell), dark oak scale, Dripping Springs, NM, on *Quercus* sp., IV-?-1898.

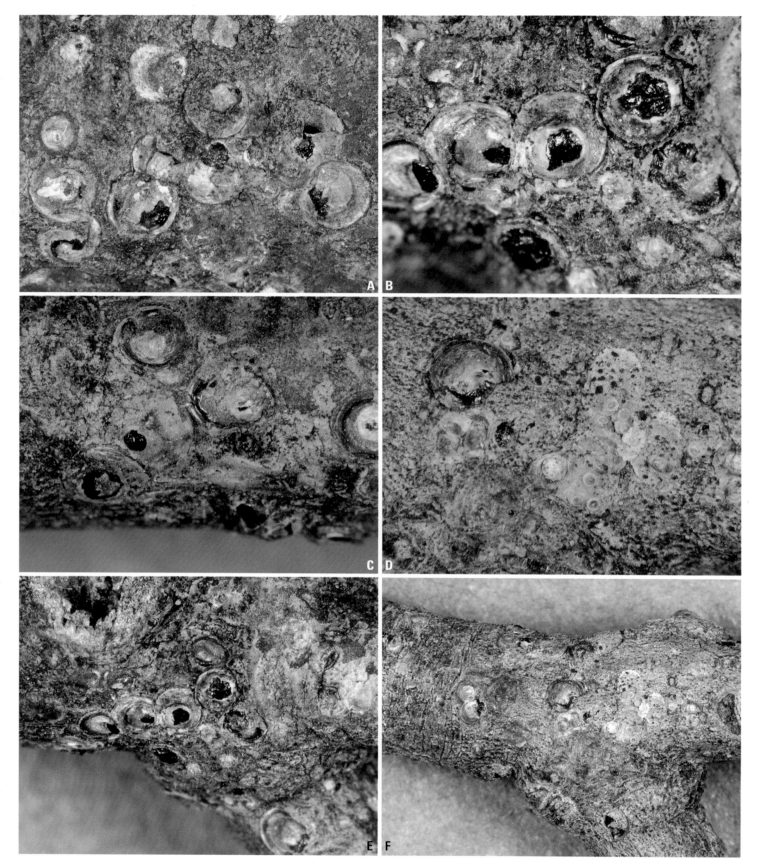

Plate 78. *Melanaspis lilacina* (Cockerell), Dark oak scale

A. Adult female covers encrusting bark of oak (J. A. Davidson).
B. Rubbed female covers showing black shed skins on oak (J. A. Davidson).
C. Covers on oak (J. A. Davidson).

D. Clustered male covers on oak (J. A. Davidson).
E. Mostly rubbed covers on oak (J. A. Davidson).
F. Distance view of cryptic covers on oak (J. A. Davidson).

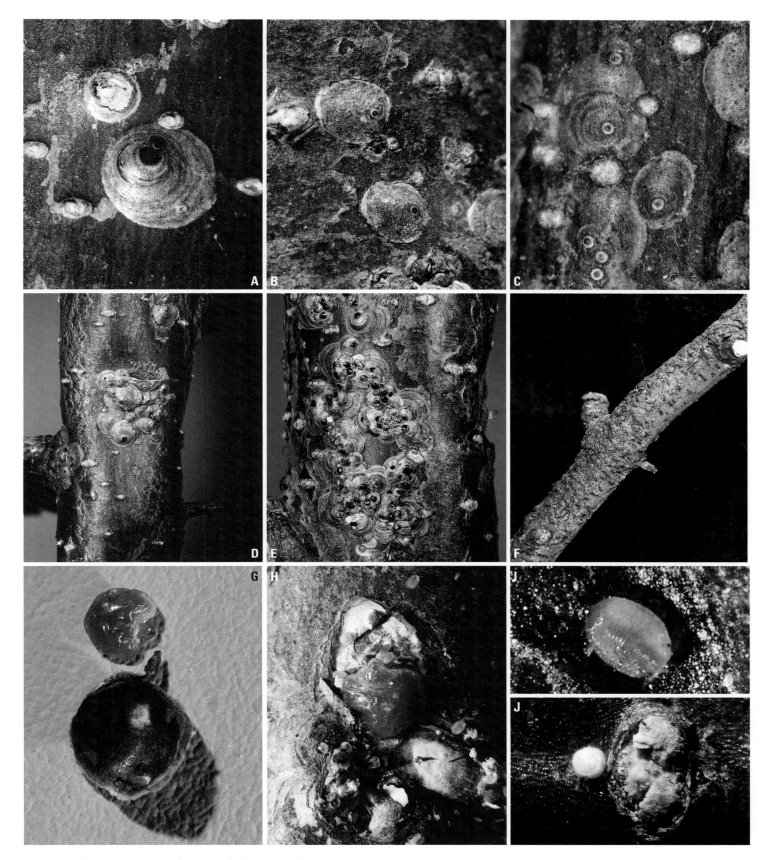

Plate 79. *Melanaspis obscura* (Comstock), Obscure scale

A. Mature female cover on pin oak (J. A. Davidson).
B. Male covers on pin oak (J. A. Davidson).
C. Male and female covers on chestnut oak (J. A. Davidson).
D. Characteristic cluster of covers in depressions caused by feeding of scale on pin oak (J. A. Davidson).
E. Distance view of heavy infestation on pin oak (J. A. Davidson).
F. Encrustation of old covers on dead limb of pin oak (J. A. Davidson).
G. Adult female prior to egg laying from pin oak (J. A. Davidson).
H. Adult female with eggs and crawlers on pin oak (J. A. Davidson).
I. Settled crawler on pin oak (J. A. Davidson).
J. Capped crawler on pin oak (J. A. Davidson).

281

Melanaspis obscura (Comstock)

ESA Approved Common Name Obscure scale.

Common Synonyms and Combinations *Aspidiotus obscurus* Comstock, *Aspidiotus (Melanaspis) obscurus* Comstock, *Chrysomphalus obscurus* (Comstock), *Chrysomphalus (Melanaspis) obscurus* (Comstock).

Field Characters (Plate 79) Adult female cover relatively flat, circular, gray; shed skins subcentral, shiny black when rubbed. Male cover similar in texture and color to female cover, elongate; shed skin submarginal, black when rubbed. Body of adult female, eggs, and crawler faintly pink or purple. On bark, occasionally in heavy encrustations.

Slide-mounted Characters Adult female (Fig. 83) with 4 pairs of well-developed lobes, occasionally with 2 or 3 projections in area of fifth lobes; paraphysis formula normally 2-3-3, occasionally 2-4-3, 2-3-4, or 2-3-5. Median lobes separated by space 0.2–0.3 (0.2) times width of median lobe, with broad basal sclerosis, without medial paraphyses and yoke, medial margin with axes parallel or diverging apically, lateral margins converging apically, with 1–2 (1) lateral notches, 0–1 (0) medial notches; second lobes simple, wider than median lobes, with 1–3 (2) lateral notches, 0–1 (0) medial notches; third lobes simple, about same size as second lobes, with 2–4 (3) lateral notches, 0–1 (1) medial notches; fourth lobes usually with more than 1 lobule, with 0–4 (3) lateral notches, without medial notches; fifth lobes usually absent, occasionally represented by several small points. Gland spines difficult to see, when visible, with gland-spine formula of 2-2-2; sometimes with 2 or 3 gland-spine-like structures anterior of segment 5, absent elsewhere; with small inconspicuous gland spines between median lobes. Macroducts decreasing in size anteriorly, 2 macroducts between median lobes 115–176 (160) μ long, 0.9–1.2 (1.0) times as long as distance between base of median lobe and posterior apex of anal opening; macroduct in first space 124–178 (163) μ long, with 28–47 (37) macroducts on each side of pygidium on segments 5 to 8, marginal ducts on segments 5 to 8, submarginal on segments 5 to 7; macroduct orifices not occurring anterior of anal opening; prepygidial macroducts on margin and submargin of segment 4; total of 32–48 (40) macroducts on each side of body. Pygidial microducts on venter of segment 5 and sometimes 6, with 2–9 (5) microducts on each side of pygidium; prepygidial microducts on venter from head to segment 4 in submarginal areas, from anterior thorax to anterior abdominal segments in submedial areas; pygidial microducts absent from dorsum, prepygidial microducts on dorsum from head to segment 3 in submarginal areas, on posterior thoracic and anterior abdominal segments in submedial areas. Perivulvar pores in 5 indistinct groups, 17–23 (19) pores on each side of pygidium. Perispiracular pores absent. Anal opening located 6–15 (9) times length of anal opening from base of median lobes, anal opening 12–17 (15) μ long. Minute dorsal seta laterad of median lobes. Eyes rarely present, represented by small sclerotized area. Antennae normally each with 1 long seta and 1 small seta. Cicatrices usually near body margin of prothorax and segment 2, occasionally on segment 4. Pygidium with sclerotized areas and with at least 1 interlobular paraphysis longer than other paraphyses. Macroducts restricted to medial margin of sclerotized area arising from base of third lobe.

Affinities Of the economic species, obscure scale resembles *Melanaspis tenebricosa*, gloomy scale, but differs by possessing perivulvar pores that gloomy scale lacks. Of the 4 *Melanaspis* species with perivulvar pores in the United States, only *M. obscura* is common on oaks. *Melanaspis obscura* resembles *M. nigropunctata* but differs by having the dorsal seta of lobes 3 and 4 longer than the lobes, lobes 2 and 3 usually with 2 or more lateral notches, a broad basal sclerosis on each median lobe, and macroducts restricted to medial margin of sclerotized area arising from third lobes. *M. nigropunctata* has the dorsal seta of lobes 3 and 4 shorter than these lobes, lobes 2 and 3 usually with 1 lateral notch each, narrow basal sclerosis on each median lobe, and macroducts on medial and lateral margin of sclerotized area arising from third lobes.

Hosts Borchsenius (1966) records this species from *Acer*, *Carya*, *Castanea*, *Cornus*, *Juglans*, *Prunus*, *Prosopis*, *Quercus*, and *Vitis*. In addition to the above, Johnson and Lyon (1988) report this scale on *Fagus*, *Fraxinus*, *Celtis*, *Myrtus*, and *Spondias*. We have seen material from *Acer*, *Carya*, *Castanea*, *Fagus*, *Quercus*, and *Ulmus*. This species prefers *Quercus*, *Castanea*, and *Carya*. In our field experience we have seen this scale only on the latter 3 hosts. We suspect many of the other hosts recorded for this species, particularly in the South, represent misidentifications of *Melanaspis nigropunctata*, which resembles *M. obscura* but has a wider host range.

Distribution This species mainly occurs east of the Mississippi River. We have seen U.S. material from AL, AR, CT, DC, FL, GA, IL, IN, KS, KY, LA, MD, MO, MS, NC, NJ, OH, OK, PA, SC, TN, TX, VA. It has been reported from CA, DE, IA, NY, WV, and Ontario, Canada. This pest is known only from the U.S. and Canada.

Biology The obscure scale has one generation per year in Louisiana (Baker 1933), Maryland (Stoetzel and Davidson 1971), Alabama (Hendricks and Williams 1992), Kentucky (Potter et al. 1989), and California (Ehler 1995). In Maryland (Stoetzel and Davidson 1971) males and females overwintered as second instars, matured and mated in May, and most eggs were laid in July. Most crawlers appeared in July, and by early October only second instars were present. In Alabama (Hendricks and Williams 1992) the same pattern was present, although specific dates varied somewhat. Potter et al. (1989) developed a day-degree model to predict obscure scale development. Stoetzel and Davidson (1973) found evidence for sibling species between infestations on hosts in the red and white oak groups in Maryland. On white oak they found a nearly one-month lag in development time of most stages resulting in settled crawlers forming the overwintering stage. On pecan in Louisiana, Baker (1933) found eggs and crawlers from mid-May to mid-July. By mid-September the population was composed mainly of second instars. Adult females appeared between early March and mid-April. The peak adult male emergence period varied from mid-March to mid-April.

Economic Importance Obscure scale is considered to be one of the most damaging insect pests of shade trees in many parts of the eastern United States (Miller and Davidson 1990). According to Dekle (1977) the obscure scale occasionally is a pest of transplanted oaks in Florida. It is considered among the most serious pests of shade trees in Pennsylvania (Nixon 1968) and Maryland (Stoetzel and Davidson 1971). We believe that pin oak and willow oak are among the most susceptible hosts to be damaged by this pest. The bark of heavily infested young trees often develops a knurled appearance after extended periods of infestation. Oak trees commonly exhibit branch dieback. Kosztarab (1963) noted serious damage to oaks in Ohio. Sprays timed for crawlers generally have been unsuccessful in controlling this pest. This is due to: (1) the habit of crawlers settling under the protective covers of dead scales making contact sprays ineffective, and (2) the long crawler emergence period which obviates the effectiveness of 1 or 2 sprays of nonresidual insecticides. Control is further complicated on white oaks because, as noted above, crawlers appear in August instead of July as on red oaks. In our experience, horticultural oil sprays in the dormant season give the best control of the obscure scale on oak. This scale is

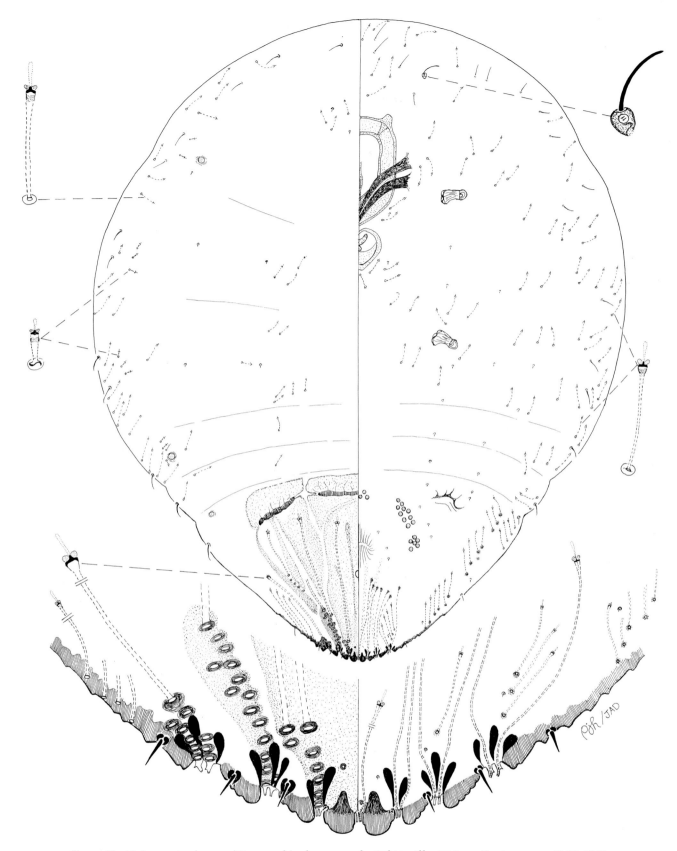

Figure 83. *Melanaspis obscura* (Comstock), obscure scale, Whiteville, TN, on *Castanea* sp., II-25-1955.

most abundant on yard and park oaks despite relatively high parasitism rates (Pinto 1980). It has not been reported to be a problem on forest oaks. Ehler (1995) implemented a successful biological control program by introducing a single parasite, *Encarsia aurantii*. This scale species is being tested as a host of the recently introduced lady beetle predator, *Chilocorus kuwanae* Silvestri (Bull et al. 1993). Both Woodard (1925) in Texas and Mackie (1933) in California consider the obscure scale as potentially the worst pest of pecan in their respective states.

Selected References Deitz and Davidson (1986); Ehler (1995); Potter et al. (1989); Stoetzel and Davidson (1971, 1973).

Melanaspis tenebricosa (Comstock)

ESA Approved Common Name Gloomy scale.

Common Synonyms and Combinations *Aspidiotus tenebricosus* Comstock, *Aspidiotus (Chrysomphalus) tenebricosus* (Comstock), *Aonidiella tenebricosa* (Comstock), *Chrysomphalus tenebricosus* (Comstock), *Chrysomphalus (Melanaspis) tenebricosus* (Comstock).

Field Characters (Plate 80) Adult female cover highly convex, circular, brown to gray; shed skins subcentral, shiny black when rubbed. Male cover smaller, similar in texture and color to female cover, elongate; shed skin submarginal, black when rubbed. Body of adult female pink to purple; eggs faint pink; crawler cream. On bark.

Slide-mounted Characters Adult female (Fig. 84) with 4 pairs of well-developed lobes, fifth lobe sometimes represented by series of small projections; paraphysis formula normally 2-3-2, occasionally 2-3-3 or 2-2-2. Median lobes separated by space 0.2–0.3 (0.2) times width of median lobe, with conspicuous basal sclerosis, without medial paraphysis and yoke, medial margin axes normally parallel, rarely diverging apically, lateral margins converging apically, with 1 lateral notch, 0–1 (0) medial notches; second lobes simple, smaller than median lobes, with 1–3 (2) lateral notches, 0–1 (0) medial notches; third lobes slightly larger than second lobes, with 1–3 (2) lateral notches, 0–1 (1) medial notches; fourth lobes usually with more than 2 lobules, with 2–5 (4) lateral notches, without medial notches; fifth lobes represented by series of 3–8 (6) points. Gland spines difficult to see, when visible, with gland-spine formula of 2-2-2 or 1-2-2; sometimes with 2–4 (3) gland-spine-like structures anterior of segment 5, absent elsewhere; with 2 inconspicuous gland spines between median lobes. Macroducts approximately equal in size; 2 macroducts between median lobes 100–145 (117) μ long, 2.8–3.9 (3.3) times as long as distance between base of median lobe and posterior apex of anal opening; macroduct in first space 109–157 (131) μ long, macroducts difficult to see, with about 19–28 (23) macroducts on each side of pygidium on segments 5 to 8, marginal ducts on segments 5 to 8, submarginal on segments 5 and 6; some macroduct orifices occurring anterior of anal opening; without prepygidial macroducts. Pygidial microducts on venter of segment 5, with 7–12 (9) microducts on each side pygidium; prepygidial ducts on venter over most of head, thorax, and prepygidial abdominal segments; on dorsum pygidial microducts absent, prepygidial microducts of two sizes, smaller size restricted to marginal areas of thorax and anterior abdominal segments, larger size in marginal areas of head and thorax, in submedial areas of thorax and anterior abdominal segments. Perivulvar and perispiracular pores absent. Anal opening located 3–5 (4) times length of anal opening from base of median lobes, anal opening 7–15 (11) μ long. Minute dorsal seta laterad of median lobes. Eyes represented by small sclerotized area, on lateral margin of pro- or mesothorax. Antennae each with 1 long seta. Cicatrices usually near body margin of prothorax and segment 2, sometimes on 4. Pygidium with sclerotized areas with at least 1 interlobular paraphysis longer than other paraphyses. Paraphysis attached to medial margin of second lobe longer than paraphysis located between median lobe and second lobe.

Affinities Of the economic species, gloomy scale is most similar to *Melanaspis obscura*, obscure scale, but differs by lacking perivulvar pores, which are present on obscure scale. Of the *Melanaspis* species without perivulvar pores, gloomy scale most resembles *M. indurata*, but the former has the first paraphysis in the first interlobular space short, while the latter has this paraphysis long. Gloomy scale mainly infests *Acer* while *M. indurata* occurs on *Pinus*.

Hosts Borchsenius (1966) records this species from *Acer*, *Carya*, *Celtis*, *Fraxinus*, *Maclura*, *Morus*, *Nyssa*, and *Populus*; Dekle (1977) reports it from 12 species in 8 genera in Florida. We have examined specimens from *Acer*, *Carya*, *Catalpa*, *Celtis*, *Cornus*, *Juglans*, *Maclura*, *Morus*, *Pithecellobium*, *Populus*, *Robinia*, *Salix*, *Sapindus*, and *Vitis*. It has been reported on *Acacia*, *Aesculus*, *Castanea*, *Gardenia*, *Gleditsia*, *Ilex*, *Liriodendron*, *Malus*, *Nerium*, *Philadelphus*, *Platanus*, *Prunus*, *Ribes*, *Thevetia*, and *Ulmus*. Maple seems to be a favored host.

Distribution This species occurs primarily in the eastern United States. We have seen U.S. material from AL, AR, DC, GA, IL, KY, LA, MD, MS, NC, OK, SC, TN, TX, VA, WV. It has been reported from CA, DE, FL, MI, MO, NJ, NY, OH, PA. Gill (1997) questioned the authenticity of the California record as it was collected only once in this state. This pest also is known from Mexico and Panama (Deitz and Davidson 1986).

Biology In Maryland, Stoetzel (1975) reports 1 generation per year with fertilized females overwintering. Egg laying began in early July and continued through August. The eggs hatched immediately after being laid, and crawlers were present through this period. Winged adult males were found from mid-August to mid-September. Immatures were noted to be cream colored while adult females were pink to purplish. In South Carolina, Metcalf (1922) also found 1 generation per year with fertilized females overwintering, but in early May females gave 'birth to living young.' Adult males appeared from mid-August to mid-September. Crawlers were described as 'pale straw-yellow' and adult females 'chitin yellow.'

Economic Importance According to Baker (1972) gloomy scale often is quite destructive on soft maples from Maryland to Florida, and Texas. Metcalf (1922) reports that heavy infestations of this pest killed terminal tree growth, which promoted basal sucker growth. Johnson and Lyon (1976) found gloomy scale to be a pest primarily of silver maple and red maple. They note that it may kill entire trees if not controlled, and that control measures are only partially effective. Miller and Davidson (1990) consider this species to be an occasional pest.

Selected References Deitz and Davidson (1986); Metcalf (1922); Stoetzel (1975).

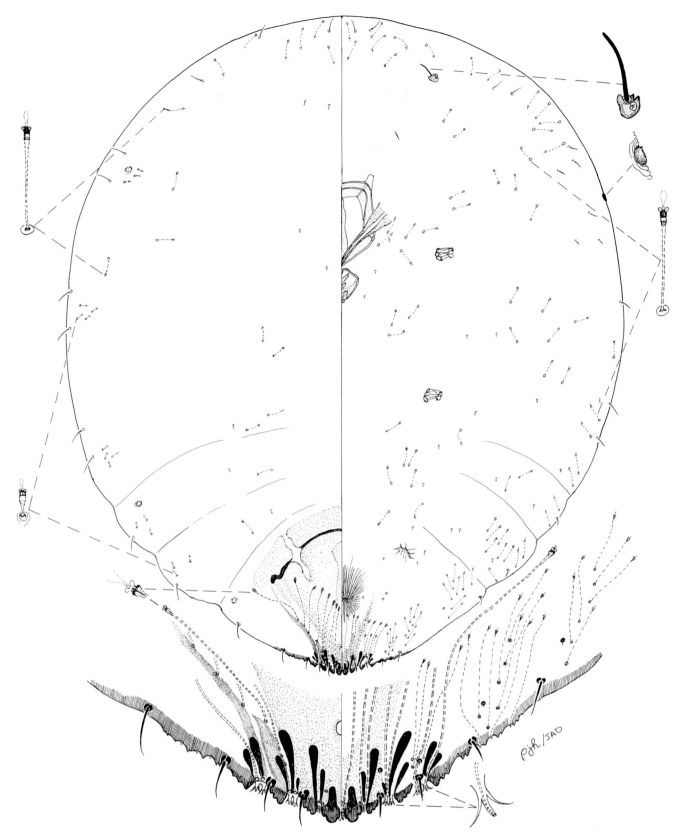

Figure 84. *Melanaspis tenebricosa* (Comstock), gloomy scale, College Park, MD, on *Acer saccharinum*, VI-7-1977.

Plate 80. *Melanaspis tenebricosa* (Comstock), Gloomy scale

A. Female covers (top) and male covers (bottom) obscured by bark flakes on red maple (J. A. Davidson).

B. Enlarged male cover with bark flakes removed from red maple (J. A. Davidson).

C. Mostly characteristically humped-up female covers on red maple (J. A. Davidson).

D. Covers largely obscured by sooty mold on sugar maple (J. A. Davidson).

E. Distance view of heavy infestation on red maple (J. A. Davidson).

F. Body of adult female with egg and crawler on red maple (J. A. Davidson).

G. Capped crawler on red maple (J. A. Davidson).

H. Branch dieback on heavily infested silver maple (M. L. Williams).

286

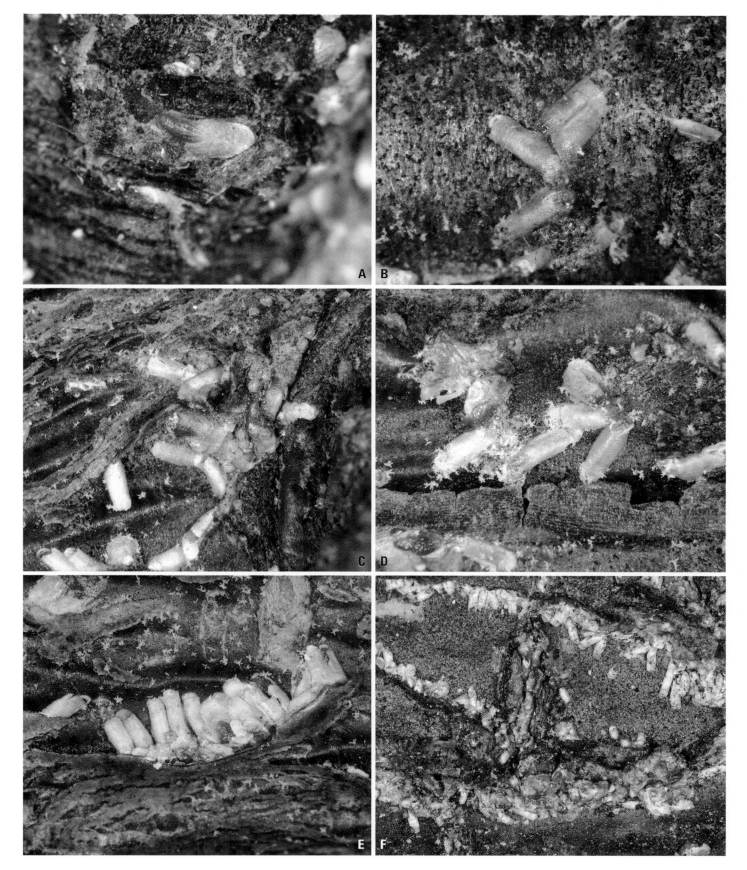

Plate 81. *Mercetaspis halli* (Green), Hall scale

A. Adult female cover on almond (J. A. Davidson).
B. Adult male covers on almond (J. A. Davidson).
C. Male and female covers on almond (J. A. Davidson).
D. Infestation on almond (J. A. Davidson).

E. Female cover (center) and male covers in bark cracks on almond (J. A. Davidson).
F. Distance view of mostly male covers on almond (J. A. Davidson).

Mercetaspis halli (Green)

Common Synonyms and Combinations *Lepidosaphes (Coccomytilus) halli* Green, *Lepidosaphes (Coccomytilus) zlocistii* Bodenheimer, *Chionaspis zlocistii* (Bodenheimer), *Lepidosaphes halli* Green, *Mytilococcus halli* (Green), *Coccomytilus zlocistii* (Bodenheimer), *Coccomytilus halli* (Green), *Nilotaspis halli* (Green).

Field Characters (Plate 81) Adult female cover unusually small, elongate, oyster-shell shaped, often irregular in shape due to irregularities of bark, yellow brown, often covered with whitish wax, slightly convex; shed skins marginal, reddish brown or bright yellowish brown. Male cover smaller, felted, white; shed skin marginal, yellow brown. Body of adult female and crawlers yellow. In cracks of bark, cryptic; when infestations heavy, also on fruit.

Slide-mounted Characters Adult female (Fig. 85) with 1 or 2 pairs of definite lobes; paraphyses variable usually with 1 pair between median lobes. Median lobes separated by space 0.7–1.4 (1.0) times width of median lobe, with elongate, weakly sclerotized paraphysis-like sclerotization attached to each of medial and lateral margins, without basal sclerotization or yoke, medial margins parallel, lateral margins converging or parallel, with 0–1 (0) lateral notch, 0–1 (0) medial notch; second lobes usually present, bilobed, slightly smaller than median lobes, medial lobule largest, medial and lateral lobules usually without notches, sometimes with a series of small lateral lobes on medial lobule. Gland-spine formula 1-1-1, gland spine in second and third spaces frequently difficult to discern, with 1–2 (1) gland spines near each body margin on segment 4 and sometimes segment 3; with 2 gland spines between median lobes. Macroducts of 2 or 3 variable sizes, largest in marginal areas of pygidium and segment 4, with 4–6 (5) such ducts, medium size in submedial and submarginal areas of pygidium and segment 4, small size on dorsum in marginal and submarginal areas from mesothorax or metathorax to segment 4, in submedial areas from metathorax, segment 1, or segment 2 to segment 4, sometimes present in medial areas of any or all of segments 2 to 4, on venter in marginal and submarginal areas from head, prothorax, or mesothorax to segments 3 or 4, sometimes in submedial areas of segments 2 or 3, without duct between median lobes, macroduct in first space 20–27 (22) μ long, with 11–20 (14) macroducts on each side of body on segments 5 to 7, total of 91–154 (116) macroducts on each side of body, some macroduct orifices anterior of anal opening. Pygidial microducts on venter in submarginal areas of segments 5 to 7 or 8, with 3–15 (10) ducts; prepygidial ducts of 1 size, in medial areas of any or all of segments 1 to 4 and on head, in small group near each spiracle and posterior of labium, in submedial areas of segments 1 or 2 to segment 4; pygidial microducts absent from dorsum; prepygidial microducts, when present, near submedial macroducts on mesothorax and/or metathorax. Perivulvar pores absent. Perispiracular pores with 3 loculi, anterior spiracles with 2–5 (3) pores, posterior spiracles with 0–4 (1) pores. Anal opening located 8–14 (12) times length of anal opening from base of median lobes, anal opening 9–18 (12) μ long. Dorsal seta laterad of median lobes 1.5–3.4 (2.0) times length of median lobe. Eyes usually absent, when present, represented by small lightly sclerotized area, on body margin at level of anterior margin of mouthparts. Antennae usually with 2 short setae. Cicatrices absent. Body elongate. Posterior end of body appears slightly rolled giving appearance of fold over lobes and causing median lobes to appear slightly sunken with 2 or more small knobs between median lobes. Older adult females with part of head sclerotized.

Affinities Hall scale is unique among the economic species of the United States by having 1 or 2 pairs of lobes, no perivulvar pores, small body size, and macroducts of 2 sizes on the pygidium. Hall scale shows some similarity to certain economic species of *Lepidosaphes*, but the latter differ by having perivulvar pores and a relatively large body size.

Hosts Borchsenius (1966) records Hall scale from Rosaceae only, including *Amygdalus, Cydonia, Malus, Persica, Prunus, Pyrus,* and *Spiraea*; Bazarov and Shmelev (1971) also report it on *Caragana*. In California, McKenzie (1956) reports it from *Cydonia, Prunus,* and *Spiraea*. We have examined specimens from *Amygdalus, Prunus,* and *Spiraea*.

Distribution This scale probably is native to the near East. We have examined U.S. specimens from CA where it is reported to be eradicated (Gill 1997). Nakahara (1982) reports Hall scale from Egypt, Greece, the former Soviet Union, and several countries in the near East. This species occasionally is taken in quarantine from Afghanistan, Egypt, Iran, Syria, and Turkey.

Biology In Israel there are population peaks of crawlers in April, June–July, and September–October suggesting there are 3 generations per year (Berlinger et al. 1984). The gravid adult females are the overwintering stage and males are reported (Fallek et al. 1988). Fosen et al. (1953) found non-gravid females as the overwintering stage in California. They indicated that crawlers began to emerge in April or late March; the first molt took place in mid-May, adult female and males were present in late June, gravid females were present in late July, and by mid-October all stages could be located. Based on this information they concluded there was at least a partial second generation.

Economic Importance This species was considered to be a serious enough pest that when it was first detected in California a major eradication program was implemented (Fosen et al. 1953). Ebeling (1959) believed that Hall scale would become more serious on almonds than either San Jose scale or European fruit lecanium if it were not eradicated from California. Fosen et al. (1953) suggested that the most important injury by this pest results from infestation of the fruit, where it makes conspicuous blotches, especially on peaches. In Israel, settling sites of Hall scale develop pits and red spots on the skin of ripe fruit and are an economic problem on peaches and nectarines (Berlinger et al. 1984). Hall scale is considered to be an important pest in the southern part of the Palearctic region, especially on stone fruits (Kozár 1990a). It is mentioned as a serious pest in Tadzhikistan by Bazarov (1962). Miller and Davidson (1990) consider it to be a serious pest of a small area of the world.

Selected References Berlinger et al. (1984); Fosen et al. (1953); McKenzie (1956).

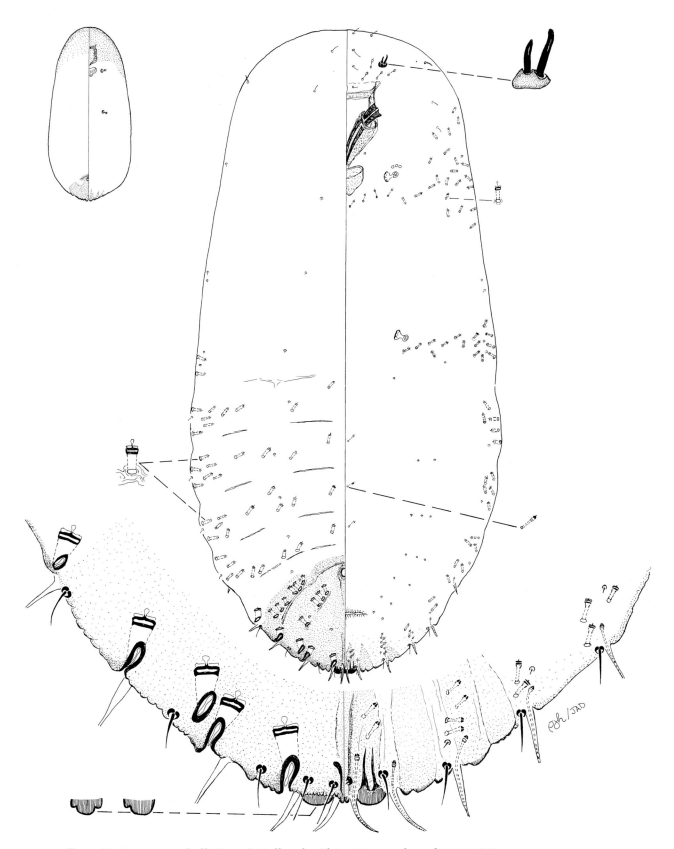

Figure 85. *Mercetaspis halli* (Green), Hall scale, Chico, CA, on almond, V-12-1950.

Morganella longispina (Morgan)

Suggested Common Name Plumose scale (also called Maskell scale).

Common Synonyms and Combinations *Aspidiotus longispina* Morgan, *Aspidiotus (Morganella) maskelli* Cockerell, *Aspidiotus (Morganella) longispina* Morgan, *Hemiberlesia longispina* (Morgan), *Aspidiotus longispina* var. *ornata* Maskell, *Hemiberlesia maskellii* (Cockerell), *Aspidiotus maskelli* Cockerell, *Morganella maskelli* (Cockerell), *Aspidiotus ornatus* Maskell.

Field Characters (Plate 82) Adult female cover convex, circular to slightly oval, gray to black; unusually thick, with ventral cover as thick as dorsal cover, bivalved (Foldi 1990); shed skins central to subcentral, black when rubbed. Male cover similar in texture and color to female cover, elongate; shed skin submarginal, black when rubbed. Body of young adult female white, older female light yellow; eggs yellow; crawlers yellow. On bark and fruit, rarely on leaves and roots. Crawler flap of adult female cover distinctive, turned up.

Slide-mounted Characters Adult female (Fig. 86) with 1 pair of lobes; with small paraphysis laterad of median lobe and in space where lobe 2 would be. Median lobes closely appressed medially, with conspicuous basal sclerosis about equal to length of lobe, without yoke, lateral margins of lobes convergent apically, medial margins parallel, with 1 lateral notch, 0–1 (0) medial notch. Plate formula difficult to interpret without intervening lobes but using dorsal setae as landmarks usually 3-3-4 but also 3-3-5, 3-4-5, 3-4-4, microducts associated with plates, with 1 or more protruding microducts on lateral margin of segments 3 and/or 4; plates broadly fimbriate, increasing in length posteriorly, unusually long for diaspidid; without plates between median lobes. Macroducts about equal in size, except duct between median lobes and nearest anal opening which is longer, with 2 or 3 ducts between median lobes, longest extending 1.9–3.2 (2.4) times distance between posterior apex of anal opening and base of median lobes, 38–52 (44) μ long, macroduct in first space difficult to measure, when visible about 45 μ long, with 7–12 (9) macroducts on each side of body of segments 5 or 6 to 8, without prepygidial macroducts, some orifices occurring anterior of anal opening. Pygidial microducts on venter approximately same size as macroducts in submarginal clusters on segments 5, 6, and sometimes 7, with 3–8 (5) ducts on each side of body; prepygidial ducts on venter in submedial and submarginal areas from head to pygidium; pygidial microducts on dorsum absent; prepygidial ducts on dorsum of 2 sizes, shorter size in submarginal clusters from prothorax or mesothorax to segment 4, longer size in submedial areas from near anterior spiracle to pygidium. Perivulvar and perispiracular pores absent. Anal opening located 1.1–2.6 (1.8) times length of anal opening from base of median lobes, anal opening 7–15 (10) μ long. Dorsal seta laterad of median lobes 0.4–0.7 (0.6) times length of lobe. Eyes normally represented by small sclerotized spot on body margin laterad of anterior spiracle. Antennae each with 1 large seta. Cicatrices usually submarginal on prothorax, segment 1, and 3, sometimes with submedial cicatrix-like structure anterior of anterior spiracle and submedial on segment 3. Dorsal and ventral setae on lateral margins of segments 4 to 7 unusually long for armored scale. Sclerotized areas on pygidium on dorsal and ventral surfaces.

Affinities Of the economic species, plumose scale is most similar to *Hemiberlesia palmae*, palm scale, as both species have a pygidial fringe formed by long, highly branched plates. They can be distinguished because plumose scale lacks perivulvar pores, lacks second and third lobes, has the setae of segments 5 and 6 longer than the median lobes, and has a small anal opening between the median lobe scleroses. Palm scale has perivulvar pores, has second and third lobes, has the setae of segments 5 and 6 shorter than the median lobes, and has a large anal opening anterior of the median lobe scleroses.

Hosts Borchsenius (1966) records this species from 36 genera in 26 families. We have examined specimens from *Aleurites*, *Artocarpus*, *Aucoumea*, *Camellia*, *Carica*, *Citrus*, *Ficus*, *Hibiscus*, *Ligustrum*, *Loranthus*, *Mangifera*, *Michelia*, *Nerium*, *Orania*, and *Pelagodoxa*. In addition, it is reported on *Averrhoa*, *Bauhinia*, *Eugenia*, *Jasminum*, *Lagerstroemia*, *Persea*, *Psidium*, and *Tecoma* (Williams and Watson 1988). Hamon (1981) found it on *Callistemon* and *Severinia*, and Hodgson and Hilburn (1990) reported it on *Celtis*.

Distribution Plumose scale probably is a native of South America. In the United States it occurs in FL and HI. In 1980 it was discovered to be established in the United States on *Ligustrum sinense* and *Nerium oleander* at Dodge Island Seaport, Miami, Florida, and other localities in the Miami region (Hamon 1981). Through the years it often has been intercepted in quarantine on hosts such as *Citrus* and *Nerium* from various Caribbean islands and South America. This is a cosmopolitan species found throughout the tropics.

Biology Plumose scale normally occurs on the branches and fruit of its host; although Ogilvie (1926) reports it as occurring several feet underground on the roots of *Nerium* in Bermuda. We have seen adult females with eggs and active crawlers on material from Miami, Florida, collected in early January. According to Balachowsky (1926), in Algeria the species does not lay eggs but is viviparous and has several generations per year. Balachowsky describes the scales of the adult female and male and provided a partial illustration and description of the first instar.

Economic Importance Ebeling (1959) treats it as a minor pest of *Citrus* in Brazil. Cohic (1958) notes this pest sometimes killed *Bauhinia* and *Jasminum* (jasmine) and caused 'cankerous' swellings on fig in New Caledonia. Cunningham (1929) reports that plumose scale caused much injury to fig and killed a number of papaya plants in Bermuda. Webster (1920) cites it as a pest of mango in the Philippines. It is considered to be a pest of: fig in Algeria (Balachowsky 1926); citrus in Haiti (Dekle 1964); fig and oleander in Bermuda (Hodgson and Hilburn 1991); fig in Florida (Mead 1991); tea in India (Nagarkatti and Sankaran 1990); grapefruit, lemon, and fig in the South Pacific (Williams and Watson 1988); citrus in Brazil (Wolff and Corseuil 1993); and mango in Japan (Kinjo et al. 1996). Swezey (1950) found it to be very abundant on avocado leaves in Hawaii. Miller and Davidson (1990) consider this species to be an occasional pest.

Selected References Balachowsky (1926); Ferris (1938a).

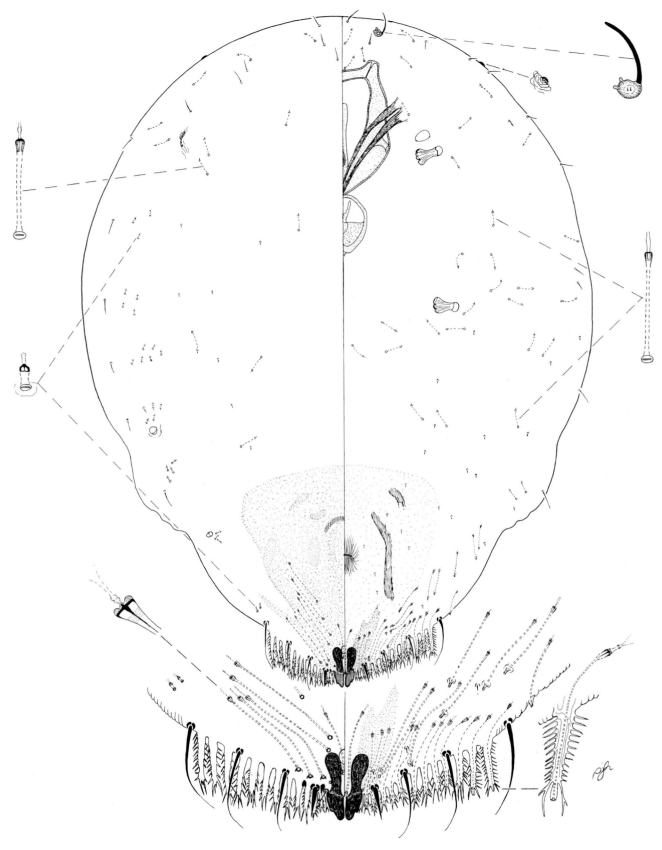

Figure 86. *Morganella longispina* (Morgan), plumose scale, St. Croix, Virgin Islands, on *Citrus sinensis*, VI-2-1977.

Plate 82. *Morganella longispina* (Morgan), Plumose scale

A. Circular adult female scale cover on *Ligustrum* (J. A. Davidson).
B. Female scale covers under bark debris on *Ligustrum* (J. A. Davidson).
C. Elongate adult male cover on *Ligustrum* (J. A. Davidson).
D. Opened male cover on *Ligustrum* (J. A. Davidson).
E. Two halves of male cover with dead second instar on *Ligustrum* (J. A. Davidson).

F. Male (bottom) and female (top) covers on *Ligustrum* (J. A. Davidson).
G. Distance view of covers on *Ligustrum* (J. A. Davidson).
H. Distance view of heavily infested *Ligustrum* twig (J. A. Davidson).
I. Cover turned to show body of adult female and yellow crawlers on *Ligustrum* (J. A. Davidson).

292

Plate 83. *Neopinnaspis harperi* McKenzie, Harper scale

A. Possibly adult male cover on island cherry (D. R. Miller).
B. Adult female cover on carob (J. A. Davidson).
C. Adult female cover on toyon (J. A. Davidson).
D. Distance shot of infestation on bark of island cherry (J. A. Davidson).
E. Distance shot of infested bark on island cherry (J. A. Davidson).
F. Distance shot of infestation on toyon (J. A. Davidson).
G. Pink body of adult female (D. R. Miller).

293

Neopinnaspis harperi McKenzie

Suggested Common Name Harper scale.

Common Synonyms and Combinations *Africaspis harperi* (McKenzie).

Field Characters (Plate 83) Adult female cover elongate, narrow, generally with irregular margins, white to light brown; shed skins marginal; bright bronze when rubbed. Males not observed and apparently nonexistent. Body of adult female light purple or red; eggs light purple. On bark of twigs and branches.

Slide-mounted Characters Adult female (Fig. 87) with 3 pairs of well-developed lobes; fourth- and fifth-lobe areas occasionally sclerotized and represented by series of small points; inconspicuous paraphysis-like sclerotization attached to medial margin of median lobes. Median lobes nearly touching, less than 1 μ apart, without basal sclerosis and yoke, medial margin axes parallel, lateral margins diverging apically, with 1–2 (2) lateral notches, without medial notches; second lobes simple, smaller than median lobes, with 2 lateral notches, without medial notches; third lobes simple, smaller than second lobes, with 1–3 (2) notches; fourth and fifth lobes usually absent, rarely represented by series of small sclerotized points. Gland-spine formula normally 1-2-3, occasionally 0-2-3, 2-2-3, or 1-2-2; gland spines in first space inconspicuous, 16–26 (22) gland spines near each body margin anterior of lobe 4 area, present to metathorax or segment 1; with 2 inconspicuous gland spines between median lobes. Macroducts of 2 sizes, smaller size similar in size to microducts, larger size on margin from segments 4 to 7, smaller size on submargin of segments 6 to 8, on submedial area of segment 6, forming segmental bands across segments 3 to 5, restricted to marginal areas on mesothorax to segment 2, with inconspicuous duct slightly anterior of base of median lobes 20–27 (25) μ long, macroduct in first space 17–30 (19) μ long, with 14–23 (19) macroducts on each side of body on segments 5 to 8, total of 81–132 (109) macroducts on each side of body, some orifices occurring anterior of anal opening, ventral macroducts in band across metathorax and on marginal areas of segments 1 and 2 and sometimes 3. Pygidial microducts on venter of segments 5 to 6 or 7 with total of 7–9 (8) ducts; prepygidial ducts in submedial areas of segments 1 or 2 to 4 and near mouthparts; dorsal microducts indistinguishable from macroducts. Perivulvar pores in 5 groups, 9–12 (10) pores. Perispiracular pores with 3 loculi, anterior spiracles each with 1–2 (2) pores, posterior spiracles without pores. Anal opening located 10–12 (10) times length of anal opening from base of median lobes, anal opening 10–12 (12) μ long. Dorsal seta laterad of median lobes 1.1–1.7 (1.2) times length of lobes. Eyes represented by weakly sclerotized spot laterad of antennae. Antennae each with 2 conspicuous setae. Cicatrices absent. Body shape elongate. Occasionally with microduct near antennae. Conspicuous sclerotized areas attached to first, second, and third lobes; sclerosis near anal ring.

Affinities Harper scale is unlike any other species reported in the United States. The occurrence of closely appressed median lobes, an apically acute pygidium, 2 sizes of macroducts, and an elongate body are a distinctive combination of characters. The species is most similar to species of *Africaspis*, a genus unknown in the United States. *Africaspis* species differ by having only 1 size of macroduct on the pygidial dorsum.

Hosts Polyphagous, Borchsenius (1966) records it from 44 genera in 26 families; Dekle (1977) reports it from 15 genera; McKenzie (1956) reports it from 38 genera; and Beshear et al. (1973) found it on 20 genera in Georgia. We have examined specimens from *Ceratonia, Eriobotrya, Macadamia, Persea*, and *Prunus*. It is common on *Ilex* in the southeastern United States (McComb 1986).

Distribution We have examined U.S. specimens from CA, FL, GA, HI. It has been reported from outside the United States in Japan and Taiwan and probably is native to the Old World.

Biology We have been unable to find an account of the biology of this species. According to McKenzie (1956), Harper scale primarily occurs on the bark of twigs and branches; in cases of heavy infestations, it may be found on the trunk, occasionally on petioles, and rarely on leaf surfaces.

Economic Importance Harper scale is reported as a pest of apple in the eastern part of the former Soviet Union (Konstantinova 1976). In the United States it infests a wide range of host plants indicating the potential for economic importance. Miller and Davidson (1990) consider this species to be an occasional pest.

Selected References Gill (1997); McKenzie (1956).

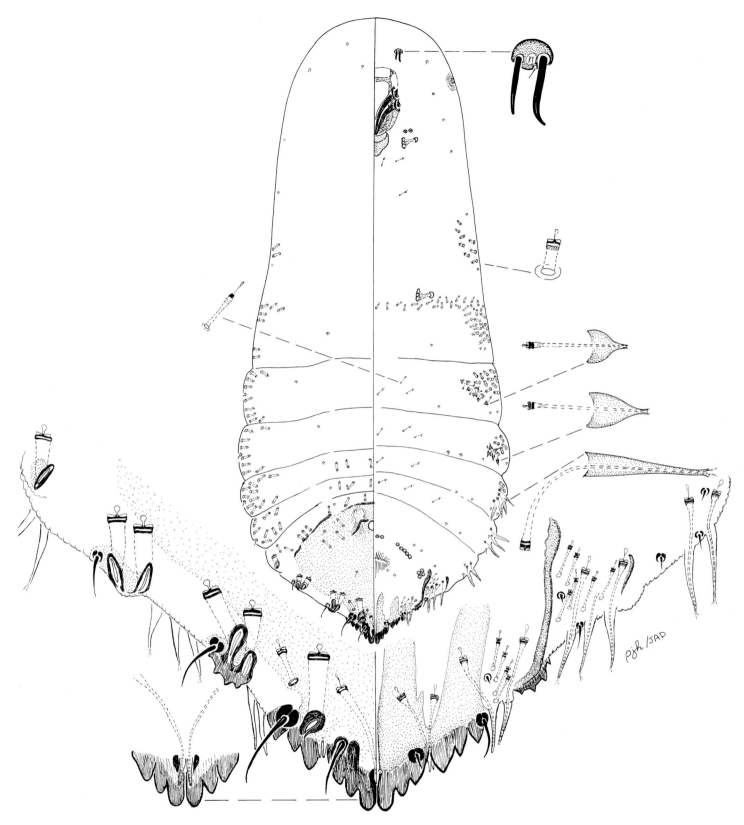

Figure 87. *Neopinnaspis harperi* McKenzie, Harper scale, Gainesville, FL, on *Eriobotrya japonica*, V-22-1946.

Nuculaspis californica (Coleman)

ESA Approved Common Name Black pineleaf scale.

Common Synonyms and Combinations *Aspidiotus pini* Comstock, *Aspidiotus californicus* Coleman, *Chrysomphalus californicus* (Coleman), *Aspidiotus (Nuculaspis) californicus* Coleman.

Field Characters (Plate 84) Adult female cover shape influenced by diameter of needle, usually elongate oval, gray to black or blackish brown with lighter margin; shed skins central to subcentral, yellow when rubbed. Male cover similar in texture and color to female cover, elongate; shed skin central to subcentral, yellow. Body of newly mature adult females yellow, older females dark brown with yellow pygidium and pygidium retracted into body, reniform; eggs yellow; crawlers yellow. On needles.

Slide-mounted Characters Adult female (Fig. 88) with 4 pairs of lobes; without paraphyses. Median lobes separated by space 0.8–1.5 (1.0) times width of median lobe, without sclerosis or yoke, medial margin axes parallel or diverging apically, lateral margins parallel or diverging apically, with 1–2 (1) lateral notches, 1 medial notch; second lobes simple, normally same shape as median lobes, slightly smaller, occasionally wider, with 0–3 (2) lateral notches, 0–2 (1) medial notches; third lobes simple, wider than second lobes, with 3–6 (5) lateral notches, 0–1 (1) medial notches; fourth lobes usually simple, rarely with 2 lobules, wider than third lobes, with 3–11 (5) lateral notches, 0–1 (0) medial notches. Plate formula normally 2-3-3 or 2-3-4, occasionally 2-2-4 or 2-3-5, without associated microducts; plates broadly fimbriate, generally becoming thinner anteriorly; with 2 broad plates between median lobes, about as long as median lobes. Macroducts on pygidium approximately same size, those on anterior abdominal segments and thorax slightly smaller, 1 macroduct between median lobes 42–53 (53) µ long, extending 0.8–1.1 (0.9) times distance between posterior apex of anal opening and base of median lobes, macroduct in first space 40–49 (43) µ long, with 20–33 (27) macroducts on each side of pygidium on segments 5 to 8, marginal ducts on segments 5 to 8, submarginal ducts on segments 5 to 7; total of 63–110 (79) macroducts on each side of body; some macroduct orifices occurring anterior of anal opening; prepygidial macroducts on margin and submargin of meso- and metathorax to segment 4. Pygidial microducts on venter on margin or submarginal areas of segments 5 and 6, with 2–6 (4) ducts on each side of body; prepygidial ducts on venter in submedial areas near mouthparts, on segment 1, sometimes also submedial between spiracles and on segments 2 and 3, submarginal ducts on segments 2 to 3 or 3, marginal ducts on segments 1 to 4; pygidial microducts absent from dorsum; prepygidial ducts on dorsum in marginal or submarginal areas from head to metathorax, occasionally absent from head and anterior thorax. Perivulvar pores unusually variable, in 2–9 (4) clusters, with 9–32 (22) pores on each side of pygidium. Perispiracular pores absent. Anal opening located 2–3 (2) times length of anal opening from base of median lobes, anal opening 18–28 (23) µ long. Dorsal seta laterad of median lobes 0.8–1.5 (1.1) times length of median lobe. Eyes represented by small, dome-shaped area laterad of antenna. Antennae normally each with 1 conspicuous seta. Cicatrices often absent, occasionally on prothorax and segments 1 and 2. Body shape elongate oval. Space between median lobes with 2 small setae. Macroducts forming marginal band from meso- or metathorax to pygidial apex. At maturity body sclerotized and pygidium retracted into body.

The western and eastern 'forms' of this species are somewhat different. The western form typically has 5 or more clusters of perivulvar pores and few or no dorsomarginal microducts on the thorax and head. The eastern form has 2 or 3 clusters of perivul-vars and a band of dorsomarginal microducts on the head and thorax. These characters are not always consistent.

Affinities Black pineleaf scale does not resemble other economic diaspidids. Of the pine infesting armored scales, only *Nuculaspis* species retract the pygidium at maturity. Unretracted black pineleaf scale females superficially resemble females of *Aspidaspis florenciae* (Coleman), which also occur on pine needles, but the latter has only 2 pairs of lobes that are closely appressed while the former has 4 pairs of widely spaced lobes. Within *Nuculaspis*, black pineleaf scale resembles *N. apacheca* Ferris, but the former has a more posteriorly placed anal opening, small lobe 3 and 4, and appears to have an incompletely retracted pygidium.

Hosts Borchsenius (1966) records this species from *Pinus*, *Pseudotsuga*, and *Tsuga*; McKenzie (1956) reports it from *Pinus* and *Pseudotsuga* in California; and Dekle (1977) notes that it had only been collected on *Pinus* in Florida. We have examined specimens from *Pinus* and *Pseudotsuga*.

Distribution We believe black pineleaf scale is native to North America. We have examined U.S. material from AZ, CA, CO, CT, FL, GA, ID, IL, MA, ME, MI, MN, MO, NC, NH, NJ, NY, OH, OR, SC, TN, TX, VA, WA, WI. Specimens also are at hand from Canada and Mexico. One specimen was seen from *Pinus* in Haiti at 5000 feet.

Biology Black pineleaf scale has been reported only from conifer needles. According to Johnson and Lyon (1976) and Luck and Dahlsten (1974), this species has been associated with population increases of the pine needle scale, *Chionaspis pinifoliae* (Fitch). Johnson and Lyon (1976) state that 1 to 3 generations occur per year depending on the geographic location; immature stages overwinter, and eggs or crawlers are deposited over a long period. In a study of black pineleaf scale ecology in Washington, Edmunds (1973) found about 40 eggs internally per female. These hatched over a 2-week period beginning about mid-July when the new needles had nearly completed growth elongation. Crawlers tended to settle on new growth where they fed and grew until October. Normally feeding resumed the following April, and growth was completed by later May. Adult males emerged about mid-June and tended to be diurnal in their time of flight (Alstad et al. 1980). Edmunds (1973) notes that on individual trees, scale population size was directly proportional to tree age. He hypothesized that a genetically distinct subpopulation (deme) of the scale was selected for its fitness to combat the specific chemical defense mechanisms of each infested tree in a stand. In a series of papers after this prediction Edmunds and Alstad and their coworkers (Alstad et al. 1980; Alstad and Edmunds 1983, 1987; Alstad and Corbin 1990; Alstad 1998) provided a great deal of information on the effect of host plant variability on this scale insect. They demonstrated that there is a higher proportion of males produced in populations that are poorly adapted to a particular host, but in successive successful generations there were proportionally more females and fewer males (Alstad and Edmunds 1983). They provided evidence that supported the hypothesis that offspring from matings between individual scales from different plant specimens produced locally maladapted offspring. The net result was that densities on infested trees that were interdigitated or located close to one another had lower densities of scales than trees that were isolated (Alstad and Edmunds 1987). Young et al. (1993) provide data that female crawlers move greater distances from the mother than males and seek out younger needles. This provided support for the hypothesis that female crawlers seek nutritionally rich needles, and males remain on nutritionally inferior needles damaged by the mother's feeding. Alstad and Corbin

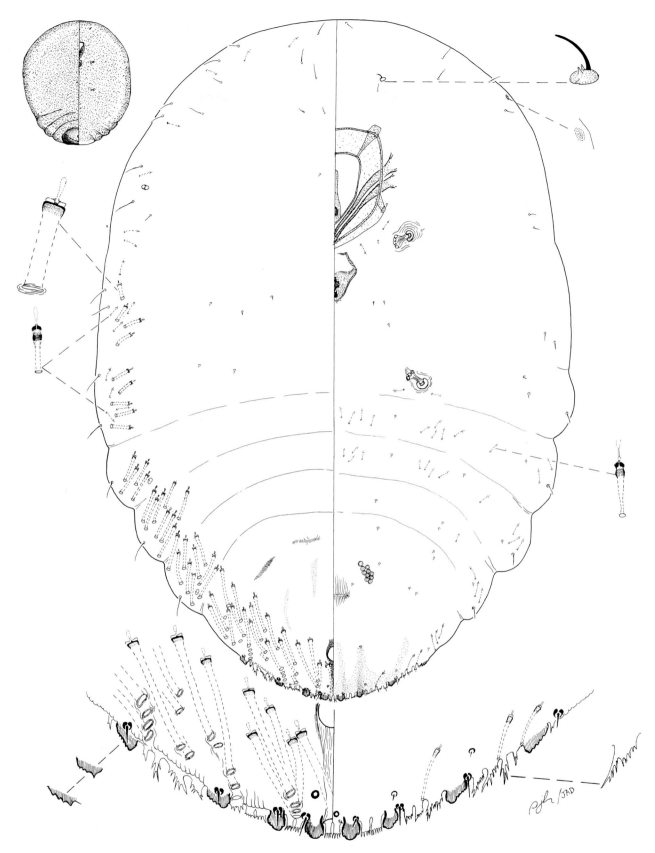

Figure 88. *Nuculaspis californica* (Coleman), black pineleaf scale, Ordell, NJ, on *Pinus resinosa*, IV-?-1961.

(1990) showed that the allelic frequencies of three enzyme systems were unique to populations of black pineleaf scale in different pine plots, on different trees, and even on different twigs of individual trees. This gave strong support to the hypothesis that demes of the scale become adapted to the host over time. A recent summary was presented by Alstad (1998).

Economic Importance Johnson and Lyon (1976) make the following comments: The black pineleaf scale is primarily an economic pest on Douglas fir and several pine species in western North America. Heavy infestations cause premature needle drop while extremely high populations prevent full needle elongation and may kill trees. Heavily infested trees exhibit thin crowns, followed by reddish discoloration, chlorosis, and finally necrosis of the needles. Trees growing under common urban conditions such as moisture deficit, soil compaction, root damage, smog, and dust appear to be particularly susceptible to attack. Edmunds (1973) notes that this scale is occasionally of considerable economic importance, causing death of various pines, especially of ponderosa pine in western North America. He commented on one remarkable infestation that has persisted at high damage levels for 20 years in Spokane County, Washington. Baker (1972) reports that this species has caused serious damage to pines in nurseries and plantations, and to natural tree reproduction in the Lake states. Beardsley and González (1975) consider this to be one of 43 major armored scale pests. Miller and Davidson (1990) consider this species to be a serious pest in a small area of the world.

Selected References Alstad (1998); Edmunds (1973); Ferris (1938a).

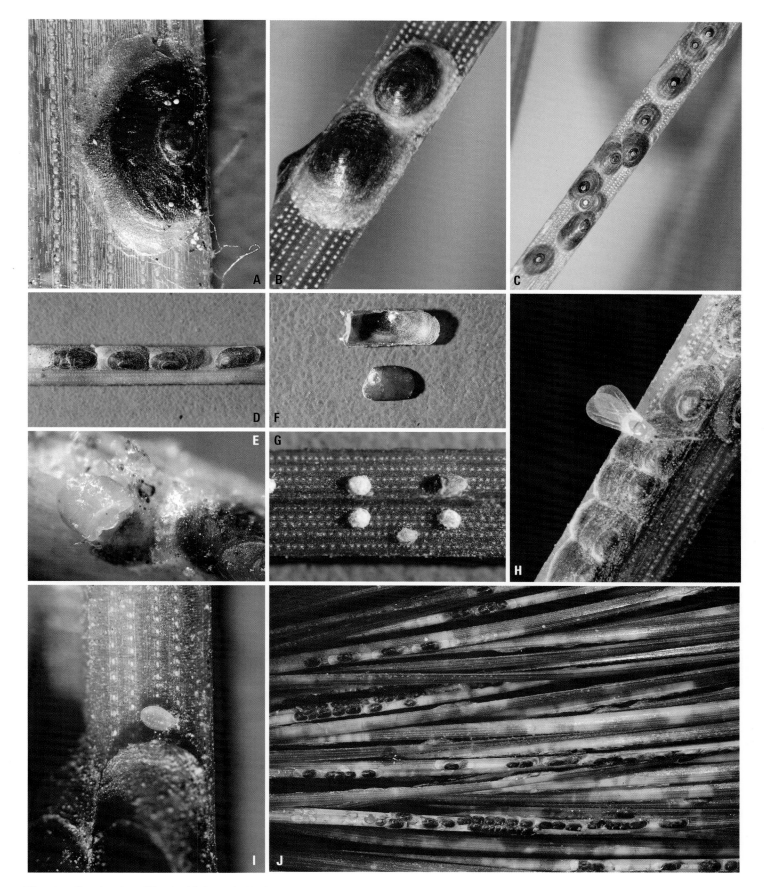

Plate 84. *Nuculaspis californica* (Coleman), Black pineleaf scale

A. Adult female cover on black pine (J. A. Davidson).
B. Adult female covers on digger pine (R. J. Gill).
C. Distance view of covers on digger pine (R. J. Gill).
D. Female covers (left) and male cover (right) on pine (J. A. Davidson).
E. Body of young adult female on pine (J. A. Davidson).
F. Body of mature asymmetrical adult female about to lay eggs (J. A. Davidson).

G. Capped crawlers on pine (J. A. Davidson).
H. Adult male mating with female under cover on pine (D. N. Alstad).
I. Crawler on pine (J. A. Davidson).
J. Distance view of damage on pine (J. A. Davidson).

Nuculaspis pseudomeyeri (Kuwana)

Suggested Common Name False Meyer scale.

Common Synonyms and Combinations *Aspidiotus pseudomeyeri* Kuwana, *Tsugaspidiotus pseudomeyeri* (Kuwana).

Field Characters (Plate 85) Adult female cover elongate oval, brown, strongly convex; shed skins central to subcentral, light orange or yellow. Male cover similar in texture and color to female cover, oval, not as convex; shed skin subcentral, yellow. Body of adult female yellow; eggs and crawlers probably yellow. On needles.

Slide-mounted Characters Adult female (Fig. 89) with 2–3 (3) pairs of lobes; usually with small paraphyses on medial and lateral margins of median and second lobes and with paraphysis on medial margin of third lobe, when present. Median lobes separated by space 0.4–0.8 (0.6) times width of median lobe, without sclerosis or yoke, medial margin axes parallel or diverging apically, lateral margins parallel or converging apically, with 1–2 (1) lateral notches, 0–1 (1) medial notch; second lobes simple, normally same shape as median lobes, slightly smaller, with 1–2 (1) lateral notches, 0–1 (1) medial notches; third lobes present in 7 of 10 specimens examined, simple, represented by sclerotized point or rounded projection, narrower and smaller than second lobes, with 0–3 (1) lateral notches, 0–1 (0) medial notches; fourth lobes usually absent, rarely with small sclerotized area. Plate formula normally 2-3-0, rarely 3-3-0, without associated microducts; plates narrowly fimbriate; with 2 narrow plates between median lobes, about as long as, or slightly shorter than median lobes. Macroducts on dorsal pygidium and prepygidium approximately same size, those on venter slightly smaller, without macroduct between median lobes, macroduct in first space 35–42 (38) μ long, with 10–15 (13) macroducts on each side of pygidium on segments 5 to 8, marginal ducts on segments 5 to 8, submarginal ducts on segments 5 to 7; total of 25–38 (30) macroducts on each side of body; some macroduct orifices occurring anterior of anal opening; dorsal prepygidial macroducts on margin and submargin of segments 2 or 3 to 4; ventral prepygidial macroducts on margin of mesothorax to segments 3 or 4. Pygidial microducts on venter on margin or submarginal areas of segments 5 and/or 6, with 2–5 (3) ducts on each side of body; prepygidial ducts variable, when abundant present on venter in submedial areas near mouthparts, submarginal near spiracles and on segments 1 to 3, marginal on head, thorax, and segments 3 to 4, when in small numbers restricted to submarginal areas near spiracles, on any of segments 1 to 3, and marginal on segment 4; pygidial microducts absent from dorsum; prepygidial ducts on dorsum in marginal or submarginal areas from head to metathorax, occasionally absent from head and anterior thorax, often present submargin-ally or submedially on 1 or more of segments 1 to 3. Perivulvar pores in 4 indistinct clusters, with 2–7 (5) pores on each side of pygidium. Perispiracular pores absent. Anal opening located 0.6–1.4 (1.0) times length of anal opening from base of median lobes, anal opening 10–16 (13) μ long. Dorsal seta laterad of median lobes 0.5–1.0 (0.8) times length of median lobe. Eyes represented by distinct sclerotized spur with 1–3 (1) projections, located on body margin laterad of anterior edge of clypeolabral shield. Antennae normally each with 1 conspicuous seta. Cicatrices generally indistinct, but usually present on prothorax and segment 1. Body shape broadly oval. Space between median lobes with 2 small setae. Macroducts forming marginal band from meso- or metathorax to pygidial apex. Body not sclerotized at maturity.

Affinities False Meyer scale is remarkably similar to *Abgrallaspis degenerata*, degenerate scale, by having the dorsal macroducts scattered along the body margin from segment 1 to segment 7 or 8, 3 pairs of lobes with the third pair much smaller, and long narrow plates between the lobes. False Meyer scale differs by lacking plates anterior of lobe 3 and no macroduct between the median lobes; *A. degenerata* has simple plates anterior of lobe 3 and a macroduct between the median lobes.

Hosts Borchsenius (1966) records this species from *Chamaecyparis* and *Juniperus*; Nakahara (1982) reports it from *Cedrus, Chamaecyparis, Juniperus, Thuja*, and *Tsuga*. We have examined specimens from the same hosts as Nakahara excluding *Tsuga*. Paik and Kim (1977) found it on *Biota, Chamaecyparis*, and *Tsuga* in Korea. The species commonly is taken in quarantine on *Chamaecyparis* and *Thuja*.

Distribution We believe that false Meyer scale is native to Japan. We have examined U.S. material from DC and NY; it also occurs in PA (Nakahara 1982; Stimmel 2001, personal communication). Numerous specimens are at hand from Japan. Paik and Kim (1977) reported the species in Korea. Single collections of material are available from New Zealand (adult female) and China (second instar), but these need confirmation.

Biology We have been unable to locate published information on the life history of this species. Stimmel (2002, personal communication) suggested that the species overwinters as second instars.

Economic Importance The only record of this species as a pest is in Pennsylvania in 2001 and 2002 where it has caused economic damage to conifers in nurseries (Stimmel 2002, personal communication). These may be isolated incidents but could forebode more widespread problems in the future.

Selected References Kosztarab (1996); Paik and Kim (1977).

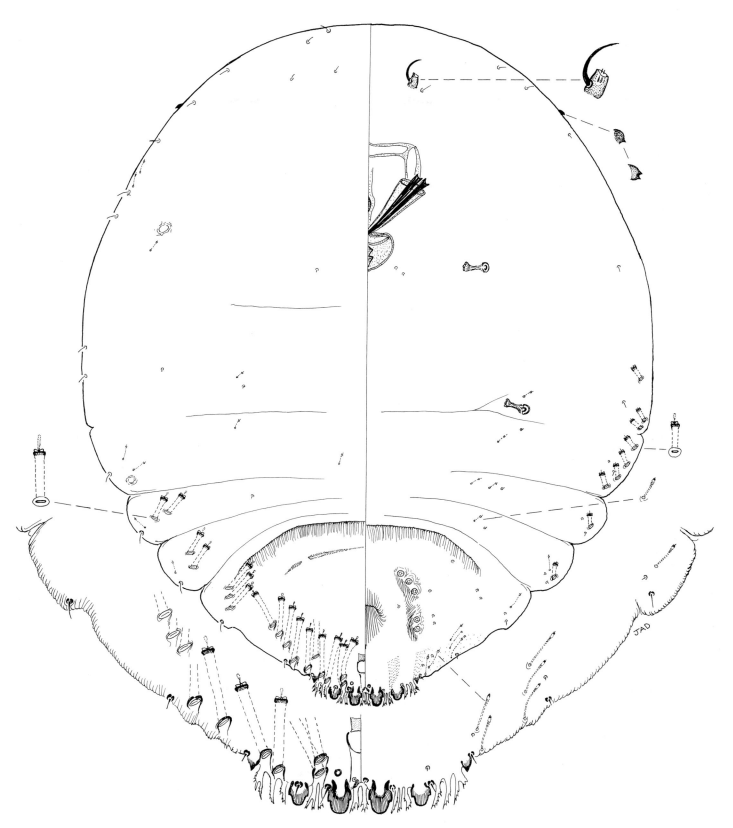

Figure 89. *Nuculaspis pseudomeyeri* (Kuwana), false Meyer scale, Japan, intercepted at Anchorage, AK, on *Thuja* sp., VIII-3-1984.

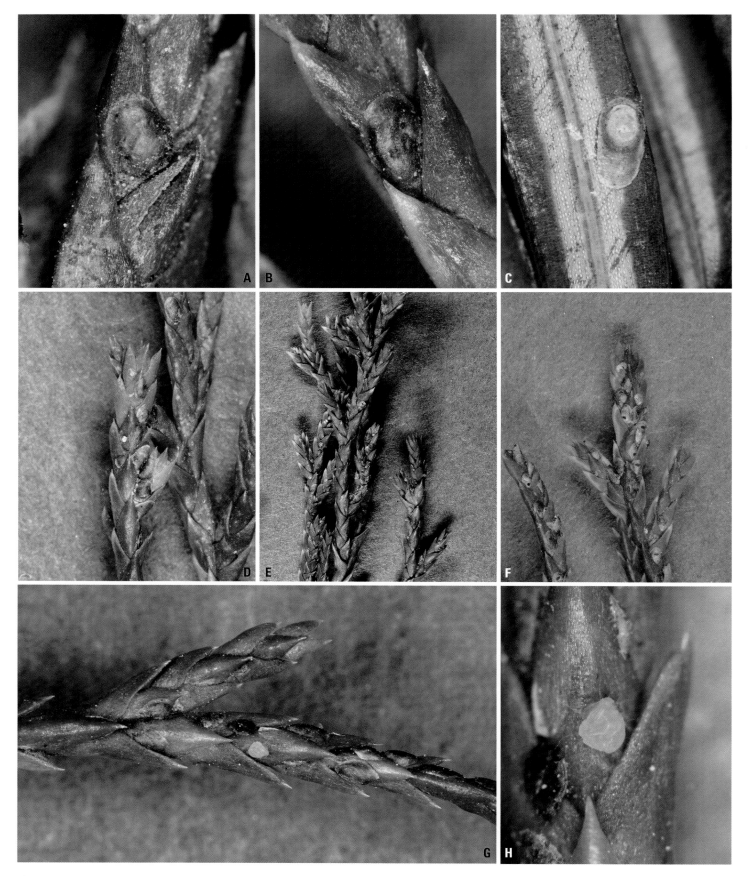

Plate 85. *Nuculaspis pseudomeyeri* (Kuwana), False Meyer scale

A. Adult female cover on juniper (J. A. Davidson).
B. Adult female cover on juniper (J. A. Davidson).
C. Male cover on hemlock (J. A. Davidson).
D. Distance view on juniper (J. A. Davidson).
E. Distance view on juniper (J. A. Davidson).

F. Distance view with parasite emergence holes in covers on juniper (J. A. Davidson).
G. Distance view showing female body on juniper (J. A. Davidson).
H. Body of adult female (J. A. Davidson).

Plate 86. *Nuculaspis tsugae* (Marlatt), Shortneedle conifer scale

A. Adult female cover on white spruce (J. A. Davidson).
B. Male covers on white spruce (J. A. Davidson).
C. Distance view of infestation on white spruce (J. A. Davidson).

D. Yellowing of needles on white spruce (R. A. Casagrande).
E. Necrosis of needles on white spruce (R. A. Casagrande).
F. Dieback on heavily infested white spruce (R. A. Casagrande).

Nuculaspis tsugae (Marlatt)

Suggested Common Name Shortneedle conifer scale (also called shortneedle evergreen scale).

Common Synonyms and Combinations *Aspidiotus* (*Diaspidiotus*) *tsugae* Marlatt, *Aspidiotus tsugae* Marlatt, *Furcaspis tsugae* (Marlatt), *Tsugaspidiotus tsugae* (Marlatt).

Field Characters (Plate 86) Adult female cover convex, circular to oval, gray; shed skins subcentral, outlined by white wax, brown. Male cover oval, same color and texture as female cover; shed skin submarginal, brown. Body of adult female yellow; eggs and crawlers yellow. On hemlock on underside of needles.

Slide-mounted Characters Adult female (Fig. 90) with 3 pairs of definite lobes, lobes usually heavily worn, notches small and inconspicuous; paraphyses especially noticeable in first space, paraphysis formula usually 2-3-1, occasionally 2-2-2, 3-3-1, or 2-3-2. Median lobes separated by space 0.7–1.1 (0.9) times width of median lobe, with small paraphysis-like sclerotization on medial margin of each lobe, without yoke, medial margins parallel or diverging slightly, lateral margins parallel, with 0–1 (1) lateral notch, 0–1 (0) medial notch; second lobes simple, about same size as median lobes, with 0–3 (1) lateral notches, with 0–1 (0) medial notch; third lobes simple, about one-half size of second lobes, with 1–3 (2) notches. Plates without associated microducts except often with 1 gland-spine-appearing structure anterior of plates, plates broad, with 2 to 4 tines, plates anterior of third lobe, when present, low, with many small tines; plate formula 2-3-1 or 2-3-2, rarely 2-3-3; with 2 plates between median lobes, equal to or slightly shorter than length of median lobes. Macroducts of 1 variable size, largest posteriorly, smallest on segments 1 or 2, 1 macroduct between median lobes 29–39 (36) µ long, extending 1.1–2.1 (1.5) times distance between posterior apex of anal opening and base of median lobes, marginal macroduct in first space 27–36 (34) µ long, with 14–18 (15) macroducts on each side of pygidium on segments 5 to 8, ducts in submarginal and marginal areas, total of 36–88 (54) macroducts on each side of body, some macroduct orifices anterior of anal opening; prepygidial macroducts on segments 1 to 4 in marginal and submarginal areas, occasionally in medio-lateral areas of segments 1 and/or 2. Pygidial microducts on venter in submarginal areas of segment 5 and sometimes 6, with 2–5 (3) ducts; prepygidial ducts of 1 size, in cluster posterior of mouthparts, in medial areas on 1 or more of mesothorax, metathorax, segment 1, segment 2, or segment 3, in marginal and submarginal areas of any or all of segments 2, 3, or 4; pygidial microducts absent from dorsum; prepygidial microducts of 2 sizes, smaller size, when present, restricted to marginal areas of head and anterior thorax, larger size in medial areas of any or all of mesothorax, metathorax, or segments 1 or 2, in submarginal areas on metathorax and segment 1, on marginal and submarginal areas of mesothorax or metathorax to segments 1 or 2. Perivulvar pores in 5, rarely 4, groups, with 8–15 (12) pores on each side of body. Perispiracular pores absent. Anal opening located 1.1–1.6 (1.4) times length of anal opening from base of median lobes, anal opening 15–20 (16) µ long. Dorsal seta laterad of median lobes 1.0–1.9 (1.3) times length of median lobe. Eyes represented by small, sclerotized, spurlike projection or dome, located anterior and laterad of mouthparts. Antennae each with 1 long seta. Cicatrices usually present on prothorax and segment 1. Body oval to round.

Affinities The shortneedle conifer scale is similar to the *Nuculaspis californica*, black pineleaf scale, by having the macroducts scattered along the body margin from segment 1 to segment 8, a convex scale cover, and occurring on coniferous hosts.

The shortneedle conifer scale differs by having 3 pairs of lobes, shorter paraphyses, an unsclerotized prosoma in old adult females, an unretracted pygidium, and short plates anterior of lobe 3; *N. californica* has 4 pairs of lobes, no paraphyses, a sclerotized prosoma and retracted pygidium in old adult females, and at least 1 plate longer than lobe 3 anterior of lobe 3.

Hosts Occurs on coniferous hosts only. Borchsenius (1966) records it from *Abies Picea*, and *Tsuga*. We have examined specimens from *Picea*, *Taxus*, and *Tsuga*. It has been reported from *Cedrus* (McClure and Fergione 1977), *Chamaecyparis* (Murakami 1970), *Juniperus* (Nakahara 1982), and *Pinus* (Danzig 1978).

Distribution This scale is native to Japan (Takagi 1961; McClure 1988). We have examined U.S. specimens from CT, NJ, and RI. It rarely has been intercepted in quarantine from Japan and Korea. Nakahara (1982) reports the species from Japan, Korea, and the far eastern part of the former Soviet Union.

Biology McClure (1978) has worked out the life history of this species in Connecticut. He found that it is bivoltine on eastern hemlock and overwinters as second instars. Adults appear in April and May, and eggs are first laid in early to mid-May and continue through June. Crawlers are present throughout most of June. Adults of the summer generation occur in July and August with the first eggs present in early August. Crawlers of the fall generation are present in August and September. Generation times of the summer and overwintering generations require 83 and 282 days, respectively. McClure (1980a) also studied the competitive interactions between the shortneedle conifer scale and the elongate hemlock scale, *Fiorinia externa*, on eastern hemlock. Both species preferentially colonize young needles of the lower crown, but *F. externa* appeared to have a competitive edge because of its ability to hatch 2 to 4 weeks earlier in the spring. In a later study McClure (1981) found that solitary populations of *Nuculaspis tsugae* increased their density by 68% annually, but populations coexisting with *F. externa* were reduced in size by 74% each year. Population trend indices and key factors extracted from life tables were used to identify the mortality factors responsible for the competitive decrease in populations of *N. tsugae*. Parasitism and starvation of nymphs were the key factors in the competitive dynamics of the two species. *Fiorinia externa* adversely affects the growth of *N. tsugae* as a superior competitor and also by concentrating the parasites that emerge from it onto the nymphs of *N. tsugae* that occur in the fall. Denno et al. (1995) summarize evidence provided by McClure demonstrating that feeding by *F. externa* reduced the nitrogen available in hemlock foliage, which decreased the survival rate of *N. tsugae*. Brown (1965) documents the presence of males of the shortneedle conifer scale. Danzig (1980) states that this species has 1 generation per year in the far eastern portion of the former Soviet Union.

Economic Importance According to McClure (1980b) this pest can cause hemlock needles to discolor, and heavy infestations can result in premature needle drop. McClure and Fergione (1977) observed hemlocks in all stages of maturity in both native stands and ornamental plantings that were either dead or dying from heavy infestations of this scale. According to Richard Casagrande (1983, personal communication) the shortneedle conifer scale can cause damage to spruce in Rhode Island. Miller and Davidson (1990) consider this species to be an occasional pest.

Selected References McClure (1978, 1980b, 1981, 1982); McClure and Fergione (1977); Takahashi and Takagi (1957).

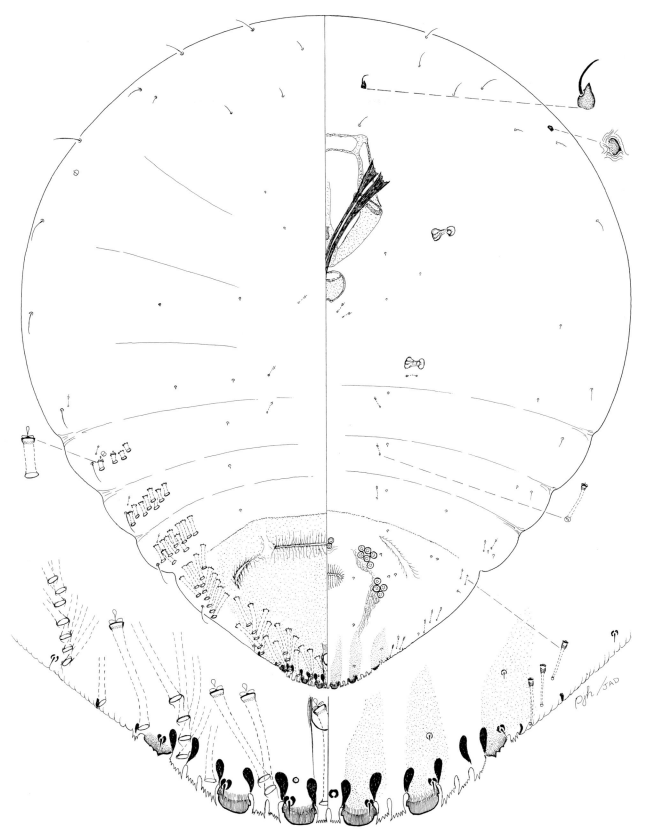

Figure 90. *Nuculaspis tsugae* (Marlatt), shortneedle conifer scale, Fairfield County, CT, on *Tsuga* sp., X-?-1975.

Odonaspis ruthae (Kotinsky)

Suggested Common Name Bermuda grass scale (also called Ruth's scale, hard grass scale, couch scale).

Common Synonyms and Combinations *Aonidia ruthae* (Kotinsky), *Dycryptaspis ruthae* (Kotinsky), *Odonaspis pseudoruthae* Mamet.

Field Characters (Plate 87) Adult female cover moderately convex, oval to cone-shaped, white; ventral cover thick and well developed; shed skins marginal, yellow or brown when rubbed. Male cover elongate oval, similar in color and texture to female cover; shed skin marginal, yellow when rubbed. Body of adult female pinkish; eggs pink or red; crawlers pink. Primarily on grasses, mainly beneath sheathing bases of leaves, and on stolons and roots, occasionally on leaves.

Slide-mounted Characters Adult female (Fig. 91), lobes indefinite; paraphyses formula normally 1-1-0, rarely 1-0-0. Median lobes fused, represented by small, central swelling, projection laterad of median lobes, without notches, yoke, or basal sclerosis; second lobe area with 1 or 2 small projections, without notches; third lobe area with or without projections. Gland spines and plates absent. Macroducts of 1 size, on dorsum in segmental rows on segments 6 to 8, in submarginal and marginal clusters on metathorax or segment 1 to segment 5, macroduct in first space 14–22 (19) μ long, with 144–204 (170) macroducts on each side of body on segments 5 to 8, total of 453–556 (498) macroducts on each side of body, some orifices occurring anterior of anal opening, ventral macroducts in marginal and submarginal clusters on mesothorax to segment 8. Pygidial microducts on venter normally difficult to see because of pygidial sclerotization, present on margins of segments 5 to 8 with total of 13–17 (15) ducts; prepygidial ducts in cluster laterad of clypeolabral shield, in segmental rows on meso- and metathorax, in medial rows of segments 1 to 4, conspicuous cluster laterad of hind spiracles; microducts absent from dorsum. Perivulvar pores in 3 groups, 36–62 (47) pores on each side of body. Perispiracular pores usually with 5 loculi, occasionally with some triloculars, anterior spiracles each with 11–26 (21) pores, posterior spiracles with 8–18 (14) pores. Anal opening located 16–22 (18) times length of anal opening from base of median lobes, anal opening 8–12 (9) μ long. Dorsal seta laterad of median lobes 1.0–1.2 (1.1) times as long as median lobes. Eyes absent. Antennae each with 1 large seta. Cicatrices absent. Setae on margin of pygidium with base set in dermal pocket. With 2 or 3 microducts on dorsum between median lobes, with 2 on venter in same area.

Affinities Of the economic species that occur on grasses, Bermuda grass scale most closely resembles *Aspidiella sacchari*, sugarcane scale; both possess small ducts on the dorsal and ventral pygidial surfaces. Bermuda grass scale lacks distinct pygidial lobes and plates, while sugarcane scale has these structures. Of the 10 *Odonaspis* species recorded from the United States (Ben-Dov 1988), Bermuda grass scale is most similar to *O. minima* Howell and Tippins (1978). Both possess large macroduct orifices but the former has intersegmental paraphyses and lacks gland spines while the latter lacks intersegmental paraphyses and has pro- and mesothoracic gland spines.

Hosts Polyphagous on graminaceous hosts. Borchsenius (1966) records Bermuda grass scale from 6 genera in the Gramineae, as well as *Juncus* and *Echeveria*; Dekle (1977) lists 16 grass genera as hosts; Ben-Dov (1988) records it from 21 grass genera as well as *Cattleya* and *Euphorbia*. We have examined material from the following genera of grasses: *Agropyron, Andropogon, Axonopus, Cynodon, Digitaria, Distichlis, Festuca, Lepturus, Setaria, Sorghum,* and *Sporobolus.* Bermuda grass (*Cynodon dactylon* [L.]) is the most commonly recorded host for this species.

Distributions We have examined U.S. specimens from AL, AZ, CA, FL, GA, KS, LA, MS, NC, SC, TX. This species rarely has been collected by quarantine officers. It has been reported from Africa, southern Asia, Australia, Caribbean islands, Central America, Christmas Island, Cook Island, French Polynesia, Hawaii, Kiribati, Mexico, New Caledonia, New Guinea, South America, Tahiti, Tuvalu, and Vanuatu.

Biology No detailed biological studies on this species could be found. Tippins and Martin (1982) sampled two Bermuda grass fields in Georgia and found population rates to be about 100 million specimens per acre. They discuss samples taken previously that were in excess of 200 individuals per square inch on turf Bermuda grass. Their sampling revealed that there are two annual peaks, one in March to May, and one in July to November or December. They suggest that the latter peak might represent two overlapping generations. Thus, there were 2 or 3 generations per year with population lows in January and February and again in June and sometimes July and August. Potter (1998) indicates that development from egg to adult took about 60 to 70 days. He reports several overlapping generations in Georgia and northern Florida and year-round reproduction farther south. Berlinger and Barak (1981) studied the phenology of the Bermuda grass scale in lawns in Israel. Brown (1965) examined the chromosomes of Bermuda grass scale collected in Oahu, Hawaii, and Riverside, California, and concluded that the populations were parthenogenetic. We have collected males of this species, and McKenzie (1956) and Dekle (1977) reported males in California and Florida, respectively. It appears there are both parthenogenetic and sexual populations.

Economic Importance Potter (1998) indicates that this pest can cause stunting of the new growth in Bermuda grass turf and even wilting and browning. When infestations are severe, large patches of turf turn brown and become very thin. Dekle (1977) reports this species as an economic pest of lawns and golf courses in Florida. Bibby (1961) records it as a serious pest of seed crop Bermuda grass in Yuma County, Arizona. Tippins and Beshear (1968) consider Bermuda grass scale to be the grass-infesting scale insect with the greatest potential for becoming a serious pest of both forage and ornamental Bermuda grass in Georgia. Tippins and Martin (1982) documented as many as 100 million scales per acre on Bermuda grass in Georgia and this was considered a moderate or low infestation. This species can become so abundant that it gives the plant a white moldy appearance; the scales can be so dense that the covers appear like overlapping shingles. Miller and Davidson (1990) consider this species to be a serious pest in a small area of the world.

Selected References Ben-Dov (1988); Tippins and Martin (1982).

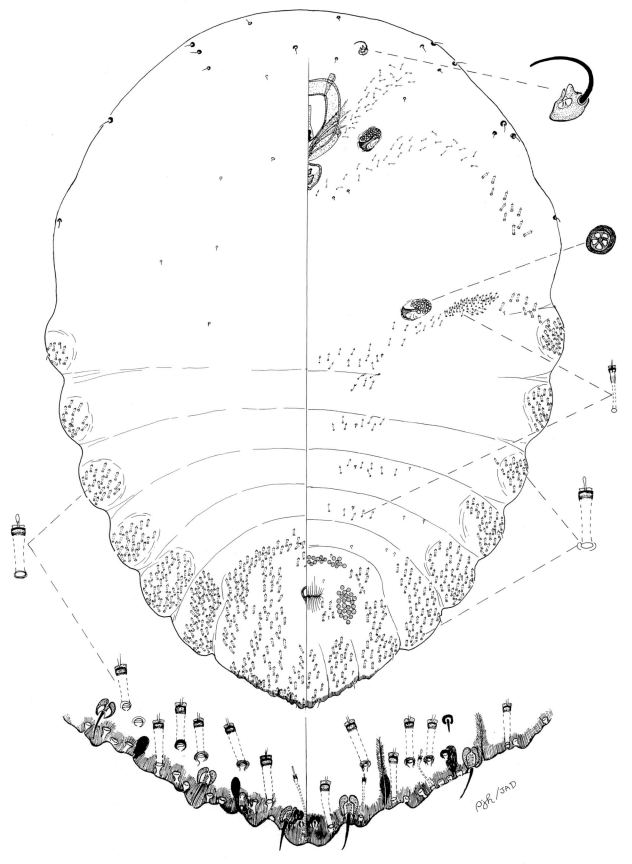

Figure 91. *Odonaspis ruthae* Kotinsky, Bermuda grass scale, Phoenix, AZ, on *Cynodon dactylon*, XII-16-1938.

Plate 87. *Odonaspis ruthae* Kotinsky, Bermuda grass scale

A. Unrubbed young adult female before elongation on grass (J. A. Davidson).

B. Rubbed young (right) and mature female (left) covers on grass (J. A. Davidson).

C. Male (left) and female (right) covers on grass (J. A. Davidson).

D. Rubbed second-instar covers on grass (J. A. Davidson).

E. Signs of infestation on grass crown (J. A. Davidson).

F. Leaves removed to show extent of grass crown infestation (J. A. Davidson).

G. Body of young adult female (J. A. Davidson).

H. Body of mature adult female with eggs (J. A. Davidson).

308

Plate 88. *Parlatoreopsis chinensis* (Marlatt), Chinese obscure scale

A. Circular adult female cover on unknown host (J. A. Davidson).
B. Elongate white male cover on lilac (J. A. Davidson).
C. Distance view of cryptic infestation on lilac (J. A. Davidson).

D. Body of adult female with eggs on lilac (J. A. Davidson).
E. Crawler on lilac (J. A. Davidson).
F. Male covers on loquat (J. A. Davidson).

Parlatoreopsis chinensis (Marlatt)

ESA Approved Common Name Chinese obscure scale (also called Chinese obscura scale).

Common Synonyms and Combinations *Parlatoria chinensis* Marlatt, *Cryptoparlatoria chinensis* (Marlatt), *Parlatoria (Parlatoreopsis) chinensis* Marlatt.

Field Characters (Plate 88) Adult female cover flat, circular, gray to white; shed skins marginal, light brown when rubbed. Male cover smaller, similar in texture and color to female cover, elongate; shed skin submarginal to marginal, light brown. Body of adult females dark purple; eggs light purple; crawlers purple. On bark, cryptic.

Slide-mounted Characters Adult female (Fig. 92) with 2 pairs of well-developed lobes; paraphyses absent. Median lobes separated by space 0.1–0.3 (0.2) times width of median lobe, with narrow paraphysis-like structures attached to medial and lateral margins of lobe, paraphysis-like structure converging but rarely touching, medial margin axes parallel, lateral margin axes diverging apically, with 4–7 (5) lateral notches, 1–2 (2) medial notches; second lobes simple, about one-half size of median lobes, about same shape, with 1–3 (2) lateral notches, without medial notches. Gland-spine formula usually 2-2-3, rarely 2-3-3 or 2-3-2, gland spines in third space frequently small, 15–27 (22) gland spines from prothorax to segment 4; with 2 small gland spines between median lobes, about one-quarter of the length of median lobes. Macroducts on pygidium of 3 sizes, marginal ducts between median lobes and in first space usually narrower and longer than remaining marginal ducts, submedial and submarginal ducts smallest, 1 macroduct between median lobes 12–18 (15) μ long, extending 0.2 of distance to posterior apex of anal opening, macroduct in first space 12–17 (14) μ long, with 8–14 (10) macroducts on each side of pygidium on segments 5 to 8, marginal macroducts on segments 5 to 8, submarginal ducts on segments 5 and 6, submedial ducts usually on segment 7, rarely absent, sometimes also on segment 6; total of 30–54 (41) macroducts on each side of body; some macroduct orifices occurring anterior of anal opening; prepygidial macroducts on dorsum of margin of metathorax or segment 1 to segment 4, on submargin of metathorax, segment 1, 2, or 3 to 4; on venter on metathorax to segments 1 or 2. Pygidial microducts on venter in submarginal and/or submedial areas of segments 5 or 6 to 7, with 6–10 (8) ducts on each side of body; prepygidial ducts on venter in submedial areas from mouthparts to metathorax and on segments 1 or 2 to 3 or 4, in submarginal or marginal areas from prothorax to mesothorax or metathorax and segments 3 and 4; pygidial microducts usually on dorsum of segment 5, rarely absent; prepygidial microducts on dorsum in submedial areas of segments 2 to 4. Perivulvar pores in 4 groups, 10–13 (11) pores on each side of body. Perispiracular pores normally with 5 loculi, anterior spiracles each with 1–3 (2) pores, posterior spiracles without pores. Anal opening located 4–7 (6) times length of anal opening from base of median lobes, anal opening 9–15 (12) μ long. Dorsal seta laterad of median lobes 0.8–1.0 (0.9) times as long as length of median lobe. Eyes usually represented by weakly sclerotized spot in center of concentric dermal pattern, occasionally with small sclerotized spot at apex of prosomal protrusion. Antennae each with 1 conspicuous seta. Cicatrices absent. Body shape broadly oval. Prepygidial segments each usually protruding laterally, usually adjacent to anterior spiracle. Openings of marginal macroducts of pygidium transverse. Orifice of marginal macroducts in spaces 1 and 2 with conspicuous globular sclerotization, remaining marginal ducts occasionally with similar but smaller sclerotization. Lateral setae on dorsum and venter unusually long for armored scale.

Affinities A combination of 3 characters distinguishes the Chinese obscure scale from all other diaspidids in the United States. Enlarged, multinotched median lobes; mesally enlarged sclerotized rims on 2 to 3 marginal pygidial macroducts; and a prothoracic marginal protuberance opposite the anterior spiracle. *Parlatoreopsis pyri* (Marlatt) occurs in DC and MD but has not been reported to be of economic importance. *Parlatoreopsis pyri* has lobes 1 and 2 without medial notches and with 1 lateral notch and lacks a prothoracic protuberance. *Parlatoreopsis longispinus* (Newstead) occurs in the Middle East and has the anal opening about equal in length to space from posterior end of anal opening to base of median lobes, while *P. chinensis* and *P. pyri* have the anal opening 5 or more times length of anal opening from base of median lobes.

Hosts Polyphagous; Borchsenius (1966) records it from 42 genera in 21 families; Dekle (1977) reports it from 21 *Ficus* species in Florida. Baker et al. (1943) list it from 67 host genera in Missouri and give the following as most commonly infested or appearing to be satisfactory hosts: *Aronia*, *Chaenomeles*, *Hibiscus*, *Koelreuteria*, *Ligustrum*, *Rosa*, and *Syringa*. We have examined specimens from *Althaea*, *Amorpha*, *Chaenomeles*, *Ficus*, *Gymnocladus*, *Hibiscus*, *Juglans*, *Mahonia*, *Malus*, *Olea*, *Periploca*, *Platanus*, *Populus*, *Prunus*, *Pyracantha*, *Pyrus*, *Syringa*, *Thuja*, *Xanthoxylum*, and *Zizyphus*.

Distribution The Chinese obscure scale probably was introduced into the United States from China and/or Japan (Takagi 1969). McKenzie (1945) believes this species to be a native of the Manchurian subregion of the Palearctic region. We have seen U.S. specimens from CA, FL, and MO. This scale most frequently has been intercepted in quarantine from China on *Pyrus*.

Biology Baker et al. (1943) found a 2:1 male to female ratio. They report 2 complete generations and a partial third generation per year in St. Louis, Missouri. Apparently only fertilized second-generation females overwintered. Each female produced a total of about 40 eggs, which were laid a few at a time and hatched 5 to 12 days after being deposited. The generations were found to overlap. In 1942, eggs were present from April 24 to October 30; first-generation crawlers were present from May 1 to July 24; second-generation crawlers were present from July 10 to near the end of September; and third-generation crawlers appeared in the last half of October.

Economic Importance Baker et al. (1943) observed a few plants believed to have been killed by this scale. They found 'young apple trees where local injury was caused by only a few specimens of the scale. On these trees, bark tissue first became red and then died at the points of infestation.' According to Sasscer (1918) the Chinese obscure scale is 'apparently a serious pest in China.' According to Hanning (1981, personal correspondence) 'In the Missouri Delta area many plantings of Rose-of-Sharon are receiving heavy damage from a scale which appears to be *P. chinensis*.' Miller and Davidson (1990) consider this species to be an occasional pest.

Selected References Baker et al. (1943); McKenzie (1956); Takagi (1969).

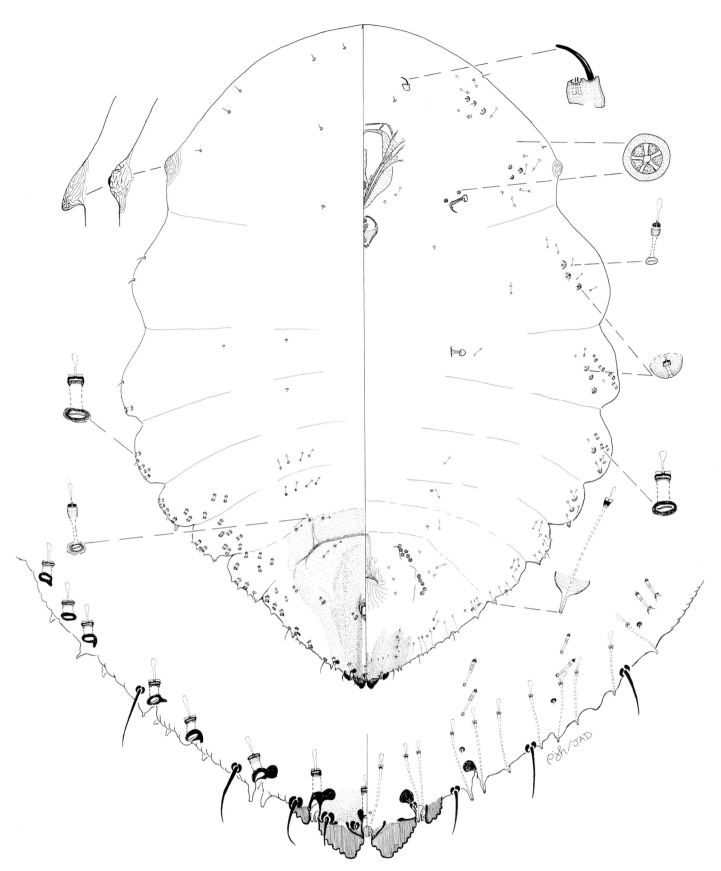

Figure 92. *Parlatoreopsis chinensis* (Marlatt), Chinese obscure scale, Yokosuka, Japan, intercepted at San Diego, CA, on *Prunus* sp., IV-28-1952.

Parlatoria blanchardi (Targioni Tozzetti)

ESA Approved Common Name Parlatoria date scale (also called white scale).

Common Synonyms and Combinations *Aonidia blanchardi* Targioni Tozzetti, *Coccus blanchardi* (Targioni Tozzetti), *Apteronidia blanchardi* (Targioni Tozzetti), *Parlatoria victrix* Cockerell, *Parlatoria proteus* var. *palmae* Maskell, *Parlatoria blanchardi* var. *victrix* Cockerell, *Parlatoria* (*Websteriella*) *blanchardi* (Targioni Tozzetti), *Parlatoria palmae* Maskell.

Field Characters (Plate 89) Adult female cover relatively flattened, elongate oval, translucent, with dark brown or black central area, light ends, covered with white powder; shed skins marginal, yellow brown to black with greenish tinge, about three-quarters of length of cover. Male cover smaller, white, elongate; shed skin marginal, light brown to black. Body of adult female light red; crawlers red; eggs pink to light red. On leaves and fruit. Difficult to detect in light infestations when hidden under overlapping leaf bases.

Slide-mounted Characters Adult female (Fig. 93) with 4 pairs of lobes, fourth lobes smaller and proportionally broader than first 3 pair, without associated microduct; paraphyses attached to first 3 lobes, formula normally 2-2-1; similar paraphysis-like structure attached to medial margin of median lobes. Median lobes separated by space 0.9–1.5 (1.2) times width of median lobe, without sclerosis or yoke, medial margin axes usually converging apically, rarely parallel, lateral margin axes diverging apically, 0–1 (0) lateral notches, without medial notches; second lobes simple, slightly smaller or equal in size to median lobes, about same shape, with 0–1 (0) lateral notches, without medial notches; third lobes simple, slightly smaller than second lobes, about same shape, with 0–2 (0) lateral notches, without medial notches; fourth lobes rarely absent, when present, simple, smaller than third lobes, proportionally wider, with 4–10 (6) notches. Plate formula usually 2-3-3, each plate or gland spine with microduct, plates with fimbriations from near base to apex, plates in first 3 spaces predominantly have 3 or more tines, with tines gradually disappearing on segments 4 or 5, changing to gland spines anteriorly, 11–15 (13) marginal plates and gland spines from metathorax, segment 1 or segment 2 to segment 4; with 2 slender plates between median lobes, usually slightly shorter than median lobes. Macroducts on pygidium of 1 size, 1 macroduct between median lobes 12–17 (14) μ long, extending 0.1–0.2 (0.2) of distance to posterior apex of anal opening, macroduct in first space 10–15 (12) μ long, with 8–17 (13) macroducts on each side of pygidium on segments 5 to 8, marginal 5 to 8, submarginal ducts usually on segments 5 to 7, sometimes absent from segment 5 or 6; total of 28–56 (40) macroducts on each side of body; some macroducts occurring anterior of anal opening; prepygidial macroducts on dorsum of margin of segments 1 or 2 to 4, absent from submargin; prepygidial macroducts present on venter from meso- or metathorax to segment 4. Pygidial microducts on venter in submarginal areas of segments 5 or 6 or 7, with 2–7 (3) ducts on each side of body; prepygidial ducts on venter in submedial areas on segments 2 or 3 to 4, in submarginal or marginal areas from prothorax or mesothorax to segment 1; pygidial ducts on dorsum usually on submedial area of segment 5, rarely absent; prepygidial microducts on dorsum in submedial areas of metathorax or segment 1 to segment 4, submarginal ducts sometimes on some of segments 1 to 4. Perivulvar pores in 4 groups, 9–18 (14) pores on each side of body. Perispiracular pores normally with 5 loculi, anterior spiracles each with 1–4 (2) pores, posterior spiracles without pores. Anal opening located 5–9 (6) times length of anal opening from base of median lobes, anal opening 10–15 (13) μ long. Dorsal seta laterad of median lobes 0.5–0.7

(0.6) of length of median lobe. Eyes absent. Antennae each with 1 conspicuous seta. Cicatrix near anterior spiracle unusually large, sclerotized, remaining cicatrices on segments 1 and 3 when present. Body shape broadly oval. Posterior spiracle without associated dermal pocket. Dermal pattern anterior of vulva transverse. Openings of marginal macroducts of pygidium transverse. First 3 pairs of lobes with rounded apices; remaining lobe with flat or acute apex.

Affinities Parlatoria date scale is not similar to other economic species of *Parlatoria* in the United States. The slender macroducts without enlarged orifices, lobes 1 to 3 spatulate, absence of gland spines on pro- and mesothorax, and large sclerotized prothoracic cicatrix make this an easily recognized *Parlatoria* species. Outside the United States, *P. blanchardi* most resembles *P. asiatica* Borchsenius as illustrated by Balachowsky (1953). Both have spatulate lobes, but the former has narrow macroducts without enlarged orifices, enlarged sclerotized prothoracic cicatrices, and no gland spines on the pro- and mesothorax. The latter has wide macroducts with enlarged orifices, no enlarged sclerotized cicatrices, has gland spines on the pro- and mesothorax, and is known only from *Ephedra*.

Hosts Known only from palms; Borchsenius (1966) reports it from *Hyphaene*, *Latania*, *Neowashingtonia*, *Phoenix*, *Pritchardia*, and *Washingtonia*. We have examined specimens from *Phoenix canariensis* and *Phoenix dactylifera*, the date palm.

Distribution The date-growing industry in the United States began in 1890 with the first successful importation of date palm offshoots from Algeria and Egypt. These palms were found infested with parlatoria date scale, and eradication procedures were undertaken. These procedures failed because the palms developed serious scale infestations in the field. Later date palm importations also were infested and this pest eventually spread to California, Arizona, and Texas. Eradication programs began in 1907 and were terminated successfully in 1932. Boyden (1941) provides a detailed history of the date-growing industry and the parlatoria date scale eradication program in the United States. Parlatoria date scale is common throughout the remaining date-growing areas of the world (Bodenheimer 1943). United States quarantine inspectors should watch for this scale from Argentina, Brazil, Bolivia, Italy, much of northern and western Africa, the Middle East (such as Israel and Saudi Arabia), southwestern Asia (such as India, Iran, and Pakistan), the coastal Mediterranean countries (such as Algeria and Tunisia), and Australia. Avidov and Harpaz (1969) believe it is indigenous to the oases of Mesopotamia. A distribution map of this species was published by CAB International (1962b).

Biology In the southwestern United States, Stickney et al. (1950) found that egg hatch required 7 to 11 days in mid-March and 2 to 7 days in late March to early April. In March the first molt took place 21 to 27 days after hatching. Reproduction occurred throughout the year with a slight slowdown during the winter months. Hussain (1974) found 4 generations per year in Iraq, as did Khoualdia et al. (1993) in Tunisia. Avidov and Harpaz (1969) note 4 to 5 generations per year in the lower Jordan Valley of Israel. Development required 16 days in late summer, 28 to 47 days in autumn and early winter, and 32 to 36 days in late winter and spring. During the same time periods, preoviposition was 96, 78, and 58 days, respectively. In late summer and autumn, males represented about 25% of the population. In winter, almost all scales were females. Crawlers were most abundant from October to November, and February and May. In Mauritania (Tourneur et al. 1975) the parlatoria date scale populations peaked in the cooler periods from April to May at

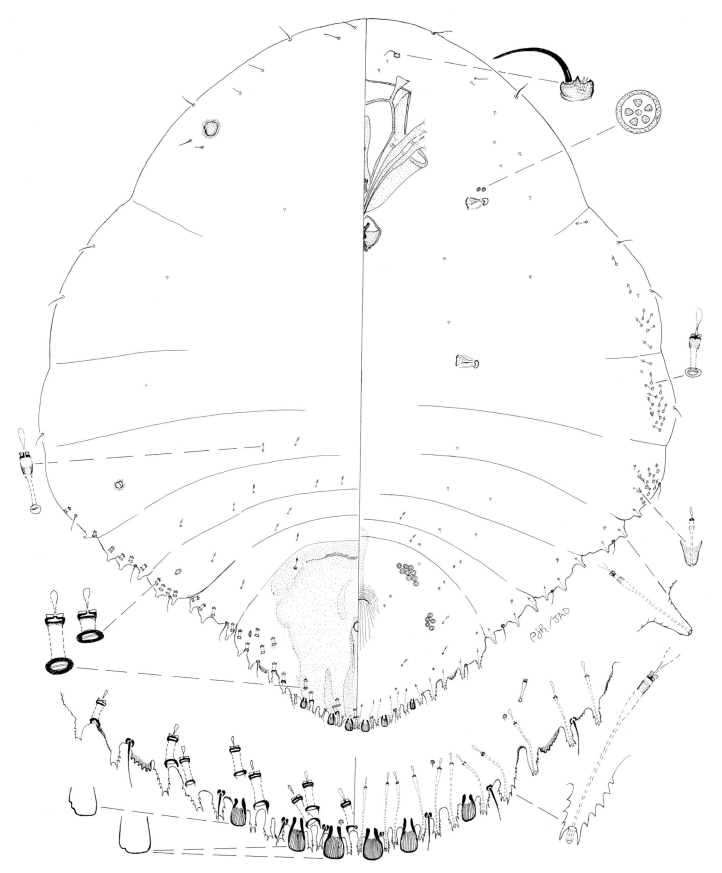

Figure 93. *Parlatoria blanchardi* (Targioni Tozzetti), parlatoria date scale, El Centro, CA, on *Phoenix dactylifera*, I-5-1924.

21 °C to 32 °C, with a secondary peak in December. Populations are lowest in August at 40 °C and again during cold weather in February. Ghauri (1962) found a 10:1 ratio of winged to wingless males in material from West Pakistan and he describes both forms. El-Kareim (1998) demonstrated that adult females release a sex pheromone when they are about 10 days old. Use of pheromone-baited traps showed that adult males are active in the morning just as the sun rises. Winged males predominated in the spring, and wingless ones were most abundant in the summer generation. Generally, wingless individuals were more abundant at the end of each generation.

Economic Importance Beardsley and González (1975) list parlatoria date scale as one of the 43 major armored scale insect pests of the world. In Algeria, a major infestation of this scale caused the death of nearly 100,000 date palms in 1920 (Rosen 1990). Carpenter and Elmer (1978) note that heavy infestations on the pinnae cause them to wither and die. This damage often renders the dates unfit for human consumption and reduces tree vigor, but seldom kills trees. Stickney (1934) considers this scale to be the most serious insect enemy of the date palm and notes that it contributed to the death of neglected palms (Stickney et al. 1950). Bénassy (1990) provides a summary of the economic importance of this scale. Chemical control is difficult because all stages of the scale are present all year-round, and there is a tendency for the scales to conceal themselves in areas where pesticides cannot efficiently reach them. The most effective method of control focuses on a pest management strategy of using few pesticides and several predators (Bénassy 1990). Biological control has been very successful in many situations (Rosen 1990). In a trial using pheromone-baited, blue sticky traps to attract and kill males, trees containing the traps had as much as a 39% reduction in scale numbers compared to control trees (El-Kareim 1998). Khoualdia et al. (1993) provide evidence that certain cultivars (particularly Kentichi) are more resistant to attack than others.

Selected References Bénassy (1990); Carpenter and Elmer (1978); McKenzie (1945); Stickney (1934).

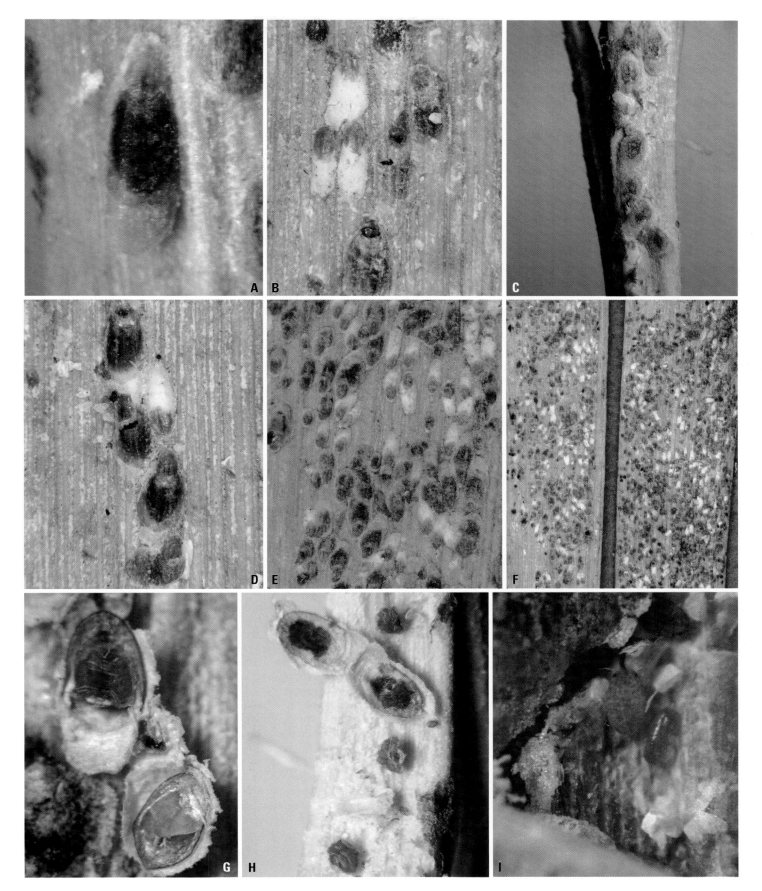

Plate 89. *Parlatoria blanchardi* (Targioni Tozzetti), Parlatoria date scale

A. Adult female cover on palm (R. J. Gill).
B. Dark female covers and white male covers on palm (J. A. Davidson).
C. Female covers on palm (R. J. Gill).
D. Female and male covers on palm (J. A. Davidson).

E. Distance view on palm (J. A. Davidson).
F. Distance view on palm (J. A. Davidson).
G. Bodies of young (light) and mature (dark) females (R. J. Gill).
H. Bodies of females after laying eggs (R. J. Gill).
I. Eggs and crawlers (R. J. Gill).

Parlatoria camelliae (Comstock)

Suggested Common Name Camellia parlatoria scale.

Common Synonyms and Combinations *Parlatoria pergandii* var. *camelliae* Comstock, *Parlatoria proteus* var. *virescens* Maskell, *Parlatoria (Euparlatoria) pergandii* var. *camelliae* Comstock, *Parlatorea pergandii* var. *camelliae* Comstock.

Field Characters (Plate 90) Adult female cover flat, broadly elongate oval, grayish to brown; shed skins marginal, faintly yellow with longitudinal, green stripe. Male cover smaller, similar in texture and color to female cover, elongate; shed skin marginal, faintly yellow, with ventral green stripe. Body of adult female purple; eggs purple; crawlers faintly purple. On leaves.

Slide-mounted Characters Adult female (Fig. 94) with 5 pairs of well-developed lobes, fourth and fifth lobes often shorter and broader than first 3 pairs of lobes; paraphyses attached to first 3 lobes, formula 2-2-1; similar paraphysis-like structure attached to medial margin of median lobes. Median lobes separated by space 1.0–1.5 (1.2) times width of median lobe, without sclerosis or yoke, medial margin axes parallel or converging apically, lateral margin axes parallel or diverging, with 1 lateral and 1 medial notch; second lobes simple, about same size and shape as median lobes, with 1 lateral and 1 medial notch; third lobes simple, about same shape as second lobes, slightly smaller, with 1 lateral notch, 0–1 (1) medial notch; fourth lobes simple, often shorter and broader than third lobes, with 2–4 (3) lateral notches, 0–2 (1) medial notches; fifth lobes simple, about same shape as fourth lobes, with 0–3 (2) lateral notches, 0–3 (1) medial notches. Plate formula 2-3-3, each plate and gland spine with at least 1 microduct, plates broadly fimbriate, plates in first 3 spaces predominantly have 3 or more tines, with tines gradually disappearing on segments 3 and 4, changing to gland spines anteriorly, 30–43 (36) marginal plates and gland spines from prothorax to segment 4; with 2 broad plates between median lobes, usually slightly shorter than median lobes. Macroducts on pygidium of 2 sizes, marginal ducts larger than submarginals, 1 macroduct between median lobes, 10–15 (12) μ long, extending 0.2–0.3 (0.2) of distance to posterior apex of anal opening, marginal macroduct in first space 10–15 (13) μ long, 14–23 (16) macroducts on each side of pygidium on segments 5 to 8, marginal ducts on segments 5 to 8, submarginal ducts on segments 5 to 7; total of 37–51 (45) macroducts on each side of body; some macroduct orifices occurring anterior of anal opening; prepygidial macroducts on dorsum of margin of metathorax to segment 4. Pygidial microducts on venter in submarginal or submedial areas of segments 5 to 7, with 3–5 (4) ducts on each side of body; prepygidial ducts on venter in submedial areas from mouthparts to segments 3 or 4, in submarginal or marginal areas on prothorax to metathorax or segment 1 and on segment 4; pygidial microducts on dorsum in submedial area of segment 5; prepygidial microducts on dorsum in submedial area of segment 4, sometimes on 3. Perivulvar pores in 4 groups, 10–17 (13) pores on each side of body. Perispiracular pores normally with 5 loculi, anterior spiracles each with 2–3 (2) pores, posterior spiracles without pores. Anal opening located 4–7 (5) times length of anal opening from base of median lobes, anal opening 12–16 (14) μ long. Dorsal seta laterad of median lobes 0.5–0.8

(0.6) times length of median lobe. Eyes variable, sometimes represented by flat, sclerotized area, sometimes in form of small, unsclerotized dome; located near body margin posterior of anterior spiracle. Antennae each with 1 conspicuous seta. Cicatrices often absent, occasionally on segments 1 and 3. Body shape broadly oval. Posterior spiracle with associated dermal pocket. Dermal pattern anterior of vulva transverse. Openings of marginal macroducts on pygidium transverse. First 3 pairs of lobes with rounded apices, lobes 4 and 5 with acute apices.

Affinities Camellia parlatoria scale is most similar to *Parlatoria crotonis*, the croton parlatoria scale, but the former lacks a sclerotized marginal tubercle on the prothorax, which is present on the latter. In addition, croton parlatoria scale has a yellow scale cover and appears to be restricted to croton (*Codiaeum*), while camellia parlatoria scale has a gray to brown scale cover and has not been reported from croton.

Hosts Polyphagous. Borchsenius (1966) records it from 26 genera in 19 families; Dekle (1977) notes it from 5 genera in Florida; and McKenzie (1956) reports 19 host genera in California. We have examined specimens from *Acer, Aucuba, Camellia, Cinnamomum, Citrus, Cleyera,* 'cycad,' *Euonymus, Eurya, Ficus, Gardenia, Gronophyllum, Laurus, Litsea, Mangifera,* 'myrtle,' *Olea, Osmanthus, Poncirus,* and *Rhododendron.* In the literature, camellia parlatoria scale is recorded from numerous hosts but not from conifers. *Camellia* is the most commonly reported host.

Distribution This species generally occurs in the United States throughout the growing range of camellias. We have seen U.S. specimens from AL, CA, DC, FL, GA, LA, MD, MS, NC, OR, SC, TX, VA. It has been reported from DE. We have seen material from China, France, Indonesia, Japan, Korea, Mexico, Portugal, and Switzerland. It has been reported from India, Pakistan, Taiwan, and the Abkhazia. Camellia parlatoria scale most often has been intercepted in U.S. quarantine on camellias from Japan.

Biology We have been unable to find an account of the biology of this species. In Gridley, California, during late July 1979 we collected the camellia parlatoria scale just at the end of the egg-laying period. Most adult females were shriveled but a few still had a few eggs under their covers. Settled crawlers were abundant. The infestation was restricted to the leaves of camellia, and both males and females were found on the top and undersides of leaves. There was a tendency for settling to occur close to the veins of the leaf.

Economic Importance McKenzie (1956) found this species to be the most important scale insect pest of camellias in California. Morrison (1946) treats it as one of the most important scale pests of camellia on the Pacific coast and notes that it is the most commonly collected species on this host. Steinweden (1942) reports that nurserymen were very cautious with this species and sprayed whenever it was found to prevent damage to infested hosts. Miller and Davidson (1990) consider this species to be an occasional pest.

Selected References McKenzie (1945); Morrison (1939).

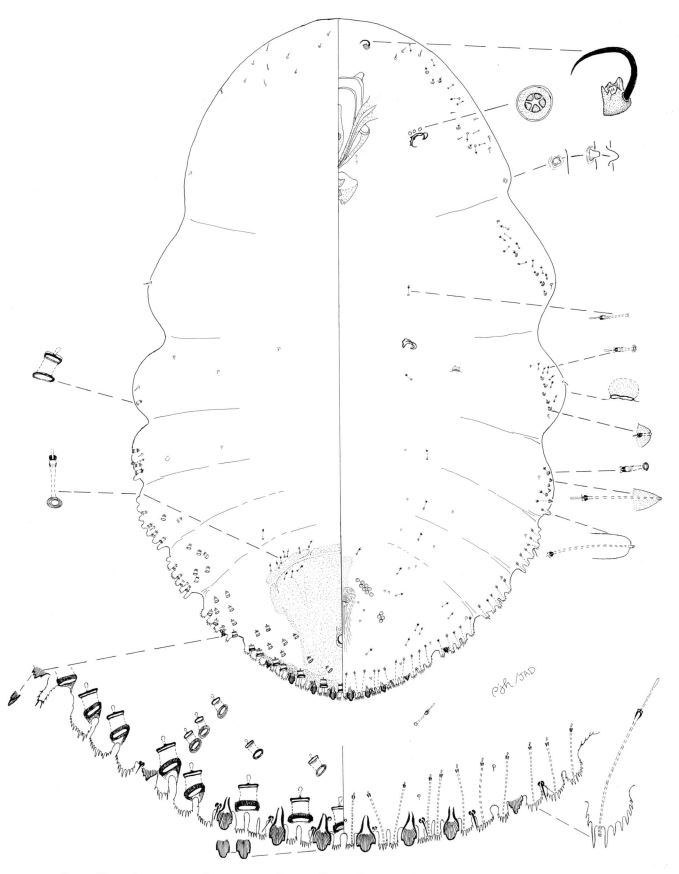

Figure 94. *Parlatoria camelliae* Comstock, camellia parlatoria scale, Leesville, LA, on *Camellia* sp., II-19-1952.

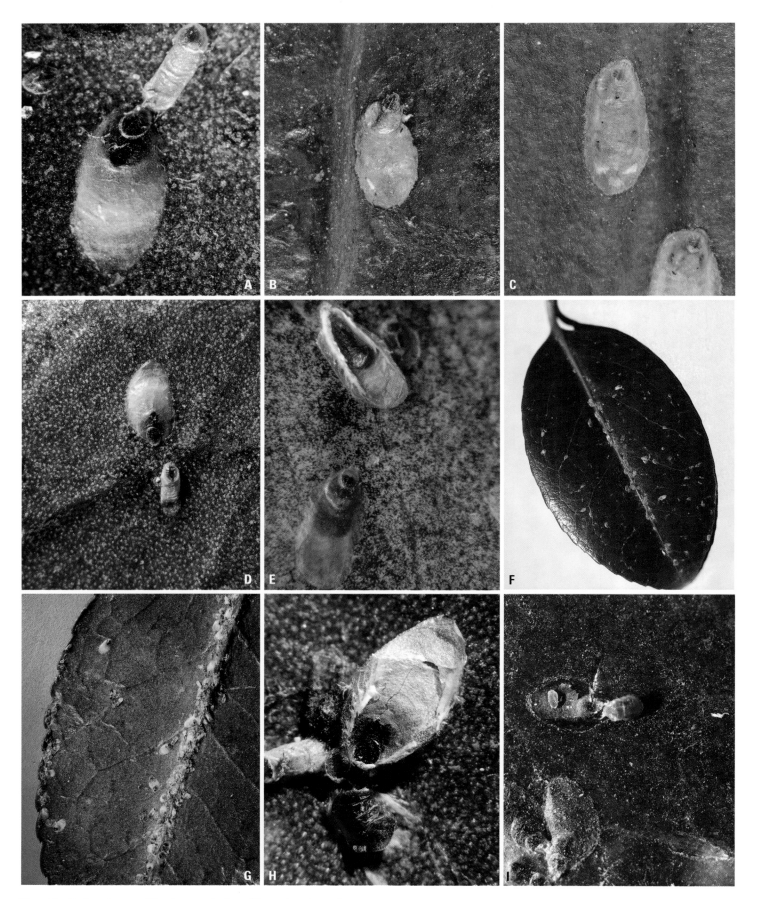

Plate 90. *Parlatoria camelliae* Comstock, Camellia parlatoria scale

A. Opaque adult female (lower left) and male (upper right) covers on camellia (J. A. Davidson).
B. Translucent adult female cover on camellia (J. A. Davidson).
C. Translucent adult female cover with white egg shells beneath on camellia (J. A. Davidson).
D. Male and female covers on camellia (J. A. Davidson).
E. Turned adult female scale cover on viburnum (R. J. Gill).

F. Distance view of mostly female covers on upper surface of camellia leaf (J. A. Davidson).
G. Lower surface of camellia leaf with mostly male covers along midvein (J. A. Davidson).
H. Body of adult female and ventral scale cover through which body was removed (J. A. Davidson).
I. Egg and adult female body (J. A. Davidson).

318

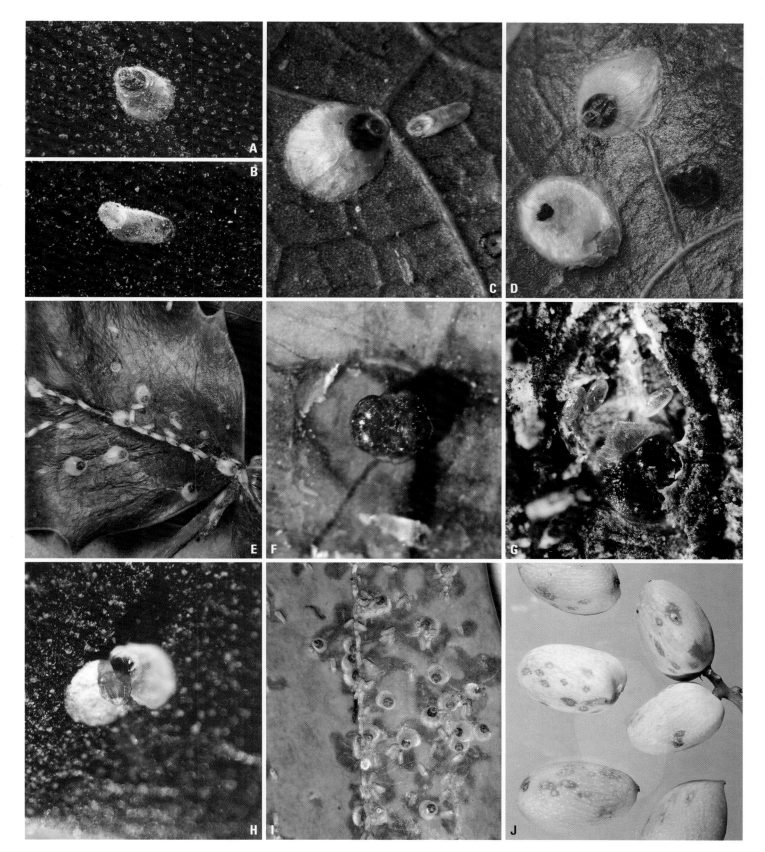

Plate 91. *Parlatoria oleae* (Colvee), Olive scale

A. Young adult female cover on olive leaf (J. A. Davidson).
B. Male cover on olive leaf (J. A. Davidson).
C. Female and male covers on *Mahonia* leaf (R. J. Gill).
D. Dorsal and ventral views of female cover and body of adult female on *Mahonia* leaf (R. J. Gill).
E. Distance view of infestation on upper surface of *Mahonia* leaf (R. J. Gill).

F. Body of young adult female on *Mahonia* leaf (J. A. Davidson).
G. Body of mature adult female with eggs on red maple (J. A. Davidson).
H. Second-instar female on olive (D. R. Miller).
I. Feeding damage on undersides of *Mahonia* leaf (R. J. Gill).
J. Feeding damage on olive fruit (R. J. Gill).

Parlatoria oleae (Colvee)

ESA Approved Common Name Olive scale (also called olive parlatoria scale, plum scale, and olive parlatoria).

Common Synonyms and Combinations *Diaspis oleae* Colvee, *Parlatoria affinis* Newstead, *Parlatoria calianthina* Berlese and Leonardi, *Parlatoria (Euparlatoria) calianthina* Berlese and Leonardi, *Diaspis squamosus* Newstead and Theobald, *Parlatorea oleae* (Colvee), *Syngenaspis oleae* (Colvee), *Parlatoria judaica* Bodenheimer, *Parlatoria morrisoni* Bodenheimer.

Field Characters (Plate 91) Adult female cover slightly convex, circular to broadly oval, white to gray; shed skins marginal, dark brown to greenish yellow. Male cover smaller, similar in texture and color to female cover, elongate; shed skin marginal, yellow with greenish tinge. Body of newly matured adult female purple, turning dark red to brown with age; eggs purple; crawlers yellow with tint of burgundy red. On all arboreal parts of host; most common on leaf midribs.

Slide-mounted Characters Adult female (Fig. 95) with 6 pairs of well-developed lobes, fifth and sixth lobes small but sclerotized; without paraphyses. Median lobes separated by space 0.3–0.5 (0.4) times width of median lobe, without sclerosis or yoke, medial margin axes parallel or converging apically, lateral margin parallel or diverging, with 1–2 (1) lateral notches, 0–1 (0) medial notches; second lobes simple, slightly smaller than median lobes, about same shape, with 1–2 (1) lateral notches, without medial notches; third lobes simple, about same size and shape as second lobes, with 1–2 (1) lateral notches, without medial notches; fourth lobes simple, smaller than third lobes, represented by 1 to several points, with 1–4 (2) lateral notches, 0–3 (1) medial notches; fifth lobes same shape as fourth lobes, usually smaller, with 0–3 (2) lateral notches, 0–1 (1) medial notches; sixth lobes same shape as fifth lobes, usually smaller, with 0–2 (0) lateral notches, 0–1 (0) medial notches. Plate formula 2-3-3, rarely 2-4-4, 2-3-4, or 3-3-5, each plate with microduct terminating through tube at apex of plate, gland spines also with microduct; plates usually have 2 or 3 tines, with tines gradually disappearing on segments 3 and 4, changing to gland spines anteriorly, 19–67 (31) marginal plates and gland spines from prothorax to segment 4; with 2 bifurcate plates between median lobes, plates about one-half to three-quarters length of median lobes. Macroducts on pygidium of 2 sizes, marginal ducts on posterior 3 or 4 segments of larger size, remaining ducts about same size, 1 macroduct between median lobes 10–16 (14) μ long, extending 0.2 of distance to posterior apex of anal opening, marginal macroduct in first space 12–16 (14) μ long, with 25–38 (30) macroducts on each side of pygidium on segments 5 to 8, marginal ducts on segments 5 to 8, submarginal ducts on segments 5 to 7, submedial ducts on segment 5, rarely on 6; total of 73–135 (106) macroducts on each side of body; some macroduct orifices anterior of anal opening; prepygidial macroducts on dorsum of margin of metathorax or segment 1 to segment 4, on submargin of segments 1 or 2 to 4, submedial on segment 4, sometimes on segment 3 on venter of margin of metathorax and segment 1. Pygidial microducts on venter in submarginal and sometimes submedial areas of segments 5 to 7, with 7–14 (11) ducts on each side of body; prepygidial ducts in submedial areas from near mouthparts to segments 4, occasionally absent from some segments, in submarginal or marginal areas of prothorax, mesothorax, and segments 3 and 4; pygidial microducts usually absent from dorsum, rarely with 1 duct associated with submedial cluster of macroducts on segment 5; prepygidial ducts often in small numbers on submedial and submarginal areas of pro- and mesothorax. Perivulvar pores in 4 or 5 groups, 28–43 (33) pores on each side of body. Perispiracular pores normally with 5 loculi, anterior

spiracles each with 4–7 (6) pores, posterior spiracles without pores. Anal opening located 4–6 (5) times length of anal opening from base of median lobes, anal opening 12–15 (14) μ long. Dorsal seta laterad of median lobes one-half to about equal length of median lobes. Eyes variable, sometimes absent, occasionally represented by flat sclerotized area or by small sclerotized point, located near body margin posterior of anterior spiracle. Antennae each with 1 conspicuous seta. Cicatrices absent. Body shape rotund to oval. Posterior spiracle without associated dermal pocket. Dermal pattern anterior of vulva transverse. Openings of marginal macroducts on pygidium transverse. First 3 pairs of lobes with rounded apices; lobes 4 to 6 with acute apices.

Affinities The olive scale is different from other species of *Parlatoria* by having single notched lobes 2 and 3, basically bifurcate posterior plates, 6 pairs of lobes, and 4 plates between lobes 3 and 4. Of the economic species in the United States, it is similar to *Parlatoria pergandii*, chaff scale, but can be separated by usually having 4 plates between lobes 3 and 4, by having notches restricted to the lateral margins of lobes 2 and 3, and by usually having the plates posterior of lobe 3 with 2 or 3 tines. The chaff scale usually has 3 plates between lobes 3 and 4, notches on the lateral and medial margins, and plates posterior of lobe 3 often have more than 3 tines. Within *Parlatoria*, olive scale is most similar to *P. boycei* McKenzie, which differs by having the anal opening located about 6.5 times length of anal opening from base of median lobes, has medial notches on lobes 2 and 3, anal opening from 20 μ long, submedial macroducts on segment 2, and 46 macroducts on each side of pygidium.

The large amount of variation in life-history behavior of this pest has prompted several authors to suggest the possibility that this species is a complex of siblings. Applebaum and Rosen (1964) and Bustshik (1960) investigated this possibility but found no evidence to support it.

Hosts Polyphagous. Borchsenius (1966) records it from 82 genera in 37 families. McKenzie (1956) states 'In California this scale has been collected . . . from about 211 host species. Many of these hosts will not . . . support the scale by themselves and must be replenished from more preferred hosts.' We have examined specimens from *Arctostaphylos, Aucuba, Berberis, Citrus, Corylus, Crataegus, Diospyros, Elaeagnus, Eriobotrya, Ficus, Fraxinus, Jasminum, Laurus, Ligustrum, Mahonia, Mespilus, Nerium, Olea, Photinia, Pinus, Pistacia, Prunus, Pyrus, Rhamnus, Ribes, Rosa, Sorbus,* and *Vitis*. This species is commonly collected on olive, various species of *Prunus*, apple, and pears. *Citrus* and *Pinus* are unusual hosts. According to Wehrle (1935) this species is unable to reproduce on citrus, although it has been considered a minor pest of citrus in Palestine (Bodkin 1928).

Distribution Olive scale probably is native to northern India, West Pakistan, and the Middle East. We have examined U.S. specimens from AZ, CA, and MD; it is reported from DE. It most commonly is taken in quarantine from the Mediterranean region, particularly Italy and Greece, on apple, pear, other rosaceous hosts, and olive. It is known from Asia, Australia, Mexico, North Africa, South America, southern Europe, and southern former Soviet Union. A distribution map of this species was published by CAB International (1962d).

Biology The olive scale has from 1 (Imamkuliev 1966) to 4 (Grandi 1951) generations per year depending on the location; most areas have 2 generations although 3 generations are not uncommon. In the United States 2 generations are reported in California (Huffaker et al. 1962) and Maryland (Ezzat 1957).

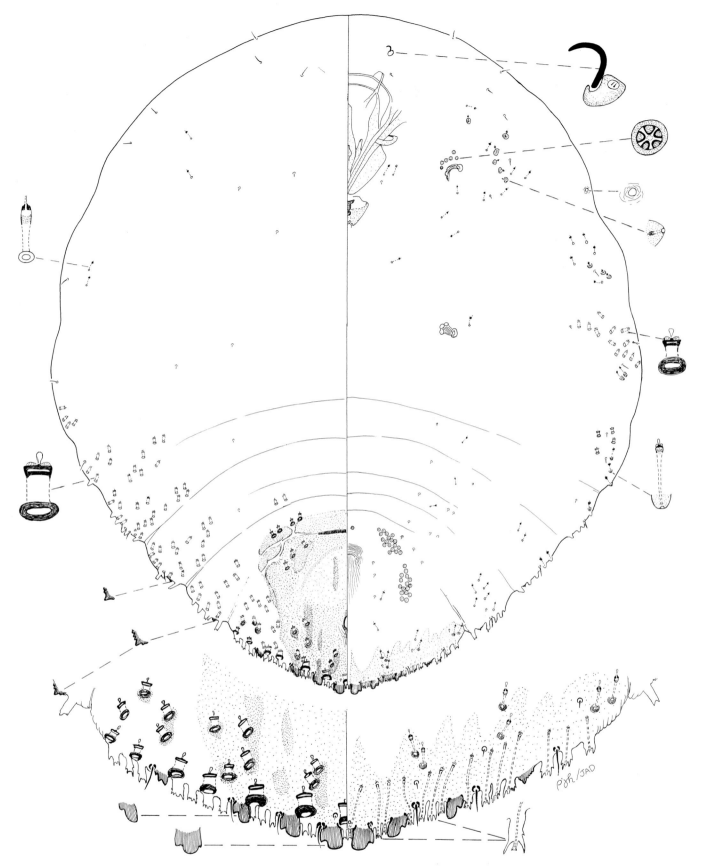

Figure 95. *Parlatoria oleae* (Colvee), olive scale, Cochise, AZ, on *Rosa* sp., III-16-1958.

Overwintering occurs as mated adult females, although a small portion of the population may overwinter as second instars (Huffaker et al. 1962). In areas where 2 generations occur, eggs are laid in early spring (mid-April to early June in California on olive and late April to early July in Maryland on *Ligustrum*). Crawlers first appear 2 or 3 weeks after the first eggs are laid (early May in California and mid-May in Maryland) and reach peak abundance 1 or 2 weeks after the first crawlers appear. Adult females of the spring generation first appear in early summer (mid-June in California and early July in Maryland) and are present until late summer. Egg laying for the summer generation begins in midsummer (late July in California and mid-July in Maryland) and crawlers reach peak abundance in late summer (mid-August in California and late August in Maryland). Adult females are present in the fall through the winter. Adult males are necessary for reproduction (Stafford 1947) and are at peak abundance in California in late June or early July and again in September (Huffaker et al. 1962). Adult females each lay a maximum of about 100 eggs although 30 is closer to average (Huffaker et al. 1962). Egg laying for each female lasts 2 or 3 weeks (Huffaker et al. 1962). In Tucson, Arizona, the generations are not in close synchrony. Eggs are found every month of the year except December and January; crawlers are present every month but January and February; and adult males occur every month except November, December, January, and February. There are 2 major egg-hatching periods in mid-April and mid-July, and 2 peak flight periods of males in early July and late October (Nichol 1935). The development threshold is about 10 °C, and approximately 1300 day-degrees are required for the development of a generation (Applebaum and Rosen 1964). According to Huffaker et al. (1962) and Stafford (1954) the life history may vary considerably depending on host and climatic conditions. Biche and Sellami (1999) demonstrated that two different hosts caused differences in duration of oviposition, sex ratio, and body size. The species apparently does not do well under coastal conditions where the climate is relatively cool during the summer (Huffaker et al. 1962). Several forecasting models have been developed (Pinhassi et al. 1996; Nestel et al. 1995).

Economic Importance This species has been reported as a pest in most areas where it occurs. It is especially important on olive (Huffaker et al. 1962; Argyriou 1990), certain stone fruits, apples (Sadeh and Gerson 1968) and pears (Kozár 1990b), and several ornamentals (Ezzat 1957). Although damage sometimes is incurred through devitalization caused by heavy scale populations, the most significant injury is sustained on the fruit. On apples, reddish spots develop around each feeding scale, and on olives, feeding spots are black. Crawlers that settle during early fruit development cause abnormalities and deformations on the fruit, making it unpalatable. According to Stafford (1948), heavily infested olives may have their oil content reduced by as much as 20%. The olive scale has been reported as a pest of peach and plum in Armenia (Babaian 1986), mango and almond in India (Chua and Wood 1990), and citrus and *Prunus* on the Canary Islands (Perez Guerra 1986). Although olive scale was once a serious economic problem in California, it now is effectively controlled as a result of the combined activities of two introduced wasps, *Aphytis paramaculicornis* DeBach and Rosen and *Coccophagoides utilis* Doutt. In most of the biological control literature *Aphytis paramaculicornis* is referred to as the Persian strain of *Aphytis maculicornis* (Masi). *Aphytis maculicornis* also is a parasite of olive scale, but it is not nearly as effective in biological control as *A. paramaculicornis* under California environmental conditions. Beardsley and González (1975) consider olive scale to be one of 43 major armored scale pests. Miller and Davidson (1990) consider it to be a serious world pest.

Selected References Ezzat (1957); Huffaker et al. (1962); McKenzie (1945).

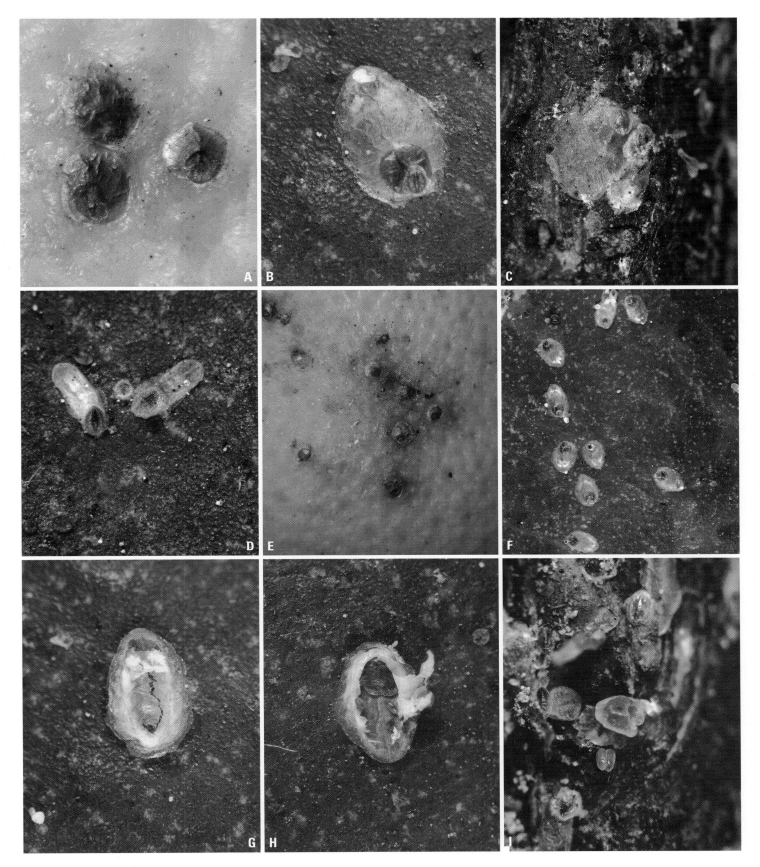

Plate 92. *Parlatoria pergandii* Comstock, Chaff scale

A. Adult female covers on orange (R. J. Gill).
B. Adult female cover on citrus leaf (J. A. Davidson).
C. Female scales on citrus bark (J. A. Davidson).
D. Male covers on citrus leaf (J. A. Davidson).
E. Distance view on orange (R. J. Gill).
F. Distance view on citrus leaf (J. A. Davidson).

G. Turned cover showing ventral cover on adult female (J. A. Davidson).
H. Ventral cover torn to reveal adult female with eggs (J. A. Davidson).
I. Adult female body with eggs on citrus bark (J. A. Davidson).

Parlatoria pergandii (Comstock)

ESA Approved Common Name Chaff scale (also called Pergande's scale).

Common Synonyms and Combinations *Parlatoria sinensis* Maskell, *Parlatoria proteus pergandei* Comstock, *Parlatoria pergandei* Comstock, *Parlatorea pergandei* (Comstock), *Parlatoria (Euparlatoria) pergandii* Comstock, *Syngenaspis pergandei* Comstock.

Field Characters (Plate 92) Adult female cover flat, round to oval, often with several wrinkles, translucent brown to gray; shed skins marginal, yellowish brown, sometimes with longitudinal green stripe. Male cover smaller, similar in texture to female cover, elongate; shed skin marginal, yellowish brown, with central, green stripe; adult male wine red with eyes black (Bodenheimer 1951). Body of adult female purple; eggs purple; crawlers purple. On twigs, leaves, and fruit.

Slide-mounted Characters Adult female (Fig. 96) with 5 rarely 6 pairs of well-developed lobes, fourth and fifth lobes smaller and proportionally broader than first 3 pairs of lobes; paraphyses attached to first 3 pairs of lobes, formula 2-2-1, 2-2-0, 2-1-1, or 2-1-0; similar paraphysis-like structure attached to medial margin of median lobes. Median lobes separated by space 0.5–1.0 (0.8) times width of median lobe, without sclerosis or yoke, medial margin axes parallel or diverging apically, lateral margin axes parallel or diverging apically, with 1 lateral and 1 medial notch; second lobes simple, slightly smaller than median lobes, about same shape, with 1 lateral and 1 medial notch; third lobes simple, slightly smaller than second lobes, about same shape, with 0–1 (1) lateral notch, 1–2 (1) medial notches; fourth lobes simple, smaller than third lobes, proportionally wider, with 1–4 (3) lateral notches, 1–2 (1) medial notches; fifth lobes simple, same size and shape as fourth lobes, with 2–5 (3) lateral notches, 0–1 (1) medial notches; sixth lobes present on 1 of 10 specimens examined, represented by single, small point. Plate formula normally 2-3-3, each plate and gland spine with microduct, plates broadly fimbriate, plates in first 3 spaces predominately have 3 or more tines, with tines gradually disappearing on segments 3 and 4 changing to gland spines anteriorly, 23–34 (30) marginal plates and gland spines from prothorax to segment 4; with 2 broad plates between median lobes, varying from slightly shorter to slightly longer than median lobes. Macroducts on pygidium of 2 sizes, marginal ducts slightly larger than submarginal ducts, 1 macroduct between median lobes 10–15 (13) μ long, extending 0.1–0.2 (0.2) of distance to posterior apex of anal opening, macroduct in first space 10–15 (12) μ long, with 15–30 (21) macroducts on each side of pygidium on segments 5 to 8, marginal ducts on segments 5 to 8, submarginal ducts on segments 5 to 7; total of 73–118 (85) macroducts on each side of body; some macroduct orifices occurring anterior of anal opening; prepygidial macroducts on dorsum of margin of metathorax or segment 1 to segment 4, on submargin of segments 2 to 4; on venter absent or on metathorax and/or segment 1. Pygidial microducts on venter in submarginal and/or submedial areas of segments 5 to 7, with 4–11 (7) ducts on each side of body; prepygidial ducts on venter in submedial areas from mouthparts to segment 4, in submarginal or marginal areas from head or prothorax to segment 4; pygidial microducts absent from dorsum; prepygidial microducts on dorsum in submedial area of segment 4. Perivulvar pores in 4 groups, 11–15 (13) pores on each side of body. Perispiracular pores normally with 5 loculi, anterior spiracles each with 2–4 (3) pores, posterior spiracles without pores. Anal opening located 7–13 (11) times length of anal opening from base of median lobes, anal opening 6–9 (8) μ long. Dorsal seta laterad of median lobes 0.7–1.0 (0.8) times length of median lobe. Eyes variable, absent,

represented by flat unsclerotized area, or represented by 2 or 3 small sclerotized points; located near body margin posterior of anterior spiracle. Antennae each with 1 conspicuous seta. Cicatrices usually on dorsal submargin of prothorax and segments 1 and 3. Body shape broadly oval. Posterior spiracle without associated dermal pocket. Dermal pattern anterior of vulva transverse. Openings of marginal macroducts on pygidium transverse. First 3 pairs of lobes with rounded apices; remaining lobes with acute apices.

Affinities Of the economic species occurring in the United States chaff scale most closely resembles *Parlatoria camelliae*, camellia parlatoria scale. The former lacks a dermal pocket near the posterior spiracle, has an anal opening 6–9 (8) μ long, has 15–30 (21) macroducts on each side of the pygidium, usually has submedial microducts on the dorsum of segment 4 only, and has submarginal microducts on segments 2 and 3. Camellia parlatoria scale has a dermal pocket near the posterior spiracle, an anal opening 12–16 (14) μ long, 14–17 (16) macroducts on the pygidium, submedial microducts on the dorsum of segments 4 and 5, and lacks submarginal microducts on segments 2 and 3.

Hosts Polyphagous. Borchsenius (1966) reports it from 25 genera in 17 families; Dekle (1977) records it from 72 genera in Florida; and McKenzie (1956) records 11 genera in California. We have examined specimens from *Aucuba, Atalantia, Citrus, Euonymus, Fortunella, Jasminum, Maclura, Mangifera, Osmanthus, Parkinsonia, Persea, Poncirus, Raphanus, Rauvolfia,* and *Yucca. Citrus* is by far the most commonly reported host.

Distribution This species probably is native to the Oriental region. In the United States it occurs out-of-doors in southern citrus-growing areas; records from northern areas are from greenhouses or from infested citrus sold in stores. We have seen U.S. specimens from AL, CA, CT, DC, FL, GA, HI, IN, KS, LA, MD, MS, NC, NY, OH, PA, SC, TX, VA, WA. It has been reported from IL, MO, OK. It commonly is taken in quarantine from tropical and subtropical areas of the world on *Citrus*. A distribution map of this species was published by CAB International (1964a).

Biology Chaff scale is reported to have 3 or 4 generations per year in Israel (Gerson 1967), 4 generations in Florida (Watson 1926), 3 generations in Alabama (English 1933), and 5 or 6 generations in Australia (Smith et al. 1997). This pest overwinters in all stages in Israel (Gerson 1967; Bodenheimer 1951), although it is reported to be predominately in the adult female form in the winter in Italy, with 14 to 15% of the population in the second instar (Lauricella 1957). Under normal circumstances, field populations have overlapping generations with all stages present at all times during the year (Bodenheimer 1951). Gerson (1967) found higher percentages of ovipositing females at 3 or 4 times during the year and therefore suggested there were 3 or 4 generations. Extensive overlap of generations probably is caused by the long oviposition period. Bodenheimer (1951) reported a single female that oviposited for a period of 76 days, laying 1 or 2 eggs each day. The average number of eggs laid by each female is about 88, and hatching may require up to 2 weeks. Watson (1926) reported that crawlers were most abundant in Florida in March and April and again in September and October. Population lows occur in the dry summer months in Israel (Bodenheimer 1951; Gerson 1967). Chaff scale prefers well-established trees that are 10 years or older. It most often is found on the inner parts of the tree and apparently prefers shady, humid areas on the host.

Economic Importance Although some consider chaff scale to be a relatively minor pest of citrus (Rosen and DeBach 1978), it is

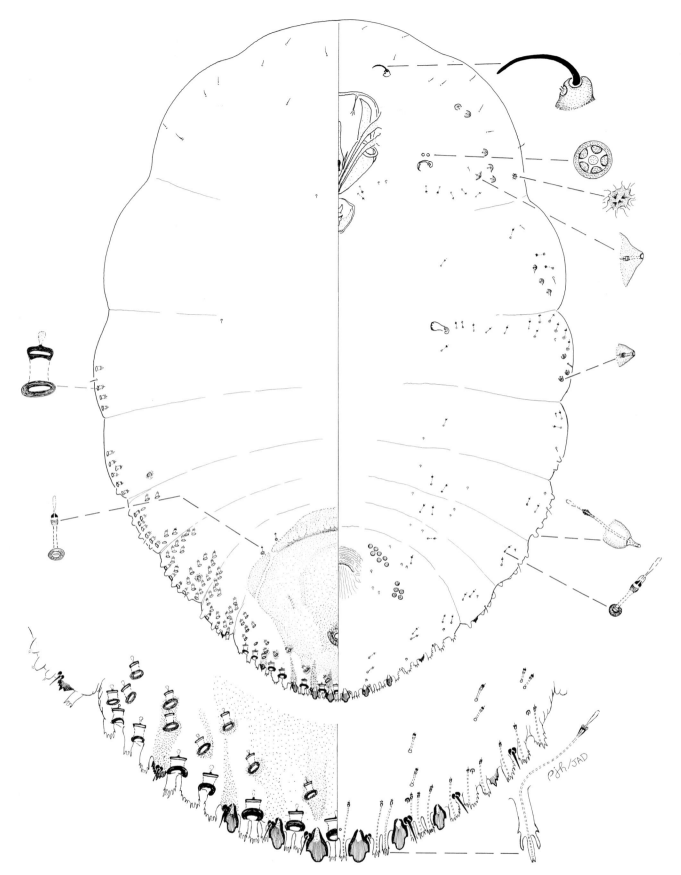

Figure 96. *Parlatoria pergandii* Comstock, chaff scale, St. Louis, MO, on *Persea* sp., I-27-1944.

included in every table presented by Rose (1990b) summarizing pests of citrus from various areas of the world. It is reported to be a very important pest in southern Japan and Italy, and an important pest in Spain, Turkey, Lebanon, Israel, Southeast Asia, Central America, Mexico, Florida, and Texas (Talhouk 1975). Under very dry conditions it causes severe twig damage in Texas (Dean 1955) and causes green areas on the ripe fruit of citrus (Gerson 1967), making it unsaleable as fresh fruit. This pest apparently is not as much of a problem in young citrus groves but usually is found on trees 10 years and older. Biological control probably will be most successful with *Aphytis comperei* and *A. paramaculicornis*; *A. hispanicus* frequently is collected from chaff scale but it is not always effective (Rosen and DeBach 1979). This species is occasionally considered a pest of hosts other than citrus. In China it causes damage to camphor trees, *Cinnamomum camphora* (Shen and Liu 1990), and magnolia, *Magnolia grandiflora* (Chao and Zeng 1997); and in India it is a pest of pepper, *Piper nigrum* (Koya et al. 1996). Beardsley and González (1975) consider this to be one of 43 major armored scale pests. Miller and Davidson (1990) treat this species as a serious world pest.

Selected References Bodenheimer (1951); Gerson (1967); Morrison (1939); Williams and Watson (1988).

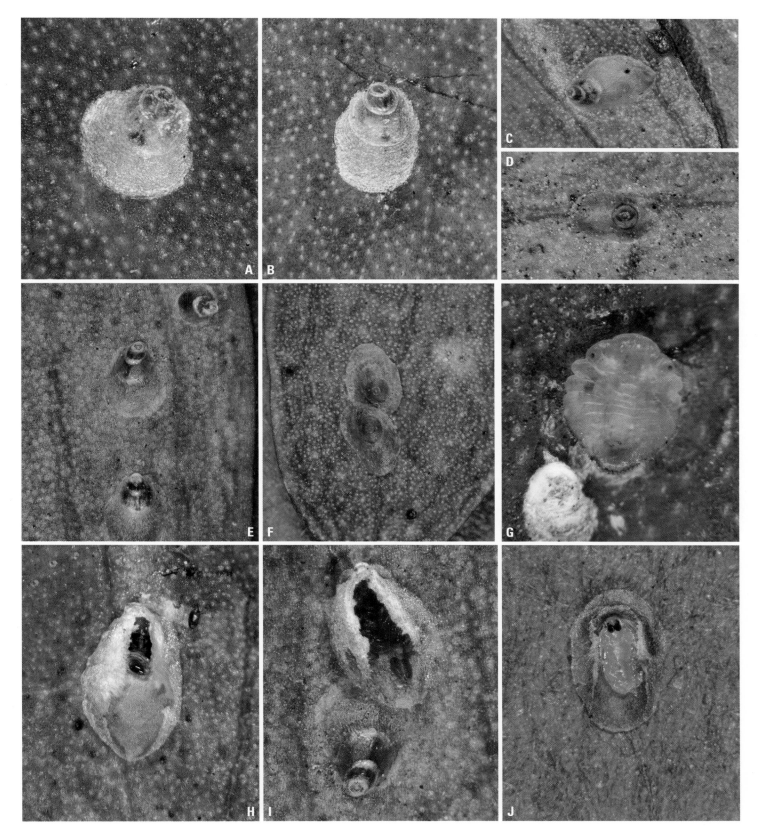

Plate 93. *Parlatoria pittospori* Maskell, Pittosporum scale

A. Young adult female cover on melaleuca leaf (J. A. Davidson).
B. Young adult female cover on melaleuca leaf (J. A. Davidson).
C. Old adult female cover on melaleuca leaf (J. A. Davidson).
D. Adult male cover on melaleuca leaf (J. A. Davidson).
E. Distance view of adult female covers on melaleuca leaf (J. A. Davidson).
F. Distance view of adult male covers on melaleuca leaf (J. A. Davidson).

G. Body of young adult female on melaleuca leaf (J. A. Davidson).
H. Mature female laying an egg on melaleuca leaf (J. A. Davidson).
I. Nearly full complement of eggs within cover on melaleuca leaf (J. A. Davidson).
J. Male pupa on melaleuca leaf (J. A. Davidson).

Parlatoria pittospori Maskell

Suggested Common Name Pittosporum scale (also called mauve pittosporum scale, pine parlatoria scale, pittosporum diaspidid).

Common Synonyms and Combinations *Parlatoria myrtus* Maskell, *Parlatoria dryandrae* Fuller, *Parlatoria petrophilae* Fuller, *Parlatoria (Euparlatoria) myrtus* Maskell, *Syngenaspis myrtus* (Maskell), *Syngenaspis dryandrae* (Fuller), *Syngenaspis petrophilae* (Fuller).

Field Characters (Plate 93) Adult female cover slightly convex, elongate oval, gray to brown; shed skins marginal, yellow-brown to green sometimes with longitudinal green stripe. Male cover smaller, similar in texture and color to female cover, elongate; shed skin marginal, dark yellow, sometimes with medial stripe. Body of adult female white before laying eggs, dark purple with eggs; crawlers and eggs light to dark purple. Normally on leaves and fruit, occasionally on smooth stems.

Slide-mounted Characters Adult female (Fig. 97) with 5 to 7 pairs of lobes, fourth to seventh lobes usually platelike in form, without associated microduct; paraphyses attached to first 3 lobes, formula normally 2-2-1, rarely 2-1-0 or 2-2-0; similar paraphysis-like structure attached to medial margin of median lobes. Median lobes separated by space 0.9–1.5 (1.1) times width of median lobe, without sclerosis or yoke, median margin axes parallel or converging apically, with 1 lateral and 1 medial notch; second lobes simple, about same size or slightly smaller than median lobes, same shape, with 1 lateral notch and 1 medial notch; third lobes simple, smaller than third lobes, same shape, with 1–3 (1) lateral notches, 1–2 (1) medial notches; fourth lobes simple, usually platelike, with 0–4 (2) lateral notches, 1–3 (2) medial notches; fifth lobes simple, platelike, with 0–4 (2) lateral notches, 0–3 (1) medial notches; sixth and seventh lobes, when present, platelike, distinguished from plate by lacking microduct. Plate formula usually 2-3-3, rarely 2-4-3, each plate and gland spine with at least 1 microduct; plates broadly fimbriate, plates in first 3 spaces generally have 3 or more tines, with tines gradually disappearing on segments 2 and 3, changing to gland spines anteriorly, 32–45 (38) plates and gland spines from prothorax to segment 4; with 2 broad plates between median lobes varying from slightly longer to slightly shorter than median lobes. Macroducts on pygidium of 2 sizes, marginal ducts longer than submarginal and submedian ducts, 1 macroduct between median lobes 12–18 (14) μ long, extending 0.2 of distance to posterior apex of anal opening, macroduct in first space 12–15 (13) μ long, with 17–21 (19) macroducts on each side of pygidium on segments 5 to 8, marginal ducts on segments 5 to 8, submarginal ducts on segments 5 to 7, submedian ducts on segment 5; total of 54–97 (72) macroducts on each side of body; some macroduct orifices occurring anterior of anal opening; prepygidial macroducts on dorsum of margin of mesothorax, metathorax, or segment 1 to segment 4, on submargin of segments 2 to 4, on submedial areas of segment 4 and sometimes 3; on venter of margin of meso- and metathorax, or metathorax only, or absent. Pygidial microducts on venter in submarginal or submedial areas of segments 5 to 7, with 2–6 (5) on each side of body; prepygidial ducts on venter in submedial areas usually near mouthparts and metathorax to segment 4, occasionally also on mesothorax, in submarginal or marginal areas on prothorax and metathorax, sometimes also on segments 2, 3, or 4; pygidial microducts absent from dorsum; prepygidial microducts in submedial areas of segment 3, sometimes on metathorax, segments 1 and 2. Perivulvar pores in 2 or 4 groups, 15–27 (20) pores on each side of body. Perispiracular pores normally with 5 loculi, anterior spiracles each with 2–4 (3) pores, posterior spiracle without pores. Anal opening located 5–8 (7) times length of anal opening from base of median lobes, anal opening 10–14 (11) μ long. Dorsal seta laterad of median

lobes 0.7–1.0 (0.8) times length of median lobes. Eyes variable, usually represented by small sclerotized area, occasionally with 1 or 2 small spurs, sometimes absent; when present located near body margin posterior of anterior spiracle. Antennae each with 1 conspicuous seta. Cicatrices often absent, sometimes on prothorax and segments 1 and 3. Body shape broadly oval. Dermal pattern anterior of vulva transverse. Openings of marginal macroducts on pygidium transverse. First 3 or 4 pairs of lobes with rounded apices; lobes 4 to 6 usually with acute apices or with platelike apices. With 3 marginal macroducts in fourth space.

Affinities Pittosporum scale is most similar to *Parlatoria pergandii*, chaff scale, but the former has submedial macroducts on segment 5 and sometimes on segments 3 and 4, has 3 submarginal macroducts in the fourth space, has the anal opening 10–14 (11) μ long, has the anal opening located 5–8 (7) times length of anal opening from base of median lobes, has 15–27 (20) perivulvar pores on each side of body, and has 32–45 (38) gland spines and plates on each side of the body from the prothorax to segment 4. Chaff scale has microducts only in submedial areas of segments 3 to 5, has 2 submarginal macroducts in the fourth space, has the anal opening 6–9 (8) μ long, has the anal opening located 7–13 (11) times length of anal opening from base of median lobes, has 11–15 (13) perivulvar pores on each side of body, and has 23–34 (30) gland spines and plates on each side of the body from the prothorax to segment 4.

Hosts Polyphagous. Borchsenius (1966) reports it from 33 genera in 17 families; Timlin (1964b) mentions 26 genera in 15 families as hosts; and McKenzie (1956) records 3 genera in California. We have examined specimens from *Banksia, Citrus, Dracaena, Eucalyptus, Hakea, Haworthia, Leucodendron, Leucopogon, Malus, Melaleuca, Mesembryanthemum, Mimosa, Nuytsia, Petrophila, Phoenix, Pimelea, Pinus, Pittosporum,* and *Xanthorrhoea.* In New Zealand pittosporum scale is collected commonly on Monterey pine, and in California it is frequently collected on pittosporum and Australian tea tree (*Leptospermum*).

Distribution This species is native to Australia. In the United States it is known only from CA. It occurs in several provinces in Australia and also is known from New Zealand and South Africa. It rarely has been taken at U.S. ports-of-entry from Australia, New Zealand, and South Africa.

Biology The life history of the pittosporum scale was studied by Timlin (1964b) in New Zealand on Monterey pine. All stages were found throughout the year. Ovipositing adult females were the predominant stage in August and September, and peak crawler emergence occurred in December. Each female laid about 46 eggs with a maximum of 62. In field experiments 1 generation required about 10 months. The preferred settling site was on the inside of the needles of the pine. Timlin also studied the pittosporum scale on apple. He was unable to collect the scale on any part of the host other than the fruit with the exception of one isolated case where he found a specimen on a smooth stem. He was able to infest smooth apple branches artificially with crawlers and development to a second generation occurred in about 10 months, but he found no evidence that this occurred in commercial orchards. In areas where apples were protected by windbreaks of Monterey pine, apple fruit commonly was contaminated with this scale. The pines apparently act as a reservoir of crawlers, which are blown by the wind onto the fruit of the apples. Pesticides are believed to eliminate development of the scale on the apples themselves. A generation may develop as rapidly as 4 months on apple fruit.

Economic Importance McKenzie (1956) reports this species to be a minor pest of certain ornamentals in southern California.

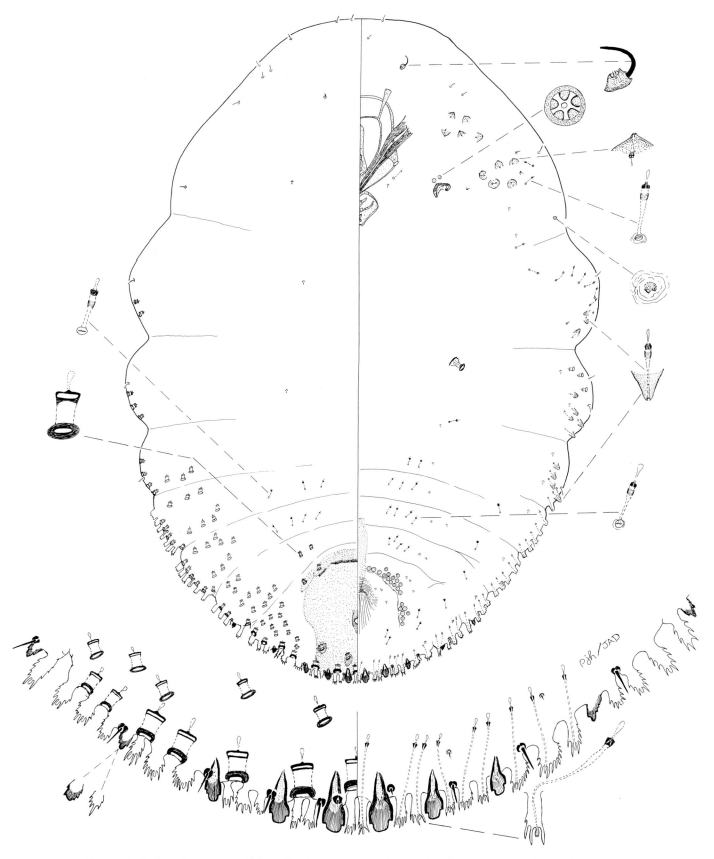

Figure 97. *Parlatoria pittospori* Maskell, pittosporum scale, South Africa, intercepted at Hoboken, NJ, on *Howorthia* sp., IV-9-1948.

In Australia (Webster 1968) and New Zealand (Timlin 1964b) infested apple fruit is rejected for shipment to certain foreign countries because of the possibility of establishment of the pest in a new area. On certain green varieties of apples, feeding sites of the scale turn red giving the fruit an unsightly appearance (Richards 1960).

Selected References McKenzie (1945); Timlin (1964b).

Parlatoria proteus (Curtis)

Suggested Common Name Proteus scale (also called sanseveria scale, small brown scale, common parlatoria scale, cattleya scale).

Common Synonyms and Combinations *Aspidiotus proteus* Curtis, *Diaspis parlatoris* Targioni Tozzetti, *Parlatoria orbicularis* Targioni Tozzetti, *Parlatoria selenipedii* Signoret, *Parlatoria* (*Euparlatoria*) *proteus* Curtis, *Parlatorea proteus* (Curtis), *Parlatoria potens* Leonardi, *Syngenaspis proteus* (Curtis).

Field Characters (Plate 94) Adult female cover flat or slightly convex, elongate oval, translucent yellow or brown; shed skins, marginal, yellow or brown with dark longitudinal stripe. Male cover smaller, white, elongate; shed skin marginal, yellowish brown, with central dark stripe. Adult female light purple or reddish purple; eggs similar in color; mature crawlers yellow with central dark area. On leaves, bark, and fruit.

Slide-mounted Characters Adult female (Fig. 98) with 3, 4, or 5 pairs of lobes, fourth and fifth lobes, when present, platelike in form, without associated microduct; paraphyses attached to first 3 pairs of lobes, formula usually 2-2-1, sometimes 2-2-0, 2-1-0; similar paraphysis-like structure attached to medial margin of median lobes. Median lobes separated by space 1.0–1.6 (1.2) times width of median lobe, without sclerosis or yoke, medial margin axes parallel or converging apically, lateral margin parallel or diverging apically, with 1 lateral and 1 medial notch; second lobes simple, about same size or slightly smaller than median lobes, about same shape, with 1 lateral notch and 0–1 (1) medial notch; third lobes simple, slightly smaller than second lobes, about same shape, with 1 lateral notch and 0–1 (1) medial notch; fourth lobes rarely replaced by plate with microduct, usually platelike in structure but without microduct, with 2–6 (5) apical notches; fifth lobes usually replaced by plate with microduct, rarely platelike in structure but without microducts, with 4–6 (5) apical notches. Plate formula normally 2-3-3, rarely 2-3-4 or 1-3-3, each plate and gland tubercle with microduct, plates broadly fimbriate, plates in first 3 spaces predominantly have 3 or more tines, with tines gradually disappearing on segments 2 and 3 changing to gland tubercles anteriorly, 23–24 (29) marginal plates and gland tubercles from head or prothorax to segment 4; with 2 broad plates between median lobes; varying from longer than to equal to length of median lobes. Macroducts on pygidium of 2 sizes, marginal ducts slightly larger than submarginals, 1 macroduct between median lobes 12–15 (13) µ long, extending 0.2 of distance to posterior apex of anal opening, macroduct in first space 10–15 (13) µ long, with 12–15 (14) macroducts on each side of pygidium on segments 5 to 8, marginal ducts on segments 5 to 8, submarginal ducts on segments 5 to 7; total of 31–48 (42) macroducts on each side of body, some macroduct orifices occurring anterior of anal opening; prepygidial macroducts on dorsum of margin of metathorax, segment 1, or segment 2 to segment 4, on submargin of segments 1 or 2 to 4; on venter usually absent, rarely on metathorax. Pygidial microducts on venter in submarginal and submedial areas of segments 5 to 7, with 3–7 (4) ducts on each side of body; prepygidial ducts on venter in submedial areas from mouthparts to segment 4, sometimes absent from metathorax, in submarginal or marginal areas from head or prothorax to metathorax and on segment 4; pygidial microducts on dorsum on submedial area of segment 5; prepygidial microducts on submedial area of segment 4. Perivulvar pores in 4 groups, 10–12 (11) pores on each side of body. Perispiracular pores normally with 5 loculi, anterior spiracles each with 2–3 (2) pores, posterior spiracles without pores. Anal opening located 5–10 (7) times length of anal opening from base of median lobes, anal opening 8–13 (11) µ long. Dorsal seta laterad of median lobes 0.5–0.8 (0.7) times length of median lobe. Eyes represented by sclerotized, spurlike process; located near body margin posterior of anterior spiracle. Antennae each with 1 conspicuous setae. Cicatrices usually on dorsal submargin of prothorax and segments 1 and 3. Body unusually broad for species of *Parlatoria*. Posterior spiracle with associated dermal pocket. Dermal pattern anterior of vulva transverse. Openings of marginal macroducts of pygidium transverse. First 3 pairs of lobes with rounded apices; remaining lobes platelike.

Affinities Of the economic species occurring in the United States, proteus scale most closely resembles *Parlatoria camelliae*, camellia parlatoria scale. The former has a sclerotized, spurlike eye, and lobes 4 and 5 platelike, when present. The camellia parlatoria scale has the eye represented by a flat sclerotized dome or a low unsclerotized dome, and lobes 4 and 5 lobelike.

Hosts Polyphagous. Borchsenius (1966) records it from 58 genera in 22 families; Dekle (1977) reports it from 122 genera in Florida. We have examined specimens from *Acalypha, Aechmea, Aerides, Agave, Aglaonema, Amomum, Angraecum, Anthurium, Arachnis, Aralia, Araucaria, Areca, Ascocentrum, Asparagus, Aspidistra, Barringtonia, Beaucarnea, Brassia, Broughtonia, Bulbophyllum, Calanthe, Calathea, Camellia, Caryota, Catasetum, Chlorophytum, Citrus, Coccothrinax, Cocos, Coelia, Coelogyne, Corypha, Croton, Cycas, Cymbidium, Cypripedium, Dendrobium, Dieffenbachia, Dracaena, Epidendrum, Eria, Ficus, Gardenia, Gongora, Grammatophyllum, Hedera, Heliconia, Hemigraphis, Jasminum, Kentia, Laelia, Lamium, Ligustrum, Lilium, Lycaste, Mangifera, Maranta, Maxillaria, Nipa, Olea, Ophiopogon, Pandanus, Paphiopedilum, Pedilanthus, Persea, Phalaenopsis, Philodendron, Phoenix, Pholidota, Pinanga, Psychotria, Quercus, Renanthera, Rhipsalis, Rhynchostylis, Sabal, Saccolabium, Sansevieria, Schefflera, Schomburgkia, Scindapsus, Sobralia, Spathoglottis, Stanhopea, Talauma, Thrinax, Thunbergia, Trichocentrum, Trichoglottis, Triphasia, Vanda,* and *Yucca.* Orchids and palms are the most commonly reported hosts (Williams and Watson 1988).

Distribution This species probably is native to the tropical areas of Asia. We have seen U.S. specimens from CA (eradicated when found), DC, FL, GA, IN, KS, LA, MA, MD, NJ, NY, OH, PA, TX. It has been reported from IL, MO, MS. It occurs out-of-doors in warmer areas and is in greenhouses elsewhere. It commonly is taken in quarantine from tropical areas of the world.

Biology We have been unable to find information pertaining to the life history of this species. On orchids it apparently reproduces slowly in the greenhouse; each female produces up to 30 eggs (Zahradník 1968). Both parthenogenetic (Nur 1971) and sexual races (Schmutterer 1952b) are reported. Scales may be found on the leaves or branches of the host.

Economic Importance Proteus scale is considered to be a serious pest of ornamentals in Florida (Dekle 1977). It is a minor pest of citrus in Argentina (Talhouk 1975), Brazil (Bondar 1914), China, Egypt, Florida, Southeast Asia, and occasionally is found on avocado (Ebeling 1959). It has been reported as a greenhouse pest of orchids and palms (Weigel and Sasscer 1923), lilies and amaryllis (Baker 1993), and bananas in the Canary Islands and Madagascar (Chua and Wood 1990). Beardsley and González (1975) consider this to be one of 43 major armored scale pests. Miller and Davidson (1990) treat it as a serious world pest.

Selected References McKenzie (1945); Williams and Watson (1988).

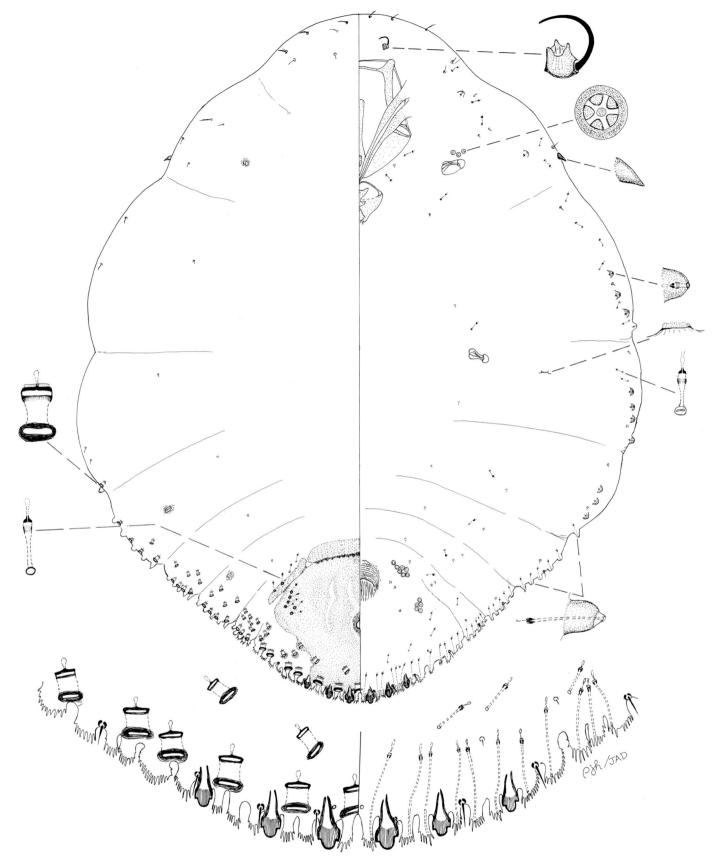

Figure 98. *Parlatoria proteus* (Curtis), proteus scale, Canal Zone, intercepted at Miami, FL, on *Anthurium* sp., II-18-1953.

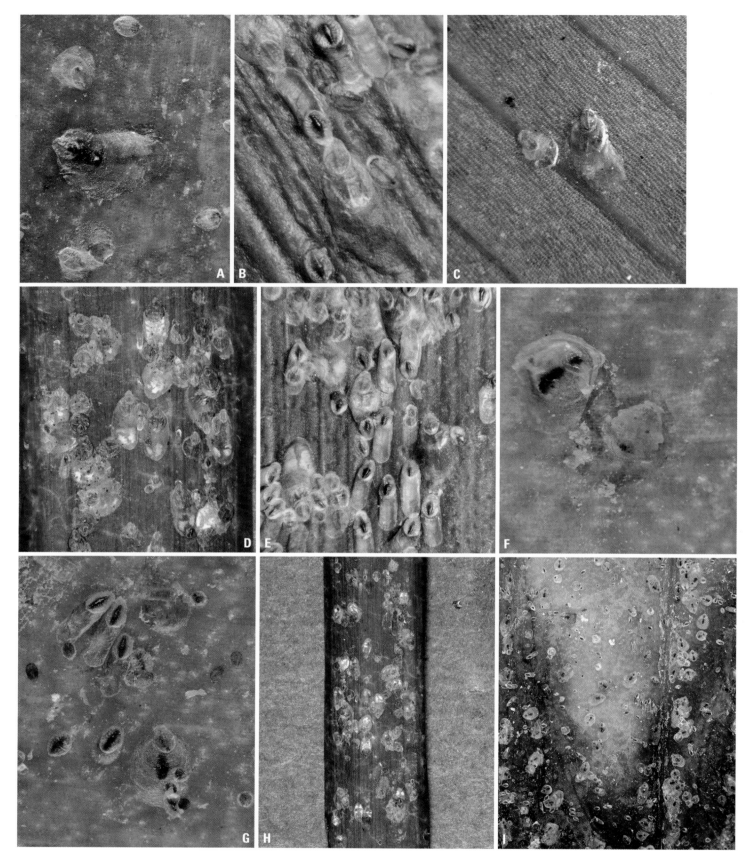

Plate 94. *Parlatoria proteus* (Curtis), Proteus scale

A. Translucent female cover with two rows of eggs on cattleya orchid (J. A. Davidson).
B. Male and female covers on bromeliad (R. J. Gill).
C. Second-instar (left) and adult female (right) covers on *Ctenanthe oppenheimeriana* (R. J. Gill).
D. Distance view of infestation on liriope (J. A. Davidson).

E. Distance view of infestation on bromeliad (R. J. Gill).
F. Body of adult female on cattleya orchid (J. A. Davidson).
G. Infestation with diverse instars including settled crawlers on cattleya orchid (J. A. Davidson).
H. Heavy infestation on liriope leaf (J. A. Davidson).
I. Damaged cattleya orchid (J. A. Davidson).

Plate 95. *Parlatoria theae* Cockerell, Tea parlatoria scale

A. Adult female cover on acuba (J. A. Davidson).
B. Adult female covers (top) and male covers on unknown leaf (J. A. Davidson).
C. Adult male covers on unknown leaf (J. A. Davidson).
D. Distance view of covers on bark of unknown plant (J. A. Davidson).
E. Distance view of covers on bark of unknown plant (J. A. Davidson).

333

Parlatoria theae Cockerell

Suggested Common Name Tea parlatoria scale (also called tea scale).

Common Synonyms and Combinations *Parlatoria theae* v. *viridis* Cockerell, *Parlatoria theae* v. *euonymi* Cockerell, *Parlatoria viridis* Cockerell, *Parlatoria (Euparlatoria) theae* Cockerell, *Syngenaspis theae* Cockerell, *Parlatoria pergandei* v. *dives* Bellio, *Parlatoria pergandei* v. *theae*.

Field Characters (Plate 95) Adult female cover flat, broadly elongate oval, grayish to brown; shed skins marginal, black, sometimes light brown or even slightly green. Male cover smaller, similar in texture and color to female cover, elongate; shed skin marginal, black or brown. Body of adult female and eggs probably light purple to red. Normally on stems and branches.

Slide-mounted Characters Adult female (Fig. 99) with 5 pairs of well-developed lobes, fourth and fifth lobes shorter and broader than first 3; paraphyses attached to first 3 lobes, formula 2-2-1 or 2-2-0; similar paraphysis-like structure attached to medial margin of median lobes. Median lobes separated by space 0.8–1.5 (1.1) times width of median lobe, without sclerosis or yoke, medial margin axes usually parallel, rarely converging apically, lateral margin axes parallel, with 1 lateral and 1 medial notch; second lobes simple, same shape as medial lobes, slightly smaller, with 1 lateral notch, 0–1 (1) medial notch; third lobes simple, same shape as second lobes, slightly smaller, with 1–2 (1) lateral notch, 0–1 (1) medial notch; fourth lobes simple, shorter and proportionally broader than third lobes, with 0–3 (1) lateral notches, 0–3 (2) medial notches; fifth lobes simple, about same size and shape as fourth lobes, with 0–2 (1) lateral notches, 0–2 (1) medial notches. Plate formula usually 2-3-3, rarely 2-3-4, each plate between medial lobes and in first 2 or 3 spaces with 1 or 2 microducts, remaining plates and gland tubercles with 1 or 2 microducts; plates broadly fimbriate, plates in first 3 spaces predominantly have 3 or more tines, with tines gradually disappearing on segments 2, 3, or 4 changing to gland tubercles anteriorly, 37–61 (49) marginal plates and gland tubercles from thorax to segment 4; with 2 broad plates between median lobes, usually about equal to length of median lobes, sometimes shorter. Macroducts on pygidium of 2 sizes, marginal ducts slightly larger than submarginals, 1 macroduct between median lobes 12–18 (15) µ long, extending 0.1–0.2 (0.2) of distance to posterior apex of anal opening, macroduct in first space 10–15 (12) µ long, with 17–27 (23) macroducts on each side of pygidium on segments 5 to 8, marginal ducts on segments 5 to 8, submarginal ducts on segments 5 to 7; total of 62–90 (77) macroducts on each side of body; some macroduct orifices occurring anterior of anal opening; prepygidial macroducts on dorsum of margin of metathorax or segment 1 to segment 4, on submargin of segments 2 to 4; on venter absent or on metathorax and/or segment 1. Pygidial microducts on venter in submedial and/or submedial areas of segments 5 to 7, with 6–18 (10) ducts on each side of body; prepygidial ducts on venter in submedial areas from mouthparts to segment 4, in submarginal or marginal areas from prothorax to segment 1, and on segment 3 and/or 4; pygidial microducts absent from dorsum or on submedial area of segment 5; prepygidial microducts on dorsum in submedial area of segment 4. Perivulvar pores in 4 or 5 groups,

22–41 (31) pores on each side of body. Perispiracular pores normally with 5 loculi, anterior spiracles each with 2–3 (3) pores, posterior spiracles without pores. Anal opening located 6–9 (7) times length of anal opening from base of median lobes, anal opening 10–17 (13) µ long. Dorsal seta laterad of median lobes 0.6–0.9 (0.8) times length of median lobe. Eyes variable, usually represented by low dome, sometimes in form of several small points, occasionally absent, located near body margin posterior of anterior spiracle. Antennae each with 1 conspicuous seta. Cicatrices, when present, on dorsal submargin of prothorax and segments 1 and 3. Body shape broadly oval. Posterior spiracle with associated dermal pocket. Dermal pattern anterior of vulva transverse. Openings of marginal macroducts of pygidium transverse. First 3 pairs of lobes with rounded apices; remaining lobes with acute apices. Usually with a band of microducts from body margin to posterior spiracle.

Affinities Of the economic species occurring in the United States, tea parlatoria scale most closely resembles *Parlatoria camelliae*, camellia parlatoria scale. The former has 2 microducts on most plates in spaces 1 to 3, with 17–27 (23) macroducts on the pygidium, and 22–41 (31) perivulvar pores on each side of body. The camellia parlatoria scale has 1 microduct on most plates in spaces 1 to 3, with 14–17 (16) macroducts on the pygidium, and 10–17 (13) perivulvar pores on each side of the body.

Hosts Polyphagous. Borchsenius (1966) records it from 36 genera in 21 families; Murakami (1970) reports it from 20 genera in Japan; McKenzie (1956) found it on 24 genera of host in California. We have examined specimens from *Acer*, *Althaea*, *Aucuba*, *Camellia*, *Citrus*, *Cornus*, *Corylopsis*, *Cotoneaster*, *Crataegus*, *Croton*, *Diospyros*, *Enkianthus*, *Euonymus*, *Euphorbia*, *Hibiscus*, *Malus*, *Prunus*, *Punica*, *Pyrus*, *Rhamnus*, *Spiraea*, *Staphylea*, *Syringa*, *Viburnum*, and *Xanthoxylum*. This species often is collected on *Acer*, *Aucuba*, and *Euonymus*.

Distribution This species probably is native to Asia. We have examined U.S. specimens from CA (eradicated), DC, HI, MD, NC, VA. It has been reported from MS and TX. The species occurs in Asia, Europe, and the southern former Soviet Union. It commonly is taken in quarantine from Japan.

Biology Very little biological information has been published about this species. It has been reported to have 2 (Hadzibejli 1983) or 2 to 3 (Borchsenius 1963) generations per year. Some biological data have been published by Chou (1984) and Kobakhidze (1954).

Economic Importance The tea parlatoria scale is considered to be a pest of apple (Kozár 1990b), pear (Clausen 1931), and tea (Ebeling 1959; Kobakhidze 1954). It has been reported as a citrus pest in Mediterranean areas (Balachowsky 1953), but these records may be based on misidentifications. The species was considered to be a serious enough threat to California agriculture that intensive control measures were taken and the species was eradicated from that state. Westcott (1973) considers it a serious pest of camellia in the southeastern United States. Beardsley and González (1975) consider this to be one of 43 armored scale pests.

Selected References McKenzie (1945); Morrison (1939).

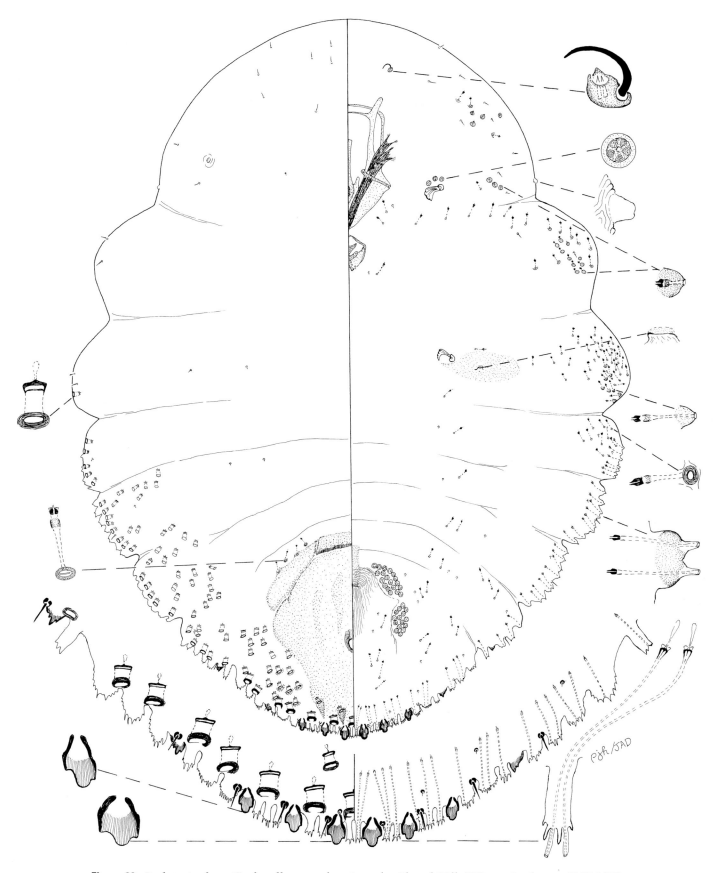

Figure 99. *Parlatoria theae* Cockerell, tea parlatoria scale, Chapel Hill, NC, on *Acuba* sp., X-27-1970.

Parlatoria ziziphi (Lucas)

Suggested Common Name Black parlatoria scale (also called citrus parlatoria, ebony scale, Mediterranean scale, black parlatoria).

Common Synonyms and Combinations *Coccus ziziphi* Lucas, *Chermes aurantii* Boisduval, *Parlatoria Lucassi* Targioni Tozzetti, *Parlatoria zizyphi* (Lucas), *Parlatoria zizyphe* (Lucas), *Parlatoria* (*Websteriella*) *Zizyphi* (Lucas), *Parlatoria zizyphus* (Lucas), *Parlatoria ziziphus* (Lucas), *Parlatoria zyziphi* (Lucas), *Parlatorea zizyphi* (Lucas), *Parlatoria zozypium* (Lucas), *Parlatoria* (*Websteriella*) *ziziphus* (Lucas), *Apteronidia ziziphi* (Lucas), *Diaspis ziziphus* (Lucas), *Parlatoreopsis ziziphi* (Lucas).

Field Characters (Plate 96) Adult female cover flat, broadly elongate oval, black with narrow white fringe, with 2 or 3 longitudinal ridges; shed skins marginal, black, primary component of cover second shed skin. Male cover smaller, white, elongate; shed skin marginal, black. Body of newly matured adult female yellow brown. Normally on leaves, also on branches and fruit. Margin of body laterad of mouthparts with small lobe.

Slide-mounted Characters Adult female (Fig. 100) with 4 pairs of well-developed lobes (3 out of 10 specimens with a fifth lobe), fourth and fifth lobes shorter and narrower than first 3; paraphyses very inconspicuous, when present attached to first 3 lobes, formula highly variable from 1-1-1 to 2-2-1; similar paraphysis-like structure attached to medial margin of median lobes. Median lobes separated by space 1.1–1.6 (1.3) times width of median lobe, without sclerosis or yoke, medial margin axes converging apically, lateral margin axes diverging, with 1 lateral and 1 medial notch; second lobes simple, same shape as median lobes, equal to or slightly smaller than median lobes, with 1 lateral notch, 1 medial notch; third lobes simple, same shape as second lobes, equal to or slightly smaller than second lobes, with 1–3 (1) lateral notch, 1 medial notch; fourth lobes simple, shorter and thinner than third lobes, with 0–2 (1) lateral notches, 0–1 (0) medial notches; fifth lobes when present simple, same size as fourth lobes, smaller than fourth lobes, without notches. Plate formula usually 2-3-3, rarely 2-2-3, each plate between medial lobes and in first 2 or 3 spaces with 1 microduct, remaining plates and gland tubercles with 1 or 2 microducts; plates broadly fimbriate, plates sometimes have as many as 15 tines, with tines gradually disappearing on segments 2, 3, or 4 changing to gland tubercles anteriorly, 37–61 (49) marginal plates and gland tubercles from thorax to segment 4; with 2 broad plates between median lobes, usually about equal to length of median lobes, sometimes shorter. Macroducts on pygidium of 2 sizes, marginal ducts slightly larger than submarginals, 1 macroduct between median lobes 10–16 (13) µ long, extending 0.1–0.2 (0.1) of distance to posterior apex of anal opening, macroduct in first space 10–15 (13) µ long, with 13–18 (14) macroducts on each side of pygidium on segments 5 to 8, marginal ducts on segments 5 to 8, submarginal ducts on segments 5 to 8; total of 36–49 (42) macroducts on each side of body; some macroduct orifices occurring anterior of anal opening; prepygidial macroducts on dorsum of margin of metathorax or segment 1 to segment 4, on submargin of segments 3 or 4; on venter absent or rarely with 1 or 2 on metathorax. Pygidial microducts on venter in submedial and/or submedial areas of segments 5 and 6, sometimes on 7, with 3–6 (4) ducts on each side of body; prepygidial ducts on venter in submedial areas from near antennae to segment 4, in submarginal or marginal areas on prothorax and mesothorax; pygidial microducts either absent from dorsum or on submedial area of segment 5; prepygidial microducts on dorsum absent. Perivulvar pores in 4 rarely 5 groups, 13–18 (15) pores on each side of body. Perispiracular pores normally with 5 loculi, anterior spiracles each with 2–4 (3) pores, posterior spiracles without pores.

Anal opening located 7–9 (8) times length of anal opening from base of median lobes, anal opening 9–13 (11) µ long. Dorsal seta laterad of median lobes 0.5–0.8 (0.7) times length of median lobe. Eyes possibly part of earlike lobe laterad of anterior spiracle. Antennae each with 1 conspicuous seta. Cicatrices usually present on dorsal submargin of prothorax and segments 1 and 3. Body shape broadly oval. Posterior spiracle without associated dermal pocket. Dermal pattern anterior of vulva transverse. Openings of marginal macroducts of pygidium transverse. First 3 pairs of lobes with rounded apices; remaining lobes with acute apices. Usually with a band of microducts from body margin to anterior spiracle. With earlike lobes laterad of anterior spiracles.

Affinities Black parlatoria scale can be distinguished from all other species of *Parlatoria* by the earlike lobes laterad of the anterior spiracles and the black rectangular scale cover with a white fringe that surrounds the posterior half of the cover.

Hosts Very limited host range. Blackburn and Miller (1984) provided evidence demonstrating that many of the records reported in the literature on hosts other than in the family Reattach are most likely erroneous. *Zizyphus* (Family Rhamnaceae) would seem to be an exception. Thus, hosts for this species probably are *Citrus*, *Murraya*, *Poncirus*, and *Severinia* with *Citrus* being the predominant host.

Distribution This species probably is native to Asia. We have examined U.S. specimens from FL and HI only. The species occurs in Africa, Asia, Central and South America, Europe, Oceania, and the West Indies. It commonly is taken in quarantine from southern Asia and Japan. A distribution map of this species was published by CAB International (1964b).

Biology In the Levant region of Spain there are 3 to 5 generations per year (Gomez Clemente 1943). All stages of development can be found throughout the year. The adult female lays from 8 to 20 eggs, which may serve as the overwintering form when they are laid in the fall (Gomez Clemente 1943). In the Caucasus the scale produces 2.5 generations per year and overwinters in the second instar (Borchsenius 1950). In Taiwan there are up to 7 generations per year; a generation requires about 42 days to develop from June to August and about 93 days during cooler weather. Depending on the season, the egg stage lasts 2 to 4 days, the first instar 6 to 13 days, the second instar 13 to 30 days, and the adult female lives 11 to 24 days. The oviposition period lasts 7 to 18 days (Chang and Tao 1963). There are 5 generations per year in Sicily, and the complete life cycle takes 30 to 40 days under favorable conditions (Monastero 1962). A generation requires 75 to 80 days in Tunisia during warm periods and up to 160 days during cool periods (Bénassy and Soria 1964).

Economic Importance The black parlatoria scale has long been considered one of the major pests of citrus in certain areas. It is reported as important in the following countries: Brazil (Gravena et al. 1992), China (Chen 1936), Egypt (Coll and Abd-Rabou 1998; Hosny 1943), Iran (Zoebelein 1966), Italy (Liotta 1970), Libya (Martin 1954), Nigeria (Boboye 1971), Puerto Rico (Cruz and Segarra 1991), Taiwan (Chang and Tao 1963), and Tunisia (Bénassy and Soria 1964). Talhouk (1975) in his treatment of citrus pests of the world, lists black parlatoria scale as economically important in Algeria, Morocco, Tunisia, and southeast Asia; he lists it as causing some damage in Greece, Italy, Spain, Israel, Egypt, and South Africa. He notes that in some countries the scale may not be considered a pest but populations occasionally become a problem in localized areas. Rose (1990b) also includes the black parlatoria scale as an important pest of citrus in many parts of the world. Heavy infestations of this scale cause chlorosis and premature drop of leaves, dieback

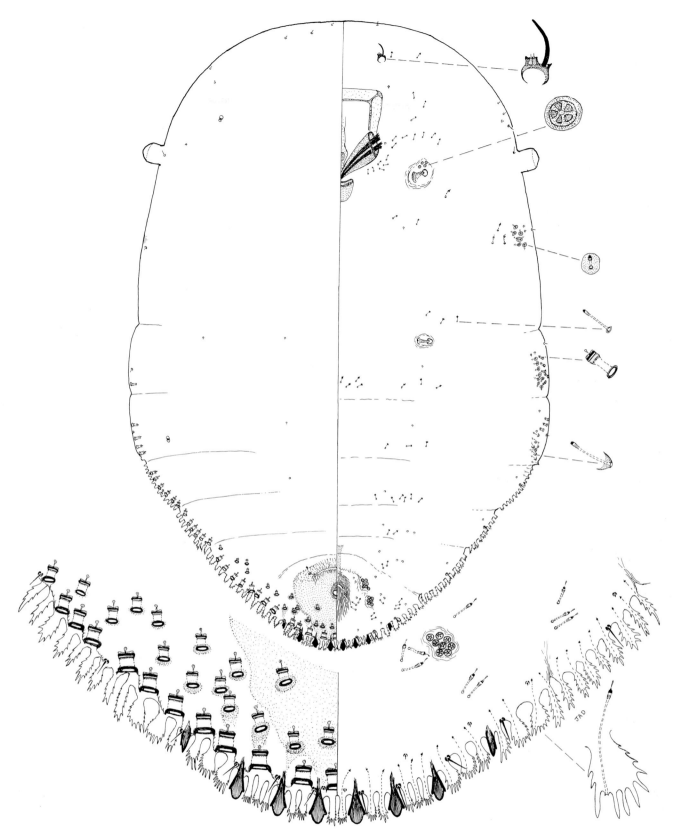

Figure 100. *Parlatoria ziziphi* (Lucas), black parlatoria scale, St. Croix, intercepted at San Juan, PR, on *Citrus* fruit, VIII-6-1979.

of twigs and branches, stunting and distortion of fruit, and premature fruit drop. Perhaps the most characteristic damage is the virtually unremoveable scale cover on the fruit. Beardsley and González (1975) consider this to be one of 43 major armored scale pests. Miller and Davidson (1990) treat it as a serious world pest.

Selected References Blackburn and Miller (1984); McKenzie (1945); Rose (1990b).

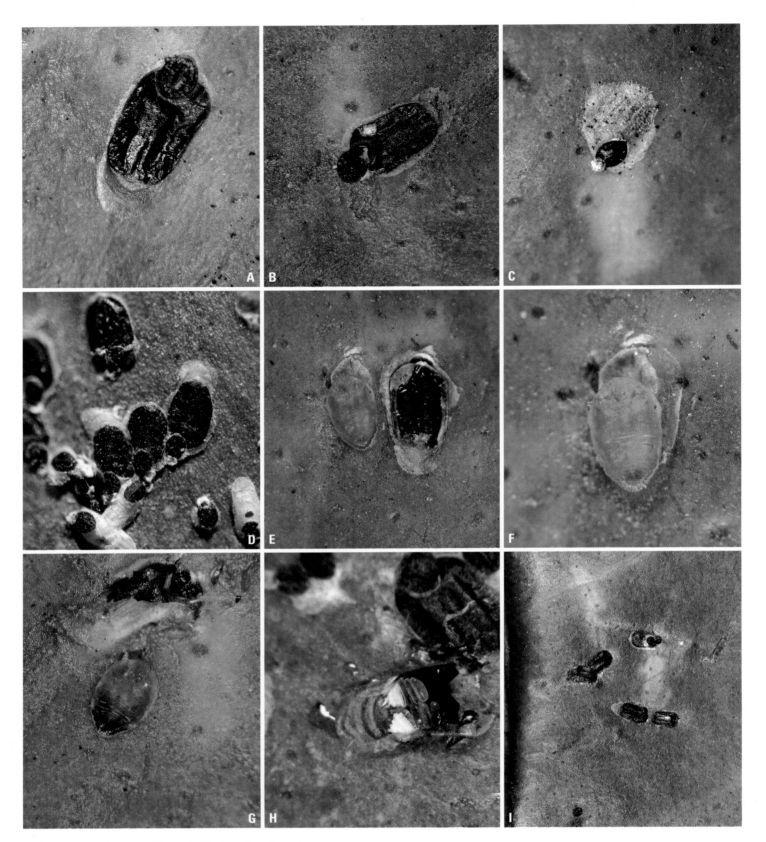

Plate 96. *Parlatoria ziziphi* (Lucas), **Black parlatoria scale**

A. Adult female cover showing large brown crawler flap on citrus leaf (J. A. Davidson).
B. Adult female cover on citrus leaf (J. A. Davidson).
C. Adult male cover on citrus leaf (J. A. Davidson).
D. Infestation on citrus leaf (D. R. Miller).
E. Body of adult female removed from scale cover (J. A. Davidson).
F. Body of young adult female showing thoracic tubercles (J. A. Davidson).

G. Body of older adult female (J. A. Davidson).
H. Turned female cover showing rows of eggs on citrus (J. A. Davidson).
I. Female covers and feeding damage on citrus leaf (J. A. Davidson).

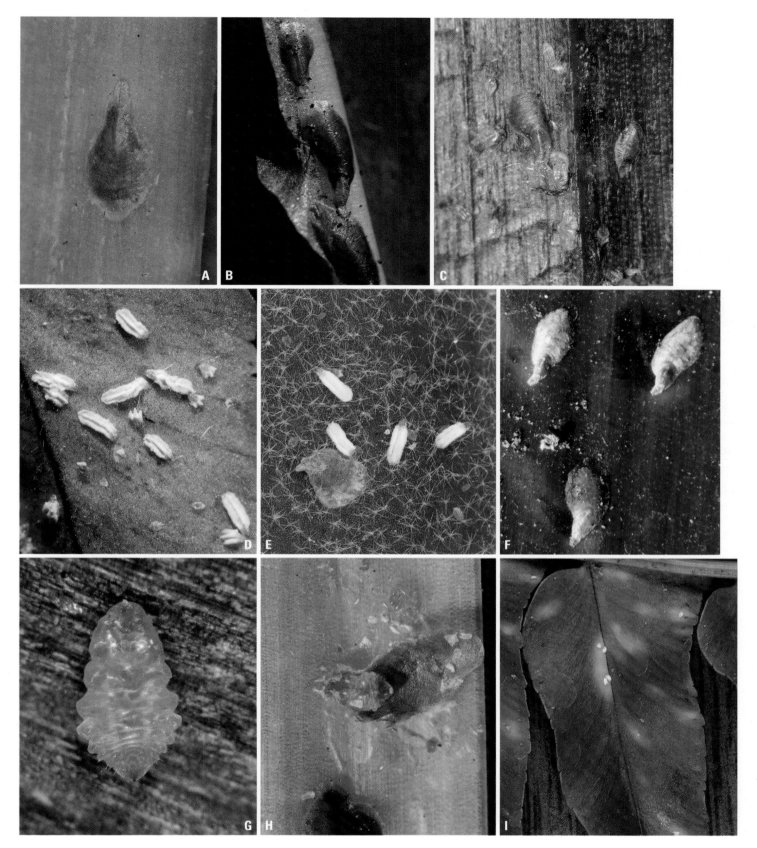

Plate 97. *Pinnaspis aspidistrae* (Signoret), **Fern scale**

A. Young adult female cover on liriope outdoors in Maryland (J. A. Davidson).

B. Old adult female covers on liriope (J. A. Davidson).

C. Immature and mature female covers on *Rhapis excelsa* (R. J. Gill).

D. Adult male covers on rabbits foot fern (J. A. Davidson).

E. Adult female and male covers on fern (J. A. Davidson).

F. Transparent covers on palm including some with eggs and others with egg shells on liriope (J. A. Davidson).

G. Body of adult female before egg laying on liriope (J. A. Davidson).

H. Body of adult female with eggs on liriope (J. A. Davidson).

I. Feeding damage on fern leaf with male covers (J. A. Davidson).

Pinnaspis aspidistrae (Signoret)

ESA Approved Common Name Fern scale (also called Brazilian snow scale, aspidistra scale, and flour scale).

Common Synonyms and Combinations *Chionaspis aspidistrae* Signoret, *Chionaspis braziliensis* Signoret, *Chionaspis latus* Cockerell, *Hemichionaspis aspidistrae* (Signoret), *Hemichionaspis aspidistrae* v. *brasiliensis* Hempel, *Hemichionaspis aspidistrae* v. *lata* Cockerell, *Chionaspis (Pinnaspis) aspidistrae* Signoret, *Pinnaspis ophiopogonis* Takahashi, *Pinnaspis caricis* Ferris.

Field Characters (Plate 97) Adult female cover narrowly to broadly oyster-shell shaped, light to dark brown; shed skins marginal, yellow brown to brown. Male cover smaller, white, felted, elongate, with 3 ridges; shed skin marginal, pale yellow. Body of adult female yellow; eggs pale brown; crawlers yellow. Mainly on leaves, occasionally on bark and fruit.

Slide-mounted Characters Adult female (Fig. 101) with 2 pairs of definite lobes, third and fourth lobes usually represented by series of points; paraphyses usually attached to medial margin of medial lobule of second lobes, occasionally attached to lateral margin of median lobule of second lobe, lateral margin of lateral lobule of second lobe, and medial margin of medial lobule of third lobes. Median lobes usually closely appressed with space less than 1μ separating them, rarely space between lobes in invert 'V' shape, then space between lobes at apex about 0.5 times width of lobe, with weakly indicated basal sclerosis, rarely with small sclerotized knob at posterior end of each lobe, medial margin axes usually parallel, lateral margins converging apically, with 2–3 (3) lateral notches, 0–1 (0) medial notches; second lobes with 2 lobules, medial lobule frequently protruding beyond median lobes, medial lobule largest with 0–1 (0) lateral notches, 0–1 (0) medial notches, lateral lobule with 0–2 (1) lateral notches, with 0–1 (0) medial notches; third lobes bilobed, represented by broad series of points, medial lobules with 3–6 (5) notches, lateral lobule with 4–10 (6) notches; fourth lobes rarely absent, usually represented by 2 low lobules, medial lobule with 0–7 (5) notches, lateral lobule with 0–8 (4) notches. Gland-spine formula 1-1-1, with 9–19 (14) gland spines near each body margin from mesothorax, metathorax, or segment 1 to segment 4; without gland spines between medial lobes. Macroducts of 2 sizes, larger size in marginal areas on segments 4 to 7, in submarginal areas on segments 3 to 5, smaller size in submarginal and marginal areas from mesothorax or metathorax to segment 4, on dorsum from metathorax, segment 1, or segment 2 to segment 4, on venter from mesothorax or metathorax to segments 1, 2, or 3, without duct between median lobes, macroduct in first space 17–25 (22) μ long, with 6–8 (7) macroducts on each side of body on segments 5 to 7, total of 26–38 (34) macroducts on each side of body, some orifices occurring anterior of anal opening. Pygidial microducts on venter, in submarginal areas of segments 5 to 6 or 7, with 2–9 (4) ducts; prepygidial ducts in submedial areas from head, prothorax, or mesothorax to segment 4, in submarginal or marginal areas of head to mesothorax or metathorax and on segments 3 and 4; pygidial microducts normally absent, rarely with 1 submedial duct on segment 5, prepygidial ducts often in submedial areas of abdomen, occasionally also on thorax. Perivulvar pores in 5 groups, 29–59 (44) pores. Perispiracular pores usually with 3 loculi, anterior spiracles each with 7–34 (16) pores, posterior spiracles with 1–12 (4) pores. Anal opening located 4–11 (7) times length of anal opening from base of median lobes, anal opening 10–18 (12) μ long. Eyes usually on small protrusion laterad of anterior margin of clypeolabral shield, eyes usually represented by flat sclerotization, rarely with 1 or 2 small points. Antennae each with 1 conspicuous seta. Cicatrices or cicatrix-like structures 5 in number on head, prothorax, segments 1, 3, and 5. Body elongate with prepygidial segments from mesothorax to segment 3 with lateral margin forming distinct protuberance. Usually with microduct near antenna. Preanal scar absent or represented by light sclerotized area. Gland spines on segments 5 to 7 with small microducts compared to those in remaining gland spine.

Affinities Of the economic species in the United States, fern scale is most similar to *Pinnaspis strachani*, lesser snow scale. In most instances these species can be separated by the combination of characters given below; a few specimens are difficult or impossible to determine using these criteria. We suspect that the *aspidistrae–strachani* complex either may be a highly variable species or more than 2 species. Fern scale has small median lobes that protrude less than or about the same distance as the large second lobes, the preanal scleroses lacking or represented by light sclerotized patches, the posterior spiracles each with 1–12 (4) pores. Lesser snow scale has large median lobes that protrude beyond or about the same distance as the small second lobes, the preanal sclerosis represented by heavily sclerotized bars, and posterior spiracles each with 0–4 (2) pores.

Hosts Polyphagous. Borchsenius (1966) records it from 60 genera in 27 families; Dekle (1977) reports it from 57 genera in Florida including 18 genera of ferns. We have examined specimens from *Atalantia*, *Anemia*, *Anthurium*, *Aspidistra*, *Asplenium*, *Camellia*, *Carica*, *Chrysalidocarpus*, *Citrus*, *Cocos*, *Cordyline*, *Cycas*, *Cypripedium*, *Diospyros*, *Eurya*, *Lagerstroemia*, *Liparis*, *Liriope*, *Livistona*, *Muehlenbeckia*, *Musa*, *Nephrolepis*, *Ophiopogon*, *Paeonia*, *Phalaris*, *Philodendron*, *Pieris*, *Piper*, *Platycerium*, *Polyalthia*, *Polystichum*, *Prunus*, *Pteris*, *Rhapis*, *Ruscus*, *Saintpaulia*, *Strelitzia*, *Strobilanthes*, *Tillandsia*, and *Ziziphus*. Based on material in the USNM Collection, commonly infested hosts include *Aspidistra*, *Citrus*, and ferns.

Distribution Fern scale probably is native to the Oriental Region. We have examined U.S. specimens from AL, AR, CA, DC, FL, GA, HI, IA, IL, IN, KS, LA, MA, MD, MI, MO, NC, NJ, NY, OH, OK, OR, PA, SC, TN, TX, UT, VA, WV. It has been reported from CT, KY, MS. The species occurs out-of-doors in the warmer southern areas and in greenhouses in the north. We have found populations that overwinter in protected areas in Maryland on *Liriope*. Fern scale is taken in quarantine from most parts of the world.

Biology According to Werner (1931) eggs hatched on fern in 18 days at 70–85 °F and 60–70% RH. The female life span averaged about 95 days, and that of the male about 37 days. Fertilized adult females lived an average of 68 days, and unfertilized females lived an average of 65.5 days. Unmated females did not lay eggs. Eggs were observed developing 8 days after mating, and oviposition began 12 days after mating. Up to 6 eggs were laid per female per day. Females produced an average of 57 eggs each, with 108 as the maximum number. The sex ratio of second instars was 75.3% males to 24.7% females. Saakyan-Baranova (1954) reports 2 generations per year on ferns in a greenhouse in Moscow, Russia. Schmutterer (1952b) found that eggs hatched about 10 days after being laid. Brown (1965) studied a fern scale population in California that had the sexual diaspidid chromosome system. Takahashi and Tachikawa (1956) report 2 strains of fern scale in Japan. One was found on *Ophiopogon*, *Liriope*, and *Rohdea* outdoors but never was found on *Aspidistra*. It lacked males and was parthenogenetic throughout the year. Another strain that had males occurred on *Nephrolepis* and *Platycerium* ferns in greenhouses, rarely on

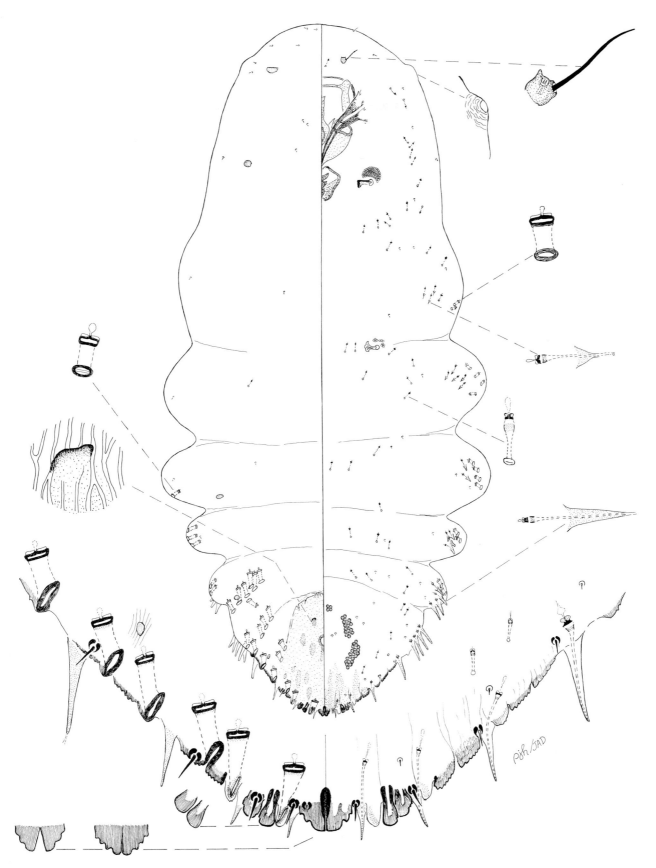

Figure 101. *Pinnaspis aspidistrae* (Signoret), fern scale, Belgium, intercepted at Hoboken, NJ, on *Aspidistra* sp., VIII-22-1949.

Aspidistra outdoors, and never on *Ophiopogon*. Ghauri (1962) described the adult male and saw only winged males from ferns and camellias.

Economic Importance Beardsley and González (1975) consider fern scale to be one of the 43 principal armored scale pests of the world. Dekle (1977) notes that this species often is a serious pest on ferns and other foliage plants in Florida. Considering only *Citrus*, Talhouk (1975) reports it as a very important pest in Japan and Venezuela; an important pest in Chile, Southeast Asia, and India; and a minor pest in Brazil and Florida. Giacometti (1983) and Nascimento (1981) both consider this scale to be a serious pest of citrus in Brazil. Chua and Wood (1990) discuss this species as a pest on the fruit of oil palm and the leaves and stems of rubber trees in Malaysia and on banana on several Pacific islands. Fern scale also has been reported to injure coconut in Sri Lanka (Rutherford 1914), bananas in Fiji (Jepson 1915), *Dracaena* in Poland (Labanowski 1999), and *Ficus* in India (Ramakrishna Ayyar 1930). Heavy infestations on ferns makes the plants unsightly because male covers turn the leaves white, and female feeding sometimes causes chlorotic spots on green foliage (Baker and Shearin 1992). In Pennsylvania this species is considered an important pest of foliage plants, particularly *Liriope* and ferns in the genera *Aspidistra* and *Nephrolepis*. When populations are heavy, severe chlorosis occurs and leaf tissue may be killed (Stimmel 1997).

Selected References Ferris (1937); Werner (1931).

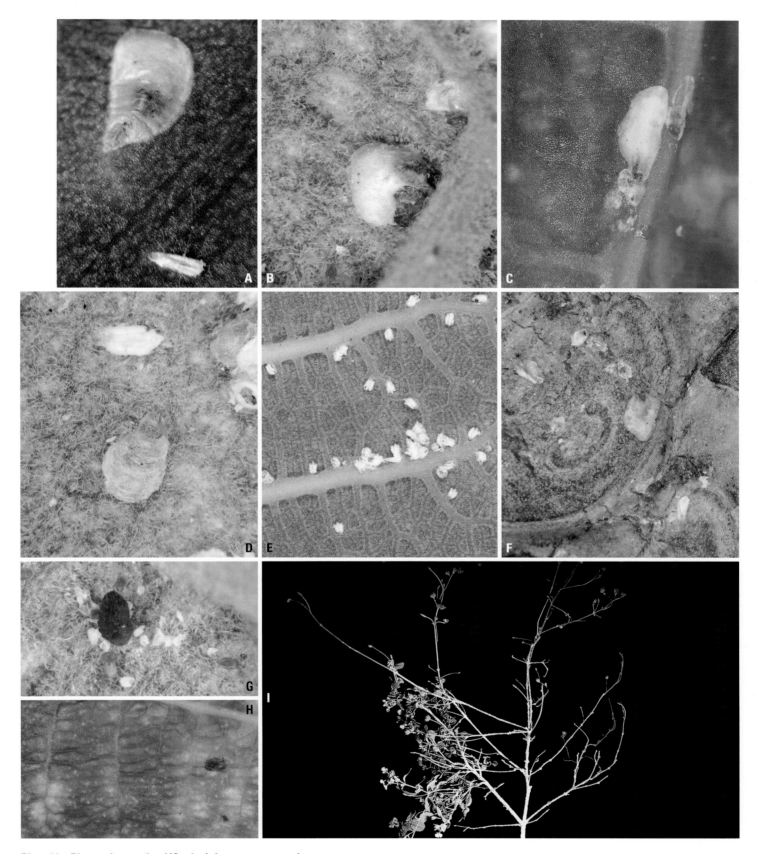

Plate 98. *Pinnaspis strachani* (Cooley), **Lesser snow scale**

A. Male cover (bottom) and adult female cover on dracaena (R. J. Gill).

B. Adult female cover on leucaena (J. A. Davidson).

C. Adult female cover on leucaena (J. A. Davidson).

D. Young adult female cover (center) on leucaena (J. A. Davidson).

E. Distance view of developing male covers on leucaena (J. A. Davidson).

F. Distance view of infestation on twig on leucaena (J. A. Davidson).

G. Adult female body with eggs on leucaena (J. A. Davidson).

H. Damage on top of leaf caused by infestation on underside on leucaena (J. A. Davidson).

I. Lantana dieback caused by heavy infestation (D. R. Miller).

Pinnaspis strachani (Cooley)

Suggested Common Name Lesser snow scale (also called cotton white scale).

Common Synonyms and Combinations *Hemichionaspis minor* v. *strachani* Cooley, *Hemichionaspis marchali* Cockerell, *Hemichionaspis townsendi* (Cockerell), *Chionaspis* (*Hemichionaspis*) *aspidistrae* v. *gossypii* Newstead, *Hemichionaspis aspidistrae gossypii* (Newstead), *Hemichionaspis proxima* Leonardi, *Chionaspis* (*Hemichionaspis*) *minor* Maskell (a common misidentification), *Hemichionaspis minor* Maskell, *Chionaspis* (*Pinnaspis*) *proxima* (Leonardi), *Pinnaspis minor* (Newstead), *Pinnaspis minor strachani* (Cooley), *Pinnaspis proxima* (Leonardi), *Pinnaspis temporaria* Ferris, *Pinnaspis marchali* (Cockerell), *Pinnaspis gossypii* (Newstead), *Chionaspis* (*Hemichionaspis*) *gossypii* (Newstead).

Field Characters (Plate 98) Adult female cover narrowly to broadly oyster-shell shaped, white to light gray; shed skins marginal, yellow brown. Male cover smaller, white, felted, elongate, with 3 ridges; shed skin marginal, yellow brown. Body of adult female, eggs, and crawlers probably yellow. On bark, fruit, and leaves.

Slide-mounted Characters Adult female (Fig. 102) with 2 pairs of definite lobes, third lobes usually present and represented by low series of points, fourth lobes absent in about half of specimens, when present similar to third lobes; paraphyses usually absent, when present, on medial and lateral margins of medial lobule of second lobe. Median lobes usually closely appressed, with interlobular space less than 1 μ, rarely space between lobes in inverted 'V' shape, normally without basal sclerosis, with yoke, medial margin axes usually parallel, lateral margins converging apically, with 2–4 (3) lateral notches, without medial notches; second lobes with 2 lobules, medial lobule usually not protruding beyond median lobes, medial lobule largest, with 0–1 (0) lateral notches, without medial notches, lateral lobule with 0–1 (0) lateral notches, 0–1 (0) medial notches; third lobes, when present, bilobed, represented by broad series of points, medial lobule with 0–4 (2) notches, lateral lobule with 0–3 (2); fourth lobes, when present represented by 1 or 2 lobules, medial lobule with 0–3 (2) notches, lateral lobule with 0–5 (2) notches. Gland-spine formula 1-1-1, with 8–17 (14) gland spines near each body margin from mesothorax, metathorax, segment 1 or 2 to segment 4; without gland spines between median lobes. Macroducts of 2 sizes, larger size on dorsum in marginal areas on segments 4 to 7, in submarginal areas on segments 3 to 5 or 6, smaller size in submarginal and marginal areas from mesothorax or metathorax to segment 4, on dorsum from metathorax, segment 1 or 2 to segment 4, on venter from mesothorax or metathorax to segments 2 and 3, without duct between median lobes, macroduct in first space 17–25 (21) μ long, with 6–9 (7) macroducts on each side of body on segments 5 to 7, total of 28–51 (38) macroducts on each side of body, some orifices occurring anterior of anal opening. Pygidial microducts on venter in submarginal areas of segments 5 to 7, with 3–7 (5) ducts; prepygidial ducts in submedial areas from head or prothorax to segment 4, occasionally absent from mesothorax, segments 4 and/or 3, in submarginal or marginal areas of head or prothorax to metathorax and on segments 4 and 3; pygidial microducts normally absent from dorsum, rarely with 1 or 2 submedial ducts on segment 5, prepygidial ducts often in submedial areas of segments 2 or 3 to 4, absent elsewhere. Perivulvar pores in 5 groups, 35–67 (46) pores. Perispiracular pores usually with 3 loculi, anterior spiracles each with 4–19 (10) pores posterior spiracles with 0–8 (2) pores. Anal opening located 7–14 (9) times length of anal opening from base of median lobes, anal opening 10–15 (12) μ long. Eyes usually on small protrusion laterad of anterior margin of clypeolabral shield, eyes usually represented by flat sclerotization, rarely absent. Antennae each with 1 conspicuous seta. Cicatrices usually on prothorax, and segments 1 and 3, rarely with cicatrix-like structures on head and segment 5. Body elongate with prepygidial segments from mesothorax to segment 3 or 4 with lateral margin forming protuberance. Usually with microduct near antennae. Preanal scar usually conspicuous. Gland spines on segments 5 to 7 with small microducts compared to those in remaining gland spines.

Affinities Of the economic species in the United States, lesser snow scale is most similar to *Pinnaspis aspidistrae*, fern scale. For a comparison see the affinities section of the treatment of fern scale.

Hosts Polyphagous. Borchsenius (1966) records it from 59 genera in 27 families; Dekle (1977) reports it from 170 genera in Florida, with *Bauhinia*, *Hibiscus*, and palms the most frequently reported hosts. We have examined specimens from *Annona*, *Aralia*, *Bauhinia*, *Bougainvillea*, *Caesalpinia*, *Cajanus*, *Casimiroa*, *Ceiba*, *Chrysalidocarpus*, *Cissus*, *Cocos*, *Cordyline*, *Crotalaria*, *Croton*, *Cucurbita*, *Cycas*, *Dioscorea*, *Dombeya*, *Dracaena*, *Euphorbia*, *Genipa*, *Gossypium*, *Heterospathe*, *Hibiscus*, *Latania*, *Licuala*, *Liriope*, *Livistona*, *Mangifera*, *Manihot*, *Melicoccus*, *Musa*, *Odontoglossum*, *Pandanus*, *Phalaenopsis*, *Plumeria*, *Prunus*, *Randia*, *Sabal*, *Salix*, *Sechium*, *Solanum*, *Strelitzia*, *Terminalia*, *Thespesia*, *Thrinax*, *Viscum*, *Vitis*, *Yucca*, *Zanthoxylum*, and *Ziziphus*. Based on material in the USNM Collection this species commonly infests hosts in the genera *Citrus*, *Cocos*, and *Hibiscus*.

Distribution Lesser snow scale probably is native to the Oriental Region. We have examined U.S. specimens from FL, HI, LA, MD, TX. It has been reported from AL, CA (eradicated [Gill 1997]), MS. This species occurs out-of-doors in southern areas and in greenhouses in the North. Lesser snow scale frequently is taken in quarantine from most tropical parts of the world.

Biology Hadzibejli (1983) indicates there is a full generation and a partial second in the warmer parts of Georgia. According to Johnson and Lyon (1976) lesser snow scale engages in continuous reproduction in the subtropics, producing several generations per year. The development in Cuba was 22.6 and 44.5 days for males and females, respectively, on grapefruit (Fernandez et al. 1996). The biology of this species probably is very similar to the life history of the fern scale. Brown (1965) sampled populations on various hosts from Mexico, Jamaica, Brazil, and Peru and found only the diaspidoid sexual chromosome system.

Economic Importance Beardsley and González (1975), Miller and Davidson (1990), and Arnett (1985) treat this species as a serious pest. According to Rosen and DeBach (1978) the cotton white scale of Peru is probably the same as the lesser snow scale. Based on the material in the USNM Collection, we agree. This pest was first noted in Peru in 1905 on perennial cotton in the Rio Piura and Chira cotton districts. A poorly documented biological control program was carried out during 1909–1912. Introduced parasites along with changes in cotton cultivation practices were credited with providing economically satisfactory control of lesser snow scale in Peru. This scale has seriously damaged coconut palm in Guam (Vandenberg 1929). Bondar (1926) considers it to be a pest of pineapple in Brazil, while Gowdey (1925) reports it as the worst pest of citrus in Jamaica. It occasionally causes serious damage to *Hibiscus* spp. in the Seychelles (Dupont 1918), New Caledonia (Cohic 1958), and Florida (Dekle 1965). It is reported as a pest on tamarind (Butani 1978), citrus (Delucchi 1975; Fernandez et al. 1998), black pepper (Koya et al. 1996), arecanut (*Areca catechu*) (Nair et al. 1963), olives (Canales Canales and Valdivieso Jara 1999), and several wild plants (Sankaran et al. 1984).

Selected References Rosen and DeBach (1978); Takagi (1970).

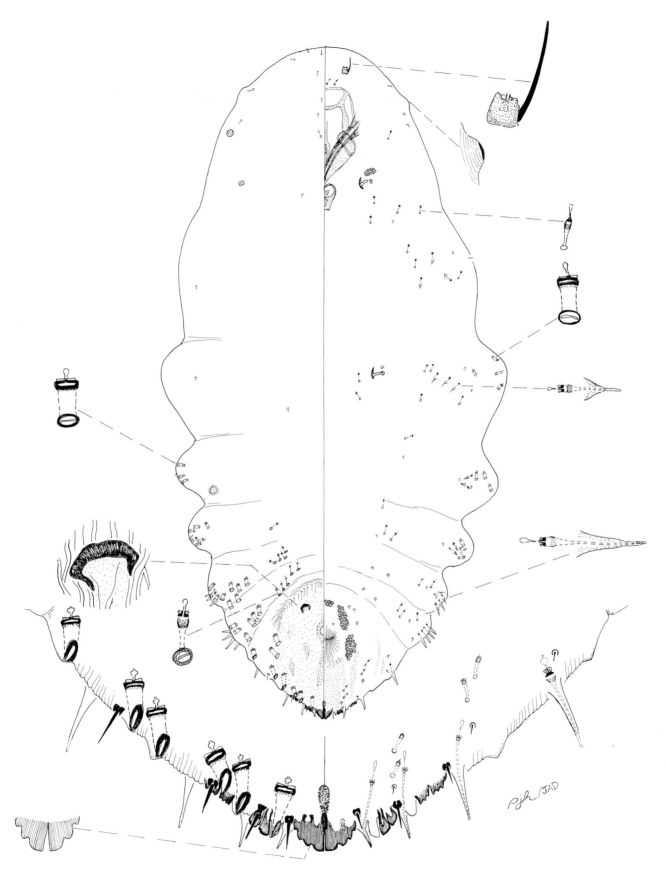

Figure 102. *Pinnaspis strachani* (Cooley), lesser snow scale, Mexico, on *Cassia* sp., I-2-1957.

Pseudaonidia duplex (Cockerell)

ESA Approved Common Name Camphor scale (also called Japanese camphor scale).

Common Synonyms and Combinations *Aspidiotus theae* Maskell (junior primary homonym of *A. theae* Green), *Aspidiotus duplex* Cockerell, *Aspidiotus* (*Pseudaonidia*) *duplex* Cockerell, *Aspidiotus* (*Pseudaonidia*) *theae* Maskell, *Aspidiotus* (*Evaspidiotus*) *theae* Maskell, *Aspidiotus* (*Evaspidiotus*) *duplex* Cockerell, *Aspidiotus theae rhododendri* Green, *Pseudaonidia rhododendrii* (Green), *Pseudaonidia rhododendri thearum* Fernald, *Pseudaonidia theae* (Maskell).

Field Characters (Plate 99) Adult female cover circular to oval, moderately to highly convex, brown, usually not overlain by bark flakes; shed skins submarginal to subcentral, orange yellow. Male cover similar in texture, smaller, elongate oval, brown; shed skin orange yellow, submarginal. Body of adult female pink, turning to purple; eggs purple; crawlers purple. Occurs on bark and leaves. Thorax and head of adult female heavily sclerotized on old adult females, with clear areas on head and prothorax.

Slide-mounted Characters Adult female (Fig. 103) with 4 pairs of lobes; paraphyses indefinite, occasionally bases of lobes with enlarged sclerotizations, rarely interlobular space with sclerotization attached to macroduct orifice. Median lobes separated by space 0.1–0.4 (0.3) times width of median lobes, often with tear-shaped sclerosis between median lobes in form of yoke, sometimes with small paraphysis-like sclerotization attached to medial margin, medial margin parallel or diverging apically, lateral margins parallel or converging apically, with 1 lateral and 1 medial notch; second lobes simple, noticeably thinner and shorter than median lobes, longer than wide, with 1 lateral notch, 0–1 (0) medial notch; third lobes same shape as second lobes, usually slightly shorter and thinner, with 1 lateral notch, 0–1 (1) medial notch; fourth lobes shorter than third, about as long as wide, with 1–2 (1) lateral notches, 0–1 (0) medial notch. Plates usually bifurcate, occasionally trifurcate; plate formula 2-3-3, without plates anterior of third space, each plate apparently with 1 microduct; with 2 plates between medial lobes slightly less or equal to length of median lobes. Macroducts on pygidium of 1 variable size, decreasing in length anteriorly, 1 macroduct between median lobes 64–122 (83) µ long, extending 0.6–1.1 (0.8) of distance to posterior apex of anal opening, longest macroduct in first space 50–88 (66) µ long, with 38–54 (47) macroducts on each side of pygidium on segments 5 to 8, ducts in marginal and submarginal areas, total of 112–249 (177) macroducts on each side of body some macroducts anterior of anal opening; prepygidial macroducts on dorsum of margin and submargin of segment 1 to segment 4. Pygidial microducts on venter in submarginal and marginal areas of segment 5 and occasionally segment 6 with 2–8 (4) ducts; pygidial microducts on dorsum absent; prepygidial microducts in submarginal areas of segments 1 to 2; prepygidial microducts in submedial areas of metathorax, usually 1 or 2 ducts on segment 1, on submarginal and marginal areas of thorax and anterior abdominal segments. Perivulvar pores in 4 lateral groups, without medial group, with 39–56 (47) pores on each side of body. Perispiracular pores normally with 5 loculi, anterior spiracles each with 8–15 (13) pores, posterior spiracles without pores. Anal opening located 6–11 (8) times length of anal opening from base of median lobes, anal opening 8–13 (11) µ long. Dorsal seta laterad of median lobes three-quarters to about equal to length of median lobes. Eyes absent. Antennae each with 1 long seta. Cicatrices absent. Body broadly oval, with distinct marginal indentation between prothorax and mesothorax, thoracic and prepygidial abdominal segments forming lateral segmental protuberances. Dorsomedial area of pygidium with conspicuous areolated pattern.

Affinities Of the economic species in the United States, camphor scale is similar to *Pseudaonidia paeoniae*, peony scale. Camphor scale differs by having lobes 2 and 3 narrow (see Fig. 103), the lateral perivulvar pore clusters arranged in 4 groups, the plates bifurcate or trifurcate, dorsal microducts in the submarginal clusters of macroducts on segments 1 and 2, no perispiracular pores associated with the posterior spiracle, 8–15 (13) perispiracular pores associated with the anterior spiracle, a single microduct in each plate, 39–56 (47) perivulvar pores, 38–57 (47) macroducts on each side of the pygidium, and the length of the longest macroducts in the first space 50–88 (66) µ. Peony scale has lobes 2 and 3 broad (Fig. 104), the lateral perivulvar pore clusters arranged in 2 groups, the plates simple or rarely with 1 or 2 tines, dorsal microducts absent from the submarginal clusters of macroducts, 1–6 (3) perispiracular pores associated with the posterior spiracle, 10–24 (17) perispiracular pores associated with the anterior spiracle, often with 2 or more microducts in each plate, 45–107 (75) perivulvar pores, 41–81 (56) macroducts on each side of the pygidium, and the length of the longest macroduct in the first space 75–140 (103) µ.

Hosts Broad host range. Borchsenius (1966) records it from 14 genera in 10 families; Dekle (1977) records it from 57 genera in Florida. Baker (1972) indicates that it occurs on more than 200 species of host plants. We have examined specimens from *Acer*, *Alnus*, *Althaea*, *Ampelopsis*, *Aucuba*, *Buxus*, *Camellia*, *Carya*, *Castanea*, *Cinnamomum*, *Citrus*, *Corylopsis*, *Croton*, *Diospyros*, *Ficus*, *Ligustrum*, *Magnolia*, *Malus*, *Myrica*, *Osmanthus*, *Parkinsonia*, *Pittosporum*, *Prunus*, *Pyracantha*, *Pyrus*, *Quercus*, *Rhododendron*, *Rosa*, *Salix*, *Solidago*, and *Vitis*.

Distribution Camphor scale probably is native to the Orient. We have examined specimens in the United States from AL, FL, LA, MS, TX. It is reported from GA (Nakahara 1982) and VA (Kosztarab 1996). This species also occurs in China, India, Japan, Korea, Sri Lanka, Taiwan, and the former Soviet Union. It most frequently has been collected by quarantine officers on *Citrus* and *Diospyros* from Japan.

Biology Cressman and Plank (1935) state that in Louisiana the minimum time to complete a generation is 51 days. They found 3 generations per year with a partial fourth generation developing in some years. The adult was the overwintering stage and oviposition began in January with crawlers appearing in February or March, depending on the temperature. Bliss et al. (1935) reports that in the laboratory 13 first-generation females produced an average of 125 crawlers while 24 second-generation females produced an average of 110 crawlers. Between 14° and 29 °C the rate of oviposition increased with rising temperature, but temperatures below and above these limits retarded egg laying. In the field, Bliss et al. found that the oviposition period ranged from 13 weeks for the overwintered females to 5 or 6 weeks for the first- and second-generation females. Crawlers were diurnal with most emerging from beneath the female cover between 6 AM and 11 AM. The highest percentage of crawlers settled between 27° and 30 °C, but settling was possible between 22° and 32 °C. More than 90% of the crawlers that survived settled in the first 6 hours after being transferred to twigs. On camphor trees most females settled around nodes and petiole bases while males usually settled along the midribs of leaves. Cressman et al. (1935) studied the biology of the camphor scale from the time of settling to the beginning of the reproductive period in the laboratory and field. They present curves for the length of the first and second stage, the period

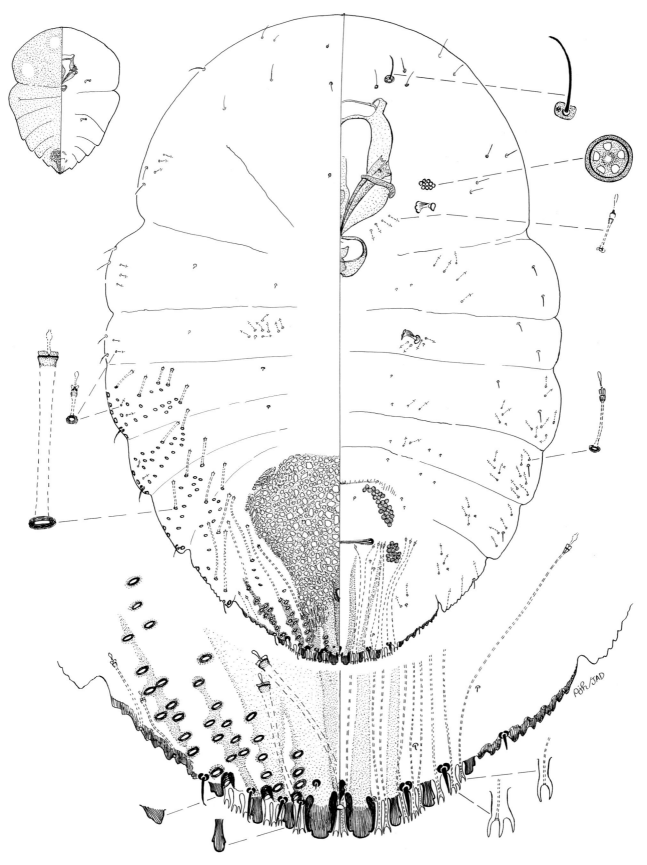

Figure 103. *Pseudaonidia duplex* (Cockerell), camphor scale, Japan, *Diospyros chinensis*, XII-22-1955.

from the second molt to egg production, and the time from the first egg to appearance of newly settled crawlers, as a function of temperature. Of the females that settle on the stem, 33.4% reproduced. Approximately equal numbers of males and females were produced, and fertilization was necessary for reproduction. Parasites and predators were found to be ineffective in the control of camphor scale in Louisiana. Dozier (1924) notes that about 40 days were required from second-generation females to develop from egg to adult in the field, while males required only 37 days. About 72 days were required for development of the third generation. According to Murakami (1970) this species has only 1 generation per year in Japan in Honshu, Shikoku, and Kyushu. This observation was made earlier by Kuwana (1933). As this area is in the same general latitude as Louisiana, where all studies indicated 3 and sometimes a partial fourth generation, it is possible that more than one species may be involved. The life history of this scale also was discussed by Shiao (1978) in Taiwan.

Economic Importance Beardsley and González (1975), Miller and Davidson (1990), and Arnett (1985) list this species as one of the principal armored scale pests in the world. Talhouk (1975) lists the camphor scale as an economically important pest of citrus in China and Japan because control measures occasionally are required. In Louisiana Cressman et al. (1935) reports this species to be capable of serious injury to camphor, Satsuma orange, fig trees (*Ficus carica*), Japanese privet, glossy privet, camellia, osmanthus, roses, jasmine, and Japanese persimmon. Because of its general use for ornamental and shade purposes, and the severity with which it is attacked, they found the camphor tree to be the most important host in Louisiana. Injury was first evident on the younger branches and twigs. Ebeling (1959) considers this scale to be a serious pest on camphor, Satsuma orange, Japanese persimmon, fig, and pecan in the gulf states except Florida. Dozier (1924) notes the reason this pest is important in citrus production is because the infestation reaches its peak just about the time the fruit is beginning to ripen. Because the scale population tended to confine itself to the twigs and leaves most of the year, large numbers of crawlers were available to infest the developing fruits. Wang et al. (1998a) list this species as a pest of red bayberry (*Myrica rubra*). Although introduced into Louisiana from Japan in 1920, this scale was not detected in Florida until 1958 where it never attained serious pest status (Dekle 1977). *Aphytis cylindratus* Compere is a very effective parasite of camphor scale in Japan and has considerable biological control potential in other areas of the world. *Aphytis longicaudus* also should be included in a biological control program (Rosen and DeBach 1979).

Selected References Cressman and Plank (1935); Ferris (1938a).

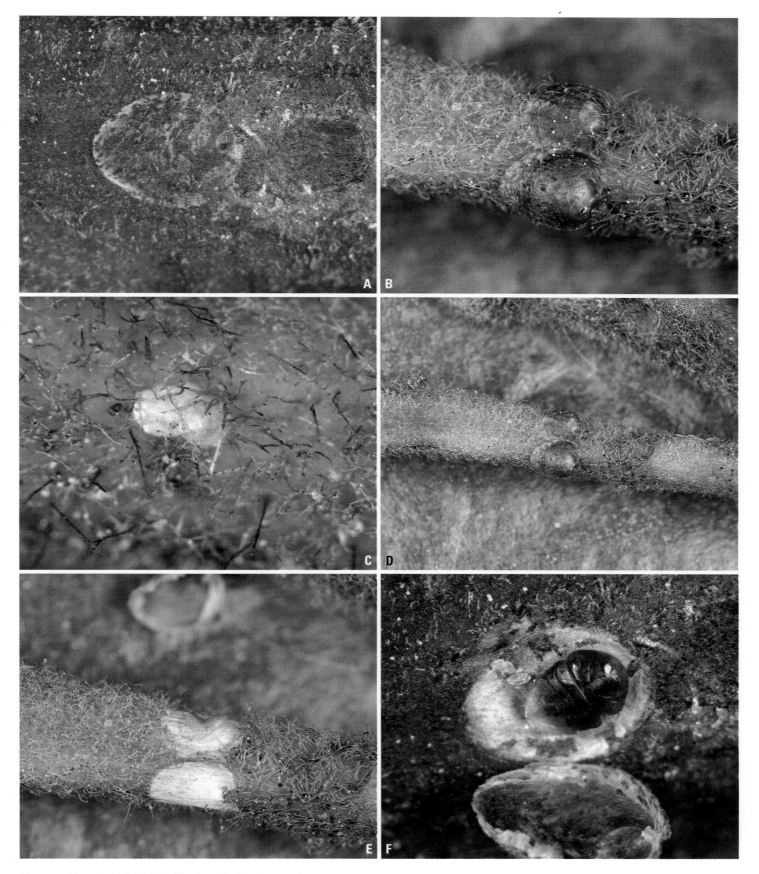

Plate 99. *Pseudaonidia duplex* (Cockerell), Camphor scale

A. Adult female cover on red bay (J. A. Davidson).
B. Adult female covers on red bay (J. A. Davidson).
C. Adult male cover on red bay (J. A. Davidson).
D. Distance view of adult female covers on red bay (J. A. Davidson).

E. Wax circles left after scale covers fall from host (J. A. Davidson).
F. Body of adult female with purple egg on red bay (J. A. Davidson).

349

Pseudaonidia paeoniae (Cockerell)

Suggested Common Name Peony scale (also called Japanese camellia scale).

Common Synonyms and Combinations *Aspidiotus duplex* v. *paeoniae* Cockerell, *Aspidiotus (Evaspidiotus) duplex* v. *paeoniae* Cockerell, *Pseudaonidiella paeoniae* (Cockerell).

Field Characters (Plate 100) Adult female cover circular to oval, moderately to highly convex, brown, commonly overlain by gray bark flakes; shed skins subcentral to submarginal, orange yellow. Male cover smaller, elongate oval, brown; shed skin orange yellow, submarginal. Body of adult female purple; eggs purple. On bark usually covered by bark flakes. Thorax and head of adult female heavily sclerotized on old adult females, with clear areas weakly indicated on head and prothorax.

Slide-mounted Characters Adult female (Fig. 104) with 4 pairs of lobes; paraphysis formula usually 1-1-1, occasionally medial margins of second, third, or fourth lobes with small paraphysis giving formulas 1-1-2, 2-2-1, or 2-2-2. Median lobes separated by space 0.2–0.3 (0.3) times width of median lobe, with inconspicuous tear-shaped sclerosis between median lobes, possibly remnant of yoke, sometimes with small paraphysis-like sclerotization attached to medial margin, medial margin and lateral margins usually parallel, with 1–3 (1) lateral notch, 1 medial notch; second lobes simple, about three-quarters of size of median lobes, same shape, with 1 lateral and 1 medial notch; third lobes simple about same size and shape as second lobes, with 1 lateral and 1 medial notch; fourth lobes simple, slightly smaller than third lobes, about same shape, with 0–1 (1) lateral notch, 0–1 (1) medial notch. Plates simple, occasionally with 1 or 2 tines; plate formula usually 2-3-3, sometimes 2-3-2 or 2-3-4, without plates anterior of third space; with 2 plates between median lobes, about one-half length of median lobe. Macroducts on pygidium of 1 variable size, decreasing anteriorly, 1–2 (1) macroducts between median lobes 66–78 (69) μ long extending 0.6–1.0 (0.7) of distance to posterior apex of anal opening, longest macroduct in first space 75–110 (103) μ long, with 41–81 (56) macroducts on each side of pygidium on segments 5 to 8, ducts in marginal and submarginal areas, total of 152–341 (243) macroducts on each side of body, some macroduct orifices anterior of anal opening; prepygidial macroducts on venter of margin and submargin of segment 1 to segment 4. Pygidial microducts on venter in submarginal and marginal areas of segment 5, with 8–12 (11) ducts; prepygidial ducts in submedial areas near mouthparts, between anterior and posterior spiracle, posterior of hind spiracle, and on segments 1 and 2; in submarginal and marginal areas of segments 1 to 4; pygidial microducts absent from dorsum; prepygidial ducts in submedial areas of metathorax to segment 2, sometimes absent from metathorax or segment 2, on marginal and submarginal areas of thorax. Perivulvar pores usually in 2 lateral groups only, occasionally with medial group composed of 1–4 (2) pores, 45–107 (75) pores on each side of body. Perispiracular pores normally with 5 loculi, anterior spiracles each with 10–24 (17) pores, posterior spiracles each with 1–6 (3) pores. Anal opening located 8–13 (10) times length of anal opening from base of median lobes, anal opening 9–15 (11) μ long. Dorsal seta laterad of median lobes three-quarters to about equal to length of median lobes. Eyes absent. Antennae each with 1 unusually long seta. Cicatrices absent. Body broadly oval, with distinct marginal indentation between prothorax and mesothorax, and between metathorax and segment 1, prepygidial abdominal segments forming lateral segmental protuberances. Plates usually with more than 1 microduct, dorsomedial area of pygidium with conspicuous areolated pattern.

Affinities Of the economic species in the United States, peony scale is related to *Pseudaonidia duplex*, the camphor scale and *P. trilobitiformis*, the trilobe scale. The former has lobes 2 and 3 times wider than long, all plates simple or unevenly bifurcate and shorter than the lobes, and most perivulvar pores in 2 clusters. The other 2 species each have lobes 2 and 3 times longer than wide, all plates equally bifurcate and as long or longer than lobes 2 and 3, and perivulvar pores in 4 clusters.

Hosts Relatively limited host range. Borchsenius (1966) records it from 5 genera in 4 families; Dekle (1977) records it from 3 genera in Florida with *Camellia* and *Rhododendron* the most frequently reported hosts. We have examined specimens from *Acer*, *Althaea*, *Aspidistra*, *Buxus*, *Cotoneaster*, *Diospyros*, *Eurya*, *Ilex*, *Ligustrum*, *Osmanthus*, *Paeonia*, *Rhododendron*, and *Rosa*. It has been reported from *Clethra* and *Cleyera*.

Distribution Peony scale probably is native to the Oriental Region. We have examined U.S. specimens from AL, AR, CA, DC, FL, GA, LA, MA, MD, NC, NJ, NY, PA, SC, TX, VA. It has been reported from MO and MS. This species also occurs in China, Georgia, Italy, and Japan. It has been collected most often by quarantine officers on *Camellia* and *Rhododendron* from Japan.

Biology Wescott reports 1 generation per year in the South. English and Turnipseed (1940) report 1 generation per year with crawlers appearing from early April to late May in Alabama and mature females acting as the overwintering stage. In Georgia, Tippins et al. (1977) found 1 generation per year with egg laying completed at Savannah by March 15, while at Experiment, egg laying had just begun in March. Egg hatch was complete in Savannah by April 12, while 100% hatch did not occur at Experiment and Tifton until May 10. It was suggested that unseasonably warm temperatures may have advanced egg laying by 1 to 2 weeks. In Maryland, McComb and Davidson (1969) found 1 generation per year with crawlers appearing from late May to mid-June outdoors. Also in Maryland, Stoetzel (1975) reports overlapping generations in the greenhouse on camellia with all stages present. Females laid 1 clutch of 30–40 eggs, and then 2–4 eggs per day until death. Johnson and Lyon (1976) report crawlers in late May in parts of Virginia, which suggests 1 generation per year. However, they also state '3 generations of the peony scale occur each year.' In Japan, Murakami (1970) noted 1 generation per year with egg laying in early April and hatching in mid-June. Males pupated in early August. Mating occurred in August, and fertilized females overwintered. Kunincka (1970) reports 1 generation per year in Russia on camellia, holly, and tea. Tikhonov (1966) also notes 1 generation per year in Russia on tea with mature females serving as the overwintering form. Shel'Deshova (1972) found that this scale could withstand cold temperatures as low as −15°C. Dzhashi (1989) indicates there is one complete generation per year and a second facultative one, with both first and second instars and also immature and mature adults overwintering on tea in Georgia; females laid up to 200 eggs.

Economic Importance Dekle (1977) reports that severe infestations of peony scale killed the branches of azalea and camellia in Florida. According to Tippins et al. (1981) this scale is a notorious pest of azaleas, camellias, and other valuable ornamentals including hollies and boxwood. English and Turnipseed (1940) note that heavily infested plants have reduced vitality to the extent that the foliage becomes thin. In Georgia (Dzhashi 1989) the species is a pest of tea, and yields are reduced by up to 30–50%. Large colonies are formed on stems and branches near

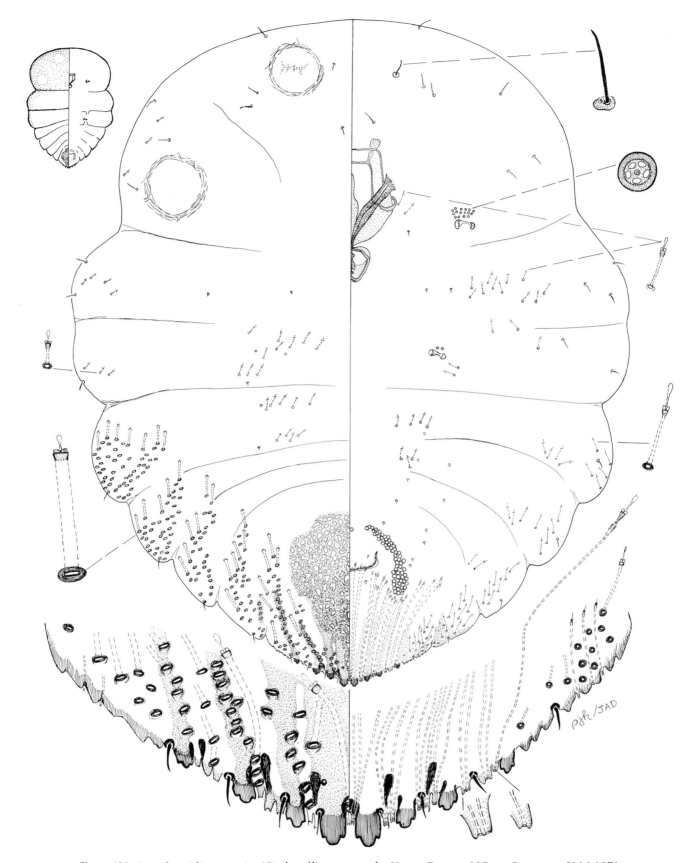

Figure 104. *Pseudaonidia paeoniae* (Cockerell), peony scale, Vance County, NC, on *Buxus* sp., V-16-1971.

the soil surface. Bark and phloem are damaged, leading to weakening and drying out of bushes and to leaf loss. Considerable scale mortality is caused by winter cold, coccinellids (including *Exochomus*), and the citrus aspidiotiphagus (*Encarsia citrina*).

Miller and Davidson (1990) consider this species to be a serious pest in a small area of the world.

Selected References Ferris (1938b); Pegazzano (1953); Tippins et al. (1977).

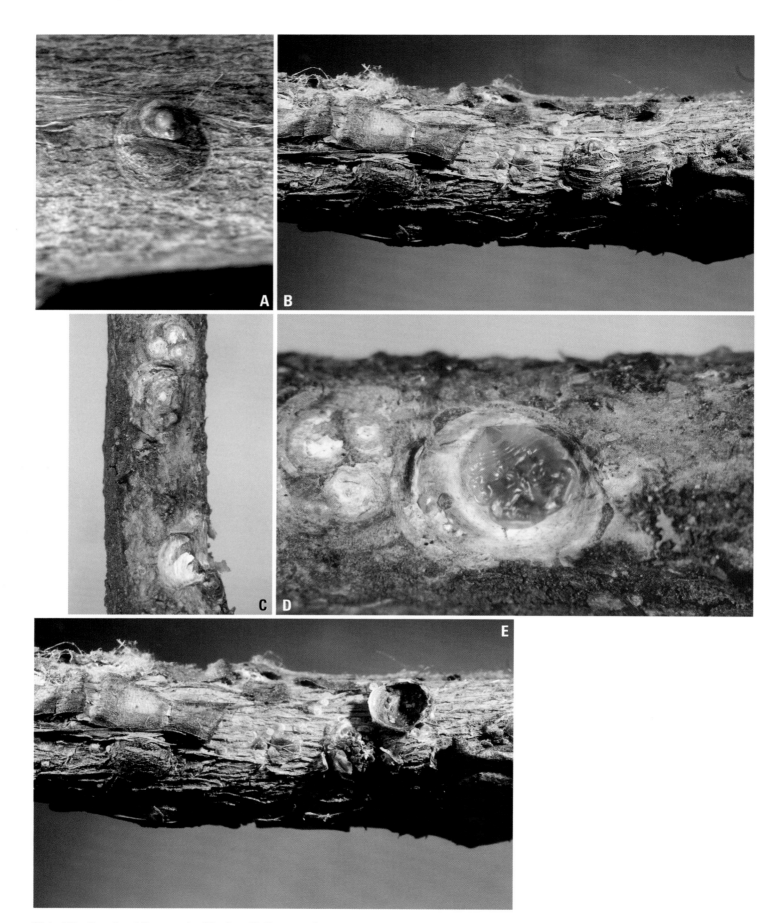

Plate 100. *Pseudaonidia paeoniae* (Cockerell), Peony scale

A. Rubbed adult female cover on peony (J. A. Davidson).
B. Female covers under bark of camellia (J. A. Davidson).
C. Female cover on azalea (J. A. Davidson).

D. Body of young adult female on azalea (J. A. Davidson).
E. Female cover turned showing dead crawlers (J. A. Davidson).

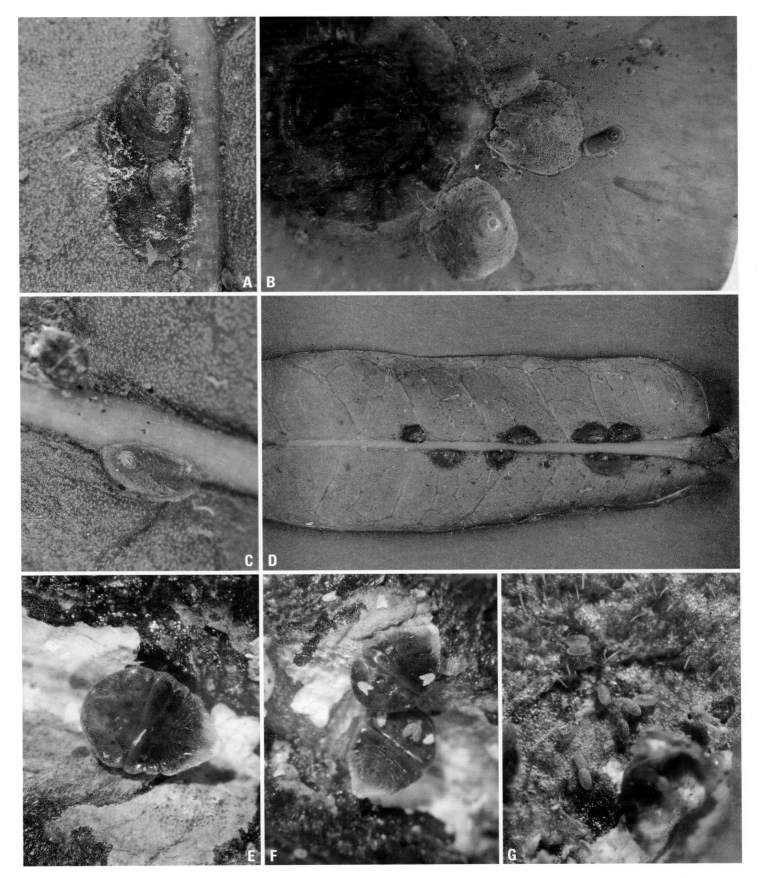

Plate 101. *Pseudaonidia trilobitiformis* (Green), Trilobe scale

A. Adult female covers on mango leaf (J. A. Davidson).
B. Female covers (left) and male cover (right) on mango fruit (J. A. Davidson).
C. Male cover on mango leaf (J. A. Davidson).
D. Distance view of female covers on mango leaves (J. A. Davidson).

E. Body of adult female on pachypodium (J. A. Davidson).
F. Female body and eggs on pachypodium (J. A. Davidson).
G. Eggs on pachypodium (J. A. Davidson).

Pseudaonidia trilobitiformis (Green)

Suggested Common Name Trilobe scale (also called trilobite scale, Gingging scale, and cashew scale).

Common Synonyms and Combinations *Aspidiotus trilobitiformis* Green, *Aspidiotus (Pseudaonidia) trilobitiformis* Green, *Aspidiotus (Evaspidiotus) trilobitiformis* Green, *Aspidiotus darvtyi* D'Emmerez de Charmoy, *Aspidiotus daruyi* D'Emmerez de Charmoy, *Pseudaonidia trilobitiformis darutyi* (D'Emmerez de Charmoy), *Pseudaonidia darutyi* (D'Emmerez de Charmoy).

Field Characters (Plate 101) Adult female cover convex, circular unless one side against plant structures such as midribs or veins, dark to light brown; shed skins subcentral to submarginal, brown with yellow tinge. Male cover smaller, elongate oval, similar in color and texture; shed skin submarginal, light yellow. Body of adult female brown, developing purplish tinge during egg production; eggs and crawlers light purple. On bark, leaves, and fruit. Thorax and head of old adult females heavily sclerotized, with clear areas on head and prothorax.

Slide-mounted Characters Adult female (Fig. 105) with 4 pairs of lobes; paraphyses indefinite, bases of lobes often with slender paraphyses. Median lobes separated by space 0.2–0.4 (0.3) times width of median lobes, with noticeable sclerosis between median lobes in form of yoke, usually with small paraphysis-like sclerotization attached to medial margin, medial margin parallel, lateral margins parallel, rarely diverging apically, with 1 lateral and 1 medial notch; second lobes simple, about same size as median lobes, medial and lateral margin diverging apically giving lobe apically swollen appearance, second lobes protruding beyond median lobes, with 1 lateral and 1 medial notch; third lobes about same size and shape as second lobes, with 0–1 (1) lateral notch, 0–1 (1) medial notch; fourth lobes same shape as third, slightly smaller, with 1–2 (1) lateral notches, 0–1 (0) medial notch. Plates usually bifurcate, middle plate in third space usually with 3 or more tines; plate formula 2-3-3, without plates anterior of third space, each plate with 1 microduct; with 2 plates between medial lobes usually noticeably longer than median lobes. Macroducts on pygidium of 1 variable size, decreasing in length anteriorly, 1 macroduct between median lobes 64–81 (76) μ long, extending 1.2–1.8 (1.5) times distance to posterior apex of anal opening, first macroduct in first space 125–187 (162) μ long, with 26–35 (30) macroducts on each size of pygidium on segments 5 to 8, ducts in marginal and submarginal areas, total of 122–192 (147) macroducts on each side of body, some macroducts anterior of anal opening; prepygidial macroducts on margin and submargin of segment 1 to segment 4, rarely on metathorax. Pygidial microducts on venter in submarginal areas of segment 5, with 0–2 (1) ducts; prepygidial ducts in submarginal areas from prothorax to segment 2, 3, or 4, in submarginal and marginal areas of segments 1 or 2 to 4; dorsal pygidial microducts absent; prepygidial microducts usually with 1 microduct in submedial area of metathorax, rarely with 1 duct on segment 1, on submarginal and marginal areas of head and thorax. Perivulvar pores usually in 4 groups, rarely with medial group, with 32–60 (38) pores on each side of body. Perispiracular pores normally with 5 loculi, anterior spiracles each with 9–26 (15), posterior spiracles without pores. Anal opening located 3–6 (4) times length of anal opening from base of median lobes, anal opening 12–17 (15) μ long. Dorsal seta laterad of median lobes one-half to nearly equal to length of median lobes. Eyes absent. Antennae each with single conspicuous seta. Cicatrices absent. Body elongate oval, with distinct marginal indentation between prothorax and mesothorax, thoracic and prepygidial abdominal segments forming lateral segmental protuberances. Dorsomedial area of pygidium with conspicuous areolated pattern.

Affinities Of the economic species in the United States, trilobe scale is similar to *Pseudaonidia duplex*, camphor scale. Trilobe scale differs by having the second lobes protruding beyond the median lobes, the second lobes about the same width as the median lobes, the second lobes with apically diverging lateral and medial margins giving an apically swollen appearance, the macroduct between the median lobes protruding anterior of the posterior apex of the anal opening, the first macroduct in the first space 125–187 (162) μ long, 26–35 (30) macroducts on each side of the pygidium, 0–2 (1) ventral microducts on each side of the pygidium, no microducts on the dorsal submargin of segments 1 and 2, no or 1 dorsal microduct submedially on the metathorax, the anal opening located 3–6 (4) times length of anal opening from base of median lobes, and the anal opening 12–17 (15) μ long. Camphor scale has the median lobes protruding beyond or the same distance as the second lobes, the second lobes noticeably thinner than the median lobes, the second lobes with the lateral and medial margins nearly parallel, the macroduct between the median lobes not reaching posterior apex of the anal opening, the first macroduct in the first space 50–88 (66) μ long, with 35–54 (47) macroducts on each side of the pygidium, 2–8 (4) ventral microducts on each side of the pygidium, several microducts on the dorsal submargin of segments 1 and 2 (mixed in with the macroduct clusters), a conspicuous cluster of dorsal microducts submedially on the metathorax, the anal opening located 6–11 (8) times the length of the anal opening from the base of the median lobes, and the anal opening 8–13 (11) μ long.

Hosts Polyphagous. Borchsenius (1966) records it from 80 genera in 42 families. Hamon (1982, personal correspondence) has seen trilobe scale from 22 genera in Florida. We have examined specimens from *Adenium*, *Anacardium*, *Annona*, *Anthurium*, *Bowdichia*, *Buxus*, *Calophyllum*, *Camellia*, *Canna*, *Cassia*, *Chrysophyllum*, *Citrus*, *Coccoloba*, *Coffea*, *Diospyros*, *Eucalyptus*, *Ficus*, *Gardenia*, *Ixora*, *Kentia*, *Lonicera*, *Loranthus*, *Mangifera*, *Monstera*, *Myrciaria*, *Nerium*, *Persea*, *Plumeria*, *Psidium*, *Quercus*, *Rosa*, *Sobralia*, *Spondias*, *Terminalia*, *Theobroma*, and *Vitis*.

Distribution The trilobe scale was discovered for the first time in the United States in Florida in 1979 (Hamon 1980). It is widely distributed in tropical areas of the world including Africa, southern Asia, Japan, Australia, the Caribbean islands, Central America, Mexico, the Pacific Islands, and South America. A distribution map of this scale was published by CAB International (1981a). It most frequently has been collected by quarantine officers on *Citrus* and *Mangifera* from tropical areas of the world.

Biology The biology of this species has not been studied. On cashew it occurs principally along the midrib and primary veins of the leaf. It is equally abundant on both leaf surfaces (Wheatley 1961).

Economic Importance As this species is a relatively recent introduction into the United States and several natural enemies attack it (Hamon 1980), there are no literature reports of its economic importance in the United States. In Brazil it has been reported as one of the more important pests of citrus (Azevedo Marques 1923) and cashews (Silva et al. 1977), and a potential pest of avocados (Ebeling 1959). Trilobe scale has been reported as a major pest of cashew in Kenya where pure stands occur over large areas. In most regions infestations on cashews are quite light apparently due to parasite effects. When infestations are heavy, chlorotic areas occur around the scale and partial defoliation often occurs (Wheatley 1961). Dupont (1926) notes it as a minor pest of coconut in the Seychelles. Cohic (1958) reports

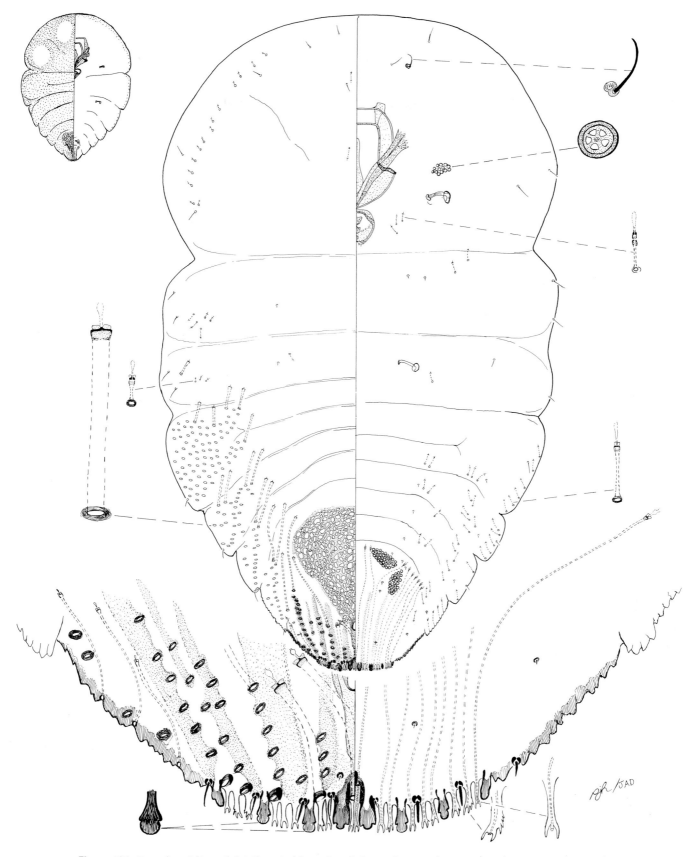

Figure 105. *Pseudaonidia trilobitiformis* (Green), trilobe scale, Brazil, intercepted at NY, on *Mangifera indica*, I-7-1944.

that this scale caused serious damage to young citrus and *Ficus pumila* in New Caledonia. Gressit and Djou (1950) record it as one of the more important citrus pests in China. Miller and Davidson (1990) consider this species to be an occasional pest.

Selected Reference Balachowsky (1958).

Pseudaulacaspis cockerelli (Cooley)

Suggested Common Name False oleander scale (also called white magnolia scale, oleander scale, magnolia white scale, Fullaway oleander scale, mango scale, oyster scale).

Common Synonyms and Combinations *Chionaspis cockerelli* Cooley, *Chionaspis aucubae* Cooley, *Chionaspis dilatata* Green, *Phenacaspis natalensis* Cockerell, *Phenacaspis aucubae* (Cooley), *Phenacaspis cockerelli* (Cooley), *Phenacaspis dilatata* (Green), *Chionaspis miyakoensis* Kuwana, *Chionaspis syringae* Borchsenius, *Chionaspis hattorii* Kanda, *Phenacaspis eugeniae* var. *sandwicensis* Fullaway, *Trichomytilus aucubae* (Cooley), *Trichomytilus cockerelli* (Cooley), *Trichomytilus natalensis* (Cockerell), *Trichomytilus dilatatus* (Green), *Chionaspis akebiae* Takahashi, *Pseudaulacaspis biformis* Takagi, *Phenacaspis ferrisi* Mamet.

Field Characters (Plate 102) Adult female cover flat, white, broadly oyster-shell shaped; shed skins marginal, yellow to brown. Male cover elongate, white, felted, with or without 3 faint, longitudinal ridges; shed skin light yellow. Body of adult female, eggs, and crawlers yellow. Mainly along major veins on leaves, occasionally on stems.

Slide-mounted Characters Adult female (Fig. 106) with 2 or 3 pairs of well-developed lobes, third lobes sometimes with lateral lobule represented by series of low points, fourth-lobe area and rarely fifth-lobe area with series of points; paraphyses usually attached to medial and lateral margins of each lobule of second lobe. Median lobes with lateral margins separated by 25–50 (38) μ at apex, with medial yoke, with narrow paraphysis-like structures attached to medial and lateral margins, medial margin strongly divergent apically, lateral margins parallel or divergent, with 4–7 (5) medial notches, 0–2 (1) lateral notches; second lobes smaller than median lobes, rounded, protruding posteriorly beyond median lobes, bilobed, medial lobule largest, medial and lateral lobules without notches; third lobes with medial lobule wider and shorter than medial lobule of second lobe, without notches, lateral lobule represented by series of low points; fourth lobes usually bilobed, represented by series of low points; fifth lobes usually absent, rarely represented by series of small points. Gland-spine formula 1-1-1, with 16–25 (21) gland spines near each body margin anterior of fourth lobe area, medial lobes without gland spines between them. Macroducts on pygidium about same size, without duct between medial lobes, macroduct in first space 18–27 (25) μ long, with 12–16 (15) macroducts on each side of body on segments 5 to 7, marginal ducts on segments 5 to 7, submarginal ducts on segment 5, submedial ducts on segments 5 and 6, segment 6 with 1–3 (2) ducts, total of 56–104 (81) macroducts on each side of body, macroduct orifices occurring anterior of anal opening; prepygidial macroducts of 2 sizes, larger size same as on pygidium, located on marginal, submarginal, and submedial areas of segments 2 to 4, smaller size on marginal or submarginal areas of mesothorax or metathorax to segments 2 or 3. Ventral macroducts of small size, on marginal or submarginal areas of mesothorax or metathorax to segment 2 or 3. Pygidial microducts on venter of segment 5, usually also on 6 and 7, with 3–8 (5) microducts on each side of pygidium; prepygidial microducts on venter in marginal or submarginal areas from head to metathorax, in submedial areas from prothorax or mesothorax to segment 3; pygidial microducts absent from dorsum; prepygidial microducts on submedial areas of metathorax to segment 3 or any part thereof. Perivulvar pores in 5 groups, 35–47 (42) pores on each side of body. Perispiracular pores primarily with 3 loculi; anterior spiracle with 7–13 (10) pores, posterior spiracles without pores. Anal opening located 7–12 (9) times length of anal opening from base of median lobes, anal opening 13–20 (15) μ long. Dorsal seta

laterad of medial lobes one-half to three-quarters length of lobe. Eyes represented by irregular clear area or slightly sclerotized area on derm laterad of anterior spiracle. Antennae each with 1 conspicuous seta. Small cicatrix on submargin of segment 1, occasionally also on prothorax. Body elongate; mesothorax, metathorax, and segments 1 to 3 or 4 with body margin forming abdominal protuberances.

Takagi (1970) has studied this species in considerable detail and has described much more variation than we found in the specimens in the USNM Collection. Essentially, the description above is of the leaf form, which seems to be far more common than the bark form. Some of the variation mentioned by Takagi is as follows: Submedian macroducts on each side of segment 2 (1–16), segment 3 (2–12), segment 4 (1–8), segment 5 (2–8), segment 6 (0–5); submarginal macroducts on each side of segment 2 (5–25), segment 3 (5–21), segment 4 (2–16), segment 5 (2–13), segment 6 (0–3); perivulvar pore numbers (based on summation of ranges in each cluster in Takagi 1970) on each side of pygidium 23–146; gland-spine formula 1-1-1, 1-2-2, 1-2-3, 1-1-2, 1-1-3, 1-1-4, 1-1-5; with 11–53 gland spines near body margin anterior of fourth lobe area. There is a tendency for the leaf form to have fewer perivulvar pores, macrotubular ducts, and gland spines; to have strongly divergent median lobes that do not protrude beyond the second lobes with lateral lobule of second lobes relatively large, and to have a slender body. The bark form has many perivulvar pores, macrotubular ducts, and gland spines; has median lobes less divergent than on leaf form; and lobes protruding well beyond apex of second lobes, lateral lobules of second lobe with lateral lobule relatively small, and has broad body.

Affinities False oleander scale can be separated from other economic species of *Pseudaulacaspis* by having an elongate body shape and by having the medial margin of the median lobes much longer than the lateral margin. *Pseudaulacaspis pentagona* and *P. prunicola* have a turbinate body shape and have the medial margin and lateral margin of the median lobes approximately equal in size.

Hosts Polyphagous. Borchsenius (1966) records it from 30 genera in 21 families; Dekle (1977) notes it from 84 genera in Florida with *Magnolia grandiflora* L. and *Bischofia* the most frequently recorded hosts. We have examined specimens from *Adenium, Aleurites, Areca, Asparagus, Aucuba, Bischofia, Camellia, Chrysalidocarpus, Citrus, Clinostigma, Cocos, Cycas, Daphniphyllum, Dianella, Diospyros, Dracaena, Helianthus, Hevea, Ilex, Livistona, Mangifera, Michelia, Musa, Myristica, Nerium, Plumeria, Rhizophora, Seaforthia, Strelitzia, Trachelospermum, Trachycarpus, Viburnum, Vriesea,* and *Yucca.*

Distribution False oleander scale is native to the Orient. We have examined U.S. material from AL, AR, FL, GA, HI, LA, TX from out-of-doors and from DC, KS, MD, MO, TN, VA in greenhouses. It has been reported as eradicated from CA and SC. This species commonly has been intercepted in quarantine inspections from the Philippines and Japan on *Mangifera* and various ornamental plants. We have seen material from Africa, Asia, Australia, Caribbean (St. Thomas), Central America (Guatemala), and several Pacific Islands.

Biology Tippins (1968) reports that false oleander scale occurred in Georgia as far north as Atlanta. Reproduction continued year-round in southern Georgia where 50 to 60 days were required for a generation. The female to male sex ratio was 61:39. The upper leaf surface was preferred for settling by 70% of the female crawlers, but by only 2% of the male crawlers; the remainder settled on the lower leaf surface. Both sexes

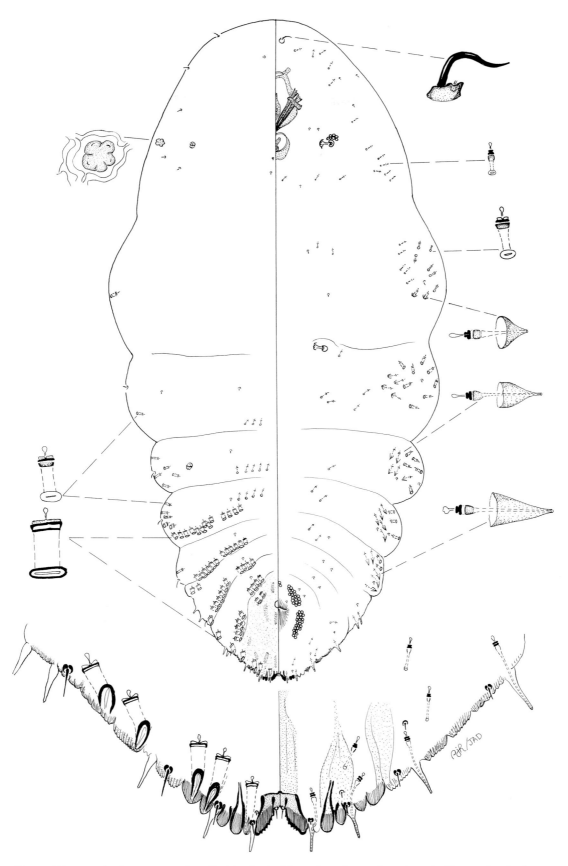

Figure 106. *Pseudaulacaspis cockerelli* (Cooley), false oleander scale, Miami, Fl, on *Bischofia* sp., VII-5-1975.

principally were found on leaves with an occasional individual on fruit or tender twigs. Greenhouse studies showed the minimum period to egg hatch was 3 days and all eggs hatched in 8 days. The first nymphal stadium averaged 9.4 days. The second stadium averaged 17.9 days. The minimum preoviposition period of adult females was 23 days. Reproducing females lived about 61 days and appeared dead within 1 week after all their crawlers had emerged. Virgin females remained alive for about 10 months. Takagi and Kawai (1967) recognized that this species produced two morphs depending on where feeding occurred: a broad median lobe form on bark, and a narrow median lobe form on leaves. Danzig (1980) reports 2 generations per year in southeastern Russia.

Economic Importance Dekle (1970) reports that false oleander scale was introduced into Florida in 1942. It rapidly spread throughout Florida on infested nursery stock. Soon it was the most serious economic pest of ornamentals in Florida, and known hosts included many of the major ornamental plants found in commercial nurseries (Dekle 1977). It has recently been reported as a pest in Italy (Russo and Mazzeo 1992) and France (Picart and Matile-Ferrero 2000). It has been reported to be a serious pest of *Mangifera* in Australia (Jarvis 1946), India (Fletcher 1921), Mauritius (Moutia 1935), and Madagascar (Frappa 1937). Dupont (1926) notes it as 1 of 10 coconut pests in the Seychelles. The false oleander scale is a pest of kiwi fruit in China (Zou and Zhou 1993). Hara et al. (1993) developed a hot water treatment that will kill all stages of this species on bird of paradise plants, *Strelitzia reginae*. Crawlers seemed to be the most resistant to the treatment. Miller and Davidson (1990) consider this species to be a serious pest in a small area of the world.

Selected References Takagi (1970); Tippins (1968); Williams and Watson (1988).

Plate 102. *Pseudaulacaspis cockerelli* (Cooley), False oleander scale

A. Adult female cover showing yellow stylet track and cluster of crawlers on bird-of-paradise plant (J. A. Davidson).
B. Adult female covers on palm (R. J. Gill).
C. Adult female covers on oleander (D. R. Miller).
D. Aggregation of male covers on cattleya orchid (J. A. Davidson).
E. Female with long stylet track on palm (R. J. Gill).

F. Distance view on oleander (J. A. Davidson).
G. Body of adult female before laying eggs on bird-of-paradise plant (J. A. Davidson).
H. Body of female with eggs on bird-of-paradise plant (J. A. Davidson).
I. Crawler on bird-of-paradise plant (J. A. Davidson).

Pseudaulacaspis pentagona

(Targioni Tozzetti)

ESA Approved Common Name White peach scale (also called mulberry scale, West Indian peach scale, oleander scale, and white scale).

Common Synonyms and Combinations *Diaspis pentagona* Targioni Tozzetti, *Diaspis amygdali* Tryon, *Diaspis lanatus* Morgan, *Diaspis patelliformis* Sasaki, *Aspidiotus vitiensis* Maskell, *Diaspis geranii* Maskell, *Aulacaspis (Diaspis) pentagona* (Targioni Tozzetti), *Aulacaspis pentagona* (Targioni Tozzetti), *Diaspis (Aulacaspis) pentagona* (Targioni Tozzetti), *Sasakiaspis pentagona* (Targioni Tozzetti), *Aspidiotus lanatus* (Morgan). Kreiter et al. (1999) examined specimens reared from potato for more than 100 generations and found specimens that fit both the white peach scale (*P. pentagona*) and the white prunicola scale (*P. prunicola*). They conclude that the two are simply variants of the same species.

Field Characters (Plate 103) Adult female cover convex, circular white (with a tinge of yellow on mulberry); shed skins usually subcentral, yellowish orange (on mulberry). Male cover smaller, felted, white, elongate, sometimes with slight median carina completely enclosing developing male; shed skin white, sometimes tinged with yellow. Body of adult female light yellow, developing pink tinge during egg production; eggs of males white, of females yellow or pink. Primarily on bark and fruit, occasionally on leaves. Male covers characteristically in conspicuous masses on the bottoms of branches.

Slide-mounted Characters Adult female (Fig. 107) with 4 or 5 pairs of lobes, fifth lobes absent or small; paraphyses usually restricted to medial lobule of second lobe, rarely absent. Median lobes separated by space 0.3–0.5 (0.4) times width of median lobe, each margin with thin paraphysis-like sclerosis, with yoke, medial margin axes diverging apically, lateral margin axes converging apically, with 1–5 (2) lateral notches, 0–4 (2) medial notches; second lobes with 2 lobules, noticeably smaller than median lobes, medial lobule largest with 0–1 (0) lateral notches, without medial notches, lateral lobule without notches; third lobes bilobed, lateral margins rarely with 1 notch, remaining lobule without notches; fourth lobe bilobed or simple, medial lobule with 0–3 (1) notches, lateral lobule without notches; fifth lobes simple, bilobed, or absent, when present, with 0–6 (1) notches. Gland-spine formula usually 1-1-1 (of 74 specimens examined, 2 specimens had 2 gland spines in first space, 3 had 2 gland spines in second space, and 30 had 2 or 3 gland spines in third space), with 23–45 (32) gland spines near each body margin from mesothorax or metathorax to segment 4; without gland spines between median lobes. Macroducts of 2 sizes, larger size on dorsum in marginal areas on mesothorax, metathorax, segment 1, or segment 2 to segment 7, in submarginal areas of segments 2 to 5 or rarely to 6, in submedial areas of segments 2 to 5 or rarely to 6, smaller size in submarginal and marginal areas on mesothorax, metathorax, or segment 1 to segments 2, 3, or 4, on dorsum on segments 1, 2, or 3, on venter from mesothorax, metathorax, or segment 1 to segments 2, 3, or 4, without duct between median lobes, macroduct in first space 18–25 (22) μ long, with 12–27 (17) macroducts on each side of body on segments 5 to 7, total of 40–106 (88) large macroducts on each side of body, some orifices occurring anterior of anal opening. Pygidial microducts on venter in submarginal areas of segments 5 to 7 with 4–8 (6) ducts; prepygidial ducts in submedial areas from prothorax, mesothorax, or metathorax to segment 2 or 3, in submarginal or marginal areas of head or prothorax to mesothorax, metathorax or segment 1 and on segments 3 and/or 4; pygidial microducts absent from dorsum,

prepygidial ducts usually present on segment 1, frequently also on mesothorax and metathorax, rarely absent. Perivulvar pores in 5 groups, 51–124 (77) pores on each side of body. Perispiracular pores usually with 3 loculi, anterior spiracles each with 11–23 (17), posterior spiracles without associated pores. Anal opening located 8–14 (11) times length of anal opening from base of median lobes, anal opening 12–20 (14) μ long. Eyes usually represented by small, flat sclerotization near body margin laterad of anterior edge of clypeolabral shield, eyes rarely absent. Antenna each with 1 conspicuous seta. Cicatrices on segment 1, rarely also on prothorax. Body turbinate, with prepygidial segments from mesothorax or metathorax to segment 3 or 4 with lateral margin forming protuberance. Of 75 specimens, all had at least 1 gland spine with bifurcate or trifurcate apex, with 1–5 (3) apically divided gland spines in first four spaces on each side of pygidium. With 5–22 (11) marginal and submarginal, small macroducts on each side of metathorax and segment 1 combined.

Affinities The white peach scale has been confused with *Pseudaulacaspis prunicola*, the white prunicola scale (Davidson et al. 1983; Rhoades et al. 1985). The 2 can be separated only on the basis of a combination of characters; no single feature is diagnostic of all specimens of one species. Features used in combination for the white peach scale are: the third space usually has 1 gland spine; at least 1 bifurcate or trifurcate gland spine in the second, third, or fourth spaces; the combined number of small macroducts on each side of the metathorax and segment 1 is 5–22 (11); the total number of large macroducts on each side of the body is 40–106 (68); the total number of perivulvar pores on each side of the pygidium is 51–124 (77); the most commonly recorded host is mulberry; and the distribution is tropical and subtropical. In contrast, the white prunicola scale characteristics are: the third space usually has 2 or more gland spines; the gland spines rarely have a bifurcate or trifurcate apex; the combined number of small macroducts on each side of the metathorax and segment 1 is 0–15 (5); the total number of large macroducts on each side of the pygidium is 38–86 (58); the total number of perivulvar pores on each side of the pygidium is 33–99 (64); the most commonly recorded hosts are species of *Prunus*, particularly the Japanese flowering cherry; and the distribution of the species is temperate, with the exception of Alabama, Louisiana, and Okinawa.

Hosts One of the most polyphagous armored scale insect species in the world, but we have seen none from coniferous hosts. Borchsenius (1966) records it from 108 genera in 55 families; Dekle (1977) reports it from 115 genera in Florida with *Callicarpa, Diospyros, Melia,* and *Prunus* the most frequently reported. We have examined specimens from *Acacia, Acanthus, Actinidia, Allamanda, Amygdalus, Aralia, Boehmeria, Broussonetia, Bryophyllum, Buddleia, Callicarpa, Camellia, Capsicum, Carica, Cassava, Castanea, Castilla, Catalpa, Cedrela, Cinchona, Citrus, Clematis, Cornus, Crossandra, Crotalaria, Croton, Cucurbita, Cycas, Cydonia, Diospyros, Dombeya, Euonymus, Euphorbia, Ficus, Fraxinus, Geranium, Gomphrena, Gossypium, Hedera, Hibiscus, Hydrangea, Iberis, Ilex, Ipomoea, Juglans, Jasminum, Jatropha, Kalanchoe, Lantana, Ligustrum, Magnolia, Mahonia, Malachra, Mangifera, Manihot, Melia, Morus, Nerium,* 'orchid,' *Osmanthus, Palicourea, Passiflora, Pelargonium, Philodendron, Piper, Pittosporum, Plumeria, Populus, Prunus, Pueraria, Rhamnus, Rhus, Ribes, Ricinus, Salix, Sassafras, Schinus, Sedum, Solanum, Spondias, Stanhopea, Sterculia, Tecoma, Theobroma, Trema,*

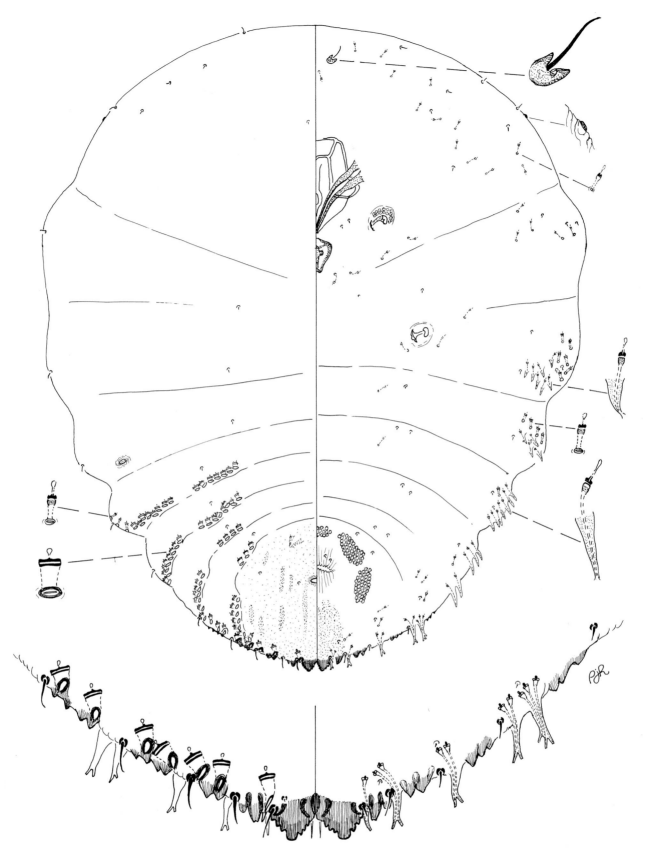

Figure 107. *Pseudaulacaspis pentagona* (Targioni Tozzetti), white peach scale, Hertford County, NC, on umbrella tree, IX-29-1958.

Vitis, and *Ziziphus*. Based on material in the USNM Collection, white peach scale generally infests the same hosts as white prunicola scale except *Syringa*, which is a common host of white prunicola scale but is not a host of white peach scale.

Distribution White peach scale is native to eastern Asia. We have examined U.S. material from AL, CA, DC, FL, GA, HI, IN, LA, MD, MS, NC, NM, SC, TN, TX, VA. The white peach scale has not been recorded from CA for more than 50 years. It also was absent from Hawaii for more than 50 years but recently has become a serious pest there. We have material in the USNM Collection from papaya in 1997. CAB International (1996b) presented a map of the world distribution of this species. This species commonly has been intercepted in quarantine inspections from Caribbean islands, Japan, and South America. We have seen material from Africa, Asia, Australia, Caribbean islands, most countries in South America, southern Central America, western Europe, and many Pacific Islands.

Biology The white peach scale is easy to rear on potatoes (Dustan 1953). As a result, much work has been done on its biology. Bennett and Brown (1958) report that in Trinidad and Bermuda females deposited all their eggs in 8 to 9 days. Individual eggs hatched in 3 to 4 days and female crawlers were more active than male crawlers. Both sexes molted to second instars in 7 to 8 days. Females completed the final molt to the third instar in 19 to 20 days. Males molted to third instars in 12 days and the fifth instars (adults) appeared in 19 to 22 days. Males emerged in late afternoon and immediately began to inseminate females. Fertilization occurred while the eggs were in the egg follicles. Oviposition began 14 to 16 days after mating. A generation was completed in 36 to 40 days during the summer at about 25 °C average temperature and in 80 to 90 days during the winter. Van Duyn and Murphey (1971) report a generation time from 49 to 51 days in the lab at 21 °C and 65% RH on potatoes. Ball (1980) found a minimum generation time of 40.4 days at 26.4 °C and 110.8 days at 13.3 °C in the lab on potatoes. Bennett and Brown (1958) report that mating was required for reproduction. They note that the eggs and crawlers exhibited sexual dimorphism by color differences; eggs containing female embryos were coral when laid while male embryo-containing eggs were pinkish white. This color difference persisted up to the end of the first instar. They also note dichronism; eggs containing females were laid first, and male-containing eggs were laid last. They pointed out that the latter trait was a valuable tool because potatoes could be infested exclusively with either female or male crawlers. Monti (1955) records orange female eggs and white male eggs in Italy, and Bedford (personal communication in Bennett and Brown 1958) also saw 2 color groups of eggs in South Africa. Brown and Bennett (1957) report *P. pentagona* to be a haplodiploid species with 16 chromosomes in the female and 8 in the male. The male haploidy resulted from chromosome elimination during early embryogeny. They found the sex ratio to be close to 1:1, but this varied widely, with some individual females producing nearly all males or females. They also found this to be true when old virgins were mated. Brown (1960) notes an instance of tetraploidy in this species. Nelson-Rees (1960) reports that exposure of egg-bearing females to X-rays stimulated higher male production. Morere and Seuge (1976) found reduced fecundity in females reared on previously irradiated potato tubers and reduction was dosage dependent. Tremblay (1969) discovered that the symbionts (mycetocytes) are carried by maternal cells that penetrate the developing oocytes in the ovary. The number of white peach scale generations in the United States has been reported to be 4 in central Florida (Kuitert 1967), 3.5 to 4 in

northern Florida (Van Duyn and Murphey 1971), 3 in North Carolina (Smith 1969), 3 in Virginia (Bobb et al. 1973), and the authors have observed 3 in Maryland. We have not reported literature dealing with *P. pentagona* north of Maryland because the species studied was probably *P. prunicola* (Maskell), white prunicola scale. Reported generation times in other countries were 4 to 5 in Bermuda (Bennett and Brown 1958), 3 in the Mediterranean areas of France and Italy (Bénassy 1958; Monti 1955; Battaglia et al. 1994), 3 in Turkey (Bodenheimer 1953), 4 in Turkey (Erkilic and Uygun 1998), 3 in Russia (Kunincka 1970), 3 in Japan (Murakami 1970), 3 in Romania (Brailoiu Tanase 1998). Crawler appearance times in the United States were reported as follows: Central Florida (Kuitert 1967)—first generation about mid-February, second generation about mid-May, third generation about mid-July, fourth generation about early October; North Carolina (Smith 1969)—first generation early May, second generation late July, third generation late August; Virginia (Bobb et al. 1973)—first generation early May, second generation early August, third generation early September; central Maryland (Davidson 1981, personal observation)—first generation mid-May, second generation early July, third generation late August. Murakami (1970) summarized 5 papers treating this pest in Japan as follows: Overwintering females began to enlarge in mid-February and their ovaries matured in April. From April to mid-May 100 to 150 eggs were deposited by each female on mulberry, with about 50 eggs less per female on tea plants. First generation crawlers appeared in early May, those of the second generation in late July, and those of the third generation from September to October. Males of the last generation appeared in early November to mate with the females that overwintered. Better results were obtained rearing this pest on squash than on potato.

Economic Importance Beardsley and González (1975) list white peach scale as one of the 43 principal armored scale pests of the world. Dekle (1977) reports it as an economic pest in Florida. Johnson and Lyon (1976) note that this scale was very destructive to ornamental trees and shrubs such as those in the following genera: *Catalpa*, *Diospyros*, *Hibiscus*, *Ligustrum*, *Prunus*, and *Syringa*. They list Kwansan cherry (*Prunus serrulata*) as a favorite host, which is highly susceptible to damage by this pest. This observation probably pertains to white prunicola scale. They also note it as a serious stone fruit pest in the Carolinas and Virginia. Snapp (1954) reports white peach scale to be as injurious to peach trees as San Jose scale. Bertels (1956) records it as a pest of great economic importance in Brazil on *Prunus* and *Pyrus*. Milaire (1962) claims that white peach scale was fatal to peaches in an area on the Mediterranean coast and adjacent areas. Vashadze (1955) reports it to be highly injurious to Japanese quince, decorative cherry, *Prunus avium*, and mulberry along the Black Sea Coast of western Georgia, SSR. Yasumatsu and Watanabe (1965) list it in a parasite–predator catalog of injurious insects in Japan. In Bermuda it nearly eradicated oleander (Simmonds 1958). The white peach scale has been reduced to subeconomic levels in most of South America and Europe, principally as a result of the introduced parasite *Encarsia berlesei*. In Bermuda control has been sustained by *Aphytis diaspidis* and *Aspidiotiphagus* spp. A detailed study of the life history, host plants, and natural enemies of white peach scale is presented by Hanks and Denno (1993a). They also present additional ecological information on this species in a series of other papers (Hanks and Denno 1993b, 1993c, 1994).

Selected References Bennett and Brown (1958); Davidson et al. (1983); Hanks and Denno (1993a).

Plate 103. *Pseudaulacaspis pentagona* (Targioni Tozzetti), **White peach scale**

A. Adult female covers under mulberry bark flakes (J. A. Davidson).
B. Adult female covers on yellow-twig dogwood (J. A. Davidson).
C. Male covers on white mulberry (J. A. Davidson).
D. Distance view of young adult female covers on basswood (J. A. Davidson).
E. Distance view of adult female covers (top) and male covers (bottom) on white mulberry (J. A. Davidson).

F. Laboratory colony on potato showing tendency of males to aggregate (J. A. Davidson).
G. Body of young adult female on basswood (J. A. Davidson).
H. Salmon-colored female eggs (bottom) and white male eggs (top) on white mulberry (J. A. Davidson).
I. Clusters of first-instar males on white mulberry (J. A. Davidson).

Pseudaulacaspis prunicola (Maskell)

Suggested Common Name White prunicola scale.

Common Synonyms and Combinations *Chionaspis prunicola* Maskell, *Diaspis amygdali* var. *rubra* Maskell (in part), *Howardia prunicola* (Maskell), *Aulacaspis pentagona rubra* (Maskell), *Diaspis auranticolor* Cockerell, *Aulacaspis pentagona auranticolor* (Cockerell), *Diaspis rubra* Maskell. It is important to point out that Kreiter et al. (1999) examined specimens reared from potato for more than 100 generations and found specimens that fit both the white peach scale (*P. pentagona*) and the white prunicola scale (*P. prunicola*). They conclude that the two are simply variants of the same species.

Field Characters (Plate 104) Adult female cover convex, circular, white; shed skins usually subcentral, light yellow (on *Prunus*). Male cover smaller, felted, white, sometimes with faint median carina, dorsal and ventral covers enclosing developing male; shed skin yellow. Body of adult female light yellow, developing pink tinge during egg production; eggs and first instars pink. Primarily infests bark and fruit, occasionally on leaves. In heavy infestations, male covers in conspicuous white masses on bottoms of branches.

Slide-mounted Characters Adult female (Fig. 108) with 4 or 5 pairs of lobes, fifth lobes absent or small; paraphyses usually restricted to medial lobule of second lobe. Median lobes separated by space 0.3–0.5 (0.4) times width of median lobe, each margin with thin paraphysis-like sclerosis, with yoke, medial margin axes diverging apically, lateral margin axes converging apically, with 1–5 (3) lateral notches, 0–3 (2) medial notches; second lobes with 2 lobules, noticeably smaller than median lobes, medial lobule largest with 0–1 (0) lateral notches, without notches on remainder of lobe; third lobes bilobed, lateral margins rarely with 1 notch, notches absent elsewhere; fourth lobe bilobed or simple, medial lobule with 0–3 (1) notches, lateral lobule without notches; fifth lobe simple or absent, when present, with 0–3 (2) notches. Gland-spine formula usually 1-2-2 (of 70 specimens examined, 3 had 2 gland spines in first space, 30 had 2 or 3 gland spines in second space, 9 had 1 in third space, 16 had 3 or more in third space), with 20–47 (31) gland spines near body margin from mesothorax or metathorax to segment 4; without gland spines between median lobes. Macroducts of 2 sizes, larger size on dorsum in marginal areas on segment 1 or 2 to segment 7, in submarginal areas of segments 2 to 5 or 6, in submedial areas of segments 1 or 2 to 5 or rarely to 6, smaller size in submarginal and marginal areas on metathorax or segment 1 to segments 2, 3, or 4, on dorsum on segments 1, 2, 3 or 4, on venter from metathorax or segment 1 to segments 2, 3, or 4, without duct between median lobes, macroduct in first space 17–24 (21) μ long, with 11–17 (14) macroducts on each side of body on segments 5 to 7, total of 38–86 (58) large macroducts on each side of body, some orifices occurring anterior of anal opening. Pygidial microducts on venter in submarginal areas of segments 5 to 7 with 3–8 (6) ducts; prepygidial ducts in submedial areas from prothorax or mesothorax to segments 1 or 2, sometimes absent from segment 2, in submarginal or marginal areas of head, prothorax, or mesothorax to metathorax or segment 1, also on segment 4 and/or 3; pygidial microducts absent from dorsum, prepygidial ducts usually on metathorax and segment 1, occasionally on segment 2 also. Perivulvar pores in 5 groups, 33–99 (64) pores on each side of body. Perispiracular pores usually with 3 loculi, anterior spiracles each with 9–22 (17), posterior spiracles without associated pores. Anal opening located 6–12 (8) times length of anal opening from base of median lobes, anal opening 12–17 (14) μ long. Eyes usually represented by small, flat, sclerotization on body margin laterad of anterior edge of clypeolabral shield. Antenna each with conspicuous seta. Cicatrix on segment 1, rarely also on prothorax. Body turbinate, with prepygidial segments from mesothorax or metathorax to segment 3 or 4 with lateral margin forming protuberance. Of 70 specimens, 12 had at least 1 gland spine with bifurcate or trifurcate apex, with 0–2 (0) apically divided gland spines in first four spaces on each side of pygidium. With 0–12 (5) marginal and submarginal small macroducts on each side of metathorax and segment 1 combined.

Affinities The white prunicola scale is most similar to *Pseudaulacaspis pentagona*, the white peach scale. For a comparison see the affinities section of the treatment of the white peach scale.

Hosts Polyphagous. Because this pest was considered a junior synonym of *Pseudaulacaspis pentagona*, white peach scale, until Kawai's (1980) publication, recorded host records for these species generally are unreliable. White prunicola scale has a more restricted host range than white peach scale. We have examined specimens from *Acer*, *Alnus*, *Aucuba*, *Buxus*, *Catalpa*, *Celtis*, *Croton*, *Cucurbita*, *Forsythia*, *Fraxinus*, *Ilex*, *Ligustrum*, *Magnolia*, *Malus*, *Nerium*, *Osmanthus*, *Populus*, *Prunus*, *Rhododendron*, *Salix*, and *Syringa*. Throughout its range white prunicola scale is most common on *Prunus*. In the United States *Ligustrum* and *Syringa* also are common hosts.

Distribution This species probably is native to the north temperate areas of China or Japan. In the United States we have seen specimens from AL, CA, CT, DC, FL, HI, LA, MA, MD, MS, NC, NJ, NY, OH, OR, PA, RI, VA, WV. We also have seen material from China, Japan, Korea, Okinawa, and Taiwan. It has been most commonly collected in quarantine on *Prunus* spp. from Japan.

Biology White prunicola scale long has been confused with white peach scale. Most published biologies probably deal with white peach scale because they usually report work done in warm areas of the world on hosts other than *Prunus*. White prunicola scale prefers a north temperate climate. Stimmel (1982) studied the seasonal history of *P. prunicola* on *Prunus serrulata* Lindl., Japanese flowering cherry, in northeastern Pennsylvania. We have examined voucher specimens from his study. To our knowledge this is the only published life history of this species. He found that the scale was bivoltine with mated adult females as the overwintering stage. In each generation of 1981, oviposition began May 15 and July 22, females produced an average of 27.2 and 78 eggs, crawlers first appeared May 20 and July 28, and adults first appeared July 8 and September 3. We observed the species in College Park, Maryland, in 1981, on *Prunus serrulata* where it had 3 generations per year. First-generation crawlers were present in early May; these became adults in early June and egg laying began the last week of June. Second-generation crawlers were present in early July. Adults were seen the first week of August. Egg laying began in mid-August and the first crawlers of the third generation were found in late August. Only mated adult females overwintered.

Economic Importance White prunicola scale is a serious pest of *Prunus*, especially in temperate areas. Moderate infestations have caused defacing dieback to the flowering cherry trees growing around the scenic tidal basin area of Washington, D.C., and heavy infestations have killed many of these priceless trees (personal observations of authors). Maskell (1895) in the original description reports that it was causing significant damage to Japanese plum in Hawaii. Miller and Davidson (1990) consider this species to be an occasional pest.

Selected References Davidson et al. (1983); Kawai (1980); Stimmel (1982).

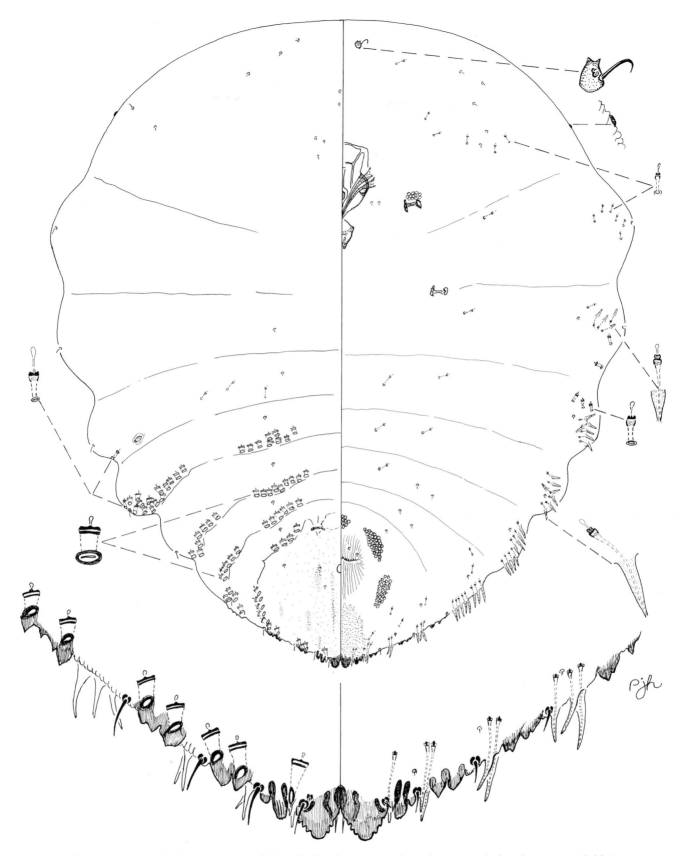

Figure 108. *Pseudaulacaspis prunicola* (Maskell), white prunicola scale, Hawaii (other data not available).

Plate 104. *Pseudaulacaspis prunicola* (Maskell), **White prunicola scale**

A. Adult female cover (top) and adult male cover on cherry laurel leaf (J. A. Davidson).
B. Adult female covers and crawlers on cherry laurel (J. A. Davidson).
C. Adult male covers in aggregation on Japanese flowering cherry (J. A. Davidson).
D. Female covers on red-twig dogwood (J. A. Davidson).
E. Young adult female covers on Japanese flowering cherry (J. A. Davidson).

F. Body of young adult female on Japanese flowering cherry (J. A. Davidson).
G. Body of adult female with eggs on Japanese flowering cherry (J. A. Davidson).
H. Heavily infested branch of Japanese flowering cherry (J. A. Davidson).
I. Japanese flowering cherry tree killed after four years of heavy infestation (J. A. Davidson).

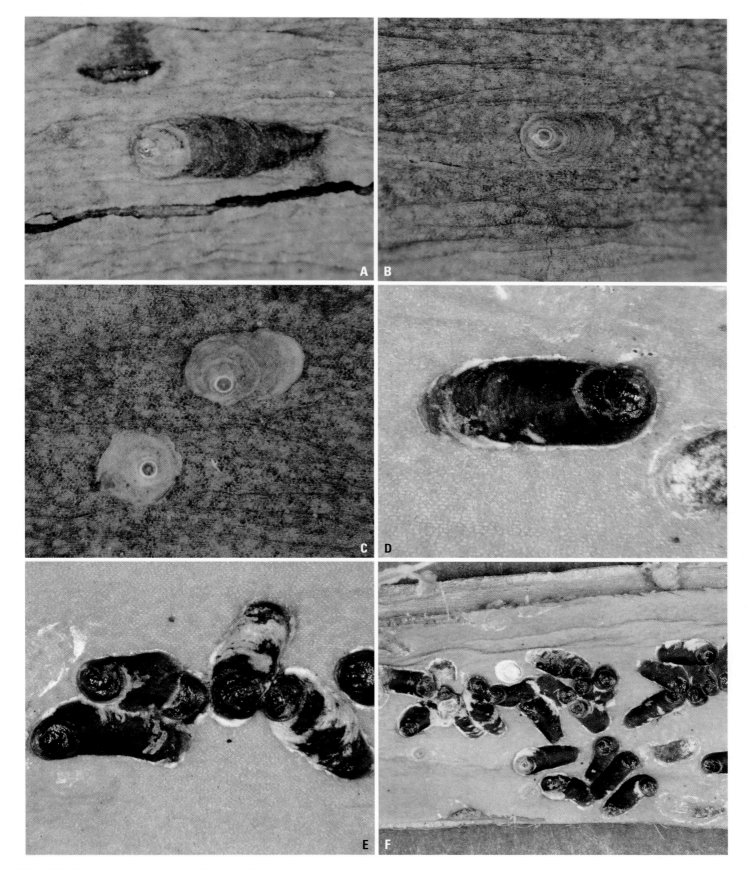

Plate 105. *Pseudischnaspis bowreyi* (Cockerell), Bowrey scale

A. Adult female cover on agave (J. A. Davidson).
B. Adult male cover on agave (J. A. Davidson).
C. Young (bottom) and mature adult male covers on agave (J. A. Davidson).

D. Rubbed adult female cover on agave (J. A. Davidson).
E. Old female covers on agave (J. A. Davidson).
F. Distance view of infestation on agave (J. A. Davidson).

367

Pseudischnaspis bowreyi (Cockerell)

Suggested Common Name Bowrey scale.

Common Synonyms and Combinations *Aspidiotus bowreyi* Cockerell, *Aspidiotus (Chrysomphalus) bowreyi* (Cockerell), *Chrysomphalus bowreyi* (Cockerell), *Pseudischnaspis linearis* Hempel, *Aspidiotus linearis* (Hempel), *Aspidiotus (Chrysomphalus) longisimma* Cockerell, *Aspidiotus longisimma* Cockerell, *Chrysomphalus longisimma* (Cockerell), *Pseudischnaspis longisimma* (Cockerell).

Field Characters (Plate 105) Newly molted adult females with circular cover, older adult female cover, elongate oval, moderately convex, black with blue to purple tinge; shed skins marginal to submarginal; brownish black. Ventral cover well developed. Male cover similar in texture, smaller, elongate oval, brown to black; shed skin brown to black. On bark and leaves.

Slide-mounted Characters Adult female (Fig. 109) with 3 pairs of definite lobes, area anterior of lobe 3 with series of 4 or 5 lobelike projections; paraphysis formula usually 2-2-1, occasionally 2-3-1 with paraphysis attached to medial margin of lobe 3. Median lobes separated by space 0.3–0.7 (0.5) times width of median lobe, without basal sclerosis and yoke, with distinct paraphysis attached to medial margin, medial margin axes parallel, lateral margins rounded, with 0–1 (0) lateral notches, without medial notches; second lobes simple, wider or equal to median lobes, with 2–3 (2) lateral notches, without medial notches; third lobes simple, slightly wider than second lobes, with 3–5 (4) lateral notches, 0–1 (0) medial notches. Gland spines difficult to see, with gland-spine formula usually 2-2-3, sometimes 2-2-1, 2-3-4, 2-2-2, with 2 or 3 gland spines anterior of segment 5, with 2 gland spines between median lobes one-half to equal length of lobe. Macroducts of 2 distinct sizes, larger size located posterior of anal opening becoming slightly smaller anteriorly, smaller size located anterior of posterior end of anal opening near to or attached to narrow sclerotized area on segment 6, macroduct between median lobes 92–218 (158) μ long, 1.0–1.2 (1.0) times as long as distance between base of median lobe and posterior apex of anal opening; macroduct in first space 107–170 (149) μ long, with 13–25 (20) macroducts on each side of pygidium on segments 5 to 8, marginal ducts on segments 5 to 8, submarginal on segments 5 to 7; total of 17–34 (23) macroducts on each side of body; some small-sized macroducts with orifices occurring anterior of anal opening; many small-sized macroducts on lateral margin of segments 4 and 5. Pygidial microducts on venter of segment 5, with 8–13 (10) microducts on each side of pygidium; prepygidial microducts on venter in marginal or submarginal areas of head or head and prothorax, and on mesothorax, metathorax, segment 2, or segment 2 to segment 4, in submedial areas near spiracles, around mouthparts, and on mesothorax or metathorax to segments 1 or 2; pygidial microducts absent from dorsum; prepygidial microducts on dorsum in submarginal areas of prothorax or mesothorax to segment 1, 2, or 3. Perivulvar pores usually in 4 or 5 groups, rarely with 1 or 2 pores comprising fifth group, 7–15 (11) pores on each side of body. Perispiracular pores absent. Anal opening located 6–11 (9) times length of anal opening from base of median lobes, anal opening 11–20 (15) μ long. Dorsal seta laterad of median lobes one-half to three-quarters length of lobe. Eyes usually absent or represented by sclerotized patch on body margin anterior of antenna. Antennae each with 1 long seta. Cicatrices near body margin of prothorax, usually also on segments 1 and 3. Pygidium with sclerotized areas and with at least 1 interlobular paraphysis longer than other paraphyses. Pygid-

ium shape may vary from having body margins concave to slightly convex depending on body length. Body 818–1407 (1111) μ long apparently depending on age of adult female. Posterior plate in third space usually very slender, seta-like. Sclerotized area conspicuously swollen adjacent to seta marking segment 5. Anal opening located 97–160 (133) μ from base of median lobes.

Affinities The Bowrey scale is similar to *Pseudischnaspis acephala* Ferris, the flatheaded scale, but can be separated by having 7–15 (11) perivulvar pores on each side of pygidium; 4–8 (6) macroduct orifices in row beginning between medial margin of third lobe and interlobular paraphysis in second space; 2–6 (4) macroduct orifices in same row located anterior of imaginary line drawn between anterior apex of paraphysis attached to medial margin of lobe 3 and anterior apex of interlobular paraphysis in second space; 13–19 (15) large macroducts on each side of pygidium; elongate, ventral microducts in band on marginal or submarginal areas of mesothorax, metathorax, segment 1, or segment 2 to segment 5; distance from anal opening to base of median lobes 97–160 (133) μ; enlarged sclerotized area adjacent to seta marking segment 5; 6–14 (9) microducts in each cluster on sublateral area of segment 5 posterior of ventral seta marking segment; *P. acephala* has 4–7 (5) perivulvar pores on each side of pygidium; 4–6 (4) macroducts in row mentioned above; 1–3 (2) macroducts anterior of imaginary line described above; 9–13 (11) large macroducts on each side of pygidium; elongate, ventral microducts usually on marginal or submarginal areas of segments 4 and 5, rarely with 1 or 2 on segment 3; distance from anal opening to base of median lobes 78–122 (91) μ; narrow sclerotized area adjacent of seta marking segment 5; 5–9 (7) microducts in each cluster on sublateral area of segment 5 posterior of ventral seta marking segment.

Hosts Polyphagous. Borchsenius (1966) records it from 12 genera in 10 families; Dekle (1977) records it from 19 genera in Florida, with *Coccoloba* the most frequently reported host. We have examined specimens from *Aechmea*, *Agave*, *Aloe*, *Annona*, *Asparagus*, *Beaucarnea*, *Bromelia*, *Calanthe*, *Cannabis*, *Carya*, *Cattleya*, *Ceiba*, *Citrus*, *Coccoloba*, *Cocos*, *Epidendrum*, *Eucalyptus*, *Hibiscus*, *Hylocereus*, *Jasminum*, *Mangifera*, *Musa*, *Nerium*, *Oncidium*, *Pachira*, *Passiflora*, *Persea*, *Phoenix*, *Pithecellobium*, *Poinciana*, *Prunus*, *Psidium*, *Pyrus*, *Rosa*, *Spondias*, and *Tillandsia*.

Distribution Bowrey scale is a native of the New World tropics. We have examined U.S. specimens from FL out-of-doors and MO and NY in greenhouses. This species occurs throughout the new world tropics of several Caribbean islands, Central America, Mexico, and South America. It most often has been collected by quarantine officers on *Agave* and *Rosa* from the Caribbean islands.

Biology We could not find a life-history study of this species. Brown (1965) reports that Bowrey scale has the diaspidid (sexual) chromosome system.

Economic Importance Dekle (1965) notes that Bowrey scale occasionally was a serious pest on rose in Key West, Florida. He also reports that it was most frequently collected on *Rosa* and orchids in the state. Ballou (1934) found this species to be a harmful pest of *Prunus*. Gowdey (1921) reports it as a serious pest on *Agave* during dry years. Otero (1935) considers it to be a pest of guava in Cuba. Miller and Davidson (1990) consider this species to be an occasional pest.

Selected References Ferris (1941); Miller et al. (1984).

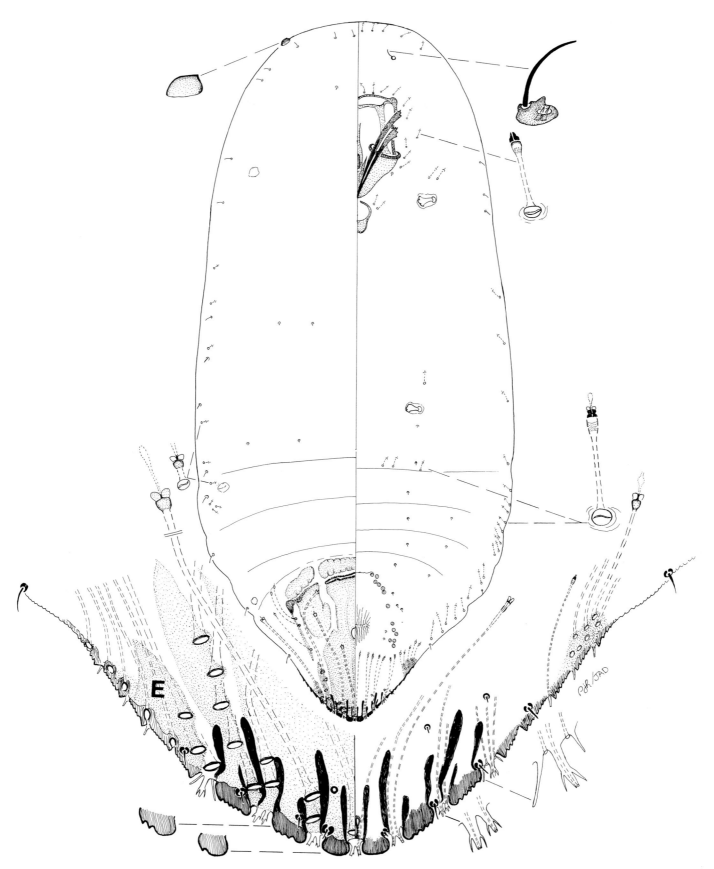

Figure 109. *Pseudischnaspis bowreyi* (Cockerell), Bowrey scale, Guatemala, intercepted at New Orleans, LA, on bromeliad, VII-19-1976.

Pseudoparlatoria ostreata (Cockerell)

Suggested Common Name Gray scale (also called acalypha scale).

Common Synonyms and Combinations *Diaspis tricuspidata* Leonardi, *Pseudoparlatorea tricuspidata* (Leonardi).

Field Characters (Plate 106) Adult female cover on leaves irregularly circular, flat, thin, parchment-like, usually gray, occasionally yellow brown, newly formed adult female scales probably translucent; shed skins marginal. Adult female covers on bark thicker, dark brown, opaque. Male cover similar in texture, smaller, elongate oval, gray; shed skin marginal to submarginal. Body of adult female probably yellow. On bark and leaves.

Slide-mounted Characters Adult female (Fig. 110) with 3 pairs of definite lobes, fourth lobes represented by series of small points; paraphyses absent. Median lobes separated by space 0.4–1.0 (0.6) times width of median lobe, with conspicuous transverse sclerotization attached to medial and lateral margin, axis of medial margin diverging apically, lateral margin converging apically, with 1 lateral notch, 1–2 (1) medial notches; second lobes with 2 lobules, medial lobule not protruding beyond median lobes, medial lobule largest, with 0–1 (1) lateral notches, 0–1 (0) medial notches, lateral lobule with 0–1 (0) lateral notch, without medial notch; third lobes with 2 lobules, often small, medial lobule largest, with 0–2 (1) lateral notches, 0–1 (1) medial notch, lateral lobule with 0–2 (1) lateral notches, without medial notches; fourth lobe represented by series of points. Gland-spine formula usually 1-1-1, rarely 1-1-0 or 2-1-1, with 7–13 (9) gland spines near each body margin from segment 2 or 3 to segment 4; with conspicuous 'fish-tail' shaped gland spines between medial lobes. Macroducts of 2 sizes, larger size in marginal areas on segments 5 to 7, with 5, 6, or 7 such ducts on each side of pygidium, with 1–4 (2) large ducts submarginally on segment 7 and 6, smaller size in submarginal and marginal areas from segments 3 to 7, increasing in size posteriorly, without duct between median lobes, macroduct in first space 17–25 (19) µ long, with 15–22 (19) macroducts on each side of body on segments 5 to 7, total of 24–52 (35) macroducts on each side of body, some orifices occurring anterior of anal opening. Pygidial microducts on venter in submarginal areas of segments 5 to 7, with 2–12 (6) ducts; prepygidial ducts in submedial areas from head or prothorax to segments 1, 2, 3, or 4 in submarginal or marginal areas of head, prothorax, or mesothorax to metathorax, segment 1, or segment 2 and on segment 3 and 4; pygidial microducts absent from dorsum, prepygidial ducts present on head and mesothorax, metathorax, or segment 1 to segment 2, 3, or 4. Perivulvar pores usually in 5 groups, rarely in 4, with 27–44 (33) pores on each side of body. Perispiracular pores absent. Anal opening located 1.3–3.0 (2.0) times length of anal opening from base of median lobes, anal opening 17–27 (22) µ long. Eyes laterad of antenna, usually in form of slightly convex dome, occasionally with acute apex. Antennae each with 2–5 (4) conspicuous setae. Cicatrices usually on prothorax, near intersegmental area of segments 2 and 3, and on segment 4. Body turbinate. Median cluster of perivulvars with 0–7 (4) pores.

Affinities Of the economic species in the United States, gray scale is similar to *Pseudoparlatoria parlatorioides*, false parlatoria scale. Both have similar pygidial margin structures and have distinctive 'fish-tail' shaped gland spines between the median lobes. The gray scale differs by having: 24–52 (35) macroducts on each side of body; 15–22 (19) macroducts on each side of pygidium; and by usually having an anteromedial cluster of perivulvar pores. The false parlatoria scale differs by having: 14–22 (17) macroducts on each side of body; 8–15 (12) macroducts on each side of pygidium; and by usually lacking an anteromedial cluster of perivulvar pores. Several other species of *Pseudoparlatoria* are similar to gray scale including *P. serrulata* Townsend and Cockerell, but the identity of most of these species is so confused that it currently is impossible to make useful comparisons.

Hosts Polyphagous. Borchsenius (1966) records it from 10 genera in 6 families; Dekle (1977) reports it from 7 genera in Florida, with *Acalypha* and *Carica* the most frequently reported hosts. We have examined specimens from *Acalypha, Agave, Antigonon, Asparagus, Barkeria, Bryophyllum, Carica, Catasetum, Cattleya, Cestrum, Chrysobalanus, Cocos, Cycnoches, Dieffenbachia, Dioscorea, Euonymus, Ionopsis, Jasminum, Laelia, Lonicera, Oncidium, Opuntia, Pelargonium, Persea, Psychotria, Senecio, Solanum, Verbena,* and *Vitis.*

Distribution Gray scale probably is native to the Caribbean islands. We have examined U.S. specimens from FL. This species also occurs in Central America, several Caribbean islands, Venezuela, and Africa. It most often has been collected by quarantine officers on *Carica* and orchids from the Caribbean islands.

Biology We have been unable to find a published account detailing the life cycle of the gray scale. Brown (1965) reports that this species has the diaspidid (sexual) chromosome system. The material he examined was from Jamaica and Brazil.

Economic Importance This species is considered to be an occasional pest by Miller and Davidson (1990). Dekle (1977) notes that gray scale occasionally was a serious pest on copperleaf, *Acalypha wilkesiana,* in Florida. It also has been reported as an important pest of papaya, *Carica papaya,* in Puerto Rico (Wolcott 1937; Martorell 1945). Wolcott and Martorell (1943, 1944) indicate that the scale was under reasonable control with the lady beetle *Chilocorus cacti* (Linnaeus). It is an occasional pest of avocado (McKenzie 1935b).

Selected References Balachowsky (1954); Ferris (1942).

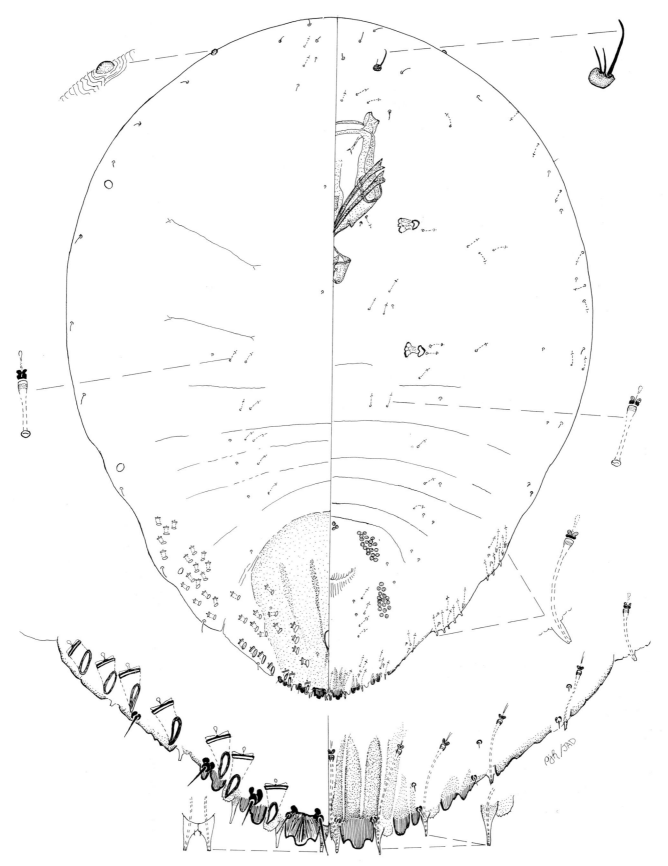

Figure 110. *Pseudoparlatoria ostreata* Cockerell, gray scale, Honduras, intercepted at Los Angeles, CA, on orchid, IV-6-1978.

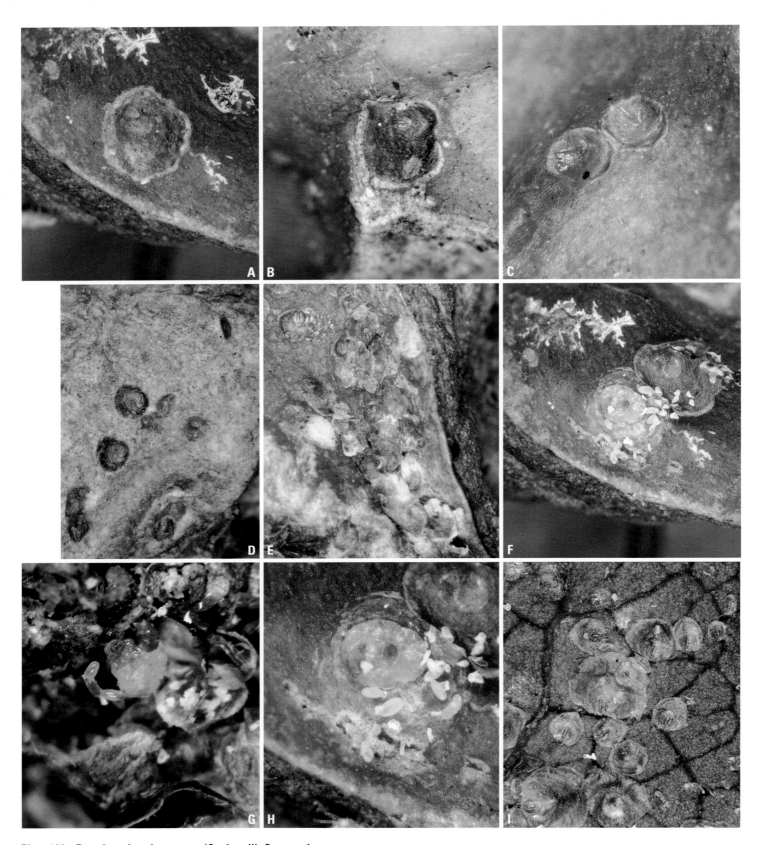

Plate 106. *Pseudoparlatoria ostreata* (Cockerell), Gray scale

A. Adult female cover on papaya (J. A. Davidson).
B. Adult female cover on papaya (J. A. Davidson).
C. Mature second-instar covers on papaya (J. A. Davidson).
D. Adult female and male covers on bark of papaya (J. A. Davidson).
E. Heavy infestation on shoot of papaya (J. A. Davidson).
F. Body of adult female with hatching eggs on papaya (J. A. Davidson).

G. Body of young adult female with newly laid eggs on papaya (J. A. Davidson).
H. Body of adult female showing distinctive shape of body (J. A. Davidson).
I. Distance view of aggregation on honeysuckle leaf (J. A. Davidson).

372

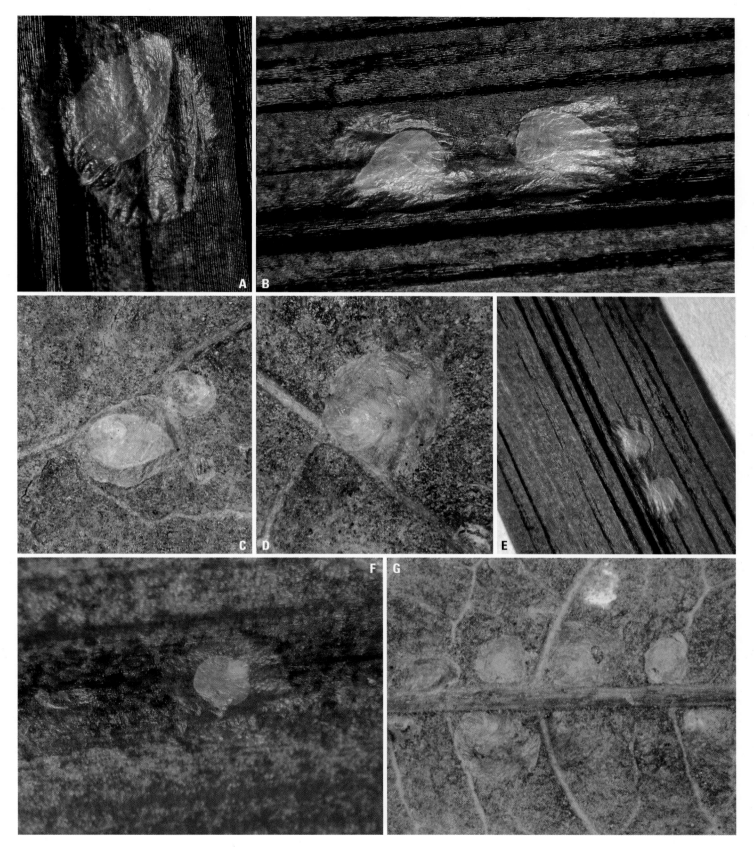

Plate 107. *Pseudoparlatoria parlatorioides* (Comstock), False parlatoria scale

A. Transparent adult female cover showing live body of adult female (characteristically shaped) on *Dracaena marginata* (R. J. Gill).

B. Body of live adult female visible through cover on *D. marginata* (R. J. Gill).

C. Body of dead adult female under cover on *Inga* (J. A. Davidson).

D. Close-up view of cover along midvein of *Inga* (J. A. Davidson).

E. Distance view of bodies of live females under covers on *Dracaena* (R. J. Gill).

F. Body of live young adult female under cover on *D. marginata* (R. J. Gill).

G. Distance view of infestation on *Inga* (J. A. Davidson).

Pseudoparlatoria parlatorioides (Comstock)

Suggested Common Name False parlatoria scale (also called parlatoria-like scale and false scale).

Common Synonyms and Combinations *Aspidiotus parlatorioides* Comstock, *Pseudoparlatorea parlatoreoides* (Comstock).

Field Characters (Plate 107) Adult female cover circular, slightly convex, thin, parchment-like, yellow or yellow brown, probably transparent when young female; shed skins usually marginal, reddish yellow. Male cover similar in texture, smaller, elongate oval, yellow brown; shed skin marginal, yellow. Body of adult female yellow; eggs and crawlers probably yellow. On bark and leaves.

Slide-mounted Characters Adult female (Fig. 111) with 3 pairs of definite lobes, fourth lobes represented by series of small points; paraphyses absent. Median lobes separated by space 0.6–1.2 (0.8) times width of median lobe, with conspicuous transverse sclerotization originating from setal bases, sometimes incomplete, axis of medial margin diverging apically, lateral margin converging apically, with 1 lateral and 1 medial notch; second lobes with 2 lobules, medial lobule not protruding beyond median lobes, medial lobule largest, with 0–1 (1) lateral notch, 0–1 (0) medial notch, lateral lobule with 0–1 (0) lateral notch, without medial notch; third lobes with 2 lobules, medial lobule largest, 0–2 (1) lateral notches, 0–1 (0) medial notch, lateral lobule with 0–1 (0) lateral notch, without medial notch; fourth lobe represented by series of points. Gland-spine formula usually 1-1-1, occasionally 1-1-0, with 4–10 (8) gland spines near each body margin on segments 3 and 4; with 'fish-tail' shaped gland spines between medial lobes. Macroducts of 2 intergrading sizes, larger size in marginal areas on segments 5 to 7, with 5, 6, or 7 such ducts on each side of pygidium, with 1 large duct submarginally on each of segments 7 and 6, smaller size in submarginal and marginal areas from segment 3 or 4 to 7, increasing in size posteriorly, without duct between median lobes, macroduct in first space 17–22 (20) µ long, with 8–15 (12) macroducts on each side of body on segments 5 to 7, total of 14–22 (17) macroducts on each side of body, some orifices occurring anterior of anal opening. Pygidial microducts on venter in submarginal areas of segments 5 to 7, with 1–9 (5) ducts; prepygidial ducts in submedial areas from head or prothorax to segments 2 or 3, in submarginal or marginal areas of head or prothorax to metathorax and on segments 1, 2, 3 and/or 4; pygidial microducts absent from dorsum; prepygidial ducts present on head and on mesothorax or metathorax to segment 1 or 2. Perivulvar pores usually in 4 groups, rarely in 5, with 18–39 (28) pores. Perispiracular pores absent. Anal opening located 1.1–3.1 (2.0) times length of anal opening from base of median lobes, anal opening 18–30 (23) µ long. Eyes laterad of antenna, usually in form of convex dome, rarely with acute apex. Antennae each with 1–4 (1) conspicuous setae. Cicatrices usually on prothorax, near intersegmental area of segments 2 and 3, and on segment 4. Body turbinate. Median cluster of perivulvar pores with 0–3 (1) pores.

Affinities Of the economic species in the United States false parlatoria scale is most similar to *Pseudoparlatoria ostreata*, gray scale. For a comparison of these species see the affinities section of the gray scale.

Hosts Polyphagous. Borchsenius (1966) records it from 47 genera in 30 families; Dekle (1977) reports it from 56 genera in Florida, with *Cestrum* and *Magnolia* the most frequently reported hosts. We have examined specimens from *Amyris*, *Anthurium*, *Artocarpus*, *Batemannia*, *Begonia*, *Borrera*, *Brassavola*, *Broughtonia*, *Cattleya*, *Cephalocereus*, *Chamaedorea*, *Cinchona*, *Cinnamomum*, *Citrus*, *Cocos*, *Comparettia*, *Cypripedium*, *Dracaena*, *Elaeodendron*, *Epidendrum*, *Gerbera*, *Hedera*, *Huntleya*, *Inga*, *Jatropha*, *Laelia*, *Lycaste*, *Mangifera*, *Maxillaria*, *Melicoccus*, *Musa*, *Nephthytis*, *Odontoglossum*, *Oncidium*, *Peperomia*, *Peristeria*, *Persea*, *Philodendron*, *Phoenix*, *Pimenta*, *Psidium*, *Punica*, *Reinhardtia*, *Reynosia*, *Rosa*, *Schomburgkia*, *Sobralia*, *Spiranthes*, *Theobroma*, *Tillandsia*, *Trachycarpus*, *Vanilla*, *Washingtonia*, and *Zygopetalum*.

Distribution False parlatoria scale probably is native to tropical America. We have examined U.S. specimens from CA (eradicated), FL, GA, TX. It is reported from AL (Nakahara 1982) and has been found in greenhouses in MD and NJ. The species occurs in most of Central and South America, Europe, Hawaii, Mexico, New Zealand, Russia, and Sri Lanka. It most frequently is taken in quarantine from Mexico and South America.

Biology We could find no definitive life-history study of this species. Brown (1965) notes that specimens from São Paulo, Brazil, on *Eugenia jaboticaba* had a haploid complement of 5 chromosomes and had both male and female embryos. Schmutterer (1952b) indicates that both sexes occurred in populations in greenhouses in Germany, but because of the paucity of males he believed the species might be at least partially parthenogenetic. Under greenhouse conditions in middle Europe adult females lay 30–130 eggs, and the larvae apparently hatch in about 10 days.

Economic Importance This species is considered to be a serious pest in a small area of the world by Miller and Davidson (1990) and Arnett (1985). Merrill and Chaffin (1923) report false parlatoria scale in Florida as an occasionally very serious pest, especially on *Acalypha* and palms. Steinweden (1945) lists *P. parlatorioides* as one of the more important orchid pests in California; it has since been eradicated from this state. These reports are possibly misidentifications of *P. ostreata*.

Selected References Balachowsky (1954); Ferris (1942); Schmutterer (1959).

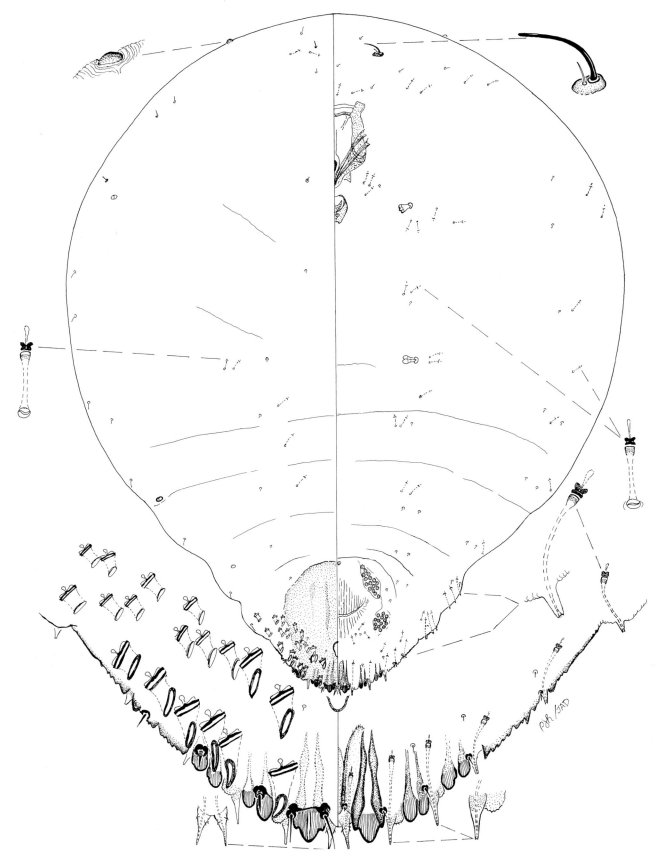

Figure 111. *Pseudoparlatoria parlatorioides* (Comstock), false parlatoria scale, Mexico, intercepted at San Antonio, TX, on *Chamaedorea* sp., XII-23-1974.

Quernaspis quercus (Comstock)

Suggested Common Name Oak scale.

Common Synonyms and Combinations *Chionaspis quercus* Comstock, *Fundaspis quercus* (Comstock), *Jaapia quercus* (Comstock), *Pinnaspis quercus* (Comstock), *Chionaspis* (*Quernaspis*) *quercus* Comstock.

Field Characters (Plate 108) Adult female cover oyster-shell shaped, anterior end usually curved, convex, white to gray; shed skins marginal, brown when unrubbed, yellowish brown when rubbed. Male cover smaller, elongate, felted, distinctly tricarinate, white; shed skin marginal, yellow. Body of adult female reddish orange. On leaves and bark. Based on our observations in California, males on bark and leaves but preferred leaves. Females primarily on bark, often concealed beneath loose epidermal layer.

Slide-mounted Characters Adult female (Fig. 112) with 3 pairs of definite lobes, fourth lobes absent or represented by series of small points; paraphyses absent. Median lobes fused medially, without sclerosis or yoke, medial margin absent, lateral margin converging or diverging apically, with 1–3 (2) lateral notches, absent medially; second lobes bilobed, noticeably smaller than median lobes, medial lobule largest with 0–2 (1) lateral notches, 0–1 (0) medial notches, lateral lobule pointed, without notches; third lobes bilobed or simple, medial lobule largest with 0–2 (1) notch, lateral lobule, when present, without notches; fourth lobes, when present, simple, represented by 1 or more small points. Gland-spine formula usually 1-2-2, rarely with 2 in first space, 1 in second space, 3 in third space, with 18–41 (28) gland spines near each body margin from prothorax or mesothorax to segment 4; without gland spines between median lobes. Macroducts of 2 variable sizes, larger size on dorsum in marginal areas on segments 4 or 5 to 7, smaller size on dorsum on submedial areas of segments 2 or 3 to 4 or 5, on marginal or submarginal areas of pro-, meso-, or metathorax to segment 4, on venter on submargin from mesothorax to segment 4, without duct between median lobes, macroduct in first space 15–21 (17) μ long, with 6–12 (8) macroducts on each side of pygidium on segments 5 to 7, total of 50–131 (78) macroducts on each side of body, some orifices occurring anterior of anal opening. Pygidial microducts on venter in submarginal areas of segments 5 to 7, with 3 to 7 (5) ducts; prepygidial ducts in submedial areas on head and metathorax or segment 1 to segment 3 or 4, in submarginal or marginal areas of head and segment 4; pygidial microducts absent from dorsum, prepygidial ducts usually present on head and anterior thorax. Perivulvar pores in 5 groups, 18–59 (40) pores. Perispiracular pores with 3 loculi, anterior spiracles with 2–9 (5), posterior spiracles with 1–2 (1) pores. Anal opening located 10–25 (14) times length of anal opening from base of medial lobes, anal opening 6–15 (11) μ long. Eyes rarely absent, usually represented by small starfish-shaped sclerotization on body margin laterad of anterior edge of clypeolabral shield. Antenna with conspicuous seta. Cicatrices absent. Body elongate, mesothorax, metathorax, and segments 1 to 4 with body margin forming slight abdominal protuberances. There is a disturbingly large amount of variation within the confines of this species. It is possible that more than a single entity is involved.

Affinities The oak scale differs from all other economic scales that occur in the United States by having definitely fused median lobes. It is very similar to *Quernaspis insularis* Howell but differs by having a broad pygidium and significantly different ($P = .975$) but overlapping ranges of perivulvar pores. *Quernaspis quercus* has 18–59 (40) perivulvar pores and an acute pygidium. Takagi and Howell (1977) distinguish between *Q. quercus* and *Q. insularis* by the above characters and by the numbers of gland spines. Although their counts appear to be significantly different, our data do not support this difference.

Hosts Because of the relatively narrow host range of oak scale, we are presenting species names of all of the plant hosts that we have authenticated. In addition to *Quercus*, Borchsenius (1966) records it from *Lithocarpus*. Dekle (1977) notes it only from *Quercus* in Florida. Takagi and Howell (1977) studied specimens from *Quercus lobata*, *Q. douglasii* and *Q. velutina*. We have seen material from *Quercus agrifolia*, *Q. alba*, *Q. chrysolepis*, *Q. garryana*, *Q. kelloggii*, *Q. lobata*, *Q. pungens*, and *Q. stellata*.

Distribution Oak scale is native to North America. We have examined U.S. specimens from AZ, CA, FL, NC, TX, WA. Nakahara (1982) records it from GA, LA, NM, OH, and Mexico.

Biology According to Riley (1903) the oak scale overwinters as adult females or partly grown females in California; eggs are laid; male 'larvae' and pupae have been collected as late as August. We believe that the overwintering stage must be adult females as males were not found in winter.

Economic Importance Herbert (1936) reports the oak scale as causing damage to oaks in California and New Mexico. Essig (1915) gave control suggestions, indicating economic importance in California. Apparently it rarely reaches pest proportions. Miller and Davidson (1990) consider this species to be an occasional pest.

Selected References Gill (1997); Takagi and Howell (1977).

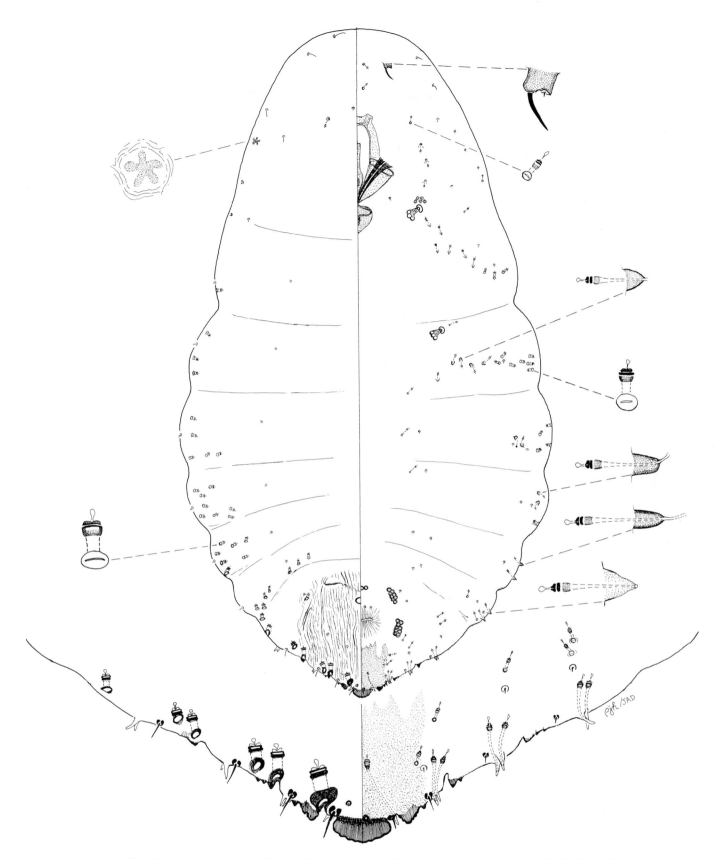

Figure 112. *Quernaspis quercus* (Comstock), oak scale, Palo Alto, CA, on *Quercus agrifolia*, X-31-1914.

Plate 108. *Quernaspis quercus* (Comstock), Oak scale

A. Adult female (top) and male (right bottom) covers on oak (J. A. Davidson).
B. Covers partially hidden by live oak bark flakes (R. J. Gill).
C. Covers among bark cracks on oak (J. A. Davidson).
D. Male covers on oak (J. A. Davidson).
E. Distance view of heavy infestation on oak (R. J. Gill).
F. Distance view of male covers clustered on oak gall (J. A. Davidson).
G. Orange body of young adult female on oak (D. R. Miller).
H. Red body of older adult female on oak in center of infestation (R. J. Gill).

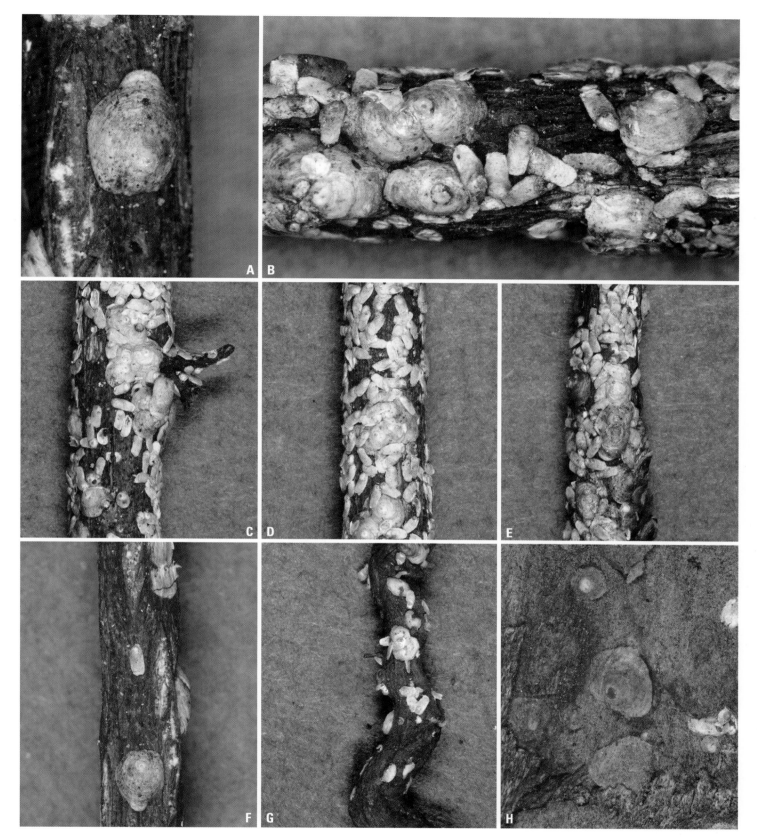

Plate 109. *Rhizaspidiotus dearnessi* (Cockerell), Dearness scale

A. Adult female cover on unknown shrub (J. A. Davidson).
B. Small male covers (central cluster) and larger female covers on unknown shrub (J. A. Davidson).
C. Infestation mostly with male covers on unknown shrub (J. A. Davidson).
D. Distance view of heavy infestation on unknown shrub (J. A. Davidson).

E. Distance view of infestation on unknown shrub (J. A. Davidson).
F. Adult female cover (bottom) and male cover (center) on unknown shrub (J. A. Davidson).
G. Distance view of infestation on unknown shrub (J. A. Davidson).
H. Adult female cover overlain by soil (J. A. Davidson).

Rhizaspidiotus dearnessi (Cockerell)

Suggested Common Name Dearness scale.

Common Synonyms and Combinations *Aspidiotus dearnessi* Cockerell, *Aspidiotus (Targionia) helianthi* Parrott, *Aspidiotus (Targionia) gutierreziae* Cockerell and Parrott, *Targionia dearnessi* (Cockerell), *Targionia gutierreziae* (Cockerell and Parrott), *Targionia helianthi* (Parrott), *Rhizaspidiotus helianthi* (Parrott), *Chorizaspidiotus gutierreziae* (Cockerell and Parrott), *Remotaspidiotus dearnessi* (Cockerell), *Pseudodiaspis helianthi* (Parrott), *Aspidiotus gutierreziae* Cockerell and Parrott, *Aspidiotus helianthi* Parrott.

Field Characters (Plate 109) Adult female cover convex, circular, light brown or gray, ventral scale thick; shed skins subcentral, yellow or tan, sometimes covered with white secretion. Male cover similar to female except shed skin submarginal. Body of adult female dark yellowish. On bark, usually well hidden; on composite hosts usually found on subterranean crown and primary root system; on ericaceous hosts usually found on trunk and branches. Marginal areas of head and thorax sclerotized on old adult females.

Slide-mounted Characters Adult female (Fig. 113) with 2–3 (3) pairs of definite lobes, rarely with small fourth lobe; paraphyses absent. Median lobes closely appressed, separated by space less than 1 μ in width, usually with conspicuous basal sclerotization approximately as long as median lobes, without yoke, medial margins parallel, lateral margins parallel or converging apically, with 0–1 (1) lateral notch, 0–1 (0) medial notch; second lobes usually bilobed, rarely simple, medial lobule largest, without notches, lateral lobule without notches; third lobes usually simple, rarely bilobed, without notches; fourth lobe normally absent, when present represented by small sclerotized lobe. Plates usually with associated microduct but difficult to discern without oil immersion, with up to 3 tines, often simple; plate formula 2-3-0, rarely 3-3-0 or 2-2-0; without plates between median lobes. (The distinction between macroducts and microducts in this species is vague at best. We consider macroducts to be the largest of the tubular ducts with a dark sclerotized inner duct. Microducts are smaller with a more lightly sclerotized inner duct.) Macroducts of 1 size, with 1–3 (2) between median lobes (not necessarily with orifice on body margin), 35–50 (43) μ long, extending 0.6–0.7 (0.6) times distance between posterior apex of anal opening and base of median lobes, longest duct in first space 32–60 (42) μ long, with 51–110 (73) macroducts on each side of pygidium on segments 5 to 8, ducts in submarginal and marginal areas on dorsal and ventral surfaces, total of 143–265 (203) macroducts on each side of body, some macroduct orifices anterior of anal opening; prepygidial dorsal macroducts occasionally in medial areas on any or all of segments 1 to 4, in submedial areas from metathorax or segment 1 to segment 4, in submarginal and marginal areas from metathorax or segment 1 to segment 4; prepygidial ventral macroducts in submedial areas of segments 1 and/or 2, in submarginal and marginal areas from metathorax or segment 1 to segment 4. Pygidial microducts on venter absent; prepygidial ducts of 1 size, in band anterolateral of mouthparts, in marginal and submarginal areas of head to segments 1, 2, or 3, in submedial areas of head to segments 1, 2, or 3, in clusters near spiracles; pygidial microducts absent from dorsum; prepygidial microducts of 1 size, in submarginal areas from head to segment 1, in submedial areas from head, prothorax, or mesothorax to segment 1 or 2. Perivulvar and perispiracular pores absent. Anal opening located 2.7–6.6 (4.1) times length of anal opening from base of median lobes, anal opening 10–27 (18) μ long. Dorsal seta laterad of median lobes 1.0–2.0 (1.3) times length of median lobe. Eyes absent or small and represented by small sclerotized area, located anterior and laterad of mouthparts. Antennae each with 1 long seta. Cicatrices absent. Body pear shaped. Area posterior and mesad of anterior spiracle with lightly sclerotized, oval area containing several microducts. Venter with 13–23 (18) macroducts on each side of pygidium on segments 5 to 8, ducts in submarginal and marginal areas, prepygidial ventral macroducts in submarginal and marginal areas from metathorax or segment 1 to segment 4. Body margin of segments 2 or 3 to pygidium distinctly crenulate. Marginal areas of head and thorax sclerotized on old adult females. Occasionally with oval clear area in medial area on dorsum anterior of mouthparts.

Affinities The Dearness scale is very different from all other economic scales of the United States. Superficially it shows some resemblance to *Odonaspis ruthae*, Bermuda grass scale, by having numerous macroducts on the dorsum and the venter, reduced second and third lobes, and crenulate posterior body margin. Dearness scale differs by having conspicuous, separate median lobes, each with a lateral notch, one-barred macroducts, no perivulvar or perispiracular pores, and by not occurring on grasses. Bermuda scale has small, partially fused median lobes that do not possess notches, two-barred macroducts, perivulvar and perispiracular pores, and by occurring on grasses. It also is similar to *Targionia bigeloviae* (Cockerell), bigelovia scale, by having 3 pairs of lobes, similarly shaped median lobes, and no perivulvar or perispiracular pores. Dearness scale differs by having marginal crenulations on the posterior part of the abdomen, relatively short and wide macroducts, and a basal sclerotization on the median lobes. Bigelovia scale has no marginal crenulations, relatively long and narrow macroducts, and no basal sclerotization.

Hosts Borchsenius (1966) records this species from 21 genera in 5 families, most of which are composites; Dekle (1977) reports it from 5 genera in Florida; and McKenzie (1956) reports it on 15 genera from California. We have examined specimens from *Achillea*, *Ageratum*, *Ambrosia*, *Aplopappus*, *Arctostaphylos*, *Artemisia*, *Aster*, *Astragalus*, *Baccharis*, *Bahia*, *Ceanothus*, *Corethrogyne*, *Erigeron*, *Gutierrezia*, *Haplopappus*, *Helianthus*, *Heliotropium*, *Hudsonia*, *Oenothera*, *Parthenium*, *Polygala*, *Pterocaulon*, *Solidago*, *Thymus*, and *Vaccinium*. This scale most commonly is collected on perennial hosts in the Compositae and Ericaceae.

Distribution Dearness scale is native to North America. We have examined U.S. material from AZ, CA, CO, FL, GA, IN, KS, MA, MD, MO, NB, NC, NJ, NM, OH, OR, SC, TX, VA, WI. Nakahara (1982) records it from AL, NY, OK, UT. Outside of the United States the species is known from Canada, Cuba, and Mexico, and it occasionally is taken in quarantine from Mexico.

Biology According to Lacroix (1926), this species has 1 generation per year and overwinters as adult females in Massachusetts on cranberry. Eggs first appear within the body of the adult in spring but rarely can be found in midwinter. Crawlers were first observed in mid-June, and settling took up to 3 days. Female crawlers tended to settle singly on upright stems at the base of leaf petioles or bracts. Males tended to aggregate on upright stems. Second stages first appeared in mid-July and pupal males were observed in early August. Adult males emerged in mid- to late August, and adult females were prevalent in early September. Approximately 75 days elapsed between the time that female crawlers first appeared and adult females were present.

Economic Importance Based on information provided by Charles F. Brodel, Cranberry Experiment Station, East Wareham, Massachusetts, this species is a pest of cranberries in Massachusetts and Quebec. Feeding by the scale is reported to

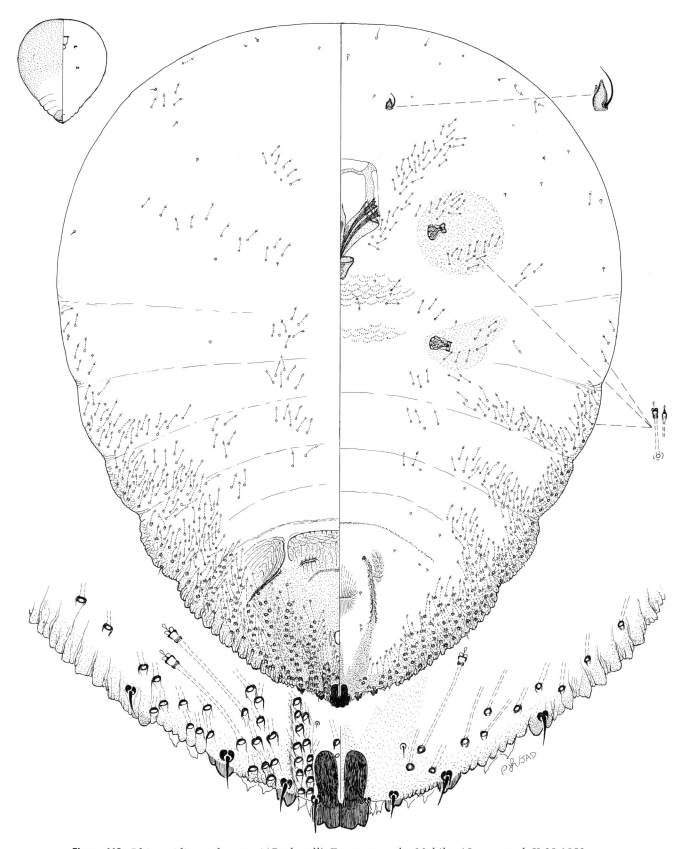

Figure 113. *Rhizaspidiotus dearnessi* (Cockerell), Dearness scale, Mobile, AL, on weed, X-30-1952.

weaken new growth, causing infested branches to break when subjected to heavy winds or an insect net. Lacroix (1926) mentions that feeding by the scale caused swelling of the stem. It is considered a pest of guayule (*Parthenium argentatum*) (Cassidy et al. 1950; Lange 1944; and Romney 1946). Miller and Davidson (1990) consider this species to be an occasional pest.

Selected References Ferris (1938a); Lacroix (1926).

Selenaspidus albus (McKenzie)

Suggested Common Name White euphorbia scale.
Common Synonyms and Combinations None.
Field Characters (Plate 110) Adult female cover flat, circular, white or yellowish white; shed skins central, orange or yellowish brown. Body of adult female yellowish orange, heavily sclerotized, with indentation between meso- and metathorax. Male reported as unknown (Gill 1997), but photographs depict male scale covers (Plate 110C). Male cover elongate oval, lighter and thinner than female cover; shed skin light yellow or white, subcentral. On green areas of host including stems and leaves.

Slide-mounted Characters Adult female (Fig. 114) with 3 pairs of definite lobes; paraphyses small and inconspicuous or absent. Median lobes separated by space 0.4–0.7 (0.6) times width of median lobe, median lobes without paraphyses, basal sclerotization, or yoke, medial margins parallel, lateral margins parallel or diverging slightly, without notches; second lobes simple, about same size and shape or slightly smaller than median lobes, without notches; third lobes simple, slender, spurlike, without notches. Plates with associated microduct, plates in first space slender, with 2 or 3 tines, plates in second space broad, with many tines, plates anterior of third lobe expanded apically, becoming shorter anteriorly; plate formula 2-3-3; with 2 plates between median lobes, slightly longer or equal to length of median lobes. Macroducts of 2 variable sizes, largest in marginal and submarginal areas of segments 2 or 3 to 8, smallest in marginal areas of metathorax or segment 1 to segment 2 or 3; 2 macroducts between median lobes 42–59 (50) μ long, extending 0.5–0.8 (0.7) times distance between posterior apex of anal opening and base of median lobes, longest macroduct in first space 41–58 (49) μ long, with 38–58 (49) macroducts on each side of pygidium on segments 5 to 8, ducts in submarginal and marginal areas, total of 74–98 (86) macroducts on each side of body, some macroduct orifices anterior of anal opening. Pygidial microducts on venter in submarginal areas of segments 5 and sometimes 6 and 7, with 0–5 (2) ducts; prepygidial ducts of 1 size, in cluster posterior of mouthparts, with 1 or 2 laterad of each spiracle, in marginal and submarginal areas of any or all of head, mesothorax, and segments 1, 2, 3, or 4; pygidial microducts absent from dorsum; prepygidial microducts of 1 size, in submarginal areas from head to segments 2, 3, or 4, usually in row on metathorax. Perivulvar pores absent. Perispiracular pores absent. Anal opening located 1.9–4.4 (3.2) times length of anal opening from base of median lobes, anal opening 18–31 (24) μ long. Dorsal seta laterad of median lobes 0.6–1.2 (0.9) times length of median lobe. Eyes visible in unsclerotized specimens, represented by small unsclerotized or weakly sclerotized dome, located laterad of anterior margin or clypeolabral shield. Antennae each with 1 short seta. Cicatrices absent. Body pear shaped, with conspicuous constriction between mesothorax and metathorax. Sclerotized tubercle or spine at posterolateral apex of mesothorax; tubercle usually acute apically, occasionally bifurcate. Body of mature females sclerotized throughout, with a few spotty clear areas.

Affinities The white euphorbia scale is most similar to *Selenaspidus articulatus*, the rufous scale, by having a spur and constriction located between the meso- and metathorax and by having the third lobe developed into a spur. They are easily separated by the absence of perivulvar pores in white euphorbia scale and the presence of these structures in rufous scale.

Hosts Monophagous. McKenzie (1956), Borchsenius (1966), Nakahara (1982), and Gill (1997) record it only from *Euphorbia*, and we have not examined specimens from hosts other than *Euphorbia*. A few specimens are labeled as being collected on cactus, but because they originated from South Africa where euphorbia species are common and look like cacti, we suspect that these are misidentifications.

Distribution This scale probably is native to the southern parts of Africa. In the United States it is known only from California. Worldwide it has been reported from Eritrea, Germany, Italy, Netherlands, South Africa, and South-West Africa. It commonly is taken at U.S. ports-of-entry from South Africa on various species of *Euphorbia*.

Biology The life history of this species has not been studied. Schmutterer (1959) indicates that the female produces crawlers rather than eggs (ovoviviparous) and is parthenogenetic. The species can develop large populations on ornamental plants.

Economic Importance The white euphorbia scale is a serious pest of ornamental euphorbias in California (Gill 1997). In some instances it has built to such large populations in greenhouses that it has been necessary to destroy infested plants by burning (McKenzie 1956). It is considered to be a pest by Marotta and Garonna (1992) in Italy. They indicate that the species can kill plants if control measures are not implemented.

Selected References Mamet (1958); Marotta and Garonna (1992); McKenzie (1956); Schmutterer (1959).

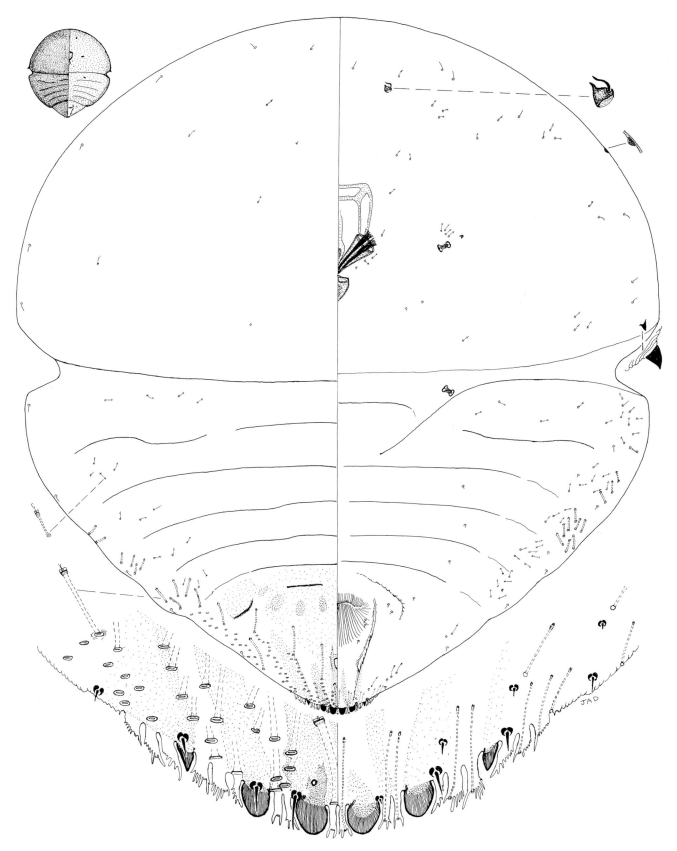

Figure 114. *Selenaspidus albus* (McKenzie), white euphorbia scale, South Africa, intercepted at Hoboken, NJ, on *Euphorbia* sp., II-29-1946.

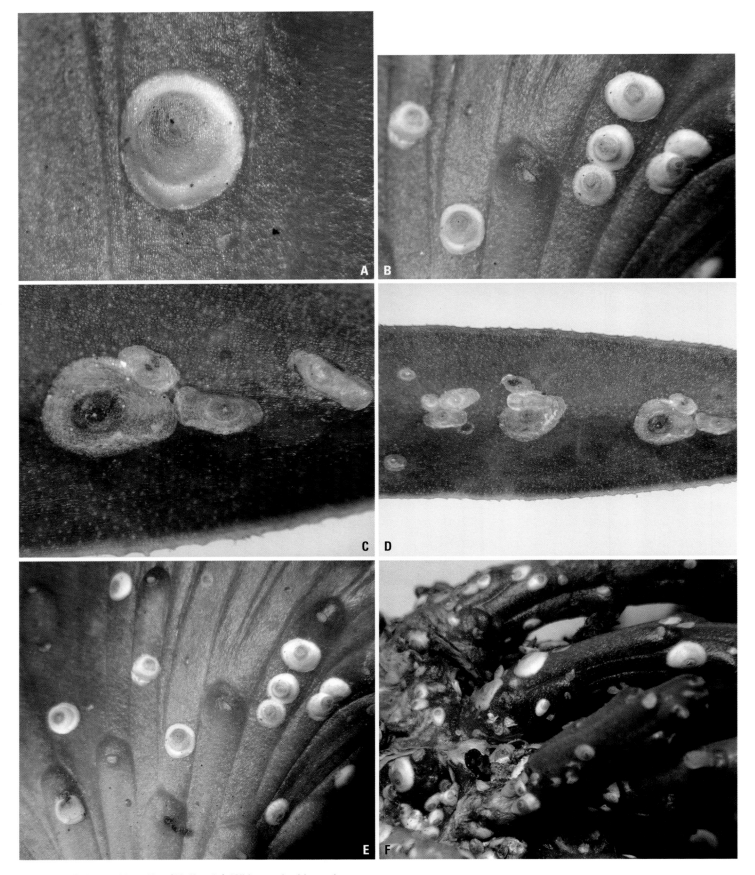

Plate 110. *Selenaspidus albus* (McKenzie), White euphorbia scale

A. Adult female cover on euphorbia (R. J. Gill).
B. Adult female covers on euphorbia (R. J. Gill).
C. Old adult female cover (left) and three male covers (right) on euphorbia (R. J. Gill).

D. Distance view of male and female covers on euphorbia (R. J. Gill).
E. Distance view of adult female covers on euphorbia (R. J. Gill).
F. Heavy infestation on euphorbia (R. J. Gill).

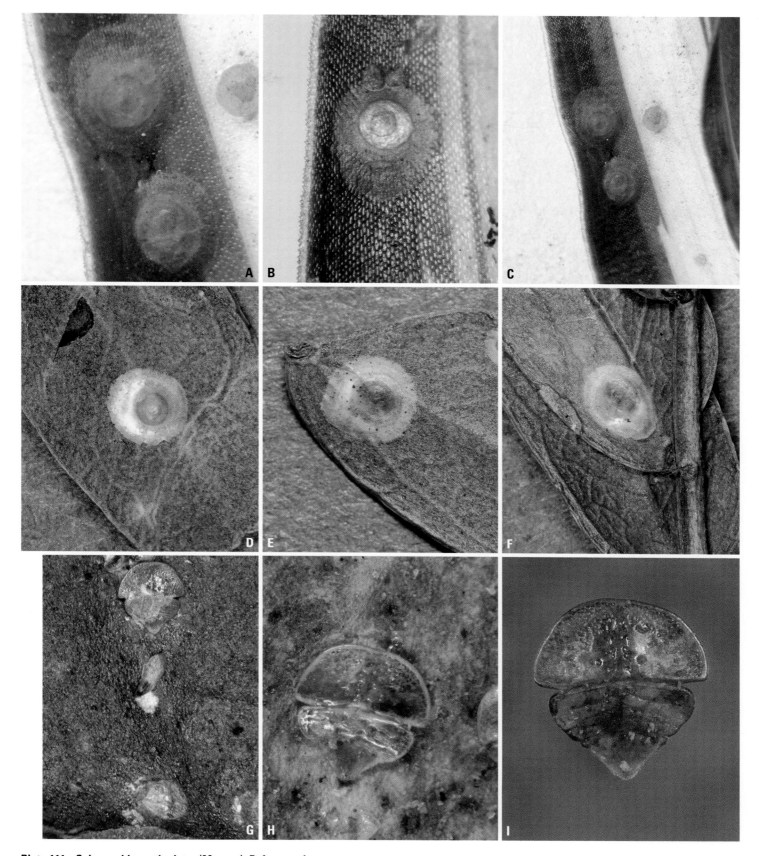

Plate 111. *Selenaspidus articulatus* (Morgan), Rufous scale

A. Translucent covers showing constrictions on body of adult female on spider plant (R. J. Gill).
B. Adult female cover on spider plant (R. J. Gill).
C. Distance view of female covers on spider plant (R. J. Gill).
D. Old adult female cover on tamarind (J. A. Davidson).
E. Old adult female cover on tamarind (J. A. Davidson).
F. Old adult female cover on tamarind (J. A. Davidson).
G. Distance shot of adult female body showing unique body shape on citrus (J. A. Davidson).
H. Close-up of dead adult female on citrus (J. A. Davidson).
I. Body of adult female removed from citrus host (J. A. Davidson).

Selenaspidus articulatus (Morgan)

Suggested Common Name Rufous scale (also called West Indian red scale).

Common Synonyms and Combinations *Aspidiotus articulatus* Morgan, *Aspidiotus (Selenaspidus) articulatus* Morgan, *Aspidiotus (Selenaspidus) articulatus* var. *simplex* D'Emmerez de Charmoy, *Pseudaonidia articulata* (Morgan).

Field Characters (Plate 111) Adult female cover nearly flat, circular, semitransparent, gray or white to light brown, margin lighter; shed skins subcentral, yellow, second shed skin about one-third of diameter of adult cover. Male cover elongate oval, lighter than female cover; shed skin dark yellow, subcentral. Body of adult female yellow; eggs and crawlers probably yellow. Primarily on leaves and fruit, rarely on bark.

Slide-mounted Characters Adult female (Fig. 115) with 3 pairs of definite lobes; paraphyses small and inconspicuous, paraphysis formula 0-1-0 or 1-1-0. Median lobes separated by space 0.5–1.0 (0.7) times width of median lobe, with small paraphysis-like sclerotization on medial margin of each lobe, sclerotization of lobe ending abruptly giving appearance of transverse bars at base of lobe, without yoke, medial margins parallel or diverging slightly, lateral margins parallel, with 1–2 (1) lateral notches, 0–1 (1) medial notch; second lobes simple, about same size as median lobes, with 0–2 (1) lateral notches, without medial notches; third lobes simple, slender, spurlike, without notches. Plates with associated microduct, plates in first space slender, with 2 or 3 tines, plates in second space broad, with many tines, plate anterior of third lobe elongate, somewhat expanded apically, remaining plates variable; plate formula 2-3-4 or 2-3-5, rarely 2-3-6; with 2 plates between median lobes, slightly longer or slightly shorter than length of median lobes. Macroducts of 1 variable size, widest at posterior end of pygidium, 1 macroduct between median lobes 55–70 (63) μ long, extending 0.7–1.0 (0.8) times distance between posterior apex of anal opening and base of median lobes, longest macroduct in first space 54–102 (65) μ long, with 33–63 (52) macroducts on each side of pygidium on segments 5 to 8, ducts in submarginal and marginal areas, some macroduct orifices anterior of anal opening; prepygidial macroducts absent. Pygidial microducts on venter in submarginal areas of segments 5 and 6, with 1–8 (5) ducts; prepygidial ducts of 1 size, in cluster posterior of mouthparts, with 1 or 2 laterad of each spiracle, in marginal and submarginal areas of any or all of mesothorax and segments 1, 2, 3, or 4; pygidial microducts absent from dorsum; prepygidial microducts of 1 size, in submarginal areas from head to segments 2, 3, or 4. Perivulvar pores in 1 group on each side of pygidium, with 4–9 (6) pores on each side of body. Perispiracular pores absent. Anal opening located 4.5–6.7 (5.5) times length of anal opening from base of median lobes, anal opening 14–17 (16) μ long. Dorsal seta laterad of median lobes 1.8–5.0 (3.9) times length of median lobe. Eyes absent or small and represented by small sclerotized area, located anterior and laterad of mouthparts. Antennae each with 1 long seta. Cicatrices inconspicuous or absent, when present on segments 1 and 3. Body pear shaped, with conspicuous constriction between mesothorax and metathorax, pygidial margin slightly concave. Vulva with noticeable V-shaped flap. Sclerotized tubercle at posterolateral apex of mesothorax; tubercle usually acute apically, occasionally bifurcate or trifurcate. Anterior margin of head with small serrations on most specimens.

Affinities The rufous scale is most similar to *Selenaspidus albus*, white euphorbia scale. For a comparison of these species see the affinities section of the latter. Two species of *Selenaspidus* occur in California, but they are not easily confused with rufous scale. Both lack perivulvar pores, have an indefinite constriction between the mesothorax and metathorax, and lack the V-shaped flap associated with the vulva.

Hosts Polyphagous. Borchsenius (1966) records it from 60 genera in 31 families; Dekle (1977) reports it from 30 genera in Florida and considers *Citrus*, *Codiaeum*, and *Tamarindus* to be the most common hosts. Mamet (1958) lists 35 host genera. We have examined specimens from the following: *Acalypha*, *Achras*, *Afzelia*, *Agave*, *Aiphanes*, *Alpinia*, *Amherstia*, *Annona*, *Anthurium*, *Antidesma*, *Artocarpus*, *Arundinaria*, *Astrocaryum*, *Attalea*, *Bauhinia*, *Blighia*, *Brassia*, *Brunfelsia*, *Bursera*, *Calathea*, *Calophyllum*, *Camellia*, *Canna*, *Carludovica*, *Ceiba*, *Chamaedorea*, *Chrysalidocarpus*, *Chrysophyllum*, *Cinnamomum*, *Citrus*, *Cocos*, *Codiaeum*, *Coffea*, *Cola*, *Congea*, *Cordyline*, *Corozo*, *Dendrobium*, *Diospyros*, *Dracaena*, *Duranta*, *Elaeis*, *Eucalyptus*, *Eugenia*, *Euonymus*, *Eupritchardia*, *Euterpe*, *Ficus*, *Funtumia*, *Gardenia*, *Geonoma*, *Gongora*, *Heliconia*, *Huernia*, *Hymenaea*, *Hyphaene*, *Ixora*, *Jasminum*, *Lagerstroemia*, *Litchi*, *Livistona*, *Lockhartia*, *Loranthus*, *Macrozamia*, *Magnolia*, *Malpighia*, *Mangifera*, *Masdevallia*, *Melicoccus*, *Miltonia*, *Murraya*, *Musa*, *Myristica*, *Narcissus*, *Nerium*, *Pandanus*, *Passiflora*, *Persea*, *Phoenix*, *Pimenta*, *Plumeria*, *Poinciana*, *Pothomorphe*, *Pothos*, *Prunus*, *Pyrenoglyphis*, *Quisqualis*, *Robinia*, *Rosa*, *Roystonea*, *Sabal*, *Pritchardia*, *Schinus*, *Spondias*, *Sterculia*, *Strychnos*, *Synechanthus*, *Tabernaemontana*, *Taraktogenos*, *Theobroma*, *Thevetia*, *Thrinax*, *Tillandsia*, *Vitis*, *Xanthoxylum*, and *Zizyphus*.

Distribution This scale probably is native to the tropical-subtropical areas of Africa. At the present time it is known only from Florida in the United States. It was discovered in a nursery in California in 1947 and was eradicated. The rufous scale is commonly intercepted in quarantine from the Caribbean islands and Central and South America. We have examined specimens from Bolivia, Dominican Republic, Ecuador, Ethiopia, Mexico, Nicaragua, Panama, Peru, and Turkey. Nakahara (1982) reports it from the Philippines and Fiji. Mamet (1958) examined many specimens from tropical Africa. A distribution map of this species was published by CAB International (1981b).

Biology In the coastal areas of Peru during the months of January and February development from the egg to the adult requires about 45 days for the female and 30 days for the male. Oviposition begins soon after female maturation and continues for over 30 days. A single female may lay more than 120 eggs. First instars hatch from the eggs within hours of when they are laid. Males apparently are required for reproduction (Beingolea 1969; Herrera 1964). Bartra (1974) indicates that a generation required 48 days at about 27 °C and 67 days at about 17 °C. Suris (1999) demonstrated that the species showed an aggregated distribution in citrus trees in Cuba. Brown (1965) reports this species to have a typical diaspidoid chromosome system.

Economic Importance The rufous scale is considered to be a serious pest by Beardsley and González (1975) and Miller and Davidson (1990). It has been reported by Ebeling (1959) to be economically important on citrus according to the following pest rankings: seventh in Mexico, ninth in Trinidad, second in Ecuador, first in Peru (tied with purple scale), first in Bolivia (along with 4 other scale species). Among other listings of world citrus pests, Talhouk (1975) considers this armored scale to be a citrus pest of major economic importance requiring regular control measures in Bolivia, Peru, Venezuela, and tropical Central America. He also found it to be a pest requiring occasional treatment in Southeast Asia and present but of little importance in Brazil. Since that time it has elevated its status in Brazil to an important citrus pest (Santos and Gravena 1995; Perruso and Cassino 1997). It is considered important in Cuba (Suris 1999). LePelley (1968) considers rufous scale to be an important pest of robusta coffee in Ecuador, causing defoliation

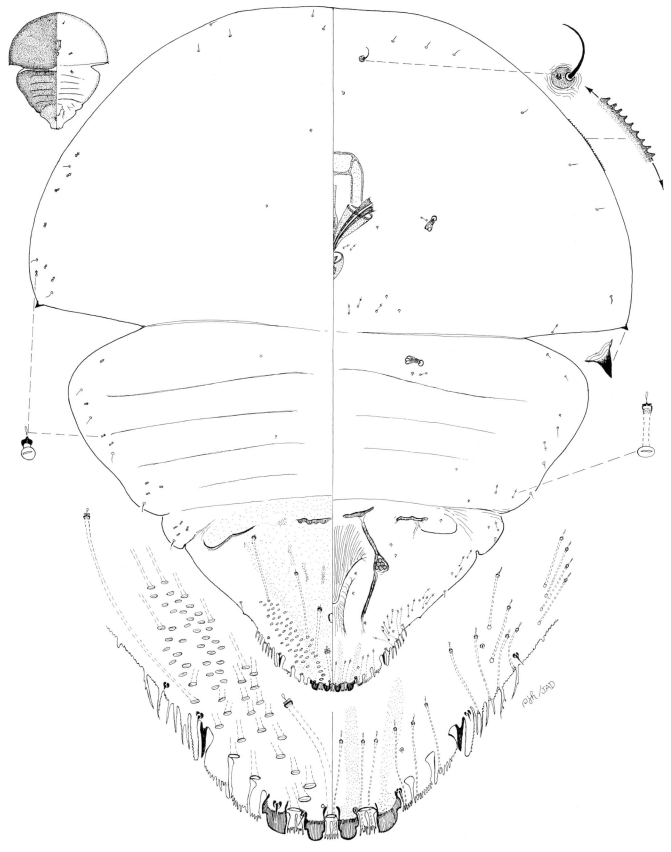

Figure 115. *Selenaspidus articulatus* (Morgan), rufous scale, Roseau, Dominica, on tamarind tree, III-25-1964.

and greatly reducing production. Arabica coffee was not as severely attacked. Dekle (1977) did not consider it to be an economic pest in Florida. *Aphytis roseni* DeBach and Gordh is considered to be an effective biological control agent in Peru (Bartra 1974).

Selected References Beingolea (1969); Mamet (1958).

Situlaspis yuccae (Cockerell)

Suggested Common Name Yucca scale (also called small situlaspis scale, celtis scale).

Common Synonyms and Combinations *Aspidiotus yuccae* Cockerell, *Aspidiotus (Chrysomphalus) yuccae* Cockerell, *Xerophilaspis parkinsoniae* Cockerell, *Hemiberlesia yuccae* (Cockerell), *Hemiberlesia yuccae* var. *neomexicanus* Cockerell and Parrott, *Diaspis celtidis* Cockerell, *Pseudodiaspis parkinsoniae* (Cockerell), *Targionia parkinsoniae* (Cockerell), *Targionia yuccae* (Cockerell), *Targionia yuccae neomexicana* (Cockerell and Parrott), *Aspidiotus celtidis* (Cockerell), *Aspidiotus neomexicanus* (Cockerell and Parrott), *Pseudodiaspis yuccae* (Cockerell), *Neosignoretia yuccae* (Cockerell).

Field Characters (Plate 112) Adult female cover unusually small, circular, or tear-drop shaped, moderately convex, dirty white; shed skins subcentral, reddish brownish. Male cover elongate, smaller than female cover, dirty white; shed skin marginal, tan. Body of adult female light to dark purple; crawlers presumably same color. Primarily on stems and leaves.

Slide-mounted Characters Adult female (Fig. 116) with 2 or 3 pairs of definite lobes, third lobes usually present but small, fourth lobes absent in about half of specimens examined; without paraphyses. Median lobes separated by space 0.1–0.4 (0.2) times width of median lobe, without basal sclerotization, without yoke, medial margins parallel or diverging apically, lateral margins parallel or converging apically, with 0–2 (1) lateral notches, 0–1 (0) medial notch; second lobes bilobed, about one-quarter to one-half size of median lobes, usually with acute or slightly rounded medial apex, with 0–1 (0) lateral and medial notch; third lobes when present simple or bilobed, about one-half size as second lobe, usually apically acute, without notches; fourth lobes usually represented by small swelling. Gland spines usually simple, spine in first space simple, inconspicuous, with associated microduct, spine in second space usually simple, rarely with a lateral projection, with associated microduct; gland-spine formula usually 1-1-0, sometimes 1-2-0, without plates anterior of second space; gland spines between median lobes 0.5–0.8 (0.6) times as long as median lobes. Macroducts of 2 sizes, gradually decreasing in size anteriorly, larger size present in marginal areas of segments 5 to 8, smaller size in marginal and/or submarginal areas of segments 2 or 3 to segments 6 or rarely 7; duct between median lobes 15–20 (17) µ long, extending 0.4–0.6 (0.5) times distance to posterior apex of anal opening, longest duct in first space 12–22 (18) µ long, with 8–17 (11) macroducts on each side of pygidium on segments 5 to 8, ducts in submarginal and marginal areas, total of 15–38 (23) macroducts on each side of body, some macroduct orifices anterior of anal opening; prepygidial macroducts present submarginally on segments 2 or 3 to 4. Pygidial microducts on venter in submarginal and marginal areas of segments 5 to 7 or 8, with 3–8 (6) ducts; prepygidial ducts of 1 size, in marginal and submarginal areas of segments 3 and 4, in submedial areas of any or all of metathorax to segment 4, absent elsewhere; on dorsum pygidial microducts absent, marginal ducts absent on 5 of 10 specimens examined, highly variable when present, some specimens with only 1 or 2 ducts on each segment, others with 10 or more on each segment, present from head or prothorax to segment 2 or 3. Perivulvar pores very small and abortive but always present usually with 1 or 2 pores in any or all of medial, anterolateral or posterolateral groups, 2–4 (3) pores on each side of body. Perispiracular pores absent. Anal opening located 2.6–4.9 (3.6) times length of anal opening from base of median lobes, anal opening 8–12 (10) µ long. Dorsal seta laterad of median lobes 0.9–1.6 (1.3) times length of median lobe. Eyes present on 5 of 10 specimens examined, when present, represented by small dome or sclerotized spot located near body margin at level of antenna. Antennae each with 1 robust seta and 1 or 2 short setae. Cicatrices usually absent, rarely on segment 2 and 4. Body pear shaped. Older specimens with head and thorax partially sclerotized. With conspicuous sclerotized spot on dorsum in triangular intersection of segments 4, 5, and 6.

Affinities The yucca scale is very different from all other economic scales of the United States by having reduced numbers of perivulvar pores that often are very small and inconspicuous. It is most similar to *Situlaspis daleae* Ferris and *S. atriplicis* (Ferris) but differs from both by the presence of perivulvar pores; it also differs from *S. daleae* by having many fewer macroducts and from *S. atriplicis* by having 3 or 4 pairs of lobes.

Hosts Polyphagous. Borchsenius (1966) records it from 14 genera in 9 families. We have examined specimens from the following: *Acacia, Agave, Cassia, Cercidium, Cercis, Cereus, Coursetia, Dasylirion, Forsythia, Fraxinus, Gleditsia, Hedera, Nolina, Olea, Parkinsonia, Pedilanthus, Phoenix, Prosopis,* and *Yucca.*

Distribution This scale probably is native to Mexico and the southwestern United States. It commonly is intercepted in quarantine from Mexico. We have examined U.S. specimens from AZ, CA, NM, OK, TX, and Mexico. Dekle (1977) reports it from Florida.

Biology Brown (1960) reports this species to have a typical diaspidoid chromosome system with a 2n chromosome number of 10.

Economic Importance The yucca scale is considered to be of occasional economic importance by Miller and Davidson (1990). Gill (1997) indicates that it is a troublesome pest of ornamentals in Arizona and is abundant on a number of yard plants, especially ivy and yucca. He indicates that the pest is transported on nursery stock into the San Joaquin and Sacramento valleys.

Selected Reference Gill (1997).

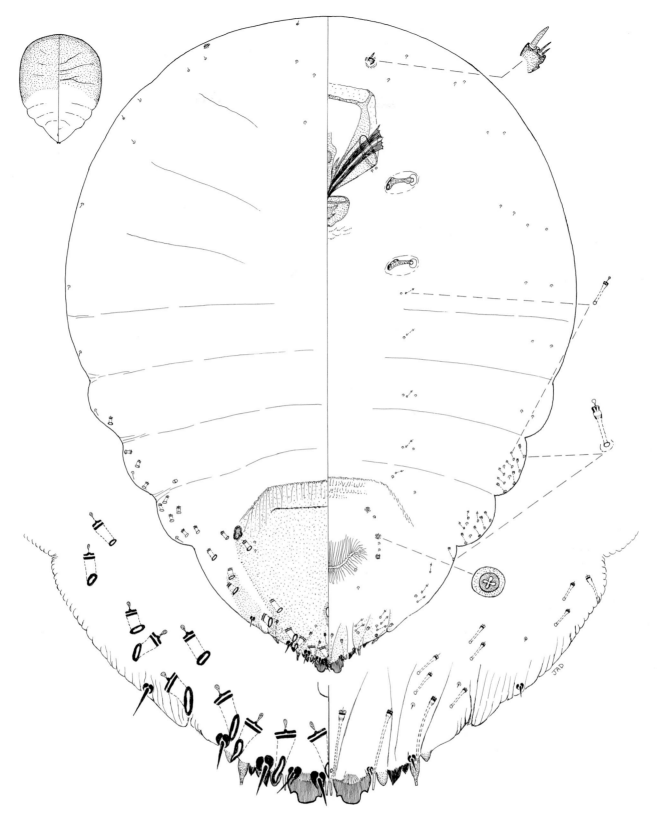

Figure 116. *Situlaspis yuccae* (Cockerell), yucca scale, Palo Verde, Mexico, intercepted at Nogales, AZ, on *Cercidium floridum*, IV-10-1952.

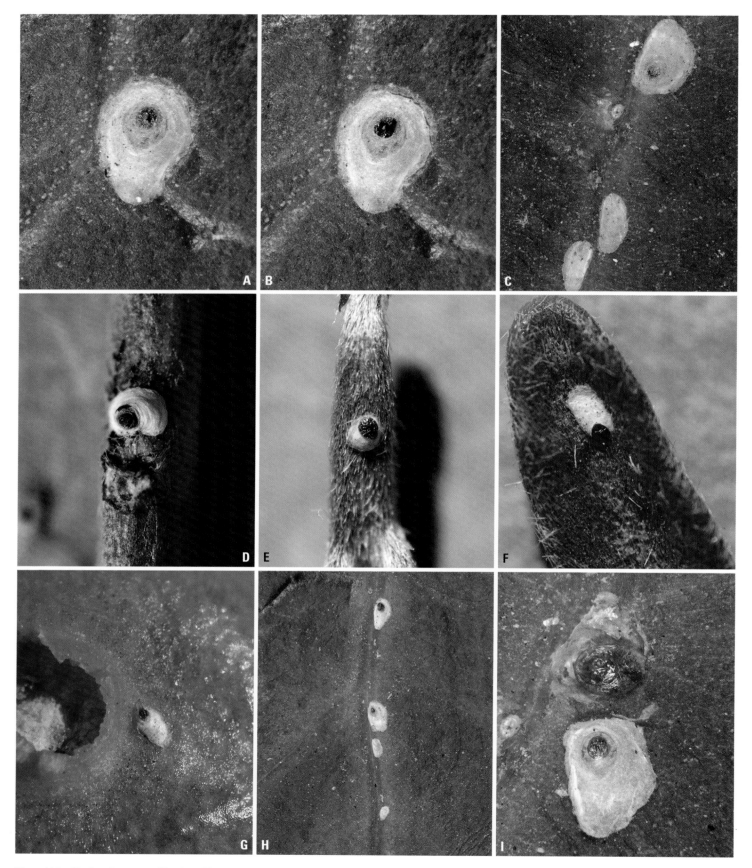

Plate 112. *Situlaspis yuccae* (Cockerell), Yucca scale

A. Unrubbed adult female cover on ivy (J. A. Davidson).
B. Rubbed adult female cover showing black shed skin on ivy (J. A. Davidson).
C. Female (top) and male covers on ivy (J. A. Davidson).
D. Female cover on twig of *Prunus fremontiae* (J. A. Davidson).
E. Second instar cover cover on leaf of *P. fremontiae* (J. A. Davidson).

F. Second instar cover on leaf of *P. fremontiae* (J. A. Davidson).
G. Female cover on fruit of *P. fremontiae* (J. A. Davidson).
H. Distance view of infestation on ivy (J. A. Davidson).
I. Body of mature adult female on ivy (J. A. Davidson).

390

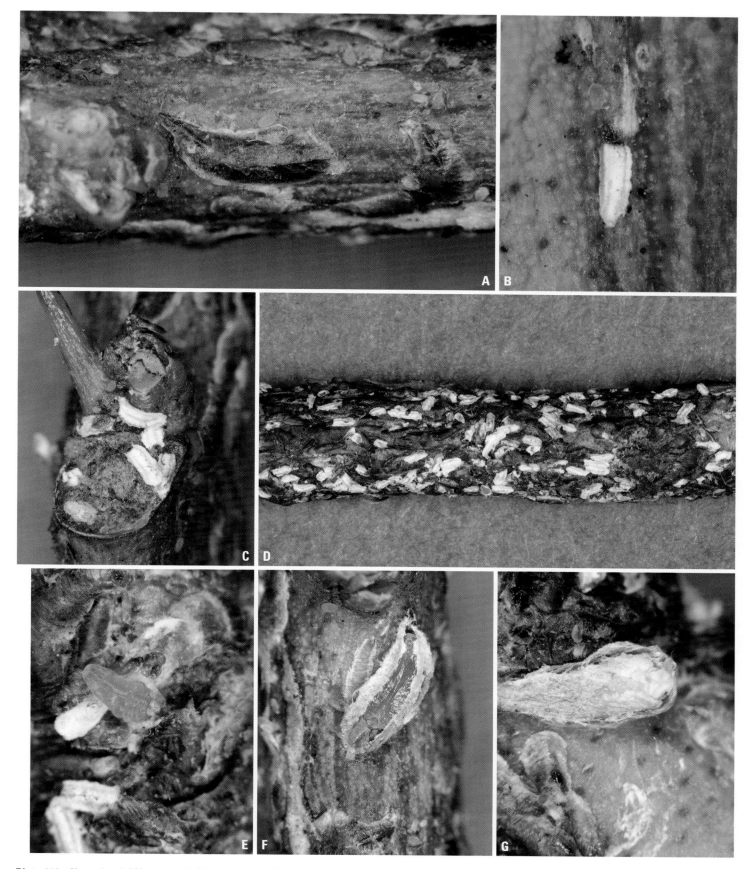

Plate 113. *Unaspis citri* (Comstock), **Citrus snow scale**

A. Adult female covers with crawlers on citrus (J. A. Davidson).
B. Male cover and crawlers on citrus (J. A. Davidson).
C. Male covers on citrus (J. A. Davidson).
D. Distance view of heavy infestation on citrus (J. A. Davidson).
E. Young adult female body on citrus (J. A. Davidson).

F. Mature adult female body partially enclosed in ventral cover on citrus (J. A. Davidson).
G. White eggs barely visible inside ventral cover on citrus (J. A. Davidson).

Unaspis citri (Comstock)

ESA Approved Common Name Citrus snow scale (also called orange snow scale, snow scale, orange chionaspis, white louse, and white louse scale).

Common Synonyms and Combinations *Chionaspis citri* Comstock, *Howardia citri* (Comstock), *Prontaspis citri* (Comstock), *Dinaspis veitchi* Green and Laing, *Trichomytilus veitchi* (Green and Laing), *Dinaspis annae* Malenotti, *Diaspis annae* (Malenotti).

Field Characters (Plate 113) Adult female cover elongate, oyster-shell shaped to rectilinear, brown to blackish brown with lighter margins, moderately convex, sometimes with median longitudinal ridge; shed skins marginal, brownish yellow. Male cover elongate, shorter than female cover, lateral margins usually parallel, white, felted, tricarinate; shed skin marginal, brownish yellow. Body of adult female orange or yellow; eggs and crawlers yellow. Primarily on trunk and branches, occasionally on leaves and fruit.

Slide-mounted Characters Adult female (Fig. 117) with 4 pairs of definite lobes; paraphyses thin and inconspicuous, of 2 general types, either with small paraphysis attached to medial margin of medial lobule of lobe 2, or with 2 long converging paraphyses attached to medial and lateral margins of medial lobules of lobes 2 and 3, paraphysis formula 1-0-0, 1-2-1, or 1-1-0, excluding paraphysis attached to lateral margin of medial lobule of lobes 2 and 3. Median lobes separated by space 0.1–0.3 (0.2) times width of median lobe, with diagonal paraphysis-like sclerotization on each margin of each lobe, without basal sclerotization or yoke, medial margins diverging apically, lateral margins converging or parallel, with 1–6 (3) lateral notches, 4–8 (6) medial notch; second lobes bilobed, slightly smaller than median lobes, medial lobule largest with 0–4 (2) lateral notches, 0–1 (0) medial notches, lateral lobule with 0–2 (1) lateral notches, 0–1 (0) medial notches; third lobes bilobed, smaller than second lobes, medial lobule largest with 0–5 (2) lateral notches, 0–2 (0) medial notches, lateral lobule with 0–5 (2) lateral notches, 0–1 (0) medial notches; fourth lobes bilobed, represented by low series of notches, medial lobule with 3–9 (6) notches, lateral lobule with 4–8 (5) notches. Gland-spine formula usually 1-2-2, sometimes 1-1-2, 1-2-1, or 1-1-1, second gland spine in second and third spaces frequently difficult to discern, with 27–39 (33) gland spines near each body margin from mesothorax to segment 4, occasionally with 1 gland spine on prothorax; without gland spines between median lobes. Macroducts of 2 or 3 variable sizes, largest in marginal areas of pygidium, medium size in marginal areas from metathorax or segment 1 to segment 4, in mediolateral and submedial areas from segment 2 or 3 to segment 7, small size in marginal areas of mesothorax or metathorax to segment 1 or 2, without duct between median lobes, macroduct in first space 18–24 (21) µ long, with 18–35 (25) macroducts on each side of body on segments 5 to 7, total of 66–154 (98) macroducts on each side of body, some macroduct orifices anterior of anal opening. Pygidial microducts on venter in submarginal areas of segments 5 and sometimes 6, with 0–4 (3) ducts; prepygidial ducts of 1 size, sometimes in medial areas of segment 2 and/or 3, in small group near each spiracle, in submedial areas of metathorax to segment 3, in marginal and submarginal areas of any or all of segments between head and segment 4; pygidial microducts absent from dorsum; prepygidial microducts, when present, near submedial macroducts on segments 2 and/or 3. Perivulvar pores present in 5 of 10 specimens examined, when present, with 1–3 (2) pores on each side of body. Perispiracular pores with 3 loculi, anterior spiracles with 9–16 (11) pores, posterior spiracles with 1–7 (4) pores. Anal opening located 8–12 (10) times length of anal opening from base of median lobes, anal opening 9–15 (12) µ long. Dorsal seta laterad of median lobes 0.5–0.9 (0.7) times length of median lobe. Eyes represented by small sclerotized area sometimes on small tubercle, located on body margin at level of anterior margin of mouthparts. Antennae each with 1, rarely 2, long seta. Cicatrices usually present on segment 1, sometimes visible on segment 3. Body elongate. Median lobes of 2 general types, either not protruding beyond second lobes or extending farther than second lobes. Older adult females with most of prosoma sclerotized.

There appear to be 2 intergrading forms of this species. One form usually has perivulvar pores, median lobes that do not protrude beyond the second lobes, long slender paraphyses attached to medial and lateral margins of medial lobules of lobes 2 and 3, and marginal row of microducts from head to mesothorax. The other form usually has no perivulvar pores, median lobes that protrude beyond the second lobes, short paraphysis attached to medial margin of medial lobule of lobe 2 only, and microducts in marginal areas of head and thorax not forming distinct row.

Affinities The citrus snow scale is similar to *Unaspis euonymi*, the euonymus scale, but differs by having less than 4 perivulvar pores on each side of body, 1 gland spine in the first space, no microtubular ducts posterior of labium, and prosoma sclerotized on old adult females. The euonymus scale has more than 4 perivulvar pores on each side of body, usually has 2 gland spines in the first space, cluster of microtubular ducts posterior of labium, and prosoma not sclerotized on old adult females.

Hosts Borchsenius (1966) records this species from *Citrus*, *Osmanthus*, and *Severinia*. Dekle (1977) reports it in Florida from *Ananas*, *Citrus*, *Fortunella*, *Hibiscus*, and *Poncirus*; he considers *Citrus* to be the most common host. We have examined specimens from the following: *Acacia*, *Citrus*, *Mangifera*, and *Pittosporum*.

Distribution This scale probably is native to the Orient. We have examined U.S. specimens from FL and LA. Nakahara (1982) reports citrus snow scale from Mexico, Central and South America, Europe, Africa, Asia, Australia, and Oceania. It generally appears to be distributed throughout the humid citrus-growing areas of the world. This species frequently is taken in quarantine from tropical Central America and the Caribbean islands. A distribution map of this species was published by CAB International (1962c).

Biology Citrus snow scale has several unsynchronized generations per year in Florida (Bullock and Brooks 1975). Brooks (1977) reports 4 widely overlapping generations per year as common in typical citrus-growing areas of the world. Borchsenius (1950) records 2 to 3 generations in Japan with eggs as the overwintering stage. Tikhonov (1966) and Kuhmina (1970) state that there are 2 or 3 generations in the citrus-growing areas of the former Soviet Union. In Australia, Smith et al. (1997) indicate that there are 5 to 6 generations in Queensland and 3 to 4 in New South Wales. They also state that each female produces up to 150 eggs over 2 or 3 months, and a life cycle takes about 8 weeks during warm months.

Economic Importance This species is considered a serious pest by Beardsley and González (1975) and Miller and Davidson (1990). Talhouk (1975) suggests that it is a citrus pest of major economic importance with control measures usually required more than once each season in Australia, Argentina, Venezuela, Colombia, Florida, tropical Central America, Mexico, and China. He considers it to be an important pest requiring occasional control measures in Peru, Chile, and Brazil. Ebeling (1959) suggests that this pest is not a problem in temperate areas

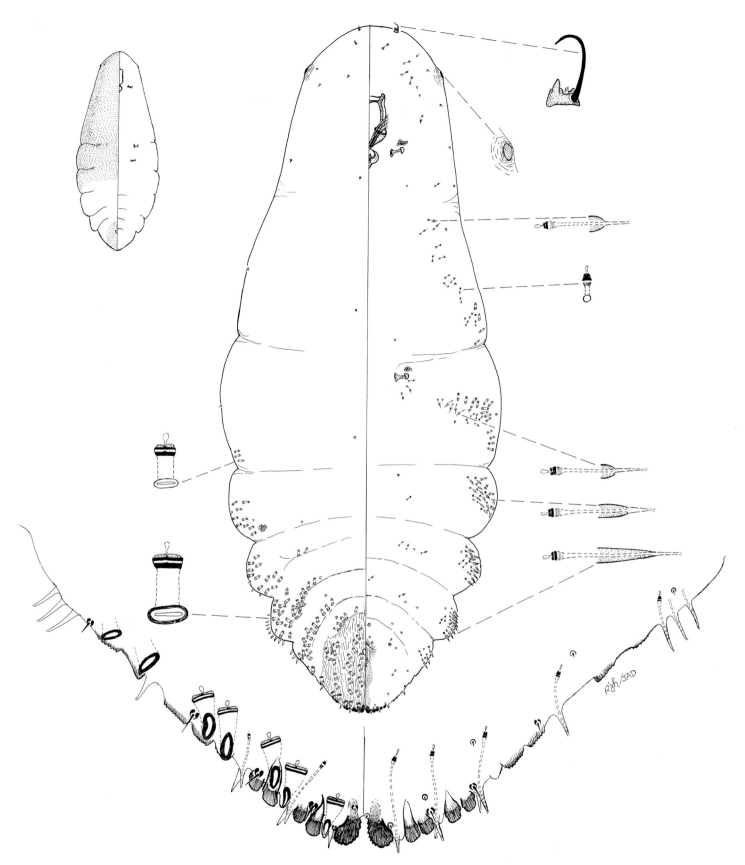

Figure 117. *Unaspis citri* (Comstock), citrus snow scale, Machala, Ecuador, on orange, VIII-21-1953.

such as California. In Florida the spread of citrus snow scale began in 1963 after a major citrus kill, and it apparently was dispersed throughout the state on infested nursery stock (Simanton 1976). Dekle (1977) and Bullock and Brooks (1975) consider this species to be a serious pest in Florida. The latter authors believe that the pest was exceeded in importance in Florida only by citrus rust mite and greasy spot disease. They support their point with the following information: Prior to 1960 less than 1% of Florida's citrus groves contained this scale, but in 1972 50% of the groves were infested and 25% of these required extra chemical treatments specifically for this species. Infestations generally were confined to the trunk, scaffold limbs, and smaller branches, with only occasional infestations on fruit and leaves. Heavy infestations cause bark splitting, loss of large limbs, and sometimes tree death. Before the introduction of *Aphytis lingnanensis*, two thorough scalacide applications were required each year for adequate control. Studies conducted by Muma (1970) in Florida showed that orange and grapefruit trees were severely attacked by the citrus snow scale, while tangerine trees were nearly immune. With the introduc-

tion of the Hong Kong variety of *A. lingnanensis* many thought that this parasite had achieved major success in controlling this devastating pest (Mead 1975). However, it is now evident that reduction of citrus snow scale populations to subeconomic levels has not been achieved; the situation in Florida is difficult to evaluate because of the introduction and natural occurrence of so many different strains of *A. lingnanensis* that are morphologically identical (Browning 1994). The lady beetle, *Chilocorus circumdatus* Gyllenhal, is reported to be a reasonably successful biological control agent in Australia (Smith et al. 1997). Balachowsky (1959) reported that orange, mandarin, and lime trees were severely attacked in Colombia and sometimes had premature leaf drop. Hearn (1979) found that 'Sunburst,' a moderately vigorous hybrid of *Citrus reticulata* and *C. paradisi*, was tolerant of this pest. Hangartner et al. (1976) demonstrated that a growth regulator reduced population sizes of this scale by interfering with larval molts, male metamorphosis, and female fertility.

Selected References Rao (1949); Smith et al. (1997).

Plate 114. *Unaspis euonymi* (Comstock), Euonymus scale

A. Adult female cover (center) with developing white covers of males on euonymus (J. A. Davidson).

B. Adult female and male covers on euonymus (J. A. Davidson).

C. Close-up of male covers with males inside; note black eyes visible through front of cover (J. A. Davidson).

D. Infestation on euonymus stem (J. A. Davidson).

E. Distance view of mostly male covers, which predominate on undersides of euonymus leaves (J. A. Davidson).

F. Yellow feeding damage visible on top of euonymus leaves (J. A. Davidson).

G. Body of mature adult female next to turned cover on euonymus (J. A. Davidson).

H. Eggs on euonymus (J. A. Davidson).

I. Crawlers with male and female covers on euonymus stem (R. J. Gill).

J. Dieback after 2 or 3 years of infestation of *Euonymus japonica* (J. A. Davidson).

395

Unaspis euonymi (Comstock)

ESA Approved Common Name Euonymus scale.

Common Synonyms and Combinations *Chionaspis euonymi* Comstock, *Chionaspis evonymi* Comstock, *Chionaspis nemausensis* Signoret, *Unaspis nakayamai* Takahashi and Kanda.

Field Characters (Plate 114) Adult female cover elongate, broadly oyster-shell shaped, dark brown; shed skins marginal, brownish yellow. Male cover elongate, shorter than female cover, lateral margins usually parallel, white, felted, tricarinate; shed skin marginal, brownish yellow. Body of adult female orange or yellow; eggs and crawlers yellow. On all aboveground portions of host. Males predominate on leaves and females most abundant on stems and branches.

Slide-mounted Characters Adult female (Fig. 118) with 4 pairs of definite lobes, fifth lobes sometimes indicated by series of points; paraphyses small and inconspicuous, usually with small paraphysis attached to medial margin of medial lobule of lobe 2, sometimes with paraphyses absent, paraphysis formula 1-0-0 or 1-1-0, excluding paraphysis attached to lateral margin of medial lobule of lobe 2. Median lobes separated by space 0.1–0.7 (0.2) times width of median lobe, with small, diagonal paraphysis-like sclerotization on each margin of each lobe, without basal sclerotization or yoke, medial margins diverging apically, lateral margins converging or parallel, with 1–5 (3) lateral notches, 0–4 (3) medial notch; second lobes bilobed, slightly smaller than median lobes, medial lobule largest with 1–4 (2) lateral notches, 0–2 (1) medial notches, lateral lobule with 0–2 (1) lateral notches, 0–1 (0) medial notches; third lobes bilobed, equal to or smaller than second lobes, medial lobule largest with 0–3 (2) lateral notches, 0–2 (1) medial notches, lateral lobule with 0–2 (1) lateral notches, 0–2 (1) medial notches; fourth lobes bilobed, represented by low series of notches, medial lobule with 2–6 (4) notches, lateral lobule with 0–5 (2) notches; fifth lobe usually absent. Gland-spine formula usually 2-2-2, sometimes 1-2-2, 1-3-2, or 1-2-3, with 20–40 (30) gland spines near each body margin from prothorax, mesothorax, or metathorax to segment 4; without gland spines between median lobes. Macroducts of 2 or 3 variable sizes, largest in marginal areas of pygidium, medium size in marginal areas from metathorax or segment 1 to segment 4, in mediolateral and submedial areas from segment 2 or 3 to segment 7, small size in marginal areas of prothorax, mesothorax, or metathorax to segment 1 or 2, without duct between median lobes, macroduct in first space 17–23 (20) μ long, with 25–38 (30) macroducts on each side of body on segments 5 to 7, total of 95–148 (116) macroducts on each side of body, some macroduct orifices anterior of anal opening. Pygidial microducts on venter in submarginal areas of segments 5 and 6, with 4–9 (6) ducts; prepygidial ducts of 1 size, sometimes in medial areas of any or all of segments 2, 3, or 4, in small group near each spiracle and posterior of mouthparts, in submedial areas of metathorax to segment 3 or 4, in marginal and submarginal areas of any or all of segments between prothorax and segment 4; pygidial microducts absent from dorsum; prepygidial microducts, when present, near submedial macroducts on any or all of segments 2, 3, or 4. Perivulvar pores with 8–14 (12) pores on each side of body. Perispiracular pores with 3 loculi, anterior spiracles with 10–29 (18) pores, posterior spiracles with 3–8 (5) pores. Anal opening located 10–14 (11) times length of anal opening from base of median lobes, anal opening 9–15 (12) μ long. Dorsal seta laterad of median lobes 0.6–1.5 (0.9) times length of median lobe. Eyes absent or represented by small sclerotized area sometimes on small tubercle, located on body margin at level of anterior margin of mouthparts. Antennae each with 1, rarely 2 or 3, long seta. Cicatrices usually present on segment 1, sometimes visible on prothorax and segment 3. Body elongate. Median lobes generally protruding beyond second lobes. Older adult females without prosoma sclerotized.

Affinities The euonymus scale is similar to *Unaspis citri*, the citrus snow scale. For a comparison of these species see the affinities section of the latter.

Hosts Borchsenius (1966) records this species from 14 plant genera in 7 families; Dekle (1977) reports it in Florida from *Buxus, Camellia, Eugenia,* and *Euonymus.* We have examined specimens from the following: *Acer, Althaea, Buxus, Camellia, Celastrus, Citrus, Daphne, Euonymus, Ficus, Fraxinus, Ilex, Ligustrum, Magnolia, Pachysandra, Philadelphus, Primula, Symphoricarpos,* and *Tripterygium.* The scale most commonly is collected on *Euonymus;* we have examined specimens from 11 species of this genus.

Distribution This scale probably is native to the Orient. We have examined U.S. material from AL, AZ, CT, DC, DE, FL, GA, IL, IN, KY, LA, MA, MD, MO, MS, NC, NJ, NM, NY, OH, OK, PA, RI, SC, TN, TX, VA. We have examined specimens from outside of the United States from Argentina, Asia, Canada, Europe, Honduras, and Mexico. Nakahara (1982) reports it also from the former Soviet Union and North Africa. The euonymus scale occasionally has been taken in quarantine from Japan, Italy, and Spain. A distribution map of this species was published by CAB International (1970b).

Biology An extensive treatment of the life history of this species was given by Gill et al. (1982). The following summarizes that paper. In the northern areas of the United States there are 2 generations per year and in the South there are 3. There are 1 or 2 reports that are inconsistent with these generalities. The overwintering stage is the adult female; immatures apparently are unable to survive the winter. In Maryland egg laying began in early May and ended in late May or early June. Crawlers were present in early May to early July. Adult males occurred from mid- to late June, and adult females of the first generation first appeared in mid-June. Eggs of the second generation began in late June and crawlers appeared in mid-July. Adult males of this generation were observed in early to mid-September and adult females began to occur in late August. Savopoulou-Soultani (1997) conducted a detailed analysis of the life history of this species under controlled conditions in the laboratory. Mussey and Potter (1997) pointed out that plant phenology was a better predictor of the developmental stages of this and several other species of scale than were calendar dates.

Economic Importance This scale has been reported as a serious pest of *Euonymus* in most areas of the world where it occurs (e.g., U. S.—Gill 1997; Cockfield and Potter 1990; France—Chauvel 1999; Germany—Schmutterer 1998; Hungary—Asef 1999; Poland—Labanowski and Soika 1998; Yugoslavia—Kozarzhevskaya and Vlainic 1982). Van Driesche et al. (1998) calculated annual economic losses by this scale on *Euonymus fortunei* in Massachusetts at approximately $355,000 and for all of New England about $711,000. If these costs were extended to the entire United States the annual cost would be multiple millions of dollars, and these figures do not consider the cost of control strategies. Gill et al. (1982) discussed the economic importance of euonymus scale in detail and presented a lengthy literature search. They noted that insecticides were the major control method but suggested that plant resistance and biological control were alternatives worth investigation. They have been proved correct, since in recent years several biological control agents have been introduced into the United States with some success (Drea and Carlson 1987; Alvarez and Van Driesche 1998; Hendrickson et al. 1991; Jefferson et al. 1995; Van Dri-

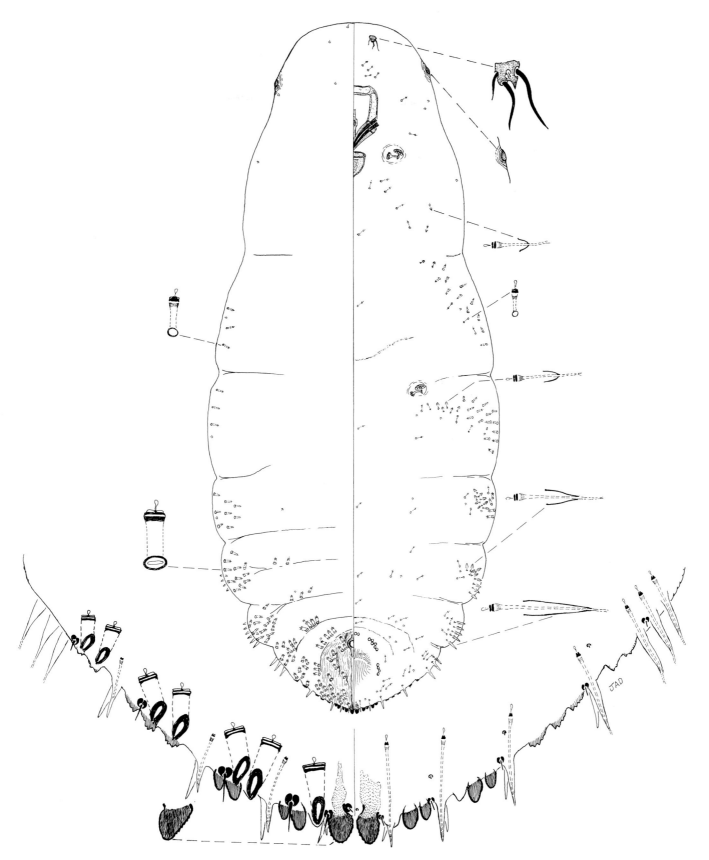

Figure 118. *Unaspis euonymi* (Comstock), euonymus scale, Washington, DC, on ?, IX-20-1944.

esche et al. 1998). Host-plant resistance also has been examined, and it is clear that certain species and varieties are more susceptible than others (Brewer and Oliver 1987; Jefferson and Schultz 1995; Sadof and Raupp 1991). In Maryland this scale is rarely found on hosts other than *Euonymus*. Although dieback has been observed on several species of *Euonymus* including decumbent species, only *E. japonica* cultivars are consistently killed by the pest. In a study area used by Gill et al. (1982) for their research, specimens of *E. japonica* and *E. kiautschovica* (= *E. sieboldiana*) were growing together in a hedge and were observed for more than 2 years. Several of the *E. japonica* plants were killed by the scale during the study and the others were heavily infested and showed obvious symptoms of serious scale damage. The *E. kiautschovica* plants, however, showed no signs of damage and had only light infestations. *Euonymus fortunei* also is reported to sustain severe damage. Warner (1949) found similar evidence of resistance in the Arnold Arboretum. Among heavily infested plants, he found specimens of *E. alata* cv. Compacta, *E. sachalinensis*, and *E. sanguinea* that were uninfested. Beardsley and González (1975) consider this to be one of 43 major armored scale pests. Miller and Davidson (1990) treat this species as a serious world pest.

Selected References Gill et al. (1982); Rao (1949).

References Cited

Abdel-Fattah, M. I., A. M. El-Minshawy, and E. T. Darwish. 1978. The seasonal abundance of two scale insects *Lepidosaphes beckii* (New.) and *Aonidiella aurantii* Mask. infesting citrus trees in Egypt. Proceedings of the fourth Conference on Pest Control, Part I. Cairo, 1978 June 7–12. 74–84.

Abdelrahman, I. 1974. Growth, development and innate capacity for increase in *Aphytis chrysomphali* Mercet and *A. melinus* DeBach, parasites of California red scale, *Aonidiella aurantii* (Mask.), in relation to temperature. Australian Journal of Zoology 22:213–230.

Ackerman, A. J. 1923. Preliminary report on control of San Jose scale with lubricating-oil emulsion. United States Department of Agriculture, Circular No. 263. 18 pp.

Ahmad, R. and M. A. Ghani. 1971. Laboratory studies on the biology of *Lepidosaphes conchiformis* (Gmel.) (Hem., Diaspididae) and of its parasite *Aphytis maculicornis* (Masi) (Hym.: Aphelinidae). Bulletin of Entomological Research 61:69–74.

Alam, M. Z. and A. Sattar. 1965. On the biology of citrus yellow scale *Aonidiella citrina* Coquillett in East Pakistan. Review of Research, Division of Entomology, Agricultural Information Service. 167–171.

Aleksidze, G. 1995. Armored scale insects (Diaspididae), pests of fruit orchards and their control in Republic of Georgia. Israel Journal of Entomology 29:187–190.

Alexandrakis, V. 1983. Biological data on *Aonidiella aurantii* Mask. on citrus in Crete. Fruits 38:831–838.

Alexandrakis, V. and P. Neuenschwander. 1980. Le role d'*Aphytis chilensis*, parasite d'*Aspidiotus nerii* sur olivier en Crete. Entomaphaga 15:61–71.

Alexandrakis, V., T. Neuenschwander, and S. Michelaskis. 1977. The effect of *Aspidiotus nerii* Bouché on the yield of olive trees. Fruits 32:412–417.

Alfieri, A. 1929. Les principaux insectes nuisibles infestant le jardin de Nouzha. Bulletin de la Société Royale Entomologique d'Egypte 13:7–9.

Allen, W. H. and H. B. Weiss. 1953. Insects of importance in New Jersey Nurseries. New Jersey Circular No. 390. 175 pp.

Almeida, D. M. de. 1972. Pests of mango. Gazeta Agricultura 24:2–5.

Alstad, D. 1998. Population structure and the conundrum of local adaptation. Pages 3–21 in S. Moppet and S. Y. Strauss, eds., Genetic structure and local adaptation in natural insect populations: Effects of ecology, life history, and behavior. Chapman & Hall, New York.

Alstad, D. N. and K. W. Corbin. 1990. Scale insect allozyme differentiation within and between host trees. Evolutionary Ecology 4:43–56.

Alstad, D. N. and G. R. Edmunds. 1983. Selection, outbreeding depression, and the sex ratio of scale insects. Science 220:93–95.

Alstad, D. N. and G. F. Edmunds. 1987. Black pineleaf scale population density in relation to interdemic mating. Annals of the Entomological Society of America. 80:652–654.

Alstad, D. N., G. F. Edmunds, and S. C. Johnson. 1980. Host adaptation, sex ratio, and flight activity in male black pineleaf scale. Annals of the Entomological Society of America. 73:665–667.

Alvarez, J. M. and R. Van Driesche. 1998. Biology of *Cybocephalus* sp. nr. *nipponicus* (Coleoptera: Cybocephalidae), a natural enemy of euonymus scale (Homoptera: Diaspididae). Environmental Entomology 27:130–136.

Andres, F. 1979. Lucha integrada en el olivar. Agricultura 48:711–713, 716–717.

Anonymous. 1977. Scale insects on fruit trees. Ministry of Agriculture, Fisheries and Food, Advisory Leaflet No. 88. 7 pp.

Anonymous. 1981. San Jose scale: *Quadraspidiotus perniciosus* (Comstock). Agriculture Canada, Information Service, Insect Identification Sheet No. 38. 2 pp.

Anthon, E. W. 1960. Insecticidal control of San Jose scale on stone fruits. Journal of Economic Entomology 53:1085–1087.

Antonelli, A., E. Elsner, and C. Shanks. 1992. Arthropod management. Pages 55–75 in M. P. Pitts, J. F. Hancock, and B. Strik, eds., Highbush Blueberry Production Guide. Northeast Region Agricultural Engineering Service Bulletin 55.

Applebaum, S. W. and D. Rosen. 1964. Ecological studies on the olive scale, *Parlatoria oleae*, in Israel. Journal of Economic Entomology 57:847–850.

Arancibia O., C., L. Sazo R., and R. Charlin C. 1990. Observaciones de la biologia de la escama del acacio *Diaspidiotus ancylus* (Putnam) en acacia blanca (*Robinia pseudoacacia*). Simiente 60:106–108.

Argyriou, L. C. 1976. Some data on biology, ecology, and distribution of *Aspidiotus nerii* Bouché in Greece. Annals of the Institute of Phytopathology. 11:209–218.

Argyriou, L. C. 1977. Recherches sur un programme de lutte biologique et intégrée contre les ravageurs des agrumes en Grèce. Fruits 32:630–634.

Argyriou, L. C. 1990. Olive. Pages 579–583 in D. Rosen, ed., Armored scale insects. Vol. 4B. Elsevier, Amsterdam, The Netherlands.

Argyriou, L. C. and A. L. Kourmadas. 1981. Contribution to the timing for the control of Diaspididae scales of olive trees. Annals of the Institute of Phytopathology. 13:65–72.

Arnett, R. H. 1985. American insects: A handbook of the insects of America north of Mexico. Van Nostrand Reinhold Co., New York. 850 pp.

Asef, F. M. 1999. Kemiai vedekezesi kiserletek a kecskerago-es boroka-pajzstetvek (*Unaspis euonymi*, *Carulaspis juniperi*). [Chemical control trials against the euonymus and juniper scales (*Unaspis euonymi* and *Carulaspis juniperi*, Homoptera: Coccoidea).] (In Hungarian.) Novenyvedelem 35:567–569.

Asplanato, G. and F. Garcia Mari. 1998. Distribución del piojo rojo de California *Aonidiella aurantii* (Maskell) (Homoptera: Diaspididae) en árboles de naranjo. Boletín de Sanidad Vegetal, Plagas 24:3, 637–646.

Avidov, Z. and I. Harpaz. 1969. Plant pests of Israel. Israel University Press, Jerusalem. 549 pp.

Azevedo Marques, L. A. De. 1923. Insectos nocivos as principaes culturas do Estado da Bahia. Anno Correio Agricultura 1:151–155.

Azim, A. 1961. Mass-production of *Chrysomphalus bifasciculatus* Ferris and its hymenopterous parasites. Mushi 35:97–109.

Babaian, G. A. 1986. Scale-insects of stone fruit crops and control measures against them. Proceedings of the Fifth International Symposium of Scale Insect Studies. Portici, Italy, 1986 June 24–28. Bollettino del Labortorio di Entomologia Agraria "Filippo Silvestri." 43:299–306.

Baccetti, B. 1960. Le coccineglie italiane delle cupressacee. Estratto da Redia 45:23–111.

Badawi, A. and A. M. Al-Ahmed. 1990. The population dynamics of the oriental scale insect, *Aonidiella orientalis* (Newstead) and factors affecting its seasonal abundance. Arab Gulf Journal of Scientific Research 8:81–89.

Baker, H. 1933. The obscure scale on the pecan and its control. United States Department of Agriculture, Circular No. 295. 20 pp.

Baker, H., N. Stahler, A. C. Johnson, L. Adams, and R. O. Froeschner. 1943. *Parlatoria chinensis* Marlatt. United States Department of Agriculture, Circular E-595. 16 pp.

Baker, J. R. 1993. Insects. Pages 101–153 in A. De Hertogh and M. Le Nard, eds., The physiology of flower bulbs: A comprehensive treatise on the physiology of utilization of ornamental flowering bulbous and tuberous plants. Elsevier, Amsterdam, The Netherlands.

Baker, J. R. 1994. Insect and related pests of flowers and foliage plants: Some important, common, and potential pests in the southeastern

United States. North Carolina Cooperative Extension Service, Raleigh. 106 pp.

Baker, J. R. and E. A. Shearin. 1992. Fern scale insects. North Carolina Flower Growers 37:1–3.

Baker, W. L. 1972. Eastern forest insects. United States Department of Agriculture, Forest Service, Miscellaneous Publications 1175. 642 pp.

Balachowsky, A. S. 1926. Note sur un coccide de la faune Neo-tropicale recemment aclimate et nuisible au figuier en Algerie. Bulletin de la Societe d'Histoire Naturelle de l'Afrique du Nord 17:63–69.

Balachowsky, A. S. 1929. Contribution à l'étude de la faune du Congo Belge. Diaspines nuisibles au caféier et au cacaoyer. Revue de Pathologie Végétale et d'Entomologie Agricole de France 16:141–145.

Balachowsky, A. S. 1950. Les cochenilles de France, d'Europe, du Nord de l'Afrique et du Bassin Méditerranéen. V.—Monographie des Coccoidea; Diaspidinae (deuxième partie) Aspidiotini. Entomologique Applicata Actualités Sciences et Industrielles 1087:397–557.

Balachowsky, A. S. 1953. Les cochenilles de France d'Europe, du Nord de l'Afrique, et du Bassin Méditerranéen. VII.—Monographie des Coccoidea; Diaspidinae-IV, Odonaspidini-Parlatorini. Entomologique Applicata Actualités Sciences et Industrielles 1202:725–929.

Balachowsky, A. 1954. Les cochenilles Palearctic de la tribu des Diaspidini. Paris, Memoires Scientifiques de l'Institut Pasteur. 450 pp.

Balachowsky, A. S. 1956. Les cochenilles du continent African Noir. Vol. 1—Aspidiotini. Annales du Musée Royal du Congo Belge (Sciences Zoologiques) 3:1–142.

Balachowsky, A. S. 1958. Les cochenilles du continent African Noir. Vol. II—Aspidiotini, Odonaspidini et Parlatorini. Annales du Musée Royal du Congo Belge (Sciences Zoologiques) 4:149–429.

Balachowsky, A. S. 1959. Nuevos cochinillas de Colombia. Revista Academia Colombiana 10:337–361.

Balás, G. and G. Sáringer. 1982. Horticultural pests. Akadémiai Kiadó. 1069 pp.

Ball, J. C. 1980. Development and fecundity of the white peach scale at two constant temperatures. Florida Entomologist 63:188–194.

Ballou, H. A. 1922. Report on the prevalence of some pests and diseases of crops in the West Indies during 1920. West Indian Bulletin 19:239–271.

Ballou, C. H. 1934. Informe anual 1932: Informe de la Sección de Entomologiá. Boletín Centro Nacional de Agricultura 17:19–25.

Baranowski, R. M. and H. B. Glenn. 1999. Establishment of Cybocephalus binotatus Grouvelle and Coccobius fulvus (Compere & Annecke) for suppression of cycad scale, Aulacaspis yasumatsui Takagi (Homoptera: Diaspididae) in South Florida. Florida Entomological Society, 82nd Annual Meeting (Abstracts). San Juan, Puerto Rico 1999 July 25–29. Display 3.

Barnes, M. M. 1959. Deciduous fruit insects and their control. Annual Review of Entomology 4:343–362.

Bartra, P. C. E. 1974. Biología de Selenaspidus articulatus Morgan y sus principales controladores biológicos. Revista Peruana de Entomología 17:60–68.

Bartra, P. C. E. 1978. Biological observations on the oleander scale (Aspidiotus hederae Vallot). Revista Peruana Entomologica 19:43–48.

Batiashvili, I. D. 1954. Pests of citrus and other subtropical fruit crops. Tbilisi, Georgia. 250 pp.

Battaglia, D., A. di Leo, P. Malinconico, G. Rotundo, and A. di Leo. 1994. Osservazioni sulla cocciniglia bianca del pesco e del gelso in Basilicata. Informatore Agrario 50:3, 77–80.

Battaglia, D. and G. Viggiani. 1982. Observation on the distribution and phenology of Aonidiella aurantii and on its natural enemies in Campania. Annali della Facolta di Scienze Agrarie della Universita degli Studi di Napoli 16:125–132.

Bazarov, B. 1962. Agricultural importance and means of control of coccids in Tadzhikistan. Izvestiya Akademii Nauk Tadzhikskoi SSR. Otdelenie Biologicheskikh Nauk 3:48–57.

Bazarov, B. B. and G. P. Shmelev. 1971. Scale insects of Tadzhikistan and the adjoining regions of middle Asia. In "Fauna of Tadzhikistan SSR." Akademii Nauk Tadzhikskol SSR. Instituta Zoologicheskogo I Parasitologicheskogo 11:1–238.

Beardsley, J. W. 1970. Aspidiotus destructor Signoret, an armored scale pest new to the Hawaiian Islands. Proceedings of the Hawaiian Entomology Society 20:505–508.

Beardsley, J. W. and R. H. González. 1975. The biology and ecology of armored scales. Annual Review of Entomology 20:47–73.

Bedford, E. C. G. 1968. The biological control of red scale, Aonidiella aurantii (Mask.) on citrus in South Africa. Journal of the Entomological Society of Southern Africa 31:1–15.

Bedford, E. C. G. 1978. Other citrus insects of minor importance. Pages 129–133 in E. C. G. Bedford, ed., Citrus pests in the Republic of South Africa. Science Bulletin 391, Department of Agricultural Technical Services, Republic of South Africa.

Bedford, E. C. G. 1996. Problems we face in bringing red scale, Aonidiella aurantii (Maskell), under biological control in citrus in South Africa. Proceedings of the International Society of Citriculture. 1:485–492.

Bedford, E. C. G. and C. J. Cilliers. 1994. The role of Aphytis in the biological control of armored scale insects on citrus in south Africa. Pages 143–179 in D. Rosen, ed., Advances in the study of Aphytis. Intercept Ltd., Andover, UK.

Bedford, E. C. G. and M. B. Georgalia. 1978. Red scale Aonidiella aurantii (Mask.) Pages 109–118 in E. C. G. Bedford, ed., Citrus pests in the Republic of South Africa. Science Bulletin 391, Department of Agricultural Technical Services, Republic of South Africa.

Beingolea, G. O. 1969. Notas sobre la biologia de Selenaspididus articulatus Morgan (Hom: Diaspididae), "Queresa redonda de los citricos."Revista Peruana de Entomología 12:119–129.

Bénassy, C. 1958. Influence de l'accouplement et de la fécondation sur la differenciation du sexe et la formation du bouclier protecteur chez Pseudaulacaspis pentagona Targ. (Homoptera: Diaspidiae). Comptes Rendus des Séances de l'Academie des Sciences 246:649–652.

Bénassy, C. 1959. Remarques écologiques sur la cochenille du rosier Aulacaspis rosae Bouché attaquant le frambioisie dans le sud-est de la Provence. Compte Rendu de l'Academie d'Agriculture de France 45:421–424.

Bénassy, C. 1969. La lutte biologique contre le pou de San Jose. Publications de l'OEPP (Ser. A) No. 48:33–39.

Bénassy, C. 1990. Date palm. Pages 585–591 in D. Rosen, ed., Armored scale insects. Vol. 4B. Elsevier, Amsterdam, The Netherlands.

Bénassy, C. and G. Euverte. 1967. Notes sur Chrysomphalus dictyospermi au Maroc. Al Awamia 24:95–111.

Bénassy, C., E. Franco, and J. C. Onillon. 1975. Utilisation en France D'Aphytis lepidosaphes Comp. (Chalcidien, Aphelinidae), parasite specifique de la cochenille virgule des Citrus (Lepidosaphes beckii Newm.) 1. Evolution de la cochenille. Fruits 30:185–189.

Bénassy, C. and F. Soria. 1964. Observations ecologiques sur les cochenilles diaspines nuisibles aux agrumes en Tunisie. Annales de l'Institut National Agronomique de Tunisie 37:193–222.

Ben-Dov, Y. 1974. A revision of Ischnaspis Douglas with a description of a new allied genus. Journal of Entomology (B) 43:19–32.

Ben-Dov, Y. 1988. A taxonomic analysis of the armored scale tribe Odonaspidini of the World. United States Department of Agriculture, Agricultural Research Service, Technical Bulletin No. 1723. 142 pp.

Bennett, F. D. and S. W. Brown. 1958. Life history and sex determination in the diaspine scale Pseudaulacaspis pentagona (Targ.) (Coccoidea). Canadian Entomologist 90:317–324.

Bennett, F. D. and I. W. Hughes. 1959. Biological control of insect pests in Bermuda. Bulletin of Entomological Research 50:423–436.

Bentley, G. M. and I. L. Bartlett. 1931. Insects and allied pests of greenhouse plants with recommendations for their control. Tennessee Department of Agriculture, Division of Insects and Plant Disease, Control Bulletin 57. 70 pp.

Beran, F. 1943. Untersuchungen zur Bekampfung der San Jose-Schildlaus. Zeitschrift fur Pflanzenkrankheiten 53:79–80.

Berlinger, M. J. and R. Barak. 1981. The phenology of Antonina graminis and Odonaspis ruthae on lawn grass in Israel. Zeitschrift für Angewandte Entomologie 91:62–67.

Berlinger, M. J., Y. Dahan, Y. Ben-Dov, and M. Cohen. 1984 (1983). The phenology, distribution and control in Israel of the Hall scale, Nilotaspis halli (Green). Hassadeh 64:722–725.

Berlinger, M. J., L. Segre, H. Podoler, and R. A. J. Taylor. 1999. Distribution and abundance of the oleander scale (Homoptera: Diaspididae) on jojoba. Journal of Economic Entomology 92:1113–1119.

Bertels, M. A. 1956. Entomologia Agrícola Sul—Brasileira. Série Didáctica, No. 16. Ministerio da Agricultura, Serviço de Informaçao Agrícola, Rio de Janeiro. 458 pp.

Beshear, R. J., H. H. Tippins, and J. O. Howell. 1973. The armored scale insects of Georgia and their hosts. University of Georgia, College of Agriculture Experiment Station, Research Bulletin 146. 15 pp.

Bethune, C. J. S. 1908 (1907). Injurious insects in Ontario in 1907. Annual Report of the Entomological Society of Ontario 38:95–99.

Bianchi, H. and C. Bénassy. 1979. La cochenille rouge du poirier, Epidiaspis leperii Sign. (Homoptera, Coccoidea) ravageur en France du prunier. Annales de Zoologie—Ecologie Animale 11:493–511.

Bianchi, A., A. Pacchiacucchi, L. Guarino, and E. Maffeo. 1994. Segnalata una nuova cocciniglia su actinidia nel Lazio. Informatore Agrario 50:46, 73–75.

Bibby, F. F. 1931. Coccoids collected on wild plants in semi-arid regions of Texas and Mexico. Journal of the New York Entomological Society 39:587–591.

Bibby, F. F. 1961. Notes on miscellaneous insects of Arizona. Journal of Economic Entomology 54:324–444.

Biche, M. and M. Sellami. 1999. Étude de quelques variations biologiques possibles chez *Parlatoria oleae* (Colvee) (Hemiptera, Diaspididae). Bulletin de la Société Entomologique de France 104:287–292.

Bichina, T. I. and E. S. Gatina. 1976. Varietal susceptibility of apple to *Aspidiotus perniciosus* and integrated measures for its control. Zashchita Rastenii 30:124–131.

Bienkowski, A. O. 1993. Morphology and systematics of the adult male of *Lopholeucaspis japonica* (Cockerell) (Coccinea Diaspididae). Russian Entomological Journal 2:25–29.

Bitancourt, A., J. P. da Fonseca, and M. Autuori. 1933. Doenças, Pragas e Tratamentos. Manual de Citricultura. Chacaras e Quintaes, Sao Paulo. 212 pp.

Blackburn, V. L. and D. R. Miller. 1984. Pests not known to occur in the United States or of limited distribution, No. 44: Black parlatoria scale. United States Department of Agriculture, Plant Protection and Quarantine, Animal and Plant Health Inspection Service 81-45:1–13.

Blaisinger, P. 1979. Évolution d'une population de la cochenille rouge du poirier *Epidiaspis leperii* Sign. (Homoptera, Diaspidoidea) dans un verger de mirabelliers en Alsace. Sa repercussion sur le rendement du verger. Annales de Zoologie—Ecologie Animale 11:487–492.

Blackman, M. W. 1916. Sap-eating insects. Bulletin of the New York State College 16:93–112.

Blank, R. H., G. S. C. Gill, and B. W. Dow. 1999. Armoured scale (Hemiptera: Diaspididae) distribution in kiwifruit blocks with reference to shelter. New Zealand Journal of Crop and Horticultural Science 27:1–12.

Blank, R. H., G. S. C. Gill, M. H. Olson, and M. P. Upsdell. 1995. Greedy scale phenology on Taraire based on Julian days and degree-day accumulations. Environmental Entomology 24:1569–1575.

Blank, R. H., M. H. Olson, J. B. Clark, and G. S. C. Gill. 1993. Investigating two bee-safe materials for controlling latania scale on avocados during pollination. Proceedings of the Forty-Sixth New Zealand Plant Protection Conference. Christ Church, New Zealand, 1992 August 10–12. 80–85.

Blank, R. H., M. H. Olson, and G. S. C. Gill. 1992. Armoured scale, *Hemiberlesia lataniae* and *H. rapax*, infestation of kiwifruit rejected for export at two packhouses from 1987 to 1991. New Zealand Journal of Crop and Horticultural Science 20:397–405.

Blank, R. H., M. H. Olson, and P. L. Lo. 1990. Armoured scale (Hemiptera: Diaspididae) aerial invasion into kiwifruit orchards from adjacent host plants. New Zealand Journal of Crop and Hort Science 18:81–87.

Bliss, C. I., A. W. Cressman, and B. M. Broadbent. 1935. Productivity of the camphor scale and the biology of its egg and crawler stages. Journal of Agricultural Research 50:243–266.

Bobb, M. L., J. A. Weidhaas Jr., and L. F. Ponton. 1973. White peach scale: Life history and control studies. Journal of Economic Entomology 66:1290–1292.

Boboye, S. O. 1971. Scale insects on citrus and their distribution in Western Nigeria. Journal of Economic Entomology 64:307–309.

Bodenheimer, F. S. 1924. The Coccidae of Palestine. First report on this family. Zionist Organization Institute of Agriculture and Natural History, Agricultural Experiment Station Bulletin 1. 100 pp.

Bodenheimer, F. S. 1943. A first survey of the Coccoidea of Iraq. Government of Iraq, Ministry of Economics, Directorate General of Agriculture, Bulletin 28. 33 pp.

Bodenheimer, F. S. 1951. Citrus entomology in the Middle East with special reference to Egypt, Iran, Iraq, Palestine, Syria, Turkey. W. Junk, The Hague. 663 pp.

Bodenheimer, F. S. 1953. The Coccoidea of Turkey III. Revue de la Faculté des Sciences de l'Université d'Istanbul (Ser. B) 18:91–164.

Bodenheimer, F. S. and H. Steinitz. 1937. Studies in the life history of citrus mussel scale (*Lepidosaphes pinnaeformis* Bouché) in Palestine. Hadar 10:152–219.

Bodkin, G. E. 1928. Fumigation of citrus trees in Palestine. Citrus Industry 9:10–11, 30, 34.

Boecklen, W. J. and S. Mopper. 1998. Local adaptation in specialist herbivores: Theory and evidence. Pages 64–88 *in* S. Mopper and S. Y. Strauss, eds., Genetic structure and local adaptation in natural insect populations: Effects of ecology, life history, and behavior. Chapman & Hall, New York.

Bognár, S. and G. Vinis. 1979. Scale insects in the parks and avenues of Budapest. Acta Agronomica Academiae Scientiarum Hungaricae 28:13–26.

Bohart, R. M. 1942. Life history of *Diaspis boisduvalii* and its control on cattleya with calcium cyanide. Journal of Economic Entomology 35:365–368.

Bondar, G. 1914. Praga das laranjeiras e outras auranciaceas. Boletim da Agricultura 15:1064–1106.

Bondar, G. 1924. Relatoria apresentado por Gr. Bondar sobre a viagem nos municipios de Areia e Jequié, em estudo das condicoes de diversas lavouras. Boletim Laboratoria Pathologia Vegetal Bahia 1:3–16.

Bondar, G. 1926. Molestias das fructeiras da familia das Amonaceas. Boletim Laboratoria Pathologia Vegetal 3:75–83.

Boratynski, K. L. 1953. Sexual dimorphism in the second instars of some Diaspididae. Transactions of the Royal Entomological Society of London 104:451–479.

Boratynski, K. 1957. On the two species of the genus *Carulaspis* MacGillivray in Britain. Entomologists Monthly Magazine 93:246–251.

Borchsenius, N. S. 1950. Mealybugs and Scale Insects of USSR (Coccoidea). Akademii Nauk SSSR, Zoological Institute, Moscow, 32. 250 pp.

Borchsenius, N. S. 1963. Practical guide to the determination of scale insects of cultivated plants and forest trees of the USSR. Akademii Nauk SSSR, Zoological Institute, Leningrad. 311 pp.

Borchsenius, N. S. 1966. A catalogue of the armoured scale insects (Diaspidoidea) of the world. (In Russian.) Nauka, Moscow and Leningrad. 449 pp.

Borg, J. 1932. Scale insects of the Maltese Islands. Malta Government Printing Office. 20 pp.

Boyden, B. L. 1941. Eradication of the parlatoria date scale in the United States. United States Department of Agriculture, Miscellaneous Publication No. 433. 62 pp.

Boyer, F. D. and P. H. Ducrot. 1999a. Total synthesis of the enantiomers of *Aspidiotus nerii* sex pheromone. Comptes Rendus de l'Academie des Sciences Serie II Fascicule C-Chimie 2:29–33.

Boyer, F. D. and P. H. Ducrot. 1999b. Syntheses of cyclobutane derivatives: Total synthesis of (+) and (–) enantiomers of the oleander scale *Aspidiotus nerii* sex pheromone. European Journal of Organic Chemistry 5:1201–1211.

Brailoiu Tanase, D. 1998. White Peach Scale *Pseudaulacaspis pentagona* (Targioni-Tozzetti) (Homoptera: Diaspididae) damage, identification, biology and control in Bucharest–Baneasa area. Zhivotnov"dni Nauki Supplement No. 10:11.

Brain, C. K. 1915. The Coccidae of South Africa. Transactions of the Royal Society of South Africa 5:65–194.

Bray, D. F. 1974. The fieldman's guide to entomology. Special Bulletin, Cooperative Extension Service, University of Delaware. 33 pp.

Brewer, B. S. and A. D. Oliver. 1987. Euonymus scale, *Unaspis euonymi* (Comstock) (Homoptera: Diaspididae): Effects of host cultivar age, and location on infestation levels. Journal of Entomological Science 22:119–122.

Brimblecombe, A. R. 1956. The pineapple scale in Queensland. Queensland Department of Agriculture and Stock, Division of Plant Industry, Leaflet 415. 2 pp.

Brimblecombe, A. R. 1961. Scale insects on papaws. Queensland Agricultural Journal 87:1–2.

Brimblecombe, A. R. 1962. Studies of the Coccoidea 12. Species occurring on deciduous fruit and nut trees in Queensland. Queensland Journal of Agricultural Science 19:219–229.

Britton, W. E. 1903. Two common scale-insects of the orchard. Connecticut Agricultural Experiment Station Bulletin, No. 143. 10 pp.

Britton, W. E. 1908. Tests of various gases for fumigating nursery trees. Journal of Economic Entomology 1:110–112.

Britton, W. E. and R. B. Friend. 1935. Insect pests of elms in Connecticut. Bulletin of the Connecticut Agricultural Experiment Station 369:265–307.

Brooks, R. F. 1977. Leafhoppers, whiteflies, aphids, and scale insects. Pages 364–367 *in* J. Kranz, H. Schmutterer, and W. Koch, eds., Diseases, pests and weeds in tropical crops. John Wiley & Sons, New York.

Brown, L. R. and C. O. Eads. 1967. Insects affecting ornamental conifers in southern California. Bulletin of the California Agricultural Experiment Station 834. 72 pp.

Brown, S. W. 1960. Spontaneous chromosome fragmentation in the armored scale insects. Journal of Morphology 106:159–185.

Brown, S. W. 1965. Chromosomal survey of the armored and palm scale insects (Coccoidea: Diaspididae and Phoenicococcidae). Hilgardia 36:189–294.

Brown, S. W. and F. D. Bennett. 1957. On sex determination in the diaspine scale *Pseudaulacaspis pentagona* (Targ.) (Coccoidea). Genetics 42:510–523.

Browning, H. W. 1994. Biological control of the citrus snow scale, *Unaspis citri*, in Florida: evaluation of *Aphytis* and other natural enemy species. Pages 119–142 in D. Rosen, ed., Advances in the study of *Aphytis* (Hymenoptera: Aphelinidae). Intercept Limited, Andover, UK.

Bruwer, I. J. 1998. Long mussel scale: *Lepidosaphes gloverii* (Packard). Pages 153–157 in E. C. G. Bedford, M. A. Van den Berg, and E. A. De Villiers, eds., Citrus pests in the Republic of South Africa. Institute for Tropical and Subtropical Crops, Nelspruit.

Bruwer, I. J. and A. S. Schoeman. 1990. Key factor analysis of two populations of the long mussel scale, *Insulaspis gloverii* (Packard) (Hemiptera: Diaspididae). Journal of the Entomological Society of Southern Africa 53:101–105.

Bull, B. C., M. J. Raupp, M. R. Hardin, and C. S. Sadof. 1993. Suitability of five horticulturally important armored scale insects as hosts for an exotic predaceous lady beetle. Journal of Environmental Horticultural 11:28–30.

Bullington, S. W., M. Kosztarab, and G. Jiang. 1989. II. Adult Males of the Genus *Chionaspis* (Homoptera: Coccoidea: Diaspididae) in North America. Virginia Agricultural Experiment Station, Blacksburg, Bulletin 88:127–184.

Bullock, R. O. and R. F. Brooks. 1975. Citrus pest control in the U.S.A. Pages 35–37 in E. Hafliger, ed., Citrus. Ciba-Geigy Agrochemicals Technology Monograph No. 4. 88 pp.

Burden, D. J. and E. R. Hart. 1989. Degree-day model for egg eclosion of the Pine Needle Scale (Hemiptera: Diaspididae). Environmental Entomology 18:223–227.

Burger, H. C. and S. A. Ulenberger. 1990. Quarantine problems and procedures. Pages 313–327 in D. Rosen, ed., Armored scale insects. Vol. 4B. Elsevier, Amsterdam, The Netherlands.

Bustshik, T. N. 1960. Coccid fauna of western Kopet-Dagh. (In Russian.) Trudy Akademii Nauk SSR Zoologicheskogo Instituta 27:167–182.

Busuiok, M. N. 1972. Damageability of apple varieties by San Jose scale in relation to the geographic factor. Trudov Kishinev S-kh. Instituta 88:26–31.

Butani, D. K. 1974. Insect pests of fruit crops and their control II. Guava. Pesticides 8:26–30.

Butani, D. K. 1978. Insect pests of tamarind and their control II. Pesticides 8:34–41.

CAB International. 1951. *Chrysomphalus dictyospermi* (Morg.). Distribution maps of pests, series A, Agricultural Map No. 3. 2 pp.

CAB International. 1957. *Howardia biclavis* (Comstock). Distribution maps of pests, series A, Agricultural Map No. 80. 2 pp.

CAB International. 1958. *Lepidosaphes ulmi* (L.). Distribution maps of pests, series A, Agricultural Map No. 85. 2 pp.

CAB International. 1962a. *Lepidosaphes gloverii* (Packard). Distribution maps of pests, series A, Agricultural Map No. 146. 2 pp.

CAB International. 1962b. *Parlatoria blanchardi* Targ. Distribution maps of pests, series A, Agricultural Map No. 148. 2 pp.

CAB International. 1962c. *Unaspis citri* (Comst.). Distribution maps of pests, series A, Agricultural Map No. 149. 2 pp.

CAB International. 1962d. *Parlatoria oleae* (Colvee). Distribution maps of pests, series A, Agricultural Map No. 147. 2 pp.

CAB International. 1964a. *Parlatoria pergandii* (Cost.). Distribution maps of pests, series A, Agricultural Map No. 185. 2 pp.

CAB International. 1964b. *Parlatoria ziziphus* (Lucas). Distribution maps of pests, series A, Agricultural Map No. 186. 2 pp.

CAB International. 1966. *Aspidiotus destructor* (Sign.). Distribution maps of pests, series A, Agricultural Map No. 218. 2 pp.

CAB International. 1967. *Ischnaspis longirostris* (Signoret). Distribution maps of pests, series A, Agricultural Map No. 235. 2 pp.

CAB International. 1970a. *Aspidiotus nerii* Bouche. Distribution maps of pests, series A, Agricultural Map No. 268. 2 pp.

CAB International. 1970b. *Unaspis euonymi* (Comst.). Distribution maps of pests, series A, Agricultural Map No. 269. 2 pp.

CAB International. 1973. *Diaspis bromeliae* (Kerner). Distribution maps of pests, series A, Agricultural Map No. 307. 2 pp.

CAB International. 1976. *Hemiberlesia lataniae* (Sign.). Distribution maps of pests, series A, Agricultural Map No. 360. 2 pp.

CAB International. 1978. *Aonidiella orientalis* (Newstead). Distribution maps of pests, series A, Agricultural Map No. 386. 2 pp.

CAB International. 1981a. *Pseudaonidia trilobitiformis* (Green). Distribution maps of pests, series A, Agricultural Map No. 418. 2 pp.

CAB International. 1981b. *Selenaspidus articulatus* (Morg.). Distribution maps of pests, series A, Agricultural Map No. 419. 2 pp.

CAB International. 1982. *Lepidosaphes beckii* (Newmn.). Distribution maps of pests, series A, Agricultural Map No. 49 (revised). 3 pp.

CAB International. 1986. *Quadraspidiotus perniciosus* (Comstock). Distribution maps of pests, series A, Agricultural Map No. 7 (revised). 3 pp.

CAB International. 1987. *Hemiberlesia rapax* (Comstock). Distribution maps of pests, series A, Agricultural Map No. 49 (revised). 3 pp.

CAB International. 1988. *Chrysomphalus aonidum* (Linnaeus). Distribution maps of pests, series A, Agricultural Map No. 4 (revised). 3 pp.

CAB International. 1993. *Aulacaspis tubercularis* Newstead. Distribution maps of pests, series A, Agricultural Map No. 540. 2 pp.

CAB International. 1996a. *Aonidiella aurantii*. Distribution maps of pests, series A, Agricultural Map No. 2 (revised). 5 pp.

CAB International. 1996b. *Pseudaulacaspis pentagona*. Distribution maps of pests, series A, Agricultural Map No. 58 (2nd revision). 5 pp.

CAB International. 1997. *Aonidiella citrina* (Coquillett). Distribution maps of pests, series A, Agricultural Map No. 349 (revised). 3 pp.

CAB International. 1998. *Lopholeucaspis japonica* Cockerell. Distribution maps of pests, series A, Agricultural Map No. 582. 2 pp.

CAB International. 2000. *Aulacaspis yasumatsui* Takagi. Distribution maps of pests, series A, Agricultural Map No. 610. 2 pp.

Canales Canales, A. and L. Valdivieso Jara. 1999. Manual de control biológico para la conducción del cultivo del olivo. Servicio Nacional de Sanidad Agraria, Jesus María, Peru. 37 pp.

Carbonell Bruhn, J. and J. Briozzo Beltrame. 1975. San Jose scale, *Quadraspidiotus perniciosus*. Boletín Divulgación Ministerio de Ganadería y Agricultura Central Investigacion Agricultura Alberto Boer German 30:1–11.

Carpenter, J. B. and H. S. Elmer. 1978. Pests and diseases of the date palm. United States Department of Agriculture Handbook No. 527. 42 pp.

Carrillo L., R., C. Cifuentes C., and N. Mundaca B. 1995. Ciclo estacional de *Lepidosaphes ulmi* (L.) (Hemiptera: Diaspididae). Revista Chilena de Entomología 22:5–8.

Carroll, D. P. and R. F. Luck. 1984. Bionomics of California red scale, *Aonidiella aurantii* (Maskell), on orange fruits, leaves and wood in California's San Joaquin Valley. Environmental Entomology 13:847–853.

Cassidy, T. P., V. E. Romney, W. D. Buchanan, and G. T. York. 1950. Damage to guayule by insects and mites with notes on control. United States Department of Agriculture, Circular No. 842. 19 pp.

Caswell, G. H. 1962. Agricultural entomology in the tropics. Edward Arnold, London. 152 pp.

Chai, X. M., R. D. Hu, Z. M. Ruan, and Y. C. Xu. 1999. [Coccids endangering pines in Zhejiang.] (In Chinese.) Journal of Zhejiang Forestry Science and Technology 19:1–6.

Chambliss, C. E. 1898. Scale insects: San Jose and other species. University of Tennessee, Agricultural Experiment Station Bulletin 19:141–146.

Chang, L. C. and C. C. Tao. 1963. Black parlatoria, *Parlatoria zizyphus* Lucas. Taiwan Agricultural Research Institute 12:1–47.

Chao, J. and X. H. Zeng. 1997. [Control of *Parlatoria pergandii* Comstock with buprofezin.] (In Chinese.) Grassland of China No. 5. 46 pp.

Chatterjee, S. W. 1961 (1960). Some insect pests of mulberry in Kalimpong, Darjeeling District, West Bengal. Indian Journal of Entomology 22:258–260.

Chauvel, G. 1999. Quel est votre diagnostic? PHM Revue Horticole 408:44–46.

Chen, F. G. 1936. [Notes on the scale insects of citrus in several districts of East Chekiang, with description of one new species.] (In Chinese.) Entomology and Phytopathology. Hangchow 4:208–228.

Chi, D., J. C. Miao, H. Qu, W. J. Xiang, C. Y. Li, C. J. Lu, and L. S. Yi. 1997a. [Control of *Quadraspidiotus gigas* and *Lepidosaphes salicina* using insect growth regulator, RH-5849]. (In Chinese.) Journal of Northeast Forestry University 25:10–14.

Chi, D. F., F. B. Zhang, Y. Y. Hu, and Y. Sun. 1997b. [The influence of the kairomone of *Quadraspidiotus gigas* and oviposition deterring pheromone in parasitoids on the control ability of these parasitoids]. (In Chinese.) Journal of Northeast Forestry University 5:15–21.

Chiesa Molinari, O. 1948. Las plagas de la agricultura, manual práctico de procedimientos modernos para combatirlas. "El Ateneo," Buenos Aires. 497 pp.

Chorley, J. K. 1939. Report of the Division of Entomology for the year ending 31st December, 1938. Rhodesia Agricultural Journal 36:598–622.

Chou, I. 1984. Monograph of the Diaspididae of China. Shaanxi Publication House of Science and Technology. 432 pp.

Chua, T. H. and B. J. Wood. 1990. Other tropical fruit trees and shrubs. Pages 543–552 *in* D. Rosen, ed., Armored scale insects. Vol. 4B. Elsevier, Amsterdam, The Netherlands.

Chumakova, B. M. 1965. The role of the parasite *Aspidiotiphagus citrinus* (Craw.) in reducing the numbers of injurious scale insects in the subtropical regions of the Russian Soviet Federated Socialist Republics. Entomologicheskoe Obozrenye 44:520–526; Entomological Review 44:305–308.

Claps, L. E. 1987. Caracteristicas de ciclo biologico de *Cornuaspis beckii* (Newman, 1896) e *Insulaspis gloverii* (Packard, 1896) en condiciones de insectario. CIPRON—Revista de Investigación 5:1–4.

Claps, L. E. 1991. Morfologia de estados immaduros y adultos de *Cornuaspis beckii* (Newman, 1896) e *Insulaspis gloverii* (Packard, 1896). Revista de la Sociedad Entomologica Argentina 90:137–149.

Clausen, C. P. 1931. Insects injurious to agriculture in Japan. United States Department of Agriculture, Circular No. 168. 115 pp.

Clausen, C. P. 1958. The biological control of insect pests in the continental United States. Proceedings of the Tenth International Congress of Entomology. Montreal, Canada 1956 June 10–15. 4:443–447.

Clausen, C. P. 1978. Introduced parasites and predators of arthropod pests and weeds: A world review. United States Department of Agriculture, Washington, D.C. 545 pp.

Clift, A. D. and G. A. C Beattie. 1993. SCALEMAN: A computer program to help citrus growers manage red scale, *Aonidiella aurantii* (Homoptera: Diaspididae). Pages 470–472 *in* S. A. Corey, D. J. Dall, and W. M. Milne, eds., Pest Control and Sustainable Agriculture. CSIRO Division of Entomology, Canberra.

Cochereau, P. 1969. Biological control of *Aspidiotus destructor* Signoret on the Island of Vaté (New Hebrides) by means of *Rhizobius pulchellus* Montrouzier. Cahiers Orstom Series Biologia 8:57–100.

Cockerell, T. D. A. 1922. Some Coccidae found on orchids (Hom.). Entomological News 33:149.

Cockfield, S. D. and D. A. Potter. 1990. Euonymus scale patterns of damage to woody plants. Journal of Arboriculture 16:239–241.

Cohen, E., H. Podoler, and M. El-Hamlauwi. 1987. Effects of the malathion-bait mixture used on citrus to control *Ceratitis capitata* (Wiedemann) (Diptera: Tephritidae) on the Florida red scale, *Chrysomphalus aonidum* (L.) (Hemiptera: Diaspididae), and its parasitoid *Aphytis holoxanthus* DeBach. Bulletin of Entomological Research 77:303–307.

Cohen, E., H. Podoler, and M. El-Hamlauwi. 1994. Delayed effects of malathion on hymenopteran parasites: The role of the scale cover of diaspidid insects. Pages 183–190 *in* D. Rosen, ed., Advances in the study of *Aphytis* (Hymenoptera: Aphelinidae). Intercept Limited, Andover, UK.

Cohic, F. 1958. Contribution a l'etude des cochenilles d'interet economique de Nouvelle-Caledonie et dependaces. Documents Techniques de la Commission du Pacifique Sud 116:1–35.

Coll, M. and S. Abd-Rabou. 1998. Effect of oil emulsion sprays on parasitoids of the black parlatoria, *Parlatoria ziziphi*, in grapefruit. Biocontrol 43:1, 29–37.

Cooley, R. A. 1900. Injurious fruit insects. Montana Experiment Station Bulletin 23:64–114.

Cooper, R. M. and R. D. Oetting. 1989. Life history and field development of the camellia scale. Annals of the Entomological Society of America 82:730–736.

Costilla, M. A., V. M. Osores, and H. J. Basco. 1970. Biologia y dando de la cochinella roja Australiana (*Aonidiella aurantii* Masks.) bajo las condiciones de Tucuman. Revista Industrial y Agrícola de Tucumán 47:57–65.

Couch, J. N. 1931. The biological relationship between *Septobasidium retiforme* (B&C) Pat. and *Aspidiotus Osborni* New. and Ckll. Quarterly Journal of Microscopical Science 74:383–437.

Couch, J. N. 1935. *Septobasidium* in the United States. Elisha Mitchell Science Society Journal 51:1–75.

Couch, J. N. 1938. The genus *Septobasidium*. University of North Carolina Press, Chapel Hill. 473 pp.

Cressman, A. W. 1933. Biology and control of *Chrysomphalus dictyospermi* (Morg.). Journal of Economic Entomology 26:696–706.

Cressman, A. W., C. I. Bliss, L. T. Kessels, and J. O. Dumestre. 1935. Biology of the camphor scale and a method for predicting the time of appearance of stages in the field. Journal of Agricultural Research 50:267–283.

Cressman, A. W. and H. K. Plank. 1935. The camphor scale. United States Department of Agriculture, Circular No. 365. 20 pp.

Cruz, C. and A. Segarra. 1991. Recent biological control experiences in Puerto Rico. Pages 7–9 *in* C. Pavis and A. Kermarrec, eds., Caribbean Meetings on Biological Control (Colloques de l'INRA). Guadeloupe, France: November 5–7, 1990. Institut National de la Recherche Agronomique, Paris, France. 35 pp.

Cumming, M. E. P. 1953. Notes on the life history and seasonal development of the pine needle scale *Phenacaspis pinifoliae* (Fitch) (Diaspididae: Homoptera). Canadian Entomologist 85:347–352.

Cunningham, H. S. 1929. Report of plant pathologist. Report of the Department of Agriculture of Bermuda. 2–28.

Danzig, E. M. 1959. [On the scale insect fauna of the Leningrad.] (In Russian.) Entomologicheskoe Obozrenye 38:443–455.

Danzig, E. M. 1964. [Suborder Coccinea-Coccids or mealy bugs and scale insects.] (In Russian.) Pages 800–850 *in* G. Y. Bei-Bienko, ed., Keys to the insects of the European USSR. Vol. I. Keys Fauna USSR, 84. 1214 pp.

Danzig, E. M. 1970. [Synonymy and some polymorphic species of coccids.] (In Russian.) Zoologicheskii Zhurnal 49:1015–1024.

Danzig, E. M. 1972. [Insects and ticks.] Pages 189–221 *in* [Pests of Forest.] (In Russian.) Akademii Nauk (SSR) Zoologicheskogo Instituta, Leningrad.

Danzig, E. M. 1978. [Scale insect fauna of South Sakhalin and Kunashir.] (In Russian.) Trudy Biologo-Pochvennogo, Akademii Nauk SSR, Vladivostok 50:3–23.

Danzig, E. M. 1980. [Coccoids of the Far East USSR (Homoptera, Coccinea) with phylogenetic analysis of scale insects fauna of the world.] (In Russian.) Nauka, Leningrad. 367 pp.

Danzig, E. M. and G. Pellizzari. 1998. Diaspididae. Pages 172–370 *in* F. Kozár, ed., Catalogue of Palaearctic Coccoidea. Plant Protection Institute, Hungarian Academy of Sciences, Budapest, Hungary.

Darvas, B. and L. Varjas. 1990. Insect growth regulators. Pages 393–408 *in* D. Rosen, ed., Armored scale insects. Vol. 4B. Elsevier, Amsterdam, The Netherlands.

Das, G. M. and S. C. Das. 1962. On the biology of *Fiorinia theae* occurring on tea in north-east India. Indian Journal of Entomology 24:27–35.

Das, G. M. and R. N. Ganguli. 1961. Coccoids on tea in North-East India. Indian Journal of Entomology 23:245–256.

Davidson, J. A. 1964. The genus *Abgrallaspis* in North America (Homoptera: Diaspididae). Annals of the Entomological Society of America 57:638–643.

Davidson, J. A., C. F. Cornell, and D. C. Alban. 1988. The untapped alternative. American Nurseryman 167:99–109.

Davidson, J. A. and C. W. McComb. 1958. Notes on the biology and control of *Fiorinia externa* Ferris. Journal of Economic Entomology 51:405–406.

Davidson, J. A. and D. R. Miller. 1990. Ornamental plants. Pages 603–632 *in* D. Rosen, ed., Armored scale insects. Vol. 4B. Elsevier, Amsterdam, The Netherlands.

Davidson, J. A., D. R. Miller, and S. Nakahara. 1983. The white peach scale, *Pseudaulacaspis pentagona* (Targioni-Tozzetti) (Homoptera: Diaspididae): Evidence that current concepts include two species. Proceedings of the Entomological Society of Washington 85:753–761.

Davidson, J. A. and M. J. Raupp. 1999. Landscape IPM. Guidelines for integrated pest management of insect and mite pests on landscape trees and shrubs. University of Maryland, Extension Service, College Park. 109 pp.

Davidson, N. A., J. E. Dibble, M. L. Flint, P. J. Marer, and A. Guye. 1991. Managing insects and mites with spray oils. IPM Education and Publications, Statewide Integrated Pest Management Project, University of California, Division of Agriculture and Natural Resources, Publication 3347. 47 pp.

Davis, J. J. 1933. Justifying expenditures for entomological research. Journal of Economic Entomology 26:75.

Dean, H. A. 1955. Factors affecting biological control of scale insects on Texas citrus. Journal of Economic Entomology 48:444–447.

Dean, H. A. 1975. Complete biological control of *Lepidosaphes beckii* on Texas citrus with *Aphytis lepidosaphes.* Environmental Entomology 4:110–114.

DeBach, P. 1962. Biological control of the California red scale *Aonidiella aurantii* (Maskell) on citrus around the world. Proceedings of the International Congress on Entomology. Wien, Germany 1960 August 17–25. 2:749–753.

DeBach, P. 1969. Biological control of diaspine scale insects on citrus in California. Proceedings of the First International Citrus Symposium. Riverside, California, 1968 March 16–26. 2:801–815.

DeBach, P. and T. W. Fisher. 1956. Experimental evidence for sibling species in the oleander scale, *Aspidiotus hederae* (Mallot). Annals of the Entomological Society of America 49:235–239.

DeBach, P. and D. Rosen. 1976. Armoured scale insects. Pages 139–178 *in* V. Delucchi, ed., Studies in Biological Control. (International Biological Programme, No. 9.) Cambridge University Press, Cambridge, England.

DeBach, P., D. Rosen, and C. E. Kennett. 1971. Biological control of coccids by introduced natural enemies. Pages 165–194 *in* C. B. Huffaker, ed., Biological control. Plenum Press, New York.

DeBach, P. and R. A. Sundby. 1963. Competitive displacement between ecological homologues. Hilgardia 34:105–166.

Deitz, L. L. and J. A. Davidson. 1986. Synopsis of the armored scale genus *Melanaspis* in North America. North Carolina Agricultural Research Service Technical Bulletin, No. 279. 91 pp.

Dekle, G. W. 1954. Some lychee insects of Florida. Proceedings of the Florida State Horticultural Society 67:226–228.

Dekle, G. W. 1964. An armored scale (*Morganella longispina* [Morgan]) on Haitian citrus. Entomology Circular, Florida Department of Agriculture and Consumer Services, Division of Plant Industry, No. 226. 2 pp.

Dekle, G. W. 1965. Florida armored scale insects. Arthropods of Florida and Neighboring Land Areas. Florida Department of Agriculture, Division of Plant Industry Library, Gainesville, 3. 265 pp.

Dekle, G. W. 1966. Aglaonema scale (*Temnaspidiotus excisus* [Green]). Florida Department of Agriculture, Division of Plant Industry, Entomology Circular No. 49. 2 pp.

Dekle, G. W. 1970. Oleander scale (*Phenacaspis cockerelli* [Cooley]). Florida Department of Agriculture, Division of Plant Industry, Entomology Circular No. 95. 2 pp.

Dekle, G. W. 1977 (1976). Florida armored scale insects. Florida Department of Agriculture, Division of Plant Industry Library, Gainesville. 345 pp.

Dekle, G. W. and Kuitert, L. C. 1975. Camellia mining scale, *Duplaspidiotus claviger* (Cockerell) (Diaspididae: Homoptera). Florida Department of Agriculture and Consumer Services, Division of Plant Industry. Entomology Circular 152:1–4.

Del Bene, G. 1984. Osservazioni sulla biologia di *Dynaspidiotus britannicus* (Newstead) (Homoptera: Diaspididae) e sui suoi nemici naturali in Toscana. Redia 67:323–336.

Delucchi, V. 1965. Notes sur le pou de Californie (*Aonidiella aurantii* Maskell). Annales de la Société Entomologique de France 1:739–788.

Delucchi, V. 1975. Scale insects and whiteflies of citrus fruit. Ciba-Geigy Agrochemicals, Technical Monograph, Supplement B. 4:24–27.

Denno, R. F., M. S. McClure, and J. R. Ott. 1995. Interspecific interactions in phytophagous insects: Competition reexamined and resurrected. Annual Review of Entomology 40:297–331.

Devasahayam, S. 1992. Insect pests of black pepper and their control. Planters Chronicle 155:153.

Dhileepan, K. 1992. Insect pests of oil palm (*Elaeis guineensis*) in India. Planter 68:183–191.

Dickson, R. C. 1941. Inheritance of resistance to hydrocyanic acid fumigation in the California red scale. Hilgardia 13:515–522.

Dinabandhoo, C. L. and O. P. Bhalla. 1980. Bionomics of the San Jose scale, *Quadraspidiotus perniciosus* infesting temperate fruits. Entomological Research Journal 4:63–67.

Dodge, B. O. and H. W. Rickett. 1943. Diseases and pests of ornamental plants. Jaques Cattell Press, Lancaster, Pennsylvania. 638 pp.

Don, D., A. Miyanoshita, and S. Tatsuki. 1995. [Host-associated differences in *Aspidiotus cryptomeriae* Kuwana (Homoptera: Coccoidea: Diaspididae): II. Morphology of *Torreya nucifera* race reared on non-host plant, *Cryptomeria japonica*.] (In Japanese.) Japanese Journal of Applied Entomology and Zoology 39 (2):159–162.

Downing, R. S. and D. M. Logan. 1977. A new approach to San Jose scale control (Hemiptera: Diaspididae). Canadian Entomologist 10:1249–1252.

Dozier, H. L. 1924. Insect pests and diseases of the Satsuma orange. Gulf coast citrus exchange. Educational Bulletin No. 1. 101 pp.

Drake, C. J. 1934. Report of the State Entomologist. Iowa Yearbook in Agriculture 4:71–77.

Drake, C. J. 1935. Report of the State Entomologist. Iowa Yearbook in Agriculture 5:83–91.

Drea, J. J. 1990a. Other Coleoptera. Pages 41–49 *in* D. Rosen, ed., Armored scale insects. Vol. 4B. Elsevier, Amsterdam, The Netherlands.

Drea, J. J. 1990b. Neuroptera. Pages 51–59 *in* D. Rosen, ed., Armored scale insects. Vol. 4B. Elsevier, Amsterdam, The Netherlands.

Drea, J. J. and R. W. Carlson. 1987. Establishment of *Cybocephalus* sp. (Coleoptera: Nitidulidae) from Korea on *Unaspis euonymi* (Homoptera: Diaspididae) in the eastern United States. Proceedings of the Entomological Society of Washington 90:307–309.

Drea, J. J. and R. D. Gordon. 1990. Coccinellidae. Pages 19–40 *in* D. Rosen, ed., Armored scale insects. Vol. 4B. Elsevier, Amsterdam, The Netherlands.

Dreistadt, S. H., J. K. Clark, and M. L. Flint. 1994. Pests of landscape trees and shrubs: An integrated pest management guide. Division of Agriculture and Natural Resources, University of California, Publication 3359. 327 pp.

Duda, E. J. 1959. Notes on scale insects. Science Tree Topics 2:9–10.

Dupont, P. R. 1918. Insect notes. Seychelles Annual Report Agriculture, Crown Lands for 1917. 22 pp.

Dupont, P. R. 1926. Report of the director of agriculture. Seychelles Annual Report Agriculture, Crown Lands for 1925. 8 pp.

Dustan, A. G. 1953. A method of rearing the oleander scale, *Pseudaulacaspis pentagona* (Targ.) on potato tubers. Bulletin of the Bermuda Department of Agriculture 27:1.

Dutta, S. and C. L. Baghel. 1991. On the morphology of mature female *Aonidiella orientalis* (Newstead). Uttar Pradesh Journal of Zoology 11:31–35.

Duzgunes, Z. 1969. Position in Turkey of the San Jose scale (*Quadraspidiotus perniciosus* Comst.). Rapport de la Conférence Internationale OEPP-OILB sur le Pou de San José (Milan-Vérone, 1968). European Mediterranean Plant Protection Organization, Paris. 111 pp.

Dzhashi, V. S. 1989. [The Japanese camellia scale—a specialised pest of the branches and shoots of the tea plant.] (In Russian.) Subtropicheskie Kul'tury 3:93–101.

Dziedzicka, A. 1989. Scale insects Coccinea occurring in Polish greenhouses. I. Diaspididae. Acta Biologica Cracoviensia, Series Zoologica 31:93–114.

Ebeling, W. 1945. Citrus in South America. Citrus Leaves 25:8–9, 29–30.

Ebeling, W. 1950. Subtropical entomology. University of California Press, Berkeley. 747 pp.

Ebeling, W. 1959. Subtropical fruit pests. University of California Press, Berkeley. 436 pp.

Ebstein, R. P. and U. Gerson. 1971. The non-waxy component of the armored-scale insect shield. Biochimica et Biophysica Acta 237: 550–555.

Edmunds, G. F., Jr. 1973. Ecology of black pineleaf scale (Homoptera: Diaspididae). Environmental Entomology 2:765–777.

Efimoff, A. 1937. List of insect pests in Spain and Portugal. Central Laboratory of Plant Quarantine Administration, Moscow. 110 pp.

Eguagie, W. E. 1972. Observations on the biology of some armoured scale insects (Homoptera: Diaspididae) on citrus in Ibadan, Nigeria. Bulletin of the Entomological Society of Nigeria 3:99–107.

Ehler, L. E. 1995. Biological control of obscure scale in California: An experimental approach. Environmental Entomology 24:779–795.

Einhorn, J., A. Guerrero, P. H. Ducrot, F. D. Boyer, and M. Gieselmann. 1998. Sex pheromone of the oleander scale, *Aspidiotus nerii*: Structural characterization and absolute configuration of an unusual functionalized cyclobutane. Proceedings of the National Academy of Science 95:9867–9872.

Elder, R. J. and K. L. Bell. 1998. Establishment of *Chilocorus* spp. (Coleoptera: Coccinellidae) in a *Carica papaya* L. orchard infested by *Aonidiella orientalis* (Newstead) (Hemiptera: Diaspididae). Australian Journal of Entomology 37:362–365.

Elder, R. J., D. Gultzow, D. Smith, and K. L. Bell. 1997. Oviposition by *Comperiella lemniscata* Compere and Annecke (Hymenoptera: Encyrtidae) in *Aonidiella orientalis* (Newstead) (Hemiptera: Diaspididae). Australian Journal of Entomology 36:299–301.

Elder, R. J. and D. Smith. 1995. Mass rearing of *Aonidiella orientalis* (Newstead) (Hemiptera: Diaspididae) on butternut gramma. Journal of the Australian Entomological Society 34:253–254.

Elder, R. J., D. Smith, and K. L. Bell. 1998. Successful parasitoid control of *Aonidiella orientalis* (Newstead) (Hemiptera: Diaspididae) on *Carica papaya* L. Australian Journal of Entomology 37:74–79.

Eliason, E. A. and D. G. McCullough. 1997. Survival and fecundity of three insects reared on four varieties of Scotch pine Christmas trees. Journal of Economic Entomology 90:1598–1608.

El-Kareim, A. I. A. 1998. Swarming activity of the adult males of parlatoria date scale in response to sex pheromone extracts and sticky color traps. Archives of Phytopathology and Plant Protection 31:3, 301–307.

El-Kareim, A. I. A., B. Darvas, and F. Kozár. 1988. Effects of the juvenoids fenoxycarb, hydroprene, kinoprene and methoprene on first instar larvae of *Epidiaspis leperii* Sign. (Hom., Diaspididae) and on its ectoparasitoid, *Aphytis mytilaspidis* (Le Baron) (Hym., Aphelinidae). (Hungary) Journal of Applied Entomology 106:270–275.

El-Kareim, A. I. A. and F. Kozár. 1988a. The host plant, a key factor in the population dynamics of *Epidiaspis leperii* (Homoptera: Coccoidea). Novenyvedelem 34:17–20.

El-Kareim, A. I. A. and F. Kozár. 1988b. Extraction and bioassay of the sex pheromone of the red pear scale, *Epidiaspis leperii*. Entomologia Experimentalis et Applicata 46:79–84.

El-Minshawy, A. M., S. K. El-Sawaf, S. M. Hammad, and A. Donia. 1971. The biology of *Hemiberlesia lataniae* (Sign.) in Alexandria District. Bulletin de la Société Entomologique d'Egypte 55:461–467.

El-Minshawy, A. M., S. K. El-Sawaf, S. M. Hammad, and A. Donia. 1974. Survey of the scale insects attacking fruit trees in Alexandria district (Part 1: Fam. Diaspididae; subfamily Diaspidinae, tribe Aspidiotini). Alexandria Journal of Agricultural Research 22:223–232.

Ellis, H. C. and J. O. Howell. 1982. Scales. Pages 3, 26, 27 *in* E. F. Suber, ed., Summary of economic losses due to insect damage and costs of control in Georgia, 1980. University of Georgia, College of Agriculture, Experiment Stations, Special Publications 20. 27 pp.

English, L. L. 1933. Life history and control of citrus insects. Alabama Polytechnic Institute, Agricultural Experiment Station Bulletin, 44th Annual Report. 27 pp.

English, L. L. 1955. The need for common sense in the control of insect pests. Journal of Economic Entomology 48:279–282.

English, L. L. 1970. Illinois trees and shrubs: Their insect enemies. Illinois Natural History Survey Circular No. 47. 47 pp.

English, L. L. 1990. Camellia pests. Pages 2–5 *in* A. B. Blair, ed., American Camellia Yearbook. American Camellia Society, Fort Valley, Georgia.

English, L. L. and G. F. Turnipseed. 1940. Control of the major pests of the Satsuma orange in south Alabama. Alabama Agricultural Experiment Station Bulletin 248. 48 pp.

Erkilic, L. B. and N. Uygun. 1998. Observations on the population development of *Pseudaulacaspis pentagona* (Hemiptera: Coccoidea: Diaspididae) under semi-field conditions in the east Mediterranean region of Turkey. Eighth International Symposium on Scale Insect Studies. Wye, UK, 1998 August 31–Sept 6. 16 pp.

Erol, T. and B. Yasar. 1996. Van ili elma bahcelerinde bulunan zarari turler ile dogal dusmanlari. [Studies on determination of harmful and beneficial fauna in the apple orchards in Van province.] (In Turkish.) Turkiye Entomoloji Dergisi 20:4, 281–293.

Essig, E.O. 1915. Injurious and beneficial insects of California. California State Printing Office, Sacramento. 541 pp.

Evans, G. A. and P. A. Pedata. 1997. Parasitoids of *Comstockiella sabalis* (Homoptera: Diaspididae) in Florida and description of a new species of the genus *Coccobius* (Hymenoptera: Aphelinidae). Florida Entomologist 80:328–334.

Ezzat, Y. M. 1957. Biological studies on the olive scale, *Parlatoria oleae* (Colvee). Bulletin de la Société Entomologique d'Egypte 41:351–363.

Fabres, G. 1980. Analyse structurelle et fonctionelle de la biocoenose d'un Homoptère (*Lepidosaphes beckii*). Office Recherche Scientifique Technique Outre-mer, Paris. 291 pp.

Fallek, C., G. Yablonka, R. Dahan, S. Mordechai, and M. J. Berlinger. 1988. Winter sprays to control the Hall scale and the olive scale on deciduous fruit trees in Israel. Hassadeh 69:454–455.

Felt, E. P. 1901. Scale insects of importance and list of species in New York State. Bulletin of the New York State Museum of Natural History 46:289–377.

Felt, E. P. 1907. Notes on insects of the year 1906 in New York State. Bureau of Entomology, United States Department of Agriculture, Bulletin No. 67. 39 pp.

Fennah, R. G. 1947. Scale insects of the British West Indies. Proceedings of the Agricultural Society of Trinidad and Tobago 47:63–77.

Fernald, H. J. 1899. The San Jose scale, and other scale insects. Commonwealth of Pennsylvania Department of Agriculture Bulletin 43:5–16.

Fernandes, I. M. 1987. Contribuicao para o conhecimento da quermorfauna da Guine-Bissau. Garcia de Orta, Serie de Zoologia 14:31–37.

Fernandes, I. M. 1989. On the presence in Portugal of the cochineal insect *Aulacaspis tubercularis*, new record Newst. (Homoptera, Coccoidea). García de Orta, Serie de Zoologia. Lisboa 14:27–30.

Fernandez, M., T. Burgos, I. de Val, and M. A. Proenza. 1998. Causas de mortalidad de *Pinnaspis strachani* C. en el cultivo del toronjo en la Isla de la Juventud. Parte II. Revista de Protección Vegetal 13:3, 179–188.

Fernandez, M., I. de Val, M. A. Proenza, and G. Garcia. 1996 (1993). Diagnóstico de la guagua nevada (Homoptera: Diaspididae) en el municipio especial Isla de la Juventud. (In Spanish; Summary in English.) Revista de Protección Vegetal 8:17–22.

Ferris, C. 1981. San Jose scale. Insect identification handbook, University of California, Berkeley, Division of Agricultural Sciences, Publication No. 4099:13–15.

Ferris, G. F. 1921. Report upon a collection of Coccidae from Lower California. Stanford University Publications, Biological Sciences, Palo Alto 1:61–132.

Ferris, G. F. 1937. Atlas of the scale insects of North America (Ser. 1) [Vol. 1], Nos. 1–136. Stanford University Press, Palo Alto, CA.

Ferris, G. F. 1938a. Atlas of the scale insects of North America (Ser. 2) [Vol. 2], Nos. 137–268. Stanford University Press, Palo Alto, CA.

Ferris, G. F. 1938b. Contribution No. 9. Contributions to the knowledge of the Coccoidea VII. Illustrations of fifteen genotypes of the Diaspididae. Microentomology 3:37–70.

Ferris, G. F. 1941. Atlas of the scale insects of North America (Ser. 3) [Vol. 3], Nos. 269–384. Stanford University Press, Palo Alto, California.

Ferris, G. F. 1942. Atlas of the scale insects of North America (Ser. 4) [Vol. 4], Nos. 385–448. Stanford University Press, Palo Alto, California.

Fisher, T. W. and P. DeBach. 1976. Principles, strategies, and tactics of pest population regulation and control in citrus ecosystems. Proceedings of the Tall Timbers Conference. Gainesville, Florida, 1974 Feb 28–March 1. 6:43–50.

Fjelddalen, J. 1996. Skjoldlus (Coccinea, Hom.) I Norge. [Scale insects (Coccinea, Hom.) in Norway.] (In Norwegian.) Insekt-Nytt 21:4–24.

Flanders, S. E. 1948. Biological control of yellow scale. California Citrograph 34:56, 76–77.

Fletcher, T. B. 1921. Additions and corrections to the list of Indian croppests. Proceedings of the Fourth Entomological Meeting. Pusa, India. 2:14–20.

Flint, M. L. 1984. Integrated pest management for citrus. University of California Statewide IPM Project, Division of Agriculture and Natural Resources, Publication 3303. 144 pp.

Flint, M. L. and P. Gouveia. 2001. IPM in practice: Principles and methods of integrated pest management. University of California Statewide IPM Project, Division of Agriculture and Natural Resources, Publication 3418. 296 pp.

Flint, W. P. and M. D. Farrar. 1940. Protecting shade trees from insect damage. University of Illinois, Agricultural Experiment Station and Extension, Circular 509. 59 pp.

Foldi, I. 1990. Moulting and scale-cover formation. Pages 257–265 *in* D. Rosen, ed., Armored scale insects. Vol. 4A. Elsevier, Amsterdam, The Netherlands.

Foldi, I. and S. J. Soria. 1989. Les cochenilles nuisibles a la vigne en amerique du sud (Homoptera: Coccoidea). Annales de la Societe entomologique de France 25:411–430.

Fonseca, J. P. da. 1963. Uma nova praga da mangueira recentemente introduzida no Brasil. O Biológico 29:32–35.

Fosen, E. H., A. W. Cressman, and H. M Armitage. 1953. The Hall scale eradication project. United States Department of Agriculture, Circular No. 92. 16 pp.

Frappa, C. 1937. Les principaux insectes nuisibles aux cultures de Madagascar. Revue de botanique appliqué 17:513–516.

Freitas, A. 1966 (1964). O comportamento bio-ecologico da Cochonilha-de-Sao-Jose (*Quadraspidiotus perniciosus* [Comst.]) em Portugal Continental. Agronomia Lusitana 26:289–335.

Fryer, J. C. S. 1936. Report on insect pests of crops in England and Wales: 1932–1934. Bulletin of the Ministry of Agriculture and Forestry, No. 99. 50 pp.

Fullaway, D. T. 1928. In "Notes and Exhibits." Proceedings of the Hawaiian Entomological Society 7:11–12.

Fullaway, D. T. 1946. Coccidae of Guam. Bernice P. Bishop Museum Bulletin 189:157–162.

Furniss, R. L. and V. M. Carolin. 1977. Western forest insects. United States Department of Agriculture, Miscellaneous Publications No. 1339. 654 pp.

Gallardo-Covas, F. 1983. Mangoes (*Mangifera indica* L.) susceptibility to *Aulacaspis tubercularis* Newstead (Homoptera: Diaspididae) in Puerto Rico. Journal of Agriculture of the University of Puerto Rico 67:179.

Gan, Z. Y., X. Q. Liu, and S. J. Zhang. 1993. [Occurrence and forecasting of the orange brown scale, *Chrysomphalus aonidum* (L.)] (In Chinese.) Entomological Knowledge 30:6, 347–348.

Garcia, D. 1998. Interaction between juniper *Juniperus communis* L. and its fruit pest insects: Pest abundance, fruit characteristics and seed viability. Acta Oecologica, Oecologia Applicata 19:6, 517–525.

Garrett, W. T. 1972. Biosystematics of the oystershell scale, *Lepidosaphes ulmi* (L.) (Homoptera: Diaspididae) in Maryland. Unpublished Dissertation, University of Maryland. 108 pp.

Gatina, E. S. 1973. [Plum in the flood plain of the Dnester Selekt.] (In Russian.) I Sortoizuch. I Yagod. Kul'tur. Kishinev, Moldavian SSR. 121–192.

Geier, P. 1949. Contribution à l'étude de la cochenille rouge du poirer (*Epidiaspis leperii* Sign.) en Suisse. Revue de Pathologie Végétale et d'Entomologie Agricole de France 28:177–266.

Gellatley, J. G. 1968. Citrus scale insects. Agricultural Gazette New South Wales 79:9–15.

Gentile, A. G. and F. M. Summers. 1958. The biology of San José scale on peaches with special reference to the behavior of males and juveniles. Hilgardia 27:269–284.

Geoffrion, R. 1976. La cochenille rouge du poirier (pear scale). Arboriculture Fruitière 265:23–31.

Geoffrion, R. 1979. Traitements d'hiver des arbres fruitiers a pepins. Winter treatment of pip-fruit trees. Phytoma 313:89.

Gerson, U. 1967. Studies of the chaff scale on citrus in Israel. Journal of Economic Entomology 60:1145–1151.

Gerson, U. and A. Hazan. 1979. A biosystematic study of *Aspidiotus nerii* Bouché, with the description of one new species. Journal of Natural History 13:275–284.

Gerson, U., B. M. O'Connor, and M. A. Houk. 1990. Acari. Pages 77–97 *in* D. Rosen, ed., Armored scale insects. Vol. 4A. Elsevier, Amsterdam, The Netherlands.

Gerson, U. and Y. Zor. 1973. The armoured scale insects of avocado in Israel. Journal of Natural History 7:513–533.

Ghauri, M. S. K. 1962. The morphology and taxonomy of male scale-insects (Homoptera: Coccoidea). British Museum (Natural History). 221 pp.

Giacometti, D. C. 1983. Present situation and outlook of the Brazilian citrus industry. Proceedings of the International Society of Citriculture. 2:947–950.

Gieselmann, M. J. 1990. Pheromones and mating behaviour. Pages 221–224 *in* D. Rosen, ed., Armored scale insects. Vol. 4A. Elsevier, Amsterdam, The Netherlands.

Gieselmann, M. J., D. S. Moreno, J. Fargerlund, H. Tashiro, and W.L. Roelofs. 1979a. Identification of the sex pheromone of the yellow scale. Journal of Chemical Ecology 5:27–33.

Gieselmann, M. J. and R. E. Rice. 1990. Use of pheromone traps. Pages 349–352 *in* D. Rosen, ed., Armored scale insects. Vol. 4A. Elsevier, Amsterdam, The Netherlands.

Gieselmann, M. J., R. E. Rice, R. A. Jones, and W. L. Roelofs. 1979b. Sex pheromone of the San Jose scale. Journal of Chemical Ecology 5:891–900.

Gill, R. J. 1982. Color-photo and host keys to the armored scales of California. State of California Department of Food and Agriculture, Environmental Monitoring and Pest Management. Scale and Whitefly Key No. 5. 8 pp.

Gill, R. J. 1997. The Scale Insects of California: Part 3. The Armored Scales (Homoptera: Diaspididae). California Department of Food and Agriculture, Sacramento, CA. 307 pp.

Gill, S. A. 1996. Maryland nursery nutrient management survey report. Cooperative Extension Service, University of Maryland, College Park. 17 pp.

Gill, S. A., D. R. Miller, and J. A. Davidson. 1982. Bionomics and taxonomy of the euonymus scale, *Unaspis euonymi* (Comstock), and detailed biological information on the scale in Maryland (Homoptera: Diaspididae). University of Maryland, Agricultural Experiment Station, College Park, Miscellaneous Publication 969. 36 pp.

Glendenning, R. 1923. Insects of economic importance in the Fraser Valley in 1921. Proceedings of the Entomological Society of British Columbia 17:167–172.

Glick, P. A. 1922. Insects injurious to Arizona crops during 1922. Fourteenth Annual Report of the Arizona Commissoner of Agriculture and Horticulture. 55–77.

Glover, P. M. 1933. *Aspidiotus (Furcaspis) orientalis* Newstead (Coccidae), its economic importance in lac cultivation and its control. Bulletin of the Indian Lac Research Institute No. 16. 23 pp.

Gobbato, C. R. 1936. Principaes pragas e molestias dasvides cultivadas no Rio Grande do Sul. Rodriguesia 2:186–190.

Goberdhan, L. C. 1962. Scale insects of the coconut palm with special reference to *Aspidiotis destructor*. Journal of the Agricultural Society of Trinidad and Tobago 62:49–70.

Gomez Clemente, F. 1943. Cochinillas que atacan a los agrios en la region de Levante. Boletín de Patología Vegetal y Entomología Agrícola 12:299–328.

Gomez-Menor Ortega, J. 1939. La plagas de la vid (*Vitis vinifera*) en la Republica Dominicana. Revista de Agricultura de Republica Dominicana 30:374–378.

Gomez-Menor Ortega, J. 1958. Un insecto plaga del cultivo de la pina de America (*Ananas ananas* L.) existente en las Islas Canarias. Anales de Estudios Atlánticos 4:1–11.

González, R. H. 1969. Biological control of citrus pests in Chile. Pages 839–847 *in* H. D. Chapman, ed., Proceedings of the first international citrus symposium. Riverside, California, 1968 March 16–26. Vol. 2. University of California, Riverside.

González, R. H. 1981. Biologia, ecologia y control de la escama de San Jose en Chile, *Quadraspidiotus perniciosus* (Comst.). Publicaciones en Ciencia Agricultura, Universidad de Chile, Facultad Ciencia Agrarias, Veterinarias, y Forestales, No. 9. 64 pp.

González, R. H. 1989. Manejo de plagas del kiwi en Chile: 1. Degradacion de residuos de lso insecticidas Chloropyrifos y Phosmet. Revista Fruticola 10:35–43.

González, C. and D. Hernandez. 1988. El genero *Lepidosaphes* sobre citricos en Cuba. Ciencia y tecnica en la Agricultura, Citricos y Otros Frutales 11:7–46.

Gordon, F. C. and D. A. Potter. 1988. Seasonal biology of the walnut scale, *Quadraspidiotus juglansregiae* (Homoptera: Diaspididae), and associated parasites on red maple in Kentucky. Journal of Economic Entomology 81:1181–1185.

Gordon, R. D. 1985. The Coccinellidae of America north of Mexico. Journal of the New York Entomological Society 93:1–912.

Gowdey, C. C. 1921. Report of the Government Entomologist. Jamaica Department of Agriculture Report. 1920:25–27.

Gowdey, C. C. 1925. Report of the Government Entomologist. Jamaica Department of Science and Agriculture, Annual Report. 1924:17–20.

Grafton-Cardwell, E., Y. Ouyang, R. Striggow, and S. Vehrs. 2001. Armored scale insecticide resistance challenges San Joaquin Valley citrus growers. California Agriculture 55:20–25.

Grandi, G. 1951. Introuzione allo studio della entomologia. Edizion Agricole, Bologna. 916 pp.

Gravena, S., P. T. Yamamoto, O. D. Fernandes, and I. Benetoli. 1992. Effect of ethion and aldicarb against *Selenaspidus articulatus* (Morgan), *Parlatoria ziziphus* (Lucas) (Hemiptera: Diaspididae) and influence on beneficial fungus. Anais da Sociedade Entomologica do Brasil 21:101–111.

Greaves, A. J., S. W. Davys, B. W. Dow, A. R., Tomkins, C. Thomson, and D. J. Wilson. 1994. Seasonal temperatures and the phenology of greedy scale populations on kiwifruit vines in New Zealand. New Zealand Journal of Crop and Horticultural Science 22:7–16.

Gressit, J. L. and Y. W. Djou. 1950. [Introduction to the study of citrus scale insects and their biological control in China.] (In Chinese; Summary in English.) Lingnan Natural History Survey and Museum. Special Publication 13:1–19.

Grillo, M., N. del Valle, and M. Alvarez. 1983. *Howardia biclavis* (Comstock) on citrus. Centro Agricola 10:47–53.

Griswold, G. H. 1925. A study of the oyster shell scale, *Lepidosaphes ulmi* (L.), and one of its parasites, *Aphelinus mytilaspidis* Le B. Part I. Biology and morphology of the two forms of the oyster-shell scale. Part II. Biology of a parasite of the oyster-shell scale. Cornell University Agricultural Experiment Station, Memoir 93. 67 pp.

Guilleminot, R. A. and J. U. Apablaza. 1986. Insects and arachnids associated with raspberry (*Rubus idaeus*) in the metropolitan region, Chile. Ciencia e Investigación Agraria 13:251–256.

Gumus, M. and N. Uygun. 1992. [Determining a better sampling scheme of the *Aonidiella aurantii* (Maskell) (Homoptera: Diaspididae) which is an important pest of citrus.] (In Turkish.) Turkiye Entomoloji Dergisi 16:209–216.

Habib, A., H. S. Salama, and A. H. Amin. 1971. Population studies on scale insects infesting citrus trees in Egypt. Zeitschrift für Angewandte Entomologie 69:318–330.

Habib, A., H. S. Salama, and A. H. Amin. 1972. The build up of populations of the red scale *Aonidiella aurantii* (Maskell) on citrus in Egypt. Zeitschrift für Angewandte Entomologie 70:378–385.

Hadzibejli, Z. K. 1983. The coccids of the subtropical zone of the Georgian SSR. Akademii Nauk Gruzinskoii SSR "Metswiereba" Tbilisi. 296 pp.

Halbert, S. E. 1996. Entomology section. Tri-ology 35:4–9.

Halbert, S. E. 1998. Entomology section. Tri-ology 37:5–9.

Hall, R. W., L. E. Ehler, and B. Bisabri-Ershadi. 1980. Rate of success in classical biological control of arthropods. Bulletin of the Entomological Society of America 26:280–282.

Halteren, P. van. 1970. Notes on the biology of the scale insect *Aulacaspis mangiferae* Newstead (Diaspididae, Hemiptera) on mango. Ghana Journal of Agricultural Science 3: 83–85.

Hamilton, D. W. and S. A. Summerland. 1953. Distribution and control of Forbes scale in the midwest. Journal of Economic Entomology 46:494–498.

Hammer, O. H. 1938. The scurfy scale and its control. Journal Economic Entomology 31:244–249.

Hamon, A. B. 1980. *Pseudaonidia trilobitiformis* (Green) (Homoptera: Diaspididae). Thirty-third Biennial Report, Florida Department of Agriculture and Consumer Services, Division of Plant Industry 38–39.

Hamon, A. B. 1981. Plumose scale, *Morganella longispina* (Morgan). Entomology Circular, Florida Department of Agriculture and Consumer Services, Division of Plant Industry, No. 266. 2 pp.

Hamon, A. B. 2002. White mango scale, *Aulacaspis tubercularis* Newstead. Pest Alert. <http://doacs.state.fl.us/~pi/enpp/ento/aulacaspis-tubercularis.html>.

Handa, S. and K. K. Dahiya. 1999. Chemical management of mango scales. Agricultural Science Digest 19:112–114.

Hangartner, W. W., M. Suchy, H. K. Wipf, and K. Zurflueh. 1976. Synthesis and laboratory and field evaluation of a new, highly active and stable insect growth regulator. Agriculture and Food Chemical Journal 24:169–175.

Hanks, L. M. and R. F. Denno. 1993a. The white peach scale, *Pseudaulacaspis pentagona* (Targioni-Tozzetti) (Homoptera: Diaspididae): Life history in Maryland, host plants, and natural enemies. Proceedings of the Entomological Society of Washington 95:79–98.

Hanks, L. M. and R. F. Denno. 1993b. The role of demic adaptation in colonization and spread of scale insect populations. Pages 393–422 *in* K. C. Kim and B. A. McPheron, eds., Evolution of insect pests: patterns of variation. Wiley, New York.

Hanks, L. M. and R. F. Denno. 1993c. Natural enemies and plant water relations influence the distribution of an armored scale insect. Ecology 74:1081–1091.

Hanks, L. M. and R. F. Denno. 1994. Local adaptation in the armored scale insect *Pseudaulacaspis pentagona* (Homoptera: Diaspididae). Ecology 75:2301–2310.

Hara, A. H., T. Hata, B. K. S. Hu, and V. L. Tenbrink. 1993. Hot-water immersion as a potential quarantine treatment against *Pseudaulacaspis cockerelli* (Homoptera: Diaspididae). Journal of Economic Entomology 86:1167–1170.

Hare, J. D. and R. F. Luck. 1991. Indirect effects of citrus cultivars on life history parameters of a parasitic wasp. Ecology 72:1576–1585.

Hare, J. D., J. G. Millar, and R. F. Luck. 1993. A caffeic acid ester mediates host recognition by a parasitic wasp. Naturwissenschaften 80:92–94.

Hargreaves, E. 1937. Some insects and their food-plants in Sierra Leone. Bulletin of Entomological Research 28:505–520.

Harris, K. M. 1990. Cecidomyiidae and other Diptera. Pages 61–66 *in* D. Rosen, ed., Armored scale insects. Vol. 4B. Elsevier, Amsterdam, The Netherlands.

Havron, A. and D. Rosen. 1994. Selection for organophosphorus pesticide resistance in two species of *Aphytis*. Pages 209–220 *in* D. Rosen, ed., Advances in the study of *Aphytis* (Hymenoptera: Aphelinidae). Intercept, Andover, U.K.

Hawthorne, R. M. 1975. Estimated damage and crop loss caused by insect/mite pests 1974. California Department of Food and Agriculture. E-82-14:7–8.

He, G. F., W. M. Bao, A. P. Lu, and G. X. Zhang. 1998. [A biological study of armored scale, *Abgrallaspis cyanophylli* with emphasis on temperature and humidity relations.] (In Chinese.) Chinese Journal of Biological Control 14:1–3.

Hearn, C. J. 1979. "Sunburst" citrus hybrid. Hortscience 14:761–762.

Heller, P. R. 1977. Oak Kermes scale. Ornamentals and turf. Cooperative Extension Service, Pennsylvania State University, March, 1977. 2 pp.

Hely, P. C. and J. G. Gellatley. 1961. The purple or mussel scale. Agricultural Gazette of New South Wales 72:14–18, 39.

Hendricks, H. J. and M. L. Williams. 1992. Life history of *Melanaspis obscura* infesting pin oak in Alabama. Annals of the Entomological Society of America. 85:452–457.

Hendrickson, R. M., J. J. Drea, and M. Rose. 1991. A distribution and establishment program for *Chilocorus kuwanae* (Silvestri) (Coleoptera: Coccinellidae) in the United States. Proceedings of the Entomological Society of Washington 93:197–200.

Herbert, F. B. 1936. Insect pests of western oaks and their control. Proceedings of the Annual Meeting of the Western Shade Tree Conference. Berkeley, California, 1930 March 19–20. 3:32–43.

Herms, D. A. and W. J. Mattson. 1992. The dilemma of plants: To grow or defend. Quarterly Review of Biology 67:283–335.

Herrera, J. M. 1964. Ciclos biológicos de las queresas de los cítricos en la costa central. Métodos para su control. Revista Peruana de Entomología 7:1–8.

Heu, R. A. and M. E. Chun. 2000. Sago palm scale *Aulacaspis yasumatsui* Takagi. New pest advisory No. 99–01. <http://www.hawaiiag.org/hdoa/npa/npa99–01>. 2 pp.

Hibbs, E. T. 1956 (1955). Insect pests of roses and their control. Transactions and Proceedings of the Iowa State Horticultural Society 90:86–90.

Hill, C. H. 1952. The biology and control of the scurfy scale on apples in Virginia. Agricultural Experiment Station, Virginia Polytechnic Institute, Blacksburg, Virginia, Technology Bulletin 119. 39 pp.

Hill, M. G., D. J. Allan, R. C. Henderson, and J. G. Charles. 1993. Introduction of armoured scale predators and establishment of the predatory mite *Hemisarcoptes coccophagus* on Latania scale *Hemiberlesia lataniae* in kiwifruit shelter trees in New Zealand. Bulletin of Entomological Research 83:369–376.

Hiltabrand, W. F. 1961. Scale insect report. California Department of Agriculture. Nursery Service Bulletin 50:140.

Hippe, C., F. Schwaller, E. Mani, H. Kull, and F. Kozár. 1995. Bekampfung einheimischer Austernschildlause. Obst- und Weinbau 131:4, 84–85.

Hix, R. L., C. D. Pless, D. E. Deyton, and C. E. Sams. 1999. Management of San Jose scale on apple with soybean-oil dormant sprays. HortScience 34:106–108.

Hodgson, C. J. and D. J. Hilburn. 1990. List of plant hosts of Coccoidea recorded in Bermuda up to 1989. Bulletin No. 39. Department of Agriculture, Fisheries and Parks, Hamilton, Bermuda. 22 pp.

Hodgson, C. J. and D. J. Hilburn. 1991. An annotated checklist of the Coccoidea of Bermuda. Florida Entomologist 74:133–146.

Hodson, A. C. and J. A. Lofgren. 1970. Control of scale insects on trees and shrubs. Agricultural Extension Service Fact Sheet, University of Minnesota, Entomology, No. 34. 2 pp.

Hoffman, R. W. and C. E. Kennett. 1985. Tracking CRS (California red scale) development by degree-days. California Agriculture 39:19–20.

Hole, U. B. and G. N. Salunkhe. 1999. Relationship between the population build up of *Aonidiella aurantii* (Maskell) on rose and weather parameters. Indian Journal of Agricultural Research 33:93–102.

Hollinger, A. H. 1923. Scale insects of Missouri. Research Bulletin. Missouri Agricultural Experiment Station, Columbia 58. 71 pp.

Holmes, J. J. and J. A. Davidson. 1984. Integrated pest management for arborists: Implementation of a pilot program. Journal of Arboriculture 10:65–70.

Hosny, M. 1943. Coccidae new to Egypt, with notes on some other species. Bulletin de la Société Fouad 1er d'Entomologie 27:113–123.

Houser, J. S. 1908. The more important insects affecting Ohio trees. Bulletin of the Ohio Agricultural Experiment Station 194:173–181.

Houser, J. S. 1918. Destructive insects affecting Ohio shade and forest trees. Ohio Agricultural Experiment Station Bulletin 332:165–487.

Howard, F. W., A. Hamon, M. McLaughlin, T. Weissling, and S. L. Yang. 1999. *Aulacaspis yasumatsui* (Hemiptera: Sternorrhyncha: Diaspididae), a scale insect pest of cycads recently introduced into Florida. Florida Entomologist 82:14–26.

Howell, J. O. 1975. Descriptions of some immature stages in two *Diaspis* species (Homoptera: Coccoidea: Diaspididae). Annals of the Entomological Society of America 68:409–416.

Howell, J. O. 1979. The adult male of *Comstockiella sabalis*; morphology and systematic significance. Annals of the Entomological Society of America 72:556–558.

Howell, J. O. and H. H. Tippins. 1973. Immature stages of *Kuwanaspis howardi* (Homoptera: Diaspididae). Journal of the Georgia Entomological Society 8:241–244.

Howell, J. O. and H. H. Tippins. 1977. Descriptions of first instars of nominal type-species of eight diaspidid tribes. Annals of the Entomological Society of America 70:119–135.

Howell, J. O. and H. H. Tippins. 1978. Morphology and systematics of *Odonaspis minima* n. sp. Annals of the Entomological Society of America 71:762–766.

Hoyt, S. C., P. H. Westigard, and R. E. Rice. 1983. Development of pheromone trapping techniques for male San Jose scale (Homoptera: Diaspididae). Environmental Entomology 12:371–375.

Hu, Y. Y., H. G. Dai, and C. S. Hu. 1982. [A preliminary study on poplar scale-insect *Quadraspidiotus gigas* (Thiem et Gerneck).] (In Chinese.) Scientia Silvae Sinicae 18:160–169.

Huba, A. 1962. [The prognosis of the spread and harmfulness of the San Jose scale (*Quadraspidiotus perniciosus* Comst.) in Europe.] (In Czech; Summary in English.) Polnohospodarstvo. Bratislava 9:415–422.

Huffaker, C. B., C. E. Kennett, and G. L. Finney. 1962. Biological control of olive scale, *Parlatoria oleae* (Colvee), in California by imported *Aphytis maculicornis* (Masi). Hilgardia 32:541–636.

Hulley, P. E. 1962. On the behaviour of crawlers of the citrus mussel scale, *Lepidosaphes beckii* (Newm.) (Homoptera: Diaspididae). Journal of the Entomological Society of South Africa 25:56–72.

Hussain, A. A. 1974. Date palms and dates with their pests in Iraq. University of Baghdad, Baghdad, Iraq. 166 pp.

Hutson, R. 1933. Insect pests of stone fruits in Michigan. Michigan State College, Agricultural Experiment Station Special Bulletin No. 244. 40 pp.

Hutson, R. 1936. Sucking insects infesting apples and pears in Michigan. Michigan State College, Extension Bulletin 161. 19 pp.

Ibarra-Nunez, G. 1990. Arthropods associated with coffee trees in mixed plantations in Soconusco, Chiapas, Mexico. I. Variety and abundance. Folia Entomologica Mexicana 79:207–231.

Imamkuliev, A. G. 1966. Coccids most injurious to fruit and subtropical cultures in the Lenkoran Zone of Azerbaidzhan. (In Russian.) Izvestiya Akademii Nauk Azerbaidzan SSR, Seriya Biologii 4:45–51.

Inserra, S. 1969. La cocciniglia rossa forte degli agrumi (*Aonidiella aurantii* Maskell) in Sicilia. Bollettino del Laboratorio di Entomologia Agraria 'Filippo Silvestri.' Portici 27:1–26.

Inserra, S. 1970. Acclimatizzation, diffusion et notes sur la biologic *defeatist Malawians* DeBach en Sicily. Al Awamia 37:39–46.

James, D. G., M. M. Stevens, and K. J. Amalia. 1997. The impact of foraging on populations of *Coccus hesperidum* L. (Hem., Coccidae) and *Aonidiella aurantii* (Maskell) (Hem., Diaspididae) in an Australian citrus grove. Journal of Applied Entomology 121:257–259.

Jarvis, H. 1946. Pests of the mango. Queensland Agricultural Journal 62:10–14.

Jefferson, D. K. and P. B. Schultz. 1995. Differential susceptibility of six *Euonymus* species and cultivars to Euonymus scale, *Unaspis euonymi* (Comstock). Journal of Environmental Horticulture 13:140–142.

Jefferson, D. K., P. B. Schultz, and M. D. Bryan. 1995. Distribution of natural enemies of Euonymus scale, *Unaspis euonymi* (Comstock), in Virginia. Journal of Entomological Science 30:273–278.

Jepson, F. P. 1915. III. Division of Entomology report. Fiji Island Agriculture Annual Report for 1914, Council Paper No 24:17–27.

Johnson, D. T., B. A. Lewis, and J. D. Whitehead. 1999. Grape scale (Homoptera: Diaspididae) biology and management on grapes. Journal of Entomological Science 34:161–170.

Johnson, W. G. 1896. Notes on new and old scale insects. United States Department of Agriculture, Division of Entomology Bulletin 6:74–78.

Johnson, W. G. 1898. Notes on external characters of the San Jose scale, cherry scale, and Putnam's scale. Canadian Entomologist 30:82–83.

Johnson, W. T. and H. H. Lyon. 1976. Insects that feed on trees and shrubs. Comstock Publishing Associates, a division of Cornell University Press, Ithaca, New York. 464 pp. (First edition).

Johnson, W. T. and H. H. Lyon. 1988. Insects that feed on trees and shrubs. Cornell University Press, Ithaca, New York. 556 pp. (Second edition).

Jones, P. R. 1910. Tests of sprays against the European fruit lecanium and the European pear scale. Bureau of Entomology, United States Department of Agriculture. Bulletin 80.147–160.

Jorgensen, C. D., R. E. Rice, S. C. Hoyt, and P. H. Westigard. 1981. Phenology of the San Jose scale (Homoptera: Diaspididae). Canadian Entomologist 113:149–159.

Kalabekov, A. A. 1974. Varietal resistance of fruit species to *Aspidiotus perniciosus*. Referativnyi Zhuranl. 655–665.

Karaca, I. and N. Uygun. 1992. [Population development of *Aonidiella aurantii* (Maskell) (Homoptera: Diaspididae) on different citrus species and varieties.] (In Turkish.) Proceedings of the Second Turkish National Congress of Entomology. Izmir, Turkey. 9–19.

Karaca, I. and N. Uygun. 1993. [Life tables of oleander scale, *Aspidiotus nerii* Bouche on different host plants.] (In Turkish.) Turkiye Entomoloji Dergisi 17:217–224.

Kathiresan, K. 1993. Dangerous pest on nursery seedlings of *Rhizopora*. Indian Forester 119:1026.

Katsoyannos, P. 1992. Olive pests and their control in the Near East. Food and Agriculture Plant Production and Protection Paper No. 115. 178 pp.

Kawai, S. 1980. [Scale insects of Japan in colors.] (In Japanese.) National Agricultural Education Association, Tokyo. 455 pp.

Kawecki, Z. 1956. Studies on the genus *Lecanium* Burim. III. *Lecanium sericeum* Ldgr. (Homoptera, Lecaniidae). Polskie Pismo Entomologiczne 25:213–226.

Kenneth, C. E. 1981. California red scale. Insects identification handbook. University of California Berkeley, Division of Agricultural Science, Publication No. 4099. 2 pp.

Khoualdia, O., A. Rhouma, and M. S. Hmidi. 1993. Contribution à l'étude bio-ecologique de la cochenille blanche *Parlatoria blanchardi* Targ. (Homoptera, Diaspididae) du palmier dattier dans le Djerid (Sud Tunisien). (In French.) Annales de l'institute Nationale de la Recherche Agronomique de Tunisie 66:89–108.

Kimouilo, H. and A. S. Costa. 1987. Symptoms of yellow net of grapevine associated with scale insect feeding. Fitopatologia Brasileira 12:104–106.

Kinjo, M., F. Nakasone, Y. Higa, M. Nagamine, S. Kawai, and T. Kondo. 1996. [Scale insects on mango in Okinawa Prefecture.] (In Japanese.) Proceedings of the Association for Plant Protection of Kyushu 42:125–127.

Knipscher, R. G., D. R. Miller, and J. A. Davidson. 1976. Biosystematics of *Chionaspis nyssae* Comstock (Homoptera: Diaspididae) with evidence supporting leaf and bark dimorphism of the scale. Melanderia 25:1–30.

Knowlton, G. F. and C. F. Smith. 1936. Rose insects and strawberry insects. Proceedings of the Utah Academy of Sciences, Arts and Letters 13:263–267, 289–292.

Kobakhidze, D. N. 1954. Injurious insects of the tea plantations of USSR. Akademii Nauk SSSR Moscow. 100 pp.

Koehler, C. S. 1964. Insect pest management guidelines for California landscape ornamentals. Cooperative Extension, University of California, Division of Agriculture and Natural Resources, Oakland, California. 82 pp.

Köhler, G., V. Tenckhoff, and G. Motte. 1980. Krankheiten und Schälinge an Kaffee in Kuba. Institute Tropicale Landwirten, Leipzig. 166 pp.

Komosinska, H. 1968. Investigations on the scale insects (Homoptera, Coccoidea, Diaspididae) living in greenhouses in Poland. Part I. Polskie Pismo Entomologiczne 38:205–208.

Komosinska, H. 1969. Studies of scale insects (Homoptera, Coccoidea, Diaspididae) of Poland, I. Fragmenta Faunistica 15:267–271.

Komosinska, H. 1975. Studies on the biology of *Mytilaspis conchiformis* forma *conchiformis* (Gmelin) (Homoptera, Coccoidea, Diaspididae) in Poland. Annales Zoologici 33:127–148.

Konar, A. and M. R. Ghosh. 1990. Important pests of orange *Citrus reticulata* in the Darjeeling District. West Benegal Environment and Ecology 8:11–18.

Kondo, T. and S. Kawai. 1995. Scale insects (Homoptera: Coccoidea) on mango in Colombia. Japanese Journal of Tropical Agricultural 39:57–58.

Konstantinova, G. M. 1976. [Coccids—pests of apple.] (In Russian.) Zashchita Rastenii 12:49–50.

Konstantinova, G. M. 1992. [The Japanese scale.] (In Russian.) Zashchita Rastenii 7:43–45.

Koronéos, J. 1934. Les Coccidae de la Gréce Surtout. Du Pélion (Thessalie). I. Diaspididae. Athens. 95 pp.

Kosztarab, M. 1963. The armored scale insects of Ohio (Homoptera: Coccoidea: Diaspididae). Bulletin of the Ohio Biological Survey 2. 120 pp.

Kosztarab, M. 1977. The current state of coccoid systematics. Bulletin of the Virginia Polytechnic Institute and State University, Research Division 127:1–4.

Kosztarab, M. 1990. Economic importance. Pages 307–311 *in* D. Rosen, ed., Armored scale insects, Vol. 4B. Elsevier, Amsterdam, The Netherlands.

Kosztarab, M. 1996. Scale insects of northeastern North America. Identification, biology, and distribution. Virginia Museum of Natural History, Martinsburg, Virginia. 650 pp.

Kosztarab, M. and F. Kozár. 1978. Scale insects—Coccoidea. Fauna Hungariae No. 131, Vol. 17, Part 22. 192 pp.

Kosztarab, M. and F. Kozár. 1988. Scale insects of Central Europe. Akademiai Kiado, Budapest. 456 pp.

Koteja, J. 1990. Life history. Pages 243–254 *in* D. Rosen, ed., Armored scale insects. Vol. 4A. Elsevier, Amsterdam, The Netherlands.

Kotinsky, J. 1909. Report of superintendent of entomology. Division of Entomology, Hawaii Board of Agriculture and Forestry, Report 1909:97–113.

Koya, K. M. A., S. Devasahayam, S. Selvakumaran, and M. Kallil. 1996. Distribution and damage caused by scale insects and mealy bugs associated with black pepper (*Piper nigrum* Linnaeus) in India. Journal of Entomological Research 20:129–136.

Kozár, F. 1990a. Forecasting. Pages 335–340 *in* D. Rosen, ed., Armored scale insects. Vol. 4B. Elsevier, Amsterdam, The Netherlands.

Kozár, F. 1990b. Deciduous fruit trees. Pages 593–602 *in* D. Rosen, ed., Armored scale insects. Vol. 4B. Elsevier, Amsterdam, The Netherlands.

Kozár, F., M. Tzalev, R.A. Viktorin, and J. Horváth. 1979. New data to the knowledge of the scale insects of Bulgaria (Homoptera: Coccoidea). Folia Entomologica Hungarica 32:129–132.

Kozarzhevskaya, E. F. 1956. Biology of *Leucaspis japonica* in Abkhazia (Homoptera, Coccoidea). Review of Applied Entomology 45:472.

Kozarzhevskaya, E. and A. Vlainic. 1982. Bioecological survey of scale insects (Homoptera: Coccoidea) on cultivated flora in Belgrade. Zastita Bilja 33:183–202.

Kreiter, P., A. Panis, and R. Tourniaire. 1999. Morphological variability in a population of *Pseudaulacaspis pentagona* (Targioni Tozzetti) (Hemiptera : Diaspididae) from the Southeastern France. Annales de la Société Entomologique de France 35 (Suppl. S):33–36.

Kr'steva, L. 1977. [Scale insects on vegetable crops.] (In Bulgarian.) Zashchita Rastenii 25:40–44.

Kuhmina, A. V. 1970. *Unaspis citri*. Pages 89–90 *in* N. N. Shutova, ed., Handbook of quarantine and other dangerous pests. Diseases and Weedy Plants. Kolos, Moscow.

Kuitert, L. C. 1967. Observations on the biology, bionomics, and control of white peach scale, *Pseudaulacaspis pentagona* (Targ.) Proceedings of the Florida State Horticultural Society 80:376–381.

Kunincka, G. M. 1970. *Pseudaulacaspis pentagona, Pseudaonidia paeoniae*. Pages 47–51 *in* N. N. Shutova, ed., Handbook of quarantine and other dangerous pests. Diseases and Weedy Plants. Kolos, Moscow.

Kuwana, I. 1911. [Coccidae of Japan.] (In Japanese.) Vol. I., publisher unknown. 157 pp.

Kuwana, I. 1925. I. The Diaspine Coccidae of Japan. III. The genus *Fiorinia*. Technical Bulletin No. 3, Department of Finance, Japan Imperial Plant Quarantine Service. 20 pp.

Kuwana, I. 1928. The Diaspine Coccidae of Japan. V. Genera *Chionaspis, Tsukushiaspis* [n. gen.], *Leucaspis, Nikkoaspis* [n. gen.]. Ministry of Agriculture and Forestry, Japan, Scientific Bulletin No. 1. 39 pp.

Kuwana, I. 1933. The Diaspine Coccidae of Japan. VII. Ministry of Agriculture and Forestry, Japan, Scientific Bulletin No. 3. 42 pp.

Kuznetsov, N. N. 1971. Scales as pests of conifers and some prospects of the integrated control. Proceedings of the 13th International Congress of Entomology, Moscow, 1968. 3: 63–64.

Labanowski, G. 1999. [Scale insects—dangerous pests of *Dracaena*.] Tarczniki—grozne szkodniki draceny. (In Polish.) Ochrona Roslin 43:14–15.

Labanowski, G. and G. Soika. 1998. [Euonymus scale—a potential pest of euonymus in Poland.] Tarcznik trzemielinowiec—potencjalny szkodnik trzmieliny w Polsce. (In Polish.) Ochrona Roslin 42:12–13.

Labuschagne, T. I., M. S. Daneel, and M. De Beer. 1996. Establishment of *Aphytis* sp. (Hymenoptera: Aphelinidae) and *Cybocephalus binotatus* Grouvelle (Coleoptera: Nitidulidae) in mango orchards in South Africa for control of the mango scale, *Aulacaspis tubercularis* Newstead (Homoptera: Diaspididae). Yearbook South African Mango Growers' Association. 16:20–22.

Labuschagne, T. I., H. van Hamburg, and I. J. Froneman. 1995. Population dynamics of the mango scale, *Aulacaspis tubercularis* (Newstead) (Coccoidea: Diaspididae), in South Africa. Israel Journal of Entomology 29:207–217.

Lacroix, D. S. 1926. Miscellaneous observations on a cranberry scale *Targionia dearnessi* (Ckll.). Entomological News 37:249–251.

LaFollette, J. R. 1949. Pest control information. California Citrograph 34:390–392.

Laing, F. 1929. Descriptions of new, and some notes on old species of Coccidae. Annals and Magazine of Natural History 4:465–501.

Lale, N. E. S. 1998. Neem in the Conventional Lake Chad Basin area and the threat of Oriental yellow scale insect (*Aonidiella orientalis* Newstead) (Homoptera: Diaspididae). Journal of Arid Environments 40:2, 191–197.

Lambdin, P. L. 1990. Development of the black willow scale, *Chionaspis salicisnigrae* (Homoptera: Diaspididae), in Tennessee. Entomological News 101:288–292.

Lambdin, P. L., D. Paulsen, and J. D. Simpson. 1993. Development and behavior of the walnut scale on flowering dogwood in Tennessee. Tennessee Farm and Home Science 166:26–29.

Lambdin, P. L. and J. K. Watson. 1980. New collection records for scale insects of Tennessee. Journal of the Tennessee Academy of Science 55:77–81.

Lange, W. H. 1944. Insects affecting guayule with special reference to those associated with nursery plants in California. Journal of Economic Entomology 37:392–399.

Langford, G. S. 1926. The life history of the willow scale (*Chionaspis salicis-nigrae* Walsh). Circular of the Colorado State Entomologist 51:50–58.

Langford, G. S. and E. N. Cory. 1939. Common insects of lawns, ornamental shrubs and shade trees. University of Maryland Extension Service, Bulletin 84:3–54.

Lauricella, S. 1957. Reperti sull' ibernazione di alcuni coccid-diaspidini in Campania. Bollettino del Laboratorio di Entomologia Agraria 'Filippo Silvestri.' 15:197–220.

Lelláková-Dusková, F. 1963. The morphology, metamorphosis, and life-cycle of the scale insect *Quadraspidiotus gigas* (Thiem and Gerneck). Acta Entomologica Musei Nationalis Prague 5:611–648.

Lelláková-Dusková, F. 1969. [Coccoidea-pests of fruit trees.] (In Czech.) Prirodovedne vyricovani, veda a vyroba 19:449–459.

Lengerken, H. G. von. 1932. Das schadlingsbuch. Brehm, Berlin. 194 pp.

LePelley, R. H. 1968. Pests of coffee. Longmans, Green and Co., Ltd. London and Harlow. 590 pp.

Li, C. D. 1996. [Two new species of Aphelinidae (Hymenoptera: Chalcidoidea) from northeastern region of China.] (In Chinese; Summary in English.) Journal of Northeast Forestry University 24:98–101.

Lindgren, D. and R. Dickson. 1941. Fumigation of purple scale with hydrocyanic acid. Journal of Economic Entomology 34:59–64.

Liotta, G. 1970. Diffusion des cochenilles des agrumes in Sicily et introduction d'une nouvelle espece en Sicily occidentale. Al Awamia 37:33–38.

Liu, J. X., K. Y. Liu, S .C. Yan, T. Lin, D. F. Chi, Q. Y. Wu, F. R. Li, and C. S. Li. 1997. [Regulations of the outbreak of *Quadraspidiotus gigas*.] (In Chinese.) Journal of Northeast Forestry University 25:5–9.

Liu, T. 1994 (1993). [Pest control in loquat.] (In Chinese.) Pages 197–203 *in* J. H. Lin and L.-R. Chang, eds., Techniques of loquat production. Taichung District Agricultura Improvement Station Special Publication, No. 34.

Liu, T.-X., R. M. Cooper, R. D. Oetting, and J. O. Howell. 1989a. Description of some immature stages of *Lepidosaphes camelliae*. Annals of the Entomological Society of America 82:9–13.

Liu, T.-X., M. Kosztarab, and M. Rhoades. 1989b. I. Biosystematics of the adult females of the genus *Chionaspis* of North America, with emphasis on polymorphism. Virginia Agricultural Experiment Station, Bulletin 88-2:1–126.

Longo, S., G. Mazzeo, A. Russo, and G. Siscaro. 1994. *Aonidiella citrina* (Coquillett) nuovo parassita degli agrumi in Italia. Informatore Fitopatologia 44:12,19–25.

Loucif, Z., and P. Bonafonte. 1977. Observation des populations du pou de San Jose, *Quadraspidiotus perniciosus* (Comst.) (Hom. Diaspididae) dans la plaine de la Mitidja (Algerie) D'Octobre 1975 to Mai 1976. Fruits 32:253–261.

Lozzia, G. C. 1985. Su alcune cocciniglie rinvenute in Lombardia e zone Limitrofe su piante ornamentali. Notiziario Malattie delle Piante. 106 (III series, No. 33):122–124.

Lu, Y. 1989. [The probability of interspecific competition between two citrus armoured scales: red scale and circular black scale.] (In Chinese.) Studies on the integrated management of citrus insect pests. Academic Book and Periodical Press, Guangzhou, Guangdong, China. 218–223.

Luck, R. F. and D. L. Dahlsten. 1974. Bionomics of the pine needle scale, *Chionaspis pinifoliae*, and its natural enemies at South Lake Tahoe, Calif. Annals of the Entomological Society of America 67:309–316.

Lugger, O. 1900. Bugs injurious to our cultivated plants. University of Minnesota, Agricultural Experiment Station, Bulletin 69:208–245.

Lupo, V. 1943. II *Mytilococcus ficifoliae* (Berlese) e uma forma estiva del *M. conchifomis* (Gmelin). Bolletino del R. Laboratorio di Entomologia Agraria 5:196–205.

Ma, L., C. D. Li, J. Q. Liu, Y. Sun, X. L. Wang, and Y. J. Ji. 1997. [Predatory function of *Chilocorus kuwanae* Silvestri on *Quadraspidiotus gigas* (Thiem et Gerneck).] (In Chinese.) Journal of Northeast Forestry University 25:64–67.

MacGillivray, A. D. 1921. The Coccidae. Scarab, Urbana, Illinois. 502 pp.

MacGowan, J. B. 1983. Insects affecting ornamentals, tropical and subtropical fruits, forest and shade trees; insect detection. Tri-ology 21:1–9.

MacGowan, J. B. 1987. Insects affecting ornamentals, tropical and subtropical fruits, forest and shade trees; insect detection. Tri-ology 26:1–9.

Mackie, D. B. 1933. Entomological service. Monthly Bulletin of the California Department of Agriculture 22:466.

Madsen, H. F. and J. C. Arrand. 1971. The recognition and life history of the major orchard insects and mites of British Columbia. British Columbia Department of Agriculture, Vernon, British Columbia, Canada. 32 pp.

Mague, D. L. and W. H. Reissig. 1983. Airborne dispersal of San Jose scale, *Quadraspidiotus perniciosus* (Comst.) (Homoptera: Diaspididae), Crawlers infesting apple. Journal of Environmental Entomology 12:692–696.

Maheswari, T. U. and K. Purushotham. 1999. Widespread occurrence of scale insects on papaya. Insect Environment 5:17.

Mahmood, R. and A. I. Mohyuddin. 1986. Integrated control of mango pests. Pakistan Agricultural Research Council, Islamaba. 11 pp.

Malipatil, M. B., K. L. Dunn, and D. Smith. 2000. An illustrated guide to the parasitic wasps associated with citrus scale insects and mealybugs in Australia. Natural Resources and Environment, Agriculture Resources Conservation Land Management, Agriculture, Victoria. 152 pp.

Mallamaire, M. A. 1954. Catalog of the main pest insects, Nematodes, Myriapodes and Acariens of cultivated plants of French West Africa and Togo. Inspection Générale de l'Agriculture, Bulletin de la Protection des Végétaux, No. $\frac{1}{2}$ Institute Français d'Outre-mer, Marseille. 24–60.

Mamet, J. R. 1953. The authorship of the species of Coccoidea (Hemiptera) described from Mauritius in 1899. Proceedings of the Royal Entomological Society of London (A) 28:149–152.

Mamet, R. 1958. The *Selenaspidus* complex (Homoptera: Coccoidea). Annales du Musée Royal du Congo Belge, Zoologiques, Miscellanea Zoologica, Tervuren 4:359–429.

Mann, J. 1969. Cactus-feeding insects and mites. Smithsonian Institution, United States National Museum, Bulletin 256. 158 pp.

Mariau, D. and J. F. Julia. 1977. Nouvelles recherches sur la cochenille du cocotier *Aspidiotus destructor* (Sign.). Oleagineux 32:217–224.

Marlatt, C. L. 1899. Temperature control of scale insects. United States Department of Agriculture, Division of Entomology Bulletin 20:73–76.

Marlatt, C. L. 1900. *Aspidiotus diffinis*. Another scale insect of probable European origin recently found in North America. Entomological News 6:425–427.

Marlatt, C. L. 1906. The San Jose or Chinese scale. United States Department of Agriculture, Bureau of Entomology Bulletin No. 62. 89 pp.

Marotta, S. and A. P. Garonna. 1992 (1991). Homoptera Coccoidea nuovi e poco conosciuti delle piante grasse in Italia. (In Italian; Summary in English.) Proceedings of the Sixteenth Italian National Congress of Entomology. (Atti XVI Congresso Nazionale italiano di Entomologia.) 741–746.

Martin, H. 1954. Scale insects on citrus in Tripolitania. FAO Plant Protection Bulletin 11:113–116.

Martorell, L. F. 1940. Some notes on forest entomology. Caribbean Forester 1:23–24.

Martorell, L. F. 1945. A survey of the forest insects of Puerto Rico. Part II. Journal of the Puerto Rico Department of Agriculture 29:355–608.

Marucci, P. E. 1966. Insects and their control. Pages 199–236 in P. Eck and N. F. Childers, eds., Blueberry culture. Rutgers University Press, New Brunswick, New Jersey.

Maskell, W. M. 1895 (1894). Synoptic list of Coccidae reported from Australasia and the Pacific Islands up to December 1894. Transactions and Proceedings of the New Zealand Institute 27:1–75.

Mathys, G. and E. Guignard. 1967. Some aspects of biological control which serve to assist the aphelinid *Prospaltella perniciosi* Tow. against the San Jose scale (*Quadraspidiotus perniciosus* Comst.). Entomophaga 12:223–224.

Mazzoni, E. and P. Cravedi. 1999. White peach scale phenology and climatic parameters. International Organisation for Biological Control—West Palaearctic Regional Section, Bulletin 22:101–106.

McBride, D. 1975. Insect pests of trees and shrubs. Cooperative Extension Service, North Dakota State University, Circular E-296. 4 pp.

McClain, D. C., G. C. Rock, and R. E. Stinner. 1990a. Thermal requirements for development and simulation of the seasonal phenology of *Encarsia perniciosi* (Hymenoptera: Aphelinidae), a parasitoid of the San Jose scale (Homoptera: Diaspididae) in North Carolina orchards. Environmental Entomology 19:1396–1402.

McClain, D. C., G. C. Rock, and R. E. Stinner. 1990b. San Jose scale (Homoptera: Diaspididae): Simulation of seasonal phenology in North Carolina orchards. Environmental Entomology 19:916–925.

McClure, M.S. 1977a. Dispersal of the scale *Fiorinia externa* (Homoptera: Diaspididae) and effects of edaphic factors on its establishment on hemlock. Environmental Entomology 6:539–544.

McClure, M. S. 1977b. Resurgence of the scale, *Fiorinia externa*, on hemlock following insecticide application. Environmental Entomology 6:480–484.

McClure, M. S. 1977c. Parasitism of the scale insects, *Fiorinia externa* by *Aspidiotiphagus citrinus* in a hemlock forest: Density dependence. Environmental Entomology 6:551–555.

McClure, M. S. 1977d. Dispersal of the scale *Fiorinia externa* and effects of edaphic factors on its establishment on hemlock. Environmental Entomology 6:539–544.

McClure, M. S. 1978. Seasonal development of *Fiorinia externa*, *Tsugaspidiotus tsugae*, and their parasite, *Aspidiotiphagus citrinus*: Importance of parasite-host synchronism to the population dynamics of two scale pests of hemlock. Environmental Entomology 7:863–869.

McClure, M. S. 1980a. Competition between exotic species: Scale insects on hemlock. Ecology 61:1391–1401.

McClure, M. S. 1980b. Foliar nitrogen: A basis for host suitabililty for elongate hemlock scale, *Fiorinia externa*. Ecology 61:72–79.

McClure, M. S. 1981. Effect of voltinism, interspecific competition and parasitism on the population dynamics of the hemlock scales, *Fiorinia externa* and *Tsugaspidiotus tsugae* (Homoptera: Diaspididae). Ecological Entomology 6:47–54.

McClure, M. S. 1982. Distribution and damage of two *Pineus* species (Homoptera: Adelgidae) on red pine in New England. Annals of the Entomological Society of America 75:150–157.

McClure, M. S. 1988. The armored scales of hemlock. Pages 46–65 in A. A. Berryman, ed., Dynamics of forest insect populations: Patterns, causes, implications. Plenum Press, New York.

McClure, M. S. 1990a. Impact of host plants. Pages 289–291 in D. Rosen, ed., Armored scale insects. Vol. 4A. Elsevier, Amsterdam, The Netherlands.

McClure, M. S. 1990b. Seasonal history. Pages 315–317 *in* D. Rosen, ed., Armored scale insects. Vol. 4A. Elsevier, Amsterdam, The Netherlands.

McClure, M. S. and M. B. Fergione. 1977. *Fiorinia externa* and *Tsugaspidiotus tsugae*: Distribution, abundance, and new hosts of two destructive scale insects of eastern hemlock in Connecticut. Environmental Entomology 6:807–811.

McComb, C. W. 1986. A field guide to insect pests of holly. Holly Society of America, Baltimore, Maryland. 122 pp.

McComb, C. W. and J. A. Davidson. 1969. Armored scale insects. A checklist of the Diaspididae of Maryland and the District of Columbia. Maryland University, Entomology Leaflet 50. 3 pp.

McDaniel, B. 1969. The armored scale insects of Texas. Part II. Southwestern Naturalist 14:89–113.

McDowell, D. N. 1960. Pests and diseases of trees and shrubs. Wisconsin State Department of Agriculture, Bulletin No. 351. 87 pp.

McKenzie, H. L. 1935a. Life history and control of latania scale on avocado. California Avocado Association, Year Book, 1935:80–82.

McKenzie, H. L. 1935b. Biology and control of avocado insects and mites. University of California, College of Agriculture, Agricultural Experiment Station, Bulletin No. 592. 48 pp.

McKenzie, H. L. 1937. Morphological differences distinguishing California red scale, yellow scale, and related species. University of California Publications in Entomology 6:323–335.

McKenzie, H. L. 1939. A revision of the genus *Chrysomphalus* and supplementary notes on the genus *Aonidiella*. Microentomology 4:51–77.

McKenzie, H. L. 1945. A revision of *Parlatoria* and closely allied genera. Microentomology 10:47–121.

McKenzie, H. L. 1956. The armored scale insects of California. Bulletin of the California Insect Survey, Vol. 5. 209 pp.

McKeown, K. C. 1945. Australian insects. 23. Homoptera 4—snow-flies and scale insects. Australian Museum Magazine 8:336–340.

McLaren, G. F. 1989. Control of oystershell scale *Quadraspidiotus ostreaeformis* (Curtis) on apples in Central Otago. New Zealand Journal of Crop and Horticultural Science 17:221–227.

McLean, R. R. 1931. Insect pests and plant diseases of the avocado in California. California Department of Agriculture Monthly Bulletin 20:443.

Mead, F. W. 1975. Economic insects of Florida. Florida Department of Agriculture and Consumer Services, Division of Plant Industry. 22 pp.

Mead, F. W. 1982. Insects affecting ornamentals, tropical and subtropical fruits, forest and shade trees; insect detection. Tri-ology 21:1–11.

Mead, F. W. 1984. Insects affecting ornamentals; insect detection. Tri-ology 23:1–6.

Mead, F. W. 1991. Technical report. Tri-ology 30:1–2.

Melander, A. L. 1914. Can insects become resistant to sprays? Journal of Economic Entomology 7:167–173.

Melander, A. L. 1915. Varying susceptibility of San Jose scale to sprays. Journal of Economic Entomology 8:475–480.

Melis, A. 1949. Elenco dei principali specie di Insetti che hanno prodotto infestazioni degne dinota in Italia durante l'anno 1949. Redia 34:17–25.

Mendel, Z., D. Blumberg, A. Zehavi, and M. Weissenberg. 1992. Some polyphagous Homoptera gain protection from their natural enemies by feeding on the toxin plants *Spartium junceum* and *Erythrina corallodendrum* (Leguminosae). Chemoecology 3:118–124.

Merrill, G. B. 1953. A revision of the scale-insects of Florida. Bulletin of the State Plant Board of Florida. 1. 143 pp.

Merrill, G. B. and J. Chaffin. 1923. Scale insects of Florida. Florida State Plant Board Quarterly Bulletin. 7:177–298.

Metcalf, C. L., W. P. Flint, and R. L. Metcalf. 1962. Destructive and useful insects. Their habits and control. McGraw-Hill Book Company, New York. 520 pp.

Metcalf, Z. P. 1922. The gloomy scale. North Carolina Agricultural Experiment Station Technical Bulletin No. 21. 23 pp.

Michener, C. D., L. A. Calkins, and R. E. Beer. 1957. Summary of activities in the Southern Division. Biennial Report of the Kansas State Entomological Commission. 25:5–8.

Milaire, H. 1962. La lutte contre les cochenilles des arbres fruitiers. Pomologie Francaise 4:6.

Miles, H. W. and M. Miles. 1935. Insect pests of glasshouse crops. H. C. Lings, Surrey, England. 174 pp.

Miller, D. R. and J. A. Davidson. 1990. A list of armored scale pests. Pages 299–306 *in* D. Rosen, ed., Armored scale insects. Vol. 4B. Elsevier, Amsterdam, The Netherlands.

Miller, D. R. and J. A. Davidson. 1998. A new species of armored scale (Hemiptera: Coccoidea: Diaspididae) previously confused with *Hemiberlesia diffinis* (Newstead). Proceedings of the Entomological Society of Washington 100:193–201.

Miller, D. R., J. A. Davidson, and M. B. Stoetzel. 1984. A taxonomic study of the armored scale *Pseudischnaspis bowreyi* Hempel (Homoptera: Coccoidea: Diaspididae). Proceedings of the Entomological Society of Washington 86:94–109.

Miller, D. R. and M. E. Gimpel. 2002. Diaspidinae and related subfamilies *in* Y. Ben-Dov, D. R. Miller, and G. A. P. Gibson. ScaleNet. <http://www.sel.barc.usda.gov/scalenet/scalenet.htm>.

Miller, D. R. and F. W. Howard. 1981. A new species of *Abgrallaspis* from Louisiana. Annals of the Entomological Society of America 74:164–166.

Miller, D. R. and M. Kosztarab. 1979. Recent advances in the study of scale insects. Annual Review of Entomology 24:1–27.

Milne, W. M. 1974. The relationship between ants and scale insects on citrus. Commonwealth Science and Industrial Research Organization, Division of Entomology, Annual Report. 1973:78.

Miskimen, G. W. and R. W. Bond. 1970. The insect fauna of St. Croix, United States Virgin Islands. Scientific Survey of Puerto Rico and the Virgin Islands. 13:29–33.

Miyanoshita, A., S. Kawai, and K. Fujii. 1993. Host-associated differences in *Aspidiotus cryptomeriae* Kuwana (Homoptera: Coccoidea: Diaspididae) I. Adult morphology and host preference. Applied Entomology and Zoology 28:71–80.

Miyanoshita, A. and S. Tatsuki. 1995. [Host-associated differences in *Aspidiotus cryptomeriae* Kuwana (Homoptera: Coccoidea: Diaspididae).] (In Japanese.) Japanese Journal of Applied Entomology and Zoology 39:159–162.

Miyanoshita, A., S. Tatsuki, T. Kusano, and F. Koichi. 1991. [Variation of esterase isozymes in *Aspidiotus cryptomeriae*.] (In Japanese.) Japanese Journal of Applied Entomology and Zoology 35:317–321.

Mohammad, Z. K., M. W. Gabbour, and M. H. Tawfit. 2001 (1999). Population dynamics of *Aonidiella orientalis* Newstead (Coccoidea: Diaspididae) and its parasite *Habrolepis aspidioti* Annecke (Hymenoptera: Encyrtidae). Entomologica 33:413–418.

Monastero, S. 1962. Le cocciniglie degli agrumi in Sicilia (*Mytilococcus beckii* New.—*Parlatoria zizyphus* Lucas—*Coccus hesperidum* L.—*Pseudococcus adonidum* L.—*Coccus oleae* Bern.—*Ceroplastes rusci* L.). Bollettino dell' Istituto di Entomologia Agraria dell Università Palmero 4:65–151.

Montgomery, J. H. 1921. Third Biennial Report. Quarantine inspector's summary for year ending April 30, 1917. Quarterly Journal of the Florida State Plant Board 2:49–65.

Monti, L. 1955. Ricerche etologiche sev due Coccidi Diaspini: "*Diaspis pentagona*" Targ. e "*Mytilococcus ulmi*" L., nella regione romagnola. Bollettino dell' Istituto di Entomologia della Università (degli Studi) di Bologna 21:141–165.

Morere, J. L. and J. Seuge. 1976. Effects of natural or artificial Preirradiate d food on the life history of two unirradiated insects; mealybugs (*Pseudaulacaspis pentagona* Targ.) and Indian Meal Moths (*Plodia interpunctella* Hubner). Radiation Research 67:120–127.

Morgan, C. V. G. 1967. Fate of the San Jose scale and the European fruit scale on apples and prunes held in standard cold storage and controlled atmosphere storage. Canadian Entomologist 99:650–659.

Morgan, C. V. G. and B. J. Angle. 1968. Notes on the habits of the San Jose scale and the European fruit scale on harvested apples in British Columbia. Canadian Entomologist 100:499–503.

Morgan, D. J. W. and J. D. Hare. 1998. Innate and learned cues: Scale cover selection by *Aphytis Malawians* (Hymenoptera: Aphelinidae). Journal of Insect Behavior 11:463–479.

Morris, H. M. 1927. Injurious insects of Cyprus. Cyprus Agriculture Department Bulletin 4:3–33.

Morrison, A. E. 1946. Insects infesting camellias. California Department of Agriculture Bulletin 35:131–135.

Morrison, H. 1939. Taxonomy of some scale insects of the genus *Parlatoria* encountered in plant quarantine inspection work. United States Department of Agriculture, Miscellaneous Publication No. 344. 34 pp.

Morse, J. G., M. J. Arbaug, and D. S. Moreno. 1985. California red scale—computer simulation on CRS population. California Agriculture 39:8–10.

Mosquera, F. 1976. Escamas protegidas mas frecuentes en Colombia. Boletin Tecnico Ministerio de Agricultura, Instituto Colombiano Agronomico Pecuario Division, Sanidad Vegetal 38. 103 pp.

Moussa, M. E. 1986. Bionomics, relative abundance and natural mortality rates of the oriental red scale, *Aonidiella orientalis* (Newstead) in Riyadh, Saudi Arabia (Homoptera, Diaspididae). (In English; Summary in Arabic.) Journal of the College of Agriculture, King Saud University 8:227–234.

Moutia, L. A. 1935. The commoner insect pests of orchards, vegetables and flower gardens in Mauritius. Mauritius Department of Agriculture General Services Bulletin 44. 39 pp.

Moutia, L. A. and R. Mamet. 1946. A review of twenty-five years of economic entomology in the Island of Mauritius. Bulletin of Entomological Research 36:439–472.

Moutia, L. A. and R. J. Mamet. 1947. An annotated list of insects and Acarina of economic importance in Mauritius. Mauritius Department of Agriculture, Bulletin No. 29. 43 pp.

Muma, M. H. 1970. Preliminary studies on environmental manipulation to control injurious insects and mites in Florida citrus groves. Proceedings of the Tall Timbers Conference on Ecology and Animal Control by Habitat Management. Tallahasee, Florida, 1968 March 10–12. 2:23–40.

Muma, M. H. and D. W. Clancy. 1961. Parasitism of purple scale in Florida citrus groves. Florida Entomologist 44:159–165.

Munir, B. and R. I. Sailer. 1985. Population dynamics of the tea scale, *Fiorinia theae* (Homoptera: Diaspididae), with biology and life tables. Environmental Entomology 14:742–748.

Munoz Ginarte, B. 1937. Cultivo del cocotero. Revista de Agricultura 2:3–9.

Munro, H. K. and F. A. Fouche. 1936. A list of the scale insects and mealy bugs (Coccidae) and their host-plants in South Africa. Bulletin of the Department of Agriculture and Forestry, Union of South Africa, Pretoria. 158. 104 pp.

Murakami, Y. 1970. A review of biology and ecology of diaspine scale in Japan (Homoptera, Coccoidea). Mushi 43:65–114.

Murdoch, W. W., S. L. Swarbrick, R. F. Luck, S. Walde, and D. S. Yu. 1996. Refuge dynamics and metapopulation dynamics: An experimental test. American Naturalist 147:424–444.

Murlidharan, C. M. 1993. Scale insects of date palm (*Phoenix dactylifera*) and their natural enemies in the date groves of Kachchh (Gugarat). Plant Protection Bulletin (Faridabad) 45:31–33.

Murray, D. A. H. 1980. Pineapple scale. Queensland Agricultural Journal 106:271–274.

Murray, D. A. H. 1982. Pineapple scale (*Diaspis bromeliae* [Kerner]) distribution and seasonal history. Queensland Journal of Agricultural and Animal Sciences 39:125–130.

Mussey, G. J. and D. A. Potter. 1997. Phenological correlations between flowering plants and activity of urban landscape pests in Kentucky. Journal of Economic Entomology 90:1615–1627.

Myburgh, A. C. 1990. Crop pests in southern Africa, Vol. 5: Flowers and other ornamentals. Plant Protection Institute, Bulletin No. 419. 98 pp.

Nagarkatti, S. and T. Sankaran. 1990. Tea. Pages 553–562 *in* D. Rosen, ed., Armored scale insects. Vol. 4B. Elsevier, Amsterdam, The Netherlands.

Nair, R., M. M. Balakrishnan, and R. Menon. 1963. Major and minor pests in the arecanut crop *Areca catechu* Linn. Arecanut Journal 14:139–147.

Nakahara, S. 1982. A checklist of the armored scale of the conterminous United States. Plant Protection and Quarantine, Animal and Plant Health Inspection Service, United States Department of Agriculture. 110 pp.

Nascimento, A. S. do. 1981. Pests of citrus and their control. Boletim da Pesquisa 1. 41 pp.

Navrozidis, E. I., Z. D. Zartaloudis, S. H. Papadopoulou, and I. Karayiannis. 1999. Biology and control of San Jose scale, *Quadraspidiotus perniciosus* (Comstock) (Hemiptera, Diaspididae) on apricot trees in northern Greece. Acta Horticulturae 488:695–698.

Negron, J. F. and S. R. Clarke. 1995. Scale insect outbreaks associated with new pyrethroids in a loblolly pine seed orchard. Journal of Entomological Science 30:149–153.

Neiswander, R. B. 1951. Investigations of nursery insects during 1950. Ohio Agricultural College Extension Nursery Notes. 20:1–8.

Neiswander, R. B. 1966. Insect and mite pests of trees and shrubs. Ohio Agricultural Research and Development Center, Research Bulletin No. 983. 50 pp.

Nel, O. G. 1933. A comparison of *Aonidiella aurantii* and *Aonidiella citrina* including a study of the internal anatomy of the latter. Hilgardia 7:417–466.

Nelson-Rees, W. A. 1960. A study of sex predetermination in the mealybug *Planococcus citri* (Risso). Journal of Experimental Zoology 144:111–137.

Nestel, D., N. Pinhassi, H. Reuveny, D. Oppenheim, and D. Rosen. 1995. Development of a predictive phenological model for the spring generation of the olive scale, *Parlatoria oleae* (Colvée), in Israel: Preliminary results. Israel Journal of Entomology 29:227–235.

Neuenschwander, P., S. Michelakis, and V. Alexandrakis. 1977. Biologic et ecologie d'aspidiotus nerii Bouché sur olivier en crete Occidentale. Fruits 32:418–427.

Nichol, A. A. 1935. The olive parlatoria, *Parlatoria oleae* Colvée, in Arizona. Part I. Life history and ecology. University of. Arizona Agricultural Experiment Station, Technical Bulletin 56:201–221.

Nielsen, D. G. 1970. Host impact, population dynamics and chemical control of the pine needle scale, *Phenacaspis pinifoliae* (Fitch) in Central New York. University Microfilms, Ann Arbor, Michigan. 122 pp.

Nielsen, D. G. and N. E. Johnson. 1973. Contribution to the life history and dynamics of pine needle scale *Phenacaspis pinifoliae*, in central New York. Annals of the Entomological Society of America 66:34–43.

Nixon, H. F. 1968. Report of activity in 1968. Pennsylvania Department of Agriculture, Bureau of Plant Industry. 28 pp.

Nur, U. 1971. Parthenogenesis in coccids. American Zoologist 11:301–308.

Oetting, R. D. 1984. Biology of the cactus scale, *Diaspis echinocacti* (Bouché). Annals of the Entomological Society of America 77:88–92.

Ofek, G., G. Huberman, Y. Yzhar, M. Wysoki, W. Kuzlitzky, and S. Reneh. 1997. [The control of the oriental red scale, *Aonidiella orientalis* Newstead and the California red scale, *A. aurantii* (Maskell) (Homoptera: Diaspididae) in mango orchards in Hevel Habsor (Israel).] (In Hebrew.) Alon Hanotea 51:212–218.

Ogilvie, L. 1926. Report of the plant pathologist for the year 1925. Bermuda Board and Department of Agriculture Report. 1925:36–63.

Ohgushi, R., T. S. Nishino, Y. Tazoe, and Y. Gamo. 1967. Ecological observations and the control of California red scale, *Aonidiella aurantii* Maskell, in Nagaski Prefecture. Proceedings of the Kyushu Association Plant Protection 13:118–121.

Onder, E. P. 1982. Investigations on the biology, food-plants, damage and factors affecting seasonal population fluctuations of *Aonidiella* species injurious to citrus trees in Izmir and its surroundings. Arastirma Eserleri Serisi, Turkey 43. 171 pp.

Orphanides, G. M. 1982. Biology of the California red scale *Aonidiella aurantii* Homoptera Diaspididae and its seasonal availability for parasitization by *Aphytis* on Cyprus. Bollettino del Laboratorio di Entomologia Agraria "Filippo Silvestri." 39:203–212.

Otero, A. R. 1935. Insectos del Guayabo en Cuba. Estación Experimental Agronómica, Santiago de Las Vegas, Cuba, Circular No. 78. 26 pp.

Paik, W. H. and J. H. Kim. 1977. [Study on the scale insects of conifers.] (In Korean; Summary in English.) Forestry Bulletin, Seoul National University. 13:41–54.

Palmer, J. M. and L. A. Mound. 1990. Thysanoptera. Pages 67–76 *in* D. Rosen, ed., Armored scale insects. Vol. 4B. Elsevier, Amsterdam, The Netherlands.

Panis, A. and C. Pinet. 1998. On the effectiveness of the diaspidid parasitoid *Coccidencyrtus malloi* Blanchard (Hymenoptera, Encyrtidae) under glasshouse conditions in southeastern France. Entomologica 33:423–428.

Pegazzano, F. 1953. Contributo alla conoscenza della *Pseudaonidia paeoniae* Cockerell. (Hemiptera, Coccidae, Diaspinae). Redia 38:281–315.

Pellizzari-Scaltriti, G. and P. Camporese. 1991. Contributo alla conoscenza delle cocciniglie (Homoptera, Coccoidea) delle querce in Italia. Atti del Convegno: Problematiche fitopatologiche del genere *Quercus* in Italia, Florence, Italy. 193–209.

Penella, J. S. 1942. El cultivo del mango. El Agricultor Venezolano 6:8–13.

Perez Guerra, G. 1986. Coccids of horticultural crops in the Canary Islands. Pages 127–130 *in* Proceedings of the Fifth International Symposium of Scale Insect Studies. Portici (Naples), Italy: June 24–28, 1986. Bollettino del Laboratorio di Entomologia Agraria "Filippo Silvestri." 43 (Suppl.).

Perruso, J. C. and P. C. R. Cassino. 1997. Presence-absence sampling plan for *Selenaspidus articulatus* (Morg.) (Homoptera: Diaspididae) on citrus. Anais da Sociedade Entomologica do Brasil 26:321–326.

Petschen, I., M. P. Borsh, and A. Guerrero. 2000. Enzyme-catalyzed synthesis and absolute configuration of (1S,2R,5S)- and (1R,2S,5R)-2-(1-

hydroxyethyl)-1-(methoxymethyloxyethyl) cyclobutane-1-carbonitrile, key intermediates for the preparation of chiral cyclobutane-containing pheromones. Tetrahedron: Assymetry 11:1691–1695.

Petschen, I., A. Parrilla, M. P. Bosch, C. Amela, and A. A. Botar. 1999. First total synthesis of the sex pheromone of the oleander scale *Aspidiotus nerii*: An unusual sesquiterpenic functionalized cyclobutane. Chemistry—A European Journal 5:3299–3309.

Pettit, R. H. 1928. Report of the Section of Entomology. Experiment Station Reports, Michigan State Board of Agriculture. 188–200.

Picart, J. L. and D. Matile-Ferrero. 2000. Cochenilles en serres de collections. Phytoma 524:44–46.

Pinhassi, N., D. Nestel, and D. Rosen. 1996. Oviposition and emergence of olive scale (Homoptera: Diaspididae) crawlers: Regional degree-day forecasting model. Environmental Entomology 25:1–6.

Pinto, L. J. 1980. Resource utilization patterns of a complex of hymenopterous parasitoids associated with obscure scale (*Melanaspis obscura*) on pin oak (*Quercus palustris*). Master's thesis, University of Maryland, College Park. 79 pp.

Pirone, P. P. 1970. Diseases and pests of ornamental plants. Ronald Press Co., New York. 546 pp.

Plank, H. K. and H. F. Winters. 1949. Insect and other animal pests of Circhona and their control in Puerto Rico. Federal Experiment Station, Puerto Rico, Bulletin 46. 16 pp.

Pless, C. D., D. E. Deyton, and C. E. Sams. 1995. Vegetable oil applications control scale insects on dormant apple and peach trees. HortScience 28:231.

Polavarapu, S., J. A. Davidson, and D. R. Miller. 2000. Life history of the Putnam scale, *Diaspidiotus ancylus* (Putnam) (Hemiptera: Coccoidea: Diaspididae) on blueberries (*Vaccinium corymbosum, ericaceae*) in New Jersey, with a world list of scale insects on blueberries. Proceedings of the Entomological Society of Washington 102:549–560.

Porter, H. L., O. W. Spilker, and J. T. Walker. 1959. The control of insects and plant diseases in the nursery. Ohio Department of Agriculture, Division of Plant Industry. 158 pp.

Potter, D. A. 1998. Destructive turfgrass insects: Biology, diagnosis and control. Ann Arbor Press (a Division of Sleeping Bear Press), Chelsea, Michigan. 344 pp.

Potter, D. A., M. P. Jensen, and F. C. Gordon. 1989. Phenology and degree-day relationships of the obscure scale and associated parasites on pin oak in Kentucky. Journal of Economic Entomology 82:551–555.

Prints, E. Y. 1971. [The type of damage to branches and the resistance of plum and apple to San Jose scale.] (In Russian.) Pages 82–86 *in* M. F. Yaroshenko, ed., The insect fauna of Moldavia. Entomofauna Moldavii. 372 pp.

Putnam, J. D. 1880. Biological and other notes on Coccidae. Proceedings of Davenport Academy of Natural Sciences 2:346–347.

Qin, H. Z., S. J. Tang, and J. R. Feng. 1997. [Preliminary study on *Aonidiella taxus* Leonardi.] (In Chinese.) Journal of Shanghai Agricultural College 15:109–113.

Quayle, H. J. 1912. The purple scale. California Agricultural Experiment Station Bulletin No. 226:319–340.

Quayle, H. J. 1932. Biology and control of citrus insects and mites. California Agricultural Experiment Station Bulletin No. 542. 87 pp.

Quayle, H. J. 1938. Insects of citrus and other subtropical fruits. Comstock Publications Co., Ithaca, New York. 583 pp.

Quiroga, D., P. Arretz, and J. E. Araya. 1991. Sucking insects damaging jojoba, *Simmondsia chinensis* (Link) Schneider, and their natural enemies, in the North Central and Central regions of Chile. Crop Protection 10:469–472.

Rajagopal, D. and A. Krishnamoorthy. 1996. Bionomics and management of oriental yellow scale, *Aonidiella orientalis* (Newstead) (Homoptera: Diaspididae): An overview. Agricultural Research 17:139–146.

Ramakrishna Ayyar, T. V. 1930 (1929). A contribution to our knowledge of South Indian Coccidae (scales and mealy-bugs). Agricultural Research Institute, Pusa, Bulletin No. 197. 73 pp.

Rao, V. P. 1949. The genus *Unaspis* Mac Gillivray. Microentomology 14:59–72.

Rao, V. P. and G. F. Ferris. 1952. The genus *Andaspis* MacGillivray (Insecta: Homoptera: Coccoidea). (Contrib. No. 77). Microentomology 17:17–32.

Rao, V. P. and T. Sankaran. 1969. The scale insects of sugarcane. Pages 325–342 *in* J. R. Williams, J. R. Metcalfe, eds., Pests of sugar cane. Elsevier Publishing Co., New York. 568 pp.

Raupp, M. J., J. J. Holmes, C. Sadof, P. Shrewsbury, and J. A. Davidson. 2001. Effects of cover sprays and residual pesticides on scale insects and natural enemies in urban forests. Journal of Arboriculture 27:203–213.

Rebek, E. J. and C. S. Sadof. 2003. Effects of pesticide applications on the euonymus scale and its parasitoid, *Encarsia citrina*. Journal of Economic Entomology 96:446–452.

Rehman, S. U., H. W. Browning, H. N. Nigg, and J. M. Harrison. 1999. Residual effects of carbaryl and dicofol on *Aphytis holoxanthus* DeBach (Hymenoptera: Aphelinidae). Biological Control 16:252–257.

Reyne, A. 1948 (1946). Studies on a serious outbreak of *Aspidiotus destructor rigidus* in the cocoanut-palms of Sangi (North Celebes). Tijdschrift voor Entomologie 89:83–123.

Rhoades, M. H., M. Kosztarab, and E. G. Rajotte. 1985. Identification, hosts and distribution of *Pseudaulacaspis pentagona* (Targioni-Tozzetti) and *P. prunicola* (Maskell) in Virginia (Homoptera: Diaspididae). Proceedings of the Entomological Society of Washington 87:545–553.

Rice, R. E. 1974. San Jose scale: Field study with a sex pheromone. Journal of Economic Entomology 67:561–562.

Rice, R. E., D. L. Flaherty, and R. A. Jones. 1982. Monitoring and modeling San Jose scale. California Agriculture, 13–14.

Rice, R. E., S. C. Hoyt, and P. H. Westigard. 1979. Chemical control of male San Jose scale in apples, pears, and peaches. Canadian Entomologist 111:827–831.

Rice, R. E. and R. A. Jones. 1977. Monitoring flight patterns of male San Jose scale (Homoptera: Diaspididae). Canadian Entomologist 109:1403–1404.

Rice, R. E. and D. S. Moreno. 1970. Flight of male California red scale. Annals of the Entomological Society of America 63:91–96.

Richards, A. M. 1960. Scale insect survey on apples 1959–1960. New Zealand Journal of Agricultural Research 3:693–698.

Richards, A. M. 1962. The oyster-shell scale, *Quadraspidiotus ostreaeformis* (Curtis), in the Christchurch District of New Zealand. New Zealand Journal of Agricultural Research 5:95–100.

Riedl, H., M. M. Barnes, and C. S. Davis. 1979. Walnut pest management: Historical perspective and present status. Pages 15–80 *in* D. J. Boethel and R. D. Eikenbary, eds., Pest management programs for deciduous tree fruits and nuts. Plenum Press, New York. 256 pp.

Riley, C. V. 1892. The kerosene emulsion: Its origin, nature, and increasing usefulness. Proceedings of the Washington Meeting of Society for the Promotion of Agricultural Science. 83–98.

Riley, C. V. 1903. Notes on Coccidae. Mode of hibernation—effects of severe cold—viviparity—remedies. Proceedings of the Entomological Society of Washington 3:65–71.

Roaf, J. R. and D. C. Mote. 1935. The holly scale, *Aspidiotus britannicus* Newstead and other insects pests of English holly in Oregon. Journal of Economic Entomology 28:1041–1049.

Rock, G. C. and D. C. McClain. 1990. Effects of constant photoperiods and temperatures on the hibernating life stages of the San Jose Scale (Homoptera: Diaspididae) in North Carolina. Journal of Entomological Science 25:615–621.

Rodrigo, E. and F. Garcia-Mari. 1990. Comparacion del ciclo biologico de los diaspinos *Parlatoria pergandii*, *Aonidiella aurantii* y *Lepidosaphes beckii*. Boletin de Sanidad Vegetal, Plagas 16:25–35.

Rodrigo, E. and F. Garcia-Mari. 1994. Estudio de la abundancia y distribucion de algunos coccidos diaspididos de citricos. Boletin de Sanidad Vegetal, Plagas. 20:151–164.

Roelofs, W., M. Gieselman, M. Carde, A. Tashiro, D. S. Moreno, C. A. Henrick, and R. S. Anderson. 1978. Identification of the California red scale sex pheromone. Journal of Chemical Ecology 9:211–224.

Romney, V. A. 1946. Insects found on guayule in northern Mexico. Journal of Economic Entomology 39:670–671.

Rose, M. 1990a. Periodic Colonization of Natural Enemies. Pages 433–440 *in* D. Rosen, ed., Armored scale insects. Vol. 4B. Elsevier, Amsterdam, The Netherlands.

Rose, M. 1990b. Citrus. Pages 535–541 *in* D. Rosen, ed., Armored scale insects. Vol. 4B. Elsevier, Amsterdam, The Netherlands.

Rose, M. and P. DeBach. 1990a. Conservation of natural enemies. Pages 461–472 *in* D. Rosen, ed., Armored scale insects. Vol. 4B. Elsevier, Amsterdam, The Netherlands.

Rose, M. and P. DeBach. 1990b. Foreign exploration and importation of natural enemies. Pages 417–431 *in* D. Rosen, ed., Armored scale insects. Vol. 4B. Elsevier, Amsterdam, The Netherlands.

Rosen, D. 1990. Biological control. Pages 497–505 *in* D. Rosen, ed., Armored scale insects. Vol. 4B. Elsevier, Amsterdam, The Netherlands.

Rosen, D. and P. DeBach. 1978. Diaspididae. Pages 78–128 *in* C. P. Clausen, ed., Introduced parasites and predators of arthropod pests and weeds: A world review. United States Department of Agriculture Handbook No. 480, Washington, D.C.

Rosen, D. and P. DeBach. 1979. Species of *Aphytis* of the world. W. Junk BV Publ., The Hague. 801 pp.

Rosen, D. and P. DeBach. 1990. Ectoparasites. Pages 99–120 *in* D. Rosen, ed., Armored scale insects. Vol. 4B. Elsevier, Amsterdam, The Netherlands.

Ruiz Castro, A. 1944. Un coccido ampelófago, nuevo en España [*Diaspidiotus uvae* (Comstock)]. Boletín de Patología Vegetal y Entomología Agrícola 13:55–74.

Russo, A. and G. Mazzeo. 1992. *Rhizoecus americanus* (Hambleton) and *Pseudaulacaspis cockerelli* (Cooley) (Homoptera Coccoidea): Pest of ornamental plants in Italy. Bollettino di Zoologia Agraria e di Bachicoltura, Serie II. 24:215–221.

Russo, A. and G. Siscaro. 1994. *Diaspis echinocacti* fitomizo del fico d'India in Sicilia. Informatore Agrario 50:37, 73–76.

Rutherford, A. 1914. Insect pests of some leguminous plants. Tropical Agriculture (Ceylon) 43:319–323.

Ryan, H. J. 1935. More citrus acres in Los Angeles County treated for pests, but costs down. California Citrograph 20:114.

Ryan, H. J. 1946. Some Los Angeles County experiences with new insect pest and insect eradication projects. Bulletin of the California Department of Agriculture 35:124–125.

Saakyan-Baranova, A. A. 1954. Pest of greenhouse plants. Akademii Nauk SSSR, Glavnogo Botanicheskogo Sada Trudy. 4:7–41.

Sadeh, D. and U. Gerson. 1968. On the control of *Parlatoria oleae* in apple orchards in the upper Galilee. (In Hebrew.) Hassadeh 48:810–822, 955–957.

Sadof, C. S. 1992. Scale insects on shade trees and shrubs. Department of Entomology, Purdue University Extension Service. 6 pp.

Sadof, C. S. and J. J. Neal. 1993. Use of host plant resources by the euonymus scale, *Unaspis euonymi* (Homoptera: Diaspididae). Annals of the Entomological Society of America 86:614–620.

Sadof, C. S. and M. J. Raupp. 1991. Effect of variegation in *Euonymus japonica* var. *aureus* on two phloem feeding insects, *Unaspis euonymi* (Homoptera: Diaspididae) and *Aphis fabae* (Homoptera: Aphidiidae). Environmental Entomology 20:83–89.

Sadof, C. S. and C. Sclar. 2000. Effects of horticultural oil and foliar- or soil-applied systemic insecticides on euonymus scale in pachysandra. Journal of Arboriculture 26:120–125.

Salama, H. S. 1970. Ecological studies on the scale insect, *Chrysomphalus dictyospermi* (Morgan) in Egypt. Zeitschrift für Angewandte Entomologie 65:427–430.

Salama, H. S. and M. K. Hamdy. 1974a. On the occurrence of *Aspidiotus hederae* on *Cassia* trees in Egypt. Anzieger fur Schadlingskunde, Pflanzen und Umweltschutz 47:138–139.

Salama, N. S. and M. K. Hamdy. 1974b. Studies on populations on two scale insects infesting fig trees in Egypt. Zeitschrift für Angewandte Entomologie 75:200–204.

Salama, N. S. and M. Saleh. 1984. Components of the essential oil of three citrus species in correlation to their infestation with scale insects. Zeitschrift für Angewandte Entomologie 97:393–397.

Samarasinghe, S. and E. J. LeRoux. 1966. The biology and dynamics of the oystershell scale, *Lepidosaphes ulmi* (L.) (Homoptera: Coccidae), on apple in Quebec. Annales de la Société Entomologique de Quebec 11:206–292.

Samways, W. J., W. Nel, and J. A. Prins. 1983. Ant foraging in citrus trees and attending honeydew producing homoptera. Phytophylactica 14:155–158.

Sankaran, T., H. Nagaraja, and A. U. Narasimham. 1984. On some south Indian armoured scales and their natural enemies. Proceedings of the Tenth International Symposium Central European Entomofaunistics. Budapest, Hungary, 1983 August 15–20. 409–411.

Santos, A. C. and S. Gravena. 1995. Controle da cochonilha pardinha *Selenaspidus articulatus* Morgan (Homoptera: Diaspididae) com óleo mineral e dimetoato. Anais da Sociedade Entomologica do Brasil 24:411–414.

Sasscer, E. R. 1918. Important foreign insect pests collected on imported nursery stock in 1917. Journal of Economic Entomology 11:125–129.

Savescu, A. D. 1961. Album de Protectia Plantelor. Vol 2. Daunatorii plantelor de ornament si legumelor. Ministry of Agriculture, Bucarest, Romania. 90 pp.

Savopoulou-Soultani, M. 1997. Laboratory rearing of euonymus scale (Homoptera, Diaspididae) at different temperatures. Journal of Economic Entomology 90:955–960.

Schaub, L., B. Bloesch, C. Hippe, C. Keimer, A. Schmid, and R. Brunetti. 1999. Validation d'un modele de la phenologie du pou de San Jose. Revue Suisse de Viticulture, d'Arboriculture et d'Horticulture 31:253–257.

Schmidt, G. 1940. Gebräuchliche Namen von Schadinsekten in verschiedenen Ländern. Entomologische Beihefte aus Berlin-Dahlem 7:161–364.

Schmutterer, H. 1951. Zur Lebensweise der nadelholz-Diaspidinen und ihrer parasiten in den Nadelwaldern Frankens. Zeitschrift für Angewandte Entomologie 33:111–136.

Schmutterer, H. 1952a. Die ökologie der cocciden (Homoptera, Coccoidea) Frankens [pts. 1–3]. Zeitschrift für Angewandte Entomologie 33:369–420, 544–584.

Schmutterer, H. 1952b. Die ökologie der cocciden Frankens. 3. Abschnitt. Zeitschrift für Angewandte Entomologie 33:65–100.

Schmutterer, H. 1959. Schildlause order Coccoidea. 1. Deckelschildlause Diaspididae Die Tierwelt Deutschlands und der Angrenzenden Meeresteile. 260 pp.

Schmutterer, H. 1998. Die Spindelstrauch-Deckelschildlaus *Unaspis euonymi* (Comst.) als neuer Zierpflanzenschädling in Deutschland. Nachrichtenblatt für den Deutschen Pflanzenschutzdienst. Berlin. 50:170–172.

Schmutterer, H., W. Kloft, and M. Lüdicke. 1957. Coccoidea, Schildläuse, scale insects, cochenilles. Pages 403–520 *in* P. Sorauer, ed., Handbuch der Pflanzendrankeiten. Vol. 5, Part 2. Paul Parey, Berlin. 602 pp.

Schread, J. C. 1955. Aphids and scale insects on ornamentals. Bulletin of the Connecticut Agricultural Experiment Station No. 588. 21 pp.

Schread, J. C. 1970. Controls of scale insects and mealybugs on ornamentals. Bulletin of the Connecticut Agricultural Experiment Station No. 710. 27 pp.

Schuh, J. and D. C. Mote. 1948. Insect pests of nursery and ornamental trees and shrubs in Oregon. Oregon State Experiment Station Bulletin No. 449. 164 pp.

Seabra, A. F. de. 1918. Observations sur quelques especes de cochenilles du Portugal. Bulletin de la Sociedad de Portugal de Sciencia Natural 8:72–81.

Sen, H. K. 1937. Indian lac industry. Science and Culture 2:454–459.

Severin, H. C. and H. H. Severin. 1909. A preliminary list of the Coccidae of Wisconsin. Journal of Economic Entomology 2:296–298.

Shel'Deshova, G. G. 1972. [The importance of frost resistance in coccids (Homoptera, Coccoidea).] Pages 51–74 *in* N. I. Goryshin, ed., [Problems of photoperiodism and diapause in insects]. (In Russian, Summary in English.) Leningrad, USSR, IZD, Leningradskoga Universiteta.

Shen, G. P. and H. M. Liu. 1990. Research on scale insect species in camphor trees in the Nanchang area. Acta Agriculturae Universitatis Jiangxiensis 12:10–17.

Shetlar, D. 2002. Scale control. The P.E.S.T. Newsletter for 22 April 2002, 1–7.

Shiao, S. N. 1978. Bionomics of the camellia scale *Pseudaonidia duplex* (Cockerell) in northern parts of Taiwan. Plant Protection Bulletin (Taiwan) 20:210–223.

Shiao, S. N. 1979. Morphology life history and bionomics of the palm scale *Hemiberlesia cyanophylli* in northern parts of Taiwan. Plant Protection Bulletin (Taiwan) 21:267–276.

Shiao, S. N. 1981. Natural enemies and population fluctuation of the palm scale *Hemiberlesia cyanophylli* Signoret. Chinese Journal of Entomology 1:69–76.

Shour, M. H. 1986. Life history studies of the pine scale *Chionaspis heterophyllae* Cooley and the pine needle scale *C. pinifoliae* (Fitch). Ph.D. dissertation, Purdue University, West Lafayette, Indiana. 269 pp.

Shour, M. H. and D. L. Schuder. 1987. Host range and geographic distribution of *Chionaspis heterophyllae* Cooley and *C. pinifoliae* (Fitch). Indiana Academy of Science. 96:297–304.

Sigwalt, B. 1971. Demographic studies of Diaspine scales. Applications to three species noxious for orange trees in Tunisia. Particular case of a species with overlapping generations *Parlatoria ziziphi* Lucas. Annales de Zoologie—Ecologie Animale 3:5–15.

Sikharulidze, A. M. 1958. The insect pests of the subtropical culture budlings and control measures. Vsesoiu. Institute Chaia 1 Subtropical. Kulture Bulletin 3:3–17.

Silva, P. 1950. The coccids of cacao in Bahia, Brazil. Bulletin of Entomological Research 41:119–120.

Silva, Q. M. A. E., R. D. Cavalcante, M. L. S. Cavalcante, and Z. B. De Castro. 1977. The cashew scale—*Pseudaonidia trilobitiformis* Green in the state of Ceara, Brazil. Fitossanidade 2:1–19.

Simanton, W. A. 1976. Occurrence of insects and mite pests of citrus, their predators and parasitism in relation to spraying operations. Proceedings of the Tall Timbers Conference on Ecological Animal Control by Habitat Management. Gainesville, Florida, 1974 Feb 28–March 1. 6:135–163.

Simmonds, F. J. 1958. The oleander scale, *Pseudaulacaspis pentagona* (Targ.), (Homoptera, Coccidae) in Bermuda. Bulletin of the Bermuda Department of Agriculture 31. 44 pp.

Simmonds, H. W. 1921. The transparent coconut scale *Aspidiotus destructor* and its enemies in southern Pacific. Department of Agriculture, Fiji. Agricultural Circular 2:14–17.

Smee, C. 1936. Report of the Entomologist. Myasaland Department of Agriculture, Annual Report. 22 pp.

Smirnoff, W. 1960. *Lepidosaphes beckii* Newm. parasite of citrus in Morocco, with a description of a method of study of coccids of the family Diaspididae, Morocco Service Recherche Agronomique Enseign, Cahias Recherche Agronomique 10:35–67.

Smit, B. 1937. The cyanide fumigation of citrus trees in the Eastern Cape Provice, South Africa. Union of South Africa, Department of Agriculture and Forestry, Bulletin No. 171. 39 pp.

Smith, C. F. 1969. Controlling peach scale. Research and farming. North Carolina Agricultural. Experiment Station Bulletin No. 28. 12 pp.

Smith, D. 1981. Red scale—a serious pest. Queensland Agricultural Journal 107:11.

Smith, D., G. A. C. Beattie, and R. H. Broadley. 1997. Citrus pests and their natural enemies: Integrated pest management in Australia. State of Queensland, Department of Primary Industries and Horticultural Research and Development Corporation. Brisbane, Australia. 263 pp.

Smith, R. I. 1905. Peach insect. Georgia State Board of Entomology Bulletin 17:59–80.

Smith-Fiola, D. C. 1994. Control of Maskell scale, a new pest of Japanese Black Pine, Ocean County, NJ, 1991. Page 348 *in* A. K. Burditt Jr., ed., Arthropod management tests, Vol. 19. Entomological Society of America, Lanham, Maryland.

Smol'yannikov, V. V. 1980. Homopterous insects as pests of orchards. Zashchita Rastenii 4:62–63.

Snapp, O. I. 1954. Insect pests of the peach in the eastern states. United States Department of Agriculture, Farmer's Bulletin No. 1861. 32 pp.

Snowball, G. J. 1971. Scale insects. Australian Natural History 17:93–96.

Spackman, E. W. 1980. Oyster shell scale and its control. Agricultural Extension Service, University of Wyoming, Bulletin 449R. 2 pp.

Srinivasan, P. M., R. Govindarajan, and A. V. N. Paul. 1974. Control of the rose scale, *Lindingaspis rossi* (Maskell) by granular insecticides. South Indian Horticulture 22:124.

Stafford, E. M. 1947. Possible control of some insects by killing the males. Journal of Economic Entomology 40:278.

Stafford, E. M. 1948. Olive scale. California Agriculture 2:8–9.

Stafford, E. M. 1954. Use of parathion to control olive scale. Journal of Economic Entomology 47:287–295.

Stafford, E. M. and D. F. Barnes. 1948. Biology of the fig scale in California. Hilgardia 18:567–598.

Stannard Jr., J. L. 1965. Polymorphism in the Putnam's scale, *Aspidiotus ancylus* (Homoptera: Coccoidea). Annals of the Entomological Society of America 58:573–576.

Steinberg, S., H. Podoler, and D. Rosen. 1987. Competition between two parasites of the Florida red scale in Israel. Ecological Entomology 12:299–310.

Steiner, M. Y. 1987. Mealybugs and scales in greenhouses and interior plantscapes. Ohio Florists' Association, Bulletin No. 694. 6 pp.

Steiner, M. Y. and D. P. Elliott. 1983. Biological pest management for interior landscapes. Alberta Environmental Centre, Vegreville, Canada. 30 pp.

Steinweden, J. B. 1942. Pest control in the nursery. American Nurseryman 76:8.

Steinweden, J. B. 1945. Identification and control of orchid pests. Orchid Digest 9:264–267.

Steinweden, J. B. 1948. Identification and control of orchid pests. Orchid Digest 12:105–111.

Sterrett, W. D. 1915. The ashes: Their characteristics and management. United States Department of Agriculture Bulletin No. 299. 88 pp.

Stevens, D., A. R. Tompkins, R. H. Blank, and J. G. Charles. 1994. A first stage integration pest management system for kiwifruit. Proceedings of Brighton Crop Protection Conference Pests and Diseases. Brighton, UK, 1993 November 13–17. 135–142.

Steyn, J. J. 1958. The effect of ants on citrus scale at Letaba, South Africa. Proceedings of the Tenth International Congress of Entomology. 4:589–594.

Stickney, F. S. 1934. The external anatomy of the parlatoria date scale, *Parlatoria blanchardi* Targioni Tozzetti, with studies of the head and associated parts. United States Department of Agriculture, Technical Bulletin No. 421. 68 pp.

Stickney, F. S., D. F. Barnes, and P. Simmons. 1950. Date palm insects in the United States. United States Department of Agriculture, Circular No. 846. 57 pp.

Stimmann, M. W. 1969. Seasonal history of a unisexual population of the pine needle scale, *Phenacaspis pinifoliae*. Annals of the Entomological Society of America 62:930–931.

Stimmel, J. F. 1976. Putnam scale, *Diaspidiotus ancylus* (Putnam), Homoptera: Diaspididae. Regulatory Horticulture, Entomology Circular No. 11, Pennsylvania Department of Agriculture, Bureau of Plant Industry. 2:19–20.

Stimmel, J. F. 1979. Seasonal history and distribution of *Carulaspis minima* (Targ.-Tozz.) in Pennsylvania. Proceedings of the Entomological Society of Washington 81:222–229.

Stimmel, J. F. 1980. Seasonal history and occurrence of *Fiorinia externa* Ferris in Pennsylvania. Proceedings of the Entomological Society of Washington 82:700–706.

Stimmel, J. F. 1982. Seasonal history of the white peach scale, *Pseudaulacaspis pentagona* in northeastern Pennsylvania. Proceedings of the Entomological Society of Washington 84:128–133.

Stimmel, J. F. 1986. *Aspidiotus cryptomeriae* Kuwana, an armored scale pest of conifers. Regulatory Horticulture, Entomology Circular No. 108, Pennsylvania Department of Agriculture, Bureau of Plant Industry. 12:21–22.

Stimmel, J. F. 1994. *Lepidosaphes pini* (Maskell), an armored scale on pines. Regulatory Horticulture, Entomology Circular No. 153, Pennsylvania Department of Agriculture, Bureau of Plant Industry. 20:19–20.

Stimmel, J. F. 1995. "Japanese Maple Scale," *Lopholeucaspis japonica* (Cockerell). Regulatory Horticulture, Entomology Circular No. 176, Pennsylvania Department of Agriculture, Bureau of Plant Industry. 21:33–34.

Stimmel, J. F. 1997. Fern scale, *Pinnaspis aspidistrae* (Signoret) Homoptera: Diaspididae. Regulatory Horticulture, Pennsylvania Department of Agriculture. 23:29–30.

Stimmel, J. F. 1999. Maskell scale, *Lepidosaphes pallida* (Maskell) Homoptera: Diaspididae. Regulatory Horticulture, Pennsylvania Department of Agriculture. 25:23–24.

Stoetzel, M. B. 1975. Seasonal history of seven species of armored scale insects of the Aspidiotini (Homoptera: Diaspididae). Annals of the Entomological Society of America 68:489–492.

Stoetzel, M. B. 1976. Scale-cover formation in the Diaspididae. Proceedings of the Entomological Society of Washington 78:323–332.

Stoetzel, M. B. and J. A. Davidson. 1971. Biology of the obscure scale, *Melanaspis obscura* (Homoptera: Diaspididae), on pin oak in Maryland. Annals of the Entomological Society of America 64:45–50.

Stoetzel, M. B. and J. A. Davidson. 1973. Life history variations of the obscure scale (Homoptera: Diaspididae), on pin oak and white oak in Maryland. Annals of the Entomological Society of America 66:308–311.

Stoetzel, M. B. and J. A. Davidson. 1974. Biology, morphology and taxonomy of immature stages of 9 species in the Aspidiotini. Annals of the Entomological Society of America 67:475–509.

Surface, H. A. 1907. Facts and treatment of San Jose scale. Missouri Bulletin of the Division of Zoology 4:365–380.

Suris, M. 1999. Disposición espacial de *Selenaspidus articulatus* Morg. (Coccoidea: Diaspididae) en naranjo Valencia (*Citrus sinensis* L.). Revista de Protección Vegetal 14:17–22.

Swailem, S. M., K. T. Awadallah, and A. A. Shaheen. 1976. Abundance of *Lindingaspis rossi* Mask. on ornamental host plants in Giza and Zagazig region, Egypt (Hemiptera-Homoptera: Diaspididae). Bulletin de la Société Entomologique d'Egypte 60:257–263.

Swaine, J. M. and C. B. Hutchings. 1926. The more important shade tree insects of Eastern Canada and their control. Dominion of Canada, Department of Agriculture, Bulletin No. 63. 58 pp.

Swan, L. A. and C. S. Papp. 1972. The common insects of North America. Harper and Row Publishers, New York. 750 pp.

Swezey, O. H. 1936. A preliminary report on an entomological survey of Guam. Hawaii. Planters' Record 40:307–314.

Swezey, O. H. 1950 (1949). *Morganella longispina* (Morgan) on avocado in Hawaii (Homoptera: Diaspididae). Proceedings of the Hawaiian Entomological Society 14:185–186.

Swirski, E. 1985. Integrated control of arthropods of subtropical fruit trees in the Mediterranean region. *In* Atti XIV Congresso Nazionale Italiano di Entomologia. Palermo, Erice, Bagheria. 781–799, 782–783, 789.

Szeremlei, B., E. Kohary, I. Frank, and K. Kienitz. 1979. [Control techniques of greenhouse plants.] (In Hungarian.) Budapest Fov Novenyved and Agrokemiai Allomas. 59 pp.

Tabatadze, E. S. and V. A. Yasnosh. 1997. Control measures of *Lopholeucaspis japonica* Cockerell (Homoptera: Coccinea) through integrated citrus pest management. Bulletin OILB/SROP (Section Regular Ouest Palearctique) 20:45–51.

Tabatadze, E. S. and V. A. Yasnosh. 1998. The population dynamics and biocontrol of the Japanese scale, *Lopholeucaspis japonica* (Cockerell) in Georgia. Entomologica 33:429–434.

Takagi, S. 1961. A contribution to the knowledge of the Diaspidini of Japan (Homoptera: Coccoidea) Pt. II. Insecta Matsumurana 24:4–42.

Takagi, S. 1969. Diaspididae of Taiwan based on material collected in connection with the Japan—U.S. co-operative science programme, 1965 (Homoptera: Coccoidea) Part I. Insecta Matsumurana. Journal of the Faculty of Agriculture, Hokkaido University, Series Entomology 32. 110 pp.

Takagi, S. 1970. Diaspididae of Taiwan based on material collected in connection with Japan—U.S. co-operative science programme, 1965. Part II. Insecta Matsumurana 3:1–146.

Takagi, S. 1977. Appendix: A new species of *Aulacaspis* associated with a cycad in Thailand (Homoptera: Coccoidea) Insecta Matsumurana (New Series) 11:68–72.

Takagi, S. 1992. *Mitulaspis* and *Sclopetaspis*: Their distributions and taxonomic positions. Insecta Matsumurana 47:44–45.

Takagi, S. and J. O. Howell. 1977. The genus *Quernaspis*, a possible Asio-American element in scale insect biogeography. Insecta Matsumurana (New Series) 11:31–59.

Takagi, S. and S. Kawai. 1967. The genera *Chionaspis* and *Pseudaulacaspis* with a criticism on *Phenacaspis*. Insecta Matsumurana 30:29–43.

Takahashi, R. 1936. A new scale insect causing galls in Formosa (Homoptera). Transactions of the Formosa Natural History Society 26:426–428.

Takahashi, R. and T. Tachikawa. 1956. Scale insects of Shikoku (Homoptera: Coccoidea). Transactions of the Shikoku Entomological Society 5:1–17.

Takahashi, R. and S. Takagi. 1957. A new genus of Diaspididae from Japan (Coccoidea, Homoptera). Kontyu 25:102–104.

Talhouk, A. M. S. 1969. Insects and mites injurious to crops in Middle Eastern Countries. Verlaq Paul Parey, Berlin. 239 pp.

Talhouk, A. S. 1975. Citrus pests throughout the world. Pages 21–23 *in* E. Hafliger, ed., Citrus. Ciba-Geigy Agrochemicals Tech. Monograph No. 4. 88 pp.

Tang, F.-T. 1984. Observations on the scale insects injurious to forestry of North China I. Shanxi Agricultural University Press, Taigu, Shanxi, China, Research Publication 2:122–133.

Tatara, A. 1999. Determination of optimum spraying time for chemical control of mulberry scale, *Pseudaulacaspis pentagona* (Targioni) (Hemiptera: Diaspididae) in tea fields. Jarq-Japan Agricultural Research Quarterly 33:155–161.

Taylor, E. P. 1908. Scale insects of the orchards of Missouri. Missouri State Fruit Experiment Station Bulletin No. 18. 87 pp.

Taylor, T. H. C. 1935. The campaign against *Aspidiotus destructor* Sign., in Fiji. Bulletin of Entomological Research 26:1–102.

Tereznikova, E. M. 1963. Trophic associations of coccids of Transcarpathia. Flora and Fauna Karpat Vol. 2, Akademii Nauk SSSR, Akademii Nauk, USSR Komissila. 182–191.

Tereznikova, E. M. 1969. Coccoidea—pests of agricultural plants in western regions of the Ukraine. Vestnik Zoologii 1:60–65.

Ter-Grigorian, M. A. 1954. Coccids of greenhouse plantings in Erevan and Leninakan. Erevan Akademii Nauk Armian SSR, Izvestiya Biologicheskii 7:61–72.

Thompson, W. L. 1940. Notes on dictyospermum scale infesting citrus. Citrus Industry 21:5, 9.

Thompson, W. L. and J. T. Griffiths Jr. 1949. Purple scale and Florida red scale as insect pests of citrus in Florida. University of Florida, Agricultural Experiment Station, Bulletin 462. 40 pp.

Tikhonov, N. P. 1966. Quarantine pests of fruit-berry crops and vineyards in the RSFSR. (In Russian.) Vol. 3. Rossel'Khozidat, Moscow. 100 pp.

Timlin, J. S. 1964a. The distribution and relative importance of some armoured scale insects on pip fruit in the Nelson/Marlborough orchards during 1959/60. New Zealand Journal of Agricultural Research 7:531–535.

Timlin, J. S. 1964b. The biology, bionomics, and control of *Parlatoria pittospori* Mask.: A pest on apples in New Zealand. New Zealand Journal of Agricultural Research 7:536–550.

Tinker, M. E. 1957. Effect of DDT on the bionomics of Putnam scale, *Aspidiotus ancylus* (Putn.) (Hemiptera, Coccidae). Ph.D. dissertation in Entomology Graduate College of the University of Illinois. 72 pp.

Tippins, H. H. 1968. Observations on *Phenacaspis cockerelli* (Cooley), a pest of ornamental plants in Georgia. Journal of the Georgia Entomological Society 3:13–15.

Tippins, H. H. and R. J. Beshear. 1968. Scale insects (Homoptera: Coccoidea) from grasses in Georgia. Journal of the Georgia Entomological Society 3:134–136.

Tippins, H. H. and R. J. Beshear. 1970. The armored scale insects of Cumberland Island, Georgia. Journal of the Georgia Entomological Society 5:7–12.

Tippins, H. H., H. Clay, and R. M. Barry. 1977. Peonyscale: A new host and biological information. Journal of the Georgia Entomological Society 12:68–71.

Tippins, H. H., J. O. Howell, and R. J. Beshear. 1981. Some immature stages of *Pseudaonidia paeoniae* (Homoptera: Coccoidea, Diaspididae). Journal of the Georgia Entomological Society 16:356–361.

Tippins, H. H. and P. B. Martin. 1982. Seasonal occurrence of bermudagrass scale. Journal of the Georgia Entomological Society 17:319–321.

Tomkins, A. R., C. Thomson, D. J. Wilson, and A. J. Greaves. 1992. Armoured scale insects on unsprayed kiwifruit vines in Waikato. New Zealand Entomologist 15:58–63.

Tourneur, J. C., A. Pham, and R. Hugues. 1975. Evolution des infestations de *Parlatoria blanchardi* (Targioni-Tozzetti) (Homoptera—Diaspididae) au cours de L'Annee dans. L'Adrar Mauritanien. Fruits 30:681–685.

Tranfaglia, A. and G. Viggiani. 1986. Scale insects of economic importance and their control in Italy. Proceedings of the Fifth International Symposium of Scale Insect Studies. Bollettino del Laboratorio di Entomologia Agraria "Filippo Silvestri." Portici, Italy, 1986 June 24–28. 43:215–221.

Tremblay, E. 1969. I casi di macro-trasmissione ereditaria dell'endosimbiosi negli insetti. Memorie della Società Entomologica Italiana 48:17–24.

Trimble, F. M. 1929. Scale insects injurious in Pennsylvania. Pennsylvania Department of Agriculture, Bulletin 12. 21 pp.

Trujillo Peluffo, A. 1942. Insects and other parasites of crops and agricultural products of Uruguay. Montevideo, Uruguay. 323 pp.

Tuncyürek, M. and E. Erkin. 1981. [Studies on the population fluctuations of the citrus armored-scale insects and the activity of their parasites.] (In Turkish; Summary in English.) Bitki Koruma Bulteni. Ankara (Plant Protection Bulletin) 21:173–196.

Turnipseed, G. F. and C. F. Smith. 1953. Life history and control of scales on apples in North Carolina. Journal of Economic Entomology 46:969–972.

Tzalev, M. 1964. [New scale insects established on plants in greenhouses in this country (Bulgaria).] (In Bulgarian.) Zashchita Rastenii 12:15–20.

Uematsu, H. 1978. [Bionomics of *Aonidiella taxus* Leonardi (Homoptera: Diaspididae).] (In Japanese; Summary in English.) Science Bulletin of the Faculty of Agriculture, Kyushu University 33:25–31.

Urich, F. W. 1893. Notes on some insect pests of Trinidad, British West Indies. Insect Life 6:196–198.

Vandenberg, S. R. 1929. Report of the Entomologist. Guam Agricultural Experiment Station Report. 1927:12–17.

Van Driesche, R. G. and T. S. Bellows Jr. 1996. Biological control. Chapman & Hall, New York. 539 pp.

Van Driesche, R. G., P. Kingsley, M. Rose, and M. Bryan. 1998. Effect of euonymus scale (Homoptera: Diaspididae) on *Euonymus* spp. survival in southern New England, with estimates of economic costs of pest damage. Environmental Entomology 7:217–220.

Van Duyn, J. and M. Murphy. 1971. Life history and control of white peach scale, *Pseudaulacaspis pentagona*. Florida Entomologist 54:91–95.

Van Harten, A. 1992. Biological pest control in the Republic of Yemen: Current status and future prospects. Boletim da Sociedade Portuguesa de Entomologia 139:127.

Vashadze, V. N. 1955. Review of pests of woody shrub, and flowering decorative plantings of the black sea coast of western Georgia. Akademii Nauk Gruzinskoi SSR, Sukhumi Botanicheskikh Sadov Trudy 8:387–396.

Vesey-FitzGerald, D. 1940. The control of Coccidae on coconuts in Seychelles. Bulletin of Entomological Research 31:253–285.

Vesey-FitzGerald, D. 1953. Review of the biological control of coccids on coconut palms in the Seychelles. Bulletin of Entomological Research 44:405–413.

Viggiani, F. and F. Iannaccone. 1973. Osservazioni sulla biologia e sui parassiti del Diaspini *Chrysomphalus dictyospermi* (Morg.) e *Lepidosaphes beckii* (Newm.) svolte in Campania nel triennio 1969–1971. Bollettino del Laboratorio di Entomologia Agraria "Filippo Silvestri" 30:104–116.

Villardebo, A. 1974. Les cochenilles des agrumes, dans L'quest African. Repartition et developpment en relation avec la climatologie. SROP Bulletin 3:67–78.

Wainhouse, D. and R. S. Howell. 1983. Intraspecific variation in beech scale populations and in susceptibility of their host *Fagus sylvatica*. Ecological Entomology 8:351–359.

Wallner, W. E. 1964. Fiorinia hemlock scale-pest of ornamental hemlock. Farming Research 1:2–3.

Wang, C. and T. Su. 1989. The effect of temperature on population parameters of the latania scale, *Hemiberlesia lataniae* Signoret. Chinese Journal of Biological Control 9:151–156.

Wang, G. Y., Q. Shen, Z. L. Xu, S. B. Zhang, and G. Q. Jiang. 1998a. [The important pests occurring in the main arbutus producing areas in Zhejiang province.] (In Chinese.) South China Fruits 27:2, 29–30.

Wang, H., R. V. Varma, and T. Xu. 1998b. Insect pests of bamboos in Asia. International Network for Bamboo and Rattan, Beijing. 200 pp.

Warner, C. N. 1949. A preliminary study of the biology and control of the euonymus scale, *Unaspis euonymi* (Comstock). Master's thesis, Department of Entomology, University of Massachusetts. 82 pp.

Waterston, J. M. 1949. The pests of juniper in Bermuda. Tropical Agriculture 26:5–15.

Watson, J. R. 1926. Citrus insects and their control. Florida University, Experiment Station Bulletin 183:293–423.

Wearing, C. H. 1976. San Jose scale, *Quadraspidiotus perniciosus* (Comstock), life cycle. Department of Scientific and Industrial Research Information Service No. 105/21. 11 pp.

Webster, P. J. 1920. The mango. Bureau of Agriculture, the Philippines, Bulletin No. 18. 70 pp.

Webster, W. J. 1968. A record of *Parlatoria pittospori* Mask. on apples. Journal of the Australian Entomological Society 7:85–86.

Wehrle, L. P. 1935. The olive parlatoria, *Parlatoria oleae* Colvee, in Arizona. Part II. Economic significance and control. University of Arizona, Agricultural Experiment Station, Technical Bulletin 56:222–235.

Weigel, C. A. and L. G. Baumhofer. 1948. Handbook on insect enemies of flowers and shrubs. United States Department of Agriculture, Miscellaneous Publication No. 626. 115 pp.

Weigel, C. A. and E. R. Sasscer. 1923. Insects injurious to ornamental greenhouse plants. United States Department of Agriculture, Farmer's Bulletin No. 1362. 80 pp.

Werner, W. H. R. 1931. Observations on the life-history and control of the fern scale, *Hemichionaspis aspidistrae* Signoret. Michigan Academy of Science, Arts, and Letters. 13:517–541.

Westcott, C. 1973. The gardener's bug book, Ed. IV. Doubleday and Company, Garden City, New York. 689 pp.

Westigard, P. H. 1979. Integrated pest management of insects and mites for pear. Pages 25–38 *in* D. C. Boethel and R. D. Eikenbary, eds., Pest management programs for deciduous tree fruits and nuts. Plenum Press, New York.

Westigard, P. H., P. B. Lombard, D. W. Berry. 1979. Integrated pest management of insects and mites attacking pears in southern Oregon. Oregon State University, Agricultural Experiment Station, Bulletin No. 634. 41 pp.

Wheatley, P. E. 1961. The insect pests of agriculture in the coast provinces of Kenya. 2—cashew. East African Agricultural and Forestry Journal 26:178–181.

Whitehead, J. D. 1963. The biology and control of the grape scale, *Aspidiotus uvae* (Comst.) in northwestern Arkansas. Master's thesis, University of Arkansas, Fayetteville. 99 pp.

Wilcox, E. V. and V. S. Holt. 1913. Ornamental hibiscus in Hawaii. Hawaii Agricultural Experiment Station Bulletin No. 29. 16 pp.

Willard, J. R. 1973. Wandering time of the crawlers of California red scale, *Aonidiella aurantii* (Mask.) (Homoptera: Diaspididae), on citrus. Australian Journal of Zoology, Melbourne 21:217–229.

Williams, D. J. 1970. Coccoidea. Pages 425–431 *in* T. E. Woodward, J. W. Evan, and V. F. Eastop, eds., Insects of Australia (Hemiptera). Melbourne University Press, Australia.

Williams, D. J. 1988. *Fiorinia externa* Ferris (Hemiptera: Diaspididae) found in Surrey infesting *Abies koreana*. Entomologist's Gazette 39:151–152.

Williams, D. J. and G. W. Watson. 1988. The scale insects of the tropical South Pacific region. Pt. 1: The armoured scales (Diaspididae). London: CAB International Institute of Entomology. 290 pp.

Williams, J. R. and D. J. Greathead. 1990. Sugar cane. Pages 563–578 *in* D. Rosen, ed., Armored scale insects. Vol. 4B. Elsevier, Amsterdam, The Netherlands.

Williams, R. N. 1991. Rose scale. Pages 77–78 *in* M. A. Ellis, R. H. Converse, R. N. Williams, and B. Williamson, eds., Compendium of raspberry and blackberry diseases and insects. American Pathological Society, APS Press, St. Paul, Minnesota.

Willoughby, P. A. and M. Kosztarab. 1974. Morphological and biological studies on two species of *Chionaspis* (Homoptera: Coccoidea: Diaspididae). Bulletin of the Virginia Polytechnic Institute and State University Research Division 92. 83 pp.

Wilson, C. E. 1923. Virgin Islands Agricultural Experiment Station Report for 1922. 9 pp.

Wilson, G. F. 1938. Pests of commercial ornamental plants. Scientific Horticulture 6:102–115.

Wolcott, G. N. 1933. An economic entomology of the West Indies. Entomological Society of Puerto Rico, San Juan. 520 pp.

Wolcott, G. N. 1937. University of Puerto Rico, Agricultural Experiment Station Annual Report 1935–36. 57 pp.

Wolcott, G. N. 1948. The insects of Puerto Rico. Journal of Agriculture of the University of Puerto Rico 32:1–224.

Wolcott, G. N. 1960. Efficiency of lady beetles in insect control. Journal of Agriculture of the University of Puerto Rico 44:166–172.

Wolcott, G. N. and L. F. Martorell. 1943. The seasonal cycle of insect abundance in Puerto Rican cane fields. Journal of Agriculture of the University of Puerto Rico 27:85–104.

Wolcott, G. N. and L. F. Martorell. 1944. Introduced lady beetles on Mona Island. Journal of Economic Entomology 37:451.

Wolfenbarger, D. O. 1951. Dictyospermum scale control on avocados. Florida Entomologist 34:54–58.

Wolfenbarger, D. O. 1955. Mango insect pest control. Proceedings of the Florida Mango Forum, Miami. 27–34.

Wolfenbarger, D. O. 1963. Insect pests of the avocado and their control. University of Florida, Agricultural Experiment Station Bulletin 605a. 52 pp.

Wolff, V. R. S. and E. Corseuil. 1993. Espécies de Diaspididae (Hom.: Coccoidea) ocorrentes em plantas cítricas no Rio Grande do Sul, Brasil: I—Aspidiotinae. Biociências. 1:5–60.

Woodard, J. S. 1925. Pecan insects: The pecan in Texas. Texas Department of Agriculture Bulletin No. 81. 169 pp.

Wysoki, M. 1977. Insect pests of macadamia in Israel. Phytoparasitica 5:187–188.

Wysoki, M. 1997. Present status of arthropod fauna in mango orchards in Israel. Acta Horticulturae 455:805–811.

Xie, Y. P., J. L. Xue, X. Q. Liu, J. P. Li, and M. Tang. 1995. [The epidemic of city forest scales and its management strategy.] (In Chinese.) Forest Research 8:114–118.

Xu, G. T. 1981. Study on *Insulaspis pini* (Maskell). Scientia Silvae Sinicae 17:314–316.

Yan, J. Y. and M. B. Isman. 1986. Environmental factors limiting emergence and longevity of male California red scale, *Aonidiella aurantii* Homoptera Diaspididae. Environmental Entomology 15:971–975.

Yan Aojin, G. L. Xie, and Z. X. Feng. 1985. Studies on the bamboo scale *Kuwanaspis pseudoleucaspis* (Kuwana). Journal of the Nanjing Institute of Forestry 4:41–50.

Yasnosh, V. A. 1986. Integrated control of scale insects in citrus groves in USSR. Bollettino del Laboratorio di Entomologia Agraria "Filippo Silvestri." 43(Suppl.):229–234.

Yasnosh, V. A. 1995. Coccids of economic importance and their control in the republic of Georgia. Israel Journal of Entomology 29:247–251.

Yasumatsu, K. and C. Watanabe. 1965. A tentative catalogue of insect natural enemies of injurious insects in Japan. Part 2. Host parasite—predator catalogue. Entomology Laboratory, Kyushu University. 47 pp.

Young, B. L., R. H. Johnson, and D. N. Alstad. 1993. Sex-ratio variation in scale insects on ponderosa pine: Effects of needle age class. Bulletin of the Ecological Society of America 74:497.

Zahradník, J. 1968. Schildlause unserer Gewachshauser. Neue Brehm-Bucherei, Germany. 42 pp.

Zahradník, J. 1990a. Scale insects (Coccinea) on greenhouse plants and house plants in the Czech provinces. (In German.) Acta Universitatis Carolinae—Biologica 34. 151 pp.

Zahradník, J. 1990b. Conifers. Pages 633–644 *in* D. Rosen, ed., Armored scale insects. Vol. 4B. Elsevier, Amsterdam, The Netherlands.

Zheng, J. T., Y. Q. Xia, Y. Y. Zhou, J. T. Zheng, Y. Q. Xia, and Y. Y. Zhou. 1993. [Forecasting the optimum control date for the first generation of pear white scale with the comprehensive correlation method.] (In Chinese.) China Tea 15:1, 14–15.

Zhou, C. A., J. J. Zou, and J. C. Peng. 1993. [Bionomics of coconut scale—a main pest insect on *Actinidia* and its control.] (In Chinese.) Entomological Knowledge 30:1, 18–20.

Zimmer, J. F. 1912. The grape scale. United States Department Agriculture, Bureau of Entomology, Bulletin No. 97. 124 pp.

Zimmerman, E. C. 1948. Insects of Hawaii, Vol. 5. Homoptera: Sternorhyncha. University of Hawaii Press, Honolulu. 464 pp.

Zocchi, R. 1960. Die in Italien derzeit wichtigsten tierischen shädlinge an olive, weinrebe und ostrbäumen. Anzeiger für Schädlingskunde wereinigt mit schädlingsbekämpfung 33:161–164.

Zoebelein, G. 1966. Probleme der Bekämpfung schädlicher Insekten in der Landwirtschaft im Iran. Anzeiger für Schädlingskunde wereinigt mit schädlingsbekämpfung 39:3–8.

Zou, J. J. and C. A. Zhou. 1993. [Predation of *Pseudaulacaspis cockerelli* (Cooley) by *Chilocorus chalybeatus* Gorham.] (In Chinese.) Entomological Knowledge 30:3, 174–176.

Zuniga, E. 1971. Biologia de la conchuela morada, *Lepidosaphes beckii* (Newm.) en trea areas citricolas de chile (Homoptera: Diaspididae). Revista Pervana Entomologia 14:285–290.

Index of Host Plants of Armored Scales

This diagnostic index provides an alphabetical list of plants by their genus names, and the pests associated with each plant.

Abies: Abgrallaspis ithacae (hemlock scale), 44; *Aspidiotus cryptomeriae* (cryptomeria scale), 68; *Chionaspis heterophyllae* (pine scale), 112; *C. pinifoliae* (pine needle scale), 116; *Fiorinia externa* (elongate hemlock scale), 200; *F. japonica* (coniferous fiorinia scale), 206; *Hemiberlesia lataniae* (latania scale), 224; *Lepidosaphes pini* (pine oystershell scale), 260; *L. ulmi* (oystershell scale), 266; *Nuculaspis tsugae* (shortneedle conifer scale), 304

Acacia: Abgrallaspis cyanophylli (cyanophyllum scale), 38; *Aspidiotus nerii* (oleander scale), 78; *Chrysomphalus dictyospermi* (dictyospermum scale), 130; *Hemiberlesia lataniae* (latania scale), 224; *H. rapax* (greedy scale), 232; *Howardia biclavis* (mining scale), 234; *Ischnaspis longirostris* (black thread scale), 238; *Lindingaspis rossi* (black araucaria scale), 272; *Melanaspis tenebricosa* (gloomy scale), 284; *Pseudaulacaspis pentagona* (white peach scale), 360; *Situlaspis yuccae* (yucca scale), 388; *Unaspis citri* (citrus snow scale), 392

Acalypha: Parlatoria proteus (proteus scale), 330; *Pseudoparlatoria ostreata* (gray scale), 370; *Selenaspidus articulatus* (rufous scale), 386

Acanthocereus: Diaspis echinocacti (cactus scale), 186

Acanthus: Pseudaulacaspis pentagona (white peach scale), 360

Acer: Aspidiotus destructor (coconut scale), 72; *Aulacaspis tubercularis* (white mango scale), 90; *Chionaspis salicis* (willow scale), 118; *Clavaspis ulmi* (elm armored scale), 136; *Diaspidiotus ancylus* (Putnam scale), 148; *D. forbesi* (Forbes scale), 152; *D. juglansregiae* (walnut scale), 158; *D. liquidambaris* (sweetgum scale), 162; *D. uvae* (grape scale), 176; *Hemiberlesia lataniae* (latania scale), 224; *Lepidosaphes conchiformis* (fig scale), 250; *L. ulmi* (oystershell scale), 266; *L. yanagicola* (fire bush scale), 270; *Lopholeucaspis japonica* (Japanese maple scale), 276; *Melanaspis obscura* (obscure scale), 282; *M. tenebricosa* (gloomy scale), 284; *Parlatoria camelliae* (camellia parlatoria scale), 316; *P. theae* (tea parlatoria scale), 334; *Pseudaonidia duplex* (camphor scale), 346; *P. paeoniae* (peony scale), 350; *Pseudaulacaspis prunicola* (white prunicola scale), 364; *Unaspis euonymi* (euonymus scale), 396

Achillea: Rhizaspidiotus dearnessi (Dearness scale), 380

Achras: Diaspis boisduvalii (Boisduval scale), 180; *Hemiberlesia rapax* (greedy scale), 232; *Howardia biclavis* (mining scale), 234; *Ischnaspis longirostris* (black thread scale), 238; *Selenaspidus articulatus* (rufous scale), 386

Acineta: Diaspis boisduvalii (Boisduval scale), 180

Acoelorrhaphe: Comstockiella sabalis (palmetto scale), 140

Acokanthera: Hemiberlesia lataniae (latania scale), 224

Acorus: Aspidiotus destructor (coconut scale), 72; *Chrysomphalus dictyospermi* (dictyospermum scale), 130

Acrocomia: Chrysomphalus aonidum (Florida red scale), 122; *Hemiberlesia lataniae* (latania scale), 224; *Ischnaspis longirostris* (black thread scale), 238

Actinidia: Diaspidiotus ancylus (Putnam scale), 148; *D. perniciosus* (San Jose scale), 172; *Hemiberlesia lataniae* (latania scale), 224; *H. rapax* (greedy scale), 232; *Pseudaulacaspis pentagona* (white peach scale), 360

Adansonia: Duplaspidiotus tesseratus (tesserate scale), 192

Adenium: Aspidiotus excisus (aglaonema scale), 76; *Hemiberlesia lataniae* (latania scale), 224; *Pseudaonidia trilobitiformis* (trilobe scale), 354; *Pseudaulacaspis cockerelli* (false oleander scale), 356

Aechmea: Aspidiotus excisus (aglaonema scale), 76; *A. nerii* (oleander scale), 78; *Diaspis boisduvalii* (Boisduval scale), 180; *D. bromeliae* (pineapple scale), 182; *Gymnaspis aechmeae* (flyspeck scale), 222; *Parlatoria proteus* (proteus scale), 330; *Pseudischnaspis bowreyi* (Bowrey scale), 368

Aegle: Aonidiella citrina (yellow scale), 54; *Aspidiotus destructor* (coconut scale), 72; *Duplaspidiotus tesseratus* (tesserate scale), 192

Aerides: Parlatoria proteus (proteus scale), 330

Aesculus: Clavaspis ulmi (elm armored scale), 136; *Diaspidiotus juglansregiae* (walnut scale), 158; *Lepidosaphes ulmi* (oystershell scale), 266; *Melanaspis tenebricosa* (gloomy scale), 284

Afzelia: Clavaspis herculeana (herculeana scale), 134; *Selenaspidus articulatus* (rufous scale), 386

Agathis: Hemiberlesia rapax (greedy scale), 232

Agave: Aonidiella orientalis (Oriental armored scale), 58; *Aspidiotus destructor* (coconut scale), 72; *A. nerii* (oleander scale), 78; *Chrysomphalus aonidum* (Florida red scale), 122; *C. dictyospermi* (dictyospermum scale), 130; *Comstockiella sabalis* (palmetto scale), 140; *Fiorinia fioriniae* (palm fiorinia scale), 204; *Furcaspis biformis* (orchid scale), 216; *Hemiberlesia lataniae* (latania scale), 224; *Parlatoria proteus* (proteus scale), 330; *Pseudischnaspis bowreyi* (Bowrey scale), 368; *Pseudoparlatoria ostreata* (gray scale), 370; *Selenaspidus articulatus* (rufous scale), 386; *Situlaspis yuccae* (yucca scale), 388

Ageratum: Rhizaspidiotus dearnessi (Dearness scale), 380

Aglaia: Aulacaspis tubercularis (white mango scale), 90; *Lepidosaphes gloverii* (Glover scale), 254

Aglaonema: Aspidiotus destructor (coconut scale), 72; *A. excisus* (aglaonema scale), 76; *Chrysomphalus aonidum* (Florida red scale), 122; *Parlatoria proteus* (proteus scale), 330

Agonis: Hemiberlesia rapax (greedy scale), 232

Agropyron: Odonaspis ruthae (Bermuda grass scale), 306

Ailanthus: Diaspidiotus ancylus (Putnam scale), 148

Aiphanes: Selenaspidus articulatus (rufous scale), 386

Albizia: Hemiberlesia lataniae (latania scale), 224; *Lepidosaphes yanagicola* (fire bush scale), 270

Alectryon: Aspidiotus destructor (coconut scale), 72

Aleurites: Aspidiotus nerii (oleander scale), 78; *A. spinosus* (spinose scale), 82; *Chrysomphalus dictyospermi* (dictyospermum scale), 130; *Diaspidiotus perniciosus* (San Jose scale), 172; *Fiorinia fioriniae* (palm fiorinia scale), 204; *Hemiberlesia lataniae* (latania scale), 224; *Morganella longispina* (plumose scale), 290; *Pseudaulacaspis cockerelli* (false oleander scale), 356

Allamanda: Chrysomphalus aonidum (Florida red scale), 122; *Howardia biclavis* (mining scale), 234; *Pseudaulacaspis pentagona* (white peach scale), 360

Alnus: Chionaspis salicis (willow scale), 118; *Diaspidiotus juglansregiae* (walnut scale), 158; *Lepidosaphes yanagicola* (fire bush scale), 270; *Lopholeucaspis japonica* (Japanese maple scale), 276; *Pseudaonidia duplex* (camphor scale), 346; *Pseudaulacaspis prunicola* (white prunicola scale), 364

Alocasia: Aspidiotus excisus (aglaonema scale), 76; *Lepidosaphes gloverii* (Glover scale), 254

Aloe: Aspidiotus nerii (oleander scale), 78; *Chrysomphalus aonidum* (Florida red scale), 122; *C. dictyospermi* (dictyospermum scale), 130; *Furcaspis biformis* (orchid scale), 216; *Hemiberlesia lataniae* (latania scale), 224; *Pseudischnaspis bowreyi* (Bowrey scale), 368

Alpinia: Chrysomphalus dictyospermi (dictyospermum scale), 130; *Selenaspidus articulatus* (rufous scale), 386

Alternanthera: Howardia biclavis (mining scale), 234

Althaea: Parlatoreopsis chinensis (Chinese obscure scale), 310; *Parlatoria theae* (tea parlatoria scale), 334; *Pseudaonidia duplex* (camphor scale), 346; *P. paeoniae* (peony scale), 350; *Unaspis euonymi* (euonymus scale), 396

Alyxia: Aspidiotus nerii (oleander scale), 78; *Chrysomphalus dictyospermi* (dictyospermum scale), 130

Ambrosia: Hemiberlesia lataniae (latania scale), 224; *Rhizaspidiotus dearnessi* (Dearness scale), 380

Amelanchier: Chionaspis furfura (scurfy scale), 110; *C. salicis* (willow scale), 118; *Lepidosaphes ulmi* (oystershell scale), 266

Amherstia: Chrysomphalus dictyospermi (dictyospermum scale), 130; *Hemiberlesia lataniae* (latania scale), 224; *Ischnaspis longirostris* (black thread scale), 238; *Selenaspidus articulatus* (rufous scale), 386

Amomum: Parlatoria proteus (proteus scale), 330

Amorpha: Diaspidiotus forbesi (Forbes scale), 152; *Lepidosaphes ulmi* (oystershell scale), 266; *Parlatoreopsis chinensis* (Chinese obscure scale), 310

Ampelopsis: Pseudaonidia duplex (camphor scale), 346

Amygdalus: Hemiberlesia lataniae (latania scale), 224; *Mercetaspis halli* (Hall scale), 288; *Pseudaulacaspis pentagona* (white peach scale), 360

Amyris: Pseudoparlatoria parlatorioides (false parlatoria scale), 374

Anacardium: Aspidiotus destructor (coconut scale), 72; *Fiorinia fioriniae* (palm fiorinia scale), 204; *Pseudaonidia trilobitiformis* (trilobe scale), 354

Ananas: Aspidiella sacchari (sugarcane scale), 66; *Aspidiotus destructor* (coconut scale), 72; *A. nerii* (oleander scale), 78; *Chrysomphalus dictyospermi* (dictyospermum scale), 130; *Diaspis boisduvalii* (Boisduval scale), 180; *D. bromeliae* (pineapple scale), 182; *Gymnaspis aechmeae* (flyspeck scale), 222; *Lepidosaphes beckii* (purple scale), 244; *Unaspis citri* (citrus snow scale), 392

Ancistrocactus: Diaspis echinocacti (cactus scale), 186

Andromeda: Aspidiotus nerii (oleander scale), 78; *Chionaspis salicis* (willow scale), 118

Andropogon: Aspidiella sacchari (sugarcane scale), 66; *Odonaspis ruthae* (Bermuda grass scale), 306

Anemia: Pinnaspis aspidistrae (fern scale), 340

Angraecum: Diaspis boisduvalii (Boisduval scale), 180; *Parlatoria proteus* (proteus scale), 330

Annona: Abgrallaspis cyanophylli (cyanophyllum scale), 38; *Aonidiella aurantii* (California red scale), 50; *A. orientalis* (Oriental armored scale), 58; *Aspidiella sacchari* (sugarcane scale), 66; *Aspidiotus destructor* (coconut scale), 72; *A. nerii* (oleander scale), 78; *Chionaspis furfura* (scurfy scale), 110; *Chrysomphalus aonidum* (Florida red scale), 122; *C. bifasciculatus* (bifasciculate scale), 126; *Clavaspis herculeana* (herculeana scale), 134; *Hemiberlesia rapax* (greedy scale), 232; *Howardia biclavis* (mining scale), 234; *Ischnaspis longirostris* (black thread scale), 238; *Lepidosaphes gloverii* (Glover scale), 254; *Pinnaspis strachani* (lesser snow scale), 344; *Pseudaonidia trilobitiformis* (trilobe scale), 354; *Pseudischnaspis bowreyi* (Bowrey scale), 368; *Selenaspidus articulatus* (rufous scale), 386

Anodendron: Howardia biclavis (mining scale), 234

Anthurium: Aspidiotus nerii (oleander scale), 78; *Chrysomphalus aonidum* (Florida red scale), 122; *C. dictyospermi* (dictyospermum scale), 130; *Hemiberlesia lataniae* (latania scale), 224; *Parlatoria proteus* (proteus scale), 330; *Pinnaspis aspidistrae* (fern scale), 340; *Pseudaonidia trilobitiformis* (trilobe scale), 354; *Pseudoparlatoria parlatorioides* (false parlatoria scale), 374; *Selenaspidus articulatus* (rufous scale), 386

Antidesma: Selenaspidus articulatus (rufous scale), 386

Antigonon: Hemiberlesia lataniae (latania scale), 224; *Howardia biclavis* (mining scale), 234; *Pseudoparlatoria ostreata* (gray scale), 370

Anubias: Ischnaspis longirostris (black thread scale), 238

Aplopappus: Rhizaspidiotus dearnessi (Dearness scale), 380

Arabis: Aspidiotus nerii (oleander scale), 78; *Lepidosaphes ulmi* (oystershell scale), 266

Arachnis: Parlatoria proteus (proteus scale), 330

Aralia: Chrysomphalus aonidum (Florida red scale), 122; *Furchadaspis zamiae* (zamia scale), 218; *Hemiberlesia lataniae* (latania scale), 224; *H. rapax* (greedy scale), 232; *Parlatoria proteus* (proteus scale), 330; *Pinnaspis strachani* (lesser snow scale), 344;

Pseudaulacaspis pentagona (white peach scale), 360

Araucaria: Aspidiotus nerii (oleander scale), 78; *Chrysomphalus dictyospermi* (dictyospermum scale), 130; *Hemiberlesia lataniae* (latania scale), 224; *Lepidosaphes pallida* (Maskell scale), 256; *Lindingaspis rossi* (black araucaria scale), 272; *Parlatoria proteus* (proteus scale), 330

Arbutus: Aonidiella citrina (yellow scale), 54; *Aspidiotus nerii* (oleander scale), 78; *Diaspidiotus juglansregiae* (walnut scale), 158

Arctostaphylos: Aspidiotus nerii (oleander scale), 78; *Chionaspis salicis* (willow scale), 118; *Parlatoria oleae* (olive scale), 320; *Rhizaspidiotus dearnessi* (Dearness scale), 380

Ardisia: Aspidiotus nerii (oleander scale), 78; *Hemiberlesia lataniae* (latania scale), 224

Areca: Aonidiella citrina (yellow scale), 54; *A. orientalis* (Oriental armored scale), 58; *Aspidiotus destructor* (coconut scale), 72; *A. nerii* (oleander scale), 78; *Chrysomphalus aonidum* (Florida red scale), 122; *C. dictyospermi* (dictyospermum scale), 130; *Diaspis boisduvalii* (Boisduval scale), 180; *Hemiberlesia lataniae* (latania scale), 224; *Ischnaspis longirostris* (black thread scale), 238; *Parlatoria proteus* (proteus scale), 330; *Pseudaulacaspis cockerelli* (false oleander scale), 356

Arecastrum: Comstockiella sabalis (palmetto scale), 140

Aregelia: Gymnaspis aechmeae (flyspeck scale), 222

Arenga: Aspidiotus spinosus (spinose scale), 82; *Chrysomphalus aonidum* (Florida red scale), 122

Argania: Aspidiotus nerii (oleander scale), 78; *A. spinosus* (spinose scale), 82; *Hemiberlesia lataniae* (latania scale), 224

Ariocarpus: Diaspis echinocacti (cactus scale), 186

Aristolochia: Hemiberlesia lataniae (latania scale), 224

Aronia: Diaspidiotus perniciosus (San Jose scale), 172; *Parlatoreopsis chinensis* (Chinese obscure scale), 310

Artemisia: Lepidosaphes ulmi (oystershell scale), 266; *Rhizaspidiotus dearnessi* (Dearness scale), 380

Artocarpus: Andaspis punicae (litchi scale), 48; *Aonidiella orientalis* (Oriental armored scale), 58; *Aspidiotus destructor* (coconut scale), 72; *Chrysomphalus dictyospermi* (dictyospermum scale), 130; *Clavaspis herculeana* (herculeana scale), 134; *Fiorinia fioriniae* (palm fiorinia scale), 204; *Ischnaspis longirostris* (black thread scale), 238; *Morganella longispina* (plumose scale), 290; *Pseudoparlatoria parlatorioides* (false parlatoria scale), 374; *Selenaspidus articulatus* (rufous scale), 386

Arum: Ischnaspis longirostris (black thread scale), 238

Arundina: Chrysomphalus aonidum (Florida red scale), 122; *Hemiberlesia lataniae* (latania scale), 224

Arundinaria: Froggattiella penicillata (penicillate scale), 212; *Kuwanaspis pseudoleucaspis* (bamboo diaspidid), 240; *Selenaspidus articulatus* (rufous scale), 386

Arundo: Aonidiella orientalis (Oriental armored scale), 58

Asclepias: Diaspidiotus perniciosus (San Jose scale), 172; *Hemiberlesia rapax* (greedy scale), 232; *Ischnaspis longirostris* (black thread scale), 238

Ascocentrum: Parlatoria proteus (proteus scale), 330

Asimina: Diaspidiotus ancylus (Putnam scale), 148

Asparagus: Aonidiella orientalis (Oriental armored scale), 58; *Aspidiotus nerii* (oleander scale), 78; *Chrysomphalus aonidum* (Florida red scale), 122; *C. dictyospermi* (dictyospermum scale), 130; *Hemiberlesia*

lataniae (latania scale), 224; *Ischnaspis longirostris* (black thread scale), 238; *Parlatoria proteus* (proteus scale), 330; *Pseudaulacaspis cockerelli* (false oleander scale), 356; *Pseudischnaspis bowreyi* (Bowrey scale), 368; *Pseudoparlatoria ostreata* (gray scale), 370

Aspidistra: Aspidiotus destructor (coconut scale), 72; *A. nerii* (oleander scale), 78; *Chrysomphalus aonidum* (Florida red scale), 122; *C. bifasciculatus* (bifasciculate scale), 126; *C. dictyospermi* (dictyospermum scale), 130; *Parlatoria proteus* (proteus scale), 330; *Pinnaspis aspidistrae* (fern scale), 340; *Pseudaonidia paeoniae* (peony scale), 350

Asplenium: Pinnaspis aspidistrae (fern scale), 340

Aster: Rhizaspidiotus dearnessi (Dearness scale), 380

Astianthus: Hemiberlesia lataniae (latania scale), 224

Astragalus: Rhizaspidiotus dearnessi (Dearness scale), 380

Astrocaryum: Diaspis boisduvalii (Boisduval scale), 180; *Selenaspidus articulatus* (rufous scale), 386

Astrophytum: Diaspis echinocacti (cactus scale), 186

Atalantia: Aonidiella orientalis (Oriental armored scale), 58; *Parlatoria pergandii* (chaff scale), 324; *Pinnaspis aspidistrae* (fern scale), 340

Atalaya: Aonidiella orientalis (Oriental armored scale), 58; *Hemiberlesia lataniae* (latania scale), 224

Atriplex: Aspidiotus nerii (oleander scale), 78

Attalea: Aonidiella orientalis (Oriental armored scale), 58; *Chrysomphalus dictyospermi* (dictyospermum scale), 130; *Diaspis boisduvalii* (Boisduval scale), 180; *Hemiberlesia lataniae* (latania scale), 224; *Selenaspidus articulatus* (rufous scale), 386

Aucoumea: Morganella longispina (plumose scale), 290

Aucuba: Aonidiella citrina (yellow scale), 54; *Aspidiotus destructor* (coconut scale), 72; *A. nerii* (oleander scale), 78; *Chrysomphalus bifasciculatus* (bifasciculate scale), 126; *C. dictyospermi* (dictyospermum scale), 130; *Hemiberlesia lataniae* (latania scale), 224; *Parlatoria camelliae* (camellia parlatoria scale), 316; *P. oleae* (olive scale), 320; *P. pergandii* (chaff scale), 324; *P. theae* (tea parlatoria scale), 334; *Pseudaonidia duplex* (camphor scale), 346; *Pseudaulacaspis cockerelli* (false oleander scale), 356; *P. prunicola* (white prunicola scale), 364

Averrhoa: Hemiberlesia lataniae (latania scale), 224; *Morganella longispina* (plumose scale), 290

Avicennia: Clavaspis herculeana (herculeana scale), 134

Axonopus: Aspidiella sacchari (sugarcane scale), 66; *Odonaspis ruthae* (Bermuda grass scale), 306

Azadirachta: Aonidiella orientalis (Oriental armored scale), 58

Baccharis: Chrysomphalus dictyospermi (dictyospermum scale), 130; *Hemiberlesia lataniae* (latania scale), 224; *H. rapax* (greedy scale), 232; *Rhizaspidiotus dearnessi* (Dearness scale), 380

Bactris: Aspidiotus spinosus (spinose scale), 82; *Diaspis boisduvalii* (Boisduval scale), 180; *Hemiberlesia lataniae* (latania scale), 224

Bahia: Rhizaspidiotus dearnessi (Dearness scale), 380

Balsamocitrus: Lepidosaphes beckii (purple scale), 244

Bambusa: Chrysomphalus aonidum (Florida red scale), 122; *Froggattiella penicillata* (penicillate scale), 212; *Kuwanaspis pseudoleucaspis* (bamboo diaspidid), 240

Banksia: Aspidiotus nerii (oleander scale), 78; *Hemiberlesia rapax* (greedy scale), 232;

Lepidosaphes pinnaeformis (cymbidium scale), 262; *Lindingaspis rossi* (black araucaria scale), 272; *Parlatoria pittospori* (pittosporum scale), 328

Barkeria: *Pseudoparlatoria ostreata* (gray scale), 370

Barringtonia: *Aonidiella aurantii* (California red scale), 50; *Aspidiotus destructor* (coconut scale), 72; *Chrysomphalus aonidum* (Florida red scale), 122; *C. dictyospermi* (dictyospermum scale), 130; *Hemiberlesia lataniae* (latania scale), 224; *Parlatoria proteus* (proteus scale), 330

Bassia: *Chrysomphalus dictyospermi* (dictyospermum scale), 130

Batemannia: *Pseudoparlatoria parlatorioides* (false parlatoria scale), 374

Batis: *Hemiberlesia lataniae* (latania scale), 224

Bauhinia: *Aonidiella orientalis* (Oriental armored scale), 58; *Chrysomphalus aonidum* (Florida red scale), 122; *C. dictyospermi* (dictyospermum scale), 130; *Hemiberlesia lataniae* (latania scale), 224; *H. rapax* (greedy scale), 232; *Howardia biclavis* (mining scale), 234; *Ischnaspis longirostris* (black thread scale), 238; *Morganella longispina* (plumose scale), 290; *Pinnaspis strachani* (lesser snow scale), 344; *Selenaspidus articulatus* (rufous scale), 386

Beaucarnea: *Hemiberlesia lataniae* (latania scale), 224; *Parlatoria proteus* (proteus scale), 330; *Pseudischnaspis bowreyi* (Bowrey scale), 368

Begonia: *Abgrallaspis cyanophylli* (cyanophyllum scale), 38; *Pseudoparlatoria parlatorioides* (false parlatoria scale), 374

Berberis: *Aspidiotus nerii* (oleander scale), 78; *Hemiberlesia lataniae* (latania scale), 224; *Lepidosaphes ulmi* (oystershell scale), 266; *Parlatoria oleae* (olive scale), 320

Bertholletia: *Hemiberlesia lataniae* (latania scale), 224

Beta: *Hemiberlesia lataniae* (latania scale), 224; *Lepidosaphes ulmi* (oystershell scale), 266

Betula: *Chionaspis furfura* (scurfy scale), 110; *C. salicis* (willow scale), 118; *Diaspidiotus ancylus* (Putnam scale), 148; *D. juglansregiae* (walnut scale), 158; *D. ostreaeformis* (European fruit scale), 168; *D. perniciosus* (San Jose scale), 172; *Lepidosaphes conchiformis* (fig scale), 250; *L. ulmi* (oystershell scale), 266

Bignonia: *Duplaspidiotus tesseratus* (tesserate scale), 192; *Hemiberlesia lataniae* (latania scale), 224

Billbergia: *Diaspis boisduvalii* (Boisduval scale), 180; *D. bromeliae* (pineapple scale), 182; *Gymnaspis aechmeae* (flyspeck scale), 222

Biota: *Carulaspis juniperi* (juniper scale), 96; *Nuculaspis pseudomeyeri* (false Meyer scale), 300

Bischofia: *Aonidiella orientalis* (Oriental armored scale), 58; *Pseudaulacaspis cockerelli* (false oleander scale), 356

Bixa: *Howardia biclavis* (mining scale), 234

Bladhia: *Aonidiella taxus* (Asiatic red scale), 62

Bletia: *Diaspis boisduvalii* (Boisduval scale), 180

Blighia: *Hemiberlesia lataniae* (latania scale), 224; *Selenaspidus articulatus* (rufous scale), 386

Boehmeria: *Pseudaulacaspis pentagona* (white peach scale), 360

Borrera: *Pseudoparlatoria parlatorioides* (false parlatoria scale), 374

Borrichia: *Hemiberlesia lataniae* (latania scale), 224

Bougainvillea: *Chrysomphalus dictyospermi* (dictyospermum scale), 130; *Duplaspidiotus tesseratus* (tesserate scale), 192; *Hemiberlesia lataniae* (latania scale), 224; *Howardia biclavis* (mining scale), 234; *Pinnaspis strachani* (lesser snow scale), 344

Bowdichia: *Pseudaonidia trilobitiformis* (trilobe scale), 354

Brahea: *Comstockiella sabalis* (palmetto scale), 140

Brassavola: *Diaspis boisduvalii* (Boisduval scale), 180; *Furcaspis biformis* (orchid scale), 216; *Pseudoparlatoria parlatorioides* (false parlatoria scale), 374

Brassia: *Diaspis boisduvalii* (Boisduval scale), 180; *Furcaspis biformis* (orchid scale), 216; *Parlatoria proteus* (proteus scale), 330; *Selenaspidus articulatus* (rufous scale), 386

Bromelia: *Diaspis bromeliae* (pineapple scale), 182; *Furcaspis biformis* (orchid scale), 216; *Gymnaspis aechmeae* (flyspeck scale), 222; *Hemiberlesia lataniae* (latania scale), 224; *Ischnaspis longirostris* (black thread scale), 238; *Pseudischnaspis bowreyi* (Bowrey scale), 368

Brosimum: *Clavaspis herculeana* (herculeana scale), 134

Broughtonia: *Chrysomphalus dictyospermi* (dictyospermum scale), 130; *Diaspis boisduvalii* (Boisduval scale), 180; *Parlatoria proteus* (proteus scale), 330; *Pseudoparlatoria parlatorioides* (false parlatoria scale), 374

Broussaisia: *Chrysomphalus aonidum* (Florida red scale), 122

Broussonetia: *Pseudaulacaspis pentagona* (white peach scale), 360

Brownea: *Ischnaspis longirostris* (black thread scale), 238

Bruguiera: *Chrysomphalus aonidum* (Florida red scale), 122

Brunfelsia: *Howardia biclavis* (mining scale), 234; *Selenaspidus articulatus* (rufous scale), 386

Bryanthus: *Duplaspidiotus tesseratus* (tesserate scale), 192

Bryophyllum: *Hemiberlesia lataniae* (latania scale), 224; *Pseudaulacaspis pentagona* (white peach scale), 360; *Pseudoparlatoria ostreata* (gray scale), 370

Buddleia: *Pseudaulacaspis pentagona* (white peach scale), 360

Bulbophyllum: *Parlatoria proteus* (proteus scale), 330

Bupleurum: *Aspidiotus nerii* (oleander scale), 78

Bursera: *Hemiberlesia lataniae* (latania scale), 224; *Selenaspidus articulatus* (rufous scale), 386

Butia: *Comstockiella sabalis* (palmetto scale), 140

Buxus: *Chrysomphalus aonidum* (Florida red scale), 122; *C. bifasciculatus* (bifasciculate scale), 126; *C. dictyospermi* (dictyospermum scale), 130; *Diaspidiotus perniciosus* (San Jose scale), 172; *Dynaspidiotus britannicus* (holly scale), 194; *Fiorinia fioriniae* (palm fiorinia scale), 204; *Hemiberlesia lataniae* (latania scale), 224; *Lepidosaphes ulmi* (oystershell scale), 266; *Lindingaspis rossi* (black araucaria scale), 272; *Pseudaonidia duplex* (camphor scale), 346; *P. paeoniae* (peony scale), 350; *P. trilobitiformis* (trilobe scale), 354; *Pseudaulacaspis prunicola* (white prunicola scale), 364; *Unaspis euonymi* (euonymus scale), 396

Caesalpinia: *Clavaspis herculeana* (herculeana scale), 134; *Hemiberlesia lataniae* (latania scale), 224; *Howardia biclavis* (mining scale), 234; *Pinnaspis strachani* (lesser snow scale), 344

Cajanus: *Aonidiella orientalis* (Oriental armored scale), 58; *Aspidiotus spinosus* (spinose scale), 82; *Pinnaspis strachani* (lesser snow scale), 344

Caladium: *Chrysomphalus aonidum* (Florida red scale), 122

Calanthe: *Chrysomphalus aonidum* (Florida red scale), 122; *Parlatoria proteus* (proteus scale), 330; *Pseudischnaspis bowreyi* (Bowrey scale), 368

Calathea: *Chrysomphalus dictyospermi* (dictyospermum scale), 130; *Diaspis boisduvalii* (Boisduval scale), 180; *Parlatoria proteus* (proteus scale), 330; *Selenaspidus articulatus* (rufous scale), 386

Callicarpa: *Pseudaulacaspis pentagona* (white peach scale), 360

Callistemon: *Aspidiotus destructor* (coconut scale), 72; *Chrysomphalus aonidum* (Florida red scale), 122; *Fiorinia fioriniae* (palm fiorinia scale), 204; *Hemiberlesia lataniae* (latania scale), 224; *H. rapax* (greedy scale), 232; *Morganella longispina* (plumose scale), 290

Callitris: *Carulaspis juniperi* (juniper scale), 96; *C. minima* (minute cypress scale), 100; *Hemiberlesia lataniae* (latania scale), 224

Calluna: *Lepidosaphes ulmi* (oystershell scale), 266

Calocarpum: *Aspidiotus destructor* (coconut scale), 72; *Clavaspis herculeana* (herculeana scale), 134; *Duplaspidiotus tesseratus* (tesserate scale), 192; *Hemiberlesia lataniae* (latania scale), 224

Calodendrum: *Hemiberlesia lataniae* (latania scale), 224

Calophyllum: *Aspidiotus destructor* (coconut scale), 72; *Chrysomphalus aonidum* (Florida red scale), 122; *Duplaspidiotus claviger* (camellia mining scale), 188; *Lepidosaphes gloverii* (Glover scale), 254; *Pseudaonidia trilobitiformis* (trilobe scale), 354; *Selenaspidus articulatus* (rufous scale), 386

Calycanthus: *Howardia biclavis* (mining scale), 234

Calycophyllum: *Hemiberlesia lataniae* (latania scale), 224

Calycotome: *Aspidiotus nerii* (oleander scale), 78

Calyptrogyne: *Hemiberlesia lataniae* (latania scale), 224

Camellia: *Abgrallaspis degenerata* (degenerate scale), 42; *Aspidiotus destructor* (coconut scale), 72; *A. nerii* (oleander scale), 78; *A. spinosus* (spinose scale), 82; *Chrysomphalus aonidum* (Florida red scale), 122; *C. bifasciculatus* (bifasciculate scale), 126; *C. dictyospermi* (dictyospermum scale), 130; *Diaspidiotus perniciosus* (San Jose scale), 172; *Duplaspidiotus claviger* (camellia mining scale), 188; *D. tesseratus* (tesserate scale), 192; *Fiorinia fioriniae* (palm fiorinia scale), 204; *F. theae* (tea scale), 210; *Hemiberlesia lataniae* (latania scale), 224; *H. rapax* (greedy scale), 232; *Howardia biclavis* (mining scale), 234; *Ischnaspis longirostris* (black thread scale), 238; *Lepidosaphes beckii* (purple scale), 244; *L. camelliae* (camellia scale), 248; *Lopholeucaspis japonica* (Japanese maple scale), 276; *Morganella longispina* (plumose scale), 290; *Parlatoria camelliae* (camellia parlatoria scale), 316; *P. proteus* (proteus scale), 330; *P. theae* (tea parlatoria scale), 334; *Pinnaspis aspidistrae* (fern scale), 340; *Pseudaonidia duplex* (camphor scale), 346; *P. paeoniae* (peony scale), 350; *P. trilobitiformis* (trilobe scale), 354; *Pseudaulacaspis cockerelli* (false oleander scale), 356; *P. pentagona* (white peach scale), 360; *Selenaspidus articulatus* (rufous scale), 386; *Unaspis euonymi* (euonymus scale), 396

Cananga: *Aonidiella citrina* (yellow scale), 54; *Howardia biclavis* (mining scale), 234

Canavalia: *Hemiberlesia lataniae* (latania scale), 224

Canna: *Aspidiella sacchari* (sugarcane scale), 66; *Chrysomphalus aonidum* (Florida red scale), 122; *Hemiberlesia lataniae* (latania scale), 224; *Ischnaspis longirostris* (black thread scale), 238; *Pseudaonidia trilobitiformis* (trilobe scale), 354; *Selenaspidus articulatus* (rufous scale), 386

Cannabis: *Pseudischnaspis bowreyi* (Bowrey scale), 368

Capsicum: *Diaspidiotus perniciosus* (San Jose scale), 172; *Fiorinia fioriniae* (palm fiorinia scale), 204; *Pseudaulacaspis pentagona* (white peach scale), 360

Caragana: *Diaspidiotus ancylus* (Putnam scale), 148; *D. ostreaeformis* (European fruit scale), 168; *Mercetaspis halli* (Hall scale), 288

Carica: *Aonidiella orientalis* (Oriental armored scale), 58; *Aspidiotus destructor* (coconut scale), 72; *A. excisus* (aglaonema scale), 76; *Hemiberlesia lataniae* (latania scale), 224;

Carica (cont.)
Howardia biclavis (mining scale), 234; Morganella longispina (plumose scale), 290; Pinnaspis aspidistrae (fern scale), 340; Pseudaulacaspis pentagona (white peach scale), 360; Pseudoparlatoria ostreata (gray scale), 370

Carissa: Aonidiella citrina (yellow scale), 54; Chrysomphalus aonidum (Florida red scale), 122; C. dictyospermi (dictyospermum scale), 130

Carludovica: Aspidiotus destructor (coconut scale), 72; Selenaspidus articulatus (rufous scale), 386

Carpinus: Chionaspis americana (elm scurfy scale), 102; Diaspidiotus ancylus (Putnam scale), 148; Lepidosaphes conchiformis (fig scale), 250

Carya: Diaspidiotus ancylus (Putnam scale), 148; D. forbesi (Forbes scale), 152; D. juglansregiae (walnut scale), 158; D. osborni (Osborn scale), 166; D. perniciosus (San Jose scale), 172; D. uvae (grape scale), 176; Hemiberlesia lataniae (latania scale), 224; H. neodiffinis (false diffinis scale), 228; H. rapax (greedy scale), 232; Melanaspis obscura (obscure scale), 282; M. tenebricosa (gloomy scale), 284; Pseudonidia duplex (camphor scale), 346; Pseudischnaspis bowreyi (Bowrey scale), 368

Caryocar: Howardia biclavis (mining scale), 234

Caryota: Aspidiotus excisus (aglaonema scale), 76; A. spinosus (spinose scale), 82; Chrysomphalus dictyospermi (dictyospermum scale), 130; Ischnaspis longirostris (black thread scale), 238; Parlatoria proteus (proteus scale), 330

Casearia: Howardia biclavis (mining scale), 234

Casimiroa: Hemiberlesia lataniae (latania scale), 224; H. rapax (greedy scale), 232; Pinnaspis strachani (lesser snow scale), 344

Cassava: Pseudaulacaspis pentagona (white peach scale), 360

Cassia: Aonidiella orientalis (Oriental armored scale), 58; Aspidiotus destructor (coconut scale), 72; A. nerii (oleander scale), 78; Chrysomphalus aonidum (Florida red scale), 122; C. dictyospermi (dictyospermum scale), 130; Clavaspis herculeana (herculeana scale), 134; Hemiberlesia lataniae (latania scale), 224; Lepidosaphes beckii (purple scale), 244; Pseudaonidia trilobitiformis (trilobe scale), 354; Situlaspis yuccae (yucca scale), 388

Castanea: Abgrallaspis cyanophylli (cyanophylli scale), 38; Chrysomphalus dictyospermi (dictyospermum scale), 130; Diaspidiotus ancylus (Putnam scale), 148; D. osborni (Osborn scale), 166; Hemiberlesia lataniae (latania scale), 224; Lepidosaphes ulmi (oystershell scale), 266; Lopholeucaspis japonica (Japanese maple scale), 276; Melanaspis obscura (obscure scale), 282; M. tenebricosa (gloomy scale), 284; Pseudaonidia duplex (camphor scale), 346; Pseudaulacaspis pentagona (white peach scale), 360

Castanopsis: Abgrallaspis cyanophylli (cyanophylli scale), 38; Chrysomphalus bifasciculatus (bifasciculate scale), 126; C. dictyospermi (dictyospermum scale), 130

Castilla: Chrysomphalus aonidum (Florida red scale), 122; Pseudaulacaspis pentagona (white peach scale), 360

Casuarina: Duplaspidiotus tesseratus (tesserate scale), 192; Hemiberlesia lataniae (latania scale), 224; Howardia biclavis (mining scale), 234

Catalpa: Aspidiotus nerii (oleander scale), 78; Clavaspis ulmi (elm armored scale), 136; Diaspidiotus uvae (grape scale), 176; Lepidosaphes ulmi (oystershell scale), 266; Melanaspis tenebricosa (gloomy scale), 284; Pseudaulacaspis pentagona (white peach scale), 360; P. prunicola (white prunicola scale), 364

Catasetum: Duplaspidiotus tesseratus (tesserate scale), 192; Parlatoria proteus (proteus scale), 330; Pseudoparlatoria ostreata (gray scale), 370

Catopsis: Diaspis boisduvalii (Boisduval scale), 180

Cattleya: Aonidiella orientalis (Oriental armored scale), 58; Chrysomphalus dictyospermi (dictyospermum scale), 130; Diaspis boisduvalii (Boisduval scale), 180; D. bromeliae (pineapple scale), 182; Duplaspidiotus tesseratus (tesserate scale), 192; Furcaspis biformis (orchid scale), 216; Hemiberlesia lataniae (latania scale), 224; Ischnaspis longirostris (black thread scale), 238; Lepidosaphes pinnaeformis (cymbidium scale), 262; Odonaspis ruthae (Bermuda grass scale), 306; Pseudischnaspis bowreyi (Bowrey scale), 368; Pseudoparlatoria ostreata (gray scale), 370; P. parlatorioides (false parlatoria scale), 374

Caularthron: Diaspis boisduvalii (Boisduval scale), 180; Ischnaspis longirostris (black thread scale), 238

Ceanothus: Aspidiotus nerii (oleander scale), 78; Diaspidiotus ancylus (Putnam scale), 148; D. forbesi (Forbes scale), 152; D. perniciosus (San Jose scale), 172; Hemiberlesia lataniae (latania scale), 224; H. rapax (greedy scale), 232; Lepidosaphes ulmi (oystershell scale), 266; Rhizaspidiotus dearnessi (Dearness scale), 380

Cedrela: Hemiberlesia lataniae (latania scale), 224; Howardia biclavis (mining scale), 234; Pseudaulacaspis pentagona (white peach scale), 360

Cedrus: Aspidiotus cryptomeriae (cryptomeria scale), 68; A. nerii (oleander scale), 78; Carulaspis minima (minute cypress scale), 100; Chionaspis pinifoliae (pine needle scale), 116; Fiorinia externa (elongate hemlock scale), 200; Hemiberlesia lataniae (latania scale), 224; Nuculaspis pseudomeyeri (false Meyer scale), 300; N. tsugae (shortneedle conifer scale), 304

Ceiba: Aspidiotus destructor (coconut scale), 72; Hemiberlesia lataniae (latania scale), 224; Pinnaspis strachani (lesser snow scale), 344; Pseudischnaspis bowreyi (Bowrey scale), 368; Selenaspidus articulatus (rufous scale), 386

Celastrus: Lepidosaphes ulmi (oystershell scale), 266; L. yanagicola (fire bush scale), 270; Lopholeucaspis japonica (Japanese maple scale), 276; Unaspis euonymi (euonymus scale), 396

Celtis: Chrysomphalus aonidum (Florida red scale), 122; Diaspidiotus ancylus (Putnam scale), 148; D. forbesi (Forbes scale), 152; D. juglansregiae (walnut scale), 158; Hemiberlesia lataniae (latania scale), 224; H. neodiffinis (false diffinis scale), 228; H. rapax (greedy scale), 232; Lepidosaphes conchiformis (fig scale), 250; L. ulmi (oystershell scale), 266; Melanaspis obscura (obscure scale), 282; M. tenebricosa (gloomy scale), 284; Morganella longispina (plumose scale), 290; Pseudaulacaspis prunicola (white prunicola scale), 364

Centaurea: Howardia biclavis (mining scale), 234

Centaurium: Hemiberlesia lataniae (latania scale), 224; Lepidosaphes ulmi (oystershell scale), 266

Centella: Howardia biclavis (mining scale), 234

Cephaelis: Ischnaspis longirostris (black thread scale), 238

Cephalanthus: Diaspidiotus forbesi (Forbes scale), 152; Hemiberlesia neodiffinis (false diffinis scale), 228

Cephalocereus: Diaspis echinocacti (cactus scale), 186; Hemiberlesia rapax (greedy scale), 232; Pseudoparlatoria parlatorioides (false parlatoria scale), 374

Cephalotaxus: Aonidiella taxus (Asiatic red scale), 62; Aspidiotus cryptomeriae (cryptomeria scale), 68; Chrysomphalus dictyospermi (dictyospermum scale), 130; Fiorinia japonica (coniferous fiorinia scale), 206; Lepidosaphes pallida (Maskell scale), 256

Ceratonia: Aonidiella aurantii (California red scale), 50; Aspidiotus nerii (oleander scale), 78; Chrysomphalus dictyospermi (dictyospermum scale), 130; Hemiberlesia lataniae (latania scale), 224; H. rapax (greedy scale), 232; Lepidosaphes ulmi (oystershell scale), 266; Neopinnaspis harperi (Harper scale), 294

Ceratozamia: Furchadaspis zamiae (zamia scale), 218

Cercidiphyllum: Lepidosaphes beckii (purple scale), 244; L. pinnaeformis (cymbidium scale), 262

Cercidium: Situlaspis yuccae (yucca scale), 388

Cercis: Diaspidiotus juglansregiae (walnut scale), 158; Lepidosaphes ulmi (oystershell scale), 266; Situlaspis yuccae (yucca scale), 388

Cereus: Aspidiotus nerii (oleander scale), 78; Diaspis echinocacti (cactus scale), 186; Hemiberlesia lataniae (latania scale), 224; Situlaspis yuccae (yucca scale), 388

Cestrum: Pseudoparlatoria ostreata (gray scale), 370; P. parlatorioides (false parlatoria scale), 374

Chaenomeles: Chionaspis furfura (scurfy scale), 110; Lopholeucaspis japonica (Japanese maple scale), 276; Parlatoreopsis chinensis (Chinese obscure scale), 310

Chaetachme: Chrysomphalus dictyospermi (dictyospermum scale), 130

Chalcas: Lepidosaphes beckii (purple scale), 244

Chamaecyparis: Aspidiotus cryptomeriae (cryptomeria scale), 68; Carulaspis juniperi (juniper scale), 96; C. minima (minute cypress scale), 100; Hemiberlesia rapax (greedy scale), 232; Lepidosaphes pallida (Maskell scale), 256; L. ulmi (oystershell scale), 266; Nuculaspis pseudomeyeri (false Meyer scale), 300; N. tsugae (shortneedle conifer scale), 304

Chamaedorea: Aspidiotus destructor (coconut scale), 72; A. spinosus (spinose scale), 82; Chrysomphalus aonidum (Florida red scale), 122; C. dictyospermi (dictyospermum scale), 130; Diaspis boisduvalii (Boisduval scale), 180; Hemiberlesia lataniae (latania scale), 224; Ischnaspis longirostris (black thread scale), 238; Pseudoparlatoria parlatorioides (false parlatoria scale), 374; Selenaspidus articulatus (rufous scale), 386

Chamaerops: Aspidiotus nerii (oleander scale), 78; Comstockiella sabalis (palmetto scale), 140; Diaspis boisduvalii (Boisduval scale), 180; Hemiberlesia lataniae (latania scale), 224

Cheirostemon: Hemiberlesia rapax (greedy scale), 232

Cherimoya: Aonidiella orientalis (Oriental armored scale), 58

Chevalieria: Diaspis bromeliae (pineapple scale), 182

Chionanthus: Lepidosaphes ulmi (oystershell scale), 266

Chloris: Aspidiella sacchari (sugarcane scale), 66

Chlorophytum: Aspidiotus excisus (aglaonema scale), 76; Chrysomphalus dictyospermi (dictyospermum scale), 130; Hemiberlesia lataniae (latania scale), 224; Parlatoria proteus (proteus scale), 330

Choisya: Aspidiotus nerii (oleander scale), 78

Chrysalidocarpus: Aonidiella orientalis (Oriental armored scale), 58; Aspidiotus destructor (coconut scale), 72; A. spinosus (spinose scale), 82; Chrysomphalus aonidum (Florida red scale), 122; C. dictyospermi (dictyospermum scale), 130; Diaspis boisduvalii (Boisduval scale), 180; Ischnaspis longirostris (black thread scale), 238; Pinnaspis aspidistrae (fern scale), 340; P. strachani (lesser snow scale), 344; Pseudaulacaspis cockerelli (false oleander scale), 356; Selenaspidus articulatus (rufous scale), 386

Chrysanthemum: Hemiberlesia lataniae (latania scale), 224

Chrysobalanus: Pseudoparlatoria ostreata (gray scale), 370

Chrysophyllum: Hemiberlesia lataniae (latania scale), 224; Howardia biclavis (mining scale), 234; Pseudaonidia trilobitiformis (trilobe scale), 354; *Selenaspidus articulatus* (rufous scale), 386

Chysis: Hemiberlesia lataniae (latania scale), 224

Cinchona: Hemiberlesia rapax (greedy scale), 232; *Howardia biclavis* (mining scale), 234; *Pseudaulacaspis pentagona* (white peach scale), 360; *Pseudoparlatoria parlatorioides* (false parlatoria scale), 374

Cinnamomum: Aspidiotus destructor (coconut scale), 72; *Aulacaspis tubercularis* (white mango scale), 90; *Chrysomphalus aonidum* (Florida red scale), 122; *C. dictyospermi* (dictyospermum scale), 130; *Hemiberlesia lataniae* (latania scale), 224; *H. neodiffinis* (false diffinis scale), 228; *H. rapax* (greedy scale), 232; *Howardia biclavis* (mining scale), 234; *Lepidosaphes pinnaeformis* (cymbidium scale), 262; *Parlatoria camelliae* (camellia parlatoria scale), 316; *Pseudaonidia duplex* (camphor scale), 346; *Pseudoparlatoria parlatorioides* (false parlatoria scale), 374; *Selenaspidus articulatus* (rufous scale), 386

Cissus: Pinnaspis strachani (lesser snow scale), 344

Citharexylum: Hemiberlesia lataniae (latania scale), 224

Citropsis: Lepidosaphes gloverii (Glover scale), 254

Citrus: Aonidiella aurantii (California red scale), 50; *A. citrina* (yellow scale), 54; *A. orientalis* (Oriental armored scale), 58; *Aspidiotus destructor* (coconut scale), 72; *A. excisus* (aglaonema scale), 76; *A. nerii* (oleander scale), 78; *A. spinosus* (spinose scale), 82; *Aulacaspis tubercularis* (white mango scale), 90; *Chrysomphalus aonidum* (Florida red scale), 122; *C. bifasciculatus* (bifasciculate scale), 126; *C. dictyospermi* (dictyospermum scale), 130; *Clavaspis herculeana* (herculeana scale), 134; *Comstockiella sabalis* (palmetto scale), 140; *Diaspidiotus perniciosus* (San Jose scale), 172; *Duplaspidiotus claviger* (camellia mining scale), 188; *Fiorinia theae* (tea scale), 210; *Hemiberlesia lataniae* (latania scale), 224; *H. rapax* (greedy scale), 232; *Howardia biclavis* (mining scale), 234; *Ischnaspis longirostris* (black thread scale), 238; *Lepidosaphes beckii* (purple scale), 244; *L. gloverii* (Glover scale), 254; *L. pallida* (Maskell scale), 256; *L. pinnaeformis* (cymbidium scale), 262; *L. ulmi* (oystershell scale), 266; *Lindingaspis rossi* (black araucaria scale), 272; *Lopholeucaspis japonica* (Japanese maple scale), 276; *Morganella longispina* (plumose scale), 290; *Parlatoria camelliae* (camellia parlatoria scale), 316; *P. oleae* (olive scale), 320; *P. pergandii* (chaff scale), 324; *P. pittospori* (pittosporum scale), 328; *P. proteus* (proteus scale), 330; *P. theae* (tea parlatoria scale), 334; *P. ziziphi* (black parlatoria scale), 336; *Pinnaspis aspidistrae* (fern scale), 340; *P. strachani* (lesser snow scale), 344; *Pseudaonidia duplex* (camphor scale), 346; *P. trilobitiformis* (trilobe scale), 354; *Pseudaulacaspis cockerelli* (false oleander scale), 356; *P. pentagona* (white peach scale), 360; *Pseudischnaspis bowreyi* (Bowrey scale), 368; *Pseudoparlatoria parlatorioides* (false parlatoria scale), 374; *Selenaspidus articulatus* (rufous scale), 386; *Unaspis citri* (citrus snow scale), 392; *U. euonymi* (euonymus scale), 396

Cladrastis: Diaspidiotus ancylus (Putnam scale), 148

Clematis: Aonidiella aurantii (California red scale), 50; *Lepidosaphes ulmi* (oystershell scale), 266; *Pseudaulacaspis pentagona* (white peach scale), 360

Clerodendron: Clavaspis herculeana (herculeana scale), 134

Clethra: Hemiberlesia lataniae (latania scale), 224; *Pseudaonidia paeoniae* (peony scale), 350

Cleyera: Abgrallaspis degenerata (degenerate scale), 42; *Hemiberlesia lataniae* (latania scale), 224; *Lepidosaphes camelliae* (camellia scale), 248; *Parlatoria camelliae* (camellia

parlatoria scale), 316; *Pseudaonidia paeoniae* (peony scale), 350

Clinostigma: Hemiberlesia lataniae (latania scale), 224; *Pseudaulacaspis cockerelli* (false oleander scale), 356

Clusia: Howardia biclavis (mining scale), 234

Cneoridium: Aspidiotus nerii (oleander scale), 78

Cobaea: Hemiberlesia rapax (greedy scale), 232

Coccoloba: Clavaspis herculeana (herculeana scale), 134; *Hemiberlesia lataniae* (latania scale), 224; *H. rapax* (greedy scale), 232; *Pseudaonidia trilobitiformis* (trilobe scale), 354; *Pseudischnaspis bowreyi* (Bowrey scale), 368

Coccothrinax: Aspidiotus destructor (coconut scale), 72; *Ischnaspis longirostris* (black thread scale), 238; *Parlatoria proteus* (proteus scale), 330

Cocculus: Aonidiella aurantii (California red scale), 50

Cocos: Abgrallaspis cyanophylli (cyanophyllum scale), 38; *Aonidiella orientalis* (Oriental armored scale), 58; *A. sacchari* (sugarcane scale), 66; *Aspidiotus destructor* (coconut scale), 72; *A. excisus* (aglaonema scale), 76; *A. nerii* (oleander scale), 78; *A. spinosus* (spinose scale), 82; *Aulacaspis tubercularis* (white mango scale), 90; *Chrysomphalus aonidum* (Florida red scale), 122; *C. dictyospermi* (dictyospermum scale), 130; *Comstockiella sabalis* (palmetto scale), 140; *Diaspis boisduvalii* (Boisduval scale), 180; *Fiorinia fioriniae* (palm fiorinia scale), 204; *Hemiberlesia lataniae* (latania scale), 224; *H. rapax* (greedy scale), 232; *Ischnaspis longirostris* (black thread scale), 238; *Lepidosaphes gloverii* (Glover scale), 254; *Parlatoria proteus* (proteus scale), 330; *Pinnaspis aspidistrae* (fern scale), 340; *P. strachani* (lesser snow scale), 344; *Pseudaulacaspis cockerelli* (false oleander scale), 356; *Pseudischnaspis bowreyi* (Bowrey scale), 368; *Pseudoparlatoria ostreata* (gray scale), 370; *P. parlatorioides* (false parlatoria scale), 374; *Selenaspidus articulatus* (rufous scale), 386

Codiaeum: Aonidiella aurantii (California red scale), 50; *Aspidiotus nerii* (oleander scale), 78; *Diaspidiotus perniciosus* (San Jose scale), 172; *Diaspis boisduvalii* (Boisduval scale), 180; *Hemiberlesia lataniae* (latania scale), 224; *Lepidosaphes beckii* (purple scale), 244; *L. gloverii* (Glover scale), 254; *Lindingaspis rossi* (black araucaria scale), 272; *Selenaspidus articulatus* (rufous scale), 386

Coelia: Chrysomphalus aonidum (Florida red scale), 122; *Parlatoria proteus* (proteus scale), 330

Coelogyne: Chrysomphalus aonidum (Florida red scale), 122; *C. dictyospermi* (dictyospermum scale), 130; *Parlatoria proteus* (proteus scale), 330

Coffea: Aspidiotus excisus (aglaonema scale), 76; *Chrysomphalus aonidum* (Florida red scale), 122; *Hemiberlesia lataniae* (latania scale), 224; *Howardia biclavis* (mining scale), 234; *Ischnaspis longirostris* (black thread scale), 238; *Lepidosaphes beckii* (purple scale), 244; *Pseudaonidia trilobitiformis* (trilobe scale), 354; *Selenaspidus articulatus* (rufous scale), 386

Cola: Selenaspidus articulatus (rufous scale), 386

Colocasia: Aonidiella aurantii (California red scale), 50; *Aspidiotus destructor* (coconut scale), 72

Comparettia: Pseudoparlatoria parlatorioides (false parlatoria scale), 374

Congea: Aspidiotus destructor (coconut scale), 72; *Ischnaspis longirostris* (black thread scale), 238; *Selenaspidus articulatus* (rufous scale), 386

Conium: Aspidiotus nerii (oleander scale), 78

Conocarpus: Hemiberlesia lataniae (latania scale), 224

Copernicia: Abgrallaspis cyanophylli (cyanophyllum scale), 38

Copiapoa: Diaspis echinocacti (cactus scale), 186

Cordia: Aonidiella citrina (yellow scale), 54; *Chrysomphalus aonidum* (Florida red scale), 122; *C. dictyospermi* (dictyospermum scale), 130; *Clavaspis herculeana* (herculeana scale), 134; *Hemiberlesia lataniae* (latania scale), 224; *Howardia biclavis* (mining scale), 234

Cordyline: Aonidiella aurantii (California red scale), 50; *Aspidiotus nerii* (oleander scale), 78; *Chrysomphalus bifasciculatus* (bifasciculate scale), 126; *Fiorinia fioriniae* (palm fiorinia scale), 204; *Hemiberlesia lataniae* (latania scale), 224; *Pinnaspis aspidistrae* (fern scale), 340; *P. strachani* (lesser snow scale), 344; *Selenaspidus articulatus* (rufous scale), 386

Corethrogyne: Rhizaspidiotus dearnessi (Dearness scale), 380

Cornus: Chionaspis corni (dogwood scale), 106; *C. salicis* (willow scale), 118; *Clavaspis ulmi* (elm armored scale), 136; *Diaspidiotus ancylus* (Putnam scale), 148; *D. forbesi* (Forbes scale), 152; *D. juglansregiae* (walnut scale), 158; *D. osborni* (Osborn scale), 166; *D. ostreaeformis* (European fruit scale), 168; *D. perniciosus* (San Jose scale), 172; *Hemiberlesia rapax* (greedy scale), 232; *Lepidosaphes ulmi* (oystershell scale), 266; *Lopholeucaspis japonica* (Japanese maple scale), 276; *Melanaspis obscura* (obscure scale), 282; *M. tenebricosa* (gloomy scale), 284; *Parlatoria theae* (tea parlatoria scale), 334; *Pseudaulacaspis pentagona* (white peach scale), 360

Corokia: Hemiberlesia lataniae (latania scale), 224

Coronilla: Aspidiotus nerii (oleander scale), 78; *Ischnaspis longirostris* (black thread scale), 238

Corozo: Diaspis boisduvalii (Boisduval scale), 180; *Selenaspidus articulatus* (rufous scale), 386

Coryanthes: Duplaspidiotus tesseratus (tesserate scale), 192

Corylopsis: Parlatoria theae (tea parlatoria scale), 334; *Pseudaonidia duplex* (camphor scale), 346

Corylus: Aspidiotus nerii (oleander scale), 78; *Chionaspis salicis* (willow scale), 118; *Diaspidiotus ancylus* (Putnam scale), 148; *D. forbesi* (Forbes scale), 152; *Hemiberlesia lataniae* (latania scale), 224; *Lepidosaphes conchiformis* (fig scale), 250; *Parlatoria oleae* (olive scale), 320

Corypha: Aonidiella orientalis (Oriental armored scale), 58; *Parlatoria proteus* (proteus scale), 330

Coryphantha: Diaspis echinocacti (cactus scale), 186

Costus: Aspidiella sacchari (sugarcane scale), 66

Cotinus: Lepidosaphes ulmi (oystershell scale), 266

Cotoneaster: Chionaspis furfura (scurfy scale), 110; *C. salicis* (willow scale), 118; *Diaspidiotus ancylus* (Putnam scale), 148; *D. forbesi* (Forbes scale), 152; *D. perniciosus* (San Jose scale), 172; *Lepidosaphes ulmi* (oystershell scale), 266; *Lopholeucaspis japonica* (Japanese maple scale), 276; *Parlatoria theae* (tea parlatoria scale), 334; *Pseudaonidia paeoniae* (peony scale), 350

Cotyledon: Diaspis echinocacti (cactus scale), 186

Coursetia: Situlaspis yuccae (yucca scale), 388

Crataegus: Chionaspis americana (elm scurfy scale), 102; *C. furfura* (scurfy scale), 110; *Diaspidiotus forbesi* (Forbes scale), 152; *D. juglansregiae* (walnut scale), 158; *D. perniciosus* (San Jose scale), 172; *Epidiaspis leperii* (Italian pear scale), 198; *Hemiberlesia lataniae* (latania scale), 224; *H. rapax* (greedy scale), 232; *Lepidosaphes ulmi* (oystershell scale), 266; *Parlatoria oleae* (olive scale), 320; *P. theae* (tea parlatoria scale), 334

Crateva: Hemiberlesia lataniae (latania scale), 224

Crinodendron: *Aspidiotus nerii* (oleander scale), 78

Crossandra: *Pseudaulacaspis pentagona* (white peach scale), 360

Crotalaria: *Hemiberlesia lataniae* (latania scale), 224; *Pinnaspis strachani* (lesser snow scale), 344; *Pseudaulacaspis pentagona* (white peach scale), 360

Croton: *Chrysomphalus dictyospermi* (dictyospermum scale), 130; *Hemiberlesia lataniae* (latania scale), 224; *Howardia biclavis* (mining scale), 234; *Lepidosaphes beckii* (purple scale), 244; *L. camelliae* (camellia scale), 248; *L. pinnaeformis* (cymbidium scale), 262; *Parlatoria proteus* (proteus scale), 330; *P. theae* (tea parlatoria scale), 334; *Pinnaspis strachani* (lesser snow scale), 344; *Pseudaonidia duplex* (camphor scale), 346; *Pseudaulacaspis pentagona* (white peach scale), 360; *P. prunicola* (white prunicola scale), 364

Cryosophila: *Aspidiotus destructor* (coconut scale), 72

Cryptocarya: *Aspidiotus destructor* (coconut scale), 72

Cryptocoryne: *Chrysomphalus dictyospermi* (dictyospermum scale), 130

Cryptomeria: *Aspidiotus cryptomeriae* (cryptomeria scale), 68; *Carulaspis juniperi* (juniper scale), 96; *C. minima* (minute cypress scale), 100; *Chrysomphalus dictyospermi* (dictyospermum scale), 130; *Lepidosaphes pallida* (Maskell scale), 256

Cucumis: *Aulacaspis tubercularis* (white mango scale), 90

Cucurbita: *Aulacaspis tubercularis* (white mango scale), 90; *Pinnaspis strachani* (lesser snow scale), 344; *Pseudaulacaspis pentagona* (white peach scale), 360; *P. prunicola* (white prunicola scale), 364

Cupania: *Howardia biclavis* (mining scale), 234

Cupressocyparis: *Carulaspis juniperi* (juniper scale), 96; *C. minima* (minute cypress scale), 100

Cupressus: *Aonidiella aurantii* (California red scale), 50; *Carulaspis juniperi* (juniper scale), 96; *C. minima* (minute cypress scale), 100; *Cupressaspis shastae* (redwood scale), 142; *Fiorinia japonica* (coniferous fiorinia scale), 206; *Hemiberlesia lataniae* (latania scale), 224; *H. rapax* (greedy scale), 232; *Lepidosaphes pallida* (Maskell scale), 256

Curcurbita: *Hemiberlesia lataniae* (latania scale), 224

Cussonia: *Furchadaspis zamiae* (zamia scale), 218; *Hemiberlesia lataniae* (latania scale), 224

Cyanotis: *Aspidiotus excisus* (aglaonema scale), 76

Cyathodes: *Fiorinia fioriniae* (palm fiorinia scale), 204; *Hemiberlesia lataniae* (latania scale), 224

Cycas: *Aonidiella aurantii* (California red scale), 50; *A. orientalis* (Oriental armored scale), 58; *Aspidiotus destructor* (coconut scale), 72; *A. excisus* (aglaonema scale), 76; *A. nerii* (oleander scale), 78; *Aulacaspis yasumatsui* (cycad aulacaspis scale), 94; *Chrysomphalus aonidum* (Florida red scale), 122; *C. dictyospermi* (dictyospermum scale), 130; *Fiorinia fioriniae* (palm fiorinia scale), 204; *Furchadaspis zamiae* (zamia scale), 218; *Hemiberlesia lataniae* (latania scale), 224; *H. rapax* (greedy scale), 232; *Ischnaspis longirostris* (black thread scale), 238; *Lepidosaphes beckii* (purple scale), 244; *L. pini* (pine oystershell scale), 260; *L. pinnaeformis* (cymbidium scale), 262; *Parlatoria proteus* (proteus scale), 330; *Pinnaspis aspidistrae* (fern scale), 340; *P. strachani* (lesser snow scale), 344; *Pseudaulacaspis cockerelli* (false oleander scale), 356; *P. pentagona* (white peach scale), 360

Cyclamen: *Chrysomphalus aonidum* (Florida red scale), 122

Cycnoches: *Diaspis boisduvalii* (Boisduval scale), 180; *Pseudoparlatoria ostreata* (gray scale), 370

Cydonia: *Aspidiotus nerii* (oleander scale), 78; *Chionaspis furfura* (scurfy scale), 110; *Diaspidiotus forbesi* (Forbes scale), 152; *D. perniciosus* (San Jose scale), 172; *Hemiberlesia lataniae* (latania scale), 224; *Lopholeucaspis japonica* (Japanese maple scale), 276; *Mercetaspis halli* (Hall scale), 288; *Pseudaulacaspis pentagona* (white peach scale), 360

Cymbidium: *Aonidiella citrina* (yellow scale), 54; *A. orientalis* (Oriental armored scale), 58; *Aspidiotus destructor* (coconut scale), 72; *A. nerii* (oleander scale), 78; *Chrysomphalus aonidum* (Florida red scale), 122; *C. dictyospermi* (dictyospermum scale), 130; *Diaspis boisduvalii* (Boisduval scale), 180; *Furcaspis biformis* (orchid scale), 216; *Hemiberlesia lataniae* (latania scale), 224; *H. rapax* (greedy scale), 232; *Lepidosaphes beckii* (purple scale), 244; *L. pinnaeformis* (cymbidium scale), 262; *Parlatoria proteus* (proteus scale), 330

Cymbopogon: *Aspidiella sacchari* (sugarcane scale), 66

Cynodon: *Aspidiella sacchari* (sugarcane scale), 66; *Chrysomphalus aonidum* (Florida red scale), 122; *Kuwanaspis pseudoleucaspis* (bamboo diaspidid), 240; *Odonaspis ruthae* (Bermuda grass scale), 306

Cyperus: *Aspidiotus nerii* (oleander scale), 78; *Hemiberlesia rapax* (greedy scale), 232

Cypripedium: *Aspidiotus nerii* (oleander scale), 78; *Chrysomphalus dictyospermi* (dictyospermum scale), 130; *Hemiberlesia lataniae* (latania scale), 224; *Parlatoria proteus* (proteus scale), 330; *Pinnaspis aspidistrae* (fern scale), 340; *Pseudoparlatoria parlatorioides* (false parlatoria scale), 374

Cyrtopodium: *Aspidiotus destructor* (coconut scale), 72

Cytisus: *Aspidiotus nerii* (oleander scale), 78; *Chionaspis salicis* (willow scale), 118; *Lepidosaphes ulmi* (oystershell scale), 266; *Lopholeucaspis japonica* (Japanese maple scale), 276

Dahlia: *Hemiberlesia lataniae* (latania scale), 224

Dalbergia: *Aspidiotus destructor* (coconut scale), 72; *Hemiberlesia lataniae* (latania scale), 224

Daphne: *Aspidiotus nerii* (oleander scale), 78; *Chrysomphalus bifasciculatus* (bifasciculate scale), 126; *Hemiberlesia rapax* (greedy scale), 232; *Unaspis euonymi* (euonymus scale), 396

Daphniphyllum: *Lepidosaphes pinnaeformis* (cymbidium scale), 262; *Pseudaulacaspis cockerelli* (false oleander scale), 356

Dasylirion: *Situlaspis yuccae* (yucca scale), 388

Datura: *Hemiberlesia lataniae* (latania scale), 224

Davidsonia: *Chrysomphalus dictyospermi* (dictyospermum scale), 130

Delonix: *Aonidiella orientalis* (Oriental armored scale), 58; *Clavaspis herculeana* (herculeana scale), 134

Delostoma: *Howardia biclavis* (mining scale), 234

Dendrobium: *Aspidiotus destructor* (coconut scale), 72; *A. nerii* (oleander scale), 78; *Chrysomphalus aonidum* (Florida red scale), 122; *C. dictyospermi* (dictyospermum scale), 130; *Diaspis boisduvalii* (Boisduval scale), 180; *Furcaspis biformis* (orchid scale), 216; *Hemiberlesia lataniae* (latania scale), 224; *Lepidosaphes pinnaeformis* (cymbidium scale), 262; *Lindingaspis rossi* (black araucaria scale), 272; *Parlatoria proteus* (proteus scale), 330; *Selenaspidus articulatus* (rufous scale), 386

Dendrocalamus: *Froggattiella penicillata* (penicillate scale), 212

Dendrocereus: *Diaspis echinocacti* (cactus scale), 186

Derris: *Howardia biclavis* (mining scale), 234

Desfontainea: *Aspidiotus nerii* (oleander scale), 78

Desmos: *Aulacaspis tubercularis* (white mango scale), 90

Dianella: *Pseudaulacaspis cockerelli* (false oleander scale), 356

Dianthus: *Aspidiotus nerii* (oleander scale), 78; *Chrysomphalus aonidum* (Florida red scale), 122; *Hemiberlesia lataniae* (latania scale), 224

Diaphananthe: *Hemiberlesia lataniae* (latania scale), 224

Dictyosperma: *Aspidiotus destructor* (coconut scale), 72; *Chrysomphalus dictyospermi* (dictyospermum scale), 130; *Ischnaspis longirostris* (black thread scale), 238

Didymosperma: *Ischnaspis longirostris* (black thread scale), 238

Dieffenbachia: *Aspidiotus destructor* (coconut scale), 72; *Parlatoria proteus* (proteus scale), 330; *Pseudoparlatoria ostreata* (gray scale), 370

Dietes: *Aulacaspis tubercularis* (white mango scale), 90

Digitalis: *Clavaspis ulmi* (elm armored scale), 136

Digitaria: *Aspidiella sacchari* (sugarcane scale), 66; *Odonaspis ruthae* (Bermuda grass scale), 306

Dillenia: *Aspidiotus destructor* (coconut scale), 72; *Chrysomphalus dictyospermi* (dictyospermum scale), 130

Dimocarpus: *Aulacaspis tubercularis* (white mango scale), 90

Dioon: *Aspidiotus nerii* (oleander scale), 78; *A. spinosus* (spinose scale), 82; *Aulacaspis yasumatsui* (cycad aulacaspis scale), 94; *Furchadaspis zamiae* (zamia scale), 218; *Hemiberlesia lataniae* (latania scale), 224

Dioscorea: *Aspidiella sacchari* (sugarcane scale), 66; *Aspidiotus destructor* (coconut scale), 72; *Pinnaspis strachani* (lesser snow scale), 344; *Pseudoparlatoria ostreata* (gray scale), 370

Diospyros: *Abgrallaspis cyanophylli* (cyanophyllum scale), 38; *Aonidiella orientalis* (Oriental armored scale), 58; *Aspidiotus nerii* (oleander scale), 78; *A. spinosus* (spinose scale), 82; *Chrysomphalus aonidum* (Florida red scale), 122; *C. dictyospermi* (dictyospermum scale), 130; *Diaspidiotus ancylus* (Putnam scale), 148; *D. perniciosus* (San Jose scale), 172; *Fiorinia fioriniae* (palm fiorinia scale), 204; *Hemiberlesia lataniae* (latania scale), 224; *H. rapax* (greedy scale), 232; *Ischnaspis longirostris* (black thread scale), 238; *Lepidosaphes conchiformis* (fig scale), 250; *L. ulmi* (oystershell scale), 266; *L. yanagicola* (fire bush scale), 270; *Lopholeucaspis japonica* (Japanese maple scale), 276; *Parlatoria oleae* (olive scale), 320; *P. theae* (tea parlatoria scale), 334; *Pinnaspis aspidistrae* (fern scale), 340; *Pseudaonidia duplex* (camphor scale), 346; *P. paeoniae* (peony scale), 350; *P. trilobitiformis* (trilobe scale), 354; *Pseudaulacaspis cockerelli* (false oleander scale), 356; *P. pentagona* (white peach scale), 360; *Selenaspidus articulatus* (rufous scale), 386

Dipteryx: *Aspidiotus destructor* (coconut scale), 72

Dischidia: *Howardia biclavis* (mining scale), 234

Distichlis: *Odonaspis ruthae* (Bermuda grass scale), 306

Dodonaea: *Aspidiotus nerii* (oleander scale), 78

Dombeya: *Pinnaspis strachani* (lesser snow scale), 344; *Pseudaulacaspis pentagona* (white peach scale), 360

Dovyalis: *Howardia biclavis* (mining scale), 234

Dracaena: *Aspidiella sacchari* (sugarcane scale), 66; *Aspidiotus nerii* (oleander scale), 78; *Chrysomphalus aonidum* (Florida red scale), 122; *C. dictyospermi* (dictyospermum scale), 130; *Diaspis boisduvalii* (Boisduval scale), 180; *Hemiberlesia lataniae* (latania scale), 224; *H. rapax* (greedy scale), 232; *Parlatoria pittospori* (pittosporum scale), 328; *P. proteus* (proteus scale), 330; *Pinnaspis strachani* (lesser snow scale), 344; *Pseudaulacaspis cockerelli* (false oleander scale), 356; *Pseudoparlatoria parlatorioides* (false parlatoria scale), 374; *Selenaspidus articulatus* (rufous scale), 386

Dudleya: *Diaspis echinocacti* (cactus scale), 186

Duranta: *Howardia biclavis* (mining scale), 234; *Ischnaspis longirostris* (black thread scale), 238; *Selenaspidus articulatus* (rufous scale), 386

Durio: *Hemiberlesia lataniae* (latania scale), 224

Dysoxylum: *Aonidiella orientalis* (Oriental armored scale), 58

Echeveria: *Aspidiotus nerii* (oleander scale), 78; *Hemiberlesia lataniae* (latania scale), 224; *Odonaspis ruthae* (Bermuda grass scale), 306

Echinocactus: *Aspidiotus nerii* (oleander scale), 78; *Diaspis echinocacti* (cactus scale), 186

Echinocereus: *Diaspis echinocacti* (cactus scale), 186

Echinofossulocactus: *Diaspis echinocacti* (cactus scale), 186

Echinopsis: *Abgrallaspis cyanophylli* (cyanophyllum scale), 38; *Diaspis echinocacti* (cactus scale), 186

Ehretia: *Abgrallaspis cyanophylli* (cyanophyllum scale), 38

Elaeagnus: *Aonidiella aurantii* (California red scale), 50; *Chrysomphalus bifasciculatus* (bifasciculate scale), 126; *Hemiberlesia lataniae* (latania scale), 224; *H. rapax* (greedy scale), 232; *Lepidosaphes beckii* (purple scale), 244; *L. gloverii* (Glover scale), 254; *L. pinnaeformis* (cymbidium scale), 262; *L. ulmi* (oystershell scale), 266; *Lindingaspis rossi* (black araucaria scale), 272; *Parlatoria oleae* (olive scale), 320

Elaeis: *Aspidiotus destructor* (coconut scale), 72; *Diaspis boisduvalii* (Boisduval scale), 180; *Hemiberlesia lataniae* (latania scale), 224; *Ischnaspis longirostris* (black thread scale), 238; *Selenaspidus articulatus* (rufous scale), 386

Elaeocarpus: *Howardia biclavis* (mining scale), 234

Elaeodendron: *Furchadaspis zamiae* (zamia scale), 218; *Hemiberlesia rapax* (greedy scale), 232; *Pseudoparlatoria parlatorioides* (false parlatoria scale), 374

Elaphrium: *Clavaspis herculeana* (herculeana scale), 134

Encephalartos: *Aulacaspis yasumatsui* (cycad aulacaspis scale), 94; *Furchadaspis zamiae* (zamia scale), 218; *Hemiberlesia lataniae* (latania scale), 224

Enkianthus: *Lopholeucaspis japonica* (Japanese maple scale), 276; *Parlatoria theae* (tea parlatoria scale), 334

Ephedra: *Aonidiella aurantii* (California red scale), 50; *Hemiberlesia rapax* (greedy scale), 232

Epidendrum: *Chrysomphalus aonidum* (Florida red scale), 122; *C. dictyospermi* (dictyospermum scale), 130; *Diaspis boisduvalii* (Boisduval scale), 180; *Furcaspis biformis* (orchid scale), 216; *Hemiberlesia lataniae* (latania scale), 224; *Parlatoria proteus* (proteus scale), 330; *Pseudischnaspis bowreyi* (Bowrey scale), 368; *Pseudoparlatoria parlatorioides* (false parlatoria scale), 374

Epiphyllum: *Diaspis echinocacti* (cactus scale), 186

Epipremnum: *Chrysomphalus aonidum* (Florida red scale), 122; *Diaspis boisduvalii* (Boisduval scale), 180

Eremochloa: *Aspidiella sacchari* (sugarcane scale), 66

Eria: *Parlatoria proteus* (proteus scale), 330

Erica: *Chionaspis salicis* (willow scale), 118; *Hemiberlesia rapax* (greedy scale), 232

Erigeron: *Hemiberlesia lataniae* (latania scale), 224; *Rhizaspidiotus dearnessi* (Dearness scale), 380

Eriobotrya: *Aspidiotus spinosus* (spinose scale), 82; *Chrysomphalus dictyospermi* (dictyospermum scale), 130; *Clavaspis*

herculeana (herculeana scale), 134; *Diaspidiotus perniciosus* (San Jose scale), 172; *Hemiberlesia lataniae* (latania scale), 224; *Ischnaspis longirostris* (black thread scale), 238; *Neopinnaspis harperi* (Harper scale), 294; *Parlatoria oleae* (olive scale), 320

Erythea: *Aspidiotus nerii* (oleander scale), 78; *Comstockiella sabalis* (palmetto scale), 140

Erythrina: *Andaspis punicae* (litchi scale), 48; *Aonidiella aurantii* (California red scale), 50; *A. orientalis* (Oriental armored scale), 58; *Clavaspis herculeana* (herculeana scale), 134; *Howardia biclavis* (mining scale), 234; *Lepidosaphes beckii* (purple scale), 244; *L. gloverii* (Glover scale), 254

Erythroxylum: *Howardia biclavis* (mining scale), 234

Eucalyptus: *Abgrallaspis cyanophylli* (cyanophyllum scale), 38; *Aonidiella aurantii* (California red scale), 50; *A. orientalis* (Oriental armored scale), 58; *Chrysomphalus aonidum* (Florida red scale), 122; *Diaspidiotus perniciosus* (San Jose scale), 172; *Hemiberlesia lataniae* (latania scale), 224; *H. rapax* (greedy scale), 232; *Ischnaspis longirostris* (black thread scale), 238; *Lepidosaphes beckii* (purple scale), 244; *Lindingaspis rossi* (black araucaria scale), 272; *Parlatoria pittospori* (pittosporum scale), 328; *Pseudaonidia trilobitiformis* (trilobe scale), 354; *Pseudischnaspis bowreyi* (Bowrey scale), 368; *Selenaspidus articulatus* (rufous scale), 386

Euclea: *Hemiberlesia lataniae* (latania scale), 224

Eucryphia: *Aspidiotus nerii* (oleander scale), 78

Eugenia: *Aspidiotus destructor* (coconut scale), 72; *Chrysomphalus aonidum* (Florida red scale), 122; *C. bifasciculatus* (bifasciculate scale), 126; *C. dictyospermi* (dictyospermum scale), 130; *Duplaspidiotus claviger* (camellia mining scale), 188; *Fiorinia fioriniae* (palm fiorinia scale), 204; *Hemiberlesia lataniae* (latania scale), 224; *Ischnaspis longirostris* (black thread scale), 238; *Morganella longispina* (plumose scale), 290; *Selenaspidus articulatus* (rufous scale), 386; *Unaspis euonymi* (euonymus scale), 396

Euonymus: *Abgrallaspis degenerata* (degenerate scale), 42; *A. aurantii* (California red scale), 50; *Aonidiella orientalis* (Oriental armored scale), 58; *Aspidiotus nerii* (oleander scale), 78; *A. spinosus* (spinose scale), 82; *Chionaspis salicis* (willow scale), 118; *Chrysomphalus aonidum* (Florida red scale), 122; *C. bifasciculatus* (bifasciculate scale), 126; *C. dictyospermi* (dictyospermum scale), 130; *Clavaspis ulmi* (elm armored scale), 136; *Diaspidiotus ancylus* (Putnam scale), 148; *D. perniciosus* (San Jose scale), 172; *Fiorinia fioriniae* (palm fiorinia scale), 204; *F. theae* (tea scale), 210; *Hemiberlesia lataniae* (latania scale), 224; *H. rapax* (greedy scale), 232; *Lepidosaphes gloverii* (Glover scale), 254; *L. yanagicola* (fire brush scale), 13; *L. yanagicola* (fire bush scale), 270; *Lindingaspis rossi* (black araucaria scale), 272; *Lopholeucaspis japonica* (Japanese maple scale), 276; *Parlatoria camelliae* (camellia parlatoria scale), 316; *P. pergandii* (chaff scale), 324; *P. theae* (tea parlatoria scale), 334; *Pseudaulacaspis pentagona* (white peach scale), 360; *Pseudoparlatoria ostreata* (gray scale), 370; *Selenaspidus articulatus* (rufous scale), 386; *Unaspis euonymi* (euonymus scale), 12, 396

Euphorbia: *Aspidiotus destructor* (coconut scale), 72; *A. excisus* (aglaonema scale), 76; *A. nerii* (oleander scale), 78; *A. spinosus* (spinose scale), 82; *Chrysomphalus aonidum* (Florida red scale), 122; *C. dictyospermi* (dictyospermum scale), 130; *Clavaspis herculeana* (herculeana scale), 134; *Hemiberlesia lataniae* (latania scale), 224; *H. rapax* (greedy scale), 232; *Howardia biclavis* (mining scale), 234; *Odonaspis ruthae* (Bermuda grass scale), 306; *Parlatoria theae* (tea parlatoria scale), 334; *Pinnaspis strachani* (lesser snow scale), 344;

Pseudaulacaspis pentagona (white peach scale), 360; *Selenaspidus albus* (white euphorbia scale), 382

Euphoria: *Chrysomphalus dictyospermi* (dictyospermum scale), 130

Eupritchardia: *Abgrallaspis cyanophylli* (cyanophyllum scale), 38; *Aspidiotus destructor* (coconut scale), 72; *Selenaspidus articulatus* (rufous scale), 386

Eurya: *Abgrallaspis degenerata* (degenerate scale), 42; *Chrysomphalus aonidum* (Florida red scale), 122; *Fiorinia theae* (tea scale), 210; *Parlatoria camelliae* (camellia parlatoria scale), 316; *Pinnaspis aspidistrae* (fern scale), 340; *Pseudaonidia paeoniae* (peony scale), 350

Euterpe: *Hemiberlesia lataniae* (latania scale), 224; *Selenaspidus articulatus* (rufous scale), 386

Fagara: *Hemiberlesia rapax* (greedy scale), 232

Fagus: *Diaspidiotus ancylus* (Putnam scale), 148; *D. perniciosus* (San Jose scale), 172; *Lepidosaphes ulmi* (oystershell scale), 266; *Melanaspis obscura* (obscure scale), 282

Fatsia: *Aspidiotus destructor* (coconut scale), 72; *A. nerii* (oleander scale), 78; *A. spinosus* (spinose scale), 82; *Chrysomphalus dictyospermi* (dictyospermum scale), 130

Feijoa: *Hemiberlesia rapax* (greedy scale), 232

Ferocactus: *Diaspis boisduvalii* (Boisduval scale), 180; *D. echinocacti* (cactus scale), 186

Feroniella: *Howardia biclavis* (mining scale), 234

Festuca: *Odonaspis ruthae* (Bermuda grass scale), 306

Ficus: *Abgrallaspis cyanophylli* (cyanophyllum scale), 38; *Andaspis punicae* (litchi scale), 48; *Aonidiella aurantii* (California red scale), 50; *A. orientalis* (Oriental armored scale), 58; *A. sacchari* (sugarcane scale), 66; *Aspidiotus destructor* (coconut scale), 72; *A. nerii* (oleander scale), 78; *A. spinosus* (spinose scale), 82; *Chrysomphalus aonidum* (Florida red scale), 122; *C. dictyospermi* (dictyospermum scale), 130; *Diaspis boisduvalii* (Boisduval scale), 180; *Duplaspidiotus claviger* (camellia mining scale), 188; *D. tesserate* (tesserate scale), 192; *Epidiaspis leperii* (Italian pear scale), 198; *Hemiberlesia lataniae* (latania scale), 224; *H. neodiffinis* (false diffinis scale), 228; *H. rapax* (greedy scale), 232; *Howardia biclavis* (mining scale), 234; *Ischnaspis longirostris* (black thread scale), 238; *Lepidosaphes beckii* (purple scale), 244; *L. conchiformis* (fig scale), 250; *L. gloverii* (Glover scale), 254; *L. pinnaeformis* (cymbidium scale), 262; *L. ulmi* (oystershell scale), 266; *Lindingaspis rossi* (black araucaria scale), 272; *Lopholeucaspis japonica* (Japanese maple scale), 276; *Morganella longispina* (plumose scale), 290; *Parlatoreopsis chinensis* (Chinese obscure scale), 310; *Parlatoria camelliae* (camellia parlatoria scale), 316; *P. oleae* (olive scale), 320; *P. proteus* (proteus scale), 330; *Pseudaonidia duplex* (camphor scale), 346; *P. trilobitiformis* (trilobe scale), 354; *Pseudaulacaspis pentagona* (white peach scale), 360; *Selenaspidus articulatus* (rufous scale), 386; *Unaspis euonymi* (euonymus scale), 396

Firmiana: *Hemiberlesia lataniae* (latania scale), 224

Fitchia: *Lepidosaphes ulmi* (oystershell scale), 266

Flacourtia: *Aonidiella orientalis* (Oriental armored scale), 58; *Hemiberlesia lataniae* (latania scale), 224

Forestiera: *Clavaspis ulmi* (elm armored scale), 136

Forsythia: *Diaspidiotus forbesi* (Forbes scale), 152; *Lepidosaphes ulmi* (oystershell scale), 266; *Pseudaulacaspis prunicola* (white prunicola scale), 364; *Situlaspis yuccae* (yucca scale), 388

Fortunella: *Aonidiella aurantii* (California red scale), 50; *Chrysomphalus dictyospermi* (dictyospermum scale), 130; *Lepidosaphes*

Fortunella (cont.)
beckii (purple scale), 244; *L. gloverii* (Glover scale), 254; *Parlatoria pergandii* (chaff scale), 324; *Unaspis citri* (citrus snow scale), 392
Fragesia: *Kuwanaspis pseudoleucaspis* (bamboo diaspidid), 240
Francoa: *Aspidiotus nerii* (oleander scale), 78
Fraxinus: *Aonidiella aurantii* (California red scale), 50; *Aspidiotus nerii* (oleander scale), 78; *Chionaspis americana* (elm scurfy scale), 102; *C. salicis* (willow scale), 118; *Diaspidiotus ancylus* (Putnam scale), 148; *D. forbesi* (Forbes scale), 152; *D. juglansregiae* (walnut scale), 158; *D. uvae* (grape scale), 176; *Hemiberlesia neodiffinis* (false diffinis scale), 228; *H. rapax* (greedy scale), 232; *Lepidosaphes conchiformis* (fig scale), 250; *L. ulmi* (oystershell scale), 266; *L. yanagicola* (fire bush scale), 270; *Lopholeucaspis japonica* (Japanese maple scale), 276; *Melanaspis obscura* (obscure scale), 282; *M. tenebricosa* (gloomy scale), 284; *Parlatoria oleae* (olive scale), 320; *Pseudaulacaspis pentagona* (white peach scale), 360; *P. prunicola* (white prunicola scale), 364; *Situlaspis yuccae* (yucca scale), 388; *Unaspis euonymi* (euonymus scale), 396
Fuchsia: *Hemiberlesia rapax* (greedy scale), 232
Funtumia: *Selenaspidus articulatus* (rufous scale), 386
Furcraea: *Aspidiotus nerii* (oleander scale), 78; *Hemiberlesia lataniae* (latania scale), 224

Gaiadendron: *Aulacaspis tubercularis* (white mango scale), 90
Garcinia: *Aonidiella citrina* (yellow scale), 54; *A. orientalis* (Oriental armored scale), 58; *Aspidiotus destructor* (coconut scale), 72; *Chrysomphalus aonidum* (Florida red scale), 122; *C. dictyospermi* (dictyospermum scale), 130; *Fiorinia fioriniae* (palm fiorinia scale), 204; *Hemiberlesia lataniae* (latania scale), 224
Gardenia: *Aonidiella aurantii* (California red scale), 50; *A. citrina* (yellow scale), 54; *Aspidiotus excisus* (aglaonema scale), 76; *A. nerii* (oleander scale), 78; *Chrysomphalus aonidum* (Florida red scale), 122; *C. dictyospermi* (dictyospermum scale), 130; *Duplaspidiotus claviger* (camellia mining scale), 188; *Fiorinia fioriniae* (palm fiorinia scale), 204; *F. theae* (tea scale), 210; *Hemiberlesia lataniae* (latania scale), 224; *H. rapax* (greedy scale), 232; *Howardia biclavis* (mining scale), 234; *Ischnaspis longirostris* (black thread scale), 238; *Lepidosaphes camelliae* (camellia scale), 248; *Melanaspis tenebricosa* (gloomy scale), 284; *Parlatoria camelliae* (camellia parlatoria scale), 316; *P. proteus* (proteus scale), 330; *Pseudaonidia trilobitiformis* (trilobe scale), 354; *Selenaspidus articulatus* (rufous scale), 386
Garrya: *Aspidiotus nerii* (oleander scale), 78
Gaultheria: *Hemiberlesia rapax* (greedy scale), 232
Geijera: *Lindingaspis rossi* (black araucaria scale), 272
Gelonium: *Hemiberlesia lataniae* (latania scale), 224
Genipa: *Clavaspis herculeana* (herculeana scale), 134; *Howardia biclavis* (mining scale), 234; *Pinnaspis strachani* (lesser snow scale), 344
Genista: *Chionaspis salicis* (willow scale), 118; *Chrysomphalus dictyospermi* (dictyospermum scale), 130; *Hemiberlesia rapax* (greedy scale), 232
Geonoma: *Selenaspidus articulatus* (rufous scale), 386
Geranium: *Pseudaulacaspis pentagona* (white peach scale), 360
Gerbera: *Pseudoparlatoria parlatorioides* (false parlatoria scale), 374
Geum: *Lepidosaphes ulmi* (oystershell scale), 266
Gigantochloa: *Froggattiella penicillata* (penicillate scale), 212
Ginkgo: *Clavaspis ulmi* (elm armored scale), 136; *Lepidosaphes ulmi* (oystershell scale), 266

Gladiolus: *Hemiberlesia lataniae* (latania scale), 224
Gleditsia: *Aspidiotus nerii* (oleander scale), 78; *Diaspidiotus ancylus* (Putnam scale), 148; *D. juglansregiae* (walnut scale), 158; *Lepidosaphes ulmi* (oystershell scale), 266; *Melanaspis tenebricosa* (gloomy scale), 284; *Situlaspis yuccae* (yucca scale), 388
Gliricidia: *Hemiberlesia lataniae* (latania scale), 224; *Ischnaspis longirostris* (black thread scale), 238
Gnaphalium: *Hemiberlesia lataniae* (latania scale), 224
Gnetum: *Aspidiotus destructor* (coconut scale), 72; *Chrysomphalus aonidum* (Florida red scale), 122
Gnidia: *Hemiberlesia rapax* (greedy scale), 232
Gomortega: *Hemiberlesia rapax* (greedy scale), 232
Gomphrena: *Pseudaulacaspis pentagona* (white peach scale), 360
Gongora: *Furcaspis biformis* (orchid scale), 216; *Parlatoria proteus* (proteus scale), 330; *Selenaspidus articulatus* (rufous scale), 386
Gonocaryum: *Ischnaspis longirostris* (black thread scale), 238
Gossypium: *Pinnaspis strachani* (lesser snow scale), 344; *Pseudaulacaspis pentagona* (white peach scale), 360
Grammatophyllum: *Aonidiella aurantii* (California red scale), 50; *Parlatoria proteus* (proteus scale), 330
Grevillea: *Aonidiella citrina* (yellow scale), 54; *Aspidiotus destructor* (coconut scale), 72; *Chrysomphalus aonidum* (Florida red scale), 122; *Hemiberlesia lataniae* (latania scale), 224; *Howardia biclavis* (mining scale), 234
Gronophyllum: *Hemiberlesia lataniae* (latania scale), 224; *Ischnaspis longirostris* (black thread scale), 238; *Parlatoria camelliae* (camellia parlatoria scale), 316
Grossularia: *Chionaspis salicis* (willow scale), 118
Guaiacum: *Aonidiella aurantii* (California red scale), 50; *Duplaspidiotus tesseratus* (tesserate scale), 192
Guilielma: *Diaspis boisduvalii* (Boisduval scale), 180
Gutierrezia: *Rhizaspidiotus dearnessi* (Dearness scale), 380
Guzmania: *Diaspis bromeliae* (pineapple scale), 182
Gymnocladus: *Diaspidiotus juglansregiae* (walnut scale), 158; *Parlatoreopsis chinensis* (Chinese obscure scale), 310

Haematoxylum: *Duplaspidiotus tesseratus* (tesserate scale), 192
Haitia: *Chrysomphalus aonidum* (Florida red scale), 122
Hakea: *Aspidiotus nerii* (oleander scale), 78; *Parlatoria pittospori* (pittosporum scale), 328
Haplopappus: *Rhizaspidiotus dearnessi* (Dearness scale), 380
Harrisia: *Diaspis echinocacti* (cactus scale), 186
Haworthia: *Aonidiella aurantii* (California red scale), 50; *Aspidiotus nerii* (oleander scale), 78; *A. spinosus* (spinose scale), 82; *Hemiberlesia lataniae* (latania scale), 224; *H. rapax* (greedy scale), 232; *Lindingaspis rossi* (black araucaria scale), 272; *Parlatoria pittospori* (pittosporum scale), 328
Hedera: *Aspidiotus nerii* (oleander scale), 78; *Chrysomphalus aonidum* (Florida red scale), 122; *C. bifasciculatus* (bifasciculate scale), 126; *C. dictyospermi* (dictyospermum scale), 130; *Clavaspis herculeana* (herculeana scale), 134; *Diaspidiotus ancylus* (Putnam scale), 148; *Dynaspidiotus britannicus* (holly scale), 194; *Fiorinia fioriniae* (palm fiorinia scale), 204; *Hemiberlesia lataniae* (latania scale), 224; *Lepidosaphes camelliae* (camellia scale), 248; *Lindingaspis rossi* (black araucaria scale), 272; *Parlatoria proteus* (proteus scale), 330; *Pseudaulacaspis pentagona* (white peach

scale), 360; *Pseudoparlatoria parlatorioides* (false parlatoria scale), 374; *Situlaspis yuccae* (yucca scale), 388
Helianthemum: *Chionaspis salicis* (willow scale), 118
Helianthus: *Pseudaulacaspis cockerelli* (false oleander scale), 356; *Rhizaspidiotus dearnessi* (Dearness scale), 380
Heliconia: *Chrysomphalus aonidum* (Florida red scale), 122; *C. dictyospermi* (dictyospermum scale), 130; *Diaspis boisduvalii* (Boisduval scale), 180; *Hemiberlesia lataniae* (latania scale), 224; *Ischnaspis longirostris* (black thread scale), 238; *Parlatoria proteus* (proteus scale), 330; *Selenaspidus articulatus* (rufous scale), 386
Heliocereus: *Diaspis echinocacti* (cactus scale), 186
Heliotropium: *Aspidiotus nerii* (oleander scale), 78; *Rhizaspidiotus dearnessi* (Dearness scale), 380
Hemerocallis: *Aspidiotus nerii* (oleander scale), 78
Hemigraphis: *Parlatoria proteus* (proteus scale), 330
Hepatica: *Abgrallaspis cyanophylli* (cyanophyllum scale), 38; *Aspidiotus nerii* (oleander scale), 78
Hernandia: *Aspidiotus destructor* (coconut scale), 72
Heteromeles: *Epidiaspis leperii* (Italian pear scale), 198
Heterospathe: *Aspidiotus destructor* (coconut scale), 72; *Ischnaspis longirostris* (black thread scale), 238; *Pinnaspis strachani* (lesser snow scale), 344
Hevea: *Aspidiotus destructor* (coconut scale), 72; *Chrysomphalus dictyospermi* (dictyospermum scale), 130; *Hemiberlesia lataniae* (latania scale), 224; *Pseudaulacaspis cockerelli* (false oleander scale), 356
Hibiscus: *Aonidiella orientalis* (Oriental armored scale), 58; *Aspidiotus nerii* (oleander scale), 78; *Chrysomphalus aonidum* (Florida red scale), 122; *Duplaspidiotus claviger* (camellia mining scale), 188; *D. tesseratus* (tesserate scale), 192; *Hemiberlesia lataniae* (latania scale), 224; *Howardia biclavis* (mining scale), 234; *Ischnaspis longirostris* (black thread scale), 238; *Morganella longispina* (plumose scale), 290; *Parlatoreopsis chinensis* (Chinese obscure scale), 310; *Parlatoria theae* (tea parlatoria scale), 334; *Pinnaspis strachani* (lesser snow scale), 344; *Pseudaulacaspis pentagona* (white peach scale), 360; *P. bowreyi* (Bowrey scale), 368; *Unaspis citri* (citrus snow scale), 392
Holboellia: *Lepidosaphes ulmi* (oystershell scale), 266
Howea: *Chrysomphalus aonidum* (Florida red scale), 122; *Hemiberlesia rapax* (greedy scale), 232; *Ischnaspis longirostris* (black thread scale), 238
Hudsonia: *Rhizaspidiotus dearnessi* (Dearness scale), 380
Huernia: *Aonidiella aurantii* (California red scale), 50; *Aspidiotus nerii* (oleander scale), 78; *Selenaspidus articulatus* (rufous scale), 386
Huntleya: *Pseudoparlatoria parlatorioides* (false parlatoria scale), 374
Hydrangea: *Diaspidiotus ancylus* (Putnam scale), 148; *Lepidosaphes ulmi* (oystershell scale), 266; *Pseudaulacaspis pentagona* (white peach scale), 360
Hylocereus: *Aspidiotus destructor* (coconut scale), 72; *Diaspis boisduvalii* (Boisduval scale), 180; *D. echinocacti* (cactus scale), 186; *Pseudischnaspis bowreyi* (Bowrey scale), 368
Hymenaea: *Clavaspis herculeana* (herculeana scale), 134; *Howardia biclavis* (mining scale), 234; *Selenaspidus articulatus* (rufous scale), 386
Hyophorbe: *Hemiberlesia lataniae* (latania scale), 224
Hypericum: *Diaspidiotus perniciosus* (San Jose scale), 172; *Hemiberlesia lataniae* (latania

scale), 224; *Lepidosaphes ulmi* (oystershell scale), 266

Hyphaene: Comstockiella sabalis (palmetto scale), 140; *Parlatoria blanchardi* (parlatoria date scale), 312; *Selenaspidus articulatus* (rufous scale), 386

Iberis: Pseudaulacaspis pentagona (white peach scale), 360

Ilex: Abgrallaspis degenerata (degenerate scale), 42; *Aonidiella aurantii* (California red scale), 50; *A. orientalis* (Oriental armored scale), 58; *Aspidiotus nerii* (oleander scale), 78; *Chrysomphalus aonidum* (Florida red scale), 122; *C. bifasciculatus* (bifasciculate scale), 126; *C. dictyospermi* (dictyospermum scale), 130; *Diaspidiotus ancylus* (Putnam scale), 148; *D. forbesi* (Forbes scale), 152; *D. juglansregiae* (walnut scale), 158; *D. osborni* (Osborn scale), 166; *Dynaspidiotus britannicus* (holly scale), 194; *Fiorinia theae* (tea scale), 210; *Hemiberlesia lataniae* (latania scale), 224; *H. rapax* (greedy scale), 232; *Lepidosaphes beckii* (purple scale), 244; *L. camelliae* (camellia scale), 248; *Lopholeucaspis japonica* (Japanese maple scale), 276; *Melanaspis tenebricosa* (gloomy scale), 284; *Neopinnaspis harperi* (Harper scale), 294; *Pseudaonidia paeoniae* (peony scale), 350; *Pseudaulacaspis cockerelli* (false oleander scale), 356; *P. pentagona* (white peach scale), 360; *P. prunicola* (white prunicola scale), 364; *Unaspis euonymi* (euonymus scale), 396

Illicium: Lepidosaphes pinnaeformis (cymbidium scale), 262

Illigera: Chrysomphalus dictyospermi (dictyospermum scale), 130

Impatiens: Hemiberlesia rapax (greedy scale), 232

Indigofera: Hemiberlesia lataniae (latania scale), 224

Inga: Duplaspidiotus claviger (camellia mining scale), 188; *Howardia biclavis* (mining scale), 234; *Ischnaspis longirostris* (black thread scale), 238; *Pseudoparlatoria parlatorioides* (false parlatoria scale), 374

Ionopsis: Furcaspis biformis (orchid scale), 216; *Pseudoparlatoria ostreata* (gray scale), 370

Ipomoea: Aspidiotus excisus (aglaonema scale), 76; *Hemiberlesia lataniae* (latania scale), 224; *Pseudaulacaspis pentagona* (white peach scale), 360

Iriartella: Ischnaspis longirostris (black thread scale), 238

Iris: Aonidiella orientalis (Oriental armored scale), 58; *Aspidiotus nerii* (oleander scale), 78; *Hemiberlesia lataniae* (latania scale), 224

Isochilus: Chrysomphalus dictyospermi (dictyospermum scale), 130

Iva: Hemiberlesia lataniae (latania scale), 224

Ixora: Chrysomphalus dictyospermi (dictyospermum scale), 130; *Howardia biclavis* (mining scale), 234; *Ischnaspis longirostris* (black thread scale), 238; *Pseudaonidia trilobitiformis* (trilobe scale), 354; *Selenaspidus articulatus* (rufous scale), 386

Jacaranda: Hemiberlesia lataniae (latania scale), 224

Jacobinia: Aspidiotus excisus (aglaonema scale), 76

Jasminum: Abgrallaspis cyanophylli (cyanophyllum scale), 38; *Aonidiella aurantii* (California red scale), 50; *A. citrina* (yellow scale), 54; *A. orientalis* (Oriental armored scale), 58; *Aspidiotus destructor* (coconut scale), 72; *A. nerii* (oleander scale), 78; *Chionaspis salicis* (willow scale), 118; *Chrysomphalus aonidum* (Florida red scale), 122; *C. dictyospermi* (dictyospermum scale), 130; *Hemiberlesia lataniae* (latania scale), 224; *H. rapax* (greedy scale), 232; *Howardia biclavis* (mining scale), 234; *Ischnaspis longirostris* (black thread scale), 238; *Morganella longispina* (plumose scale), 290; *Parlatoria*

oleae (olive scale), 320; *P. pergandii* (chaff scale), 324; *P. proteus* (proteus scale), 330; *Pseudaulacaspis pentagona* (white peach scale), 360; *Pseudischnaspis bowreyi* (Bowrey scale), 368; *Pseudoparlatoria ostreata* (gray scale), 370; *Selenaspidus articulatus* (rufous scale), 386

Jatropha: Pseudaulacaspis pentagona (white peach scale), 360; *Pseudoparlatoria parlatorioides* (false parlatoria scale), 374

Jessenia: Chrysomphalus dictyospermi (dictyospermum scale), 130

Jubaea: Hemiberlesia lataniae (latania scale), 226

Juglans: Aonidiella aurantii (California red scale), 50; *Aspidiotus nerii* (oleander scale), 80; *Chionaspis furfura* (scurfy scale), 110; *Clavaspis ulmi* (elm armored scale), 136; *Diaspidiotus juglansregiae* (walnut scale), 158; *D. perniciosus* (San Jose scale), 172; *Epidiaspis leperii* (Italian pear scale), 198; *Hemiberlesia lataniae* (latania scale), 226; *H. neodiffinis* (false diffinis scale), 228; *H. rapax* (greedy scale), 232; *Lepidosaphes conchiformis* (fig scale), 250; *L. ulmi* (oystershell scale), 266; *Melanaspis obscura* (obscure scale), 282; *M. tenebricosa* (gloomy scale), 284; *Parlatoreopsis chinensis* (Chinese obscure scale), 310; *Pseudaulacaspis pentagona* (white peach scale), 360

Juncus: Odonaspis ruthae (Bermuda grass scale), 306

Juniperus: Carulaspis juniperi (juniper scale), 96; *C. minima* (minute cypress scale), 100; *C. pinifoliae* (pine needle scale), 116; *Chrysomphalus dictyospermi* (dictyospermum scale), 130; *Cupressaspis shastae* (redwood scale), 142; *Fiorinia japonica* (coniferous fiorinia scale), 206; *Lepidosaphes pallida* (Maskell scale), 256; *Nuculaspis pseudomeyeri* (false Meyer scale), 300; *N. tsugae* (shortneedle conifer scale), 304

Kalanchoe: Pseudaulacaspis pentagona (white peach scale), 360

Kalmia: Hemiberlesia lataniae (latania scale), 226

Karatas: Gymnaspis aechmeae (flyspeck scale), 222

Kentia: Aspidiotus destructor (coconut scale), 72; *A. nerii* (oleander scale), 80; *Chrysomphalus aonidum* (Florida red scale), 122; *C. dictyospermi* (dictyospermum scale), 130; *Diaspis boisduvalii* (Boisduval scale), 180; *Fiorinia fioriniae* (palm fiorinia scale), 204; *Parlatoria proteus* (proteus scale), 330; *Pseudaonidia trilobitiformis* (trilobe scale), 354

Keteleeria: Aspidiotus cryptomeriae (cryptomeria scale), 68

Kleinhovia: Abgrallaspis cyanophylli (cyanophyllum scale), 38

Koelreuteria: Lepidosaphes ulmi (oystershell scale), 266; *Parlatoreopsis chinensis* (Chinese obscure scale), 310

Kokia: Duplaspidiotus tesseratus (tesserate scale), 192

Kydia: Hemiberlesia lataniae (latania scale), 226

Laelia: Aonidiella aurantii (California red scale), 50; *Aspidiotus nerii* (oleander scale), 80; *A. spinosus* (spinose scale), 82; *Chrysomphalus dictyospermi* (dictyospermum scale), 130; *Diaspis boisduvalii* (Boisduval scale), 180; *Furcaspis biformis* (orchid scale), 216; *Parlatoria proteus* (proteus scale), 330; *Pseudoparlatoria ostreata* (gray scale), 370; *P. parlatorioides* (false parlatoria scale), 374

Lagerstroemia: Aonidiella aurantii (California red scale), 50; *A. orientalis* (Oriental armored scale), 58; *Aspidiotus destructor* (coconut scale), 72; *Chrysomphalus dictyospermi* (dictyospermum scale), 130; *Hemiberlesia lataniae* (latania scale), 226; *Morganella longispina* (plumose scale), 290; *Pinnaspis aspidistrae* (fern scale), 340; *Selenaspidus articulatus* (rufous scale), 386

Laguncularia: Aonidiella orientalis (Oriental armored scale), 58

Lamium: Parlatoria proteus (proteus scale), 330

Landolphia: Ischnaspis longirostris (black thread scale), 238

Lansium: Aonidiella citrina (yellow scale), 54

Lantana: Chrysomphalus aonidum (Florida red scale), 122; *C. bifasciculatus* (bifasciculate scale), 126; *C. dictyospermi* (dictyospermum scale), 130; *Comstockiella sabalis* (palmetto scale), 140; *Duplaspidiotus claviger* (camellia mining scale), 188; *Hemiberlesia lataniae* (latania scale), 226; *Howardia biclavis* (mining scale), 234; *Ischnaspis longirostris* (black thread scale), 238; *Pseudaulacaspis pentagona* (white peach scale), 360

Lapageria: Aspidiotus nerii (oleander scale), 80

Larrea: Hemiberlesia lataniae (latania scale), 226

Latania: Diaspis boisduvalii (Boisduval scale), 180; *Fiorinia fioriniae* (palm fiorinia scale), 204; *Parlatoria blanchardi* (parlatoria date scale), 312; *Pinnaspis strachani* (lesser snow scale), 344

Laurelia: Aspidiotus nerii (oleander scale), 80

Laurus: Aonidiella aurantii (California red scale), 50; *Aspidiotus destructor* (coconut scale), 72; *A. nerii* (oleander scale), 80; *A. spinosus* (spinose scale), 82; *Aulacaspis tubercularis* (white mango scale), 90; *Chrysomphalus aonidum* (Florida red scale), 122; *C. dictyospermi* (dictyospermum scale), 130; *Dynaspidiotus britannicus* (holly scale), 194; *Fiorinia fioriniae* (palm fiorinia scale), 204; *Hemiberlesia lataniae* (latania scale), 226; *H. rapax* (greedy scale), 232; *Lindingaspis rossi* (black araucaria scale), 272; *Lopholeucaspis japonica* (Japanese maple scale), 276; *Parlatoria camelliae* (camellia parlatoria scale), 316; *P. oleae* (olive scale), 320

Lavatera: Aspidiotus nerii (oleander scale), 80

Lecythis: Chrysomphalus dictyospermi (dictyospermum scale), 130

Ledum: Chionaspis salicis (willow scale), 118

Lemaireocereus: Diaspis echinocacti (cactus scale), 186

Leptospermum: Lindingaspis rossi (black araucaria scale), 272; *Parlatoria pittospori* (pittosporum scale), 328

Lepturus: Odonaspis ruthae (Bermuda grass scale), 306

Lespedeza: Hemiberlesia lataniae (latania scale), 226

Leucaena: Hemiberlesia lataniae (latania scale), 226; *Howardia biclavis* (mining scale), 234

Leuchtenbergia: Diaspis echinocacti (cactus scale), 186

Leucodendron: Hemiberlesia rapax (greedy scale), 232; *Parlatoria pittospori* (pittosporum scale), 328

Leucophyllum: Hemiberlesia lataniae (latania scale), 226

Leucopogon: Hemiberlesia rapax (greedy scale), 232; *Parlatoria pittospori* (pittosporum scale), 328

Leucospermum: Lindingaspis rossi (black araucaria scale), 272

Leucothoe: Hemiberlesia lataniae (latania scale), 226

Libocedrus: Carulaspis juniperi (juniper scale), 96; *Cupressaspis shastae* (redwood scale), 142

Licania: Howardia biclavis (mining scale), 234

Licuala: Aspidiotus destructor (coconut scale), 72; *Fiorinia fioriniae* (palm fiorinia scale), 204; *Hemiberlesia lataniae* (latania scale), 226; *Pinnaspis strachani* (lesser snow scale), 344

Ligustrum: Aonidiella aurantii (California red scale), 50; *A. orientalis* (Oriental armored scale), 58; *Aspidiella sacchari* (sugarcane scale), 66; *Aspidiotus destructor* (coconut scale), 72; *A. nerii* (oleander scale), 80; *Chionaspis americana* (elm scurfy scale), 102; *C. salicis* (willow scale), 118; *Chrysomphalus aonidum* (Florida red scale), 122; *C. bifasciculatus* (bifasciculate scale), 126; *C. dictyospermi* (dictyospermum scale), 130;

Ligustrum (cont.)
Diaspidiotus forbesi (Forbes scale), 152;
Duplaspidiotus claviger (camellia mining
scale), 188; Fiorinia fioriniae (palm fiorinia
scale), 204; Howardia biclavis (mining scale),
234; Lepidosaphes camelliae (camellia scale),
248; L. ulmi (oystershell scale), 266;
Lindingaspis rossi (black araucaria scale), 272;
Lopholeucaspis japonica (Japanese maple
scale), 276; Morganella longispina (plumose
scale), 290; Parlatoreopsis chinensis (Chinese
obscure scale), 310; Parlatoria oleae (olive
scale), 320; P. proteus (proteus scale), 330;
Pseudaonidia duplex (camphor scale), 346; P.
paeoniae (peony scale), 350; Pseudaulacaspis
pentagona (white peach scale), 360; P.
prunicola (white prunicola scale), 364; Unaspis
euonymi (euonymus scale), 396
Lilium: Aonidiella aurantii (California red scale),
50; Parlatoria proteus (proteus scale), 330
Lindera: Chionaspis corni (dogwood scale), 106;
Lepidosaphes pinnaeformis (cymbidium scale),
262
Liparis: Pinnaspis aspidistrae (fern scale), 340
Liquidambar: Diaspidiotus juglansregiae (walnut
scale), 158; D. liquidambaris (sweetgum scale),
162; D. uvae (grape scale), 176; Hemiberlesia
rapax (greedy scale), 232
Liriodendron: Chionaspis salicis (willow scale),
118; Diaspidiotus juglansregiae (walnut scale),
158; Hemiberlesia neodiffinis (false diffinis
scale), 228; Melanaspis tenebricosa (gloomy
scale), 284
Liriope: Aspidiotus nerii (oleander scale), 80;
Pinnaspis aspidistrae (fern scale), 340; P.
strachani (lesser snow scale), 344
Litchi: Andaspis punicae (litchi scale), 48;
Aonidiella orientalis (Oriental armored scale),
58; Aulacaspis tubercularis (white mango
scale), 90; Chrysomphalus dictyospermi
(dictyospermum scale), 130; Hemiberlesia
rapax (greedy scale), 232; Howardia biclavis
(mining scale), 234; Ischnaspis longirostris
(black thread scale), 238; Selenaspidus
articulatus (rufous scale), 386
Lithocarpus: Quernaspis quercus (oak scale), 376
Lithraea: Hemiberlesia rapax (greedy scale), 232
Litsea: Aonidiella citrina (yellow scale), 54;
Aspidiotus destructor (coconut scale), 72;
Aulacaspis tubercularis (white mango scale),
90; Hemiberlesia lataniae (latania scale), 226;
Lepidosaphes pinnaeformis (cymbidium scale),
262; Parlatoria camelliae (camellia parlatoria
scale), 316
Livistona: Abgrallaspis cyanophylli
(cyanophyllum scale), 38; Aspidiotus
destructor (coconut scale), 72; A. nerii
(oleander scale), 80; Chrysomphalus aonidum
(Florida red scale), 122; C. dictyospermi
(dictyospermum scale), 130; Comstockiella
sabalis (palmetto scale), 140; Diaspis
boisduvalii (Boisduval scale), 180;
Hemiberlesia lataniae (latania scale), 226;
H. rapax (greedy scale), 232; Ischnaspis
longirostris (black thread scale), 238; Pinnaspis
aspidistrae (fern scale), 340; P. strachani (lesser
snow scale), 344; Pseudaulacaspis cockerelli
(false oleander scale), 356; Selenaspidus
articulatus (rufous scale), 386
Lobivia: Diaspis echinocacti (cactus scale), 186
Lockhartia: Selenaspidus articulatus (rufous
scale), 386
Lonicera: Aspidiotus destructor (coconut scale),
72; A. excisus (aglaonema scale), 76; A. nerii
(oleander scale), 80; Chrysomphalus
bifasciculatus (bifasciculate scale), 126;
Clavaspis ulmi (elm armored scale), 136;
Diaspidiotus forbesi (Forbes scale), 152;
Dynaspidiotus britannicus (holly scale), 194;
Hemiberlesia lataniae (latania scale), 226; H.
rapax (greedy scale), 232; Howardia biclavis
(mining scale), 234; Ischnaspis longirostris
(black thread scale), 238; Lepidosaphes ulmi
(oystershell scale), 266; Pseudaonidia
trilobitiformis (trilobe scale), 354;

Pseudoparlatoria ostreata (gray scale), 370
Lophocereus: Diaspis echinocacti (cactus scale),
186
Loranthus: Aonidiella citrina (yellow scale), 54;
A. orientalis (Oriental armored scale), 58;
Aspidiotus destructor (coconut scale), 72;
Clavaspis herculeana (herculeana scale), 134;
Hemiberlesia lataniae (latania scale), 226; H.
rapax (greedy scale), 232; Howardia biclavis
(mining scale), 234; Lepidosaphes ulmi
(oystershell scale), 266; Morganella longispina
(plumose scale), 290; Pseudaonidia
trilobitiformis (trilobe scale), 354; Selenaspidus
articulatus (rufous scale), 386
Lucaena: Diaspis boisduvalii (Boisduval scale), 180
Lucuma: Hemiberlesia lataniae (latania scale),
226; Howardia biclavis (mining scale), 234
Luffa: Aulacaspis tubercularis (white mango
scale), 90
Lupinus: Lepidosaphes ulmi (oystershell scale),
266
Lycaste: Aspidiotus spinosus (spinose scale), 82;
Chrysomphalus dictyospermi (dictyospermum
scale), 130; Diaspis boisduvalii (Boisduval
scale), 180; Parlatoria proteus (proteus scale),
330; Pseudoparlatoria parlatorioides (false
parlatoria scale), 374
Lyonia: Chionaspis salicis (willow scale), 118

Maackia: Lepidosaphes yanagicola (fire bush
scale), 270
Macadamia: Aonidiella aurantii (California red
scale), 50; Aspidiotus nerii (oleander scale), 80;
Chrysomphalus dictyospermi (dictyospermum
scale), 130; Duplaspidiotus claviger (camellia
mining scale), 188; Hemiberlesia lataniae
(latania scale), 226; Howardia biclavis (mining
scale), 234; Neopinnaspis harperi (Harper
scale), 294
Machilus: Aulacaspis tubercularis (white mango
scale), 90; Lepidosaphes pinnaeformis
(cymbidium scale), 262
Maclura: Diaspidiotus ancylus (Putnam scale),
148; D. juglansregiae (walnut scale), 158; D.
uvae (grape scale), 176; Lepidosaphes gloverii
(Glover scale), 254; L. ulmi (oystershell scale),
266; Melanaspis tenebricosa (gloomy scale),
284; Parlatoria pergandii (chaff scale), 324
Macrozamia: Aonidiella aurantii (California red
scale), 50; Aspidiotus nerii (oleander scale), 80;
Aulacaspis yasumatsui (cycad aulacaspis
scale), 94; Furchadaspis zamiae (zamia scale),
218; Lindingaspis rossi (black araucaria scale),
272; Selenaspidus articulatus (rufous scale),
386
Maesa: Aspidiotus destructor (coconut scale), 72;
Hemiberlesia lataniae (latania scale), 226
Magnolia: Aspidiotus destructor (coconut scale),
72; A. nerii (oleander scale), 80;
Chrysomphalus dictyospermi (dictyospermum
scale), 130; Clavaspis ulmi (elm armored
scale), 136; Diaspidiotus ancylus (Putnam
scale), 148; D. liquidambaris (sweetgum
scale), 162; D. perniciosus (San Jose scale), 172;
Hemiberlesia lataniae (latania scale), 226;
H. neodiffinis (false diffinis scale), 228;
Lepidosaphes camelliae (camellia scale), 248;
L. pinnaeformis (cymbidium scale), 262; L.
ulmi (oystershell scale), 266; Lindingaspis rossi
(black araucaria scale), 272; Lopholeucaspis
japonica (Japanese maple scale), 276;
Pseudaonidia duplex (camphor scale), 346;
Pseudaulacaspis cockerelli (false oleander
scale), 356; P. pentagona (white peach scale),
360; P. prunicola (white prunicola scale), 364;
Pseudoparlatoria parlatorioides (false
parlatoria scale), 374; Selenaspidus articulatus
(rufous scale), 386; Unaspis euonymi
(euonymus scale), 396
Mahonia: Diaspidiotus juglansregiae (walnut
scale), 158; D. perniciosus (San Jose scale), 172;
Parlatoreopsis chinensis (Chinese obscure
scale), 310; Parlatoria oleae (olive scale), 320;
Pseudaulacaspis pentagona (white peach
scale), 360

Malachra: Pseudaulacaspis pentagona (white
peach scale), 360
Malpighia: Lepidosaphes beckii (purple scale),
244; Selenaspidus articulatus (rufous scale),
386
Malus: Abgrallaspis cyanophylli (cyanophyllum
scale), 38; Chionaspis furfura (scurfy scale),
110; Chrysomphalus aonidum (Florida red
scale), 122; C. dictyospermi (dictyospermum
scale), 130; Clavaspis ulmi (elm armored
scale), 136; Diaspidiotus ancylus (Putnam
scale), 148; D. forbesi (Forbes scale), 152; D.
juglansregiae (walnut scale), 158; D.
ostreaeformis (European fruit scale), 168; D.
perniciosus (San Jose scale), 172;
Duplaspidiotus claviger (camellia mining
scale), 188; Epidiaspis leperii (Italian pear
scale), 198; Hemiberlesia lataniae (latania
scale), 226; H. rapax (greedy scale), 232;
Howardia biclavis (mining scale), 234;
Lepidosaphes conchiformis (fig scale), 250; L.
ulmi (oystershell scale), 266; Lopholeucaspis
japonica (Japanese maple scale), 276;
Melanaspis tenebricosa (gloomy scale), 284;
Mercetaspis halli (Hall scale), 288;
Parlatoreopsis chinensis (Chinese obscure
scale), 310; Parlatoria pittospori (pittosporum
scale), 328; P. theae (tea parlatoria scale), 334;
Pseudaonidia duplex (camphor scale), 346;
Pseudaulacaspis prunicola (white prunicola
scale), 364
Malvaviscus: Duplaspidiotus tesseratus
(tesserate scale), 192; Hemiberlesia lataniae
(latania scale), 226
Mammea: Andaspis punicae (litchi scale), 48;
Aspidiotus destructor (coconut scale), 72;
Chrysomphalus aonidum (Florida red scale),
122; Clavaspis herculeana (herculeana scale),
134; Hemiberlesia lataniae (latania scale), 226;
Howardia biclavis (mining scale), 234
Mammillaria: Diaspis echinocacti (cactus scale),
186
Mangifera: Abgrallaspis cyanophylli
(cyanophyllum scale), 38; Aonidiella aurantii
(California red scale), 50; A. citrina (yellow
scale), 54; A. orientalis (Oriental armored
scale), 58; Aspidiella sacchari (sugarcane
scale), 66; Aspidiotus destructor (coconut
scale), 72; A. excisus (aglaonema scale), 76; A.
nerii (oleander scale), 80; A. spinosus (spinose
scale), 82; Aulacaspis tubercularis (white
mango scale), 90; Chrysomphalus aonidum
(Florida red scale), 122; C. dictyospermi
(dictyospermum scale), 130; Clavaspis
herculeana (herculeana scale), 134; Fiorinia
fioriniae (palm fiorinia scale), 204;
Hemiberlesia lataniae (latania scale), 226; H.
rapax (greedy scale), 232; Howardia biclavis
(mining scale), 234; Ischnaspis longirostris
(black thread scale), 238; Lepidosaphes beckii
(purple scale), 244; Morganella longispina
(plumose scale), 290; Parlatoria camelliae
(camellia parlatoria scale), 316; P. pergandii
(chaff scale), 324; P. proteus (proteus scale),
330; Pinnaspis strachani (lesser snow scale),
344; Pseudaonidia trilobitiformis (trilobe
scale), 354; Pseudaulacaspis cockerelli (false
oleander scale), 356; P. pentagona (white peach
scale), 360; Pseudischnaspis bowreyi (Bowrey
scale), 368; Pseudoparlatoria parlatorioides
(false parlatoria scale), 374; Selenaspidus
articulatus (rufous scale), 386; Unaspis citri
(citrus snow scale), 392
Manihot: Aspidiotus destructor (coconut scale),
72; Pinnaspis strachani (lesser snow scale),
344; Pseudaulacaspis pentagona (white peach
scale), 360
Manilkara: Aonidiella citrina (yellow scale), 54;
Clavaspis herculeana (herculeana scale), 134;
Duplaspidiotus tesseratus (tesserate scale),
192; Fiorinia fioriniae (palm fiorinia scale),
204; Hemiberlesia lataniae (latania scale), 226;
Howardia biclavis (mining scale), 234
Maranta: Aspidiotus destructor (coconut scale),
72; Chrysomphalus dictyospermi

(dictyospermum scale), 130; *Diaspis boisduvalii* (Boisduval scale), 180; *Hemiberlesia lataniae* (latania scale), 226; *H. rapax* (greedy scale), 232; *Ischnaspis longirostris* (black thread scale), 238; *Parlatoria proteus* (proteus scale), 330

Marcgravia: Hemiberlesia rapax (greedy scale), 232

Masdevallia: Furcaspis biformis (orchid scale), 216; *Selenaspidus articulatus* (rufous scale), 386

Matisia: Howardia biclavis (mining scale), 234

Maxillaria: Diaspis boisduvalii (Boisduval scale), 180; *Parlatoria proteus* (proteus scale), 330; *Pseudoparlatoria parlatorioides* (false parlatoria scale), 374

Maytenus: Furchadaspis zamiae (zamia scale), 218

Medicago: Aspidiotus nerii (oleander scale), 80

Melaleuca: Aonidiella citrina (yellow scale), 54; *Chrysomphalus dictyospermi* (dictyospermum scale), 130; *Parlatoria pittospori* (pittosporum scale), 328

Melia: Aonidiella aurantii (California red scale), 50; *A. orientalis* (Oriental armored scale), 58; *Aspidiotus nerii* (oleander scale), 80; *Hemiberlesia lataniae* (latania scale), 226; *Lepidosaphes beckii* (purple scale), 244; *Pseudaulacaspis pentagona* (white peach scale), 360

Melicoccus: Aspidiotus excisus (aglaonema scale), 76; *Chrysomphalus aonidum* (Florida red scale), 122; *Duplaspidiotus tesseratus* (tesserate scale), 192; *Hemiberlesia lataniae* (latania scale), 226; *Pinnaspis strachani* (lesser snow scale), 344; *Pseudoparlatoria parlatorioides* (false parlatoria scale), 374; *Selenaspidus articulatus* (rufous scale), 386

Melocactus: Diaspis echinocacti (cactus scale), 186

Melochia: Duplaspidiotus tesseratus (tesserate scale), 192

Menyanthes: Lopholeucaspis japonica (Japanese maple scale), 276

Meryta: Aspidiotus nerii (oleander scale), 80; *Hemiberlesia rapax* (greedy scale), 232

Mesembryanthemum: Parlatoria pittospori (pittosporum scale), 328

Mespilus: Lepidosaphes ulmi (oystershell scale), 266; *Parlatoria oleae* (olive scale), 320

Metrosideros: Duplaspidiotus claviger (camellia mining scale), 188; *Hemiberlesia lataniae* (latania scale), 226; *H. rapax* (greedy scale), 232

Metroxylon: Furchadaspis zamiae (zamia scale), 218

Michelia: Aspidiotus destructor (coconut scale), 72; *Lepidosaphes pinnaeformis* (cymbidium scale), 262; *L. ulmi* (oystershell scale), 266; *Morganella longispina* (plumose scale), 290; *Pseudaulacaspis cockerelli* (false oleander scale), 356

Miconia: Hemiberlesia lataniae (latania scale), 226

Microcitrus: Chrysomphalus dictyospermi (dictyospermum scale), 130

Microcos: Hemiberlesia lataniae (latania scale), 226

Microcycas: Aulacaspis yasumatsui (cycad aulacaspis scale), 94

Miltonia: Aspidiotus nerii (oleander scale), 80; *Diaspis boisduvalii* (Boisduval scale), 180; *Selenaspidus articulatus* (rufous scale), 386

Mimosa: Aspidiotus nerii (oleander scale), 80; *Chrysomphalus dictyospermi* (dictyospermum scale), 130; *Hemiberlesia lataniae* (latania scale), 226; *H. rapax* (greedy scale), 232; *Parlatoria pittospori* (pittosporum scale), 328

Mimusops: Chrysomphalus dictyospermi (dictyospermum scale), 130; *Howardia biclavis* (mining scale), 234

Monodora: Chrysomphalus dictyospermi (dictyospermum scale), 130; *Howardia biclavis* (mining scale), 234

Monstera: Aspidiotus spinosus (spinose scale), 82; *Chrysomphalus aonidum* (Florida red scale), 122; *C. dictyospermi* (dictyospermum scale), 130; *Gymnaspis aechmeae* (flyspeck scale), 222; *Howardia biclavis* (mining scale), 234; *Ischnaspis longirostris* (black thread scale), 238; *Pseudaonidia trilobitiformis* (trilobe scale), 354

Montezuma: Howardia biclavis (mining scale), 234

Moraea: Ischnaspis longirostris (black thread scale), 238

Morinda: Hemiberlesia lataniae (latania scale), 226

Moringa: Aonidiella aurantii (California red scale), 50; *A. orientalis* (Oriental armored scale), 58; *Aspidiotus destructor* (coconut scale), 72

Morus: Aonidiella aurantii (California red scale), 50; *Aspidiotus nerii* (oleander scale), 80; *Chionaspis americana* (elm scurfy scale), 102; *Chrysomphalus dictyospermi* (dictyospermum scale), 130; *Clavaspis herculeana* (herculeana scale), 134; *Diaspidiotus perniciosus* (San Jose scale), 172; *Fiorinia fioriniae* (palm fiorinia scale), 204; *Hemiberlesia lataniae* (latania scale), 226; *H. rapax* (greedy scale), 232; *Lepidosaphes yanagicola* (fire bush scale), 270; *Melanaspis tenebricosa* (gloomy scale), 284; *Pseudaulacaspis pentagona* (white peach scale), 360

Muehlenbeckia: Hemiberlesia lataniae (latania scale), 226; *Pinnaspis aspidistrae* (fern scale), 340

Murraya: Aonidiella aurantii (California red scale), 50; *A. citrina* (yellow scale), 54; *A. orientalis* (Oriental armored scale), 58; *Aspidiotus destructor* (coconut scale), 72; *A. excisus* (aglaonema scale), 76; *Lepidosaphes beckii* (purple scale), 244; *L. gloverii* (Glover scale), 254; *Parlatoria ziziphi* (black parlatoria scale), 336; *Selenaspidus articulatus* (rufous scale), 386

Musa: Aonidiella citrina (yellow scale), 54; *A. orientalis* (Oriental armored scale), 58; *Aspidiotus destructor* (coconut scale), 72; *A. excisus* (aglaonema scale), 76; *A. nerii* (oleander scale), 80; *Chrysomphalus aonidum* (Florida red scale), 122; *C. dictyospermi* (dictyospermum scale), 130; *Diaspis boisduvalii* (Boisduval scale), 180; *Fiorinia fioriniae* (palm fiorinia scale), 204; *Furchadaspis zamiae* (zamia scale), 218; *Hemiberlesia lataniae* (latania scale), 226; *H. rapax* (greedy scale), 232; *Ischnaspis longirostris* (black thread scale), 238; *Lindingaspis rossi* (black araucaria scale), 272; *Pinnaspis aspidistrae* (fern scale), 340; *P. strachani* (lesser snow scale), 344; *Pseudaulacaspis cockerelli* (false oleander scale), 356; *Pseudischnaspis bowreyi* (Bowrey scale), 368; *Pseudoparlatoria parlatorioides* (false parlatoria scale), 374; *Selenaspidus articulatus* (rufous scale), 386

Mussaenda: Howardia biclavis (mining scale), 234

Myoporum: Aspidiotus nerii (oleander scale), 80

Myrciaria: Pseudaonidia trilobitiformis (trilobe scale), 354

Myrica: Howardia biclavis (mining scale), 234; *Pseudaonidia duplex* (camphor scale), 346

Myricaria: Duplaspidiotus claviger (camellia mining scale), 188

Myristica: Aspidiotus destructor (coconut scale), 72; *Ischnaspis longirostris* (black thread scale), 238; *Pseudaulacaspis cockerelli* (false oleander scale), 356; *Selenaspidus articulatus* (rufous scale), 386

Myrtillocactus: Diaspis echinocacti (cactus scale), 186

Myrtus: Chionaspis salicis (willow scale), 118; *Fiorinia fioriniae* (palm fiorinia scale), 204; *Hemiberlesia rapax* (greedy scale), 232; *Melanaspis obscura* (obscure scale), 282

Napoleona: Ischnaspis longirostris (black thread scale), 238

Narcissus: Aonidiella citrina (yellow scale), 54; *Selenaspidus articulatus* (rufous scale), 386

Neofinetia: Diaspis boisduvalii (Boisduval scale), 180

Neoglaziovia: Diaspis boisduvalii (Boisduval scale), 180

Neolitsea: Lepidosaphes pinnaeformis (cymbidium scale), 262

Neomoorea: Hemiberlesia lataniae (latania scale), 226

Neoregelia: Diaspis bromeliae (pineapple scale), 182; *Gymnaspis aechmeae* (flyspeck scale), 222

Neowashingtonia: Parlatoria blanchardi (parlatoria date scale), 312

Nepenthes: Aspidiotus destructor (coconut scale), 72; *Hemiberlesia lataniae* (latania scale), 226

Nephelium: Andaspis punicae (litchi scale), 48; *Aspidiotus destructor* (coconut scale), 72; *Aulacaspis tubercularis* (white mango scale), 90; *Chrysomphalus dictyospermi* (dictyospermum scale), 130; *Howardia biclavis* (mining scale), 234; *Ischnaspis longirostris* (black thread scale), 238

Nephrolepis: Pinnaspis aspidistrae (fern scale), 340

Nephthytis: Hemiberlesia lataniae (latania scale), 226; *Pseudoparlatoria parlatorioides* (false parlatoria scale), 374

Nerium: Aonidiella aurantii (California red scale), 50; *A. citrina* (yellow scale), 54; *A. orientalis* (Oriental armored scale), 58; *Aspidiotus destructor* (coconut scale), 72; *A. nerii* (oleander scale), 80; *Chrysomphalus aonidum* (Florida red scale), 122; *C. dictyospermi* (dictyospermum scale), 130; *Hemiberlesia lataniae* (latania scale), 226; *H. neodiffinis* (false diffinis scale), 228; *H. rapax* (greedy scale), 232; *Lindingaspis rossi* (black araucaria scale), 272; *Melanaspis tenebricosa* (gloomy scale), 284; *Morganella longispina* (plumose scale), 290; *Parlatoria oleae* (olive scale), 320; *Pseudaonidia trilobitiformis* (trilobe scale), 354; *Pseudaulacaspis cockerelli* (false oleander scale), 356; *P. pentagona* (white peach scale), 360; *P. prunicola* (white prunicola scale), 364; *Pseudischnaspis bowreyi* (Bowrey scale), 368; *Selenaspidus articulatus* (rufous scale), 386

Neyraudia: Aspidiella sacchari (sugarcane scale), 66

Nidularium: Diaspis boisduvalii (Boisduval scale), 180; *D. bromeliae* (pineapple scale), 182; *Gymnaspis aechmeae* (flyspeck scale), 222

Nipa: Aspidiotus destructor (coconut scale), 72; *Chrysomphalus aonidum* (Florida red scale), 122; *Duplaspidiotus claviger* (camellia mining scale), 188; *Parlatoria proteus* (proteus scale), 330

Nolina: Hemiberlesia lataniae (latania scale), 226; *Situlaspis yuccae* (yucca scale), 388

Nopalxochia: Diaspis echinocacti (cactus scale), 186

Norantea: Ischnaspis longirostris (black thread scale), 238

Normanbya: Aspidiotus destructor (coconut scale), 72

Nuytsia: Lindingaspis rossi (black araucaria scale), 272; *Parlatoria pittospori* (pittosporum scale), 328

Nypa: Diaspis boisduvalii (Boisduval scale), 180; *Ischnaspis longirostris* (black thread scale), 238

Nyssa: Melanaspis tenebricosa (gloomy scale), 284

Ocimum: Aspidiotus nerii (oleander scale), 80

Odontoglossum: Aspidiotus nerii (oleander scale), 80; *Chrysomphalus aonidum* (Florida red scale), 122; *C. dictyospermi* (dictyospermum scale), 130; *Diaspis boisduvalii* (Boisduval scale), 180; *Hemiberlesia lataniae* (latania scale), 226; *Ischnaspis longirostris* (black thread scale), 238; *Pinnaspis strachani* (lesser snow scale),

Odontoglossum (cont.)
344; *Pseudoparlatoria parlatorioides* (false parlatoria scale), 374

Oenothera: *Rhizaspidiotus dearnessi* (Dearness scale), 380

Olea: *Aonidiella aurantii* (California red scale), 50; *A. citrina* (yellow scale), 54; *A. orientalis* (Oriental armored scale), 58; *Aspidiotus destructor* (coconut scale), 72; *A. nerii* (oleander scale), 80; *Chrysomphalus aonidum* (Florida red scale), 122; *C. dictyospermi* (dictyospermum scale), 130; *Diaspidiotus perniciosus* (San Jose scale), 172; *Epidiaspis leperii* (Italian pear scale), 198; *Hemiberlesia lataniae* (latania scale), 226; *H. rapax* (greedy scale), 232; *Lepidosaphes beckii* (purple scale), 244; *L. ulmi* (oystershell scale), 266; *Lindingaspis rossi* (black araucaria scale), 272; *Parlatoreopsis chinensis* (Chinese obscure scale), 310; *Parlatoria camelliae* (camellia parlatoria scale), 316; *P. oleae* (olive scale), 320; *P. proteus* (proteus scale), 330; *Situlaspis yuccae* (yucca scale), 388

Oncidium: *Chrysomphalus dictyospermi* (dictyospermum scale), 130; *Diaspis boisduvalii* (Boisduval scale), 180; *Furcaspis biformis* (orchid scale), 216; *Hemiberlesia lataniae* (latania scale), 226; *Ischnaspis longirostris* (black thread scale), 238; *Pseudischnaspis bowreyi* (Bowrey scale), 368; *Pseudoparlatoria ostreata* (gray scale), 370; *P. parlatorioides* (false parlatoria scale), 374

Oncosperma: *Hemiberlesia lataniae* (latania scale), 226

Ophiopogon: *Chrysomphalus dictyospermi* (dictyospermum scale), 130; *Hemiberlesia lataniae* (latania scale), 226; *Lepidosaphes pinnaeformis* (cymbidium scale), 262; *Parlatoria proteus* (proteus scale), 330; *Pinnaspis aspidistrae* (fern scale), 340

Opizia: *Aspidiella sacchari* (sugarcane scale), 66

Opuntia: *Aonidiella aurantii* (California red scale), 50; *Diaspis boisduvalii* (Boisduval scale), 180; *D. echinocacti* (cactus scale), 186; *Hemiberlesia lataniae* (latania scale), 226; *Pseudoparlatoria ostreata* (gray scale), 370

Orania: *Morganella longispina* (plumose scale), 290

Orbignya: *Hemiberlesia lataniae* (latania scale), 226

Oreopanax: *Hemiberlesia lataniae* (latania scale), 226

Ormosia: *Hemiberlesia lataniae* (latania scale), 226

Osbeckia: *Hemiberlesia rapax* (greedy scale), 232

Osmanthus: *Aspidiotus destructor* (coconut scale), 72; *A. nerii* (oleander scale), 80; *Chrysomphalus bifasciculatus* (bifasciculate scale), 126; *Hemiberlesia rapax* (greedy scale), 232; *Parlatoria camelliae* (camellia parlatoria scale), 316; *P. pergandii* (chaff scale), 324; *Pseudaonidia duplex* (camphor scale), 346; *P. paeoniae* (peony scale), 350; *Pseudaulacaspis pentagona* (white peach scale), 360; *P. prunicola* (white prunicola scale), 364; *Unaspis citri* (citrus snow scale), 392

Osmarea: *Aspidiotus nerii* (oleander scale), 80

Osteomeles: *Hemiberlesia lataniae* (latania scale), 226

Ostodes: *Fiorinia theae* (tea scale), 210

Oxydendrum: *Lepidosaphes ulmi* (oystershell scale), 266

Pachira: *Aonidiella orientalis* (Oriental armored scale), 58; *Howardia biclavis* (mining scale), 234; *Pseudischnaspis bowreyi* (Bowrey scale), 368

Pachycereus: *Chrysomphalus dictyospermi* (dictyospermum scale), 130; *Diaspis echinocacti* (cactus scale), 186

Pachypodium: *Hemiberlesia rapax* (greedy scale), 232

Pachysandra: *Lepidosaphes ulmi* (oystershell scale), 266; *L. yanagicola* (fire bush scale), 270; *Unaspis euonymi* (euonymus scale), 396

Pachystroma: *Hemiberlesia lataniae* (latania scale), 226

Paeonia: *Diaspidiotus forbesi* (Forbes scale), 152; *D. perniciosus* (San Jose scale), 172; *Lepidosaphes ulmi* (oystershell scale), 266; *Lopholeucaspis japonica* (Japanese maple scale), 276; *Pinnaspis aspidistrae* (fern scale), 340; *Pseudaonidia paeoniae* (peony scale), 350

Palaquium: *Aspidiotus destructor* (coconut scale), 72

Palicourea: *Pseudaulacaspis pentagona* (white peach scale), 360

Panax: *Lepidosaphes ulmi* (oystershell scale), 266

Pandanus: *Abgrallaspis cyanophylli* (cyanophyllum scale), 38; *Aonidiella aurantii* (California red scale), 50; *Aspidiotus destructor* (coconut scale), 72; *Chrysomphalus aonidum* (Florida red scale), 122; *C. dictyospermi* (dictyospermum scale), 130; *Diaspis boisduvalii* (Boisduval scale), 180; *Hemiberlesia lataniae* (latania scale), 226; *Ischnaspis longirostris* (black thread scale), 238; *Parlatoria proteus* (proteus scale), 330; *Pinnaspis strachani* (lesser snow scale), 344; *Selenaspidus articulatus* (rufous scale), 386

Panicum: *Aspidiella sacchari* (sugarcane scale), 66

Paphiopedilum: *Parlatoria proteus* (proteus scale), 330

Papyrus: *Chrysomphalus aonidum* (Florida red scale), 122

Parkia: *Duplaspidiotus tesseratus* (tesserate scale), 192

Parkinsonia: *Hemiberlesia lataniae* (latania scale), 226; *Parlatoria pergandii* (chaff scale), 324; *Pseudaonidia duplex* (camphor scale), 346; *Situlaspis yuccae* (yucca scale), 388

Parodia: *Diaspis echinocacti* (cactus scale), 186

Parthenium: *Rhizaspidiotus dearnessi* (Dearness scale), 380

Paspalum: *Aspidiella sacchari* (sugarcane scale), 66; *Kuwanaspis pseudoleucaspis* (bamboo diaspidid), 240

Passiflora: *Aspidiotus nerii* (oleander scale), 80; *Chrysomphalus dictyospermi* (dictyospermum scale), 130; *Hemiberlesia lataniae* (latania scale), 226; *Howardia biclavis* (mining scale), 234; *Lepidosaphes beckii* (purple scale), 244; *Pseudaulacaspis pentagona* (white peach scale), 360; *Pseudischnaspis bowreyi* (Bowrey scale), 368; *Selenaspidus articulatus* (rufous scale), 386

Paulownia: *Lopholeucaspis japonica* (Japanese maple scale), 276

Pedilanthus: *Aonidiella orientalis* (Oriental armored scale), 58; *Furcaspis biformis* (orchid scale), 216; *Hemiberlesia rapax* (greedy scale), 232; *Parlatoria proteus* (proteus scale), 330; *Situlaspis yuccae* (yucca scale), 388

Pelagodoxa: *Morganella longispina* (plumose scale), 290

Pelargonium: *Aonidiella aurantii* (California red scale), 50; *Hemiberlesia lataniae* (latania scale), 226; *Pseudaulacaspis pentagona* (white peach scale), 360; *Pseudoparlatoria ostreata* (gray scale), 370

Pelecyphora: *Diaspis echinocacti* (cactus scale), 186

Peltophorum: *Hemiberlesia lataniae* (latania scale), 226

Peniocereus: *Diaspis echinocacti* (cactus scale), 186

Pennisetum: *Aspidiella sacchari* (sugarcane scale), 66

Pentas: *Aspidiotus excisus* (aglaonema scale), 76

Peperomia: *Aspidiotus destructor* (coconut scale), 72; *A. nerii* (oleander scale), 80; *Pseudoparlatoria parlatorioides* (false parlatoria scale), 374

Pereskia: *Diaspis echinocacti* (cactus scale), 186

Periploca: *Parlatoreopsis chinensis* (Chinese obscure scale), 310

Peristeria: *Chrysomphalus dictyospermi* (dictyospermum scale), 130; *Diaspis boisduvalii* (Boisduval scale), 180;

Pseudoparlatoria parlatorioides (false parlatoria scale), 374

Persea: *Abgrallaspis cyanophylli* (cyanophyllum scale), 38; *Aonidiella aurantii* (California red scale), 50; *A. citrina* (yellow scale), 54; *A. orientalis* (Oriental armored scale), 58; *Aspidiotus destructor* (coconut scale), 72; *A. nerii* (oleander scale), 80; *A. spinosus* (spinose scale), 82; *Aulacaspis tubercularis* (white mango scale), 90; *Chrysomphalus aonidum* (Florida red scale), 122; *C. dictyospermi* (dictyospermum scale), 130; *Diaspis boisduvalii* (Boisduval scale), 180; *Epidiaspis leperii* (Italian pear scale), 198; *Fiorinia fioriniae* (palm fiorinia scale), 204; *Hemiberlesia lataniae* (latania scale), 226; *H. neodiffinis* (false diffinis scale), 228; *H. rapax* (greedy scale), 232; *Howardia biclavis* (mining scale), 234; *Lepidosaphes beckii* (purple scale), 244; *L. ulmi* (oystershell scale), 266; *Morganella longispina* (plumose scale), 290; *Neopinnaspis harperi* (Harper scale), 294; *Parlatoria pergandii* (chaff scale), 324; *P. proteus* (proteus scale), 330; *Pseudaonidia trilobitiformis* (trilobe scale), 354; *Pseudischnaspis bowreyi* (Bowrey scale), 368; *Pseudoparlatoria ostreata* (gray scale), 370; *P. parlatorioides* (false parlatoria scale), 374; *Selenaspidus articulatus* (rufous scale), 386

Persica: *Mercetaspis halli* (Hall scale), 288

Petrea: *Duplaspidiotus tesseratus* (tesserate scale), 192; *Hemiberlesia lataniae* (latania scale), 226; *H. rapax* (greedy scale), 232; *Howardia biclavis* (mining scale), 234

Petrophila: *Parlatoria pittospori* (pittosporum scale), 328

Peumus: *Aspidiotus nerii* (oleander scale), 80

Phalaenopsis: *Parlatoria proteus* (proteus scale), 330; *Pinnaspis strachani* (lesser snow scale), 344

Phalaris: *Pinnaspis aspidistrae* (fern scale), 340

Phellodendron: *Lepidosaphes pinnaeformis* (cymbidium scale), 262; *L. ulmi* (oystershell scale), 266

Philadelphus: *Melanaspis tenebricosa* (gloomy scale), 284; *Unaspis euonymi* (euonymus scale), 396

Philodendron: *Aonidiella aurantii* (California red scale), 50; *A. orientalis* (Oriental armored scale), 58; *Chrysomphalus aonidum* (Florida red scale), 122; *C. dictyospermi* (dictyospermum scale), 130; *Hemiberlesia lataniae* (latania scale), 226; *Ischnaspis longirostris* (black thread scale), 238; *Parlatoria proteus* (proteus scale), 330; *Pinnaspis aspidistrae* (fern scale), 340; *Pseudaulacaspis pentagona* (white peach scale), 360; *Pseudoparlatoria parlatorioides* (false parlatoria scale), 374

Phlox: *Lepidosaphes ulmi* (oystershell scale), 266

Phoebe: *Aulacaspis tubercularis* (white mango scale), 90; *Howardia biclavis* (mining scale), 234; *Lepidosaphes pinnaeformis* (cymbidium scale), 262

Phoenix: *Abgrallaspis cyanophylli* (cyanophyllum scale), 38; *Aonidiella aurantii* (California red scale), 50; *A. orientalis* (Oriental armored scale), 58; *Aspidiotus destructor* (coconut scale), 72; *A. nerii* (oleander scale), 80; *Chrysomphalus aonidum* (Florida red scale), 122; *C. dictyospermi* (dictyospermum scale), 130; *Comstockiella sabalis* (palmetto scale), 140; *Diaspis boisduvalii* (Boisduval scale), 180; *Fiorinia fioriniae* (palm fiorinia scale), 204; *Hemiberlesia lataniae* (latania scale), 226; *H. rapax* (greedy scale), 232; *Ischnaspis longirostris* (black thread scale), 238; *Parlatoria blanchardi* (parlatoria date scale), 312; *P. pittospori* (pittosporum scale), 328; *P. proteus* (proteus scale), 330; *Pseudischnaspis bowreyi* (Bowrey scale), 368; *Pseudoparlatoria parlatorioides* (false parlatoria scale), 374; *Selenaspidus articulatus* (rufous scale), 386; *Situlaspis yuccae* (yucca scale), 388

Pholidota: *Parlatoria proteus* (proteus scale), 330

Phoradendron: *Aspidiotus nerii* (oleander scale), 80; *Clavaspis herculeana* (herculeana scale), 134; *Hemiberlesia lataniae* (latania scale), 226; *H. neodiffinis* (false diffinis scale), 228; *H. rapax* (greedy scale), 232; *Lepidosaphes beckii* (purple scale), 244

Phormium: *Aspidiotus nerii* (oleander scale), 80; *Fiorinia fioriniae* (palm fiorinia scale), 204; *Hemiberlesia rapax* (greedy scale), 232

Photinia: *Diaspidiotus perniciosus* (San Jose scale), 172; *Howardia biclavis* (mining scale), 234; *Parlatoria oleae* (olive scale), 320

Phylica: *Hemiberlesia rapax* (greedy scale), 232

Phyllostachys: *Aspidiotus nerii* (oleander scale), 80; *Froggattiella penicillata* (penicillate scale), 212; *Kuwanaspis pseudoleucaspis* (bamboo diaspidid), 240

Physalis: *Hemiberlesia lataniae* (latania scale), 226

Phytelephas: *Aspidiotus nerii* (oleander scale), 80

Phytolacca: *Aspidiotus nerii* (oleander scale), 80

Picea: *Abgrallaspis ithacae* (hemlock scale), 44; *Aspidiotus cryptomeriae* (cryptomeria scale), 68; *Carulaspis minima* (minute cypress scale), 100; *Chionaspis heterophyllae* (pine scale), 112; *C. pinifoliae* (pine needle scale), 116; *Diaspidiotus juglansregiae* (walnut scale), 158; *Fiorinia externa* (elongate hemlock scale), 200; *F. japonica* (coniferous fiorinia scale), 206; *Lepidosaphes pallida* (Maskell scale), 256; *L. ulmi* (oystershell scale), 266; *Nuculaspis tsugae* (shortneedle conifer scale), 304

Pieris: *Pinnaspis aspidistrae* (fern scale), 340

Pimelea: *Parlatoria pittospori* (pittosporum scale), 328

Pimenta: *Chrysomphalus aonidum* (Florida red scale), 122; *C. dictyospermi* (dictyospermum scale), 130; *Hemiberlesia lataniae* (latania scale), 226; *Pseudoparlatoria parlatorioides* (false parlatoria scale), 374; *Selenaspidus articulatus* (rufous scale), 386

Pinanga: *Hemiberlesia lataniae* (latania scale), 226; *Parlatoria proteus* (proteus scale), 330

Pinus: *Abgrallaspis ithacae* (hemlock scale), 44; *Aspidiotus cryptomeriae* (cryptomeria scale), 68; *Carulaspis juniperi* (juniper scale), 96; *Chionaspis heterophyllae* (pine scale), 112; *C. pinifoliae* (pine needle scale), 116; *Chrysomphalus aonidum* (Florida red scale), 122; *C. dictyospermi* (dictyospermum scale), 130; *Diaspidiotus juglansregiae* (walnut scale), 158; *D. ostreaeformis* (European fruit scale), 168; *D. perniciosus* (San Jose scale), 172; *Fiorinia externa* (elongate hemlock scale), 200; *F. japonica* (coniferous fiorinia scale), 206; *Hemiberlesia lataniae* (latania scale), 226; *Lepidosaphes pallida* (Maskell scale), 256; *L. pini* (pine oystershell scale), 260; *Lindingaspis rossi* (black araucaria scale), 272; *Nuculaspis californica* (black pineleaf scale), 296; *N. tsugae* (shortneedle conifer scale), 304; *Parlatoria oleae* (olive scale), 320; *P. pittospori* (pittosporum scale), 328

Piper: *Aonidiella citrina* (yellow scale), 54; *Chrysomphalus dictyospermi* (dictyospermum scale), 130; *Hemiberlesia lataniae* (latania scale), 226; *Howardia biclavis* (mining scale), 234; *Pinnaspis aspidistrae* (fern scale), 340; *Pseudaulacaspis pentagona* (white peach scale), 360

Pistacia: *Lepidosaphes conchiformis* (fig scale), 250; *L. ulmi* (oystershell scale), 266; *Parlatoria oleae* (olive scale), 320

Pitcairnia: *Aspidiotus nerii* (oleander scale), 80; *Diaspis boisduvalii* (Boisduval scale), 180

Pithecellobium: *Clavaspis herculeana* (herculeana scale), 134; *Hemiberlesia lataniae* (latania scale), 226; *Melanaspis tenebricosa* (gloomy scale), 284; *Pseudischnaspis bowreyi* (Bowrey scale), 368

Pittosporum: *Aspidiotus nerii* (oleander scale), 80; *Aulacaspis tubercularis* (white mango scale), 90; *Chrysomphalus bifasciculatus* (bifasciculate scale), 126; *Parlatoria pittospori* (pittosporum scale), 328; *Pseudaonidia duplex* (camphor scale), 346; *Pseudaulacaspis pentagona* (white peach scale), 360; *Unaspis citri* (citrus snow scale), 392

Planera: *Lepidosaphes ulmi* (oystershell scale), 266

Platanus: *Aspidiotus nerii* (oleander scale), 80; *Chionaspis americana* (elm scurfy scale), 102; *Chrysomphalus dictyospermi* (dictyospermum scale), 130; *Diaspidiotus ancylus* (Putnam scale), 148; *D. osborni* (Osborn scale), 166; *D. ostreaeformis* (European fruit scale), 168; *D. uvae* (grape scale), 176; *Hemiberlesia lataniae* (latania scale), 226; *Melanaspis tenebricosa* (gloomy scale), 284; *Parlatoreopsis chinensis* (Chinese obscure scale), 310

Platycerium: *Pinnaspis aspidistrae* (fern scale), 340

Pleioblastus: *Kuwanaspis pseudoleucaspis* (bamboo diaspidid), 240

Pleiospilos: *Lindingaspis rossi* (black araucaria scale), 272

Pleurothallis: *Diaspis boisduvalii* (Boisduval scale), 180

Plocosperma: *Hemiberlesia lataniae* (latania scale), 226

Plumeria: *Andaspis punicae* (litchi scale), 48; *Aonidiella aurantii* (California red scale), 50; *A. citrina* (yellow scale), 54; *A. orientalis* (Oriental armored scale), 58; *Aspidiotus destructor* (coconut scale), 72; *Chrysomphalus aonidum* (Florida red scale), 122; *Clavaspis herculeana* (herculeana scale), 134; *Hemiberlesia lataniae* (latania scale), 226; *Howardia biclavis* (mining scale), 234; *Pinnaspis strachani* (lesser snow scale), 344; *Pseudaonidia trilobitiformis* (trilobe scale), 354; *Pseudaulacaspis cockerelli* (false oleander scale), 356; *P. pentagona* (white peach scale), 360; *Selenaspidus articulatus* (rufous scale), 386

Podocarpus: *Aonidiella orientalis* (Oriental armored scale), 58; *A. taxus* (Asiatic red scale), 62; *Aspidiotus destructor* (coconut scale), 72; *Chrysomphalus aonidum* (Florida red scale), 122; *C. dictyospermi* (dictyospermum scale), 130; *Fiorinia japonica* (coniferous fiorinia scale), 206; *Lepidosaphes pallida* (Maskell scale), 256; *L. pini* (pine oystershell scale), 260

Poinciana: *Aspidiotus destructor* (coconut scale), 72; *Hemiberlesia lataniae* (latania scale), 226; *Pseudischnaspis bowreyi* (Bowrey scale), 368; *Selenaspidus articulatus* (rufous scale), 386

Poinsettia: *Chrysomphalus aonidum* (Florida red scale), 122

Polianthes: *Aonidiella aurantii* (California red scale), 50

Polyalthia: *Chrysomphalus aonidum* (Florida red scale), 122; *Duplaspidiotus claviger* (camellia mining scale), 188; *Fiorinia fioriniae* (palm fiorinia scale), 204; *Hemiberlesia lataniae* (latania scale), 226; *Pinnaspis aspidistrae* (fern scale), 340

Polyandrococus: *Chrysomphalus dictyospermi* (dictyospermum scale), 130

Polygala: *Rhizaspidiotus dearnessi* (Dearness scale), 380

Polyrrhiza: *Furcaspis biformis* (orchid scale), 216

Polystichum: *Pinnaspis aspidistrae* (fern scale), 340

Pomaderris: *Lepidosaphes pinnaeformis* (cymbidium scale), 262

Poncirus: *Aspidiotus destructor* (coconut scale), 72; *Lopholeucaspis japonica* (Japanese maple scale), 276; *Parlatoria camelliae* (camellia parlatoria scale), 316; *P. pergandii* (chaff scale), 324; *P. ziziphi* (black parlatoria scale), 336; *Unaspis citri* (citrus snow scale), 392

Pongamia: *Howardia biclavis* (mining scale), 234

Populus: *Chionaspis salicis* (willow scale), 118; *Diaspidiotus ancylus* (Putnam scale), 148; *D. gigas* (poplar scale), 154; *D. juglansregiae* (walnut scale), 158; *D. osborni* (Osborn scale), 166; *D. ostreaeformis* (European fruit scale), 168; *D. uvae* (grape scale), 176; *Hemiberlesia lataniae* (latania scale), 226; *H. rapax* (greedy scale), 232; *Lepidosaphes ulmi* (oystershell scale), 266; *Lopholeucaspis japonica* (Japanese maple scale), 276; *Melanaspis tenebricosa* (gloomy scale), 284; *Parlatoreopsis chinensis* (Chinese obscure scale), 310; *Pseudaulacaspis pentagona* (white peach scale), 360; *P. prunicola* (white prunicola scale), 364

Porliera: *Abgrallaspis cyanophylli* (cyanophyllum scale), 38

Pothomorphe: *Selenaspidus articulatus* (rufous scale), 386

Pothos: *Chrysomphalus aonidum* (Florida red scale), 122; *C. dictyospermi* (dictyospermum scale), 130; *Selenaspidus articulatus* (rufous scale), 386

Pouteria: *Aspidiotus spinosus* (spinose scale), 82; *Hemiberlesia lataniae* (latania scale), 226

Primula: *Unaspis euonymi* (euonymus scale), 396

Prinsepia: *Hemiberlesia lataniae* (latania scale), 226

Pritchardia: *Aspidiotus excisus* (aglaonema scale), 76; *A. spinosus* (spinose scale), 82; *Chrysomphalus dictyospermi* (dictyospermum scale), 130; *Hemiberlesia lataniae* (latania scale), 226; *Ischnaspis longirostris* (black thread scale), 238; *Parlatoria blanchardi* (parlatoria date scale), 312; *Selenaspidus articulatus* (rufous scale), 386

Prosopis: *Aonidiella orientalis* (Oriental armored scale), 58; *Clavaspis herculeana* (herculeana scale), 134; *Duplaspidiotus tesseratus* (tesserate scale), 192; *Hemiberlesia lataniae* (latania scale), 226; *Melanaspis obscura* (obscure scale), 282; *Situlaspis yuccae* (yucca scale), 388

Protea: *Aonidiella aurantii* (California red scale), 50; *Aspidiotus nerii* (oleander scale), 80; *Hemiberlesia lataniae* (latania scale), 226; *H. rapax* (greedy scale), 232; *Lindingaspis rossi* (black araucaria scale), 272

Prunus: *Aspidiotus destructor* (coconut scale), 72; *A. nerii* (oleander scale), 80; *Aulacaspis tubercularis* (white mango scale), 90; *Chionaspis americana* (elm scurfy scale), 102; *C. furfura* (scurfy scale), 110; *Chrysomphalus dictyospermi* (dictyospermum scale), 130; *Clavaspis herculeana* (herculeana scale), 134; *Diaspidiotus ancylus* (Putnam scale), 148; *D. forbesi* (Forbes scale), 152; *D. juglansregiae* (walnut scale), 158; *D. ostreaeformis* (European fruit scale), 168; *D. perniciosus* (San Jose scale), 172; *D. uvae* (grape scale), 176; *Epidiaspis leperii* (Italian pear scale), 198; *Hemiberlesia lataniae* (latania scale), 226; *H. rapax* (greedy scale), 232; *Howardia biclavis* (mining scale), 234; *Lepidosaphes gloverii* (Glover scale), 254; *L. pinnaeformis* (cymbidium scale), 262; *L. ulmi* (oystershell scale), 266; *Lindingaspis rossi* (black araucaria scale), 272; *Lopholeucaspis japonica* (Japanese maple scale), 276; *Melanaspis obscura* (obscure scale), 282; *M. tenebricosa* (gloomy scale), 284; *Mercetaspis halli* (Hall scale), 288; *Neopinnaspis harperi* (Harper scale), 294; *Parlatoreopsis chinensis* (Chinese obscure scale), 310; *Parlatoria oleae* (olive scale), 320; *P. theae* (tea parlatoria scale), 334; *Pinnaspis aspidistrae* (fern scale), 340; *P. strachani* (lesser snow scale), 344; *Pseudaonidia duplex* (camphor scale), 346; *Pseudaulacaspis pentagona* (white peach scale), 360; *P. prunicola* (white prunicola scale), 364; *Pseudischnaspis bowreyi* (Bowrey scale), 368; *Selenaspidus articulatus* (rufous scale), 386

Pseudotsuga: *Abgrallaspis ithacae* (hemlock scale), 44; *Aspidiotus cryptomeriae* (cryptomeria scale), 68; *Chionaspis pinifoliae* (pine needle scale), 116; *Fiorinia externa* (elongate hemlock scale), 200; *Nuculaspis californica* (black pineleaf scale), 296

Psidium: *Abgrallaspis cyanophylli* (cyanophyllum scale), 38; *Aonidiella aurantii* (California red scale), 50; *A. citrina* (yellow scale), 54; *A. orientalis* (Oriental armored scale), 58; *Aspidiotus destructor* (coconut scale), 72; *A. excisus* (aglaonema scale), 76;

Psidium (cont.)
A. nerii (oleander scale), 80; Aulacaspis tubercularis (white mango scale), 90; Chrysomphalus aonidum (Florida red scale), 122; C. dictyospermi (dictyospermum scale), 130; Clavaspis herculeana (herculeana scale), 134; Fiorinia fioriniae (palm fiorinia scale), 204; Hemiberlesia lataniae (latania scale), 226; H. rapax (greedy scale), 232; Howardia biclavis (mining scale), 234; Lepidosaphes beckii (purple scale), 244; L. gloverii (Glover scale), 254; Morganella longispina (plumose scale), 290; Pseudaonidia trilobitiformis (trilobe scale), 354; Pseudischnaspis bowreyi (Bowrey scale), 368; Pseudoparlatoria parlatorioides (false parlatoria scale), 374
Psittacanthus: Andaspis punicae (litchi scale), 48
Psychotria: Ischnaspis longirostris (black thread scale), 238; Parlatoria proteus (proteus scale), 330; Pseudoparlatoria ostreata (gray scale), 370
Ptelea: Diaspidiotus ancylus (Putnam scale), 148; D. juglansregiae (walnut scale), 158
Pteris: Pinnaspis aspidistrae (fern scale), 340
Pterocarya: Hemiberlesia lataniae (latania scale), 226; H. neodiffinis (false diffinis scale), 228
Pterocaulon: Rhizaspidiotus dearnessi (Dearness scale), 380
Pterogyne: Aspidiotus nerii (oleander scale), 80
Ptychosperma: Aspidiotus destructor (coconut scale), 72; Chrysomphalus aonidum (Florida red scale), 122; Fiorinia fioriniae (palm fiorinia scale), 204; Hemiberlesia lataniae (latania scale), 226
Pueraria: Aspidiotus nerii (oleander scale), 80; Pseudaulacaspis pentagona (white peach scale), 360
Punica: Aonidiella aurantii (California red scale), 50; Aspidiotus destructor (coconut scale), 72; A. nerii (oleander scale), 80; Chrysomphalus aonidum (Florida red scale), 122; Clavaspis herculeana (herculeana scale), 134; Hemiberlesia lataniae (latania scale), 226; H. rapax (greedy scale), 232; Howardia biclavis (mining scale), 234; Parlatoria theae (tea parlatoria scale), 334; Pseudoparlatoria parlatorioides (false parlatoria scale), 374
Puya: Diaspis boisduvalii (Boisduval scale), 180
Pyracantha: Chionaspis furfura (scurfy scale), 110; Chrysomphalus dictyospermi (dictyospermum scale), 130; Diaspidiotus perniciosus (San Jose scale), 172; Hemiberlesia lataniae (latania scale), 226; Howardia biclavis (mining scale), 234; Lopholeucaspis japonica (Japanese maple scale), 276; Parlatoreopsis chinensis (Chinese obscure scale), 310; Pseudaonidia duplex (camphor scale), 346
Pyrenoglyphis: Selenaspidus articulatus (rufous scale), 386
Pyrus: Aonidiella aurantii (California red scale), 50; Aspidiotus nerii (oleander scale), 80; Chionaspis furfura (scurfy scale), 110; C. salicis (willow scale), 118; Chrysomphalus dictyospermi (dictyospermum scale), 130; Diaspidiotus juglansregiae (walnut scale), 158; D. ostreaeformis (European fruit scale), 168; D. perniciosus (San Jose scale), 172; Duplaspidiotus tesseratus (tesserate scale), 192; Epidiaspis leperii (Italian pear scale), 198; Hemiberlesia lataniae (latania scale), 226; H. rapax (greedy scale), 232; Howardia biclavis (mining scale), 234; Lepidosaphes conchiformis (fig scale), 250; L. pinnaeformis (cymbidium scale), 262; L. ulmi (oystershell scale), 266; Lopholeucaspis japonica (Japanese maple scale), 276; Mercetaspis halli (Hall scale), 288; Parlatoreopsis chinensis (Chinese obscure scale), 310; Parlatoria oleae (olive scale), 320; P. theae (tea parlatoria scale), 334; Pseudaonidia duplex (camphor scale), 346; Pseudischnaspis bowreyi (Bowrey scale), 368

Quercus: Aonidiella taxus (Asiatic red scale), 62; Chionaspis americana (elm scurfy scale), 102; C. salicis (willow scale), 118; Chrysomphalus dictyospermi (dictyospermum

scale), 130; Diaspidiotus ancylus (Putnam scale), 148; D. osborni (Osborn scale), 166; D. perniciosus (San Jose scale), 172; D. uvae (grape scale), 176; Fiorinia fioriniae (palm fiorinia scale), 204; Hemiberlesia neodiffinis (false diffinis scale), 228; H. rapax (greedy scale), 232; Lepidosaphes beckii (purple scale), 244; L. pinnaeformis (cymbidium scale), 262; L. ulmi (oystershell scale), 266; Melanaspis lilacina (dark oak scale), 278; M. obscura (obscure scale), 282; Parlatoria proteus (proteus scale), 330; Pseudaonidia duplex (camphor scale), 346; P. trilobitiformis (trilobe scale), 354; Quernaspis quercus (oak scale), 376
Quesnelia: Gymnaspis aechmeae (flyspeck scale), 222
Quisqualis: Hemiberlesia lataniae (latania scale), 226; Selenaspidus articulatus (rufous scale), 386

Randia: Abgrallaspis cyanophylli (cyanophyllum scale), 38; Howardia biclavis (mining scale), 234; Pinnaspis strachani (lesser snow scale), 344
Rapanea: Hemiberlesia lataniae (latania scale), 226
Raphanus: Parlatoria pergandii (chaff scale), 324
Raphiolepis: Lepidosaphes camelliae (camellia scale), 248
Rauvolfia: Parlatoria pergandii (chaff scale), 324
Ravenala: Aonidiella aurantii (California red scale), 50; Diaspis boisduvalii (Boisduval scale), 180
Rebutia: Diaspis echinocacti (cactus scale), 186
Reinhardtia: Duplaspidiotus tesseratus (tesserate scale), 192; Pseudoparlatoria parlatorioides (false parlatoria scale), 374
Renanthera: Diaspis boisduvalii (Boisduval scale), 180; Furcaspis biformis (orchid scale), 216; Parlatoria proteus (proteus scale), 330
Reseda: Aonidiella aurantii (California red scale), 50
Retama: Hemiberlesia lataniae (latania scale), 226
Retinospora: Chrysomphalus dictyospermi (dictyospermum scale), 130
Reynosia: Pseudoparlatoria parlatorioides (false parlatoria scale), 374
Rhamnus: Chionaspis furfura (scurfy scale), 110; C. salicis (willow scale), 118; Diaspidiotus ancylus (Putnam scale), 148; Dynaspidiotus britannicus (holly scale), 194; Parlatoria oleae (olive scale), 320; P. theae (tea parlatoria scale), 334; Pseudaulacaspis pentagona (white peach scale), 360
Rhapidophyllum: Comstockiella sabalis (palmetto scale), 140
Rhapis: Chrysomphalus bifasciculatus (bifasciculate scale), 126; Lepidosaphes pinnaeformis (cymbidium scale), 262; Pinnaspis aspidistrae (fern scale), 340
Rheedia: Chrysomphalus aonidum (Florida red scale), 122; Hemiberlesia lataniae (latania scale), 226
Rhipsalis: Diaspis boisduvalii (Boisduval scale), 180; Parlatoria proteus (proteus scale), 330
Rhizophora: Aonidiella orientalis (Oriental armored scale), 58; Pseudaulacaspis cockerelli (false oleander scale), 356
Rhododendron: Aspidiotus destructor (coconut scale), 72; A. nerii (oleander scale), 80; Chionaspis salicis (willow scale), 118; Diaspidiotus ancylus (Putnam scale), 148; Fiorinia theae (tea scale), 210; Hemiberlesia lataniae (latania scale), 226; H. rapax (greedy scale), 232; Lepidosaphes ulmi (oystershell scale), 266; Parlatoria camelliae (camellia parlatoria scale), 316; Pseudaonidia duplex (camphor scale), 346; P. paeoniae (peony scale), 350; Pseudaulacaspis prunicola (white prunicola scale), 364
Rhodomyrtus: Hemiberlesia lataniae (latania scale), 226
Rhopalostylis: Aspidiotus nerii (oleander scale), 80

Rhus: Diaspidiotus ancylus (Putnam scale), 148; D. perniciosus (San Jose scale), 172; Furchadaspis zamiae (zamia scale), 218; Pseudaulacaspis pentagona (white peach scale), 360
Rhynchostylis: Chrysomphalus dictyospermi (dictyospermum scale), 130; Hemiberlesia lataniae (latania scale), 226; Parlatoria proteus (proteus scale), 330
Ribes: Aspidiotus nerii (oleander scale), 80; Chionaspis corni (dogwood scale), 106; C. furfura (scurfy scale), 110; C. salicis (willow scale), 118; Clavaspis ulmi (elm armored scale), 136; Diaspidiotus ancylus (Putnam scale), 148; D. forbesi (Forbes scale), 152; D. ostreaeformis (European fruit scale), 168; D. perniciosus (San Jose scale), 172; Epidiaspis leperii (Italian pear scale), 198; Lepidosaphes ulmi (oystershell scale), 266; Melanaspis tenebricosa (gloomy scale), 284; Parlatoria oleae (olive scale), 320; Pseudaulacaspis pentagona (white peach scale), 360
Ricinus: Aonidiella aurantii (California red scale), 50; A. orientalis (Oriental armored scale), 58; Pseudaulacaspis pentagona (white peach scale), 360
Robinia: Aonidiella citrina (yellow scale), 54; Aspidiotus nerii (oleander scale), 80; Clavaspis ulmi (elm armored scale), 136; Diaspidiotus forbesi (Forbes scale), 152; D. juglansregiae (walnut scale), 158; D. perniciosus (San Jose scale), 172; Hemiberlesia lataniae (latania scale), 226; Lepidosaphes ulmi (oystershell scale), 266; Melanaspis tenebricosa (gloomy scale), 284; Selenaspidus articulatus (rufous scale), 386
Rodriguezia: Furcaspis biformis (orchid scale), 216
Ronnbergia: Diaspis boisduvalii (Boisduval scale), 180
Rosa: Andaspis punicae (litchi scale), 48; Aonidiella aurantii (California red scale), 50; A. citrina (yellow scale), 54; A. orientalis (Oriental armored scale), 58; Aspidiotus destructor (coconut scale), 72; A. nerii (oleander scale), 80; Aulacaspis rosae (rose scale), 86; Chionaspis salicis (willow scale), 118; Chrysomphalus aonidum (Florida red scale), 122; C. dictyospermi (dictyospermum scale), 130; Clavaspis herculeana (herculeana scale), 134; Diaspidiotus ancylus (Putnam scale), 148; D. perniciosus (San Jose scale), 172; Duplaspidiotus tesseratus (tesserate scale), 192; Fiorinia fioriniae (palm fiorinia scale), 204; Hemiberlesia lataniae (latania scale), 226; Lepidosaphes ulmi (oystershell scale), 266; Lopholeucaspis japonica (Japanese maple scale), 276; Parlatoreopsis chinensis (Chinese obscure scale), 310; Parlatoria oleae (olive scale), 320; Pseudaonidia duplex (camphor scale), 346; P. paeoniae (peony scale), 350; P. trilobitiformis (trilobe scale), 354; Pseudischnaspis bowreyi (Bowrey scale), 368; Pseudoparlatoria parlatorioides (false parlatoria scale), 374; Selenaspidus articulatus (rufous scale), 386
Rosmarinus: Aspidiotus nerii (oleander scale), 80; Hemiberlesia lataniae (latania scale), 226; H. rapax (greedy scale), 232
Roystonea: Aspidiotus spinosus (spinose scale), 82; Chrysomphalus aonidum (Florida red scale), 122; C. dictyospermi (dictyospermum scale), 130; Comstockiella sabalis (palmetto scale), 140; Diaspis boisduvalii (Boisduval scale), 180; Hemiberlesia lataniae (latania scale), 226; Ischnaspis longirostris (black thread scale), 238; Selenaspidus articulatus (rufous scale), 386
Rubia: Aspidiotus nerii (oleander scale), 80
Rubus: Aulacaspis rosae (rose scale), 86; Hemiberlesia lataniae (latania scale), 226; Ischnaspis longirostris (black thread scale), 238; Lepidosaphes ulmi (oystershell scale), 266

Chrysomphalus dictyospermi (dictyospermum scale), 130; *Dynaspidiotus britannicus* (holly scale), 194; *Fiorinia fioriniae* (palm fiorinia scale), 204; *Pinnaspis aspidistrae* (fern scale), 340

Ruta: Aonidiella aurantii (California red scale), 50; *A. orientalis* (Oriental armored scale), 58; *Aspidiotus nerii* (oleander scale), 80; *Chrysomphalus aonidum* (Florida red scale), 122; *Fiorinia fioriniae* (palm fiorinia scale), 204; *Hemiberlesia rapax* (greedy scale), 232

Sabal: Aspidiotus destructor (coconut scale), 72; *A. nerii* (oleander scale), 80; *Chrysomphalus dictyospermi* (dictyospermum scale), 130; *Comstockiella sabalis* (palmetto scale), 140; *Ischnaspis longirostris* (black thread scale), 238; *Parlatoria proteus* (proteus scale), 330; *Pinnaspis strachani* (lesser snow scale), 344; *Selenaspidus articulatus* (rufous scale), 386
Saccharum: Aspidiella sacchari (sugarcane scale), 66; *Aspidiotus destructor* (coconut scale), 72
Saccolabium: Parlatoria proteus (proteus scale), 330
Saintpaulia: Pinnaspis aspidistrae (fern scale), 340
Salacca: Chrysomphalus dictyospermi (dictyospermum scale), 130
Salix: Aonidiella aurantii (California red scale), 50; *A. orientalis* (Oriental armored scale), 58; *Chionaspis americana* (elm scurfy scale), 102; *C. salicis* (willow scale), 118; *Chrysomphalus aonidum* (Florida red scale), 122; *Diaspidiotus ancylus* (Putnam scale), 148; *D. gigas* (poplar scale), 154; *D. juglansregiae* (walnut scale), 158; *Hemiberlesia lataniae* (latania scale), 226; *H. neodiffinis* (false diffinis scale), 228; *H. rapax* (greedy scale), 232; *Lepidosaphes beckii* (purple scale), 244; *L. ulmi* (oystershell scale), 266; *L. yanagicola* (fire bush scale), 270; *Lopholeucaspis japonica* (Japanese maple scale), 276; *Melanaspis tenebricosa* (gloomy scale), 284; *Pinnaspis strachani* (lesser snow scale), 344; *Pseudaonidia duplex* (camphor scale), 346; *Pseudaulacaspis pentagona* (white peach scale), 360; *P. prunicola* (white prunicola scale), 364
Salvia: Aspidiotus nerii (oleander scale), 80; *Hemiberlesia lataniae* (latania scale), 226
Samanea: Hemiberlesia lataniae (latania scale), 226
Sansevieria: Chrysomphalus dictyospermi (dictyospermum scale), 130; *Parlatoria proteus* (proteus scale), 330
Santalum: Duplaspidiotus claviger (camellia mining scale), 188
Sapindus: Aspidiotus nerii (oleander scale), 80; *Melanaspis tenebricosa* (gloomy scale), 284
Sapium: Hemiberlesia lataniae (latania scale), 226; *Howardia biclavis* (mining scale), 234
Saraca: Hemiberlesia lataniae (latania scale), 226
Sarcococca: Chrysomphalus dictyospermi (dictyospermum scale), 130; *Lepidosaphes ulmi* (oystershell scale), 266
Sarothamnus: Chionaspis salicis (willow scale), 118; *Chrysomphalus dictyospermi* (dictyospermum scale), 130; *Lepidosaphes ulmi* (oystershell scale), 266
Sarracenia: Aspidiotus nerii (oleander scale), 80
Sasa: Kuwanaspis pseudoleucaspis (bamboo diaspidid), 240
Sasamorpha: Kuwanaspis pseudoleucaspis (bamboo diaspidid), 240
Sassafras: Lepidosaphes ulmi (oystershell scale), 266; *Pseudaulacaspis pentagona* (white peach scale), 360
Schefflera: Hemiberlesia lataniae (latania scale), 226; *Parlatoria proteus* (proteus scale), 330
Schima: Abgrallaspis degenerata (degenerate scale), 42
Schinopsis: Hemiberlesia lataniae (latania scale), 226
Schinus: Hemiberlesia rapax (greedy scale), 232; *Pseudaulacaspis pentagona* (white peach

scale), 360; *Selenaspidus articulatus* (rufous scale), 386
Schisandra: Diaspidiotus ostreaeformis (European fruit scale), 168
Schomburgkia: Diaspis boisduvalii (Boisduval scale), 180; *Furcaspis biformis* (orchid scale), 216; *Ischnaspis longirostris* (black thread scale), 238; *Parlatoria proteus* (proteus scale), 330; *Pseudoparlatoria parlatorioides* (false parlatoria scale), 374
Sciadopitys: Fiorinia japonica (coniferous fiorinia scale), 206; *Lepidosaphes pallida* (Maskell scale), 256
Scindapsus: Chrysomphalus aonidum (Florida red scale), 122; *C. dictyospermi* (dictyospermum scale), 130; *Parlatoria proteus* (proteus scale), 330
Sclerocarya: Aonidiella orientalis (Oriental armored scale), 58
Seaforthia: Pseudaulacaspis cockerelli (false oleander scale), 356
Sechium: Pinnaspis strachani (lesser snow scale), 344
Sedum: Aspidiotus nerii (oleander scale), 80; *Hemiberlesia lataniae* (latania scale), 226; *Pseudaulacaspis pentagona* (white peach scale), 360
Semiarundinaria: Kuwanaspis pseudoleucaspis (bamboo diaspidid), 240
Senecio: Pseudoparlatoria ostreata (gray scale), 370
Sequoia: Carulaspis juniperi (juniper scale), 96; *C. minima* (minute cypress scale), 100; *Cupressaspis shastae* (redwood scale), 142; *Lepidosaphes pallida* (Maskell scale), 256
Serenoa: Chrysomphalus dictyospermi (dictyospermum scale), 130; *Comstockiella sabalis* (palmetto scale), 140
Serrisa: Hemiberlesia lataniae (latania scale), 226
Setaria: Odonaspis ruthae (Bermuda grass scale), 306
Severinia: Morganella longispina (plumose scale), 290; *Parlatoria ziziphi* (black parlatoria scale), 336; *Unaspis citri* (citrus snow scale), 392
Sida: Aonidiella orientalis (Oriental armored scale), 58
Sideroxylon: Chrysomphalus aonidum (Florida red scale), 122
Silene: Aspidiotus nerii (oleander scale), 80
Sinobambusa: Kuwanaspis pseudoleucaspis (bamboo diaspidid), 240
Smilax: Aspidiotus nerii (oleander scale), 80; *Chrysomphalus dictyospermi* (dictyospermum scale), 130; *Duplaspidiotus tesseratus* (tesserate scale), 192
Sobralia: Parlatoria proteus (proteus scale), 330; *Pseudaonidia trilobitiformis* (trilobe scale), 354; *Pseudoparlatoria parlatorioides* (false parlatoria scale), 374
Solanum: Andaspis punicae (litchi scale), 48; *Aonidiella orientalis* (Oriental armored scale), 58; *Aspidiotus nerii* (oleander scale), 80; *Chrysomphalus aonidum* (Florida red scale), 122; *Hemiberlesia lataniae* (latania scale), 226; *Pinnaspis strachani* (lesser snow scale), 344; *Pseudaulacaspis pentagona* (white peach scale), 360; *Pseudoparlatoria ostreata* (gray scale), 370
Solidago: Pseudaonidia duplex (camphor scale), 346; *Rhizaspidiotus dearnessi* (Dearness scale), 380
Sonneratia: Aspidiotus destructor (coconut scale), 72; *Chrysomphalus dictyospermi* (dictyospermum scale), 130
Sophora: Aspidiotus nerii (oleander scale), 80; *Hemiberlesia rapax* (greedy scale), 232
Sophronitis: Diaspis boisduvalii (Boisduval scale), 180
Sorbaria: Diaspidiotus perniciosus (San Jose scale), 172; *Lepidosaphes ulmi* (oystershell scale), 266
Sorbus: Aspidiotus nerii (oleander scale), 80; *Chionaspis furfura* (scurfy scale), 110; *C. salicis* (willow scale), 118; *Diaspidiotus ancylus* (Putnam scale), 148; *D. juglansregiae*

(walnut scale), 158; *D. ostreaeformis* (European fruit scale), 168; *D. perniciosus* (San Jose scale), 172; *Lepidosaphes ulmi* (oystershell scale), 266; *Parlatoria oleae* (olive scale), 320
Sorghum: Odonaspis ruthae (Bermuda grass scale), 306
Spartium: Lepidosaphes ulmi (oystershell scale), 266
Spathodea: Howardia biclavis (mining scale), 234
Spathoglottis: Hemiberlesia lataniae (latania scale), 226; *Parlatoria proteus* (proteus scale), 330
Spiraea: Diaspidiotus perniciosus (San Jose scale), 172; *Lepidosaphes ulmi* (oystershell scale), 266; *Mercetaspis halli* (Hall scale), 288; *Parlatoria theae* (tea parlatoria scale), 334
Spiranthes: Pseudoparlatoria parlatorioides (false parlatoria scale), 374
Spondias: Aonidiella aurantii (California red scale), 50; *A. orientalis* (Oriental armored scale), 58; *Aspidiotus destructor* (coconut scale), 72; *Chrysomphalus dictyospermi* (dictyospermum scale), 130; *Clavaspis herculeana* (herculeana scale), 134; *Fiorinia theae* (tea scale), 210; *Hemiberlesia lataniae* (latania scale), 226; *Lepidosaphes ulmi* (oystershell scale), 266; *Melanaspis obscura* (obscure scale), 282; *Pseudaonidia trilobitiformis* (trilobe scale), 354; *Pseudaulacaspis pentagona* (white peach scale), 360; *Pseudischnaspis bowreyi* (Bowrey scale), 368; *Selenaspidus articulatus* (rufous scale), 386
Sporobolus: Aspidiella sacchari (sugarcane scale), 66; *Odonaspis ruthae* (Bermuda grass scale), 306
Stachyphrynium: Hemiberlesia lataniae (latania scale), 226
Stahlia: Hemiberlesia lataniae (latania scale), 226
Stangeria: Aulacaspis yasumatsui (cycad aulacaspis scale), 94; *Furchadaspis zamiae* (zamia scale), 218
Stanhopea: Diaspis boisduvalii (Boisduval scale), 180; *Parlatoria proteus* (proteus scale), 330; *Pseudaulacaspis pentagona* (white peach scale), 360
Stapelia: Aspidiotus nerii (oleander scale), 80
Staphylea: Diaspidiotus ancylus (Putnam scale), 148; *Hemiberlesia rapax* (greedy scale), 232; *Lepidosaphes ulmi* (oystershell scale), 266; *Parlatoria theae* (tea parlatoria scale), 334
Stauntonia: Lepidosaphes pinnaeformis (cymbidium scale), 262
Stenocarpus: Hemiberlesia lataniae (latania scale), 226
Stenotaphrum: Aspidiella sacchari (sugarcane scale), 66
Sterculia: Clavaspis herculeana (herculeana scale), 134; *Dynaspidiotus britannicus* (holly scale), 194; *Pseudaulacaspis pentagona* (white peach scale), 360; *Selenaspidus articulatus* (rufous scale), 386
Steriphoma: Howardia biclavis (mining scale), 234
Stokesia: Chrysomphalus aonidum (Florida red scale), 122
Strelitzia: Aonidiella orientalis (Oriental armored scale), 58; *Aspidiotus nerii* (oleander scale), 80; *Chrysomphalus aonidum* (Florida red scale), 122; *C. dictyospermi* (dictyospermum scale), 130; *Diaspis boisduvalii* (Boisduval scale), 180; *Furchadaspis zamiae* (zamia scale), 218; *Hemiberlesia lataniae* (latania scale), 226; *Ischnaspis longirostris* (black thread scale), 238; *Pinnaspis aspidistrae* (fern scale), 340; *P. strachani* (lesser snow scale), 344; *Pseudaulacaspis cockerelli* (false oleander scale), 356
Streptosolen: Aspidiotus nerii (oleander scale), 80
Strobilanthes: Pinnaspis aspidistrae (fern scale), 340

Strongylodon: *Hemiberlesia lataniae* (latania scale), 226

Strychnos: *Selenaspidus articulatus* (rufous scale), 386

Stylophyllum: *Hemiberlesia lataniae* (latania scale), 226

Styphelia: *Duplaspidiotus claviger* (camellia mining scale), 188; *Hemiberlesia rapax* (greedy scale), 232; *Lindingaspis rossi* (black araucaria scale), 272

Styrax: *Lopholeucaspis japonica* (Japanese maple scale), 276

Swietenia: *Chrysomphalus dictyospermi* (dictyospermum scale), 130; *Ischnaspis longirostris* (black thread scale), 238

Symphoricarpos: *Diaspidiotus perniciosus* (San Jose scale), 172; *Unaspis euonymi* (euonymus scale), 396

Symplocos: *Diaspis boisduvalii* (Boisduval scale), 180; *Hemiberlesia lataniae* (latania scale), 226; *Lepidosaphes camelliae* (camellia scale), 248

Synechanthus: *Chrysomphalus dictyospermi* (dictyospermum scale), 130; *Selenaspidus articulatus* (rufous scale), 386

Syngonium: *Hemiberlesia lataniae* (latania scale), 226

Syringa: *Chionaspis americana* (elm scurfy scale), 102; *C. salicis* (willow scale), 118; *Diaspidiotus juglansregiae* (walnut scale), 158; *D. ostreaeformis* (European fruit scale), 168; *D. perniciosus* (San Jose scale), 172; *D. uvae* (grape scale), 176; *Hemiberlesia lataniae* (latania scale), 226; *H. neodiffinis* (false diffinis scale), 228; *Lepidosaphes conchiformis* (fig scale), 250; *L. ulmi* (oystershell scale), 266; *L. yanagicola* (fire bush scale), 270; *Lopholeucaspis japonica* (Japanese maple scale), 276; *Parlatoreopsis chinensis* (Chinese obscure scale), 310; *Parlatoria theae* (tea parlatoria scale), 334; *Pseudaulacaspis prunicola* (white prunicola scale), 364

Syzygium: *Chrysomphalus aonidum* (Florida red scale), 122

Tabebuia: *Abgrallaspis cyanophylli* (cyanophyllum scale), 38; *Clavaspis herculeana* (herculeana scale), 134; *Howardia biclavis* (mining scale), 234

Tabernaemontana: *Howardia biclavis* (mining scale), 234; *Ischnaspis longirostris* (black thread scale), 238; *Selenaspidus articulatus* (rufous scale), 386

Talauma: *Parlatoria proteus* (proteus scale), 330

Tamarindus: *Abgrallaspis cyanophylli* (cyanophyllum scale), 38; *Aonidiella orientalis* (Oriental armored scale), 58; *Chrysomphalus aonidum* (Florida red scale), 122; *Selenaspidus articulatus* (rufous scale), 386

Tamarix: *Aonidiella aurantii* (California red scale), 50; *Aspidiotus nerii* (oleander scale), 80; *Diaspidiotus ancylus* (Putnam scale), 148

Taraktogenos: *Selenaspidus articulatus* (rufous scale), 386

Taxodium: *Carulaspis juniperi* (juniper scale), 96; *Lepidosaphes pallida* (Maskell scale), 256

Taxus: *Aonidiella taxus* (Asiatic red scale), 62; *Aspidiotus cryptomeriae* (cryptomeria scale), 68; *Chionaspis pinifoliae* (pine needle scale), 116; *Chrysomphalus dictyospermi* (dictyospermum scale), 130; *Fiorinia externa* (elongate hemlock scale), 200; *F. japonica* (coniferous fiorinia scale), 206; *Lepidosaphes pallida* (Maskell scale), 256; *L. pini* (pine oystershell scale), 260; *L. pinnaeformis* (cymbidium scale), 262; *L. ulmi* (oystershell scale), 266; *Nuculaspis tsugae* (shortneedle conifer scale), 304

Tecoma: *Chrysomphalus dictyospermi* (dictyospermum scale), 130; *Hemiberlesia lataniae* (latania scale), 226; *Morganella longispina* (plumose scale), 290; *Pseudaulacaspis pentagona* (white peach scale), 360

Tecomaria: *Hemiberlesia lataniae* (latania scale), 226

Terminalia: *Aspidiotus destructor* (coconut scale), 72; *Chrysomphalus aonidum* (Florida red scale), 122; *Hemiberlesia lataniae* (latania scale), 226; *Howardia biclavis* (mining scale), 234; *Pinnaspis strachani* (lesser snow scale), 344; *Pseudaonidia trilobitiformis* (trilobe scale), 354

Ternstroemia: *Lepidosaphes camelliae* (camellia scale), 248

Tetracera: *Hemiberlesia lataniae* (latania scale), 226

Tetradymia: *Lepidosaphes pinnaeformis* (cymbidium scale), 262

Teucrium: *Aspidiotus nerii* (oleander scale), 80

Thea: *Aspidiotus destructor* (coconut scale), 72; *Lepidosaphes camelliae* (camellia scale), 248; *L. yanagicola* (fire bush scale), 270

Thelocactus: *Diaspis echinocacti* (cactus scale), 186

Theobroma: *Hemiberlesia lataniae* (latania scale), 226; *Howardia biclavis* (mining scale), 234; *Pseudaonidia trilobitiformis* (trilobe scale), 354; *Pseudaulacaspis pentagona* (white peach scale), 360; *Pseudoparlatoria parlatorioides* (false parlatoria scale), 374; *Selenaspidus articulatus* (rufous scale), 386

Thespesia: *Aspidiotus excisus* (aglaonema scale), 76; *Pinnaspis strachani* (lesser snow scale), 344

Thevetia: *Furchadaspis zamiae* (zamia scale), 218; *Hemiberlesia lataniae* (latania scale), 226; *Melanaspis tenebricosa* (gloomy scale), 284; *Selenaspidus articulatus* (rufous scale), 386

Thrinax: *Comstockiella sabalis* (palmetto scale), 140; *Hemiberlesia lataniae* (latania scale), 226; *Parlatoria proteus* (proteus scale), 330; *Pinnaspis strachani* (lesser snow scale), 344; *Selenaspidus articulatus* (rufous scale), 386

Thuja: *Aspidiotus nerii* (oleander scale), 80; *Carulaspis juniperi* (juniper scale), 96; *C. minima* (minute cypress scale), 100; *Chrysomphalus dictyospermi* (dictyospermum scale), 130; *Hemiberlesia lataniae* (latania scale), 226; *H. rapax* (greedy scale), 232; *Lepidosaphes pallida* (Maskell scale), 256; *Nuculaspis pseudomeyeri* (false Meyer scale), 300; *Parlatoreopsis chinensis* (Chinese obscure scale), 310

Thunbergia: *Ischnaspis longirostris* (black thread scale), 238; *Parlatoria proteus* (proteus scale), 330

Thymus: *Rhizaspidiotus dearnessi* (Dearness scale), 380

Tilia: *Chionaspis americana* (elm scurfy scale), 102; *Clavaspis ulmi* (elm armored scale), 136; *Diaspidiotus ancylus* (Putnam scale), 148; *D. juglansregiae* (walnut scale), 158; *D. osborni* (Osborn scale), 166; *D. ostreaeformis* (European fruit scale), 168; *D. perniciosus* (San Jose scale), 172; *Hemiberlesia neodiffinis* (false diffinis scale), 228; *Lepidosaphes conchiformis* (fig scale), 250; *L. ulmi* (oystershell scale), 266; *L. yanagicola* (fire bush scale), 270

Tillandsia: *Aspidiotus nerii* (oleander scale), 80; *Chrysomphalus dictyospermi* (dictyospermum scale), 130; *Diaspis boisduvalii* (Boisduval scale), 180; *D. bromeliae* (pineapple scale), 182; *Gymnaspis aechmeae* (flyspeck scale), 222; *Hemiberlesia lataniae* (latania scale), 226; *Pinnaspis aspidistrae* (fern scale), 340; *Pseudischnaspis bowreyi* (Bowrey scale), 368; *Pseudoparlatoria parlatorioides* (false parlatoria scale), 374; *Selenaspidus articulatus* (rufous scale), 386

Tipuana: *Hemiberlesia lataniae* (latania scale), 226

Torreya: *Aspidiotus cryptomeriae* (cryptomeria scale), 68; *Chionaspis pinifoliae* (pine needle scale), 116; *Cupressaspis shastae* (redwood scale), 142; *Fiorinia japonica* (coniferous fiorinia scale), 206; *Lepidosaphes pallida* (Maskell scale), 256; *L. pini* (pine oystershell scale), 260

Trachelospermum: *Hemiberlesia lataniae* (latania scale), 226; *Pseudaulacaspis cockerelli* (false oleander scale), 356

Trachycarpus: *Aspidiotus spinosus* (spinose scale), 82; *Diaspis boisduvalii* (Boisduval scale), 180; *Furchadaspis zamiae* (zamia scale), 218; *Pseudaulacaspis cockerelli* (false oleander scale), 356; *Pseudoparlatoria parlatorioides* (false parlatoria scale), 374

Trema: *Pseudaulacaspis pentagona* (white peach scale), 360

Trichilia: *Howardia biclavis* (mining scale), 234

Trichocentrum: *Parlatoria proteus* (proteus scale), 330

Trichocereus: *Diaspis echinocacti* (cactus scale), 186

Trichoglottis: *Parlatoria proteus* (proteus scale), 330

Trichopilia: *Chrysomphalus aonidum* (Florida red scale), 122; *C. dictyospermi* (dictyospermum scale), 130; *Diaspis boisduvalii* (Boisduval scale), 180

Trigonidium: *Chrysomphalus aonidum* (Florida red scale), 122

Triphasia: *Parlatoria proteus* (proteus scale), 330

Tripsacum: *Aspidiella sacchari* (sugarcane scale), 66

Tripterygium: *Unaspis euonymi* (euonymus scale), 396

Tristania: *Hemiberlesia rapax* (greedy scale), 232

Trochocarpa: *Aspidiotus nerii* (oleander scale), 80

Trochodendron: *Aspidiotus nerii* (oleander scale), 80

Tsuga: *Abgrallaspis ithacae* (hemlock scale), 44; *Aspidiotus cryptomeriae* (cryptomeria scale), 68; *Chionaspis pinifoliae* (pine needle scale), 116; *Diaspidiotus ancylus* (Putnam scale), 148; *D. juglansregiae* (walnut scale), 158; *Fiorinia externa* (elongate hemlock scale), 200; *F. japonica* (coniferous fiorinia scale), 206; *Lepidosaphes ulmi* (oystershell scale), 266; *Nuculaspis californica* (black pineleaf scale), 296; *N. pseudomeyeri* (false Meyer scale), 300; *N. tsugae* (shortneedle conifer scale), 304

Turbinocactus: *Diaspis echinocacti* (cactus scale), 186

Ulmus: *Chionaspis americana* (elm scurfy scale), 102; *C. salicis* (willow scale), 118; *Chrysomphalus aonidum* (Florida red scale), 122; *Clavaspis ulmi* (elm armored scale), 136; *Diaspidiotus ancylus* (Putnam scale), 148; *D. juglansregiae* (walnut scale), 158; *D. osborni* (Osborn scale), 166; *D. perniciosus* (San Jose scale), 172; *Hemiberlesia lataniae* (latania scale), 226; *H. neodiffinis* (false diffinis scale), 228; *H. rapax* (greedy scale), 232; *Lepidosaphes conchiformis* (fig scale), 250; *L. ulmi* (oystershell scale), 266; *Lopholeucaspis japonica* (Japanese maple scale), 276; *Melanaspis obscura* (obscure scale), 282; *M. tenebricosa* (gloomy scale), 284

Umbellularia: *Aspidiotus nerii* (oleander scale), 80; *Hemiberlesia rapax* (greedy scale), 232

Vaccinium: *Abgrallaspis degenerata* (degenerate scale), 42; *Aspidiotus nerii* (oleander scale), 80; *Chionaspis salicis* (willow scale), 118; *Diaspidiotus ancylus* (Putnam scale), 148; *D. forbesi* (Forbes scale), 152; *D. uvae* (grape scale), 176; *Hemiberlesia lataniae* (latania scale), 226; *H. rapax* (greedy scale), 232; *Lepidosaphes ulmi* (oystershell scale), 266; *Rhizaspidiotus dearnessi* (Dearness scale), 380

Vanda: *Aonidiella orientalis* (Oriental armored scale), 58; *Chrysomphalus aonidum* (Florida red scale), 122; *C. dictyospermi* (dictyospermum scale), 130; *Diaspis boisduvalii* (Boisduval scale), 180; *Duplaspidiotus tesseratus* (tesserate scale), 192; *Furcaspis biformis* (orchid scale), 216; *Hemiberlesia lataniae* (latania scale), 226; *Parlatoria proteus* (proteus scale), 330

Vandopsis: *Chrysomphalus dictyospermi* (dictyospermum scale), 130

Vanilla: *Andaspis punicae* (litchi scale), 48; *Pseudoparlatoria parlatorioides* (false parlatoria scale), 374

Vateria: Hemiberlesia lataniae (latania scale), 226

Veitchia: Chrysomphalus aonidum (Florida red scale), 122; *Hemiberlesia lataniae* (latania scale), 226

Vellozia: Chrysomphalus dictyospermi (dictyospermum scale), 130

Verbena: Lepidosaphes beckii (purple scale), 244; *Pseudoparlatoria ostreata* (gray scale), 370

Verbesina: Hemiberlesia lataniae (latania scale), 226

Veronica: Diaspidiotus perniciosus (San Jose scale), 172; *Hemiberlesia rapax* (greedy scale), 232

Vetiveria: Aspidiella sacchari (sugarcane scale), 66

Viburnum: Chionaspis corni (dogwood scale), 106; *C. salicis* (willow scale), 118; *Chrysomphalus dictyospermi* (dictyospermum scale), 130; *Diaspidiotus forbesi* (Forbes scale), 152; *D. juglansregiae* (walnut scale), 158; *D. perniciosus* (San Jose scale), 172; *Duplaspidiotus claviger* (camellia mining scale), 188; *Hemiberlesia lataniae* (latania scale), 226; *Lepidosaphes ulmi* (oystershell scale), 266; *Parlatoria theae* (tea parlatoria scale), 334; *Pseudaulacaspis cockerelli* (false oleander scale), 356

Vicia: Aonidiella orientalis (Oriental armored scale), 58

Vigna: Hemiberlesia lataniae (latania scale), 226

Vinca: Aspidiotus nerii (oleander scale), 80; *Hemiberlesia rapax* (greedy scale), 232

Viola: Abgrallaspis cyanophylli (cyanophyllum scale), 38; *Chrysomphalus dictyospermi* (dictyospermum scale), 130; *Hemiberlesia lataniae* (latania scale), 226

Viscum: Hemiberlesia lataniae (latania scale), 226; *H. rapax* (greedy scale), 232; *Pinnaspis strachani* (lesser snow scale), 344

Vitex: Aspidiotus nerii (oleander scale), 80

Vitis: Aonidiella aurantii (California red scale), 50; *A. citrina* (yellow scale), 54; *A. orientalis* (Oriental armored scale), 58; *Aspidiotus destructor* (coconut scale), 72; *A. nerii* (oleander scale), 80; *A. spinosus* (spinose scale), 82; *Chionaspis salicis* (willow scale), 118; *Chrysomphalus aonidum* (Florida red scale), 122; *C. dictyospermi* (dictyospermum scale),

130; *Clavaspis herculeana* (herculeana scale), 134; *Diaspidiotus perniciosus* (San Jose scale), 172; *Diaspis boisduvalii* (Boisduval scale), 180; *Duplaspidiotus claviger* (camellia mining scale), 188; *Hemiberlesia lataniae* (latania scale), 226; *H. rapax* (greedy scale), 232; *Lepidosaphes ulmi* (oystershell scale), 266; *Melanaspis obscura* (obscure scale), 282; *M. tenebricosa* (gloomy scale), 284; *Parlatoria oleae* (olive scale), 320; *Pinnaspis strachani* (lesser snow scale), 344; *Pseudaonidia duplex* (camphor scale), 346; *P. trilobitiformis* (trilobe scale), 354; *Pseudaulacaspis pentagona* (white peach scale), 362; *Pseudoparlatoria ostreata* (gray scale), 370; *Selenaspidus articulatus* (rufous scale), 386

Vriesea: Diaspis boisduvalii (Boisduval scale), 180; *Gymnaspis aechmeae* (flyspeck scale), 222; *Ischnaspis longirostris* (black thread scale), 238; *Pseudaulacaspis cockerelli* (false oleander scale), 356

Wallichia: Ischnaspis longirostris (black thread scale), 238

Waltheria: Howardia biclavis (mining scale), 234

Washingtonia: Aspidiotus destructor (coconut scale), 72; *Comstockiella sabalis* (palmetto scale), 140; *Diaspis boisduvalii* (Boisduval scale), 180; *Ischnaspis longirostris* (black thread scale), 238; *Parlatoria blanchardi* (parlatoria date scale), 312; *Pseudoparlatoria parlatorioides* (false parlatoria scale), 374

Weberocereus: Diaspis echinocacti (cactus scale), 186

Wikstroemia: Hemiberlesia lataniae (latania scale), 226

Wilcoxia: Diaspis echinocacti (cactus scale), 186

Wisteria: Hemiberlesia lataniae (latania scale), 226; *Howardia biclavis* (mining scale), 234; *Lopholeucaspis japonica* (Japanese maple scale), 276

Xanthorrhoea: Parlatoria pittospori (pittosporum scale), 328

Xanthoxylum: Parlatoreopsis chinensis (Chinese obscure scale), 310; *Parlatoria theae* (tea parlatoria scale), 334; *Selenaspidus articulatus* (rufous scale), 386

Xylobium: Diaspis boisduvalii (Boisduval scale), 180

Yucca: Andaspis punicae (litchi scale), 48; *Aspidiotus nerii* (oleander scale), 80; *Chrysomphalus aonidum* (Florida red scale), 122; *C. dictyospermi* (dictyospermum scale), 130; *Furcaspis biformis* (orchid scale), 216; *Hemiberlesia lataniae* (latania scale), 226; *H. rapax* (greedy scale), 232; *Lepidosaphes ulmi* (oystershell scale), 266; *Parlatoria pergandii* (chaff scale), 324; *Pinnaspis strachani* (lesser snow scale), 344; *Pseudaulacaspis cockerelli* (false oleander scale), 356; *Situlaspis yuccae* (yucca scale), 388

Zamia: Chrysomphalus aonidum (Florida red scale), 122; *Furchadaspis zamiae* (zamia scale), 218

Zanthoxylum: Aonidiella orientalis (Oriental armored scale), 58; *Chionaspis furfura* (scurfy scale), 110; *Diaspidiotus juglansregiae* (walnut scale), 158; *Pinnaspis strachani* (lesser snow scale), 344

Zelkova: Diaspidiotus perniciosus (San Jose scale), 172; *Lepidosaphes conchiformis* (fig scale), 250; *Lopholeucaspis japonica* (Japanese maple scale), 276

Zingiber: Aspidiotus destructor (coconut scale), 72; *Chrysomphalus aonidum* (Florida red scale), 122

Ziziphus: Diaspidiotus ancylus (Putnam scale), 148; *D. perniciosus* (San Jose scale), 172; *Hemiberlesia lataniae* (latania scale), 226; *Lopholeucaspis japonica* (Japanese maple scale), 276; *Pinnaspis aspidistrae* (fern scale), 340; *P. strachani* (lesser snow scale), 344; *Pseudaulacaspis pentagona* (white peach scale), 362

Zizyphus: Aonidiella aurantii (California red scale), 50; *A. orientalis* (Oriental armored scale), 58; *Parlatoreopsis chinensis* (Chinese obscure scale), 310; *Parlatoria ziziphi* (black parlatoria scale), 336; *Selenaspidus articulatus* (rufous scale), 386

Zygocactus: Diaspis echinocacti (cactus scale), 186

Zygopetalum: Pseudoparlatoria parlatorioides (false parlatoria scale), 374

Index of Armored Scales, Natural Enemies, and General Subjects

Insect names in bold type indicate valid scientific names and most frequently used common names. Page numbers in bold type indicate major discussions of insects. Insects are also indexed by synonyms and species names. When looking up a synonym or species name, refer to the page indicated to obtain the valid name; then look up the valid name in the index to find all pages where the insect is discussed.

abdominal segments, 14
Abgrallaspis cyanophylli, 27, 28, 34, **38–40**; *degenerata*, 6, 27, 35, **41–43**, 300; *fraxini*, 42; *ithacae*, 5, 6, 13, 27, 30, **44–46**
abietus, 44
acalypha scale, 370
acephala, 368
aceris, 118
achmeae scale, 222
aechmeae, **222**
aesculi solus, 146
affinis: Aspidiotus, 78; Parlatoria, 320
Africaspis spp., 294; *harperi*, 294
aglaonema scale, 27, 34, 72, **75–77**, 78
agrumincola, 130
akebiae, 356
albiventer, 158
albus: Aspidiotus, 158; **Selenaspidus, 382**
Aleurodothrips fasciapennis, 9
Aleuropteryx juniperi, 9
alluadi, 134
alluaudi, 134
alni, 118
aloes, 78
americana, **102**
amygdali: Coccus, 266; Diaspis, 360
amygdali var. *rubra*, 364
anal opening, 14
ancylus, **146**
ancylus var. *latilobis*, 146
ancylus var. *ornatus*, 146
ancylus var. *serratus*, 146
Andaspis punicae, 22, 32, **47–49**
anguinus, 244
annae, 392
anonae, 134
Anoplaspis penicillata, 212
antenna, 14
Aonidia aonidum, 50; *aurantii*, 50; *blanchardi*, 312; *fusca*, 172; *gennadii*, 50; *juniperi*, 142; *picea*, 222; *ruthae*, 306; *shastae*, 142
Aonidiella spp., 5, 194; **aurantii**, 6, 8, 10, 12, 13, 25, 33, **50–53**, 54, 58, 62; **citrina**, 8, 25, 33, 50, **54–57**, 58, 62; *inornata*, 62; **orientalis**, 26, 33, 35, **58–61**; *perniciosa*, 172; *subrossi*, 272; **taxus**, 6, 13, 25, 29, 54, **62–64**; *tenebricosa*, 284; *ulmi*, 136
aonidum: Aonidia, 50; **Chrysomphalus, 122**; Coccus, 122
apacheca, 296
Aphelinus mytilaspidus, 268
Aphycus californicus, 160
Aphytis spp., 9, 10, 210; *chilensis*, 80; *chionaspis*, 92; *comperei*, 326; *cylindratus*, 348; *diaspidis*, 362; *hispanicus*, 10, 326; *holoxanthus*, 124; *lepidosaphes*, 246, 254; *lingnanensis*, 52, 394; *longicaudus*, 348; *maculicornis*, 322; *melinus*, 52, 56, 80, 132; *paramaculicornis*, 322, 326; *roseni*, 387
apple comma scale, 266
apple mussel scale, 266

appletree bark louse, 266
Apteronidia blanchardi, 312; *ziziphi*, 336
arecae: Chermes, 204; Chrysomphalus, 130
argentina, 232
armored scales classification of, 1; color of, 1, 4, 5; cover development, 1, 3, 4, 5; detection and identification, 7; dimorphic forms, 7; economic impact, 1, 13; feeding behavior, 4–5, 6; host-plant specificity, 6; insecticide resistance in, 6, 8; life history, 1–5; management practices, 6–9, 12–13, 174–175; morphology, 1, 13–17; overwintering, 1, 6; preservation and preparation, 18–19; reproduction, 1, 4–5; seasonal history, 5–6
articulata, 386
articulatus, 386
articulatus var. *simplex*, 386
Aschersonia fungi, 210
ash scale, 42
Asiatic red scale, 6, 13, 25, 29, 54, **62–64**
asiatica, 312
Aspidaspis florenciae, 296; *ithacae*, 44
Aspidiella hartii, 66; **sacchari**, 6, 27, 28, 30, **65–67**, 306
aspidiotine scales life-history chart, 3; morphology, 1, 15, 17; scale-cover development, 3, 4, 5
Aspidiotiphagus spp., 362; *citrinus*, 254
Aspidiotus spp., 5; *aesculi solus*, 78; *affinis*, 78; *agrumincola*, 130; *aloes*, 78; *ancylus*, 146; *ancylus* var. *latilobis*, 146; *ancylus* var. *ornatus*, 146; *ancylus* var. *serratus*, 146; *argentina*, 232; *articulatus*, 386; *aspleniae*, 224; *aurantii*, 50; *aurantii* var. *citrinus*, 54; *betulae*, 168; *biformis*, 216; *biformis* var. *cattleyae*, 216; *bipromimens*, 216; *Bouchei*, 78; *bowreyi*, 368; *britannicus*, 194; *bromeliae*, 182; *budleiae*, 78; *caldesii*, 78; *californicus*, 296; *camelliae*, 232; *celtidis*, 388; *cerasi*, 110; *ceratoniae*, 78; *circularis*, 146; *citri*, 50; *citricola*, 244; *citrinus*, 54; *coccineus*, 50; *cocotiphagus*, 58; *cocotis*, 72; *comstocki*, 146; *coniferarum* var. *shastae*, 142; *convexus*, 146; *crawii*, 224; **cryptomeriae**, 6, 7, 13, 27, 30, **68–70**, 72; *cyanophylli*, 38; *cycadicola*, 78; *cydoniae*, 224; *cydoniae crawii*, 224; *cydoniae punicae*, 224; *cydoniae* var. *tecta*, 224; *daruyi*, 354; *darvtyi*, 354; *dearnessi*, 380; *degeneratus*, 42; *denticulatus*, 78; **destructor**, 27, 34, 68, **71–74**, 76; *destructor transparens*, 72; *dictyospermi*, 130; *dictyospermi* var. *arecae*, 130; *dictyospermi* var. *jamaicensis*, 130; *diffinis*, 228; *diffinis* var. *lateralis*, 224; *duplex*, 346; *duplex* var. *paeoniae*, 350; *echinocacti*, 186; *epidendri*, 78; *ericae*, 78; **excisus**, 27, 34, 72, **75–77**, 78; *falciformis*, 266; *fallax*, 72; *fernaldi*, 158; *fernaldi hesperius*, 152; *fernaldi* var. *albiventer*, 158; *fernaldi* var. *cockerelli*, 158; *ficus*, 122; *flavescens*, 232; *forbesi*, 152; *furfurus*, 110; *genistae*, 78; *gigas*, 154; *gloverii*, 254; *greeni*, 224; *gutierreziae*, 380; *harrisii*,

110; *hederae*, 78; *hederae unisexualis*, 78; *helianthi*, 380; *herculeana*, 134; *hesperius*, 152; *hippocastani*, 168; *howardi*, 146; *hunteri*, 168; *ilicis*, 78; *jamaicensis*, 130; *juglandis*, 266; *juglandis-regiae*, 158; *juglans-regiae*, 158; *juglans-regiae* var. *albus*, 158; *juglans-regiae* var. *pruni*, 158; *juniperi*, 96; *lanatus*, 360; *lataniae*, 224; *latastei*, 194; *lateralis*, 224; *lilacinus*, 278; *linearis*, 368; *liquidambaris*, 162; *longisimma*, 368; *longispina*, 290; *longispina* var. *ornata*, 290; *lucumae*, 232; *mangiferae*, 130; *maskelli*, 290; *minimus*, 118; *multiglandulatus*, 154; *neomexicanus*, 388; *nerii*, 6, 26, 34, **78–81**, 82; *nerii unisexualis*, 78; *obscurus*, 282; *ohioensis*, 146; *oleastri*, 78; *oppugnatus*, 72; *ornatus*, 290; *osbeckiae*, 58; *osborni*, 166; *ostreaeformis*, 168; *ostreaeformis magnus*, 168; *ostreaeformis oblongus*, 168; *oxyacanthae*, 168; *oxycrataegi*, 146; *paraneri*, 78; *parlatorioides*, 374; *perniciosus*, 172; *persearum*, 82; *pini*, 296; *pinifoliae*, 116; *pinnaeformis*, 262; *piricola*, 198; *pomorum*, 266; *populi*, 118; *proteus*, 330; *pseudomeyeri*, 300; *punicae*, 224; *pyricola*, 198; *pyrus-malus*, 266; *rapax*, 232; *rigidus*, 72; *rosae*, 86; *rossi*, 272; *sabalis*, 140; *sabalis* var. *mexicana*, 140; *sacchari*, 66; *salicis-nigrae*, 118; *shastae*, 142; *simillimus translucens*, 72; **spinosus**, 27, 28, 35, 36, 72, **82–84**; *subsimilis* var. *anonae*, 134; *symbioticus*, 134; *taprobanus*, 58; *taxus*, 62; *tectus*, 224; *tenebricosus*, 284; *theae*, 234, 346; *theae rhododendri*, 346; *townsendi*, 146; *translucens*, 72; *transparens*, 72; *trilobitiformis*, 354; *tsugae*, 304; *ulmi*, 136; *uvae*, 176; *uvaspis*, 176; *vaccinii*, 118; *vastatrix*, 72; *villosus*, 78; *vitiensis*, 360; *yuccae*, 388; *yulupae*, 166
Aspidiotus (Aspidiella) forbesi, 152
Aspidiotus (Chrysomphalus) biformis, 216; *biformis* var. *cattleyae*, 216; *bowreyi*, 368; *degeneratus*, 42; *lilacinus*, 278; *longisimma*, 368; *pedronis*, 58; *taprobanus*, 58; *tenebricosus*, 284; *yuccae*, 388
Aspidiotus (Comstockaspis) perniciosus, 172
Aspidiotus (Diaspidiotus) fernaldi hesperius, 152; *forbesi*, 152; *glanduliferus*, 158; *greeni*, 224; *juglans-regiae*, 158; *osbeckiae*, 58; *osborni*, 166; *punicae*, 224; *tesseratus*, 192; *tsugae*, 304; *uvae*, 176
Aspidiotus (Dynaspidiotus) britannicus, 194
Aspidiotus (Euraspidiotus) gigas, 154; *ostreaeformis*, 168
Aspidiotus (Evaspidiotus) biformis, 216; *britannicus*, 194; *convexus*, 146; *crawii*, 224; *cydoniae*, 224; *duplex*, 346; *duplex* var. *paeoniae*, 350; *greeni*, 224; *juglans-regiae*, 158; *orientalis*, 58; *osbeckiae*, 58; *punicae*, 224; *theae*, 346; *trilobitiformis*, 354
Aspidiotus (Hemiberlesia) camelliae, 232; *crawii*, 224; *cydoniae*, 224; *lataniae*, 224; *perniciosus*, 172; *rapax*, 232; *tricolor*, 232

Aspidiotus (Melanaspis) obscurus, 282
Aspidiotus (Morganella) longispina, 290;
    maskelli, 290
Aspidiotus (Nuculaspis) californicus, 296
Aspidiotus (Pseudaonidia) duplex, 346; theae,
    346; trilobitiformis, 354
Aspidiotus (Quadraspidiotus) forbesi, 152;
    ostreaeformis, 168; perniciosus, 172
Aspidiotus (Selenaspidus) articulatus, 386;
    articulatus var. simplex, 386
Aspidiotus (Targionia) gutierreziae, 380;
    helianthi, 380
aspidistra scale, 340
aspidistrae, **340**
aspidistrae gossypii, 344
aspidistrae var. brasiliensis, 340
aspidistrae var. gossypii, 344
aspidistrae var. lata, 340
aspleniae, 224
atlantica, 142
atriplicis, 388
atunicola, 270
aucubae, 356
Aulacaspis spp., 5; boisduvalii, 180; bromeliae,
    182; cattleya, 180; cinnamomi, 90; pentagona,
    360; pentagona auranticolor, 364; pentagona
    rubra, 364; **rosae**, 6, 20, 31, **85–88**, 94;
    rosarum, 86; tegalensis, 12; **tubercularis**, 13,
    20, 31, **89–92**; *yasumatsui*, 6, 20, 21, 30, **93–95**;
    zamiae, 218
Aulacaspis (Diaspis) pentagona, 360;
    tubercularis, 90
auranticolor, 364
aurantii: Aonidia, 50; **Aonidiella, 50**; Aspidiotus,
    50; Chermes, 336; Chrysomphalus, 50
aurantii var. citrinus, 54
avocado scale, 82, 204
Axion spp., 9

bamboo diaspidid, 6, 22, 30, **240–242**
bambusae, 240
bark damage symptoms, 6, 13
basal sclerosis, 14
***beckii*, 244**
beetle predators, 9; Cybocephalus binotatus, 92,
    95. See also lady beetle predators
berberidis, 266
Bermuda cedar scale, 100
Bermuda grass scale, 6, 23, 30, 212, **306–308**, 380
betulae: Aspidiotus, 168; Epidiaspis, 198
biclavate scale, 234
biclavis, 234
biclavis detecta, 234
biclavis var. detecta, 234
bifasciculate scale, 26, 35, **126–128**
***bifasciculatus*, 127**
***biformis*, 216**
biformis cattleyae, 216
biformis var. cattleyae, 216
bigelovia scale, 380
bigeloviae, 380
biological control methods, 8–12, 396. See also
    parasites of scale insects; predators of scale
    insects
biprominens, 216
bisexualis, 266
black araucaria scale, 26, 29, 30, 34, **272–274**
black lined scale, 238
black parlatoria scale, 6, 24, 31, **336–338**
black pineleaf scale, 5, 6, 13, 27, 29, 44, **296–299**,
    304
black scale, 122
black thread scale, 21, 32, **237–239**
black willow scale, 118
***blanchardi*, 312**
blanchardi var. victrix, 312
blanchardii, 6
Boisduval scale, 24, 30, **179–181**, 182, 186
***boisduvalii*, 180**
Bouchei, 78
bourbon scale, 72
Bowrey scale, 26, 32, **367–369**
***bowreyi*, 368**
boycei, 320
Brazilian snow scale, 340
braziliensis, 340

***britannicus*, 194**
***bromeliae*, 182**
bruneri, 118
budleiae, 78
burrowing scale, 234

cacti, 186
cacti var. opuntiae, 186
cacti var. opunticola, 186
cactus scale, 13, 24, 30, **185–187**
caldesii, 78
calianthina, 320
California red scale, 6, 8, 10, 12, 13, 25, 33,
    50–53, 54, 58, 62
California yellow scale, 54
***californica*, 296**
californicus, 296
calyptroides, 186
calyptroides var. cacti, 186
calyptroides var. opuntiae, 186
camellia mining scale, 23, 34, **188–190**
camellia parlatoria scale, 24, 31, 33, **316–318**,
    324, 330, 334
camellia scale, 6, 23, 32, 204, **247–249**
camelliae: Aspidiotus, 232; Aspidiotus
    (Hemiberlesia), 232; Diaspidiotus, 232;
    Fiorinia, 204; Hemiberlesia, 232;
    Hemiberlesiana, 232; Insulaspis, 248;
    ***Lepidosaphes*, 248**; ***Parlatoria*, 316**; Uhleria,
    204
camphor scale, 23, 33, 34, **346–349**, 350, 354
candida, 266
candidus, 266
cap (settled crawler), 3, 5
caricis, 340
carueli, 100
Carulaspis spp., 5; calyptroides, 186;
    heterophyllae, 6; **juniperi**, 6, 9, 13, 24, 29,
    96–98, 100; **minima**, 6, 23, 29, 96, **99–101**, 256;
    pinifoliae, 6; visci, 96
cashew scale, 354
castigatus, 130
cattleya, 180
cattleya scale, 330
cattleyae, 216
celtidis, 388
celtis scale, 388
cerasi, 110
ceratoniae: Aspidiotus, 78; Lepidosaphes, 266;
    Mytilaspis, 266
chaff scale, 24, 31, 33, 320, **323–326**, 328
chemical control methods, 7–8, 174–175
Chemnaspidiotus liquidambaris, 162
Chermes aloes, 78; arecae, 204; aurantii, 336;
    bromeliae, 182; cycadicola, 78; echinocacti,
    186; ericae, 78; rosae, 86
cherry scale, 152
***chinensis*, 310**
Chinese lepidosaphes scale, 244
Chinese obscura scale, 310
Chinese obscure scale, 24, 33, **309–311**
Chinese scale, 172
chinesis, 244
Chionaspis spp., 5; aceris, 118; akebiae, 356;
    alni, 118; **americana**, 21, 31, **102–105**;
    aspidistrae, 340; aucubae, 356; bambusae,
    240; biclavis var. detecta, 234; braziliensis,
    340; cacti, 186; citri, 392; cockerelli, 356;
    corni, 6, 21, 31, **106–108**, 118; cryptogamus,
    118; dilatata, 356; euonymi, 396; evonymi,
    396; floridensis, 102; fraxini, 118; **furfura**, 21,
    31, **109–111**; furfurea, 110; furfurus, 110;
    furfurus var. fulvus, 110; furfurus var. ulmi,
    102; hattorii, 356; **heterophyllae**, 21, 29,
    112–114, 116; latus, 340; lintneri, 106;
    micropori, 118; miyakoensis, 356; montana,
    118; nemausensis, 396; nyssae, 1; ortholobis,
    106, 110, 118; ortholobis bruneri, 118;
    pinifoliae, 4, 5, 6, 10, 21, 29, 112, **115–117**;
    pinifoliae heterophyllae, 112; pinifoliae
    semiaureus, 116; pinifolii, 116; polypora, 118;
    prunicola, 364; pseudoleucaspis, 240; quercus,

    376; **salicis**, 21, 31, 110, **118–121**; salicisnigrae,
    118; sylvatica, 1; syringae, 356; ulmi, 102;
    zlocistii, 288
chionaspis, 92
Chionaspis? biclavis, 234
Chionaspis (Hemichionaspis) aspidistrae var.
    gossypii, 344; gossypii, 344; minor, 344
Chionaspis (Howardia) biclavis, 234
Chionaspis (Pinnaspis) aspidistrae, 340;
    proxima, 344
Chionaspis (Poliaspis) pini, 260
Chionaspis (Quernaspis) quercus, 376
chorions, 4
Chorizaspidiotus gutierreziae, 380
Chrysomphalus spp., 194, 272; alluaudi, 134;
    aonidum, 2, 6, 13, 26, 35, **122–125**, 126, 130;
    arecae, 130; aurantii, 50; **bifasciculatus**, 26,
    35, 122, **126–128**, 130; biformis, 216; biformis
    cattleyae, 216; bowreyi, 368; californicus, 296;
    castigatus, 130; citrinus, 54; degeneratus, 42;
    dictyospermi, 26, 35, 122, **129–132**; ficus, 122;
    jamaicensis, 130; lilacinus, 278; longisimma,
    368; mangiferae, 130; minor, 130; niger, 272;
    obscurus, 282; orientalis, 58; pedroniformis,
    58; pedronis, 58; rossi, 272; subrossi, 272;
    taxus, 62; tenebricosus, 284
Chrysomphalus (Melanaspis) obscurus, 282;
    tenebricosus, 284
cicatrices, 14
cinnamomi: Aulacaspis, 90; Diaspis
    (Aulacaspis), 90; Lepidosaphes, 262
cinnamomi mangiferae, 90
cinnamomum scale, 90
cinnamon scale, 90
circular black scale, 122, 272
circular purple scale, 122
circular scale, 122
circularis, 146
citri: Aspidiotus, 50; Chionaspis, 392; Howardia,
    392; Prontaspis, 392; **Unaspis**, 392
citricola, 244
citricola tasmaniae, 244
***citrina*, 54**
citrinus, 54
citrus aspidiotiphagus, 351
citrus long scale, 254
citrus mussel scale, 244
citrus parlatoria, 336
citrus snow scale, 6, 13, 20, 21, 31, **391–394**, 396
Clavaspis alluadi, 134; anonae, 134; **herculeana**,
    25, 34, 35, 36, **133–135**; symbioticus, 134;
    texana, 136; **ulmi**, 25, 26, 28, 36, **136–138**
***clavigera*, 188**
Coccidencyrtus malloi, 181
Coccidophilus spp., 9
coccineus, 50
Coccobius donatellae, 140; fulvus, 95;
    varicornis, 10
Coccoidea, 1
Coccomytilus halli, 288; zlocistii, 288
Coccophagoides utilis, 322
Coccus amygdali, 266; anguinus, 244; aonidum,
    122; beckii, 244; berberidis, 266; blanchardi,
    312; bromeliae, 182; conchiformis, 250;
    harrisi, 110; salicis, 118; ulmi, 266; ziziphi, 336
la cochenille pou rouge, 130
cochinella roja Australiana, 50
Cockerell scale, 276
cockerelli: Aspidiotus, 158; Chionaspis, 356;
    Lopholeucaspis, 276; Phenacaspis, 356;
    Pseudaulacaspis, 356; Trichomytilus, 356
cocoa-nut snow scale, 180
coconut scale, 27, 34, 68, **71–74**, 76
cocos scale, 180
cocotiphagus, 58
cocotis, 72
collection of specimens, 18
common and grass seed scale, 244
common parlatoria scale, 330
Compieriella bifasciata, 56
Comstockaspis perniciosus, 172
comstocki, 146
Comstockiella sabalis, 6, 23, 32, **139–141**
***conchiformis*, 250**
conchiformis-ulmi, 250
coniferarum var. shastae, 142

coniferous fiorinia scale, 2, 6, 22, 29, **206–208**
control practices. *See* management practices
convexus, 146
corky bark aspidiotus, 136
corni, 106
Cornuaspis beckii, 244
cotton white scale, 344
couch scale, 306
cover development, 1, 3, 4, 5
crawii, 224
crawler monitoring, 7
crawler stage, 3, 4
croton parlatoria scale, 316
crotonis, 316
Cryptaspidiotus shastae, 142
cryptogamus, 118
cryptomeria scale, 6, 7, 13, 27, 30, **68–70**, 72
cryptomeriae, 68
Cryptoparlatoria chinensis, 310
Cryptophyllaspis liquidambaris, 162
cultural control practices, 12
Cupressaspis atlantica, 142; **shastae**, 6, 25, 29, **142–144**
Curtis scale, 168
***cyanophylli*, 38**
cyanophyllum scale, 27, 28, 34, **38–40**
Cybocephalus spp., 9; *binotatus*, 92, 95
cycad aulacaspis scale, 6, 20, 21, 30, **93–95**
cycad scale, 94, 218
cycadicola: Aspidiotus, 78; Chermes, 78; Lepidosaphes, 244
cydoniae, 224
cydoniae crawii, 224
cydoniae punicae, 224
cydoniae var. *tecta*, 224
cymbidicola, 262
cymbidium scale, 22, 32, **262–264**

dactylproides, 186
daleae, 388
dark oak scale, 6, 25, 34, **278–280**
darutyi, 354
daruyi, 354
darvtyi, 354
Dearness scale, 25, 35, **379–381**
***dearnessi*, 380**
degenerata, 42
degenerate scale, 6, 27, 35, **41–43**, 300
***degeneratus*, 42**
demic adaptation, 13, 296, 298
denticulatus, 78
***destructor*, 72**
destructor transparens, 72
detecta, 234
detection and identification, 7
Diaspididae, 1
diaspidine scales morphology, 1, 16, 17; scale-cover development, 4, 5
***Diaspidiotus* spp.**, 5; **ancylus**, 1, 2, 6, 9, 10, 13, 27, 28, 35, **145–150**, 166, 176; *camelliae*, 232; *cyanophylli*, 38; *degeneratus*, 42; **forbesi**, 27, 28, 36, **151–153**; **gigas**, 13, 28, 35, **154–156**, 168; **juglansregiae**, 6, 26, 27, 29, 34, 152, **157–160**; *lataniae*, 224; **liquidambaris**, 6, 26, 33, 34, **161–164**; **osborni**, 6, 28, 35, 146, **165–167**; **ostreaeformis**, 6, 28, 35, 154, 158, **168–170**; **perniciosus**, 6, 7, 8, 13, 26, 35, 36, **171–175**; *piceus*, 148; **uvae**, 6, 28, 34, 146, **176–178**; *uviae*, 176
Diaspidiotus (Aspidiotus) lataniae, 224
Diaspis* spp.**, 5, 218; *amygdali*, 360; *amygdali* var. *rubra*, 364; *ancylus*, 13, 146; *annae*, 392; *auranticolor*, 364; **boisduvalii**, 24, 30, **179–181**, 182, 186; **bromeliae**, 13, 24, 30, 180, **182–184**; *cacti*, 186; *cacti* var. *opuntiae*, 186; *cacti* var. *opunticola*, 186; *calyptroides*, 186; *calyptroides* var. *cacti*, 186; *calyptroides* var. *opuntiae*, 186; *carueli*, 100; *cattleya*, 180; *celtidis*, 388; *conchiformis*, 250; *dactylproides*, 186; ***echinocacti, 13, 24, 30, **185–187**; *echinocacti cacti*, 186; *echinocacti opuntiae*, 186; *fallax*, 198; *fioriniae*, 204; *geranii*, 360; *harrisii*, 110; *jamiae*, 218; *juniperi*, 96; *lanatus*, 360; *leperii*, 198; *linearis*, 266; *mangiferae*, 90; *oleae*, 320; *opuntiae*, 186; *opunticola*, 186; *ostreaeformis*, 198; *parlatoris*, 330;

patelliformis, 360; *pentagona*, 360; *piricola*, 198; *pyri*, 198; *rhois*, 218; *rhusae*, 218; *rosae*, 86; *rubra*, 364; *snowii*, 166; *squamosus*, 320; *tillandsiae*, 182; *tricuspidata*, 370; *trinacis*, 180; *zamiae*, 218; *ziziphus*, 336
Diaspis (Aulacaspis) cinnamomi, 90; *cinnamomi mangiferae*, 90; *pentagona*, 360; *tubercularis*, 90
***dictyospermi*, 130**
dictyospermi var. *arecae*, 130
dictyospermi var. *jamaicensis*, 130
dictyospermum scale, 26, 35, **129–132**
dieback, 6, 13
diffinis, 228
diffinis var. *lateralis*, 224
dilatata, 356
dilatatus, 356
dimorphic forms, 1
Dinaspis annae, 392; *veitchi*, 392
Diplacaspis echinocacti, 186
dives, 334
dogwood scale, 6, 21, 31, **106–108**
dormant oil sprays, 7, 174
dry preservation of specimens, 18
dryandrae, 328
Duplaspidiotus claviger, 23, 34, **188–190**; **tesseratus**, 20, 34, **191–193**
***duplex*, 346**
duplex var. *paeoniae*, 350
dusty wing, 9
Dycryptaspis penicillata, 212; *ruthae*, 306
Dynaspidiotus britannicus, 27, 34, **194–196**; *degeneratus*, 42

ebony scale, 336
***echinocacti*, 186**
echinocacti cacti, 186
echinocacti opuntiae, 186
economic impact, 1, 13
eggs, 1, 3, 4
Egyptian black scale, 122
elegans, 218
elm armored scale, 25, 26, 28, 36, **136–138**
elm clavaspis scale, 136
elm scurfy scale, 21, 31, **102–105**
elongate hemlock scale, 6, 7, 12, 13, 22, 29, **200–202**, 204
Encarsia spp., 210; *aurantii*, 283; *berlesei*, 10, 362; *citrina*, 351; *perniciosi*, 175
English walnut scale, 158
epidendri, 78
Epidiaspis *betulae*, 198; **leperii**, 13, 24, 30, **197–199**; *peragrata*, 198
ericae, 78
escama blanca del mango, 90
escama del mango, 90
escama del terminal, 134
Eucornuaspis machili, 262
eugeniae var. *sandwicensis*, 356
euonymi: Chionaspis, 396; Parlatoria, 334; **Unaspis**, 396
euonymus scale, 5, 6, 9, 11, 12, 13, 21, 31, 392, **395–398**
European fiorinia scale, 204
European fruit scale, 6, 28, 35, 154, 158, **168–170**
European pear scale, 198
Evaspidiotus cyanophylli, 38; *excisus*, 76
evonymi, 396
***excisus*, 76**
Exochomus spp., 9, 351
***externa*, 200**
exuviae, 1
eye (scale morphology), 14
ezokihadae, 262

falciformis, 266
fallax: Aspidiotus, 72; Diaspis, 198
false diffinis scale, 25, 34, 36, **228–230**
false Florida red scale, 126
false Meyer scale, 27, 29, **300–302**
false oleander scale, 2, 13, 22, 30, **356–359**
false parlatoria scale, 26, 32, 370, **373–375**
false San Jose scale, 168
false scale, 374
feeding behavior, 4–5, 6
feeding, effect on host plants, 6, 13
fern scale, 2, 9, 10, 21, 31, **339–342**, 344

fernaldi, 158
fernaldi hesperius, 152
fernaldi var. *albiventer*, 158
fernaldi var. *cockerelli*, 158
ferrisi, 356
ficifoliae ulmicola, 250
ficifolii, 250
ficus: Aspidiotus, 122; Chrysomphalus, 122; Lepidosaphes, 250; Mytilaspis, 250
fig scale, 22, 32, 122, 248, **250–252**
filiformis, 238
***Fiorinia* spp.**, 5; *camelliae*, 204; **externa**, 6, 7, 12, 13, 22, 29, **200–202**, 204, 304; **fioriniae**, 22, 31, 32, 200, **203–205**, 206; *fioriniae* var. *japonica*, 206; **japonica**, 2, 6, 22, 29, **206–208**; *juniperi*, 206; **palmae**, 204; *pellucida*, 204; **theae**, 13, 21, 31, **209–211**
fiorinia hemlock scale, 200
fiorinia scale, 204
***fioriniae*, 204**
fioriniae var. *japonica*, 206
fire bush scale, 6, 13, 22, 32, 254, **269–271**
firebush scale, 270
first instar (crawler), 3, 4
flatheaded scale, 368
flavescens: Aspidiotus, 232; Mytilaspis, 244
flies, as scale predators, 9
florenciae, 296
Florida red scale, 2, 6, 13, 26, 35, **122–125**
floridensis, 102
flour scale, 340
flyspeck scale, 25, 34, **221–223**
Forbes scale, 27, 28, 36, **151–153**
Forbesaspis (Aspidiotus) forbesi, 152
***forbesi*, 152**
fraxini: Abgrallaspis, 42; Chionaspis, 118
***Froggattiella penicillata*, 6, 20, 30, **212–214**
fruit, damage symptoms, 13
Fullaway oleander scale, 356
fulva, 244
fulvus, 110
Fundaspis americana, 102; *quercus*, 376
fungi, 9; Aschersonia, 210; Fusarium, 210; Septobasidium, 144
***Furcaspis biformis*, 5, 23, 35, **215–217**; *cyanophylli*, 38; *juglans-regiae*, 158; *tsugae*, 304
Furchadaspis zamiae, 6, 20, 32, **218–220**
Furchadiaspis elegans, 218
***furfura*, 110**
furfurea, 110
furfuris, 110
furfurus var. *fulvus*, 110
furfurus var. *ulmi*, 102
Fusarium fungi, 210
fusca, 172

gall midges, as scale predators, 9
gas fumigation, 8, 174–175
genistae, 78
gennadii, 50
geranii, 360
gigas: Aspidiotus, 154; Aspidiotus (Euraspidiotus), 154; **Diaspidiotus**, 154; Quadraspidiotus, 168
Gingging scale, 354
gland-spine formula, 15
gland spines, 14
glanduliferus, 158
glomerata, 12
gloomy scale, 25, 34, 282, **284–286**
Glover scale, 22, 32, **253–255**, 270
***gloverii*, 254**
Gonaspidiotus ithacae, 44; *shastae*, 142
gopher scale, 158
gossypii, 344
granati, 250
grape scale, 6, 28, 34, 146, **176–178**
grass seed scale, 244
gray scale, 26, 32, **370–372**, 374
greedy scale, 6, 25, 34, 35, 36, 224, 226, **231–233**
green oyster scale, 168
greeni, 224
grey pear scale, 198
grey scale, 272
growth regulators, 8

gutierreziae, 380
Gymnaspis aechmeae, 25, 34, **221–223**

Hall scale, 6, 20, 31, **287–289**
halli, 288
hard grass scale, 306
Harper scale, 21, 32, **293–295**
harperi, 294
harrisi, **110**
harrisii, 110
hartii, 66
hattorii, 356
hederae, 78
hederae unisexualis, 78
helianthi, 380
Hemiberlesia argentina, 232; camelliae, 232;
   cyanophylli, 38; degenerata, 42; diffinis, 228;
   lataniae, 5, 6, 7, 10, 13, 28, 34, 35, **224–227**,
   232; longispina, 290; maskellii, 290;
   neodiffinis, 25, 34, 36, **228–230**; palmae, 38,
   290; **rapax**, 6, 25, 34, 35, 36, 224, 226,
   231–233; tricolor, 232; yuccae, 388; yuccae var.
   neomexicanus, 388
Hemiberlesiana camelliae, 232; perniciosa, 172
Hemichionaspis aspidistrae, 340; aspidistrae
   gossypii, 344; aspidistrae var. brasiliensis, 340;
   aspidistrae var. lata, 340; marchali, 344;
   minor, 344; minor var. strachani, 344;
   proxima, 344; townsendi, 344
Hemisarcoptes malus, 9, 268
hemlock scale, 5, 6, 13, 27, 30, **44–46**
Hendaspidiotus tricolor, 232; ulmi, 136
herculean scale, 134
herculeana, 134
herculeana scale, 25, 34, 35, 36, **133–135**
hesperius, 152
heterophyllae: Carulaspis, 6; **Chionaspis**, 112;
   Phenacaspis, 112
hippocastani, 168
holly scale, 27, 34, **194–196**
honeydew, 4, 6
horticultural oil sprays, 7, 174
host plants damage symptoms, 6, 13; overview,
   1; resistance to infestation, 12–13, 175, 296,
   298, 398; specificity of armored scales, 6. See
   also separate index of host plants
Howard scale, 146
howardi: Aspidiotus, 146; Kuwanaspis, 240
Howardia biclavis, 20, 32, 34, **234–236**; biclavis
   detecta, 234; citri, 392; elegans, 218;
   prunicola, 364; ramiae, 218
Howardia (Chionaspis) biclavis, 234
hunteri, 168
hydrangae, 276
hydrangeae, 276
hydrocyanic acid (HCN) gas, 8, 174–175
Hyperaspis spp., 9

identification, 7·
ilicis, 78
indurata, 284
inornata, 62
insect growth regulators, 8
insecticidal soaps, 7–8
insecticides, 7–8, 174–175; resistance to, 6, 8;
   synthetic organic, 8, 175
instars, 1, 3, 4, 5
insularis, 376
Insulaspis camelliae, 248; gloverii, 254;
   maskelli, 256; minima, 250; pallida, 256; pini,
   260; yanagicola, 270
Integrated Pest Management (IPM), 7, 12, 13
iota, 188
Ischnaspis filiformis, 238; **longirostris**, 21, 32,
   237–239
Italian pear scale, 13, 24, 30, **197–199**
ithacae, 44
ivy scale, 78

Jaapia americana, 102; quercus, 376
jamaicensis, 130
jamiae, 218
Japanese camellia scale, 350
Japanese camphor scale, 346
Japanese maple scale, 7, 22, 32, **275–277**

Japanese scale, 206, 276
japonica: Fiorinia, **206**; Leucaspis, 276;
   Leucodiaspis, 276; **Lopholeucaspis**, **276**
judaica, 320
juglandis, 266
juglandis-regiae, 158
juglans-regiae, 158
juglans-regiae var. albus, 158
juglans-regiae var. pruni, 158
juglansregiae, **158**
juniper fiorinia scale, 206
juniper scale, 6, 9, 13, 24, 29, **96–98**, 100
juniperi: Aonidia, 142; Aspidiotus, 96;
   Carulaspis, **96**; Diaspis, 96; Fiorinia, 206;
   Targionia, 142

Kuwanaspis howardi, 240; pseudo-leucaspis,
   240; **pseudoleucaspis**, 6, 22, 30, **240–242**

la cochenille pou rouge, 130
lady beetle predators, 9, 11, 74, 351. See also
   Chilocorus
lanatus, 360
lata, 340
latania scale, 6, 7, 10, 13, 28, 34, 35, **224–227**, 232
lataniae, 224
latastei, 194
lateralis, 224
latilobis, 146
Lattaspidiotus oreodoxae, 192; tesseratus, 192
latus, 340
laurel scale, 194
leaves, damage symptoms, 6, 13
Lecanium myrtilli, 118
lemon peel scale, 78
leperii, **198**
Lepidosaphes spp., 5, 288; atunicola, 270;
   bambusae, 240; **beckii**, 22, 32, **243–246**,
   262, 266; **camelliae**, 6, 23, 32, **247–249**;
   ceratoniae, 266; chinesis, 244; cinnamomi,
   262; citricola, 244; **conchiformis**, 22, 32, 248,
   250–252; conchiformis-ulmi, 250; cycadicola,
   244; cymbidicola, 262; ezokihadae, 262;
   ficifoliae ulmicola, 250; ficifolii, 250; ficus,
   250; **gloverii**, 22, 32, **253–255**, 270; granati,
   250; halli, 288; juglandis, 266; lilacina, 6;
   machili, 262; maskelli, 256; mesasiatica, 266;
   minima, 250; obscura, 6; oleae, 266; **pallida**, 6,
   10, 23, 29, 100, **256–258**; pallida maskelli, 256;
   pini, 6, 22, 29, **259–261**; piniformis, 262;
   pinnaeformis, 22, 32, 244, **262–264**; pomorum,
   266; populi, 266; punicae, 48; ritsema-basi,
   238; tiliae, 266; tuberculata, 262; turkmenica,
   250; **ulmi**, 6, 22, 32, 48, 244, **265–268**; ulmi
   bisexualis, 266, 268; ulmi candida, 266; ulmi-
   cotini, 266; ulmi-rosae, 266; ulmi vitis, 266;
   vulva, 266; **yanagicola**, 6, 13, 22, 32, 254,
   269–271
Lepidosaphes (Coccomytilus) halli, 288; zlocistii,
   288
Lepidosaphoides bambusae, 240
lesser snow scale, 21, 31, 340, **343–345**
Leucaspis bambusae, 240; hydrangeae, 276;
   japonica, 276
Leucodiaspis hydrangae, 276; japonica, 276
lilacina: Lepidosaphes, 6; **Melanaspis**, **278**
lilacinus, 278
lime-sulphur spray, 8, 174
Lindingaspis spp., 5; **rossi**, 26, 29, 30, 34,
   272–274
linearis: Aspidiotus, 368; Diaspis, 266;
   Mytilaspis, 250; Mytilococcus, 250;
   Pseudischnaspis, 368
lintneri, 106
liquid preservation of specimens, 18
liquidambaris, **162**
litchi scale, 22, 32, **47–49**
lobes (scale morphology), 14
long mussel scale, 254
long scale, 254
longirostris, **238**
longisimma, 368
longispina, **290**
longispina var. ornata, 290
longispinus, 310

Lopholeucaspis cockerelli, 276; **japonica**, 7, 22,
   32, **275–277**
Lucassi, 336
lucumae, 232

machili, 262
machilus oystershell, 262
macroducts, 14
magnolia white scale, 356
magnus, 168
management practices, 6–9, 12–13, 174–175
mangiferae: Aspidiotus, 130; Chrysomphalus,
   130; Diaspis, 90
mango scale, 90, 356
maple leaf aspidiotus, 146
maquarti, 118
marchali, 344
Maskell scale, 6, 10, 23, 29, **256–258**, 290
maskelli: Aspidiotus, 290; Aspidiotus
   (Morganella), 290; Insulaspis, 256;
   Lepidosaphes, 256; Morganella, 290
maskellii, 290
mauve pittosporum scale, 328
median lobes (scale morphology), 14
Mediterranean fig scale, 250
Mediterranean scale, 336
Megalodiaspis zamiae, 218
Melanaspis spp., 5; glomerata, 12; indurata, 284;
   lilacina, 25, 34, **278–280**; nigropunctata, 282;
   obscura, 2, 5, 6, 7, 9, 10, 13, 26, 34, 278,
   281–283, 284; rossi, 272; **tenebricosa**, 25, 34,
   282, **284–286**
Mercetaspis halli, 20, 31, **287–289**
mesasiatica, 266
mesothorax, 14
metathorax, 14
mexicana, 140
microducts, 14
micropori, 118
Microweisea spp., 9
minima: **Carulaspis**, **100**; Insulaspis, 250;
   Lepidosaphes, 250; Mytilaspis, 250;
   Odonaspis, 306
minimus: Aspidiotus, 118; Mytilococcus, 250
mining scale, 20, 32, 34, **234–236**
minor: Chionaspis (Hemichionaspis), 344;
   Chrysomphalus, 130; Hemichionaspis, 344;
   Pinnaspis, 344
minor strachani, 344
minor var. strachani, 344
minute cypress scale, 6, 23, 29, 96, **99–101**
mites, as scale predators, 9
miyakoensis, 356
monitoring practices, 7
montana, 118
Morgan's scale, 130
Morganella longispina, 25, 34, **290–292**;
   maskelli, 290
morphology, 1, 13–17
morrisoni, 320
mulberry scale, 360
multiglandulatus, 154
mussel scale, 266
myrtilli, 118
myrtus, 328
Mytiella sexspina, 254
Mytilaspis bambusae, 240; beckii, 244;
   ceratoniae, 266; citricola, 244; citricola
   tasmaniae, 244; conchiformis, 250; ficifolii,
   250; ficus, 250; flavescens, 244; fulva, 244;
   gloverii, 254; juglandis, 266; linearis, 250;
   longirostris, 238; machili, 262; maquarti, 118;
   minima, 250; pallida, 256; pallida var.
   maskelli, 256; pinifoliae, 116; pinnaeformis,
   262; pomicorticis, 266; pomorum, 266;
   pomorum var. candidus, 266; pomorum var.
   ulicis, 266; ritzemae basi, 238; salicis, 118;
   tasmaniae, 244; ulicis, 266; ulmi, 266;
   ulmicorticis, 266; vitis, 266
Mytilaspis (Aspidiotus) gloverii, 254
Mytilaspis (Lepidosaphes) pomorum, 266
Mytilococcus beckii, 244; conchiformis, 250;
   ficifoliae ulmicola, 250; gloverii, 254; halli,
   288; linearis, 250; minimus, 250; piniformis,
   262; pinorum, 260; tuberculatus, 262

nakayamai, 396
narrow fig scale, 250
natalensis, 356
nemausensis, 396
neodiffinis, 228
neomexicana, 388
neomexicanus, 388
neonicotinoid insecticides, 8
Neopinnaspis harperi, 21, 32, **293–295**
Neosignoretia yuccae, 388
nerii, **78**
nerii unisexualis, 78
neuropteran predators, 9
niger, 272
nigropunctata, 282
Nilotaspis halli, 6, 288
Nuculaspis *abietis*, 44; *apacheca*, 296;
  californica, 5, 6, 13, 27, 29, 44, **296–299**, 304;
  pseudomeyeri, 27, 29, 42, **300–302**; **tsugae**, 6,
  27, 30, **303–305**
nyssae, 1

oak scale, 6, 21, 31, **376–378**
oblongus, 168
obscura: Lepidosaphes, 6; **Melanaspis, 282**
obscure scale, 2, 5, 6, 7, 9, 10, 13, 26, 34, 278,
  281–283, 284
obscurus, 282
Odonaspis *minima*, 306; *penicillata*, 212;
  pseudoruthae, 306; **ruthae**, 6, 23, 30, 212,
  306–308, 380
ohioensis, 146
oil sprays, 7, 174
oleae: Diaspis, 320; Lepidosaphes, 266; Parlatorea,
  320; **Parlatoria**, 320; Syngenaspis, 320
oleander scale, 6, 26, 34, **78–81**, 82, 356, 360
oleastri, 78
olive parlatoria, 320
olive parlatoria scale, 320
olive scale, 5, 13, 24, 31, 33, **319–322**
ophiopogonis, 340
oppugnatus, 72
opuntiae, 186
opunticola, 186
orange brown scale, 122
orange chionaspis, 392
orange mussel scale, 244
orange scale, 50, 244
orange snow scale, 392
orbicularis, 330
orchid scale, 5, 23, 35, **215–217**
oreodoxae, 192
organic insecticides, synthetic, 8, 175
Oriental armored scale, 26, 33, 35, **58–61**
Oriental pine scale, 260
Oriental red scale, 58
Oriental scale, 58
Oriental yellow scale, 58
orientalis, 58
orientalis yellow scale, 58
ornata, 290
ornatus, 146, 290
ortholobis, 106, 110, 118
ortholobis bruneri, 118
Osborn scale, 6, 28, 35, 146, **165–167**
osborni, 166
ostreaeformis: Aspidiotus, 168; Aspidiotus
  (Euraspidiotus), 168; Aspidiotus
  (Quadraspidiotus), 168; **Diaspidiotus, 166**;
  Diaspis, 198; Quadraspidiotus, 168
ostreaeformis magnus, 168
ostreaeformis oblongus, 168
ostreata, 370
overwintering stage, 1, 6
oxyacanthae, 168
oxycrataegi, 146
oyster scale, 356
oyster-shell scale, 168
oystershell scale, 6, 22, 32, 48, 168, 244, **265–268**

paeoniae, 350
pallida, 256
pallida maskelli, 256
pallida var. *maskelli*, 256

palm fiorinia scale, 22, 31, 32, 200, **203–205**, 206
palm scale, 38, 290
palmae: Fiorinia, 204; Hemiberlesia, 38, 290;
  Parlatoria, 312
palmetto scale, 6, 23, 32, **139–141**
paraneri, 78
paraphyses, 14
paraphysis formula, 16
parasites of scale insects, 9, 10. *See also* wasps
parkinsoniae, 388
Parlatorea oleae, 320; *pergandei*, 324; *pergandii*
  var. *camelliae*, 316; *proteus*, 330; *zizyphi*, 336
parlatoreoides, 374
Parlatoreopsis chinensis, 24, 33, **309–311**;
  longispinus, 310; *pyri*, 310; *ziziphi*, 336
Parlatoria spp., 5; *affinis*, 320; *asiatica*, 312;
  blanchardi, 24, 31, 33, **312–315**; *blanchardi*
  var. *victrix*, 312; *blanchardii*, 6; *boycei*, 320;
  calianthina, 320; **camelliae**, 24, 31, 33,
  316–318, 324, 330, 334; *chinensis*, 310;
  crotonis, 316; *dryandrae*, 328; *judaica*, 320;
  Lucassi, 336; *morrisoni*, 320; *myrtus*, 328;
  oleae, 5, 13, 24, 31, 33, **319–322**; *orbicularis*,
  330; *palmae*, 312; *pergandei*, 324; *pergandei*
  var. *dives*, 334; *pergandei* var. *theae*, 334;
  pergandii, 24, 31, 33, 320, **323–326**, 328;
  pergandii var. *camelliae*, 316; *petrophilae*, 328;
  pittospori, 24, 31, 33, **327–329**; *potens*, 330;
  proteus, 24, 31, 33, **330–332**; *proteus*
  pergandei, 324; *proteus* var. *palmae*, 312;
  proteus var. *virescens*, 316; *selenipedii*, 330;
  sinensis, 324; **theae**, 24, 31, 33, **333–335**; *theae*
  var. *euonymi*, 334; *theae* var. *viridis*, 334;
  victrix, 312; *viridis*, 334; **ziziphi**, 6, 24, 31,
  336–338; *ziziphus*, 336; *zizyphe*, 336; *zizyphi*,
  336; *zizyphus*, 336; *zozypium*, 336; *zyziphi*,
  336
parlatoria date scale, 6, 24, 31, 33, **312–315**
Parlatoria (Euparlatoria) calianthina, 320;
  myrtus, 328; *pergandii*, 324; *pergandii* var.
  camelliae, 316; *proteus*, 330; *theae*, 334
parlatoria-like scale, 374
Parlatoria (Parlatoreopsis) chinensis, 310
Parlatoria (Websteriella) blanchardi, 312;
  ziziphus, 336; *Zizyphi*, 336
parlatorioides, 374
parlatoris, 330
patelliformis, 360
pear oyster scale, 168
pear tree oyster scale, 168, 198
pear white scale, 276
pedroniformis, 58
pedronis, 58
pellucida, 204
Pelomphala lilacinus, 278
penicillata, 212
penicillate scale, 6, 20, 30, **212–214**
pentagona, 360
pentagona auranticolor, 364
pentagona rubra, 364
peony scale, 23, 33, 34, 346, **350–352**
peragrata, 198
Pergande's scale, 324
pergandei, 324
pergandei var. *dives*, 334
pergandei var. *theae*, 334
pergandii, 324
pergandii var. *camelliae*, 316
perispiracular pores, 14
perivulvar pores, 14
permanent mounts of specimens, 18–19
perniciosa, 172
perniciosus, 172
pernicious scale, 172
persearum, 82
pesticides. *See* chemical control methods
petroleum oil sprays, 7, 174
petrophilae, 328
Phenacaspis aucubae, 356; *cockerelli*, 356;
  dilatata, 356; *eugeniae* var. *sandwicensis*, 356;
  ferrisi, 356; *heterophyllae*, 112; *natalensis*,
  356; *pinifoliae*, 116
pheromones, 5, 8, 175, 314
picea, 222
piceus, 148

pine needle scale, 4, 5, 6, 10, 21, 29, 112,
  115–117
pine oystershell scale, 6, 22, 29, **259–261**
pine parlatoria scale, 328
pine scale, 6, 21, 29, **112–114**, 116
pineapple scale, 13, 24, 30, 180, **182–184**
pini: Aspidiotus, 296; Chionaspis (Poliaspis),
  260; **Insulaspis**, 260; **Lepidosaphes, 260**;
  Poliaspis, 260
pinifoliae: Aspidiotus, 116; Carulaspis, 6;
  Chionaspis, 116; Mytilaspis, 116; Phenacaspis,
  116
pinifoliae heterophyllae, 112
pinifoliae semiaureus, 116
pinifolii, 116
piniformis, 262
pinnaeformis, 262
Pinnaspis spp., 5; **aspidistrae**, 2, 9, 10, 21, 31,
  339–342, 344; *caricis*, 340; *gossypii*, 344;
  marchali, 344; *minor*, 344; *minor strachani*,
  344; *ophiopogonis*, 340; *proxima*, 344; *quercus*,
  376; **strachani**, 21, 31, 340, **343–345**;
  temporaria, 344
pinorum, 260
piricola, 198
pittospori, 328
pittosporum diaspidid, 328
pittosporum scale, 24, 31, 33, **327–329**
Plagiomerus diaspidis, 186
plate formula, 15
plates (scale morphology), 14
plum scale, 320
plumose scale, 25, 34, **290–292**
Poliaspis pini, 260
Polyaspis pinifolii, 116
polypora, 118
pomicorticis, 266
pomorum, 266
pomorum var. *candidus*, 266
pomorum var. *ulicis*, 266
poplar scale, 13, 28, 35, **154–156**, 168
populi: Aspidiotus, 118; Lepidosaphes, 266
potens, 330
predators of scale insects, 9–12. *See also* beetle
  predators; lady beetle predators
preservation and preparation techniques, 18–19
prickly pear scale, 186
Prontaspis citri, 392
proteus, 330
proteus pergandei, 324
proteus scale, 24, 31, 33, **330–332**
proteus var. *palmae*, 312
proteus var. *virescens*, 316
prothorax, 14
proxima, 344
pruni, 158
prunicola, 364
Pseudaonidia *articulata*, 386; *clavigera*, 188;
  darutyi, 354; **duplex**, 23, 33, 34, **346–349**, 350,
  354; *iota*, 188; **paeoniae**, 23, 33, 34, 346,
  350–352; *rhododendri thearum*, 346;
  rhododendrii, 346; *tesseratus*, 192; *theae*, 346;
  trilobitiformis, 21, 23, 33, 34, 350, **353–355**;
  trilobitiformis darutyi, 354
Pseudaonidiella paeoniae, 350
Pseudaulacaspis spp., 5; *biformis*, 356;
  cockerelli, 2, 6, 13, 22, 30, **356–359**;
  pentagona, 4, 5, 6, 7, 8, 10, 11, 12, 23, 30, 356,
  360–363, 364; **prunicola**, 7, 12, 13, 23, 30, 356,
  360, 362, **364–366**
Pseudischnaspis *acephala*, 368; **bowreyi**, 26, 32,
  367–369; *linearis*, 368; *longisimma*, 368
pseudo-leucaspis, 240
Pseudodiaspis helianthi, 380; *parkinsoniae*, 388;
  yuccae, 388
pseudoleucaspis, 240
pseudomeyeri, 300
Pseudoparlatorea parlatoreoides, 374;
  tricuspidata, 370
Pseudoparlatoria ostreata, 26, 32, **370–372**, 374;
  parlatorioides, 26, 32, 370, **373–375**
pseudoruthae, 306
punicae: Andaspis, 48; Aspidiotus, 224;
  Aspidiotus (Diaspidiotus), 224; Aspidiotus
  (Evaspidiotus), 224; Lepidosaphes, 48

pupillarial scale insects, 1, 5
purple scale, 22, 32, **243–246**, 266
Putnam scale, 1, 2, 9, 10, 13, 27, 28, 35, **145–150**, 166, 176
pygidium, 14, 17
pyri: Diaspis, 198; *Parlatoreopsis*, 310
pyricola, 198
pyrus-malus, 266

Quadraspidiotus fernaldi, 158; *forbesi*, 152; *gigas*, 154, 168; *juglans-regiae*, 158; *juglansregiae*, 158; *ostreaeformis*, 168; *perniciosus*, 172
***quercus*, 376**
Quernaspis* insularis**, 376; ***quercus, 6, 21, 31, **376–378**
quince scale, 224

ramiae, 218
***rapax*, 232**
red cedar scale, 142
red cochineal, 130
red orange scale, 50
red orchid scale, 216
red pear scale, 198
red scale, 50
redwood scale, 6, 25, 29, **142–144**
Remotaspidiotus dearnessi, 380
reproduction, 1, 4–5
Rhizaspidiotus dearnessi, 25, 35, **379–381**; *helianthi*, 380
Rhizobius spp., 9, 11; *lophanthae*, 9, 11; *pulchellus*, 74
rhododendri thearum, 346
rhododendrii, 346
rhododendron scale, 146
rhois, 218
rhusae, 218
ridged scale, 204
rigidus, 72
ritsema-basi, 238
ritzemae basi, 238
rosa scale, 86
***rosae*, 86**
rosarum, 86
rose scale, 6, 20, 31, **85–88**, 94, 272
Ross scale, 272
Ross' black scale, 272
***rossi*, 272**
rubra, 364
rufous scale, 26, 34, 382, **385–387**
Ruth's scale, 306
***ruthae*, 306**

sabalis: Aspidiotus, 140; **Comstockiella**, 149
sabalis var. *mexicana*, 140
***sacchari*, 66**
sago palm scale, 94
***salicis*, 118**
salicis-nigrae, 118
salicisnigrae, 118
San Jose scale, 6, 7, 8, 13, 26, 35, 36, **171–175**
sandwicensis, 356
sanseveria scale, 330
Sasakiaspis pentagona, 360
scale cover, 1, 3, 4, 5
scale insects, classification of, 1
scales, armored. *See* armored scales
Scrupulaspis machili, 262
scurfy scale, 21, 31, **109–111**
Scymnus spp., 9
seasonal history, 5–6
second instar, 5
second lobes, 14
Selenaspidus albus, 6, 25, 33, **382–384**, 386; **articulatus**, 26, 34, 382, **385–387**
selenipedii, 330
semiaureus, 116
Septobasidium fungi, 134
serratus, 146
sexspina, 254
***shastae*, 142**
shed skins, 1
shortneedle conifer scale, 6, 27, 30, **303–305**
shortneedle evergreen scale, 304
simillimus translucens, 72

simplex, 386
sinensis, 324
Situlaspis* atriplicis**, 388; *daleae*, 388; ***yuccae, 25, 26, 30, **388–390**
slide mounting of specimens, 18–19
small brown scale, 330
small situlaspis scale, 388
snow scale, 392
snowii, 166
soaps, insecticidal, 7–8
solus, 146
Spanish red scale, 130
specimens, slide mounting of, 18–19
spined scale, 82
spinose scale, 27, 28, 35, 36, 72, **82–84**
***spinosus*, 82**
spiracles, 14
squamosus, 320
***strachani*, 344**
subrossi, 272
subsimilis var. *anonae*, 134
sugarcane scale, 6, 27, 28, 30, **65–67**, 306
sweet gum scale, 162
sweetgum scale, 1, 6, 26, 33, 34, **161–164**
sylvatica, 1
symbioticus, 134
Syngenaspis dryandrae, 328; *myrtus*, 328; *oleae*, 320; *pergandei*, 324; *petrophilae*, 328; *proteus*, 330; *theae*, 334
syringae, 356

taprobanus, 58
Targionia biformis, 216; *bigeloviae*, 380; *dearnessi*, 380; *gutierreziae*, 380; *helianthi*, 380; *juniperi*, 142; *parkinsoniae*, 388; *sacchari*, 66; *yuccae*, 388; *yuccae neomexicana*, 388
tasmaniae, 244
***taxus*, 62**
tea parlatoria scale, 24, 31, 33, **333–335**
tea scale, 13, 21, 31, **209–211**, 334
tecta, 224
tectus, 224
tegalensis, 12
Telsimia nitida, 74
Temnaspidiotus excisus, 76
temporaria, 344
temporary mounts of specimens, 18
***tenebricosa*, 284**
tenebricosus, 284
tesserate scale, 20, 34, **191–193**
***tesseratus*, 192**
texana, 136
theae: *Aspidiotus*, 234, 346; *Aspidiotus (Evaspidiotus)*, 346; *Aspidiotus (Pseudaonidia)*, 346; ***Fiorinia***, 210; ***Parlatoria***, 334; *Parlatoria (Euparlatoria)*, 334; *Pseudaonidia*, 346; *Syngenaspis*, 334
theae rhododendri, 346
theae var. *euonymi*, 334
theae var. *viridis*, 334
thearum, 346
thread scale, 238
thrips, as scale predators, 9
tiliae, 266
tillandsiae, 182
townsendi: Aspidiotus, 146; *Hemichionaspis*, 344
translucens, 72
transparens, 72
transparent scale, 72
Trichomytilus aucubae, 356; *cockerelli*, 356; *dilatatus*, 356; *natalensis*, 356; *pinifolii*, 116; *veitchi*, 392
tricolor, 232
tricuspidata, 370
trilobe scale, 21, 23, 33, 34, 350, **353–355**
trilobite scale, 354
***trilobitiformis*, 354**
trilobitiformis darutyi, 354
trinacis, 180
tropical palm scale, 38
***tsugae*, 304**
Tsugaspidiotus pseudomeyeri, 300; *tsugae*, 304
Tsukushiaspis bambusae, 240; *pseudoleucaspis*, 240
***tubercularis*, 90**
tuberculata, 262

tuberculatus, 262
turkmenica, 250

Uhleria camelliae, 204; *fioriniae*, 204
ulicis, 266
ulmi: *Aonidiella*, 136; *Aspidiotus*, 136; *Chionaspis*, 102; **Clavaspis**, 136; *Coccus*, 266; *Hendaspidiotus*, 136; **Lepidosaphes**, 266; *Mytilaspis*, 266
ulmi bisexualis, 266
ulmi candida, 266
ulmi-cotini, 266
ulmi-rosae, 266
ulmi vitis, 266
ulmicola, 250
ulmicorticis, 266
Unaspis spp., 5; ***citri***, 6, 13, 20, 21, 31, **391–394**, 396; ***euonymi***, 5, 6, 9, 11, 12, 13, 21, 31, 392, **395–398**; *nakayamai*, 396
unisexualis, 78
Utah cedar scale, 142
***uvae*, 176**
uvaspis, 176
uviae, 176

vaccinii, 118
vastatrix, 72
veitchi, 392
victrix, 312
villosus, 78
virescens, 316
viridis, 334
visci, 96
vitiensis, 360
vitis, 266
voltinism, 5
vulva, 266

walnut scale, 6, 26, 27, 29, 34, 152, **157–160**
wasps, parasitic, 9, 10; *Aphelinus mytilaspidus*, 268; *Aphycus californicus*, 160; *Aspidiotiphagus* spp., 362; *A. citrinus*, 254; *Coccidencyrtus malloi*, 181; *Coccophagoides utilis*, 322; *Comperiella bifasciata*, 56; *Plagiomerus diaspidis*, 186. See also *Aphytis*; *Coccobius*; *Encarsia*
West Indian peach scale, 360
West Indian red scale, 386
white cap, 3, 5
white euphorbia scale, 6, 25, 33, **382–384**, 386
white louse scale, 392
white magnolia scale, 356
white mango scale, 13, 20, 31, **89–92**
white peach scale, 4, 5, 6, 7, 8, 10, 11, 12, 23, 30, **360–363**, 364
white prunicola scale, 7, 12, 13, 23, 30, 360, 362, **364–366**
white scale, 78, 312, 360
willow scale, 21, 31, **118–121**, 154

Xerophilaspis parkinsoniae, 388

***yanagicola*, 270**
yanagicola oystershell scale, 270
***yasumatsui*, 94**
yellow apple scale, 168
yellow oyster scale, 168
yellow scale, 8, 25, 33, 50, **54–57**, 58, 62
yoked median lobes, 14
yucca scale, 25, 26, 30, **388–390**
***yuccae*, 388**
yuccae neomexicana, 388
yuccae var. *neomexicanus*, 388
yulupae, 166

Zagloba spp., 9
zamia scale, 6, 20, 32, **218–220**
***zamiae*, 218**
Zilus spp., 9
ziziphi, 336
ziziphus, 336
zizyphe, 336
zizyphi, 336
Zizyphi, 336
zizyphus, 336
zlocistii, 288
zozypium, 336
***zyziphi*, 336**